PHEROMONE COMMUNICATION IN MOTHS

Pheromone Communication in Moths

EVOLUTION, BEHAVIOR, AND APPLICATION

Edited by

JEREMY D. ALLISON

RING T. CARDÉ

UNIVERSITY OF CALIFORNIA PRESS

University of California Press, one of the most distinguished university presses in the United States, enriches lives around the world by advancing scholarship in the humanities, social sciences, and natural sciences. Its activities are supported by the UC Press Foundation and by philanthropic contributions from individuals and institutions. For more information, visit www.ucpress.edu.

University of California Press
Oakland, California

Library of Congress Cataloging-in-Publication Data

Names: Allison, Jeremy D., 1973- editor. | Cardé, Ring T., editor.
Title: Pheromone communication in moths : evolution, behavior, and application / edited by Jeremy D. Allison and Ring T. Cardé.
Description: Oakland, California : University of California Press [2016] | Includes bibliographical references and index.
Identifiers: LCCN 2016002213 (print) | LCCN 2016003460 (ebook) | ISBN 9780520278561 (cloth : alk. paper) | ISBN 9780520964433 (pbk.) | ISBN 9780520964433 (ebook)
Subjects: LCSH: Moths. | Pheromones. | Animal communication.
Classification: LCC QP572.P47 P437 2016 (print) | LCC QP572.P47 (ebook) | DDC 573.9/2—dc23
LC record available at http://lccn.loc.gov/2016002213

Manufactured in China

25 24 23 22 21 20 19 18 17 16
10 9 8 7 6 5 4 3 2 1

The paper used in this publication meets the minimum requirements of ANSI/NISO Z39.48-1992 (R 2002) (Permanence of Paper).♾

Cover illustration: [caption and credit].

CONTENTS

LIST OF CONTRIBUTORS

JÉRÔME ALBRE The New Zealand Institute for Plant & Food Research, New Zealand

JEREMY D. ALLISON Canadian Forest Service, Canada

THOMAS C. BAKER Pennsylvania State University

ROMINA B. BARROZO University of Buenos Aires, Argentina

RING T. CARDÉ University of California, Riverside

WILLIAM E. CONNER Wake Forest University

ALAN CORK University of Greenwich, United Kingdom

TEUN DEKKER Swedish University of Agricultural Sciences, Sweden

ELDON S. EVELEIGH Canadian Forest Service Fredericton, Canada

MAYA EVENDEN University of Alberta, Canada

STEPHEN P. FOSTER North Dakota State University

MICHAEL D. GREENFIELD Université François Rabelais deTours, France

ASTRID T. GROOT University of Amsterdam, The Netherlands

SABINE HÄNNIGER Max Planck Institute for Chemical Ecology, Germany

BILL S. HANSSON Max Planck Institute for Chemical Ecology, Germany

KENNETH F. HAYNES University of Kentucky

DAVID G. HECKEL Max Planck Institute for Chemical Ecology, Germany

N. KIRK HILLIER Acadia University, Canada

YUKIO ISHIKAWA The University of Tokyo, Japan

VIKRAM K. IYENGAR Villanova University

MARÍA LAURA JUÁREZ Estación Experimental Agroindustrial Obispo Colombres, Argentina

SILVIA KOST Max Planck Institute for Chemical Ecology, Germany

WALTER S. LEAL University of California, Davis

MARJORIE A. LIÉNARD Lund University, Sweden

CHRISTER LÖFSTEDT Lund University, Sweden

JEAN-MARC LASSANCE Harvard University

J. STEVEN MCELFRESH University of California, Riverside

JOCELYN G. MILLAR University of California, Riverside

RICHARD D. NEWCOMB The New Zealand Institute for Plant & Food Research, New Zealand

WENDELL L. ROELOFS New York State Agricultural Experiment Station

PETER J. SILK Canadian Forest Service Fredericton, Canada

BERND STEINWENDER The New Zealand Institute for Plant & Food Research, New Zealand

D.M. SUCKLING The New Zealand Institute for Plant & Food Research, New Zealand

JUN TABATA National Institute for Agro-Environmental Sciences, Japan

MELANIE UNBEHEND Max Planck Institute for Chemical Ecology, Germany

NIKLAS WAHLBERG University of Turku, Finland

PART ONE

Reminiscence of the Early Days

WENDELL L. ROELOFS

BECOMING AN ENTOMOLOGIST

CHALLENGES TO PHEROMONE IDENTIFICATIONS

Oak leafroller, *Archips semiferanus* (Tortricidae)
European corn borer, *Ostrinia nubilalis* (Crambidae)
Codling moth, *Laspeyresia pomonella* (Tortricidae)
Sex attractants used as a taxonomic tool

CHALLENGES TO BEHAVIORAL STUDIES

Is a sex attractant really an attractant?
Are all emitted compounds really pheromone components?
Blend versus individual component roles

THE NEXT PHASE

REFERENCES CITED

Becoming an Entomologist

One of the hottest topics at Entomological Society of America meetings (ESA) in the mid-1960s was anything to do with insect pheromones. The recent decoding of the silkworm moth, *Bombyx mori* (Bombycidae), pheromone by German scientists (Butenandt et al. 1959) after three decades of research showed that it was possible to unravel the mysteries of these mating messages. The term "pheromone" had recently been coined (Karlson and Lüscher 1959) to describe these chemical signals, and much discussion was centered on the exact meaning of this new term. A plea in Rachel Carson's book *Silent Spring* to develop insecticide alternatives also helped to generate funds for research on pheromones and their use in pest monitoring and management programs. The idea for the practical use of pheromones was proposed in 1882 by J. A. Lintner, the first New York state entomologist (Lintner 1882). He had observed the great attraction that female Promethea moths had for conspecific males from long distances and wrote: "Can not chemistry come to the aid of the economic entomologist, in furnishing at moderate cost, the odorous substances needed? Is the imitation of some of the more powerful animal secretions impracticable?" Paper sessions and night discussions at the ESA were packed as the few scientists involved in the pheromone field debated questions regarding pheromones and their practical use. It was my great fortune to come into this scene in 1965 as part of a new thrust by the Entomology Department of Cornell University at the New York State Agricultural Experiment Station in Geneva to develop a research program on pheromones of moth pest species.

Paul Chapman, the Chair of the department, felt that the fastest route into the pheromone field was to hire a chemist. He was a wise man and elicited the help of the renowned chemist at Cornell, Jerry Meinwald, to send out the position statement to his colleagues. My postdoctoral advisor at MIT got the statement and showed it to me. Although I had no training in entomology, I was intrigued by the possibility of conducting research on pheromones and applied for the position and got it. The search committee must have been impressed with my PhD thesis on "Cyclization of ylidenemalononitriles" in Organic Chemistry from Indiana University.

The entomologists at Geneva were eager to collaborate with me and I quickly set up a project on an apple pest, the redbanded leafroller moth, *Argyrotaenia velutinana* (Tortricidae), which had become resistant to the current pesticides and had become a major pest. Another project, which was funded by the NSF, was on the giant cecropia moth, *Hyalophora cecropia* (Saturniidae). This species was being mass reared by a fellow faculty member, Frederick Taschenberg, at the Fredonia Research Laboratory and was included in the research since it was a very large insect and thought to be an easy subject for pheromone identification. It turned out that the cecropia moth has very little stored pheromone, and the chemical structure apparently so complex that so far it has eluded all efforts on its identity.

To guide our research there was little information on pheromone structures or how to identify them. After the publication of the silkworm moth pheromone in 1959, Milt Silverstein and Dave Wood reported (Silverstein et al. 1966) on a chemical blend of three compounds for a bark beetle, and then Bob

Berger (1966) published the pheromone of the cabbage looper moth, *Trichoplusia ni* (Noctuidae) as (*Z*)-7-dodecenyl acetate (Z7-12Ac). With little information available on how to identify pheromones, I decided to take a trip to the USDA labs in Washington, DC, to get up-to-date on methodology since the scientists there had been involved for years with the pheromones of a number of moth species, such as the gypsy moth and pink bollworm moth. However, accumulating thousands of field-collected whole moths in barrels of benzene did not seem like a good pheromone purification scheme. Therefore, we jumped into the fray by setting up a mass rearing effort and clipped 50,000 female abdominal tips in ether or dichloromethane for isolation of the pheromone by column, thin-layer, and gas chromatography. In a couple of years we identified the red-banded leafroller moth pheromone as (*Z*)-11-tetradecenyl acetate (Z11-14Ac). This monounsaturated acetate was synthesized by my first postdoc, Henry (Heinrich) Arn, using a Wittig reaction, and then tested in apple orchards with baited ice cream carton sticky traps to show its great activity in trapping male moths (Roelofs and Arn 1968).

I was fortunate to associate with faculty colleagues, Paul Chapman and Sieg Lienk, who were preparing a book (Chapman and Lienk 1971) on over 50 leafroller species found in wild apple trees in New York. They collected larvae by beating branches of wild apple trees around Memorial Day and then reared the dislodged larvae for identification. I joined them on these field excursions and this allowed not only for my entrée into the world of entomology, but also for material to start cultures of several pest leafroller species from the collected larvae. We were able to identify the pheromone of a number of these leafroller species and found that many also used Z11-14Ac. Since the female extracts were very specific in attracting conspecific males, we anticipated that specific blends must be involved, similar to what had been previously reported for bark beetles. After more research on the gland extracts and numerous field tests, the postdoc chemists in my lab, Henry Arn, Ada Hill, and Jim Tette, were able to show that these leafrollers used precise ratios of Z/E isomers with various additional components added to make a specific blend for each species. Interestingly (and fortuitously), the redbanded leafroller males were initially captured in great numbers with Z11-14Ac lures because this species uses a 92:8 ratio of Z/E isomers, which is exactly what was produced using the Wittig reaction to synthesize Z11-14Ac.

Challenges to Pheromone Identifications

Oak Leafroller, *Archips semiferanus* (Tortricidae)

By the mid-1970s species-specific pheromone blends had been documented in numerous moth species. However, Larry Hendry at Penn State published results that led to a proposal of startling new concepts in insect chemical communication and evolutionary biology that seriously challenged the existence of species-specific pheromone blends. He was a bright, young scientist in the Chemistry Department investigating the pheromone of the locally abundant oak leafroller moth. His initial studies on extracts from the pheromone glands indicated activity in gas chromatography (GLC) collections at the retention time of 14-carbon acetates. At that point, they conducted field-trapping studies using large sticky vane traps. These traps were normally used for bark beetles in very dense populations and could intercept thousands of insects, male

and female, even with blank (unbaited) traps. If a single trap containing a test chemical captured more males than the 3000+ males caught on a blank trap, the chemical on that trap was concluded to be an attractant for oak leafroller males. Field tests involving various monounsaturated 14-carbon acetates throughout the woods in an array of different species of oak trees led to the conclusion that the pheromone used by oak leafroller moths differed among white, red, and black oak trees and could include acetates mainly with Z3, Z4, or Z10 double bonds. This conclusion along with anecdotal observations that male oak leafroller moths were attempting to mate with leaves of the different host trees led to the controversial conclusion that with pheromones, "You Are What You Eat." In other words, the larvae obtained the pheromone from host leaves and the sequestered chemicals could be different for each host species. As a result moth sex pheromones could vary among populations. Analyses of the different oak leaves by the Penn State scientists suggested the presence of the pheromone structures to support this conclusion (Hendry et al. 1975a, 1975b, 1975c).

This conclusion undermined the growing information on pheromone blends since it suggested that pheromone blends could be highly variable within a species, and that pheromones for monitoring and insect control programs would probably fail as the pheromone production and response were plastic. Many leading biologists and chemical ecologists readily accepted the concept of "You Are What You Eat" as a viable hypothesis. With this support, the new idea was widely promoted by the Press, discussed in symposia at national and international ecology meetings, and published in several research papers in *Science*. Jane Brody, a columnist for *The New York Times*, later summed up the threats to the pheromone field in an article that discussed the pros and cons of this new idea. She said:

> If Dr. Hendry were right, pest control that relied on man-made versions of insect sex attractants, or pheromones, would be doomed, because no one could predict which chemical the insects would respond to. In scores of research projects costing millions of dollars, synthetic pheromones are being used to detect insect invasions by luring them into pheromone-baited traps and to disrupt insect mating patterns by widely spraying the pheromone that ordinarily draws in the male insect to a receptive female. Dr. Hendry's report in April 1975 that at least one insect pest, the oak leafroller, derived its pheromone from leaves it ate and produced different chemicals from different diets threw the field of pheromone research into temporary disarray.

In the mid-1970s, Jim Miller came as a postdoc to our research group after receiving his PhD from the Entomology Department at Penn State. He was appalled by how the field data on the oak leafroller were collected by the scientists in the Chemistry Department and was eager to reveal the fallacies of the new hypothesis. He initiated a rigorous re-investigation of this pheromone. He reared the larvae on a pinto bean diet and showed that the female moths produced a precise ratio of 67:33 E/Z-11-14Ac, obviously without obtaining pheromone from their diet. He then analyzed the pheromone from female moths reared on the different oak leaves and found the same ratio of pheromone components from all oak species. Furthermore, male moths reared on the various leaves were responsive in the lab only to that precise blend of geometric isomers. The final blow to the hypothesis was delivered with data from a large field-trapping study that our research group conducted in the woods near Penn State. Treatments placed in small sticky traps were replicated 10 times, and included an array of E/Z11-14Ac isomeric blends from 0:100 to 100:0, as well as other monounsaturated acetates purported to be pheromones by the Penn State scientists. The results showed that male moths were trapped only with the female-produced blend (70:30) of E and Z11-14Ac, and none by the other suggested monounsaturated acetates or off-blends of the 11-14Ac. Thus, the oak leafroller pheromone was found to be similar to other leafroller pheromones in using a very specific E11/Z11 ratio of chemicals. These data that disproved the hypothesis were presented at a fully packed ESA meeting and published in *Science* (Miller et al. 1976).

However, the idea did not die easily, since *Science* then published a rebuttal by Hendry (1976). He attributed the differences in results between the two studies to the use of different techniques and did not retract the hypothesis. Our lab then collaborated with the postdoc and graduate student from Penn State who were involved in the project and had them use procedures developed in our laboratory to verify the pheromone identification. It resulted in a complete retraction in *Science* (Hindenlang and Wichmann 1977). Hindenlang and Wichmann stated:

> Our present results indicate that the earlier data and derived hypotheses should be reconsidered. We do not deem it appropriate to advance a hypothesis regarding a direct association between plant chemistry and insect sex pheromones. Furthermore, we retract previous reports and interpretations of data suggesting such an association . . . In our present analysis of the plant material, we found compounds that were clearly not tetradecenyl acetate but gave patterns similar to it.

In other words, the leaves did not contain pheromone compounds. These two brave scientists were under much pressure to maintain their support for the new hypothesis, but in the Jane Brody report in *The New York Times* they said that as far as they were concerned, the theory of sex pheromones in plants is dead.

European Corn Borer, *Ostrinia nubilalis* (Crambidae)

Another interesting challenge to a pheromone identification came after Jerry Klun characterized the European corn borer pheromone from 10,000 females (Klun and Brindley 1970) as Z11-14Ac. He conducted the research in collaboration with a Professor of Entomology at Iowa State who was an acknowledged leader in managing European corn borer populations in the Midwest. However, in 1972, a postdoctoral chemist in my group, Jan Kochansky (Kochansky et al. 1975), found that European corn borer females in certain field plots in New York produced and males were specifically attracted to the opposite isomer, E11-14Ac, and not to the Z isomer as found in Iowa. This was a great collaboration that we had with my colleague Chuck Eckenrode and his technician Paul Robbins. We presented these findings in a symposium at an ESA meeting to a packed ballroom audience, but they were not well received by all. The Iowa State collaborator shouted from the middle aisle of the ballroom that it had to be a wrong identification and that a "corn borer, is a corn borer, is a corn borer." The controversy was resolved in the next year by full cooperation on both sides by exchanging insect cultures obtained from Iowa and New York and conducting in-depth analyses of

the pheromone of both populations. Both laboratories found that indeed the Iowa population used a 97:3 Z/E blend and the New York population used the opposite 1:99 Z/E blend. This finding provided good evidence for pheromone polymorphism in this species and led to years of research on three genetically different European corn borer "races" in New York labeled bivoltine Z, univoltine Z, and bivoltine E as defined in the field mainly through the efforts of Paul Robbins (Roelofs et al. 1985).

Codling Moth, *Laspeyresia pomonella* (Tortricidae)

In the years leading to the early 1970s there were many sex pheromones and sex attractants reported to be monounsaturated acetates and alcohols with 12-, 14-, and 16-carbon-length chains (Roelofs and Comeau 1970). A number of apple pests were included in that list, but one glaring omission was that of the codling moth, which is a major worldwide pest of fruit. By 1970, USDA scientists had initiated a project on this species and evidently had already extracted thousands of female moths for pheromone identification. That pheromone presented an interesting challenge so we took it on with the help of a novel technique that was set up in our laboratory by a creative graduate student, André Comeau.

In the 1960s, Dietrich Schneider in Germany was conducting research on antennal responses of male silkworm moths to the newly identified pheromone (Schneider 1962). He developed a technique in which an antenna was connected between two electrodes, and responses to volatiles recorded by measuring the depolarization of the antenna as it responded to active compounds. Comeau set up this electro-antennogram (EAG) in our lab as an analytical tool. He found that the large EAG responses from male antennae to their own pheromone compounds could be used to determine GLC retention times of activity, with effluent collections from injections of crude pheromone-gland extracts. The advantage of collecting the effluent in capillary tubes compared to the later use of the combined GLC-EAD technique was that the active material could be rinsed from the tubes and used for an injection on a GLC column of different polarity or used for micro-reactions for information on the double bond position or compound functionality. The retention times could be used to determine if there were several major active compounds in the crude extract, if they were alcohols or acetates, and the length of their carbon chains.

Another key factor involving the EAG technique was to puff each compound from a library of monounsaturated standards to determine which isomeric and geometric isomers in the arrays of 12-, 14-, and 16-carbon-length compounds elicited the highest EAG responses. In many cases, the combination of these two techniques would indicate the possible pheromone structure for a particular pest species from less than 50 male and female pupae sent to our lab from around the world.

These techniques were used to investigate and identify the codling moth pheromone. Crude extracts from a few codling female abdominal tips were injected on polar and nonpolar GLC columns and collected in tubes for EAG analysis. The data revealed a single EAG-active compound. On the nonpolar column the retention time was similar to dodecanol, but it was much longer than dodecanol on a polar column. This indicated that there was probably a conjugated double bond system in the active compound. A screening of the library compounds showed that 12-carbon alcohols were more active than the corresponding 14- and 16-carbon alcohols, and more than any acetate standard. The two monounsaturated compounds eliciting the largest EAG responses were (*E*)-8- and (*E*)-10-dodecenol. Combining the above results indicated that the pheromone could be (*E,E*)-8,10-dodecadienol. Ada Hill synthesized the four geometric isomers for this double bond system and we assayed them for EAG and field-trapping activity. The E,E isomer was the most active with EAG, and was the only isomer to be extremely active in trapping males in the field.

In 1971, these results were published in *Science* (Roelofs et al. 1971) and presented at an ESA meeting as identification of the codling moth sex attractant using the EAG technique. Most of our colleagues viewed this with great skepticism, since the EAG technique was not considered an acceptable method for identification. There were many, including our friends, who set out to prove us wrong and some even published other structures. Over the years, there were retractions of the other structures and finally some scientists carried out classical methods to prove our identification to be correct. Eventually the EAG technique was recognized as a powerful tool, but, unfortunately, has erroneously been used as a substitute for behavioral evidence in characterizing new pheromones.

Sex Attractants Used as a Taxonomic Tool

In the late 1960s, a number of moth pheromone structures were identified as monounsaturated acetates and alcohols with the double bond in various positions. In order to speed up efforts to identify new sex attractants, scientists began to test these compounds and their analogs in field traps to discover species that were attracted to specific compounds. This technique has now been used for decades in many countries to define attractants for hundreds of species. In 1971, we reported on the identification of sex attractants for over 90 species of Lepidoptera (Roelofs and Comeau 1971). We found it quite interesting that the reported attractant structures supported the division of the Tortricidae Family into the subfamilies Olethreutinae and Tortricinae. The Olethreutine species were all attracted by 12-carbon acetates or alcohols, with the exception of only two species, whereas all the Tortricine species, except one, were attracted to 14-carbon chain acetates and alcohols. We suggested (Comeau and Roelofs 1973) that the attractant structures could provide valuable information for structural comparisons within and among genera, subfamilies and families. However, many taxonomists around the country rejected this notion that attractants could be used as a taxonomic tool. At Cornell, the famous lepidopteran taxonomist, John Franclemont, and his student Richard Brown, embraced the idea. Richard Brown later collaborated on an *Annual Review* chapter (Roelofs and Brown 1982) in which we discussed the role of sex attractants as specific mating signals and he related structural diversities and similarities of known attractants to several published schemes of phylogenetic relationships.

Franclemont provided excellent assistance in the identification of male moths removed from sticky traps when significant numbers were trapped by a particular chemical. In some cases, however, the trapped males provided unexpected taxonomic information. For example, in one location Comeau found that some males of the noctuid moth, then known as *Amanthes*

c-nigrum (Noctuidae), were attracted to Z7-14Ac, and other morphologically similar males to its opposite geometric isomer, E7-14Ac (Roelofs and Comeau 1969). We thought it quite strange since a combination of the two isomers caught no males. A closer investigation of the male moths showed that the males attracted to the Z isomer were always significantly larger than those attracted to the E isomer. Specimens of each type were brought to Franclemont and he found many of the larger moths in the Cornell University collection, but only a single smaller moth. Franclemont had previously seen the smaller specimens at the black light in his backyard in Ithaca, but had not added them to his massive collection, which generally is notable for long series of perfect specimens. He then carried out several years of black light and attractant trapping locally with these populations. An in-depth study of the specimens of the two populations showed that they possessed differences in size and coloration, as well as some previously overlooked differences in the genitalia of both sexes. He then named (Franclemont 1980) the smaller species *Xestia adela* (Noctuidae) and the larger species *X. dolosa*.

A similar interesting study involving another lepidopteron taxonomist, John Dugdale, was conducted in New Zealand during my sabbatical leave there in 1983. My task was to investigate the pheromones of two tortricid pest species, *Planotortrix excessana* (Tortricidae) and *Ctenopseutis obliquana* (Tortricidae). The pheromones had been described previously, but they only attracted males in select areas of the two Islands. A research program was set up with Stephen Foster at the Department of Scientific and Industrial Research in Auckland. The initial findings showed that samples of these species from different areas used different pheromone blends. Dugdale was brought into the project and he enthusiastically sampled populations of these species throughout New Zealand. In a few months it was shown (Foster et al. 1986) that each of the two previously described species consisted of at least three or four sibling species with reproductive isolation effected by different pheromone blends. The *Planotortrix* sibling species utilized Z5-14Ac, or Z8-14Ac, or combinations of Z5- and Z7- or Z7- and Z9-14Ac. The *Ctenopseutis* sibling species were found to use Z5-14Ac or Z10-16Ac, or mixtures of Z5-14Ac and Z8-14Ac.

Research in New Zealand has continued on these interesting sibling species complexes, including in-depth studies at the molecular level. However, the question of how morphologically similar moth populations evolve to generate new species with different pheromones from common ancestral populations remains unresolved. Even in a case involving *Ostrinia* corn borer species (Roelofs and Rooney 2003), in which it was shown that a particular biosynthetic gene could have been turned on to produce a new pheromone structure, and that there are rare males present in the population to respond to that new compound, the mystery remains on how these factors could have driven a portion of the population to form a new species using the new pheromone.

Challenges to Behavioral Studies

Chemical communication is common throughout the animal kingdom, but the term "pheromone" is restricted to chemical communication between individuals of the same species. Early discussions on "communication" resulted in agreement that pheromones had to elicit behavioral responses in a receiving organism of the same species. Thus, it was obvious, even to chemists, that the identification of chemical structures found in female pheromone glands had to be tied to assays for proof of male behavioral responses. Our early bioassays for pheromone activity involved crude olfactometer boxes in the laboratory or trapping studies in the field for proximate conclusions on attractant activity.

Is a Sex Attractant Really an Attractant?

Harry Shorey and his students were among the first to conduct basic studies on pheromone production and male responses relative to circadian rhythms, age, mating history, environmental effects, etc. By the early 1970s, things seemed to be progressing well on the behavioral side until a bombshell was dropped at the 1972 International Congress of Entomology meeting in Australia. The closing address entitled "The emergence of behaviour" was given by Prof. John S. Kennedy, a scholarly gentleman from Imperial College, who was a critical thinker on mechanistic behavior in insects and had an adverse opinion on the use of anthropomorphic terms to describe insect behavior. His research was primarily centered on aphid behavior, but he had developed a keen interest in the new field involving moth pheromones. He was appalled at how weak he found the field of behavior to be in entomological science and used his address to challenge the community of chemical ecology to strengthen their behavioral work (Kennedy 1972). Relative to the pheromone field, he said:

> Indeed the effort put into analyzing the behavioural mechanisms by which insects find their mates has been a pathetic fraction of the effort put into isolating, identifying and synthesizing candidate attractants, all based on ad hoc bioassays of unknown relevance to the real field behaviour. . . . By a year ago the task of chemical identification has been completed for more than 80 species of Lepidoptera (Roelofs and Comeau 1971). One is filled with admiration for the expertise and pertinacity of the organic chemists here, but they only show up the emptiness of the behavioural side. The best picture we can conjure up of how these substances actually work on the free insects' behaviour (Shorey 1970) is still very largely hypothetical and highly controversial. The whole question has been begged by the label SEX ATTRACTANT that is always applied to them and serves as a fig-leaf for our ignorance. "Sex attractant" sounds a simple, straightforward, descriptive term, but in truth is very misleading because these substances are, in the first place, not only attractants, and, in the second place, are probably not, strictly speaking, attractants at all as usually understood.

I remember standing in the airport after that meeting and expressing my profound confusion along with the other entomologists on why an attractant is not an attractant. The address served its purpose well in that behavior became a focus for several scientists in the pheromone field.

Luckily for me, Ring Cardé, a recent PhD student of Franclemont at Cornell, had just joined my research group with an interest and skills in insect behavior. He set up behavioral studies with our newly identified pheromones, and, surprisingly, even identified some pheromones himself by learning the skills of the EAG and GLC techniques along with some chemical analytical tests. He took a newly hired technician, Tom Baker, under his wing and got him involved in behavioral studies as well. One major project was conducting behavioral studies on male oriental fruit moth, *Grapholita molesta* (Tortricidae),

response from a distance to pheromone in the field. Cardé moved on to the faculty of Michigan State and took Baker on as a PhD student as they continued their pheromone behavioral studies. These studies then became career-long in-depth mechanistic studies in their individual research groups to unravel the mysteries of the optomotor upwind anemotaxis, anemomenotaxis, and all levels of intricacies of modulating the male's responses to aerial trails of pheromone (Cardé, this volume). Cardé initiated the behavioral side of pheromone identifications in my lab, which then continued on with the return of Baker as a postdoc, followed by Jim Miller and then Charlie Linn. The combination of chemistry and behavior was always a great strength in my research group and mirrored the combination of Professors Jerry Meinwald and Tom Eisner at Cornell as a great team in chemical ecology. The influence of Cardé and Baker in the pheromone field has been great, not only from their own contributions, but also from the many postdocs and students who have passed through their laboratories.

Are All Emitted Compounds Really Pheromone Components?

In the early days, it was quite straightforward to conduct field-trapping studies with a few compounds identified from female gland extracts and label them as pheromones if they proved to be active. However, this changed with the advent of capillary GLC. With its greater sensitivity and powers of compound separation, it became possible to analyze one red-banded leafroller female gland, instead of 50,000, and see the elution of seven possible pheromone components in the gland. In another example, the cabbage looper female gland exhibited six possible pheromone components in the gland instead of just two. These results became common in the field and led to the next round of challenges to pheromone identifications. Which compounds in the female pheromone gland make up the natural pheromone blend and which ones are inactive? To answer this question the behaviorists needed to demonstrate activity for each compound for it to be included in the natural pheromone blend.

Various types of olfactometers, spheres, and field-trapping studies were used to demonstrate pheromone activity, but we found that studies using a laboratory wind tunnel best suited our needs. In part this was due to the fact that the upwind-flight response of males in the odor plume is the most sensitive measure of male responses to the pheromone blends that were being identified. Olfactomotor-based activation bioassays could be used for a variety of studies, but did not adequately assay the importance of minor components in the blends. Jim Miller not only worked on the oak leafroller moth, but also set up a laboratory flight tunnel patterned after one developed by Prof. John Kennedy (Kennedy and Marsh 1974), who initiated studies on male moth behavioral responses to pheromones after decrying the lack of behavioral studies. The great value of the flight tunnel for behavioral analyses was also seen by Ring Cardé, Tom Baker, Harry Shorey and others who devoted a good portion of their careers to using the flight tunnels in creative ways to analyze male flight tracks and the male's response to the flickering wafts of pheromone. Charlie Linn joined our group in the mid-1980s and proceeded to fly thousands of males individually to various compound blends in assays to determine whether a compound had activity in the blend or not. It turned out to be a great tool to define the natural pheromone blend.

Blend versus Individual Component Roles

Many pheromones were found to be blends with one predominant compound, and these findings led to another controversial topic promoted by neurophysiologists who were studying pheromone receptors and their response to the various pheromone components. They found that some moth species under investigation had thousands of receptors for the most abundant pheromone component and these receptors were extremely sensitive to that specific compound. The obvious conclusion from these studies was that a male moth far downwind from the source would detect the most abundant compound and initiate upwind flight to that compound. The other components, often present in much lower proportions and having fewer receptors on the antenna, would then be detected as the moth flew closer to the source, and perhaps some components had the role of initiating close-range behaviors to the calling female. Our feeling was that the blend worked as a unit and there were not individual roles for pheromone components. The arguments were quite enthusiastic on both sides, but finally Charlie conducted in-depth flight-tunnel studies with redbanded leafroller, cabbage looper, and oriental fruit moth males. He recorded the initiation of flight, upwind flight, and source contact to a wide range of release rates of treatments consisting of the single most abundant component, the combination of two components, and the whole blend. It was obvious in all three cases that the blend was the most active at the lowest concentrations for all behaviors and in most cases the single most abundant component had no activity except at high concentrations (Linn et al. 1986).

These data, however, did not impress the neurophysiologists, or many entomologists, since it was not obtained with free-flying males in the field. There was still strong conviction that the sensitive receptors had to detect the main component at a long distance, although there were no data to substantiate it. In one of our signature coffee break discussions it was finally decided that Charlie had to conduct a field study to settle this argument. Others had started to use soap-bubble generators to define pheromone plumes, and Tom Baker and Wendy Meyer had used this technique in our group previously. It was decided to use this technique in the field with the oriental fruit moth, which flies at dusk instead of at night like most of the other moths we had available. We did not have a bubble machine, so Charlie became the machine and simply dipped a bubble wand in a bottle of children's bubble mix and generated the required bubbles behind a rubber septum containing one of several mixtures of pheromone components. Marlene Campbell, a technician in the group, standing downwind of the bubbles would follow the bubble stream upwind with a single male oriental fruit moth in a wire screen cage and throw a flag to the ground at the distance that the male was activated and initiated upwind flight. The distance of response to the various treatments was then determined. Data (Linn et al. 1987) with the main pheromone component, two components, or all three components in the blend were the same as the flight-tunnel results. All 30 males responded 20–23 ft downwind from the three-component source (complete blend). Only four males initiated flight close to the single-component source (2–3 ft), and the two-component blends showed only slightly higher activity at 10 ft or less from the source. These data showed that the natural blend had the greatest activity at the lowest release rate and, clearly, initiated upwind flight in males at the longest distance with this species. The conclusion was that the components do not have separate roles, but are perceived as a blend throughout the behavioral sequence. Although this also has been shown with various behavioral studies in other species, the search for individual roles for pheromone components continues to resurface as new investigators come into the field with their research on other species.

The Next Phase

The above sections on the struggles and challenges from the early days are mainly from the first decade of pheromone research in my group. "The Next Phase" was ushered into my laboratory when postdocs Lou Bjostad, Peter Ma and Russ Jurenka brought the research program on chemical communication systems to the molecular level. Bjostad first unraveled some of the mysteries of pheromone biosynthesis when he found that hundreds, if not thousands, of moth species commonly utilize specialized desaturases and limited chain-shortening reactions in their pheromone biosynthetic pathways. One interesting connection to the early days was that data on the biosynthetic pathway of the cabbage looper moth revealed unsaturated acyl intermediates of different chain lengths that predicted new pheromone components for this species. The components were then identified in gland extracts, bioassayed, and a new six-component blend was defined (Bjostad et al. 1984). These studies provided the basis for decades of research on gene characterizations and their evolution in the genome.

There have been too many excellent graduate students, postdocs and visiting scientists in my laboratory throughout decades of "The Next Phase" to discuss here, but they all contributed greatly to the expanding pheromone knowledge. As I look back at all the decades of research, an overriding theme that emerges to me involves the wonderful interactions that have occurred with pheromone children and grandchildren of those who have been in my lab. Of particular note is the expansive genealogy that has come from a fortuitous interaction with Christer Löfstedt when he visited our lab in the 1980s as a graduate student from Sweden. A collaboration was started that, over the years, has grown to include many of his students and then their students on various interesting pheromone projects.

Knowledge on pheromone production and perception is detailed in this book and showcases the tremendous advances made in understanding the basis of this chemical communication system. Many of the challenges of the past have been met, but the expanding knowledge of this system continues to generate new questions that keep challenging pioneering scientists at the frontiers to use creative new approaches and technologies for answers. It is no less exciting now than it was in the early days.

References Cited

Berger, R.S. 1966. Isolation, identification, and synthesis of the sex attractant of the cabbage looper, *Trichoplusia ni. Annals of the Entomological Society of America* 59:767–771.

Bjostad, L.B., Linn, Jr., C.E., Du, J.-W., and W.L. Roelofs. 1984. Identification of new sex pheromone components in *Trichoplusia ni*, predicted from biosynthetic precursors. *Journal of Chemical Ecology* 10:1309–1323.

Butenandt, A., Beckmann, R., Stamm, D., and E. Hecker. 1959. Über den Sexual-lockstoff des Seidenspinners *Bombyx mori* Reindarstellung und Konstitution. *Zeitschrift für Naturforschung B* 14:283–384.

Chapman, P.J., and S.E. Lienk. 1971. *Tortricid Fauna of Apple in New York*. Geneva, NY: New York State Agricultural Experiment Station.

Comeau, A., and W.L. Roelofs. 1973. Sex attraction specificity in the Tortricidae. *Entomologia Experimentalis et Applicata* 16: 191–200.

Foster, S.P., Clearwater, J.R., Muggleston, S.J., and W.L. Roelofs. 1986. Probable sibling species complexes within two described New Zealand leafroller moths. *Naturwissenschaften* 73:156–158.

Franclemont, J.G. 1980. "*Noctua c-nigrum*" in eastern North America, the description of two new species of *Xestia* Hübner (Lepidoptera: Noctuidae: Noctuinae). *Proceedings of the Entomological Society of Washington* 82:576–586.

Hendry, L.B. 1976. Insect pheromones: diet related? *Science* 192:143–145.

Hendry, L.B., Jugovich, J., Mumma, R.O., Robacker, D., Weaver, K., and M.E. Anderson. 1975a. The oak leaf roller (*Archips semiferanus* Walker) sex pheromone complex: field and laboratory evaluation of requisite behavioral stimuli. *Experientia* 31:629–631.

Hendry, L.B., Anderson, M.E., Jugovich, J., Mumma, R.O., Robacker, D., and Z. Kosarych. 1975b. Sex pheromone of the oak leaf roller: a complex chemical messenger system identified by mass fragmentography. *Science* 187:355–357.

Hendry, L.B., Wichmann, J.K., Hindenlang, D.M., Mumma, R.O., and M.E. Anderson. 1975c. Evidence for origin of insect sex pheromones: presence in food plants. *Science* 188:59–63.

Hindenlang, D.M., and J.K. Wichmann. 1977. Reexamination of tetradecenyl acetates in oak leaf roller sex pheromone and in plants. *Science* 195:86–89.

Karlson, P., and M. Lüscher. 1959. "Pheromones": a new term for a class of biologically active substances. *Nature* 183:55–56.

Kennedy, J.S. 1972. The emergence of behaviour. *Australian Journal of Entomology* 11:168–176.

Kennedy, J.S., and D. Marsh. 1974. Pheromone-regulated anemotaxis in flying moths. *Science* 184:999–1001.

Klun, J.A., and T.A. Brindley. 1970. *cis*-11-Tetradecenyl acetate, a sex stimulant of the European corn borer. *Journal of Economic Entomology* 63:779–780.

Kochansky, J., Cardé, R.T., Liebherr, J., and W.L. Roelofs. 1975. Sex pheromones of the European corn borer in New York. *Journal of Chemical Ecology* 1:225–231.

Linn, C.E., Jr., Campbell, M.G., and W.L. Roelofs. 1986. Male moth sensitivity to multicomponent pheromones: the critical role of the female released blend in determining the functional role of components and the active space of the pheromone. *Journal of Chemical Ecology* 12:659–668.

Linn, C.E., Jr., Campbell, M.G., and W.L. Roelofs. 1987. Pheromone components and active spaces: what do male moths smell and where do they smell it? *Science* 237:650–652.

Lintner, J.A. 1882. A new principle in protection from insect attack. *Western New York Horticultural Society Proceedings* 27:52–66.

Miller, J.R., Baker, T.C., Cardé, R.T., and W.L. Roelofs. 1976. Reinvestigation of oak leafroller sex pheromone components and the hypothesis that they vary with diet. *Science* 192:140–143.

Roelofs, W.L., and H. Arn. 1968. Sex attractant of the red-banded leafroller moth. *Nature* 219:513.

Roelofs, W.L., and R.L. Brown. 1982. Pheromones and evolutionary relationships of tortricidae. *Annual Review of Ecology and Systematics* 13:395–422.

Roelofs, W.L., and A. Comeau. 1969. Sex pheromone specificity – taxonomic and evolutionary. *Science* 165: 398–400.

Roelofs, W.L., and A. Comeau. 1970. Lepidopterous sex attractants discovered by field screening tests. *Journal of Economic Entomology* 63:969–974.

Roelofs, W.L., and A. Comeau. 1971. Sex attractants in Lepidoptera. Pp. 91–114. In A.S. Tahori, ed. *Chemical Releasers in Insects*. New York: Gordon & Breach.

Roelofs, W.L., and A.P. Rooney. 2003. Molecular genetics and evolution of pheromone biosynthesis in Lepidoptera. *Proceedings of the National Academy of Sciences of the United States of America.* 100:9179–9184.

Roelofs, W.L., Comeau, A., Hill, A., and G. Milicevic. 1971. Sex attractant of the codling moth: characterization with electroantennogram technique. *Science* 174:297–299.

Roelofs, W.L., Du, J.-W., Tang, X.-H., Robbins, P.S., and C.J. Eckenrode. 1985. Three European corn borer populations in New York based on sex pheromones and voltinism. *Journal of Chemical Ecology* 11:829–836.

Schneider, D. 1962. Electrophysiological investigation on the olfactory specificity of sexual attracting substances in different species of moths. *Journal of Insect Physiology* 8:15–30.

Shorey, H.H. 1970. Sex pheromones of Lepidoptera. Pp. 249–284. In D.L. Wood, R.M. Silverstein, and M. Nakajima, eds. *Control of Insect Behavior by Natural Products*. New York: Academic Press.

Silverstein, R.M., Rodin, J.O., and D.L. Wood. 1966. Sex attractants in frass produced by male *Ips confusus* in ponderosa pine. *Science* 154:509–510.

Pheromones: Reproductive Isolation and Evolution in Moths

JEREMY D. ALLISON and RING T. CARDÉ

INTRODUCTION

REPRODUCTIVE ISOLATION: A "FUNCTION" OR "EFFECT"

MODELS OF PHEROMONE EVOLUTION IN MOTHS

Stabilizing preference functions

    (i) Stasis hypothesis

    (ii) Asymmetric tracking hypothesis

    (iii) Wallflower hypothesis

Directional preference functions

    (iv) Competitive signal evolution

Selective forces

    (v) Male preference functions

    (vi) Communication interference

EXISTING LITERATURE

Male preference function shape

Communication interference

Coordination of signaler and receiver traits

Variation in signaling traits and female mating
    success

SIGNIFICANCE OF PHEROMONE EVOLUTION

SUMMARY

REFERENCES CITED

Introduction

The location or recruitment of a mate is a pivotal event in sexual reproduction. Although asexual reproduction has evolved independently several times in the Lepidoptera (see Grapputo et al. 2005), sexual reproduction is an almost universal condition among moths. It is typically preceded by long-distance attraction of males and short-range courtship of females mediated by pheromones (Cardé and Haynes 2004). Despite a lack of diversification in larval ecology (i.e., in nearly all species the larvae are phytophagous), the Lepidoptera are among the most speciose insect orders with estimates of 174,250 species (156,300 of which are moths) (Mallet 2007). It has been hypothesized that one method of mitigating an Allee effect in mate finding (and ultimately the extinction of small populations) is the evolution of traits that increase mate-finding efficiency at low densities (e.g., pheromones) (Gascoigne et al. 2009). The ability to locate or recruit a mate at low densities as mediated by volatile pheromones has likely facilitated the persistence of populations of moths in evolutionary time and contributed to their phylogenetic success.

Early models of specific mate recognition systems (SMRS) in moths emphasized the role of single compound, species-specific pheromones (e.g., Shorey [1970] who argued that for a given species only one compound "has been selected behaviorally as the sex pheromone."). This was likely a consequence of two factors: (i) the first lepidopteran sex pheromone identified (bombykol) remains one of comparatively few examples of a single component pheromone and (ii) many multi-component lepidopteran sex pheromones have one component capable of inducing attraction alone and additional components that are usually present in much lower quantities and only augment attraction to the major component. As more moth sex pheromones were identified and analytical techniques became more sensitive, it became clear that the majority of moth pheromones were multi-component blends and that sympatric species with similar phenologies and calling periodicities often share components (Byers 2002; El-Sayed 2012). Chemical specificity in moth SMRS is now hypothesized to be a product of qualitative (e.g., presence/absence of components and differences in component double bond configuration or position, enantiomer configurations, chain length, and functional moieties) and often quantitative (ratios of components) differences in the pheromone blend. Although there is little debate about whether these differences can confer reproductive isolation, there is considerable debate how these differences evolve and whether these differences are involved in driving the speciation process or whether they are sequelae that follow divergence (see Paterson 1978; Templeton 1981; Lambert et al. 1987).

This chapter describes existing models of sex pheromone evolution in moths and the associated selective forces. It explores the question: "Is reproductive isolation a function or an effect of moth sex pheromones?" The available literature is then interpreted with respect to the predictions of these models and finally the significance of pheromone evolution in moths is discussed. In some cases, we have had to interpret the literature to develop explicit statements of models and their predictions. As a result, in some cases we may have attributed predictions/assumptions not intended by the authors of the relevant literature. Also, we have developed this chapter to emphasize general principles, but given the number of moths, exceptions undoubtedly exist.

Reproductive Isolation: A "Function" or "Effect"

Ever since Darwin (1859), evolutionary biologists have debated the process by which permanent reproductive isolation develops between populations (see Dobzhansky 1940; Mayr 1963; Paterson 1978). Resolution of this controversy is complicated by the fact that speciation usually occurs on a timescale that makes experimental manipulation and ultimately identification of the mechanisms involved at best difficult. Consequently, most studies compare taxa at various stages of divergence to attempt to identify the mechanisms involved in the evolution of reproductive isolation (e.g., Coyne and Orr 1989, 1997). Ideally, these studies involve populations of incipient species or closely related taxa that are isolated by traits that can be quantified with precision (Howard et al. 1998).

Reproductive isolation between populations can be the product of physical separation or prezygotic and postzygotic isolating mechanisms. Prezygotic isolating mechanisms mitigate the costs of hybridization and because of their positive fitness effects can be strengthened by natural selection, a process described as reinforcement (Dobzhansky 1940). Reinforcement can complete the speciation process when postzygotic barriers are incomplete (Dobzhansky 1940; Howard 1993; Noor 1999; Marshall et al. 2002). One of the potential outcomes of the process of reinforcement is reproductive character displacement, a pattern of greater divergence of an isolating character in sympatry than allopatry (Brown and Wilson 1957). At one time reproductive character displacement was considered uncommon (e.g., Littlejohn 1981); however, recent literature reviews suggest otherwise (Coyne and Orr 1989, 1997; Howard 1993; Noor 1999) and empirical tests in natural populations have detected the predicted pattern (Noor 1995; Sætre et al. 1997; Higgie et al. 2000; McElfresh and Millar 2001).

Although studies of reproductive isolating mechanisms can identify factors that isolate populations, a major challenge remains determining whether or not those factors were involved in speciation itself (e.g., White 1978). Opponents of speciation by reinforcement argue that there is little theoretical reason to suppose that reinforcement of differences in pheromone signals is involved in the initial stages of speciation (Paterson 1978; Templeton 1981). Paterson (1978, 1985), among others, has promoted the Recognition Concept, an alternate allopatric model. The recognition concept states "new species arise as the incidental consequence of differential adaptive evolution in isolated subpopulations of an original parental species" (Paterson 1978). This model of speciation is predicated on the belief that selection acts on the fertiliza-tion system, including the SMRS, to promote efficient signaling between potential mates, *not* to isolate one population from another. When isolation does evolve, it is a consequence of local adaptation to the signaling environment. Ultimately, this model considers reproductive isolation an *effect* of differences in the SMRS and that the *function* of these differences is efficient signaling between the sexes (*sensu* Otte 1974).

The recognition concept places an emphasis on coadaptation of signal and response traits and predicts that stabilizing selection will act to maintain the functionality of the SMRS. As a result, variation in female moth sex pheromone signals should be limited by male preference function shape. Lambert et al. (1987) conducted a case study of sex pheromones and concluded that "pheromones are generally highly specific in action and show low compositional and proportional variability." Their conclusion that sex pheromones have low proportional variability is surprising given the abundance of lepidopteran examples in their case study. The majority of studies that have characterized variation in moth sex pheromones have reported high levels of proportional variation (Allison and Cardé, this volume). The recognition concept has been challenged (e.g., Coyne et al. 1988; Ryan and Wilczynski 1991) on the grounds that significant levels of variation have been observed in SMRS. Paterson (1993) responded to these challenges, stating that (i) "the species studied were not delineated on the criteria of the recognition concept" and (ii) "the samples of signals used to demonstrate variability were not postselection samples."

The first criticism appears to be focused on the idea that many of the examples demonstrating variation in the SMRS are not reliable because they define species in taxonomic terms and the recognition concept is based on a genetic species concept. Paterson (1993) suggests that the examples used to challenge the recognition concept involve taxonomic species and as a result may be collections of genetic species (i.e., cryptic species). If true, this would result in the overestimation of variation in SMRS characters. Given that in the majority of moths examined, abundant variation has been reported, it is unlikely that all (or even a few) of these examples involve cryptic species. Additionally, in many of these cases research colonies have been established from a limited number of field-collected individuals, making it improbable that cryptic species were present and not recognized (i.e., given the limited numbers of individuals early generation matings would likely have revealed incompatibilities).

The second criticism argues that the true measure of variability in the SMRS is not the total amount of variation present in a population but rather the variation present in those individuals that actually mate. Paterson (1993) argues that to demonstrate variability in the SMRS, future studies need to estimate the variation among individuals that actually contribute to the next generation. If this criticism is valid, then female moths with aberrant (unattractive) signals should be less likely to attract a mate (see "Variation in Signaling Traits and Female Mating Success" below). Paterson (1993) also pointed out that the recognition concept implicitly assumes variation and that receiver preference function shape will determine the amount of variation in mating signals. The available literature for moth sex pheromones provides mixed support for this prediction (see "Male Preference Functions" below).

The dichotomy between the selective forces associated with speciation by reinforcement and adaptation to the signaling environment (e.g., the recognition concept) does not appear to exist for moth sex pheromones. Paterson (1993) uses the

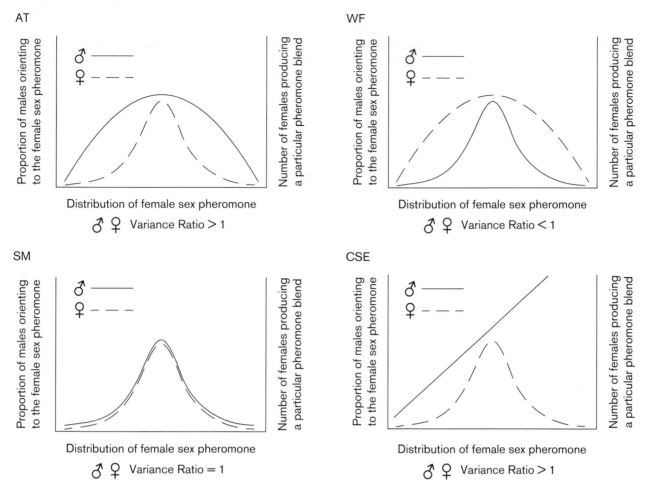

FIGURE 2.1 Graphical representation of the receiver preference function relative to the distribution of female pheromone signals for each model of moth pheromone evolution. The ratio of variance in male response and female signaling traits is >1 for the Asymmetric Tracking (AT) and Competitive Signal Evolution (CSE) models, <1 for the Wallflower Hypothesis (WF), and approximately 1 for the Stasis Model (SM).

example of the frogs *Heleophryne purcelli* and *Xenopus laevis* to illustrate adaptation to the signaling environment. The former species occurs in the rapids of streams in gorges and has a high-energy call, whereas the latter species occurs in slower moving streams and has a low-energy call (see Paterson 1993 and references therein). Paterson interprets these differences as evidence that selection has acted primarily to shape the male call to facilitate efficient signaling (i.e., *H. purcelli* has a high-energy call so that males can be heard above the background noise). In the case of moth sex pheromones, the primary factor that could generate selection for adaptation to the signaling environment would be "noise" from heterospecific (or incipient species) females that use the same or similar pheromone components. In moths it appears that the selective forces for adaptation to the signaling environment and speciation by reinforcement are the same.

Another major challenge to the recognition concept is the existence of behavioral antagonists. These compounds are released from the pheromone gland and often have no behavioral effect on conspecific males; however, their presence, in even small amounts, can eliminate responses of males to conspecific pheromone blends (e.g., Löfstedt et al. 1990). Males of some species have components of the peripheral and central nervous system that are tuned specifically to behavioral antagonists and separate from those that process pheromone components (see Cardé and Haynes 2004). As has been

pointed out by other authors (e.g., Löfstedt 1993), it is difficult to explain the evolution and maintenance of these traits without communication interference as a selective force.

Although a few studies have observed geographical patterns of variation in moth sex pheromones consistent with reproductive character displacement (e.g., Cardé et al. 1977; Thompson et al. 1991; McElfresh and Millar 1999, 2001; Gries et al. 2001), in no instance is it possible to determine if speciation and divergence in the sex pheromones occurred simultaneously or if divergence in the sex pheromones occurred after speciation. Communication interference assumes that the selective disadvantage of heterospecific attraction is not reduced hybrid viability or wasted mating effort. Rather it is the associated opportunity costs and predation risks of heterospecific courtship (Cardé and Haynes 2004).

Models of Pheromone Evolution in Moths

Several models of sex pheromone evolution in moths have been proposed. These models can be differentiated on the basis of the predicted shape of the receiver preference function relative to the distribution of female pheromone signals (i.e., the ratio of variance in male response and female signaling traits) (see figure 2.1). The Wallflower, Asymmetric Tracking, and Stasis models (hereafter WF, AT, and SM, respectively) predict that male preference functions will be stabilizing. The WF and AT hypotheses emphasize an asymmetry in the relative shapes of the preference function and signal distribution. While the AT hypothesis predicts that male preference functions will be broad relative to the signal distribution (male:female variance ratio > 1), the WF hypothesis predicts the opposite (male:female variance ratio < 1). Alternatively, the SM predicts symmetry between the preference function shape and signal distribution (male:female variance ratio ≅ 1). The principal difference between the SM and the AT and WF hypotheses is that the SM predicts that male preference function shape will determine the female distribution of pheromone signals, whereas the AT and WF hypotheses predict that the distribution of female pheromone blends will determine male preference function shape (albeit for different reasons; see below). The Competitive Signal Evolution hypothesis (hereafter the CSE) predicts directional preference functions, with receiver preferences for increasingly exaggerated signals. Like the SM, the CSE predicts that receiver preference function shape will determine the distribution of female pheromone signals.

To maintain signal function, change in communication systems requires a mechanism to facilitate parallel changes in signal and receiver traits (Butlin and Ritchie 1989). Two mechanisms that have been proposed are a genetic correlation and coevolution between signal and response traits. The former would exist because of coupled genetic control of signal production and reception traits due to either pleiotropy or linkage disequilibrium. Coevolution of signal and response traits predicts stepwise reciprocal change between genetically independent signaler and receiver traits. The initial change could be due to drift or external directional selection (e.g., communication interference) and would be followed by complementary change in appropriate signaler or receiver traits as a result of selection generated by the initial change. The WF, AT, and SM hypotheses of moth sex pheromone evolution rely on coevolution of signal and response traits to maintain signal function. The CSE hypothesis of signal evolution predicts the existence of a genetic correlation between signal and response traits.

Stabilizing Preference Functions

(I) STASIS HYPOTHESIS

This model of female sex pheromone evolution in moths is an emergent property of the hypothesis that selection should favor male preferences for (Butlin and Ritchie 1989; Groot et al. 2006) and maximum sensitivity to (Roelofs 1978; Linn et al. 1987) the median pheromone blend and the assumption that high sensitivity to one pheromone signal occurs at the expense of reduced sensitivity to other signals (see "Male Preference Function Shape"). Variation in the absolute and relative amounts of pheromone blend components could then result in variation in female mating success and ultimately stabilizing selection on the pheromone blend (e.g., Butlin and Ritchie 1989).

The SM predicts that coordination of pheromone signal and response traits occurs as a result of reciprocal stabilizing selection. This coevolutionary process would favor males most responsive to the median pheromone blend produced by females and females that produce blends that coincide with the male preference function optimum (Cardé and Baker 1984). This process can explain "fine-tuning" of the pheromone blend, but it should make the pheromone blend resis-

tant to change (e.g., Phelan 1992; Groot et al. 2006). Other mechanisms (e.g., selection for adaptation to the signaling environment as predicted by the recognition concept) are necessary to explain the evolution of novel pheromone blends and ultimately the diversity of moth pheromone signals.

(II) ASYMMETRIC TRACKING HYPOTHESIS

This model was developed as an alternative to the SM and is predicated on the belief that the existence of strong mutually stabilizing selection between signal and response traits is not consistent with the predictions of sexual selection theory. The AT hypothesis predicts an asymmetry in the strength of selection that female signal and male response traits experience as a consequence of differential investment in the zygote (Trivers 1972). Specifically, the AT hypothesis predicts the following:

1. Because of this asymmetry, the optimal reproductive strategy differs between the sexes. Assignment of signaler and receiver roles will be determined by their relative costs and the limiting sex will usually assume the less costly role. The energetic costs and predation risks of producing and releasing minute quantities of sex pheromones are likely low compared to the costs of searching for a pheromone point source and have been proposed to explain the signaling role reversal in Lepidoptera (Cardé and Baker 1984).
2. Males are expected to be maximally responsive to the median pheromone blend while maintaining a response window broad enough to include most, if not all, potential mates. This male preference function shape will impose weak or no selection on the female pheromone signal, allowing for the existence of significant amounts of variation in the pheromone signal (in the absence of other selective forces). As a result, evolution of the female sex pheromone will be driven primarily by stochastic factors and directional selection associated with external factors.
3. Male preference function shape will be determined by the distribution of female pheromone signals and will "track" changes in the distribution of female pheromone signals (Phelan 1992, 1997). The mechanism for the initial change in the signal trait remains speculative but could be drift (more likely if preference functions impose no selection than if they generate weak selection) or directional selection generated by communication interference.

A variant of the AT hypothesis, the "rare male" hypothesis, predicts rapid evolution via large changes in the pheromone blend and proposes that the mechanism for the initial change in the signal trait involves gene duplication and mutations (Roelofs et al. 2002). As a result of the gene duplications and mutations, females with novel pheromone blends arise. The persistence and evolution of these mutant lineages would be facilitated by the existence of "rare males" capable of responding to the novel pheromone blend (Roelofs et al. 2002). In support of the rare male hypothesis, the available evidence suggests that (1) the pheromone blend in moths is controlled by one or a few genes (e.g., Löfstedt et al. 1990; Haynes, this volume); (2) subtle changes (mutations) in these genes can result in significant changes in the pheromone blend produced (Lassance et al. 2013); (3) duplication events of genes involved in pheromone biosynthesis have been documented (e.g., Roelofs et al. 2002); and (4) the existence of rare males has been reported for both the European corn borer, *Ostrinia nubilalis* (Lepidoptera: Crambidae) (Roelofs et al. 2002; Linn et al. 2003) and the Asian corn borer, *O. furnacalis* (Lepidoptera: Crambidae) (Linn et al. 2007). In addition to the pheromone blend of conspecific females, these males also orient to the pheromone blend of *O. furnicalis* or *O. nubilalis*, respectively. In the case of *O. nubilalis*, selection experiments were able to increase the frequency of these rare males from 5% to 70% in 19 generations (Droney et al. 2012). In the absence of an external selective force (e.g., communication interference), it would appear that the alternate outcomes of extinction of the mutant pheromone lineage or introgression with the wild type would be more likely than the evolution of a new pheromone signal.

Like the AT hypothesis, the rare male hypothesis predicts that male preference functions will be broad enough to accommodate change in the female pheromone blend and that male response traits will track changes in female signaling traits. It differs from the AT hypothesis which predicts that variation among males in breadth of response should be low, in that it depends on intra-male variation in preference function shape (i.e., rare males). Like the WF hypothesis (see below) and unlike the SM, the AT (and rare male) hypothesis provides a mechanism to reconcile the conflict between the need to maintain coordination between signal and response traits to preserve signal function and divergence of mate recognition systems during speciation.

(III) WALLFLOWER HYPOTHESIS

This model of signal evolution emphasizes the relative variances of signal and response traits and the female fitness costs related to preference-related risks of remaining unmated (De Jong and Sabelis 1991). Unlike some other treatments of sexual selection on mating signals that predict runaway sexual selection, simulations of this model predict that, although short bouts of runaway selection are possible, persistent runaway selection is not an emergent property of the WF hypothesis. De Jong and Sabelis (1991) conclude that at equilibria (i.e., steady state conditions) the ratio of male to female variance will be less than unity and that this is due to differential male responses to variation in female sex pheromones.

Although the WF hypothesis predicts that female signals that differ significantly from the population median will be less attractive to males, sexual selection is expected to be weaker on females than males because of the wallflower effect. The wallflower effect occurs because in populations with a long enough, discrete mating season and females that mate only once or a few times, the distribution of available signal phenotypes is predicted to change during the mating season. As a result, the frequency of females with less attractive signals (i.e., wallflowers), and their likelihood of attracting a mate, will increase. Ultimately, these increases will reduce the strength of selection on female signals. This model assumes that males increase their fitness by mating multiply and that females are constrained in the number of times they can mate.

The WF hypothesis predicts that sexual selection can promote speciation via the following process: (i) a founder event populated primarily by females with outlier or rare (unattractive) signal phenotypes is derived from a source population at equilibrium (i.e., with a low male-to-female variance ratio); (ii) because of the wallflower effect, selection is stron-

ger on male response than female signal traits and male preference functions are predicted to evolve to match the founder female signal distribution; and (iii) with enough isolation, founder male preference functions may differ from the source population enough that founder males are unlikely to respond to females from the source population. Although the wallflower effect provides a mechanism for females with outlier or rare pheromone phenotypes to attract a mate, these females will experience a delay in or be less likely to attract a mate. As a result, the alternate outcome of extinction of these founder populations may be more likely than the evolution of a novel signal and speciation.

Directional Preference Functions

(IV) COMPETITIVE SIGNAL EVOLUTION

It has been argued that aspects of (i) insect SMRS are consistent with male courtship as a form of intraspecific competition for mates and (ii) male courtship signals can increase male mating success relative to conspecifics (West-Eberhard 1983, 1984). The CSE model emphasizes intraspecific social competition (West-Eberhard 1983, 1984) hypothesized to exist as a consequence of differential investment by males and females in the zygote. Individuals are predicted to compete with conspecifics for access to mates. Because males usually invest less in the zygote, they are predicted to take on the more active, costly, and competitive role in communication (Trivers 1972), usually signaling. In moths, the costs of mate searching are hypothesized to be high relative to signaling costs and likely explain the signaling role reversal.

It has been argued that competitive signal evolution is a "virtually self-perpetuating" process that occurs when (i) there is signaling competition among individuals (theoretically always the case if there is an asymmetry in parental investment by the sexes) and (ii) phenotypic variation exists among signalers (West-Eberhard 1983, 1984). Although there is abundant phenotypic variation in moth sex pheromones (Allison and Cardé, this volume), the existence of signaling competition among female moths is debatable and it is not clear how relevant the CSE model is for sex pheromones in moths. If female pheromone signaling in moths is competitive, (1) variation in female signaling traits should result in differential female mating success and (2) the ratio of variance in male response and female signal traits should be much larger than unity and male preference functions should be directional.

Selective Forces

Selection acting on the sex pheromone blend is hypothesized to be generated primarily by conspecific male preference functions and communication interference from sympatric species with similar pheromone blends. Communication interference could create directional selection on the sex pheromone signal (although the aggregate effect of communication interference from multiple species could be stabilizing) and result in reproductive character displacement (*sensu* Butlin 1987). The type of selection generated by conspecific male preference functions depends on male preference function shape relative to the distribution of female pheromone signals. The effects of the two selective forces are not necessarily independent.

Although communication interference has primarily been assumed to generate selection on the female sex pheromone (e.g., Groot et al. 2006), it may also generate selection on male preference function shape. For example, if the costs of hetero-specific courtship are high relative to mate searching costs, then communication interference should generate selection on male preference function shape.

(V) MALE PREFERENCE FUNCTIONS

Over 140 years ago, Darwin (1871) speculated that the mating preferences of the responding sex could generate selection on sexual signals. In addition to determining the type and strength of selection (e.g., Ritchie 1996; Greenfield 2002), preference function shape has been used to infer the function of sexual signals. Unimodal preferences have been interpreted as evidence for a species recognition function and directional preferences have been interpreted as evidence for Fisherian and "good gene" functions (e.g., O'Donald 1980; Ewing and Miyan 1986; Hoikkala and Aspi 1993). The reliability of this approach has been questioned (Ritchie 1996).

Open-ended preference functions favor signal values greater (or less) than the mean signal value and exert directional selection (Ryan and Keddy-Hector 1992; Ryan and Rand 1993, 2003; Gerhardt 1994). They are most common for signals with high energetic costs (e.g., acoustic signals) and because of these high energetic costs they may be "honest" (Zahavi 1975) indicators of mate quality (e.g., Brandt and Greenfield 2004; Hunt et al. 2004). Unimodal preference functions favor signal phenotypes within a particular range and select against signals outside the range. The type and strength of selection they exert depends on the relationship of the optimum to the distribution of signal phenotypes. When the maximum of the preference function and the mean signal value are coincident, selection will be stabilizing. When the mean signal value and preference function optimum do not coincide but the preference function optimum occurs within the distribution of signal phenotypes, selection will have directional and stabilizing components (Ryan and Keddy-Hector 1992). Initially selection would be directional, favoring phenotypes coincident with the preference function optimum. Once the signal mean and optimum were coincident, selection would be stabilizing. When the preference function optima occur outside the distribution of signal phenotypes, selection will be directional (Ryan and Rand 1993). In a few cases, flat preference functions have been observed (Houde 1988; Westerdahl 2004), indicating the absence of detectable mating preferences. True absence of a mating preference would result in no selection on the pheromone signal.

The WF and AT hypotheses both emphasize an asymmetry in parental investment of the two sexes. As a result, they both predict that sexual selection will be stronger on male response than female signal traits and that male preference functions will track the female signal phenotype distribution. Although both hypotheses assume that male preference functions are stabilizing, they differ in the predicted ratio of variance in male response and female signal traits. The AT hypothesis predicts that the ratio of male to female variance will be larger than unity and the WF hypothesis predicts it will be smaller than unity. In essence, the AT hypothesis predicts that male preference functions will be broad relative to the female signal phenotype distribution and the WF hypothesis that the male preference function will be narrow relative to the female sig-

nal distribution (see figure 2.1). Despite the expected differences in relative variance in male preference functions, both models predict that male preference functions generate weak or no selection acting on female sex pheromones in moths. This is so in the AT hypothesis because males are not expected to exhibit much preference over the range of female signals produced, and in the WF hypothesis because of changes in the distribution of female pheromone signals over the mating season (i.e., the wallflower effect). The SM also predicts that receiver preference functions will be stabilizing; however, unlike the WF and AT models, the SM model predicts that at equilibrium the ratio of variance in male response to female signal traits will approach unity. As a result, the SM model predicts that receiver preference functions will generate intermediate to strong selection on signal traits that will constrain signal evolution and limit variation in moth signal traits.

Like the WF and AT models, the CSE hypothesis also emphasizes differential parental investment in the zygote. Unlike the WF and AT models that predict that competition for mates will occur among receivers, the CSE model predicts that competition occurs among signalers. As a result, the CSE predicts that receiver preference functions generate intermediate to strong selection on signaler traits. Unlike the other models, the CSE model predicts directional preference functions and has the potential to generate rapid ("runaway") evolution of sexual signals (see Fisher 1930; Lande 1981). If signalers with the most developed signal traits are more successful in obtaining mates, selection should favor receiver preferences for exaggerated traits due to the advantage of producing progeny that signal more successfully. This could lead to increasingly strong selection on signal traits and a genetic correlation between signal and response traits (Fisher 1930; Lande 1981). The exaggeration of sexual signals would be opposed, and ultimately stopped by natural selection generated by costs associated with exaggerated signals (e.g., energetic, increased predation risks).

(VI) COMMUNICATION INTERFERENCE

The phenomenon of reproductive isolation is implicit to the Biological Species Concept (Mayr 1963) and can be the result of prezygotic and postzygotic isolating mechanisms (Dobzhansky 1937). Butlin (1987) suggested that reinforcement be limited to cases where hybridization can occur and is selected against. Reproductive character displacement would then be a process of enhancing prezygotic barriers between populations isolated by postzygotic mechanisms. Following this definition, reinforcement would be a part of some forms of speciation and reproductive character displacement would only occur when taxa were completely isolated by postzygotic isolating mechanisms (Butlin 1987).

For male moths, the costs associated with communication interference are hypothesized to occur as a result of the energetic and opportunity costs and increased rates of mortality associated with orienting to and courting heterospecific females. For females, selection on the pheromone signal would occur as a result of a reduced probability of, or delay in, attracting a conspecific male as a consequence of courtship by heterospecific males. Communication interference would then be a mechanism for reproductive character displacement (*sensu* Butlin 1987) of pheromone signals. As a selective force, communication interference has been hypothesized to explain qualitative and quantitative differences in the pheromone signals of moths that share components as well as the existence of behavioral antagonists. Both reinforcement and communication interference could generate selection for divergence in pheromone signals and result in patterns of pheromone variation consistent with reproductive character displacement. With the exception of incipient species undergoing speciation, in most cases it will be difficult to determine if divergence in pheromone signals occurred as a result of selection generated by hybrid disadvantage or communication interference (reinforcement vs. reproductive character displacement [*sensu* Butlin 1987]).

Existing Literature

Male Preference Function Shape

The shape of the male preference function describes the likelihood that a signal phenotype (e.g., a pheromone blend) will attract a mate (Ritchie 1996). Preference function shape could reflect variation among individual males (i.e., responses of individual males cumulatively create the breadth of response of a population) or variation among males may be low (i.e., individual male preference functions are representative of the population). Sexual selection has been hypothesized to act more strongly on male than female signaling traits, and as a result males would have lower variance in their signal responses than females in signal production (e.g., De Jong 1988). Consequently, while the distribution of female pheromone signals would reflect variation among individuals, the breadth of individual male preference functions should reflect that of the population. An attraction–mark–re-attraction assay was used to score and compare the response phenotype of individual males in natural populations of the oriental fruit moth, *Grapholita molesta* (Tortricidae) (Cardé et al. 1976) and the pink bollworm, *Pectinophora gossypiella* (Gelechiidae) (Haynes and Baker 1988). Neither study found any evidence of variation (in response phenotype) among males, suggesting that male preference function shape is similar at the individual and population level. Conversely, variation among males in preference function shape (i.e., rare males with broad preference functions) has been reported in *Ostrinia nubilalis* and *O. furnicalis* (Roelofs et al. 2002; Linn et al. 2003, 2007).

It has been suggested that a negative correlation between breadth of response and sensitivity may be a common feature of the insect peripheral nervous system and may explain the rarity of generalist natural enemies that use moth sex pheromones as foraging cues (Haynes and Yeargan 1999). Simulations of a population genetics model suggest that when a trade-off exists between sensitivity and breadth of response, that selection should favor a narrower preference function with high sensitivity to common signals and low sensitivity to uncommon signals (Butlin and Trickett 1997). In the absence of this trade-off, the optimal male preference function shape would appear to be high sensitivity to the entire range of signals defined by the male preference function. The unimodal preference functions predicted by the existence of a trade-off could generate stabilizing, directional, or no selection depending on the distribution of the female pheromone signals (see "Male Preference Functions" above). The flat preference functions predicted by the absence of a trade-off suggest that male mating preferences do not generate selection on the sex pheromone signal.

In the cabbage looper, *Trichoplusia ni* (Noctuidae), a single gene mutation with major effects on the female pheromone blend but no effects on male preference was discovered (Haynes and Hunt 1990). Selection over 49 generations of rearing resulted in mutant males equally responsive to the mutant and wild type pheromone blends (Liu and Haynes 1994). Hemmann et al. (2008) examined the relationship between sensitivity and breadth of response in mutant and wild type males and documented evidence of a trade-off between sensitivity and breadth of response. Wild type males were more sensitive than mutant males to the wild type pheromone blend but unlike mutant males did not respond well to the mutant blend. As a result, mutant males would be at a competitive advantage when competing for mutant females but at a disadvantage when competing for wild type females. Although hybrid and pure wild type males had similar blend preferences, wild type males were more sensitive than hybrid males (Hemmann et al. 2008).

The available literature on the type and strength of selection male moth preference functions generate on female sex pheromones is limited. Stabilizing male preference functions acting on the amount of pheromone (e.g., Linn et al. 1984; Valeur and Löfstedt 1996; but see Linn et al. 1986) and the number of components (e.g., Linn et al. 1984; Valeur and Löfstedt 1996) have been documented in several systems. With respect to the pheromone blend, several examples of broadly unimodal male moth preference functions exist. Unfortunately, these studies are difficult to interpret. In some cases, the male preference function is characterized over a range of pheromone blends that occur primarily outside the known distribution of female pheromone blends and/or few blends tested occur within the range of female-produced blends. In these cases although the male preference function may be unimodal for the range of pheromone blends tested, male preference functions may be open-ended or flat over the range of blends actually produced by females and generate directional or no selection, respectively. In other cases, the distribution of pheromone signals produced by females is not well characterized, making it impossible to infer the type and strength of selection generated by male preference functions.

The female pheromone blend distribution is known and male preference function adequately (i.e., many of the blends used to describe the preference function occur within the range of signals produced by females) described with wind-tunnel and field-trapping trials, in only a few systems. In the redbanded leafroller, *Argyrotaenia velutinana* (Tortricidae), and oriental fruit moth, *G. molesta*, the female pheromone blend distribution (Miller and Roelofs 1980; Lacey and Sanders 1992) and male orientation behavior in field trials and wind-tunnel bioassays (Roelofs et al. 1975; Linn and Roelofs 1983) suggest that the female pheromone signal experiences stabilizing selection as a consequence of male preference function shape. In the pink bollworm, *P. gossypiella*, comparison of the female pheromone blend distribution (Collins and Cardé 1985) and response specificity of males in wind-tunnel bioassays (Linn and Roelofs 1985) suggests that male preference functions generate no selection or directional selection on the female pheromone signal. In the almond moth, *Cadra cautella* (Pyralidae), the distribution of female pheromone signals was characterized (Allison and Cardé 2006) and the preference function shape described for this distribution of signals with choice and no-choice tests (Allison and Cardé 2008). Surprisingly, choice and no-choice tests provided different estimates of male preference function shape (unimodal in choice and flat in no-choice). The no-choice trials are likely to be more representative of natural conditions (i.e., males are unlikely to frequently have the opportunity to select among multiple receptive females), suggesting that in this system the female pheromone signal experiences little or no selection as a result of male preferences.

The AT, SM, and WF models can be differentiated from the CSE model on the basis of male preference function shape (the former predict unimodal male preference functions and the latter open-ended; see figure 2.1). To date, open-ended male moth preferences for pheromone blends have not been documented, suggesting that the CSE model has little application for the evolution of moth pheromones. As a consequence of the predicted trade-off between breadth of response and sensitivity, the AT, SM, and WF models all predict that the male preference function optima will be coincident with the most common female pheromone signal. Discrimination among the AT, WF, and SM models requires knowledge of the male preference function shape relative to the distribution of female pheromone signals. The three models predict different ratios of variance in male response and female signaling traits (AT – male:female variance ratio > 1; WF – male:female variance ratio < 1; SM – male:female variance ratio $\cong$ 1). The available literature provides mixed support for the AT hypothesis (pink bollworm and almond moth in no-choice assays) and SM hypothesis (oriental fruit moth, redbanded leafroller, and almond moth in choice assays). Additional studies that characterize the distribution of pheromone blends produced by females and then determine male preference function shape over that range of pheromone blends are needed before generalities can begin to emerge.

Communication Interference

Communication interference between sympatric species that use the same or similar pheromone components is hypothesized to generate directional selection and result in reproductive character displacement or reinforcement depending on the degree of postzygotic isolation (*sensu* Butlin 1987). For the responding sex, selection would be generated by the increased predation risks and energetic and opportunity costs of orienting to heterospecific females. The selective basis of communication interference for the signaling sex would be the attraction of heterospecific males that harass calling females, interfere with courtship by conspecific males, or result in nonfertile matings. Communication interference has been proposed as an explanation for sympatric species of moths that share some pheromone components but differ in other ways (e.g., optimal component ratios for attraction, use of exclusive constituents) and the existence of behavioral antagonists.

Patterns of variation in moth sex pheromone signals and responses that are consistent with reproductive character displacement have been reported in a few moth species (Cardé et al. 1977; Thompson et al. 1991; McElfresh and Millar 1999, 2001; Gries et al. 2001). Directional selection associated with communication interference has been proposed as a mechanism to overcome stabilizing selection associated with the SM of moth pheromone evolution. Groot et al. (2006) hybridized and backcrossed two closely related moth species to generate females with titers of minor acetate ester components representative of heterospecific females. Field-trapping experiments demonstrated that these females attracted 10 times more heterospecific males than normal wild type females. Groot et al. (2006) estimated that this level of communication interference would generate directional selection strong enough to overcome the stabilizing selection associated with the SM. The importance of communication interference as a mechanism for the divergence of prezygotic isolating mechanisms (i.e., the diversification of moth pheromone signals) remains to be determined.

Coordination of Signaler and Receiver Traits

To maintain signal function, evolutionary change in communication systems requires parallel change in receiver and signaler traits. As a result, communication systems are hypothesized to be resistant to change (e.g., Butlin and Ritchie 1989). Despite this, empirical support exists for the rapid evolution of prezygotic isolation and speciation due to divergent selection on signal and response traits (e.g., Kaneshiro and Boake 1987; Coyne and Orr 1989, 1997; Gleason and Ritchie 1998; Cain et al. 1999; Dieckmann and Doebeli 1999; Higashi et al. 1999; Kondrashov and Kondrashov 1999; Gray and Cade 2000; Svensson et al. 2006). Lacking is a consensus on the mechanism of divergence (see Butlin and Ritchie 1989).

Two mechanisms that have been proposed to maintain coordination of signal and response traits during signal evolution are a genetic correlation and coevolution of signal and response traits. The former exists as a result of coupling of genetic control of signal and response traits due to either pleiotropy or linkage disequilibrium, whereas the latter predicts stepwise reciprocal changes of genetically independent signal and response traits. The evolutionary force for the initial change could be drift or external directional selection and would be followed by complementary change in corresponding signal or response traits as a result of selection generated by the initial change. One method of testing for a genetic correlation between signal and response traits is to hybridize signal or response phenotypes and look for the predicted pattern of phenotypes in the hybrid progeny. Alternatively, selection experiments can be used to select for change in signal and response traits and look for predicted coincident change in the corresponding signal or response trait. While some studies have found evidence of a genetic correlation between signal and response traits (Collins and Cardé 1989a), others have not (Löfstedt et al. 1989; Allison et al. 2008).

To date, there is no evidence of pleiotropic effects of genes influencing signal and response traits (see Haynes, this volume). Linkage disequilibrium is maintained by assortative mating and evidence of pheromone-mediated assortative mating in moths has been reported for host races (Pashley et al. 1992; Emelianov et al. 2003; Bethenod et al. 2005; but see Takanashi et al. 2005) that are likely the by-product of a host shift (e.g., Rice 1984, 1987; Craig et al. 1993; Feder et al. 1994). Zhu et al. (1997) reported pheromone-mediated assortative mating between wild type and a mutant strain of *Trichoplusia ni*. This is a unique example involving a mutation with large effects and may not be representative of moths. Two studies have used attraction–mark–recapture field bioassays to test for pheromone-mediated assortative mating in natural populations of the oriental fruit moth (Cardé et al. 1976) and the pink bollworm (Haynes and Baker 1988). Neither found any evidence of male phenotypes favoring particular ratios.

Several hybridization studies have observed signal and response phenotypes intermediate to parental phenotypes in the F1 (Hoy and Paul 1973; Hoy et al. 1977; Klun and Maini 1979; Doherty and Gerhardt 1983). These results are necessary but not sufficient to demonstrate a genetic correlation.

The existence of F1 phenotypes intermediate to the parental phenotypes is consistent with both the existence of a genetic correlation and the alternate hypothesis of genetic independence with primarily additive genetic effects or simple intermediate dominance underlying signal and response traits. Discrimination between these alternate hypotheses would require examination of the F2 generation or higher order backcrosses or hybrids (see Butlin and Ritchie 1989). Löfstedt et al. (1989) crossed E-strain female and Z-strain male *Ostrinia nubilalis* and the F1 males were backcrossed to E-strain females to generate B1 generation hybrids. Gas chromatographic analyses of B1 female pheromone glands and electrophysiological recordings of B1 males found no evidence of a genetic correlation between signal and response traits in *O. nubilalis*.

Linkage disequilibrium would decrease by 50% per generation of random mating. As a result, random mating would interfere with an indirect response in mate preference to selection on the pheromone blend and the detection of a genetic correlation between signal and response traits. It has been suggested that in studies that use selection experiments to test for genetic correlations the use of experimental designs that impose random mating may explain the absence of a genetic correlation (Pomiankowski and Sheridan 1994). Allison et al. (2008) used a design that simulated male mate choice (i.e., females from high or low ratio selection lines were crossed to non-siblings from the same lineages) and did not observe a genetic correlation between signal and response traits in three selection lines. Conversely, Collins and Cardé (1989a) used a similar design that also simulated male mate choice and observed that in a line of pink bollworms selected for a high percentage of (Z,E)-7,11-hexadecadienyl acetate, the duration of wing-fanning in males to high percentage blends increased.

One intriguing potential example of coevolution of signal and response traits has been reported in the genus *Ostrinia* (Roelofs et al. 2002; Linn et al. 2003, 2007; Droney et al. 2012). Phylogenetic and molecular studies of pheromone biosynthesis in *Ostrinia* moths have demonstrated that differences among species in their desaturases contribute to pheromone differences and that some species possess pseudogenes for the desaturases of other species. Additionally, behavioral studies in these species have documented the existence of rare males broadly tuned and responsive to heterospecific pheromone blends. It has been suggested that the presence and subsequent expression of desaturase pseudogenes and the associated selection for rare male phenotypes would facilitate the coevolution of novel pheromone signals via asymmetric tracking (e.g., Roelofs et al. 2002). This potential mechanism for the coordination of signal and response traits during signal evolution would seem to require an external selective force favoring signal divergence. Without this selective force, the alternate outcome of extinction of the divergent signal lineage would seem far more likely.

Variation in Signaling Traits and Female Mating Success

The CSE model predicts that individual female mating success will be positively correlated with signal deviation from the median when aligned with the open-ended male preference function and negatively correlated with deviation when it opposes the open-ended male preference function. Both the AT and WF predict weak or little selection against females with rare/outlier signaling traits; however, extreme variants

(in all dimensions) will be selected against (but see the "rare" male hypothesis above). Although the WF hypothesis predicts that females with rare/outlier signaling traits will still attract a mate because of the wallflower effect, these females may experience reduced fitness despite being mated. In many moths, a negative correlation between total fecundity and female age at mating has been observed (see Mori and Evenden 2012). Additionally, when males contribute more than just sperm, there may be a negative relationship between female fitness and the age of her mate. If the wallflower effect means that females with rare/outlier signaling traits are older on average when mated or mate with older males, variation in female signaling traits could be associated with fitness costs. The SM predicts that as the female signaling trait increases in divergence (in all dimensions) from the population median, that female mating success decreases.

Although there is considerable empirical evidence of variation in female signaling traits (e.g., absolute and relative titers [Allison and Cardé 2006]), there is very little evidence that this variation impacts female mating success. Mori and Evenden (2012) performed a meta-analysis of the literature of delayed mating in female moths and observed that in general delayed mating significantly decreases fecundity. Similarly, several studies have reported mating failures in female moths (Sharov et al. 1995; Rhainds et al. 1999; Tcheslavskaia et al. 2002; Contarini et al. 2009; Régnière et al. 2012). Mating failures in female moths have been hypothesized to occur because of (1) a reduced capacity to produce sex pheromone; (2) reduced or no calling behavior; and/or (3) reduced female receptivity to mating (Proshold 1996; Torres-Vila et al. 2002; Walker and Allen 2011). To date, there is no evidence that variation in female signaling traits contributes to delayed mating or mating failure in female moths (although numerous studies have demonstrated a relationship between trap catches and the absolute and relative titers of pheromone released from pheromone-baited traps). Using caged female *Choristoneura fumiferana*, Régnière et al. (2012) observed that female mating success was lower during the period of peak flight than later in the season. Although this is not sufficient evidence to demonstrate the wallflower effect, it is consistent with it (evidence that the delay in mating was linked to the female pheromone phenotype would be necessary).

Harari et al. (2011) suggest that female moth sex pheromone may function as an honest signal of female quality and serve as the basis of male choice. As Harari et al. (2011) point out, this hypothesis requires that the female pheromone phenotype is condition-dependent. Evidence that the female pheromone phenotype is condition-dependent is limited in moths. A metabolic cost to pheromone biosynthesis in female moths has been demonstrated in mated and virgin *Heliothis virescens* (Foster, this volume). Although demonstration of a metabolic cost to pheromone biosynthesis may provide a mechanism for the hypothesis, it does not demonstrate that the pheromone phenotype is condition-dependent. A few studies have demonstrated that insecticide-resistant individuals have lower pheromone titers than susceptible individuals (Campanhola et al. 1991; Delisle and Vincent 2002), suggesting a trade-off between metabolic activities associated with resistance and pheromone biosynthesis.

Harari et al. (2011) report that the pheromone titer was significantly higher in large than small female *Lobesia botrana*, that male moths preferred the pheromone effluvia of large females, that females exposed to conspecific pheromone called significantly more than females that were not, and that when small virgin females (but not large females) were exposed to conspecific pheromone they produced significantly fewer eggs than females not exposed to pheromone. The authors of this study conclude that these results indicate that the costs of pheromone signaling are condition-dependent, at least in *L. botrana*.

Further experimental work is needed in additional species to determine if a relationship exists between variation in female signaling traits and female fitness, and if so, how common is this relationship.

Significance of Pheromone Evolution

Aside from its relevance to our conceptual understanding of the speciation process in moths, pheromone evolution has significant practical application for pest management (e.g., Borden 1990; Cardé 1990; Lanier 1990; Silverstein 1990; Howse et al. 1998). In large part, the impetus for the identification of insect sex pheromones has been generated by the expectation that they could be used in pest management programs. Historically, insect sex pheromones were used primarily to measure emergence and population levels and this information was used to "decide" when and where to apply direct management tactics. Direct pheromone-based tactics also exist, including mass-trapping and lure-and-kill formulations (Cork, this volume), and mating disruption (Evenden, this volume). Direct pheromone-based tactics often deploy pheromone analogues or incomplete pheromone blends because they are often less expensive than the natural blend to synthesize and can be more stable under field conditions (Minks and Cardé 1988).

Like chemical pesticides, the broadscale application of direct pheromone-based tactics can result in strong selection pressure. Because of the necessity of coordination of signaler and receiver, there is debate over the probability of treated populations becoming resistant. Opponents of the likelihood of the evolution of resistance argue that because coincident, complementary changes are necessary, the evolution of resistance is unlikely. The use of incomplete pheromone blends obviates the need for coincident, complementary change in signal and response traits. Resistance could evolve with no change in signal traits if responders evolved increased reliance on missing components. Despite being used operationally for over 20 years against a variety of insect pests (Cardé and Minks 1997; Howse et al. 1998), there has only been one published report consistent with the evolution of resistance to pheromone-based mating disruption (Mochizuki et al. 2002). In this case, a population of the tea pest, *Adoxophyes honmai* (Tortricidae), appears to have evolved increased reliance on the presence of minor components in response to a mating disruption program based on the major pheromone component alone.

The literature on the probability of the evolution of resistance to pheromone-based mating disruption is equivocal. It is clear that quantitative and qualitative parameters of the pheromone blend and male behavioral responses are heritable (Collins and Cardé 1985, 1989b; Sreng et al. 1989; Gemeno and Haynes 2000; Gemeno et al. 2001; Evenden et al. 2002; Svensson et al. 2002; Allison and Cardé 2006; Tabata et al. 2006) and selection experiments have demonstrated that male responses (Collins and Cardé 1989b; Liu and Haynes 1994) and the relative and absolute titers of pheromone components (Collins and Cardé 1989a; Sreng et al. 1989; Collins et al. 1990; Allison and Cardé 2007) respond to selection. An understanding of the genetics of pheromone systems and the selective forces acting on them could facilitate the preservation of direct pheromone-based pest management tactics and expansion of their use.

Summary

Comparative studies have documented that sex pheromones vary quantitatively and qualitatively among species and populations. Although there is little doubt that in moths female-produced volatile sex pheromones play a critical role in mediating mate location and reproductive isolation, there is debate over whether or not pheromone evolution plays a role in speciation. Several models of pheromone evolution have been proposed and can be differentiated on the basis of the relative variation in male response and female signal traits. Although numerous studies have characterized variation in signal and response traits in moths, few have done so at an informative scale. To date, the female pheromone blend distribution has been characterized and male preference function described using the distribution of signals actually produced by females, in four species. In two species (redbanded leafroller and oriental fruit moth), it appears that male preference functions impose stabilizing selection on the female pheromone, a result consistent with the SM model. In the other two species (pink bollworm and almond moth), it appears that male preference functions impose no or directional selection on the female pheromone, a result most consistent with the AT model.

Both hybridization and selection experiments have been used to understand the genetic basis of pheromone signal and response traits. In only one of these studies has any evidence been found of a genetic correlation between signal and response traits. Establishing generalities about how moth pheromones evolve will require additional empirical work in many more species. There is a need for studies that characterize variation in female signaling traits, the consequences (costs) of variation in signaling traits over the range of variation observed, and the genetic basis of variation in signal and response traits. Without such studies, generalities will not emerge, and it will not be possible to evaluate existing (and forthcoming) models of pheromone evolution in moths.

References Cited

Allison, J.D., and R.T. Cardé. 2006. Heritable variation in the sex pheromone of the almond moth, *Cadra cautella*. *Journal of Chemical Ecology* 32:621–641.

Allison, J.D., and R.T. Cardé. 2007. Bidirectional selection for novel pheromone blend ratios in the almond moth, *Cadra cautella*. *Journal of Chemical Ecology* 33:2293–2307.

Allison, J.D., and R.T. Cardé. 2008. Male pheromone blend preference function measured in choice and no-choice wind tunnel trials with almond moths, *Cadra cautella*. *Animal Behaviour* 75:259–266.

Allison, J.D., D.A. Roff, and R.T. Cardé. 2008. Genetic independence of female signal form and male receiver design in the almond moth, *Cadra cautella*. *Journal of Evolutionary Biology* 21:1666–1672.

Bethenod, M.T., Y. Thomas, F. Rousset, B. Frérot, L. Pélozuelo, G. Genestier, and D. Bourguet. 2005. Genetic isolation between two sympatric host plant races of the European corn borer, *Ostrinia nubilalis* Hübner. II: assortative mating and host-plant preferences for oviposition. *Heredity* 94:264–270.

Borden, J.H. 1990. Use of semiochemicals to manage coniferous tree pests in Western Canada. Pp. 281–315. In R.L. Ridgeway, R.M. Silverstein, and M.N. Inscoe, eds. *Behavior-Modifying Chemicals for*

Insect Management: Applications of Pheromones and other Attractants. New York: Marcel Dekker Inc.

Brandt, L.S.E., and M.D. Greenfield. 2004. Condition-dependent traits and the capture of genetic variance in male advertisement song. *Journal of Evolutionary Biology* 17:821–828.

Brown, W.L., Jr., and E.O. Wilson. 1957. Character displacement. *Systematic Zoology* 5:49–64.

Butlin, R.K. 1987. Speciation by reinforcement. *Trends in Ecology and Evolution* 2:8–13.

Butlin, R.K., and M.G. Ritchie. 1989. Genetic coupling in mate recognition systems: what is the evidence? *Biological Journal of the Linnaean Society* 37:237–246.

Butlin, R.K., and A.J. Trickett. 1997. Can population genetic simulations help to interpret pheromone evolution? Pp. 548–562. In R.T. Cardé and A.K. Minks, eds. *Insect Pheromone Research: New Directions*. New York: Chapman & Hall.

Byers, J.A. 2002. Internet programs for drawing moth pheromone analogs and searching literature database. *Journal of Chemical Ecology* 28:807–817.

Cain, M.L., V. Andeasen, and D.J. Howard. 1999. Reinforcing selection is effective under a relatively broad set of conditions in a mosaic hybrid zone. *Evolution* 53:1343–1353.

Campanhola, C., B.F. McCutchen, E.H. Baehrecke, and F.W. Plapp, Jr. 1991. Biological constraints associated with resistance to pyrethroids in the tobacco budworm (Lepidoptera: Noctuidae). *Journal of Economic Entomology* 84:1404–1411.

Cardé, R.T. 1990. Principles of mating disruption. Pp. 47–73. In R.L. Ridgway, R.M. Silverstein, and M.N. Inscoe, eds. *Behavior-Modifying Chemicals for Insect Management. Applications of Pheromones and other Attractants*. New York: Marcel Dekker Inc.

Cardé, R.T., and T.C. Baker. 1984. Sexual communication with pheromones. Pp. 355–384. In W.J. Bell and R.T. Cardé, eds. *Chemical Ecology of Insects*. London: Chapman & Hall.

Cardé, R.T., and K.F. Haynes. 2004. Structure of the pheromone communication channel in moths. Pp. 283–332. In R.T. Cardé and J.G. Millar, eds. *Advances in Insect Chemical Ecology*. Cambridge: Cambridge University Press.

Cardé, R.T., and A.K. Minks. 1997. *Insect Pheromone Research: New Directions*. London: Chapman & Hall.

Cardé, R.T., T.C. Baker, and W.L. Roelofs. 1976. Sex attractant responses of male oriental fruit moths to a range of component ratios: pheromone polymorphism? *Experientia* 32:1406–1407.

Cardé, R.T., A.M. Cardé, A.S. Hill, and W.L. Roelofs. 1977. Sex pheromone specificity as a reproductive isolating mechanism among the sibling species *Archips argyrospilus* and *A. mortuanus* and other sympatric tortricine moths (Lepidoptera: Tortricidae). *Journal of Chemical Ecology* 3:71–84.

Collins, R.D., and R.T. Cardé. 1985. Variation in and heritability of aspects of pheromone production in the pink bollworm moth, *Pectinophora gossypiella* (Lepidoptera: Gelechiidae). *Annals of the Entomological Society of America* 78:229–234.

Collins, R.D., and R.T. Cardé. 1989a. Selection for altered pheromone-component ratios in the pink bollworm moth, *Pectinophora gossypiella* (Lepidoptera: Gelechiidae). *Journal of Insect Behavior* 2:609–621.

Collins, R.D., and R.T. Cardé. 1989b. Heritable variation in pheromone response of the pink bollworm, *Pectinophora gossypiella* (Lepidoptera: Gelechiidae). *Journal of Chemical Ecology* 15:2647–2659.

Collins, R.D., S.L. Rosenblum, and R.T. Cardé. 1990. Selection for increased pheromone titre in the pink bollworm moth, *Pectinophora gossypiella* (Lepidoptera: Gelechiidae). *Physiological Entomology* 15:141–147.

Contarini, M., K.S. Onufrieva, K.W. Thorpe, K.F. Raffa, and P.C. Tobin. 2009. Mate-finding failure as an important cause of Allee effects along the leading edge of an invading insect population. *Entomologia Experimentalis et Applicata* 133:307–314.

Coyne, J.A., and H.A. Orr. 1989. Patterns of speciation in Drosophila. *Evolution* 43:362–381.

Coyne, J.A., and H.A. Orr. 1997. "Patterns of speciation in Drosophila" revisited. *Evolution* 51:295–303.

Coyne, J.A., H.A. Orr, and D.J. Futuyma. 1988. Do we need a new species concept? *Systematic Zoology* 37:190–200.

Craig, T.P., J.K. Itami, W.G. Abrahamson, and J.D. Horner. 1993. Behavioral evidence for host-race formation in *Eurosta solidaginis*. *Evolution* 47:1696–1710.

Darwin, C. 1859. *On the Origin of Species by Means of Natural Selection, of the Preservation of Favoured Races in the Struggle for Life*. London: John Murray.

Darwin, C. 1871. *The Descent of Man, and Selection in Relation to Sex*. London: John Murray.

De Jong, M.C.M. 1988. Evolutionary approaches to insect communication systems. Bark beetle host colonization and mate finding in small ermine moths. PhD thesis, Leiden University, 165 pp.

De Jong, M.C.M., and M.W. Sabelis. 1991. Limits to runaway sexual selection: the wallflower paradox. *Journal of Evolutionary Biology* 4:637–655.

Delisle, J., and C. Vincent. 2002. Modified pheromone communication associated with insecticidae resistance in the obliquebanded leafroller, *Choristoneura rosaceana* (Lepidoptera: Tortricidae). *Chemoecology* 12:47–51.

Dieckmann, U., and M. Doebeli. 1999. On the origin of species by sympatric speciation. *Nature* 400:354–357.

Dobzhansky, T. 1937. *Genetics and the Origin of Species*. New York: Columbia University Press.

Dobzhansky, T. 1940. Speciation as a stage in evolutionary divergence. *The American Naturalist* 74:312–321.

Doherty, J.A., and H.C. Gerhardt. 1983. Hybrid tree frogs: vocalizations of males and selective phonotaxis of females. *Science* 220:107801080.

Droney, D.C., C.J. Musto, K. Mancuso, W.L. Roelofs, and C.E. Linn, Jr. 2012. The response to selection for broad male response to female sex pheromone and its implications for divergence in close-range mating behavior in the European corn borer moth, *Ostrinia nubilalis*. *Journal of Chemical Ecology* 38:1504–1512.

El-Sayed, A.M. 2012. The Pherobase: Database of Pheromones and Semiochemicals. Available at: http://www.pherobase.com.

Emelianov, I., F. Simpson, P. Narang, and J. Mallet. 2003. Host choice promotes reproductive isolation between host races of the larch budworm *Zeiraphera diniana*. *Journal of Evolutionary Biology* 16:208–218.

Evenden, M.L., B.G. Spohn, A.J. Moore, R.F. Preziosi, and K.F. Haynes. 2002. Inheritance and evolution of male response to sex pheromone in *Trichoplusia ni* (Lepidoptera: Noctuidae). *Chemoecology* 12:53–59.

Ewing, A.W., and J.A. Miyan. 1986. Sexual selection, sexual isolation and the evolution of song in the *Drosophila repleta* group of species. *Animal Behaviour* 34:421–429.

Feder, J.L., S.B. Opp, B. Wlazlo, K. Reynolds, W. Go, and S. Spisak. 1994. Host fidelity is an effective premating barrier between sympatric races of the apple maggot fly. *Proceedings of the National Academy of Sciences of the United States of America* 91:7990–7994.

Fisher, R.A. 1930. *The Genetical Theory of Natural Selection*. Oxford: Clarendon Press.

Gascoigne, J., L. Berec, S. Gregory, and F. Courchamp. 2009. Dangerously few liaisons: a review of mate-finding Allee effects. *Population Ecology* 51:355–372.

Gemeno, C., and K.F. Haynes. 2000. Periodical and age-related variation in chemical communication system of black cutworm moth, *Agrotis ipsilon*. *Journal of Chemical Ecology* 26:329–342.

Gemeno, C., A.J. Moore, R.F. Preziosi, and K.F. Haynes. 2001. Quantitative genetics of signal evolution: a comparison of the pheromonal signal in two populations of the cabbage looper moth, *Trichoplusia ni*. Behavioral Genetics 31:157–165.

Gerhardt, H.C. 1994. The evolution of vocalization in frogs and toads. *Annual Review of Ecology, and Systematics* 25: 293–324.

Gleason, J.M., and M.G. Ritchie. 1998. Evolution of courtship song and reproductive isolation in the *Drosophila willistoni* species complex: do sexual signals diverge the most quickly? *Evolution* 52:1493–1500.

Grapputo, A., T. Kumpulainen, and J. Mappes. 2005. Phylogeny and evolution of parthenogenesis in Finnish bagworm moth species (Lepidoptera: Psychidae: Naryciinae) based on mtDNA-markers. *Annales Zoologici Fennici* 42:141–160.

Gray, D.A., and W.H. Cade. 2000. Sexual selection and speciation in field crickets. *Proceedings of the National Academy of Sciences of the United States of America* 97:14449–14454.

Greenfield, M.D. 2002. *Signalers and Receivers. Mechanisms and Evolution of Arthropod Communication*. Oxford: Oxford University Press.

Gries, G., P. W. Schaeffer, R. Gries, J. Liška, and T. Gotoh. 2001. Reproductive character displacement in *Lymantria monacha* from Northern Japan? *Journal of Chemical Ecology* 27:1163–1176.

Groot, A., J. L. Horovitz, J. Hamilton, R. G. Santangelo, C. Schal, and F. Gould. 2006. Experimental evidence for interspecific directional selection on moth pheromone communication. *Proceedings of the National Academy of Sciences of the United States of America* 103:5858–5863.

Harari, A. R., T. Zahavi, and D. Thiéry. 2011. Fitness cost of pheromone production in signaling female moths. *Evolution* 65:1572–1582.

Haynes, K. F., and T. C. Baker. 1988. Potential for evolution of resistance to pheromones: worldwide and local variation in chemical communication-system of pink bollworm moth, *Pectinophora gossypiella*. *Journal of Chemical Ecology* 14:1547–1560.

Haynes, K. F., and R. E. Hunt. 1990. Interpopulational variation in emitted pheromone blend of cabbage looper moth, *Trichoplusia ni*. *Journal of Chemical Ecology* 16:509–519.

Haynes, K. F., and K. V. Yeargan. 1999. Exploitation of intraspecific communication systems: illicit signalers and receivers. *Annals of the Entomological Society of America* 92:960–970.

Hemmann, D. J., J. D. Allison, and K. F. Haynes. 2008. Trade-off between sensitivity and specificity in the cabbage looper moth response to sex pheromone. *Journal of Chemical Ecology* 34:1476–1486.

Higashi, M., G. Takimoto, and N. Yamamura. 1999. Sympatric speciation by sexual selection. *Nature* 402:523–526.

Higgie, M., S. Chenoweth, and M. W. Blows. 2000. Natural selection and the reinforcement of mate recognition. *Science* 290:519–521.

Hoikkala, A., and J. Aspi. 1993. Criteria of female mate choice in *Drosophila littoralis*, *D. monatana*, and *D. ezoana*. *Evolution* 47:768–777.

Houde, A. E. 1988. The effects of female choice and male-male competition on the mating success of male guppies. *Animal Behaviour* 36:888–896.

Howard, D. J. 1993. Reinforcement: origin, dynamics, and fate of an evolutionary hypothesis. Pp. 46–69. In R. G. Harrison, ed. *Hybrid Zones and the Evolutionary Process*. New York: Oxford University Press.

Howard, D. J., M. Reece, P. G. Gregory, J. Chu, and M. L. Cain. 1998. The evolution of barriers to fertilization between closely related organisms. Pp. 279–288. In D. J. Howard and S. H. Berlocher, eds. *Endless Forms: Species and Speciation*. New York: Oxford University Press.

Howse, P. E., I. D. R. Stevens, and O. T. Jones. 1998. *Insect Pheromones and Their Use in Pest Management*. London: Chapman & Hall.

Hoy, R. R., and R. C. Paul. 1973. Genetic control of song specificity in crickets. *Science* 180:82–83.

Hoy, R. R., J. Hahn, R. C. Paul. 1977. Hybrid cricket auditory behavior: evidence for genetic coupling in animal communication. *Science* 195:82–84.

Hunt, J., L. F. Bussiere, M. D. Jennions, and R. Brooks. 2004. What is genetic quality? *Trends in Ecology and Evolution* 19:329–333.

Kaneshiro, K. Y., and C. R. B. Boake. 1987. Sexual selection and speciation: issues raised by Hawaiian Drosophila. *Trends in Ecology and Evolution* 2:207–212.

Klun, J. A., and S. Maini. 1979. Genetic basis of an insect chemical communication system: the European corn borer. *Environmental Entomology* 8:423–426.

Kondrashov, A. S., and F. A. Kondrashov. 1999. Interactions among quantitative traits in the course of sympatric speciation. *Nature* 400:351–354.

Lacey, M. J., and C. J. Sanders. 1992. Chemical composition of sex-pheromone of oriental fruit moth and rates of release by individual female moths. *Journal of Chemical Ecology* 18:1421–1435.

Lambert, D. M., B. Michaux, and C. S. White. 1987. Are species self-defining? *Systematic Zoology* 36:196–205.

Lande, R. 1981. Models of speciation by sexual selection on polygenic traits. *Proceedings of the National Academy of Sciences of the United States of America* 78:3721–3725.

Lanier, G. N. 1990. Principles of attraction-annihilation: mass-trapping and other means. Pp. 25–47. In R. L. Ridgway, R. M. Silverstein, and M. N. Inscoe, eds. *Behavior-Modifying Chemicals for Insect Management: Applications of Pheromones and Other Attractants*. New York: Marcel Dekker Inc.

Lassance, J.-M., M. A. Liénard, B. Antony, S. Qian, T. Fujii, J. Tabata, Y. Ishikawa, and C. Löfstedt. 2013. Functional consequences of sequence variation in the pheromone biosynthesis gene *pgFAR* for *Ostrinia* moths. *Proceedings of the National Academy of Sciences of the United States of America* 110:3967–3972.

Linn, C. E., Jr., and W. L. Roelofs. 1983. Effect of varying proportions of the alcohol component on sex pheromone blend discrimination in male oriental fruit moths. *Physiological Entomology* 8:291–306.

Linn, C. E., Jr., and W. L. Roelofs. 1985. Response specificity of male pink bollworm moths to different blends and dosages of sex pheromone. *Journal of Chemical Ecology* 11:1583–1590.

Linn, C. E., Jr., L. B. Bjostad, J. W. Du, and W. L. Roelofs. 1984. Redundancy in a chemical signal: behavioral responses of male *Trichoplusia ni* to a 6-component sex pheromone blend. *Journal of Chemical Ecology* 10:1635–1658.

Linn, C. E., Jr., M. G. Campbell, and W. L. Roelofs. 1986. Male moth sensitivity to multicomponent pheromones: critical role of female-released blend in determining the functional role of components and active space of the pheromone. *Journal of Chemical Ecology* 12:659–668.

Linn, C. E., Jr., M. G. Campbell, and W. L. Roelofs. 1987. Pheromone components and actives spaces: what do moths smell and where do they smell it? *Science* 237:650–652.

Linn, C. E., Jr., M. O'Connor, and W. L. Roelofs. 2003. Silent genes and rare males: a fresh look at pheromone blend response specificity in the European corn borer moth, *Ostrinia nubilalis*. *Journal of Insect Science* 3:15. Available at: insectscience.org/3.15.

Linn, C. E., Jr., C. Musto, and W. L. Roelofs. 2007. More rare males in *Ostrinia*: response of Asian corn borer moths to the sex pheromone of the European corn borer. *Journal of Chemical Ecology* 33:199–212.

Littlejohn, M. J. 1981. Reproductive isolation: a critical review. Pp. 298–334. In W. R. Atchley and D. S. Woodruff, eds. *Evolution and Speciation: Essays in Honor of M. J. D. White*. Cambridge: Cambridge University Press.

Liu, Y. B., and K. F. Haynes. 1994. Evolution of behavioral-responses to sex-pheromone in mutant laboratory colonies of *Trichoplusia ni*. *Journal of Chemical Ecology* 20:231–238.

Löfstedt, C. 1993. Moth pheromone genetics and evolution. *Philosophical Transactions of the Royal Society of London, Series B* 340:167–177.

Löfstedt, C., B. S. Hansson, W. Roelofs, and B. Bengtsson. 1989. No linkage between genes controlling female pheromone production and male pheromone response in the European corn borer, *Ostrinia nubilalis* Hübner (Lepidoptera; Pyralidae). *Genetics* 123:553–556.

Löfstedt, C., B. S. Hansson, H. J. Dijkerman, and W. M. Herrebout. 1990. Behavioural and electrophysiological activity of unsaturated analogues of the pheromone tetradecyl acetate in the small ermine moth *Yponomeuta rorellus*. *Physiological Entomology* 15:47–54.

Mallet, J. 2007. Taxonomy of Lepidoptera: the scale of the problem. *The Lepidoptera Taxome Project*. University College, London. Available at: http://www.ucl.ac.uk/taxome/lepnos.html (retrieved 11 July 2013).

Marshall, J. L., M. L. Arnold, and D. J. Howard. 2002. Reinforcement: the road not taken. *Trends in Ecology and Evolution* 17:558–563.

Mayr, E. 1963. *Animal Species and Evolution*. Cambridge, MA: Belknap Press.

McElfresh, J. S., and J. G. Millar. 1999. Geographic variation in sex pheromone blend of *Hemileuca electra* from southern California. *Journal of Chemical Ecology* 25:2505–2525.

McElfresh, J. S., and J. G. Millar. 2001. Geographic variation in the pheromone system of the saturniid moth *Hemileuca eglanterina*. *Ecology* 82:3505–3518.

Miller, J. R., and W. L. Roelofs. 1980. Individual variation in sex pheromone component ratios in two populations of the red-banded leafroller moth, *Argyrotaenia velutinana*. *Environmental Entomology* 9:359–363.

Minks, A. K., and R. T. Cardé. 1988. Disruption of pheromone communication in moths: is the natural blend really the most efficacious. *Entomologia Experimentalis et Applicata* 49:25–36.

Mochizuki, F., T. Fukomoto, H. Noguchi, H. Sugie, T. Morimoto, and K. Ohtani. 2002. Resistance to a mating disruptant composed of (Z)-11-tetradecenyl acetate in the smaller tea tortrix, *Adoxophyes honmai* (Yasuda) (Lepidoptera: Tortricidae). *Applied Entomology and Zoology* 37:299–304.

Mori, B.A., and M.L. Evenden. 2012. When mating disruption does not disrupt mating: fitness consequences of delayed mating in moths. *Entomologia Experimentalis et Applicata* 146:50–65.

Noor, M.A. 1995. Speciation driven by natural selection in Drosophila. *Nature* 375:674–675.

Noor, M.A. 1999. Reinforcement and other consequences of sympatry. *Heredity* 83:503–508.

O'Donald, P. 1980. *Genetic Models of Sexual Selection*. Cambridge: Cambridge University Press.

Otte, D. 1974. Effects and functions in the evolution of signaling systems. *Annual Review of Ecology and Systematics* 5:385–417.

Pashley, D.P., A.M. Hammond, and T.N. Hardy. 1992. Reproductive isolating mechanisms in fall armyworm host strains (Lepidoptera, Noctuidae). *Annals of the Entomological Society of America* 85:400–405.

Paterson, H.E.H. 1978. More evidence against speciation by reinforcement. *South African Journal of Science* 74:369–371.

Paterson, H.E.H. 1985. The recognition concept of species. Pp. 21–29. In E.S. Vrba, ed. *Species and Speciation*. Transvaal Museum Monograph No. 4, Transvaal Museum, Pretoria.

Paterson, H.E.H. 1993. The specific-mate recognition system and variation in motile animals. Pp. 209–227. In P.P.G. Bateson, P.H. Klopfer, and N.S. Thompson, eds. *Perspectives in Ethology, Vol. 10: Behavior and Evolution*. New York: Plenum Press.

Phelan, P.L. 1992. Evolution of sex pheromones and the role of asymmetric tracking. Pp. 265–314. In B.D. Roitberg and M.B. Isman, eds. *Insect Chemical Ecology: An Evolutionary Approach*. New York: Chapman & Hall.

Phelan, P.L. 1997. Evolution of mate-signaling in moths: phylogenetic considerations and predictions from asymmetric tracking hypothesis. Pp. 240–256. In J.C. Choe and B.J. Crespi, eds. *Mating Systems in Insects and Arachnids*. Cambridge: Cambridge University Press.

Pomiankowski, A., and L. Sheridan. 1994. Linked sexiness and choosiness. *Trends in Ecology and Evolution* 9:242–244.

Proshold, F.I. 1996. Reproductive capacity of laboratory-reared gypsy moth (Lepidoptera: Lymantriidae): effect of age of female at time of mating. *Journal of Economic Entomology* 89:337–342.

Régnière, J., J. Delisle, D.S. Pureswaran, and R. Trudel. 2012. Mate-finding Allee effect in spruce budworm population dynamics. *Entomologia Experimentalis et Applicata* 146:112–122.

Rhainds, M., G. Gries, and M.M. Min. 1999. Size- and density-dependent reproductive success of bagworms, *Metisa plana*. *Entomologia Experimentalis et Applicata* 91:375–383.

Rice, W.R. 1984. Disruptive selection on habitat preference and the evolution of reproductive isolation: a simulation study. *Evolution* 38:1251–1260.

Rice, W.R. 1987. Speciation via habitat specialization: the evolution of reproductive isolation as a correlated character. *Evolutionary Ecology* 1:301–314.

Ritchie, M.G. 1996. The shape of female mating preferences. *Proceedings of the National Academy of Sciences of the United States of America* 93:14628–14631.

Roelofs, W.L. 1978. Threshold hypothesis for pheromone perception. *Journal of Chemical Ecology* 4:685–699.

Roelofs, W.L., A. Hill, and R.T. Cardé. 1975. Sex pheromone components of the redbanded leafroller, *Argyrotaenia velutinana*. *Journal of Chemical Ecology* 1:83–89.

Roelofs, W.L., W. Liu, G. Hao, H. Jiao, A.P. Rooney, and C.E. Linn, Jr. 2002. Evolution of moth sex pheromones via ancestral genes. *Proceedings of the National Academy of Sciences of the United States of America* 99:13621–13626.

Ryan, M.J., and A. Keddy-Hector. 1992. Directional patterns of female mate choice and the role of sensory biases. *The American Naturalist* 139:S4–S35.

Ryan, M.J., and A.S. Rand. 1993. Species recognition and sexual selection as a unitary problem in animal communication. *Evolution* 47:647–657.

Ryan, M.J., and A.S. Rand. 2003. Sexual selection in female perceptual space: how female tungara frogs perceive and respond to complex population variation in acoustic mating signals. *Evolution* 57:2608–2618.

Ryan, M.J., and W. Wilczynski. 1991. Evolution of intraspecific variation in the advertisement call of a cricket frog (*Acris crepitans*, Hylidae). *Biological Journal of the Linnaean Society* 44:249–271.

Sætre, G.P., T. Moum, S. Bureš, M. Král, M. Adamjan, and J. Moreno. 1997. A sexually selected character displacement in flycatchers reinforces premating isolation. *Nature* 387:589–592.

Sharov, A.A., A.M. Liebhold, and F.W. Ravlin. 1995. Prediction of gypsy moth (Lepidoptera: Lymantriidae) mating success from pheromone trap counts. *Environmental Entomology* 24:1239–1244.

Shorey, H.H. 1970. Sex pheromones of Lepidoptera. Pp. 249–284. In D.L. Wood, R.M. Silverstein, and M. Nakajima, eds. *Control of Insect Behavior by Natural Products*. New York: Academic Press.

Silverstein, R.M. 1990. Practical use of pheromones and other behavior-modifying compounds: overview. Pp. 1–9. In R.L. Ridgway, R.M. Silverstein, and M.N. Inscoe, eds. *Behavior-Modifying Chemicals for Insect Management: Applications of Pheromones and Other Attractants*. New York: Marcel Dekker Inc.

Sreng, I., T. Glover, and W.L. Roelofs. 1989. Canalization of the redbanded leafroller moth sex pheromone blend. *Archives of Insect Biochemistry and Physiology* 10:73–82.

Svensson, E.I., F. Eroukhmanoff, and M. Friberg. 2006. Effects of natural and sexual selection on adaptive population divergence and premating isolation in a damselfly. *Evolution* 60:1242–1253.

Svensson, G.P., C. Ryne, and C. Löfstedt. 2002. Heritable variation of sex pheromone composition and the potential for evolution of resistance to pheromone-based control of the Indian meal moth, *Plodia interpunctella*. *Journal of Chemical Ecology* 28:1447–1461.

Tabata, J., S. Hoshizaki, S. Tatsuki, and Y. Ishikawa. 2006. Heritable pheromone blend variation in a local population of the butterbur borer moth, *Ostrinia zaguliaevi* (Lepidoptera: Crambidae). *Chemoecology* 16:123–128.

Takanashi, T., Y.P. Huang, K.R. Takahashi, S. Hoshizaki, S. Tatsuki, and Y. Ishikawa. 2005. Genetic analysis and population survey of sex pheromone variation in the adzuki bean borer moth, *Ostrinia scapulalis*. *Biological Journal of the Linnaean Society* 84:143–160.

Tcheslavskaia, K., C.C. Brewster, and A.A. Sharov. 2002. Mating success of gypsy moth (Lepidoptera: Lymantriidae) females in southern Wisconsin. *Great Lakes Entomologist* 35:1–7.

Templeton, A.R. 1981. Mechanisms of speciation; a population genetic approach. *Annual Review of Ecology and Systematics* 12:23–48.

Thompson, D.R., N.P.D. Angerilli, C. Vincent, and A.P. Gaunce. 1991. Evidence for regional differences in the response of the obliquebanded leafroller (Lepidoptera: Tortricidae) to sex pheromone blends. *Environmental Entomology* 20:935–938.

Torres-Vila, L.M., M.C. Rodriguez-Molina, and J. Stockel. 2002. Delayed mating reduces reproductive output of female European grapevine moth, *Lobesia botrana* (Lepidoptera: Tortricidae). *Bulletin of Entomological Research* 92:241–249.

Trivers, R.L. 1972. Parental investment and sexual selection. Pp. 136–179. In B. Campbell, ed. *Sexual Selection and the Descent of Man, 1871–1971*. Chicago, IL: Aldine.

Valeur, P.G., and C. Löfstedt. 1996. Behaviour of male oriental fruit moth, *Grapholita molesta*, in overlapping sex pheromone plumes in a wind tunnel. *Entomologia Experimentalis et Applicata* 79:51–59.

Walker, P.W., and G.R. Allen. 2011. Delayed mating and reproduction in the autumn gum moth *Mnesampela privata*. *Agricultural and Forest Entomology* 13:341–347.

West-Eberhard, M.J. 1983. Sexual selection, social competition, and speciation. *Quarterly Review of Biology* 58:155–183.

West-Eberhard, M.J. 1984. Sexual selection, competitive communication and species-specific signals in insects. Pp. 283–324. In T. Lewis, ed. *Proceedings of the 12th Symposium of the Royal Entomological Society of London*. London: Academic Press.

Westerdahl, H. 2004. No evidence of an MHC-based female mating preference in great reed warblers. *Molecular Ecology* 13:2465–2470.

White, M.J.D. 1978. *Modes of Speciation*. San Francisco, CA: W.H. Freeman.

Zahavi, A. 1975. Mate selection: a selection for a handicap. *Journal of Theoretical Biology* 53:205–314.

Zhu, J.W., B.B. Chastain, B.G. Spohn, and K.F. Haynes. 1997. Assortative mating in two pheromone strains of the cabbage looper moth, *Trichoplusia ni*. *Journal of Insect Behavior* 10:805–817.

CHAPTER THREE

Variation in Moth Pheromones

Causes and Consequences

JEREMY D. ALLISON and RING T. CARDÉ

Introduction

Mate finding in almost all moth lineages is dependent on males using a female-emitted pheromone. Although numerous studies have documented variation in moth pheromone traits within and among individuals (Allison and Cardé 2008), populations (Löfstedt et al. 1986), and species (Byers 2006), such variation remains incompletely documented and its causes and consequences poorly understood. From an evolutionary perspective, variation in moth pheromones is significant for multiple reasons. The extent to which genotype and environment determine moth pheromone phenotype will determine the potential for evolution by natural selection to occur on moth pheromone traits. The consequences of variation in these traits on individual reproductive success will determine the genetic contribution of an individual to the next generation. Without an understanding of these consequences, it is not possible to predict if and how the mean phenotype of pheromone traits will respond to selection. Further, to evaluate the hypothesis that phenotypic differences in pheromone traits among populations and species are the outcome of evolution by natural selection, it is necessary to understand the consequences of variation in pheromone traits in different selection regimes (i.e., environments) (see Mazer and Damuth 2001).

Variation in moth pheromone traits also has significant applications to pest management. Initially, the most common use of insect pheromones was as an indirect pest management tactic (Silverstein 1990). Emergence and population levels were monitored with traps baited with synthetic sex pheromone and trap catch data integrated into economic threshold and developmental models to determine where, when, or if to apply insecticides. Problems with resistance, nontarget effects, and environmental contamination provided the impetus for the development of direct pheromone-based pest management tactics, including mass trapping (Borden 1990; Cork, this volume), attract-and-kill formulations (Lanier 1990; Cork, this volume), and mating disruption (Cardé 1990; Evenden, this volume) (see Howse et al. 1998 and Witzgall et al. 2010a for reviews). The broadscale application of direct pheromone-based pest management strategies can generate strong selection pressure on pheromone traits in the managed population. Although there is debate over the probability of populations treated with direct pheromone-based management tactics becoming resistant, the existence of variation within populations, particularly heritable variation (e.g., Collins and Cardé 1985; Allison and Cardé 2006), is a necessary condition for the evolution of resistance. The existence of variation in pheromone traits among geographic populations also has important consequences for the use of pheromones as indirect and direct pest management tactics (i.e., the tactic would have to be modified to match the targeted population).

To date, there is only one established case of resistance to mating disruption. Mochizuki et al. (2002) documented that use of the (Z)-11-tetradecenyl acetate (Z11-14Ac) to disrupt mating of the smaller tea tortrix moth, *Adoxophyes honmai* (Tortricidae), became ineffective in one prefecture of Japan after 15 years of application. Using the full, four-component blend restored efficacy. Likely resistance was due to male moths becoming able to detect the natural blend amidst a background of Z11-14Ac and thereby orient to females.

Single Components versus Blends

The pheromone employed by some moth species evidently consists of a single compound. The hypothesis that moths might use only one compound in communication was fostered by the finding that the first pheromone to be identified, the attractant of the domestic silkworm moth, *Bombyx mori* (Bombycidae), comprises a single compound, (*E, Z*)-10,12-hexadecadienol, termed bombykol (Butenandt et al. 1959). The corresponding aldehyde, bombykal, subsequently was suggested to be a pheromone component (Kaissling et al. 1978), but a positive effect on male behavior when this compound is added to bombykol has not been established. The male gypsy moth, *Lymantria dispar* (Erebidae), as another example, only responds to 2-methyl-7R,8S-epoxy-octadecane, termed (+)-disparlure, and no other compounds are known to augment its attractivity. Shorey (1977) argued that the pheromone of all moth species should be comprised of a single component.

We now know that by far the majority of moth species use blends comprised of two or more compounds (Byers 2006). In the case of the noctuid *Trichoplusia ni* (Noctuidae), six compounds have been identified as contributing to orientation (Bjostad et al. 1984). Here, it is useful to emphasize that we are considering as true pheromones only those compounds that are released by the female (or are in the pheromone gland and presumed to be emitted) and that also have been verified to mediate mate location ("attraction") and courtship. Other pheromone-like compounds (often pheromone precursors) can be present in the female's pheromone gland and some of these may even be emitted, but, unless these compounds influence the male's behavior positively, they cannot be termed pheromones. Instead, these compounds should be labeled as components of the pheromone gland. Other compounds that are attractive or add to attractivity but have not been verified as produced by the females also cannot be properly termed a pheromone. Pheromone compounds have been classified as "primary" when their presence is required for attraction and "secondary" when they are not attractive alone but when added to the primary compound(s) they augment the attractivity (Roelofs and Cardé 1977). Many of the species now viewed as having but one component in their pheromone may upon further scrutiny turn out to have multicomponent blends, and many established blends may have yet-to-be identified components.

Challenges Associated with Defining the Complete Blend

Defining the "complete" pheromone blend has been the subject of considerable effort for many species, particularly those of economic importance. Technological advances have enabled discovery of many candidate pheromones that then can be tested in bioassays to establish behavioral activity. The most important techniques used to identify prospective pheromones are capillary gas chromatography coupled with monitoring of neurophysiological responses to individual components in the effluent by electroantennogram (EAG) and sometimes Single Sensillum Response (SSR). The issue of blend completeness eventually hinges on the diagnostic ability of the behavioral bioassay used to verify pheromonal status.

Shorey's (1977) advocacy of the moth pheromone as always being a single component led him to assert that it was only necessary to pick a single "key" response, such as wing fanning, for a bioassay, as this reaction would reflect all the behaviors that are mediated by the pheromone, from activation to mating. We now realize that attraction to the pheromone source is the *sine non qua* of bioassays and consequently blend completeness and presentation in the correct ratio and airborne concentration are usually evaluated in two ways: attraction to traps in the field and attraction to a point source in a wind tunnel.

In the case of field trapping, male behaviors are rarely observed directly and only captures are recorded. Does a field-trapping assay miss components? Moth landing behavior near traps can be influenced by the plume's structure, which in many trap designs in turn is dictated by the trap's orientation to wind flow (Lewis and Macaulay 1976). Foster et al. (1991) also have shown that retention of moths in sticky traps can vary with where in traps moths land initially and this in turn can be influenced by a trap's orientation to wind flow, details of trap configuration, and even where in the trap the lure is positioned. What we should take from these studies is that subtle but statistically distinguishable differences in patterns of capture to a range of ratios seen in different localities must be interpreted with caution, as these could be due to many factors besides the nominate lure composition: attraction levels can be influenced by differences in traps trap saturation, setup (e.g., on an isolated stake or proximate to foliage or a tree trunk), lure purity, and characteristics of dispenser release. The latter issue has not been given due attention: using rubber septa, for example, Kuenen and Seigel (2015) have shown that amount and type of solvent used to charge septa with pheromone can alter rate and ratio of compounds released.

One test that is generally considered diagnostic for determining whether the entire pheromone is known is to compare traps baited with the synthetic compound or mixture against traps baited with one or several caged females. Because synthetic lures emit continuously (when females may not be calling but males are responsive) and also often emit at a higher release rate than females, trap capture of synthetic lures simply matching or even exceeding the attractiveness of females may not verify that the full blend has been successfully characterized. High rates of release also may subsume the additive effects on attraction of secondary components that have been established in wind-tunnel assays that use rates of emission near to those from naturally calling females. Several secondary components of *Lobesia botrana* (Tortricidae), for example, increase attractivity of the main pheromone component in a wind-tunnel milieu (Witzgall et al. 2005), but these secondary components are not known to add significantly to the attractivity of traps in the field.

The influence of concentration on the perception of component ratio was considered by Roelofs (1978) and its consequences termed the "Threshold Hypothesis." This proposition states that when the release rate of a blend is in the range of concentration below or comparable to the female's natural rate of release, the most attractive ratio will be the natural ratio. If the rates are well above the natural rates (as can be the case with artificial lures), then unnatural ratios *may* become more attractive, in some species even exceeding the potency of the natural ratio. In general, lures placed in traps emit at higher rates than females and this may, in some species, distort the relative preference of ratios to which males respond.

Baker (2008) has suggested that we interpret responses to a natural blend as a "balanced antagonism," meaning that non-optimal ratios and in some cases amounts have an antagonistic effect on response, similar in terms of behavioral output to heterospecific antagonists that contribute to reproductive isolation among otherwise cross-attractive species (see Baker and Hansson, this volume; Hillier and Baker, this volume). In the case of the (Z)-strain of the European corn borer, *Ostrinia nubilalis* (Pyralidae), if in a wind-tunnel assay the ratio of the blend first encountered was the natural ratio, subsequent upwind orientation can permit orientation to unnatural ratios that otherwise would not elicit attraction. This rapid lowering of ratio specificity was correlated with a differential change in the sensitivity via selective adaptation of the receptors for the two pheromone components (Kárpáti et al. 2013). In other words, measuring of ratio after the first encounter with pheromone is compromised. Whether such generalization is applicable to other moths is unexplored.

A wind tunnel allows documentation of a repertoire of behaviors beginning with activation, upwind flight along the plume (termed "locking-on" to the plume), and source location (such as landing on the pheromone dispenser) or, when the blend does not match the natural pheromone or the release rate is abnormally high, how closely males approach the stimulus. When the blend is incomplete or is present in the wrong ratio, males may have prolonged flights, never reaching the source, and a measure how far upwind along the plume males have progressed may reflect how closely the synthetic copy matches the natural pheromone. Another indicator of the completeness of the synthetic copy may be how "precisely" males navigate along the pheromone plume, with flight paths characterized by wide zigzags and sometimes a slowed flight speed being indicative of an imperfect match to the natural blend and with a relatively straight and rapid upwind course representative of the natural blend (Witzgall and Arn 1990; Witzgall 1997). Witzgall (1997), for example, concluded that the three-component pheromone blend for the oriental fruit moth *Grapholita molesta* (Tortricidae) was incomplete, because it did not evoke as straight a path of plume following as an extract of the female's pheromone gland, with both presented at identical concentrations. This is surprising, given that this synthetic copy elicits high levels of trap capture in the field and it evokes the full range of natural courtship behaviors including hair extrusion, and otherwise would be classified as a "complete" blend (Baker and Cardé 1979a, 1979b).

A possible deficiency of the wind-tunnel bioassay is that a male is at the outset quiescent, whereas in the field the male may be engaged in ranging flight (flight presumed to be in search of the female's pheromone plume). In a quiescent state, the threshold of responsiveness may be elevated compared to a flying male and, possibly, the reaction to the completeness of the blend compromised (the former has been established for the gypsy moth [Charlton et al. 1993], but the latter remains speculative—see Cardé and Charlton 1984). Work

with the noctuid *Spodoptera littoralis* suggests that prior exposure to pheromone sensitizes males, improving their subsequent orientation to pheromone in a wind-tunnel assay (Anderson et al. 2003). Whether the males also became choosier in terms of ratio preference was not assessed.

Nearly all wind-tunnel comparisons are "no-choice," meaning that the shape of the preference function is determined by the aggregate response of multiple males, each only tested once to a single blend. Less commonly, the shape of the preference function is characterized with two-choice tests wherein moths experience a blended plume during activation and takeoff at the tunnel's downwind end and must choose between stimuli upwind where the plumes diverge, leading to separate lures near the tunnel's upwind end. In studies with the almond moth *Cadra cautella* (Pyralidae), males were able to discriminate among ratios that matched or deviated from the natural mix of its two components: (Z,E)-9,12-tetradecadienyl acetate (attractive itself) and (Z)-9-tetradecenyl acetate (Allison et al. 2008). Males were able to choose the natural ratio over off-blend variants, an unexpected response, because in no-choice tests all of these lures evoked equivalent levels of attraction. Choice tests provide insight into both a male's preferences and ability to discriminate among blends, but the relationship of wind-tunnel choices to selection in the field is tenuous. Males in the wild would not be expected to have the opportunity to choose between two equidistant females whose plumes are evenly mixed.

Another concern is that the sticky traps commonly used in field tests are subject to saturation effects. At high population densities, the trapping surface can become paved with moths and scales and therefore be less apt to retain moths. This effect will be strongest for the most attractive lures and thereby distort the relative distribution of males across treatments. Use of nonsticky, high-capacity, "no-exit" traps obviates this problem. What is clear is that in trying to document patterns of response specificity, use of differing trapping protocols (e.g., trap types, different dispensers, loading rates, charging with different solvents and solvent amounts, or sources of pheromone with differing levels of purity) in different localities, seasons, and population densities all could produce both spurious similarities and differences. A final caveat on methods is the use of ratios found in pheromone glands as a basis for formulating lures for bioassays. This is especially problematic for blends comprised of components of differing chain lengths and/or moieties. The ratios actually emitted by females can differ substantially from ratios found in pheromone glands and these can also vary during the calling cycle (e.g., Hunt and Haynes 1990). In turn, ratios emitted from dispensers can differ from the loaded ratios (Roelofs and Cardé 1977; Kuenen and Seigel 2015). The salutatory lesson is that these methodological issues can influence estimates of individual variability in the pheromone message and the response it elicits.

Variation in the Timing of Mating

EXOGENOUS AND ENDOGENOUS FACTORS

In nearly all moth lineages, females are the attractive (pheromone-emitting) sex. The time of mating in these groups is a composite of the rhythms of the male's response to and the female's release of pheromone. The behavior of a female actively protruding her abdominal tip, thereby exposing the surface of her pheromone gland, is termed "calling." When the rhythmicity of calling has been observed under constant environmental conditions (an unvarying temperature and relative humidity under either constant light or dark), generally these rhythms have been found to have a circadian (endogenous) basis (see review by Groot 2014). However, unless these rhythms have been demonstrated experimentally to persist for several days under unvarying environmental conditions, such rhythms should be simply labeled as diel.

The rhythmicity of female calling is well studied in numerous moth species, and the timing of calling in moths can be modified by many external factors, especially temperature, but also relative humidity, wind speed, light levels, odors from a host plant, and pheromone released by other nearby females. Finally, physiological factors such as a female's age and mating status can also influence the timing of calling (see McNeil 1991; Groot 2014). Although the timing of these rhythmic patterns in a given species may seem relatively fixed in given environmental regimes, these rhythms are usually variable in the field as a consequence of abiotic conditions, particularly temperature. For example, in night- or late-day-active species, mating rhythms are advanced during relatively cool days so that mating occurs during warmer intervals (e.g., Cardé et al. 1975; Comeau et al. 1976; Teal and Byers 1980), presumably to minimize energetic costs associated with flight in cooler temperatures, whereas day-active species may delay calling on cool days (e.g., Gorsuch et al. 1975; Schal and Cardé 1986; Webster and Yin 1997) for the same reason. The arctiine moth *Virbia immaculata* (Erebidae) will initiate calling within minutes of a sudden drop in temperature, provided this cue falls within an overall permissible window of calling activity (Cardé and Roelofs 1973). There can even be a carryover effect, wherein the temperature regime experienced the previous day modulates current response to temperature. In *Grapholita molesta*, for example, a cool day not only advances calling on the day it is experienced but also on a subsequent warm day (Baker and Cardé 1979c). An example of extreme flexibility is in the omnivorous leafroller *Platynota stultana* (Tortricidae), which calls exclusively in scotophase during warm conditions (above 24°C) and almost exclusively during daytime during cool conditions (13°C); this species also modulates the timing of calling based on whether temperature experienced on the previous day was warm or cool (Webster 1988). In California, this tortricid has multiple generations from spring to fall, and so these temperature regimes fall within the natural ranges that this species experiences.

Most accounts of calling assume that once it commences, it is continuous until the female retracts her pheromone gland. Several species, however, engage in calling bouts, with gland exposure of several minutes duration interspersed with gland retraction for several minutes, when there should be very little pheromone emission (Sower et al 1971; Ohbayashi et al. 1973; Marks 1976; Swier et al. 1977). It is presumed that in the field, if a given bout of calling does not attract a suitor, a female will delay calling for several minutes. The significance of calling in bouts could be to conserve pheromone, so that when release does occur the rate of emission will be higher than if calling is spread over a longer interval, and the active space can be projected farther downwind (Sower et al. 1971, 1973). But it also has been suggested that between calling bouts females might move to a new calling site, thereby increasing their chance of attracting a mate (Swier et al. 1977). Most moths, however, seem to call continuously once calling is initiated.

Besides modulating when during the permissible window a female will call, ambient temperature also can influence a female's rate of pheromone emission. A warm period elevates release in the cabbage looper *Trichoplusia ni* (Noctuidae) (Liu and Haynes 1994) and therefore extends the downwind projection of the active space; this in turn should increase a female's likelihood of attracting a male.

Another example of plasticity in the timing of calling is seen in the influence of neighboring females. Female *Utetheisa ornatrix* (Erebidae, Arctiinae) can initiate calling when they detect pheromone from another calling female (Lim and Greenfield 2007). One explanation for such "contagious" calling or as it is sometimes called "autodetection" is that males can be a limiting resource and a female may increase her chance of mating by offering a competing odor plume. Detection of a conspecific female's pheromone plume would generally indicate that the female would be downwind of her competitor and so by initiating calling she might intercept an upwind-orienting male. This phenomenon, which Lim and Greenfield termed "pheromonal chorusing," has been documented in other moth families (e.g., Tortricidae: Stelinski et al. 2006) and likely it is more widely distributed than is currently recognized. A key factor determining the value of contagious calling is the operational sex ratio (Lim and Greenfield 2007). If few males are available, then contagious calling should be favored; a male contribution to reproduction other than sperm and multiple mating of females may also promote evolution of this behavior. Whether "experience" influences this process is unclear—for example, do females that have not mated after one night of calling then have an increased propensity to call as soon as they detect another calling competitor?

There is a contrary phenomenon in which the presence of conspecific pheromone delays the onset of calling and/or also causes its early termination (e.g., Gökçe et al. 2007; Yang et al. 2009). The selective value of this response is unclear, but it could involve conservation of pheromone or waiting for female–female competition to diminish. Some tests examining the effects of autodetection use sources of synthetic pheromone and exposure regimes (high concentrations and long times of exposure) that are unlikley to mimic the demographics of female interactions under natural conditions; the relevance of such artifical conditions to naturally induced contagious calling is unclear.

It is clear with some species that a calling female in the field may not attract a male on the first night of calling, or perhaps not at all during her lifetime. Field tests with the spruce budworm, *Christoneura fumiferana* (Tortricidae) (Régnière et al. 2013), and *Lymantria dispar* (Sharov et al. 1995) in very low population levels suggest that some females never attract a mate even though males are in the vicinity. These observations suggest that female–female competition for mates when few males are available may be more prevalent than has been recognized.

SELECTIVE FORCES LIMITING THE MATING WINDOW

The rhythm of male response mostly has been documented in the field mainly by using automated "clock traps" and sometimes by a direct observation over time of males captured in traps. Some observations use synthetic lures and others virgin females. It is problematic, however, to compare periodicities of female calling, which typically are observed under unvarying controlled conditions in the lab, with the patterns of male attraction in the field. What we do expect is that patterns (i.e., onset and termination) of male response to pheromone and ranging flight (i.e., flight prior to contacting a pheromone plume) should be coincident or perhaps somewhat in advance of and after the onset and termination of female calling. Stabilizing selection, however, ought to constrain the timing of these behaviors to relatively fixed intervals, because male ranging flight when females are unavailable and female release of pheromone when males are unresponsive are both energetically wasteful. Flight also exposes males to predation. Stabilizing selection should therefore tend to act to keep mating rhythms within narrow windows, typically a duration of several hours, although exceptions can be found, as with the gypsy moth (Cardé et al. 1974; Webster and Yin 1977). Stabilizing selection should also dictate that the peaks of male flight and calling should be coincident. There are examples such as in *Virbia aurantiaca* (Erebidae, Arctiinae) in which male attraction to pheromone lures in the field occurs in daytime many hours in advance of nighttime female calling (e.g., Cardé 1974), but this is artifactual and entirely dependent on the presence of lures of synthetic pheromone, as males do not engage in daytime ranging flight in the absence of pheromone.

MATING RHYTHMS IN REPRODUCTIVE ISOLATION

Despite plasticity in the daily timing of mating, there are many cases wherein differences in timing constitute an effective isolating mechanism among closely related moth species (e.g., Collins and Weast 1961; Greenfield and Karandinos 1979; Zaspel et al. 2008). Teal and Byers (1980) documented that temperature-induced shifts in the calling rhythms of three sibling *Euoxa* species (Noctuidae) moved in parallel, preserving complete temporal partitioning.

The temporal plasticity of mating rhythms, however, suggests that partitioning of the communication channel by exclusive mating rhythms is infrequently a "fail-safe" mechanism for preventing cross-attraction of species sharing the same pheromone. Even partial separation of mating times, however, can contribute to reproductive isolation. Dopman et al. (2010), for example, have suggested that the partially offset mating times of the (Z)- and (E)-pheromone strains of the European corn borer *Ostrinia nubilalis* observed in the lab (Liebherr and Roelofs 1975) could account for an important proportion of pre-mating isolation between these strains in the field; this trait coupled with phenological differences and, of course, pheromonal specificity (see Lassance, this volume) should provide a substantial barrier to hybridization. However, a field study in Italy where both strains co-occur failed to uncover any difference in the rhythms of males attracted to the two types of pheromone lures (Camerini et al. 2015).

A parallel case of partial isolation as a consequence of the timing of mating is found in the corn and rice strains of the fall armyworm moth, *Spodoptera frugiperda* (Noctuidae) (Pashley et al. 1992; Schöfl et al. 2009; Groot et al., this volume). Here, isolation is partially by habitat (because of host-plant distribution) and this coupled with somewhat different times of male attraction to pheromone lures (the same blend) reduces gene flow where these two strains are in spatial contact. That there is some cost to hybridization in terms of hybrid sterility suggests that selection could favor further divergence by mating times among these host races (Groot

et al., this volume). The pheromone system also varies geographically, and will be considered in more detail in the "Geographic Variation" section below.

The reason why different patterns of timing exist among closely allied species is not always clear and there are likely multiple factors that dictate why particular rhythms have evolved and are maintained. Do these diverge as a consequence of the value of minimizing or eliminating cross-attraction or are there autecological factors that dictate timing? In the case of sympatric *Virbia* species, optimizing flight to favorable temperatures may have contributed to the mating rhythm of *V. lamae*. This moth is found in coastal bogs of the Maritime Provinces of Canada, Maine, and near the Great Lakes of Michigan and Wisconsin. In this habitat nighttime temperatures are low and so mating occurs during the daytime (Schal and Cardé 1986; Cardé et al. 2012). Females of two other broadly sympatric *Virbia* species, *V. aurantiaca* and *V. ferruginosa*, call after dark (Zaspel et al. 2008) and are not found in these bogs. As all of these *Virbia* share the same pheromone, 2-methylheptadecane (Roelofs and Cardé 1971), both differential habitat preference and mating times contribute to reproductive isolation of *V. lamae* among these species.

How such partitioning by rhythm first arises is unresolved. Most moths are active at nighttime, in part it is generally assumed, to avoid predation by day-active predators. Consistent with this hypothesis, many day-active species mimic well-defended species, such as aegeriid (clearwing) moths closely resembling wasps and hornets. Mating activities, furthermore, are typically relegated to only a portion of the night or day (e.g., Groot 2014, and references therein). This is suggestive that a narrow window of emission is a consequence of the limited amount of pheromone available to emit. In *Heliothis virescens*, however, the available pool of pheromone precursors seems high relative to its actual conversion to pheromone (Foster and Anderson 2015), suggesting that the cost of pheromone production is not highly limiting, but the cost of flight also may act to limit when males search for females. Flight likely is more constrained by low temperatures in small moths, given their low volume to surface area ratio and consequent dissipative heat loss (Cardé et al. 1975).

Another example of partitioning by exclusive mating rhythms is evidenced in *Callosamia promethea*, *C. angulifera*, and *C. securifera*. These saturniids are sympatric in South Carolina. *C. securifera* commences calling around 10:00, while male flight ceases by 16:00, when *C. promethea* starts its activity which continues to sunset. Finally, nocturnal *C. angulifera* mates between sunset and midnight (Collins and Weast 1961; Ferguson 1971–1972). There seems to be no obvious differentiation among the pheromones of these three *Callosamia* species, and a late-calling *C. promethea* can attract *C. angulifera* (Tuskes et al. 1966), although to date only the pheromone of *C. promethea* is known (Gago et al. 2013).

Reproductive character displacement after secondary contact of populations previously isolated could be one selective force for rhythm divergence. Such may the case in *Lymantria monacha* (Erebidae). Males were captured between 02:00 and 05:00 in Honshu, Japan, and between 18:00 and 24:00 in Bohemia, Czech Republic. Gries et al. (2001) suggested that the rhythm of attraction in Japan serves in part to minimize cross-communication between *L. monacha* and *L. fumida*, which is active early in the night but also is separated from *L. monacha* by important differences in their respective pheromone blends. Both the mating rhythms and differences in the pheromones of *L. monacha* in Europe and Japan were hypothesized to have diverged because of reproductive character displacement in the Japanese population of *L. monacha*.

Selection and Variation in Male and Female Signaling Traits

The existence of multicomponent pheromone blends creates the potential for variation in ratios and preferences among individual females and males, respectively. Although it is universally accepted that variation in these signaling traits is necessary for them to evolve by natural selection, there is considerable debate regarding the selective forces thought to act on them (see Coyne et al. 1988; Ryan and Wilczynski 1991; Paterson 1993). In essence the debate centers on whether or not selection acts on signaling traits to promote efficient signaling among potential mates (i.e., the recognition concept [Paterson 1978, 1985]) or to isolate one population from another (i.e., reinforcement or the isolation concept [Dobzhansky 1940; Howard 1993]). In general mate recognition characters are hypothesized to experience stabilizing or directional selection, which is expected to reduce the amount of genetic variation (Paterson 1978, 1985). Models of female moth sex pheromone evolution that emphasize sexual selection theory (i.e., the asymmetric tracking and wallflower hypotheses) predict that selection will be stronger on male response than female signal traits (see Allison and Cardé, this volume). As a result variation among female signals is predicted to be high, whereas variation among male preference functions is predicted to be low. In essence this means that the distribution of signals present within a population will be the aggregate effect of all females, whereas the preference function of a population will largely be an emergent property of individual males. An alternative to these models, the stasis model, predicts that male preference function shape will constrain variation in female moth pheromone traits, particularly the blend (see Allison and Cardé, this volume).

Harari and Steinitz (2013), however, in a conflicting view predict that females should compete for males and that males could select females on the basis of variation in the ratio of pheromone components, assuming the ratio of pheromone components predicts female quality. This would require, for example, component ratio or quantity to be correlated with a specific quality trait such as female age or size and therefore fecundity. It also assumes that males are more apt to reject or not be as attracted to females with a "low quality" signal.

Few studies have characterized male preference functions and fewer have studied how individual males might vary in their response to deviations from the blend optimum. In field tests with *Grapholita molesta* (Cardé et al. 1976) and *Pectinophora gossypiella* (Gelechiidae) (Haynes and Baker 1988), males were allowed to mark themselves at lures baited with optimal or suboptimal blends. Males were then captured on subsequent days with traps baited with the same blends. The resulting distributions of captures among blends failed to reveal any polymorphism. For example, males that were initially marked at a suboptimal blend distributed their subsequent attraction across the blends in precisely the same way as all males—in other words, these males were most likely to be captured in traps baited with the optimal blend. These results are consistent with the expectation that genetically based variation among male preference functions is low. It is noteworthy that in the case of *P. gossypiella*, the male response function (Haynes and Baker 1988) appears to be considerably broader than the range of ratios produced by females (Haynes et al. 1984; Collins and Cardé 1985) in agreement with the asymmetric tracking model, whereas in *G. molesta*, the male response function (Baker and Cardé 1979b) closely mirrors the ratios produced by the females (Lacey and Sanders 1992) in agreement with the stasis model.

High levels of variation have been observed in the sex pheromones of several moths (e.g., *Phthorimaea operculella* [Ono et al. 1990], *Cadra cautella* [Allison and Cardé 2006], *Heliothis virescens* (Noctuidae) [Groot et al. 2014]). Fisher's fundamental theorem predicts that selection will rapidly eliminate additive genetic variance for total fitness (Fisher 1930). Although the heritability of a trait is not defined by the amount of additive genetic variance but rather the amount relative to total phenotypic variance, Fisher's fundamental theorem has been extended to predict that characters closely related to fitness will have low heritabilities. In general, the literature supports the prediction that the heritability of a trait is negatively correlated with its fitness consequences (Roff 1997). Published estimates of narrow-sense and realized heritabilities for female moth pheromone blends are high (see Allison and Cardé 2006) and suggest that variation in the pheromone blend may have low impact on female moth fitness. Alternatively, theoretical arguments support the maintenance of additive genetic variance in traits experiencing selection expected to reduce or eliminate variation (Roff 1997).

The majority of the existing literature on moth pheromones involves laboratory colonies in which the need for males to locate females over long distances and discriminate among conspecifics and heterospecifics does not exist. Consequently, selection on the female pheromone blend is likely relaxed. Depending on how long the colony has been in culture, this could result in higher levels of variation than would be seen in natural populations. Alternatively, many of these research colonies were started from a few individuals and consequently would have passed through a population bottleneck. Although population bottlenecks are generally predicted to result in reduced levels of variation, experimental and theoretical work has demonstrated that inbreeding can increase levels of heritable variation (particularly when substantial dominance and epistasis exist) (e.g., van Buskirk and Willi 2006). Several studies have compared variation in the female pheromone blend in wild and laboratory populations and have found similar levels of variation (e.g., Miller and Roelofs 1980; Haynes et al. 1984).

Causes of Variation

Methodological Effects

Determination of the females' most common ratio and describing the breadth of variation rely principally on two methods: quantification of ratios either in gland extracts or from airborne collections from individual females. Most of what we know about variation among individuals comes from gland extracts (as this is a much simpler procedure and airborne pheromone may be released in miniscule quantities that can be difficult to measure). The ratios found in the gland, however, may not be identical to those that are released. When the components are very similar (e.g., 14-carbon acetates with one or two double bonds that differ in either geometry or position), there should be a good correlation between the gland

and emitted ratios, even when the temperatures at which samples were collected differ. For example, female *Pectinophora gossypiella* have a two-component blend of (Z,Z)- and (Z,E)-7,11-hexadecdienyl acetate and based on gland extractions of individual, laboratory-colony females originating from Arizona, the mean proportion of (Z,E)-7,11-hexadecdienyl acetate was 44 % (SD = 2.3) (Collins and Cardé 1985). Among several California field strains of *P. gossypiella*, Haynes et al. (1984) found that the emitted proportion varied between 38% and 40% Z,E isomer with SDs between 4.0 and 5.1. The means and the variability in ratios found in these two studies were very close and some (or all) of the minor discrepancies between these two studies could be due to different methods of collection and quantification. Similarly, in *Grapholita molesta* the ratio of (Z)-8-dodecenyl acetate to (E)-8-dodecenyl acetate was measured as 93.5 to 6.5 in the combined effluvium from many females (Cardé et al. 1979) and a mean of 95.8 to 4.2 in the effluvium of individual females (Lacey and Sanders 1992). Whether these slight differences are "real" is unclear, given the differences in collection and quantification methods. Generally when moths use components with the same (or similar) functional moieties and carbon chain lengths, the variability in ratio among individuals can fall within a very narrow range (e.g., with several percent). Hunt and Haynes (1990) compared the ratios of pheromone components in glands and released in *Trichoplusia ni*. As they measured volatized pheromone from forcibly extruded glands, they could compare proportions in the glands and volatilized during the time of normal calling and during early scotophase and photophase when female signaling does not occur. Throughout this cycle, proportions of the 12-carbon acetate minor components in the gland and volatilized remained constant relative to the main pheromone component (Z)-7-dodecenyl acetate (Z7-12Ac). The proportions of the 14-carbon components relative to (Z7-12Ac), however, increased substantially throughout the calling interval, while its titer in the gland remained relatively constant, suggesting a periodicity in transport to the gland's surface. Thus, use of glandular extracts to determine optimal ratio and characterize variability can be appropriate, especially in species where the components are long-chain Type I pheromones of the same chain length and moiety. In other cases, glandular extracts and airborne collections can yield significantly different values.

Environmental Effects

Tight correlations between gland content and release, however, are less expected when the components differ in chain length (e.g., 12 vs. 14 carbon-chain length), functional moiety (e.g., acetate vs. alcohol vs. aldehyde), or other characteristics affecting vapor pressure and therefore evaporation from the glandular surface (Roelofs and Cardé 1977). To achieve a constant ratio with such disparate components necessitates a precise regulation of biosynthesis and of transport of pheromone to the glandular surface. To date, however, there is no evidence of precise regulation of ratio in pheromone blends that have diverse chemistries.

Liu and Haynes (1994) observed emission rates from forcibly extruded pheromone glands in *Trichoplusia ni* and documented a 70% increase in total blend emission from 15°C to 30°C. This value was below that observed for artificial pheromone dispensers. For example, McDonough et al. (1989) observed that the half-lives of 12Ac, Z7-12Ac, and Z9-14Ac at 15°C were five times

those at 30°C, which translates into about a fourfold increase in emission rates for all three compounds. Also unsaturated and shorter acetates have greater emission rates than saturated and longer acetates from artificial substrates (Butler and McDonough 1979, 1981; McDonough et al. 1989). Liu and Haynes (1994) found that the percentages of components emitted during 10-minute intervals from extruded glands were highest for Z5-12Ac and Z7-12Ac, lowest for Z7-14Ac and Z9-14Ac, and intermediate for 12Ac and 11-12Ac, a result likely due to differences in the vapor pressure of the pheromone components.

Another example of abiotic effects on the pheromone phenotype involves the potato tuberworm moth, *Phthorimaea operculella* (Gelechiidae). The sex pheromone of *P. operculella* is a two-component blend of (E,Z,Z)-4,7,10-tridecatrienyl acetate (triene) and (E,Z)-4,7-tridecadienyl acetate (diene) (Persoons et al. 1976). In addition to high levels of variation within populations in California and Japan (Ono et al. 1990) laboratory rearing studies suggest that the ratio of the triene to diene is influenced by rearing temperatures. Biosynthesis of the triene occurs primarily during the pupal period and the diene primarily during the first few days postemergence (Ono 1993, 1994). As a result the temperature experienced during both the pupal period and shortly after emergence can affect the pheromone blend.

The release rate of pheromone components increases with increased temperature but components with different functional groups and/or chain lengths may do so differentially. Moreover, at different environmental temperatures, the emitted ratios also should differ (Roelofs and Cardé 1977). In such component mixtures physiochemical rules also dictate that male response to ratio must be flexible. An exception to this occurs in arctiines; these moths generally seem to release pheromone by atomization from paired internal abdominal reservoirs via two small pores on their abdominal tip (e.g., Yin et al. 1991). Vaporization of the resulting aerosol thus causes the composition of the plume to be identical with the gland's contents, although to date there are not any known cases of the use of precise ratios by arctiines (Cardé and Millar 2008).

Environmental effects on male response traits have also been observed. For example, temperature can modulate the breadth and shape of the male's preference function. In *Grapholita molesta* and *Pectinophora gossypiella*, Linn et al. (1988) found in wind-tunnel trials that at 20°C males were attracted to narrower range of ratios than at 26°C. As both *G. molesta* and *P. gossypiella* are a multivoltine species, active in spring and summer, these are natural temperature ranges that could be encountered during mate finding. There was also some shifting in *G. molesta* of the most attractive ratio with concentration, with a 3-µg lure causing optimal attraction with 6% (E)-12Ac (close to the natural ratio) and 30-µg lures causing optimal attraction to an unnatural 2% lure (Linn et al. 1988), a finding consistent with the Threshold Hypothesis. Another curious case is found in the multivoltine *Platynota stultana*, where the most favored ratio in field trapping shifts with season, presumably because of ambient temperature (there was no evidence that the temperatures that the female experiences during development or as an adult shifts her ratio of production) (Baker et al. 1978). This phenomenon also is consistent with the Threshold Hypothesis. During summertime flights of *P. stultana* when relatively warm conditions prevail, dispenser release rates would increase and therefore create plumes with potentially high and unnatural concentrations of pheromone; summertime is when the most favored ratio shifts away from the female's natural blend.

Another example of environmental effects on male response is seen in the (Z)-strain of *Ostrinia nubilalis*. This species is notable for its narrow range of female ratio production of Z11-14Ac to E11-14Ac in both the (Z)- and (E)-strains (Kochansky et al. 1975; see Lassance, this volume). In a wind-tunnel setting, males are attracted to a narrow range of ratios specific to their strain (Linn et al. 1997). In another wind-tunnel study, however, an initial exposure to the natural (Z)-blend early in orientation along a plume predisposed (Z)-strain males to then orient along plumes that otherwise would be unattractive. Thus, plumes from a 1:1 mixture of the (Z)- and (E)-isomers (approximating the blend of F1 hybrid females) and, most surprisingly, plumes from an (E)-strain mixture evoked good levels of plume following and source location, following an initial encounter with the (Z)-strain mixture (Kárpáti et al. 2013). Such a broadening of response specificity based on a first exposure to the natural blend may account for the unexpectedly high levels of hybridization between the two strains where they occur sympatrically in the eastern United States (Dopman et al. 2010). In dense populations (Cardé et al. 1975), plumes from the two strains occasionally may overlap, facilitating mating "mistakes" and the occurrence of hybrids. Whether such odor generalization occurs in other moth species and influences their specificity of response to ratio (or the presence of all components in a blend) has not been determined. Although the effects of concentration and temperature on male preference functions have been well studied in only two tortricids and a gelechiid, their occurrence suggests that these effects may be widespread among other moths.

Variation within Individuals

To estimate the repeatability of the measure of a trait, total phenotypic variance in a trait can be partitioned into variance within and among individuals (Falconer and Mackay 1996). Repeatability estimates range from 0 to 1, with an estimate of 1 indicating that successive measures of the trait (e.g., female pheromone blend) are identical. Few studies have estimated the repeatability of female moth pheromone traits. Du et al. (1987) observed high estimates of repeatability of the female pheromone blend in *Yponomeuta padellus* and *Y. rorellus* (0.82–0.90). Similarly, Svensson et al. (1997) observed high repeatabilities for the absolute and relative amounts of pheromone components of *Agrotis segetum* (Noctuidae), although repeatability was observed to decrease as the age of females increased. In many moth species the absolute and relative titers of pheromone components are not independent of female age. In moths that mate shortly after eclosion titers of sex pheromone components are often low at eclosion, increase for a short period of time (e.g., one day) and then decline progressively until death (e.g., Schal et al. 1987; Tang et al. 1992; Gemeno et al. 2000; Allison and Cardé 2006). In some species titers of individual components decline at different rates, resulting in a change in the pheromone blend ratio as female age increases (e.g., Allison and Cardé 2006). Similarly, the absolute and relative titers of pheromone components can vary throughout the LD cycle (e.g., Allison and Cardé 2006) and suggests that the rate of release from the pheromone gland exceeds the rate of biosynthesis. There also is evidence that adult feeding can influence absolute pheromone titers in some moths (Foster and Anderson 2015). Variation in the temperatures experienced by larvae and pupae can affect several female signaling traits in moths, including age of first calling, calling periodic-

ity, calling duration and pheromone gland content (McNeil 1991). For example, the ratio of sex pheromone components in *Phthorimaea operculella* varies with temperature (mean percent of (E,Z,Z)-4,7,10-tridecatrienyl acetate was ca. 80% at 15°C and 30% at 35°C) (Ono 1993, 1994); these males, however, are catholic in their attraction across this range of ratios (Ono and Orita 1986) and so there appears to be no preferential selection by males for a female with a particular ratio.

Variation among Individuals

It has been hypothesized that because sexual selection generally acts more strongly on males than females males should have lower variance in their signal responses than females in signal production (see Allison and Cardé, this volume). An emergent property of this hypothesis is that while the distribution of female pheromone signals reflects variation among individuals, the breadth of individual male preference functions should reflect that of the population. In support of this hypothesis in some species significant (i.e., high) amounts of variation has been observed among females and populations in the pheromone blend.

Geographic Variation

While some species have little variation in pheromone production and response traits among populations, some do. To establish true differences in male response among different populations, as noted in "Causes of Variation" above, the lures, traps, and trapping protocols must be identical; documenting variation among females in production is methodologically straightforward but requires collections of females from several localities. Many comparisons of trapping using the same blend variants in different regions do not meet all of these criteria. In general, the existing literature can be divided into two groups: those documenting species with populations that seem to have diverged and those that seem identical throughout much or all of their range. Some of the latter are species that have a widespread geographical expansion because of recent human activities and for which there is no compelling evidence of geographic divergence in their pheromone signal (e.g., *Cydia pomonella*, *Grapholita molesta*, and *Pectinophora gossypiella*) and will not be considered here. Generally, there are few cases where geographical variation in the pheromonal message and/or the response it evokes has been examined in detail by methods that allow unambiguous population comparisons. We consider here selected examples in which the methods used allow us to determine the extent of the geographical variation in the pheromone produced and/or response and also cases in which geographic divergence has been claimed. Further discussions of pheromone polymorphisms in this volume include Hillier and Baker for heliothines, McElfresh and Millar for *Hemileuca*, Lassance for *Ostrinia nubilalis*, Tabata and Ishikawa for "oriental" *Ostrinia*, and Groot et al. for *Spodoptera frugiperda*.

There are examples of widely distributed species appearing to have the same pheromone system throughout their range. For example, males of the spruce seed moth, *C. strobilella* (Tortricidae) were attracted to a blend of (E,E)- and (E,Z)-8,10-dodecadienyl acetates centered around a 6:4 ratio (Wang et al. 2010) in eight localities ranging from France and Sweden to Central Siberia and Inner Mongolia (Svensson et al. 1997). In two Swedish populations, males were attracted to a rather broad ratio of two these components (Wang et al. 2010) and this pattern also was documented in Inner Mongolia (Svensson et al. 1997). All of the populations, save that of Inner Mongolia, which has a different host plant, have little genetic differentiation based on *COI* and *EF1α* genes, suggesting recent gene flow across their spruce-forest habitat during a postglacial expansion. Because of habitat fragmentation, it is likely that this species is not currently mixing continuously across its entire range, although there seems to be no pheromonal barrier for continued gene flow.

An example of geographic segregation of pheromonal types is seen in the turnip moth, *Agrotis segetum* (Noctuidae), which is widely distributed across Eurasia and Africa, although some of its occurrence beyond Europe may be recent and human-aided. The three pheromone components are (*Z*)-5-decenyl (Z5-10Ac), (*Z*)-7-dodecenyl (Z7-12Ac), and (*Z*)-9-tetradecenyl (Z9-14Ac) acetates, and these were found to have a mean ratio of 12:59:29 in a Swedish population (Löfstedt et al. 1985), whereas a Zimbabwean population produced a 78:20:2 blend (Wu et al. 1998). Tòth et al. (1992) examined how well males were attracted to binary and ternary mixtures of these components at 11 sites in Eurasia and Africa (figure 3.1). The evidence from these field studies also points to populations in South Africa and Zimbabwe being quite distinct, as males there were attracted to the Z5 component alone with the other two compounds suppressing capture, at least at the ratios tested. Sites sampled from Europe to Egypt seemed to have a similar distribution of "types," with mixtures of Z5-10Ac and Z7-12Ac or Z7-12Ac and Z9-14Ac capturing significant numbers of males but fewer than the three-component blend. Males from Kazakhstan and Turkmenistan also were most attracted to the three-component blends, but seemed less attracted to the two-component blends than European and Egyptian males. There is some evidence that within Europe populations of *A. segetum* are further subdivided (Löfstedt et al. 1986; Hansson et al. 1990). In a French population, females produced relatively more Z5-10Ac and males possessed many receptors attuned to this compound; a Swedish population produced less Z5-10Ac and had fewer Z5-10Ac receptors; in another population from Armenia/Bulgaria there were more Z9-14Ac than Z5-10Ac receptors and little Z5-10Ac produced (Hansson et al. 1990).

Tòth et al. (1992) suggest that in Europe there is a mix of male response types, some most attracted to Z5-10Ac plus Z7-12Ac and others most attracted to Z7-12Ac plus Z9-14Ac, with both types presumably most responsive to the three-component blend. Individual females from Sweden varied widely in their ratio of these three components (Löfstedt et al. 1985), indicating that stabilizing selection by male preference has not limited signal variation. Whether the proposed different responder types and female producers tend to mate assortatively in nature remains to be determined, but, if so, this could promote disruptive selection and possibly generate divergent communication channels. There are, however, no known associations in Eurasia of loci associated with these pheromone differences with other ecological traits (such as host plant preferences) which could reinforce divergence and potentially drive these pheromone forms toward reproductive isolation.

Another factor potentially driving divergence, in this case at the geographical level, is antagonistic interactions with other species sharing some of these pheromone components. As Tòth et al. (1992) point out, in field tests Z5-10Ac alone attracted several other species in Eurasia and northern Africa, whereas in Zimbabwe and South Africa Z5-10Ac only attracted

A. segetum, a pattern consistent with *A. segetum* having tailored its pheromone in Eurasia to reduce communication interference with other species.

Another case of geographic variation in the pheromone signal is found in the obliquebanded leafroller, *Choristoneura rosaceana* (Tortricidae). Populations from Quebec, New York, Ontario, Michigan, and British Columbia were compared for field response to synthetic lures and, using lab cultures of females collected from each locality, composition of their pheromone glands (El-Sayed et al. 2003). Females from all populations contained the same four components: Z11-14Ac, E11-14Ac, Z11-14OH, and Z11-14Ald. The first three of these were found in very similar proportions in all populations, but females from Michigan and especially New York produced very little of Z11-14Ald relative to the three Canadian populations. In Ontario and Quebec, inclusion of the aldehyde at 1% of the four-component blend increased capture more than twofold and in British Columbia about fourfold. In New York and Michigan, however, adding Z11-14Ald did not elevate trap capture. An outstanding question is: which state is ancestral, the three- or the four-component system?

Some geographic divergence in signal is also evident in the black cutworm moth, *Agrotis ipsilon* (Noctuidae), a species widely distributed in Africa, Asia, Australia, Europe, and the Americas. Its pheromone is a mixture of Z7-12Ac, Z9-14Ac, and Z11-16Ac. Gemeno et al. (2000) found overall that populations from Kentucky and Kansas had Z7-12Ac as the most abundant component, whereas Z11-16Ac was most abundant in populations from Egypt and France; levels of Z9-14Ac were similar in all sampled populations. The Nearctic and Palearctic samples differed in ratio of Z7-12Ac to Z11-16Ac. Significant heritabilities for amounts and ratios were found in French and Kentucky populations (none were detected in the Egyptian strain). Geographic differences in pheromone signal (particularly in the ratio of Z7-12Ac to Z11-16Ac), however, were insufficient to create a barrier to attraction of males from any of these four strains to females from any of the four strains, as determined in wind-tunnel trials. In other words, males from geographically disparate populations accept a range of component ratios that includes females from all the populations studied (Gemeno et al. 2000).

The fall armyworm moth, *S. frugiperda*, pheromone is a mixture of Z9-14Ac and Z7-14Ac; both are required for attraction. This species also has two host forms, termed the corn and rice strains. These labels are a bit misleading in that the host range of both strains is considerably broader and the extent to which they are isolated, at least is South America, is unresolved (Juárez et al. 2014). There are two other components of the pheromone gland that have been suggested (Groot et al. 2014) to increase attraction: Z11-16OH and Z9-12Ac. Field tests of male response to two four-compound blends across North America, the Caribbean, and South America report some evidence in the corn strain of geographical divergence in response to the dose (ratio) of Z7-12Ac in the blend (see Groot et al., this volume). As with other moths using blends consisting of Type I structures (see Foster, this volume) with differing chain lengths and moieties, we expect variation in ratio of emission. This species is a good candidate for continued exploration of geographic variation in the pheromone signal and its possible linkage to other characteristics of host preference and factors that would influence the viability of progeny.

Another example of within species variation is pheromone polymorphism in the larch bud moth, *Zeiraphera diniana* (Tortricidae). There are two larval "forms," one feeding on

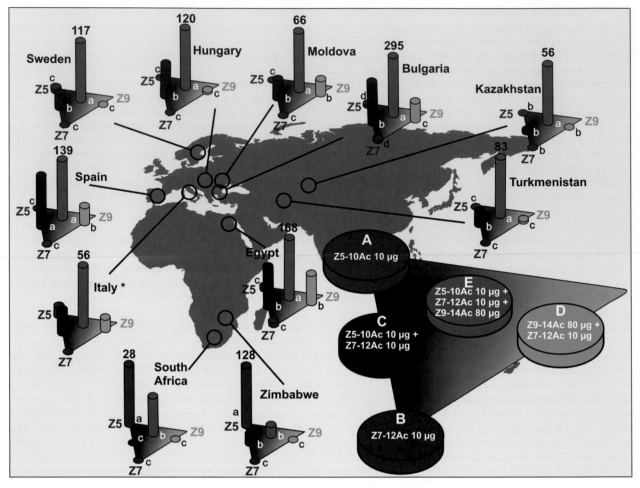

FIGURE 3.1 Capture of male *Agrotis segetum* in 11 localities in Eurasia and Africa. Each locality gives the number captured at the most attractive bait and the letters indicate significant difference among lures at $p = 0.05$. *In Italy only one trap per bait variation was set up, and data were not statistically analyzed for significance. The inset provides the lure compositions (modified from Tòth et al. [1992]; redrawn by Ron Fournier).

larch and the other on pine. In some localities these races are somewhat isolated by adult phenology and geographical distribution, but many populations are synchronous and fully sympatric. The pheromone of the larch race is E11-14Ac and that of the pine race is E9-12Ac; there are traces of the E9-12Ac in the larch race and E11-14Ac in the pine race (Baltensweiler et al. 1978), but these trace compounds do not seem to contribute to attraction in either form. In studies reported by Emelianov et al. (2001), wild-caught males were typed with high accuracy to form by allozyme markers. In an experiment to determine if proximity to host plant influenced male capture, females of both forms were set out in traps transecting a valley where both forms and hosts occur. Pine-form females calling on pine and larch generally attracted pine-form males and few larch-form males. Larch-form females on larch attracted larch males (probabilities for cross-form attraction in these three pairings ranged from 0.033 to 0.091). Larch-form females in pine attracted larch-form males but also many pine-form males (0.377 probability). Proximity of the alternative host also influenced cross-attraction: the larger the fraction of the nearby alternate host, the greater the likelihood of a female attracting an alien male. This also is consistent with the observation that both sexes collected in the field have a high propensity to rest on their own host (Emelianov et al. 2003). Attraction periodicities of males by females of the two forms were synchronous and therefore are not a

factor in partitioning of the pheromone channel. Despite differing pheromone channels and the influence of hosts on distribution, there is some gene flow between these forms (Emelianov et al. 2004).

How divergence of the larch and pine forms of *Z. diniana* arose is unclear. Jiggins et al. (2005) speculated that the initial change was via a host shift, followed by a genetic linkage with mate choice. In areas where the two forms co-occur, continued assortative mating between types of slightly divergent signals combined with the pleiotropic effects of host choice would have promoted further divergence of sexual signaling (Jiggins et al. 2005), a process that may continue in the now extant populations which are already largely reproductively isolated. The progenitor pheromone system might well have had both E11-14Ac and E9-12Ac as essential components and so the initial divergence would have been toward divergent ratios and finally in each host race use of a single component.

McElfresh and Millar (this volume) document that the pheromone of the saturniid *Hemileuca eglanterina* diverges in a population that fully overlaps with *H. nuttalli* (hybrids of *H. eglanterina* × *H. nuttalli* are infertile, see Tuskes et al. 1996). In areas where *H. eglanterina* and *H. nuttalli* do not co-occur, the optimal blend for *H. eglanterina* is a 100:0.3:03 blend of E10Z12-16Ac, E10Z12-16OH, and E10Z12-16Ald. *H. nuttalli* is most attracted to a 100:48 blend of E10Z12-16Ac and E10E12-16Ac. In a population sympatric with *H. nuttalli*, however,

H. eglanterina males are attracted to a 100:10 blend of only E10Z12-16OH and E10Z12-16Ald. Moreover, adding E10E12-16Ac, a required component for *H. nuttalli*, to this blend inhibits attraction of *H. eglanterina* only in sympatry. At another site where the two species are on the edge of contact, some *H. eglanterina* males are attracted to the three-component blend, whereas others are attracted to a two-component blend lacking E10Z12-16Ac. The selective force presumed to cause elimination of response to the acetate component in *H. eglanterina* where it co-occurs with *H. nuttalli* is that the latter species also uses this acetate component. Specificity of response of *H. eglanterina* in sympatry with *H. nuttalli* is further enhanced by an antagonistic effect of E10E12-16Ac on *H. eglanterina*. Reproductive character displacement in sympatry thus seems to be the result of *H. eglanterina* dropping a component of its pheromone formerly shared with *H. nuttalli* and adding an antagonistic response to the other component of the *H. nuttalli* pheromone. Further details on this case are given in McElfresh and Millar (this volume) and in McElfresh and Millar (2001).

Another possible example of reproductive character displacement has been suggested for *Lymantria monacha* and *L. fumida*, discussed earlier in the context of displacement of mating rhythms. Gries et al. (2001) proposed that the pheromone of *L. monacha* in Japan could have diverged because of communication interference between the Japanese populations of *L. monacha* and *L. fumida*. In Honshu, Japan, *L. monacha* has evidently substituted in its pheromone blend (7R,8S)-*cis*-epoxy-2-methyl-octadecane for (7R,8S)-*cis*-epoxy-2-methyl-octadecane which is a component of the pheromone of sympatric *L. fumida*. In Bohemia, Czech Republic, (7R,8S)-*cis*-epoxy-2-methyl-octadecane is one of three pheromone components of *L. monacha* and this blend is presumably its ancestral state. Gries et al. (2001) suggest that the divergence in Honshu is consistent with reproductive character displacement, but they also emphasized that study of other populations of these species in allopatry and sympatry in eastern Asia would be useful to substantiate the proposed mechanism of divergence.

Another case of reproductive character displacement was proposed for the sibling species *Archips argyrospila* and *A. mortunanus* (Tortricidae) (Cardé et al. 1977); both species share the same four components, but in the area of sympatry in New York, the range of attractive component ratios for Z11-14Ac and E11-14Ac appears narrower for *A. argyrospila* than for an allopatric population in British Columbia. This ratio is crucial for separation of these sibling species in New York. Again, sampling of other allopatric and sympatric populations would be important to establishing the contribution of reproductive character displacement to divergence.

Within population and geographical variation in the pheromones of *Heliothis virescens* and *H. subflexa* (Noctuidae) have been investigated extensively by Groot and colleagues (see also Hillier and Baker, this volume). One study by Groot et al. (2014) with *H. virescens* documented phenotypic variation in four widely separated populations in the ratio of 16Ald to Z11-16Ald in the pheromone gland. Selection experiments created high and low lines for the titer of 16Ald relative to Z11-16Ald and field-trapping assays with live caged females seemed to show that females with relatively high ratios of 16Ald to Z11-16Ald were less attractive to males than low-ratio females. The pheromone of *H. virescens* is comprised of Z9-14Ald and Z11-16Ald. Without knowing how these lines did or did not vary in the rate of pheromone emission and the ratio of the two attractive components, the significance of variation in 16Ald titer, if any, remains unclear.

Comparisons of male *H. subflexa* attraction to lures in the eastern United States and western Mexico were interpreted by Groot et al. (2007) to show that addition of Z7-16Ac to the pheromone blend caused an increase in capture in the eastern United States but not in western Mexico. Alternatively, Hillier and Baker (this volume) suggested that addition of Z11-16Ac may have accounted for an increase in trap capture in both regions. Another study (Groot et al. 2013) examined the amount of three acetate gland components (Z7-16Ac, Z9-16Ac, and Z11-16Ac) that contribute to blend exclusivity in *H. subflexa* in North Carolina by reducing the possibility of cross-attraction of male *H. virescens* and also *Helicoverpa zea* (Noctuidae). The pheromone of *H. subflexa* is a three-component blend of Z9-16Ald, Z11-16Ald, and Z11-16OH, and the three acetates have no known pheromonal activity (see Hillier and Baker, this volume). Groot et al. (2013) noted that the amounts of the acetates in the pheromone gland of western Mexico *H. subflexa* were reduced and suggest that this could reflect the lack of *H. virescens* in this region and therefore a reduced need for production of antagonists. Because the acetates and aldehydes are derived from the same precursor, it could be that higher acetate titers could lower aldehyde titers and suggest a pleiotropic trade-off of release of pheromone versus antagonist.

Groot et al. (2009) have proposed both geographic and between-year variation in pheromone gland composition and the associated male response in *H. virescens* and *H. subflexa*. Females from North Carolina of both species were somewhat more attractive in North Carolina to their respective males than females from Texas stocks, suggestive of some geographical differentiation, but the reciprocal comparisons of attraction in Texas were not made; other explanations (e.g., differences in the timing and duration of calling) cannot be excluded. As noted by Hillier and Baker (this volume), Groot and colleagues documented variation in *components* of the pheromone gland relative to Z9-14Ald, the minor (ca. 3%) pheromone component relative to the Z11-16Ald. The most unexpected finding was the suggestion of year-to-year variation in pheromone production and male attraction. This interpretation, however, rests on variation in gland constituents that are not known components of the pheromone. Notwithstanding, Groot et al. (2009) suggested that females alter their gland composition based on their olfactory experience either as larvae or as young adults, and that males could alter their preference based on similar influences. Whether these two *Heliothis* species varied their pheromone geographically and locally over the two-year course of this study remains unclear.

Seasonal variation in male response to pheromone component ratio, as discussed earlier, is known in the omnivorous leafroller, *Platynota stultana* (Baker et al. 1978). This example, however, seems artifactual and due to temperature-correlated differences in rates of pheromone emission from lures that shift response of males to an unnatural ratio with high levels of pheromone (Roelofs 1978), females did not vary their ratio with temperature (Baker et al. 1978).

Multiple mechanisms exist that could explain these patterns of geographic variation in moth pheromones. These include the following:

1. Reproductive character displacement, wherein divergence in the communication channel of species that are already isolated by other postmating mechanisms, is enhanced in areas of recent sympatry (Brown and Wilson 1956).

2. Reinforcement, as defined by Butlin (1987), is limited to cases where hybrids of the diverged populations are partially sterile or inviable, so that selection for a reduction in gene flow should enhance pheromone differentiation (via changes in pheromones or antagonistic heterospecific signals). Reinforcement thus enhances assortative mating and is a component of the speciation process.
3. Divergence of the channel could occur when allopatric populations change in response to drift, sexual selection, or perhaps environmental conditions (Paterson 1985).

Rarely do we have sufficient information about variation in the pheromone channel within and among populations to allow us to choose among these alternative mechanisms. For the larch bud moth and many of the other examples of geographic variation, the extent to which the proposed divergence evolved in either sympatry or in isolated populations that are now in secondary contact remains an open issue.

The extents to which females vary geographically in the ratio of their pheromone blend and local males are attuned to that blend remain outstanding questions. Relatively few moth species have been subject to sufficient scrutiny to establish such variation definitively and this limits our ability to establish the contribution of reproductive character displacement, reinforcement, or conventional allopatric divergence as potential drivers of distinct communication channels among closed related species.

Types of Variation

The study of variation within and among individuals, populations, and species is central to the fields of ecology and evolution. Comparative studies of the sex pheromones of individuals within and among populations have documented abundant variation in moth pheromone blends (see Löfstedt et al., this volume; McElfresh and Millar, this volume; Lassance, this volume; Tabata and Ishikawa, this volume; Groot et al., this volume; Hillier and Baker, this volume). Several methods have been used to estimate genetically based variation in moth pheromone traits including crossing experiments (e.g., Gemeno et al. 2001), half-sibling (Allison and Cardé 2006) and full-sibling (Collins and Cardé 1985) breeding design, and mother–daughter regression (Svensson et al. 2002). All of these studies have observed intermediate to high levels of heritable variation for pheromone component ratios (see Table 2 in Allison and Cardé 2006). Several mechanisms have been proposed to explain the maintenance of genetic variation in traits that experience strong stabilizing or directional selection (see Roff 1997). To date, it is not possible to discriminate between the hypotheses that (1) moth pheromone blends experience strong stabilizing selection but genetic variation is maintained by one or more of these mechanisms and (2) moth pheromone blends experience weak stabilizing selection. Either scenario would explain the existence of abundant genetic variation in the pheromone blend, albeit for different reasons. Discrete and continuous patterns of trait variation have been documented for the female sex pheromone blend in moths. Traits characterized by both patterns of variation can evolve; however, the evolutionary trajectories of traits with nonadditive sources of variation (i.e., discrete patterns of variation) are difficult to predict and the

effects of genetic drift can be increased (see Mazer and Damuth 2001).

Quantitative Traits

Traits with continuous variation (quantitative traits) have phenotypes that vary along a continuum and are usually determined by alleles at multiple loci. Numerous examples of moths with continuous variation in the female pheromone blend exist. Perhaps the most extreme case is the potato tuberworm, *Phtorimaea operculella* (Gelechiidae). The female produces two components, a diene (E,Z)-4,7-tridecadienyl acetate and a triene (E,Z,Z)-4,7,10-tridecatrienyl acetate. The ratio in females varies broadly, with the percentage of triene varying from 27% to 88% in California females and from 16% to 71% in Nagoya, Japan, females (Ono et al. 1990). In parallel, male *P. operculella* seem indiscriminate in their preference of ratio (Ono and Orita 1986), accepting with equal unanimity, lures ranging in ratio from 9:1 to 1:9, with both the diene and triene being somewhat attractive alone.

Discrete Traits

Traits with discrete variation have phenotypes that can be divided into distinct, nonoverlapping groups and are usually determined by one or a few genes with large effects. Discrete variation in the moth sex pheromone blend has been reported in several moth species (see Haynes, this volume). For example, the European corn borer, *Ostrinia nubilalis*, has two divergent pheromone phenotypes, the E- and Z-phenotype. Both phenotypes produce a blend of (E)-11-tetradecenyl acetate (E11-14Ac) and (Z)-11-tetradecenyl acetate (Z11-14Ac). Females from both types produce a narrow, nonoverlapping distribution of pheromone blends, E-type females produce a blend of 99:1 E11-14Ac:Z11-14Ac, while Z-type females produce a 3:97 blend of E11-14Ac:Z11-14Ac (see Lassance, this volume). A similar polymorphism has been documented in *O. scapulalis* (see Tabata and Ishikawa, this volume). One notable difference between these two pheromone polymorphisms is the prevalence of hybrids. In Japan, E- and Z-type *O. scapulalis* hybrids are the most common phenotype (52%, 33%, and 15% hybrid, E- and Z-type, respectively). Conversely, in Maryland, United States, hybrid *O. nubilalis* females were much less common (70%, 14%, and 15% for the Z-, E-, and hybrid type, respectively) (Klun and Huettel 1988). Similar to *O. nubilalis*, the distributions of pheromone blends produced by *O. scapulalis* females from both types are narrow and do not overlap (see Tabata and Ishikawa, this volume). Discrete traits controlled by one or a few loci are of interest because they are excellent model systems for testing population genetics theory predictions of allele frequency changes across generations. Discrete and continuous patterns of variation in the pheromone blend are not mutually exclusive. Both continuous and discrete patterns of variation in the pheromone blend of the cabbage looper, *Trichoplusia ni*, have been reported (see Haynes, this volume).

Consequences of Variation

Variation in the pheromone blend should only be selected against when it reduces the likelihood of attracting a male. Multiple mechanisms exist to promote reproductive isolation among sympatric species that share pheromone components including the following: (i) unique pheromone components; (ii) unique blends of shared components; (iii) behavioral antagonists; (iv) exclusive times of mating; (v) phenological and habitat separation; and (vi) male pheromones and courtship behaviors. Stabilizing selection on the pheromone blend (as a consequence of male preference functions) would only be associated with the second mechanism and reduce/limit variation in the relative amounts of components. Species are not expected to rely on a single mechanism. For example, Sasaerila et al. (2002) observed that four limacodid moths from Borneo co-occur in palm plantations and differ in their pheromone blends, with further isolation due to separation in peak times of sexual activity and the preferential height at which males locate pheromone sources.

Simulations with a population genetics model (Butlin and Trickett 1997) and wind-tunnel bioassays with wild-type and mutant male *Trichoplusia ni* (Hemmann et al. 2008) suggest that a trade-off exists between the male traits breadth of response and sensitivity—if males widen their window of response to ratio, the assumption is that their threshold of sensitivity to pheromone is diminished. As a result of this trade-off male moth preference functions are believed to be unimodal with a maximum coincident with the most common pheromone blend emitted by conspecific females (see Allison and Cardé, this volume). As a result of the shape of the male preference function, quantitative and qualitative variation in the pheromone blend may result in significant variation in female mating success (i.e., females with missing components or divergent pheromone blends may not be mated or take longer to attract a male).

In addition to species identity, pheromone signals may contain information about mate quality and have been speculated to present males with an opportunity to assess female quality (Bonduriansky 2001). Although several factors can contribute to the evolution of male mating preferences, high variance in female quality is most commonly associated with male mating preferences. The existence of male mating preferences is predicated on the ability of males to assess female quality. Among insects male mating preferences are often based on female traits that correlate positively with fecundity (e.g., female body size; see Bonduriansky 2001). In some species female size may not correlate positively with fecundity or courtship and mating may occur under conditions that prevent accurate assessment. In these cases, females may possess traits that signal female quality, i.e., "female ornamentation" (see Chenoweth et al. 2006). In many moth species the ratio of pheromone components varies predictably with female age, females are short lived and mate only once or a few times, and a significant negative relationship exists between female age and reproductive value. For example, in the almond moth *Cadra cautella*, as females age the ratio of the major to minor pheromone component decreases (Allison and Cardé 2006) and coincident with this decline in ratio is a decline in female reproductive value (i.e., as female age increases, reproductive value decreases) (Allison and Cardé, pers. comm.). In *C. cautella* it appears that the ratio of the pheromone blend can predict female age and thereby provide males with a mechanism to assess female quality before incurring the costs associated with orientation and/or courtship. In two-choice wind-tunnel trials, males preferred the ratio most apt to be produced by younger, more fecund females (Allison and Cardé 2008). In nature, however, males are most likely to have "no choice," because of the improbability of two competing females being equally distant upwind from a male and generating a single, coalesced plume. Moreover, blends within this natural range presented alone in a wind tunnel (no choice) were found to be equally attractive, and therefore it seems improbable that such mate choice in *C. cautella* could occur in nature, even though the computational machinery for selection of the natural blend can be demonstrated in a choice assay (Allison and Cardé 2008).

Summary

Distinctive pheromone channels serve as a mate-location signal among nearly all moths and in many cases serve as primary pre-mating reproductive isolating mechanisms among closely allied species (Cardé and Haynes 2004). Variation in the timing of mating serves as an effective mechanism for reproductive isolation of many pairs of closely allied species that share a pheromone channel. The most prevalent mechanism for prevention of cross-attraction, however, is separation by distinctive chemical channels. How such differentiation initially arises is an unresolved question.

Some models propose that changes in the pheromone drive the formation of distinctive channels and therefore speciation, with females possessing a new component attracting males responsive to this novelty, quickly creating a divergent population in sympatry (e.g., Baker 2002; Roelofs et al. 2002). Such a saltational scenario may have improbable assumptions about the frequency of such novelties, likelihood of such novel types mating, and perhaps this fortuitous event occurring in a small enough population to promote fixation of the new communication strain (Butlin and Trickett 1997). Saltational shifts, however, have been held to drive the evolution of new blends resulting in sibling species with dissimilar pheromones (Symonds and Elgar 2007).

The most accepted general model, however, posits gradual, allopatric divergence. In one such scenario, component ratios are assumed to shift in response to sexual selection or gradually drift to a new optimum. Other paths to change are adding new components and response, responding to an existing compound not previously sensed, dropping components, or reacting antagonistically to the emitted blend of other species. In reinforcement, allopatric populations that have partially diverged in signal traits but still produce partially sterile or inviable hybrids can after secondary contact be driven to complete signal separation by the reduced reproductive success of hybrids. Reproductive character displacement is similar in the diverging of the communication channel but the now sympatric populations are already "good" species and so the driving selective force would be reducing communication interference. So far evidence to support either reinforcement or reproductive character displacement is quite limited, perhaps because of the difficulty of identifying sympatric populations in the process of divergence.

Genetically based variation in the pheromone message and the response it evokes form a basis for evolutionary change. To understand the factors that drive evolutionary change requires documentation within and among populations of message and response variabilities and establishing the heritable nature of these traits. In moths, the pheromone typically is comprised of multiple components, which in some species must be present in a precise ratio to evoke a high level of attraction. Although the pheromones of more than 600 moth species have been identified (Löfstedt et al., this volume), for most species we know little about variation among individuals in the amounts and ratios emitted by females and the response spectrum of individual males. Given that the focus of most research on moth pheromones is to establish the most attractive "average" blend in the field, perhaps this is not surprising. Relatively, few studies, however, have used unambiguous methods to document the intrapopulation and interpopulation variabilities in the emission of these components and how such variation influences response of individual males. Even in those few cases where variation in phenotypic traits is reasonably well described, typically the genetic underpinnings of such variation have not been measured. Disentangling how environmental factors, particularly temperature, modulate emission of pheromone and response remains a considerable experimental challenge.

References Cited

Allison, J.D., and R.T. Cardé. 2006. Heritable variation in the sex pheromone of the almond moth, *Cadra cautella*. *Journal of Chemical Ecology* 32:621–641.

Allison, J.D., and R.T. Cardé. 2008. Male pheromone blend preference function measured in choice and no-choice wind tunnel trials with almond moths, *Cadra cautella*. *Animal Behaviour* 75:259–266.

Allison, J.D., D.A. Roff, and R.T. Cardé. 2008. Genetic independence of female signal form and male receiver design in the almond moth, *Cadra cautella*. *Journal of Evolutionary Biology* 21:1666–1672.

Anderson, P., M.M. Sadek, and B.S. Hansson. 2003. Pre-exposure modulates attraction to sex pheromone in a moth. *Chemical Senses* 28:285–291.

Baker, J.L., A.S. Hill, and W.L. Roelofs. 1978. Seasonal variations in pheromone trap catches of male omnivorous leafroller moths, *Platynota stultana*. *Environmental Entomology* 7:399–401.

Baker, T.C. 2002. Mechanisms for saltational shifts in pheromone communication systems. *Proceedings of the National Academy of Sciences of the United States of America* 99:13368–13370.

Baker, T.C. 2008. Balanced olfactory antagonism as a concept for understanding evolutionary shifts in moth sex pheromone blends. *Journal of Chemical Ecology* 34:971–981.

Baker, T.C., and R.T. Cardé. 1979a. Courtship behavior of the oriental fruit moth (*Grapholitha molesta*): experimental analysis and consideration of the role of sexual selection in the evolution of courtship pheromones in the Lepidoptera. *Annals of the Entomological Society of America* 72:173–188.

Baker, T.C., and R.T. Cardé. 1979b. Analysis of pheromone-mediated behavior in male *Grapholitha molesta*, the oriental fruit moth (Lepidoptera: Tortricidae). *Environmental Entomology* 8:956–968.

Baker, T.C., and R.T. Cardé. 1979c. Endogenous and exogenous factors affecting periodicities of female calling and male sex pheromone responses in *Grapholitha molesta*. *Journal of Insect Physiology* 25:943–950.

Baltensweiler, W.E., E. Priesner, H. Arn, and V. Delucci. 1978. Unterschiedliche Sexuallockstoffe bei Larchen und Arvenform des Grauen Larchenwicklers (*Zeiraphera diniana* Gn., Lep., Tortricidae). *Mitteilungen der Schweizerischen Entomologischen Gesellschaft* 51:133–142. [In German]

Bjostad, L.B., C.E. Linn, J.-W. Du, and W.L. Roelofs. 1984. Identification of new sex pheromone components in *Trichoplusia ni*, predicted from biosynthetic precursors. *Journal of Chemical Ecology* 10:1309–1323.

Bonduriansky, R. 2001. The evolution of male mate choice in insects: a synthesis of ideas and evidence. *Biological Reviews* 76:305–339.

Borden, J.H. 1990. Use of semiochemicals to manage coniferous tree pests in Western Canada. Pp. 281–315. In R.L. Ridgeway, R.M. Silverstein, and M.N. Inscoe, eds. *Behavior-Modifying Chemicals for Insect Management: Applications of Pheromones and Other Attractants*. New York: Marcel Dekker Inc.

Brown, W.L., and E.O. Wilson. 1956. Character displacement. *Systematic Zoology* 5:49–65.

Butler, L.J., and L.M. McDonough. 1979. Insect sex pheromones: evaporation rates of acetates from natural rubber septa. *Journal of Chemical Ecology* 5: 825–837.

Butler, L.J., and L.M. McDonough. 1981. Insect sex pheromones: evaporation rates of alcohols and acetates from natural rubber septa. *Journal of Chemical Ecology* 7:627–633.

Butenandt, A.R., R. Beckman, D. Stamm, and E. Hecker. 1959. Über den sexuallockstoff des seidenspinner Bombyx mori, reidarstellung und constitution. *Zeitschrift für Naturforschung B* 14:283–284. [In German]

Butlin, R. 1987. Speciation by reinforcement. *Trends in Ecology and Evolution* 2:8–11.

Butlin, R.K., and A.J. Trickett. 1997. Can population genetic simulation help to interpret pheromone evolution. Pp. 548–562. In R.T. Cardé and A.K. Minks, eds. *Insect Pheromone Research: New Directions*. New York: Chapman & Hall.

Byers, J.A. 2006. Pheromone component patterns of moth evolution revealed by computer analysis of the *Pherolist*. *Journal of Animal Ecology* 75:399–407.

Camerini, G., R. Groppali, F. Rama, and S. Maini. 2015. Semiochemicals of *Ostrinia nubilalis*: diel response to sex pheromone and phenylacetaldehyde in open field. *Bulletin of Insectology* 68:45–50.

Cardé, A.M., T.C. Baker, and R.T. Cardé. 1979. Identification of a four component sex pheromone of the female oriental fruit moth, *Grapholitha molesta* (Lepidoptera: Tortricidae). *Journal of Chemical Ecology* 5:423–427.

Cardé, R.T. 1974. Diel periodicities of female calling and male pheromone attraction in *Holomelina aurantiaca* (Lepidoptera: Arctiidae). *The Canadian Entomologist* 106:933–944.

Cardé, R.T. 1990. Principles of mating disruption. Pp. 47–73. In R.L. Ridgway, R.M. Silverstein, and M.N. Inscoe, eds. *Behavior-Modifying Chemicals for Insect Management: Applications of Pheromones and Other Attractants*. New York: Marcel Dekker Inc.

Cardé, R.T., and R.E. Charlton. 1984. Olfactory sexual communication in Lepidoptera: strategy, sensitivity and selectivity. Pp. 241–265. In T. Lewis, ed. *Insect Communication*. New York: Academic Press.

Cardé, R.T., and K.F. Haynes. 2004. Structure of the pheromone communication channel in moths. Pp. 283–332. In R.T. Cardé and J.G. Millar, eds. *Advances in the Chemical Ecology of Insects*. Cambridge: Cambridge University Press.

Cardé, R.T., and J.G. Millar. 2008. The scent of a female: sex pheromones of female tiger moths. Pp. 127–143. In W.E. Conner, ed. *Tiger Moths and Wooly Bears: Behavior, Ecology and Natural History of the Arctiidae*. New York: Oxford University Press.

Cardé, R.T., and W.L. Roelofs. 1973. Temperature modification of male sex pheromone response and factors affecting female calling *Holomelina immaculata* (Lepidoptera: Arctiidae). *The Canadian Entomologist* 105:1505–1512.

Cardé, R.T., C.C. Doane, and W.L. Roelofs. 1974. Diel rhythms of male sex pheromone response and female attractiveness in the gypsy moth. *The Canadian Entomologist* 106:479–484.

Cardé, R.T., A. Comeau, T.C. Baker, and W.L. Roelofs. 1975. Moth mating periodicity: temperature regulates the circadian gate. *Experientia* 31:46–48.

Cardé, R.T., T.C. Baker, and W.L. Roelofs. 1976. Sex attractant responses of male oriental fruit moths to a range of component ratios: pheromone polymorphism? *Experientia* 32:1406–1407.

Cardé, R.T., A.M. Cardé, A.S. Hill, and W.L. Roelofs. 1977. Sex pheromone specificity as a reproductive isolating mechanism among the sibling species *Archips argyrospilus* and *A. mortuanus* and other sympatric tortricine moths (Lepidoptera: Tortricidae). *Journal of Chemical Ecology* 3:71–84.

Cardé, R.T., A.M. Cardé, and R.D. Girling. 2012. Observations on the flight paths of the day-flying moth *Virbia lamae* during periods of mate location: do males have a strategy for contacting the pheromone plume? *Journal of Animal Ecology* 81:268–276.

Charlton, R.E., H. Kanno, R.D. Collins, and R.T. Cardé. 1993. Influence of pheromone concentration and ambient temperature on flight of the gypsy moth *Lymantria dispar* (L.), in a sustained-flight wind tunnel. *Physiological Entomology* 18:349–362.

Chenoweth, S.F., P. Doughty, and H. Kokko. 2006. Can non-directional male mating preferences facilitate honest female ornamentation? *Ecology Letters* 9:179–184.

Collins, M.M., and R.D. Weast. 1961. *Wild Silk Moths of the United States Saturniinae: Experimental Studies and Observations of Natural*

Living Habitats and Relationships. Cedar Rapids, IA: Collins Radio Company.

Collins, R.D., and R.T. Cardé. 1985. Variation in and heritability of aspects of pheromone production in the pink bollworm moth, *Pectinophora gossypiella* (Lepidoptera: Gelechiidae). *Annals of the Entomological Society of America* 78:229–234.

Comeau, A., R.T. Cardé, and W.L. Roelofs. 1976. Relationship of ambient temperature to diel periodicities of sex attraction in six species of Lepidoptera. *The Canadian Entomologist* 108:415–418.

Coyne, J.A., H.A. Orr, and D.J. Futuyma. 1988. Do we need a new species concept? *Systematic Zoology* 37:190–200.

Dobzhansky, T. 1940. Speciation as a stage in evolutionary divergence. *The American Naturalist* 74:312–321.

Dopman, E.B., P.S. Robbins, and A. Seaman. 2010. Components of reproductive isolation between North American pheromone strains of the European corn borer. *Evolution* 64:881–902.

Du, J.W., C. Löfstedt, and J. Löfqvist. 1987. Repeatability of pheromone emissions from individual female ermine moths *Yponomeuta padellus* and *Yponomeuta rorellus*. *Journal of Chemical Ecology* 13:1431–1441.

El-Sayed, A.M., J. Delisle, N. De Lury, L.J. Gut, G.J.R. Judd, S. Legrand, W.H. Reissig, W.L. Roelofs, C.R. Unelius, and R.M. Trimble. 2003. Geographic variation in pheromone chemistry, antennal electrophysiology, and pheromone-mediated trap catch of North America populations of the obliquebanded leafroller. *Environmental Entomology* 32:470–476.

Emelianov, I., M. Drès, W. Baltensweiler, and J. Mallet. 2001. Host-induced assortative mating in host races of the larch budmoth. *Evolution* 55:2002–2010.

Emelianov, I., F. Simpson, P. Narang, and J. Mallet. 2003. Host choice promotes reproductive isolation between host races of the larch budmoth. *Journal of Evolutionary Biology* 16:208–218.

Emelianov, I., F. Marec, and J. Mallet. 2004. Genomic evidence for divergence with gene flow in host races of the larch budmoth. *Proceedings of the Royal Society of London Series B* 271:95–105.

Falconer, D.S., and T.F.C. Mackay. 1996. *Introduction to Quantitative Genetics*. New York: Prentice Hall.

Ferguson, D.C. 1971–1972. Bombycoidea (Saturniidae). In R.B. Dominick, D.C. Ferguson, J.G. Franclemont, R.W. Hodges, and E.G. Munroe, eds. *The Moths of America North of Mexico*, fasc. 20.2, London: E.W. Classey Ltd. and R.B.D. Publications Inc.

Fisher, R.A. 1930. *The Genetical Theory of Natural Selection*. Oxford: Clarendon Press.

Foster, S.P., and K.G. Anderson. 2015. Sex pheromones in mate assessment: analysis of nutrient cost of sex pheromone production in females of the moth *Heliothis virescens*. *Journal of Experimental Biology* 218:1252–1258.

Foster, S.P., S.J. Muggleston, and R.D. Ball. 1991. Behavioral responses of male *Epiphyas postvittana* (Walker) to sex pheromone-baited delta trap in a wind tunnel. *Journal of Chemical Ecology* 17:1449–1468.

Gago, R., J.D. Allison, J.S. McElfresh, K.F. Haynes, J. McKenney, A. Guerrero, and J.G. Millar. 2013. A tetraene aldehyde as the major sex pheromone component of the promethea moth (*Callosamia promethea* (Drury)). *Journal of Chemical Ecology* 39:1263–1272.

Gemeno, C., A.F. Lutfallah, and K.F. Haynes. 2000. Pheromone blend variation and cross-attraction among populations of the black cutworm moth (Lepidoptera: Noctuidae). *Annals of the Entomological Society of America* 93:1322–1328.

Gemeno, C., A.J. Moore, R.F. Preziosi, and K.F. Haynes. 2001. Quantitative genetics of signal evolution: a comparison of the pheromonal signal in two populations of the cabbage looper moth, *Trichoplusia ni*. *Behavioral Genetics* 31:157–165.

Gökçe, A., L.L. Stelinski, L.J. Gut, and M.E. Whalon. 2007. Comparative behavioral and EAG responses of female obliquebanded and redbanded leafroller moths (Lepidoptera: Tortricidae) to their sex pheromone components. *European Journal of Entomology* 104:187–194.

Gorsuch, C.S., M.G. Karandinos, and C.F. Koval. 1975. Daily rhythms of *Synanthedon pictipes* (Lepidoptera: Aegeriidae) female calling behavior in Wisconsin: temperature effects. *Entomologia Experimentalis et Applicata* 18:367–376.

Greenfield, M.D., and M.G. Karandinos. 1979. Resource partitioning of the sex communication channel in clearwing moths (Lepidoptera: Sesiidae) of Wisconsin. *Ecological Monographs* 49:403–426.

Gries, G., P.W. Schaefer, R. Gries, J. Liška, and T. Gotoh. 2001. Reproductive character displacement in *Lymantria monacha* from northern Japan. *Journal of Chemical Ecology* 27:1163–1176.

Groot, A.T. 2014. Circadian rhythms of sexual activities in moths: a review. *Frontiers in Ecology and Evolution* 2:43. doi:10.3389/fevo.2014.00043.

Groot, A.T., R.G. Santangelo, E. Ricci, C. Brownie, F. Gould, and C. Schal. 2007. Differential attraction of *Heliothis subflexa* males to synthetic pheromone lures in eastern US and western Mexico. *Journal of Chemical Ecology* 33:353–368.

Groot, A.T., O. Inglis, S. Bowdridge, R.G. Santangelo, C. Blanco, J.D. Lopez, Jr., A.T. Vargas, F. Gould, and C. Schal. 2009. Geographic and temporal variation in moth chemical communication. *Evolution* 63:1987–2003.

Groot, A.T., H. Staudacher, A. Barthel, O. Inglis, G. Schöfl, R.G. Santangelo, S. Gebauer-Jung et al. 2013. One quantitative trait locus for intra- and interspecific variation in a sex pheromome. *Molecular Biology* 22:1065–1080.

Groot, A.T., G. Schöfl, O. Inglis, S. Donnerhacke, A. Classen, A. Schmalz, R.G. Santangelo et al. 2014. Within population variability in a moth sex pheromone blend: genetic basis and behavioural consequences. *Proceedings of the Royal Society B* 281:20133054.

Harari, A.R., and H. Steinitz. 2013. The evolution of female sex pheromones. *Current Zoology* 39: 569–578.

Hansson, B.S., M. Tòth, C. Löfstedt, G. Szöcs, M. Subchev, and J. Löfqvist. 1990. Pheromone variation among Eastern European and a Western Asian population of the turnip moth *Agrotis segetum*. *Journal of Chemical Ecology* 16:1611–1622.

Haynes, K.F., and T.C. Baker. 1988. Potential for evolution of resistance to pheromones: worldwide and local variation in chemical communication system of pink bollworm moth, *Pectinophora gossypiella*. *Journal of Chemical Ecology* 14:1547–1560.

Haynes, K.F., L.K. Gaston, M. Mistrot Pope, and T.C. Baker. 1984. Potential for evolution of resistance to pheromones: interindividual and interpopulation variation in chemical communication system of the pink bollworm. *Journal of Chemical Ecology* 10:1551–1565.

Hemmann, D.J., J.D. Allison, and K.F. Haynes. 2008. Trade-off between sensitivity and specificity in the cabbage looper moth response to sex pheromone. *Journal of Chemical Ecology* 34:1476–1486.

Howard, D.J. 1993. Reinforcement: origin, dynamics, and fate of an evolutionary hypothesis. Pp. 46–69. In R.G. Harrison, ed. *Hybrid Zones and the Evolutionary Process*. New York: Oxford University Press.

Howse, P.E., I.D.R. Stevens, and O.T. Jones. 1998. *Insect Pheromones and their Use in Pest Management*. London: Chapman & Hall.

Hunt, R.E., and K.F. Haynes. 1990. Periodicity in the quantity and blend ratios of pheromone components in glands and volatile emissions of mutant and normal cabbage looper moths, *Trichoplusia ni*. *Journal of Insect Physiology* 36:769–774.

Jiggins, C.D., I. Emelianov, and J. Mallet. 2005. Assortative mating and speciation as pleiotropic effects of ecological adaptation: examples in moths and butterflies. Pp. 451–473, In M. Fellowes, G. Holloway, and J. Rolff, eds. *Insect Evolutionary Ecology: Proceedings of the Royal Entomological Society's 22nd Symposium*. London: CABI Publishing.

Juárez, M.L., G. Schöfl, M.T. Vera, J.C. Vilardi, M.G. Murúa, E. Willink, S. Hänniger, D.G. Heckel, and A.T. Groot. 2014. Population structure of *Spodoptera frugiperda* maize and rice host forms in South America: are they host strains? *Entomologia Experimentalis et Applicata* 152:182–199.

Kaissling, K.E., G. Kasang, H.J. Bestmann, W. Stransky, and O. Vostrowsky. 1978. A new pheromone of the silkworm moth *Bombyx mori*: sensory pathway and behavioral effect. *Naturwissenschaften* 62:382–384.

Kárpáti, Z., M. Tasin, R.T. Cardé, and T. Dekker. 2013. Early quality assessment lessens pheromone specificity in a moth. *Proceedings of the National Academy of Sciences of the United States of America* 110:7377–7382.

Klun, J.A., and M.D. Huettel. 1988. Genetic regulation of sex pheromone production and response: interaction of sympatric pheromonal types of European corn borer, *Ostrinia nubilalis* (Lepidoptera: Pyralidae). *Journal of Chemical Ecology* 14:2047–2061.

Kochansky, J., R.T. Cardé, J. Liebherr, and W.L. Roelofs. 1975. Sex pheromone of the European corn borer, *Ostrinia nubilalis* (Lepidoptera: Pyralidae), in New York. *Journal of Chemical Ecology* 1:225–231.

Kuenen, L.P.S., and J.P. Seigel. 2015. Measure your septa releases ratios: pheromone release ratio variability affected by rubber septa and solvent. *Journal of Chemical Ecology* 41:303–310.

Lacey, M.J., and C.J. Sanders. 1992. Chemical composition of sex pheromone of oriental fruit moth and rates of release by individual female moths. *Journal of Chemical Ecology* 18: 1421–1435.

Lanier, G.N. 1990. Principles of attraction-annihilation: mass-trapping and other means. Pp. 25–47. In R.L. Ridgway, R.M. Silverstein, and M.N. Inscoe, eds. *Behavior-Modifying Chemicals for Insect Management: Applications of Pheromones and Other Attractants.* New York: Marcel Dekker Inc.

Lewis, T., and E.D.M. Macaulay. 1976. Design and evaluation of sex-attractant traps for pea moth, *Cydia nigricana* (Steph.) and the effect of plume shape on catches. *Ecological Entomology* 1:175–187.

Liebherr, J., and W. Roelofs. 1975. Laboratory hybridization and mating period studies using two pheromone strains of *Ostrinia nubilalis. Annals of the Entomological Society of America* 68:305–309.

Lim, H., and M.D. Greenfield. 2007. Female pheromonal chorusing in an arctiid moth, *Utetheisa ornatrix. Behavioral Ecology* 18:165–173.

Linn, C.E., M.G. Campbell, and W.L. Roelofs. 1988. Temperature modulation of behavioural thresholds controlling male moth pheromone specificity. *Physiological Entomology* 13:59–67.

Linn, C.E., M.S. Young, M. Gendle, T.J. Glover, and W.L. Roelofs. 1997. Sex pheromone blend discrimination in two races and hybrids of the European corn borer. *Physiological Entomology* 22:212–223.

Liu, Y.B., and K.F. Haynes. 1994. Evolution of behavioral responses to sex pheromone in mutant laboratory colonies of *Trichoplusia ni. Journal of Chemical Ecology* 20:231–238.

Löfstedt, C., B.S. Lanne, J. Löfqvist, and G. Bergström. 1985. Individual variation in the pheromone of the turnip moth *Agrotis segetum. Journal of Chemical Ecology*11:1181–1196.

Löfstedt, C., J. Löfqvist, B.S. Lanne, J.N.C. Van Der Pers, and B.S. Hansson. 1986. Pheromone dialects in European turnip moths *Agrotis segetum. Oikos* 46:250–257.

Marks, R.J. 1976. Female pheromone release and timing of male flight in the red bollworm, *Diparopsis castanea* Hmps. (Lepidoptera, Noctuidae), measured by pheromone traps. *Bulletin of Entomological Research* 66:219–241.

Mazer, S.J., and J. Damuth. 2001. Nature and causes of variation. Pp. 3–15. In C.W. Fox, D.A. Roff, and D.J. Fairbairn, eds. *Evolutionary Ecology: Concepts and Case Studies.* New York: Oxford University Press.

McDonough, L.M., D.F. Brown, and W.C. Alder. 1989. Insect sex pheromones: effect of temperature on evaporation rates of acetates from rubber septa. *Journal of Chemical Ecology* 15:779–790.

McElfresh, S.J., and J.G. Millar. 2001. Geographic variation in the pheromone system of the saturniid moth *Hemileuca eglanterina. Ecology* 82:3505–3518.

McNeil, J.N. 1991. Behavioral ecology of pheromone-mediated communication in moths and its importance in the use of pheromone traps. *Annual Review of Entomology* 36:407–430.

Miller, J.R., and W.L. Roelofs. 1980. Individual variation in sex pheromone component ratios in two populations of the red-banded leafroller. *Environmental Entomology* 9:359–363.

Mochizuki, F., T. Fukumoto, H. Noguchi, H. Sugie, T. Morimoto, and K. Ohtani. 2002. Resistance to mating disruptant composed of (Z)-11-tetradecenyl acetate in the smaller tea tortrix, *Adoxophyes honmai* (Yasuda) (Lepidoptera: Tortricidae). *Applied Entomology and Zoology* 37:299–304.

Ohbayashi, N., Y. Yushima, H. Noguchi, and Y. Tamaki. 1973. Timing of mating and sex pheromone production and release of *Spodoptera litura* (F.) Lepidoptera: Noctuidae). *Kontyû* 41:389–395.

Ono, T. 1993. Effect of rearing temperature on pheromone component ratio in potato tuberworm moth, *Phthorimaea operculella* (Lepidoptera: Gelechiidae). *Journal of Chemical Ecology* 19:71–81.

Ono, T. 1994. Effect of rearing temperature on biosynthesis of sex pheromone components in potato tuberworm moth, *Phthorimaea operculella* (Lepidoptera: Gelechiidae). *Journal of Chemical Ecology* 20:2733–2741.

Ono, T., and S. Orita. 1986. Field trapping of the potato tuber moth, *Phthorimaea operculella* (Lepidoptera: Gelechiidae), with the sex pheromone. *Applied Entomology and Zoology* 21:632–634.

Ono, T., R.E. Charlton, and R.T. Cardé. 1990. Variability on pheromone composition and periodicity of pheromone titer in potato tuberworm moth, *Phthorimaea operculella* (Lepidoptera: Gelechiidae). *Journal of Chemical Ecology* 16:531–542.

Pashley, D.P., A.M. Hammond, and T.N. Hardy. 1992. Reproductive isolating mechanisms in fall armyworm host strains (Lepidoptera: Noctuidae). *Annals of the Entomological Society of America* 79:898–904.

Paterson, H.E.H. 1978. More evidence against speciation by reinforcement. *South African Journal of Science* 74:369–371.

Paterson, H.E.H. 1985. The recognition concept of species. Pp. 21–29. In E.S. Vrba, ed. *Species and Speciation.* Transvaal Museum Monograph No. 4. Pretoria: Transvaal Museum.

Paterson, H.E.H. 1993. Variation and the specific-mate recognition system. Pp. 209–227. In P.P.G. Bateson, P.H. Klopfer, and N.S. Thompson, eds. *Perspectives in Ethology: Volume 10 Behavior and Evolution.* New York: Plenum Press.

Persoons, C.J., S. Voerman, P.E.J. Vermeil, F.J. Ritter, W.J. Nooyen, and A.K. Minks. 1976. Sex pheromone of the potato tuberworm moth, *Phthorimaea operculella*: isolation, identification and field evaluation. *Entomologia Experimentalis et Applicata* 20:289–300.

Régnière, J., J. Delisle, D.S. Pureswaran, and R. Trudel. 2013. Mate-finding allee effect in spruce budworm population dynamics. *Entomologia Experimentalis et Applicata* 146:112–122.

Roelofs, W.L. 1978. Threshold hypothesis for pheromone perception. *Journal of Chemical Ecology* 4:685–699.

Roelofs, W.L., and R.T. Cardé. 1971. Hydrocarbon sex pheromone in tiger moths (Arctiidae). *Science* 171:684–686.

Roelofs, W.L., and R.T. Cardé. 1977. Responses of Lepidoptera to synthetic sex pheromone chemicals and their analogues. *Annual Review of Entomology* 22:377–405.

Roelofs, W.L., W. Liu, G. Hao, A.P. Rooney, and C.E. Linn, Jr. 2002. Evolution of moth sex pheromones via ancestral genes. *Proceedings of the National Academy of Sciences of the United States of America* 99:13621–13626.

Roff, D.A. 1997. *Evolutionary Quantitative Genetics.* New York: Chapman & Hall.

Ryan, M.J., and W. Wilczynski. 1991. Evolution of intraspecific variation in the advertisement call of a cricket frog (*Acris crepitans,* Hylidae). *Biological Journal of the Linnaean Society* 44:249–271.

Sasaerila, Y., G. Gries, R. Gries, and T.C. Boo. 2000. Specificity of communication channels in four limacodid moths: *Darna bradleyi, Darna trima, Setothosea asigna* and *Setora nitens* (Lepidoptera: Limacodidae). *Chemoecology* 10:193–199.

Schal, C., and R.T. Cardé. 1986. Effects of temperature and light on calling behaviour in the tiger moth *Holomelina lamae* (Freeman) (Lepidoptera: Arctiidae). *Physiological Entomology* 11:75–87.

Schal, C., R.E. Charlton, and R.T. Cardé. 1987. Temporal patterns of pheromone titers and release rates in *Holomelina lamae* (Lepidoptera: Arctiidae). *Journal of Chemical Ecology* 13:1115–1129.

Schöfl, G., D.G. Heckel, and A.T. Groot. 2009. Time-shifted reproductive behaviour among fall armyworm (Noctuidae: *Spodoptera frugiperda*) host strains: evidence for different modes of inheritance. *Journal of Evolutionary Biology* 22:1447–1459.

Sharov, A.A., A.M. Liebhold, and F.W. Ravlin. 1995. Prediction of gypsy moth (Lepidoptera: Lymantriidae) mating success from pheromone trap counts. *Environmental Entomology* 31:1119–1127.

Shorey, H.H. 1977. Manipulation of insect pests of agricultural crops. Pp. 353–367. In H.H. Shorey and J.J. McKelvey, Jr., eds. *Chemical Control of Insect Behavior: Theory and Application.* New York: Wiley-Interscience.

Silverstein, R.M. 1990. Practical use of pheromones and other behavior-modifying compounds: overview. Pp. 1–9. In R.L. Ridgway, R.M. Silverstein, and M.N. Inscoe, eds. *Behavior-Modifying Chemicals for Insect Management: Applications of Pheromones and Other Attractants.* New York: Marcel Dekker Inc.

Sower, L.L., L.K. Gaston, and H.H. Shorey. 1971. Sex pheromones of noctuid moths. XXVI. Female release rate, male response threshold, and communication distance for *Trichoplusia ni. Annals of the Entomological Society of America* 64:1448–1456.

Sower, L.L., L.K. Gaston, and H.H. Shorey. 1973. Sex pheromones of noctuid moths. XLI. Factors limiting potential distance of sex pheromone communication in *Trichoplusia ni*. *Annals of the Entomological Society of America* 66:1121–1122.

Stelinski, L. L., A. L. Il'Ichev, and L. J. Gut. 2006. Antennal and behavioral responses of virgin and mated oriental fruit moth (Lepidoptera: Tortricidae) females to their sex pheromone. *Annals of the Entomological Society of America* 99:898–904.

Svensson, M. G. E., M. Bengtsson, and J. Löfqvist. 1997. Individual variation and repeatability of sex pheromone emission of female turnip moths *Agrotis segetum*. *Journal of Chemical Ecology* 23:1833–1850.

Svensson, G. P., C. Ryne, and C. Löfstedt. 2002. Heritable variation of sex pheromone composition and the potential for evolution of resistance to pheromone-based control of the Indian meal moth, *Plodia interpunctella*. *Journal of Chemical Ecology* 28:1447–1461.

Swier, S. R., R. W. Rings, and G. J. Musick. 1977. Age-related calling behavior of the black cutworm, *Agrotis ipsilon*. *Annals of the Entomological Society of America* 70:919–924.

Symonds, M. R. E., and M. A. Elgar. 2007. The evolution of pheromone diversity. *Trends in Ecology and Evolution* 23:220–228.

Tang, J. D., R. E. Charlton, R. T. Cardé, and C.-M. Yin. 1992. Diel periodicity and influence of age and mating on sex pheromone titer in gypsy moth, *Lymantria dispar*. *Journal of Chemical Ecology* 18:749–760.

Teal, P. E. A., and J. R. Byers. 1980. Biosystematics of the genus *Euxoa* (Lepidoptera: Noctuidae) XIV: effect of temperature on female calling behavior and temporal partitioning in three sibling species of the *Declarata* group. *The Canadian Entomologist* 112:113–117.

Tòth, M., C. Löfstedt, B. W. Blair, T. Cabello, A. I. Farag, B. S. Hansson, B. G. Kovalev et al. 1992. Attraction of male turnip moths *Agrotis segetum* (Lepidoptera: Noctuidae) to sex pheromone components and their mixtures at 11 sites in Europe, Asia, and Africa. *Journal of Chemical Ecology* 18:1337–1347.

Tuskes, P. M., J. P. Tuttle, and M. M. Collins. 1996. *The Wild Silk Moths of North America*. Ithaca, NY: Cornell University Press.

van Buskirk, J., and Y. Willi. 2006. The change in quantitative genetic variation with inbreeding. *Evolution* 60:2428–2434.

Wang, H.-L., G. P. Svensson, O. Rosenberg, M. Bengtsson, E. V. Jirle, and C. Löfstedt. 2010. Identification of the sex pheromone of the spruce seed moth, *Cydia strobilella* L. *Journal of Chemical Ecology* 36:305–313.

Webster, R. P. 1988. Modulation of the expression of calling by temperature in the omnivorous leafroller moth, *Platynota stultana* (Lepidoptera: Tortricidae) and other moths: a hypothesis. *Annals of the Entomological Society of America* 81:131–151.

Webster, R. P., and C.-M. Yin. 1997. Effects of temperature and photoperiod on calling behaviour of the gypsy moth *Lymantria dispar* L. (Lepidoptera: Lymantriidae). *The Canadian Entomologist* 129:843–854.

Witzgall, P. 1997. Modulation of pheromone-mediated flight in male moths. Pp. 265–274. In R. T. Cardé and A. K. Minks, eds. *Insect Pheromone Research: New Directions*. New York: Chapman & Hall.

Witzgall, P., and H. Arn. 1990. Direct measurement of the flight behaviour of male moths to calling females and synthetic sex pheromones. *Zeitschrift für Naturforschung* 45c:1067–1069.

Witzgall, P., M. Tasin, H. R. Buser, G. Wegner-Kiß, V. S. Mancebón, C. Ioriatti, A. C. Bäckman et al. 2005. New pheromone components of the grapevine moth *Lobesia botrana*. *Journal of Chemical Ecology* 31:2923–2932.

Witzgall, P., P. Kirsch, and A. Cork. 2010. Sex pheromones and their impact on pest management. *Journal of Chemical Ecology* 36:80–100.

Wu, W. Q., J. W. Zhu, J. Millar, and C. Löfstedt. 1998. A comparative study of sex pheromone biosynthesis in two strains of the turnip moth, *Agrotis segetum*, producing different ratios of sex pheromone components. *Insect Biochemistry and Molecular Biology* 11:895–900.

Yang, M.-W., S.-L. Dong, and L. Chen. 2009. Electrophysiological and behavioral responses of female beet armyworm *Spodoptera exigua* (Hübner) to the conspecific female sex pheromone. *Journal of Insect Behavior* 22:153–164.

Yin, L. R. S., C. Schal, and R. T. Cardé. 1991. Sex pheromone gland of the female tiger moth, *Holomelina lamae* (Lepidoptera: Arctiidae). *Canadian Journal of Zoology* 69:1916–1921.

Zaspel, J. M., S. J. Weller, and R. T. Cardé. 2008. A review of *Virbia* (formerly *Holomelina*) of America north of Mexico (Arctiidae: Arctiinae: Arctiini). *Bulletin of the Florida Museum of Natural History* 48:59–113.

Evolutionary Patterns of Pheromone Diversity in Lepidoptera

CHRISTER LÖFSTEDT, NIKLAS WAHLBERG, and JOCELYN G. MILLAR

Lepidoptera as a Model for Phylogenetic Reconstruction of Pheromone Evolution

The Lepidoptera, consisting of the moths and butterflies, comprise the second largest insect order, second in size only to the Coleoptera. It is also the insect order for which we have the most extensive knowledge of volatile sex pheromones and the roles and mechanisms of pheromonal communication, accumulated since the identification of the first pheromone, bombykol (($E10,Z12$)-10,12-hexadecadien-1-ol), in 1959 from females of the silkworm moth *Bombyx mori*. Since then, sex pheromones have been identified from more than 600 species of Lepidoptera ((http://web.tuat.ac.jp/~antetsu/LepiPheroList .htm); http://www.pherobase.com). In addition, sex attractants, defined as chemicals that attract males of a given species, but for which conspecific females have not yet been shown to produce the compounds (usually because no one has looked), have been identified for approximately 1200 additional species. Sex attractants are usually identified in screening trials or from serendipitous capture of novel or unexpected species in traps deployed to catch a different target species. Because of the high species specificity of lepidopteran pheromones, sex attractants with high activity are usually good "approximations" of the real pheromone. Thus, in total, more or less detailed information on female-produced sex attractant pheromone composition is available for approximately 2000 species, or slightly more than 1% of the ~160,000 described lepidopteran species (Nieukerken et al. 2011). The numbers of identified pheromones and sex attractants are constantly increasing, and the reader should bear in mind that the numbers we report in what follows primarily reflect only what was known up to 2014.

The use of volatile, female-produced sex pheromones for long-distance attraction of mates is a key component of the ground plan of Lepidoptera (Löfstedt and Kozlov 1997), with sex pheromones (or sex attractants) having been reported from 23 superfamilies that comprise the majority (ca. 135,000) of the described species. An exception is the superfamily Papilionioidea (the butterflies), which do not have female-produced sex attractant pheromones. Instead, butterflies have evolved entirely different mating systems suited to their diurnal habits, usually based on visual signals. Another exception is the superfamily Hepialoidea, which also appear to lack female-produced sex pheromones. However, this general picture of use of female-produced sex attracant pheromones throughout most of the lepidopteran superfamilies may be slightly misleading because the possible presence or absence of this type of pheromone has not yet been examined in many of the smaller superfamilies such as the Palaephatoidea or Andesianoidea.

It must also be noted that sex attractant pheromones are by no means the only type of pheromones that are employed by Lepidoptera; a number of other types of pheromones are used to convey information in different contexts. For example, male-produced chemical signals used in shorter-range interactions and courtship are widespread among Lepidoptera (Conner, this volume). The selective forces driving the evolution of courtship pheromones and the structures producing the courtship pheromones are in most cases very different from those involved in female sex pheromones. Whereas selection for efficient mate-finding coupled with the need for reproductive isolation has been the driving force for evolution of female-produced sex pheromones (Allison and Cardé, this volume), the evolution of male-produced courtship pheromones is related to female choice and sexual selection (Conner, this volume; Greenfield, this volume). Furthermore, with few exceptions, epidermal glands in the terminal abdominal segments are involved in the production and/or release of female-produced sex pheromones, whereas courtship pheromones of males are produced in and released from a variety of structures on the wings, thorax, abdomen, or legs (Birch et al. 1990).

Our knowledge of the phylogenetic relationships among the major clades of Lepidoptera has advanced remarkably during the last two decades. The milestone phylogenetic hypothesis by the late N.P. Kristensen based on morphology set the stage for our understanding of the relationships among the superfamilies (Kristensen and Skalski 1998). More recently, two extensive and independent molecular studies suggested several unexpected relationships, such as butterflies not being closely related to other "macrolepidoptera" (e.g., Noctuoidea, Bombycoidea, and Geometroidea), but provided support for most of the accepted superfamilies (Mutanen et al. 2010; Regier et al. 2013). These results were used to compile the latest family-level classification for Lepidoptera (Nieukerken et al. 2011), which we use here. Very recently, two phylogenomic studies (Bazinet et al. 2013; Kawahara and Breinholt 2014) have corroborated some of the unexpected results described above, leading to a robust phylogenetic hypothesis that is informative about the early evolution of the major lineages of Lepidoptera.

With a relatively robust phylogeny in hand and with considerable knowledge of the biosynthesis of the two major types of lepidopteran pheromones that comprise >90% of the known pheromone structures, it should be possible to trace the evolution of the patterns of pheromone use and the diversity of lepidopteran pheromone chemistry. In this chapter, focusing strictly on the female-produced sex attractant pheromones, we analyze the types and distribution of lepidopteran pheromone structures from a combined phylogenetic and biosynthetic perspective. As mentioned above, male-produced pheromones, with rare exceptions, are used in an entirely different way than female-produced sex attractant pheromones, and they are produced in morphologically different structures. The rare exceptions include a few species such as the wax moth *Galleria mellonella* (family Pyralidae), males of which produce pheromones that attract females, some arctiine species in which both males and females produce sex attractant pheromones but at different times of day (e.g., Wunderer et al. 1988), and the aforementioned hepialid moths.

We will begin with a discussion of the most current lepidopteran phylogeny, examining the relationships and phylogenetic distances between the superfamilies and in some cases families. Our discussion will then switch to an examination of the pheromone chemistry, looking at the diversity of chemical structures that are used as sex pheromones by female moths, within the framework of the biosynthetic pathways that are known or likely to be involved in their production. We will continue with a discussion of why lepidopteran sex pheromones are simultaneously structurally diverse and highly conserved both within and among different phylogenetic groups. This will include a synopsis of the chemical/biochemical mechanisms that are involved in the generation of specific individual compounds, and specific blends of individual compounds. We will conclude by discussing the extent to which lepidopteran sex pheromones might be used as reliable phylogenetic characters for taxonomic purposes.

The Phylogenetic Perspective: Lepidopteran Phylogeny (at Family/Superfamily Level)

The current classification of the Lepidoptera recognizes 53 superfamilies, 137 families, 15,648 genera, and 157,424 species (Nieukerken et al. 2011). The family-level numbers are the result of several large-scale studies, which have built upon each other to arrive at the foundation for a robust classification. The phylogenetic relationships of the superfamilies and the families within them are being further clarified and refined through the use of molecular data. Figure 4.1 presents the latest phylogenetic hypothesis at the superfamily level. The sister relationship of Lepidoptera and Trichoptera is one of the best-supported groupings among insect orders (Kristensen 1998; Beutel et al. 2011), and the reciprocal monophyly of the two orders has not been called into question (Kristensen 1998). These two lineages most likely diverged from each other in the Mid-Triassic period (estimated at 234 million years ago, henceforth Mya) (Malm et al. 2013).

Within the Lepidoptera, the early divergences took place mainly throughout the Jurassic period (Wahlberg et al. 2013) leading to 15 major lineages, of which 14 are considered superfamilies of the informal grade of nonditrysian Lepidoptera (i.e., with a single genital opening used for both mating and egg-laying; Kristensen 1998; Mutanen et al. 2010; Regier et al. 2013). The nonditrysian superfamilies are generally very species poor, with at most a few hundred species. They are often considered to be "primitive" Lepidoptera, and each superfamily is clearly defined by autapomorphies (diagnostic morphological characters).

The vast majority of lepidopteran species (98%) belong to the 15th major lineage, a clade known as Ditrysia. Although Ditrysia show an amazing variety of external morphologies, from some of the smallest gracillariid moths with wingspans of a couple of millimeters to the huge birdwing butterflies and atlas moths more than 20 cm across, the internal morphology

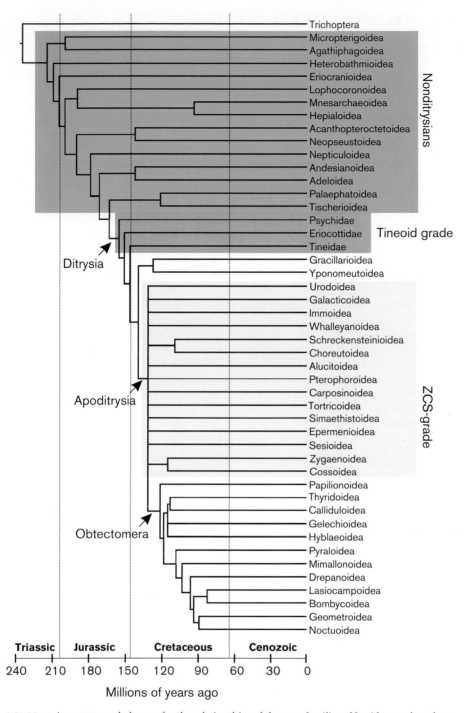

FIGURE 4.1 A consensus phylogeny for the relationships of the superfamilies of Lepidoptera based on four studies (Mutanen et al. 2010; Bazinet et al. 2013; Regier et al. 2013; Kawahara and Breinholt 2014). Relationships of superfamilies in the zygaenoid–cossoid–sesioid grade remain poorly resolved despite genomic level data.

of the ditrysians is homogenous (Kristensen and Skalski 1998), which has long been problematic for morphology-based classifications. Fortunately, DNA analyses are resolving some of the unclear relationships while also providing hard data for important revisions to relationships that had been derived on the basis of morphological characters (Regier et al. 2009, 2013; Mutanen et al. 2010). Some of the first lineages to diverge are what were formerly considered to be the superfamily Tineoidea, but which now appear to be a grade of three or more lineages, with Psychidae branching off first, followed by Erio-

cottidae, and finally by Tineidae (Mutanen et al. 2010; Regier et al. 2013, 2015). The tineoid lineages appear to have diverged from each other in the late Jurassic or early Cretaceous period (Wahlberg et al. 2013). The relationships in this region of the tree are still somewhat unstable and future transcriptomic studies should allow further clarifications.

Following the tineoid grade is a well-supported clade comprising the superfamilies Gracillarioidea and Yponomeutoidea (figure 4.1). The position of this clade as sister to the remaining Lepidoptera (a clade known as Apoditrysia) appears to be well supported, with the common ancestor diverging into the two lineages in the early Cretaceous period around 140 Mya. The two superfamilies then diverged from each other approximately 130 Mya (Wahlberg et al. 2013).

Divergences in the non-obtectomeran Apoditrysia (in figure 4.1, the 15 superfamilies from Tortricoidea to Zygaenoidea) are at present very unclear. Even a recent transcriptomic study based on 741 genes was unable to resolve their relationships (Bazinet et al. 2013). Tortricoidea tend to be found at the base of the Apoditrysia (Mutanen et al. 2010), sometimes with Galacticoidea as their sister group (Regier et al. 2013). Zygaenoidea and Cossoidea tend to be associated with each other in many molecular studies (Mutanen et al. 2010; Regier et al. 2013; Bazinet et al. 2013; Kawahara and Breinholt 2014), but rarely with any statistical support. The other superfamilies tend to cluster differently depending on the approach used. The latest phylogenomic study (Kawahara and Breinholt 2014) suggests that data from 2600 genes might help resolve these difficult nodes, but this remains speculative.

On the other hand, relationships within the Obtectomera have become clear and are well supported. Kawahara and Breinholt (2014) have shown that Papilionoidea is clearly the sister group to the rest of the Obtectomera, with a clade comprising the Gelechioidea, Thyridoidea, and Calliduloidea (possibly also including Hyblaeoidea) branching off next. Pyraloidea is clearly the sister group to Macroheterocera (Bazinet et al. 2013; Kawahara and Breinholt 2014), within which the relationships have now been well resolved (figure 4.1). Against this phylogenetic background, we then pose the questions: What patterns of distribution of pheromone structures and what evolutionary processes generating these patterns can be seen? A discussion of these questions is the major focus of this chapter.

Pheromone Types in a Biosynthetic Perspective

Any attempt to analyze and explain differences and similarities between pheromones without careful consideration of the biosynthetic pathways that underpin pheromone production immediately faces the risk of becoming speculative. For example, the monounsaturated 14-carbon acetates (Z)-8-tetradecenyl acetate (Z8-14Ac; subsequent abbreviations follow the same pattern), Z9-14Ac, Z10-14Ac, and Z11-14Ac may appear closely related by just differing in double bond position, but in species in which the biosynthesis has actually been investigated it turns out that quite different pathways are involved in the formation of these structures. Z11-16Ac and its shorter homolog Z9-14Ac produced by two-carbon chain shortening of the former may both be accounted for by the interaction of a Δ11-desaturase with palmitic acid (16:COOH). Thus, Z9-14Ac and Z11-16Ac are typically closely related biosynthetically, just as Z8-14Ac and Z10-14Ac may be closely related because the key double bond in both is introduced with a Δ10 desaturase, with or without a subsequent chain-shortening step. Alternatively Z9-14Ac and Z8-14Ac could have been produced by a Z9-desaturase and a Z8-desaturase, respectively, acting on myristic acid. Before speculating about the evolutionary history of usage of pheromone compounds in different taxa, it is thus important to know if similar compounds are produced by similar or different biosynthetic routes. Pheromone compounds may be similar because of common ancestry or as a result of independent/convergent evolution.

Roelofs and Brown (1982) published the first detailed review of lepidopteran pheromone biosynthesis within the family Tortricidae, and used that information to try to divine evolutionary relationships within the group, relating the patterns in pheromone diversity to different postulated phylogenies. For example, they identified a general trend that species in the Tortricinae use mostly 14-carbon pheromone components (acetates, alcohols, and aldehydes), whereas species in the subfamily Olethreutinae use mostly 12-carbon compounds. Furthermore, almost all tortricine species use pheromone components with double bonds in odd-numbered positions (Δ9 or Δ11 isomers), whereas pheromone components with double bonds in even-numbered positions (Δ8 and Δ10 and doubly unsaturated Δ8, Δ10 isomers) are common in the Olethreutinae. Roelofs and Brown (1982) suggested that a lepidopteran-specific Δ11-desaturase acting on 14-carbon intermediates could account biosynthetically for most of the pheromone compounds found in the Tortricinae. We will return to the details of tortricid pheromones below, but at this point we want to emphasize the fundamental merits of their approach that combined biosynthetic and phylogenetic perspectives.

As mentioned above, the first moth pheromone identified was the doubly unsaturated (E10,Z12)-10,12-hexadecadien-1-ol from the silk moth *Bombyx mori* (family Bombycidae). In hindsight, this chemical structure was not at all unexpected, in that it is a very typical example of a lepidopteran pheromone structure. The majority of identified moth sex pheromone components are monounsaturated or diunsaturated acetates, alcohols, or aldehydes with 10–18 carbon atoms in the carbon skeleton, classified as Type I pheromones by Ando et al. (2004). These types of structures constitute approximately 75% of all known moth sex pheromone components. They are synthesized de novo in the pheromone gland of the female moth, a specialized epidermal tissue usually located between the terminal abdominal segments.

In 1980, Conner et al. published the first of what have come to be known as Type II lepidopteran pheromones, from females of the artiine moth *Utetheisa ornatrix* (Conner et al. 1980). Type II pheromones typically are polyunsaturated hydrocarbons or the corresponding epoxide derivatives, with C_{17}–C_{25} unbranched carbon chains. This second major group of moth sex pheromones comprises about 15% of the reported moth pheromones. In major contrast to Type I pheromones that are biosynthesized de novo, Type II pheromones are biosynthesized from diet-derived linoleic or linolenic acid. The remaining 10% of moth pheromones are not as easily classified (Ando et al. 2004).

In the following sections we provide examples of the Type I and Type II moth pheromones and outline their biosyntheses based on postulated and confirmed pathways, and describe the enzymes involved. We then propose to extend the known classifications of lepidopteran pheromones by defining two additional pheromone types, Type 0 and Type III pheromones, based on their distinctive structural and biosynthetic features. Even with these two additional classifications, there are still a few moth sex pheromone components that do not appear to fall into any of the four major categories, and these also will be discussed.

Type I Pheromones

It is now well established that Type I pheromone compounds (figure 4.2) are biosynthesized de novo from acetate (Jones and Berger 1978; Bjostad and Roelofs 1981). Thus, studies using isotopically labeled acetate have confirmed that pal-

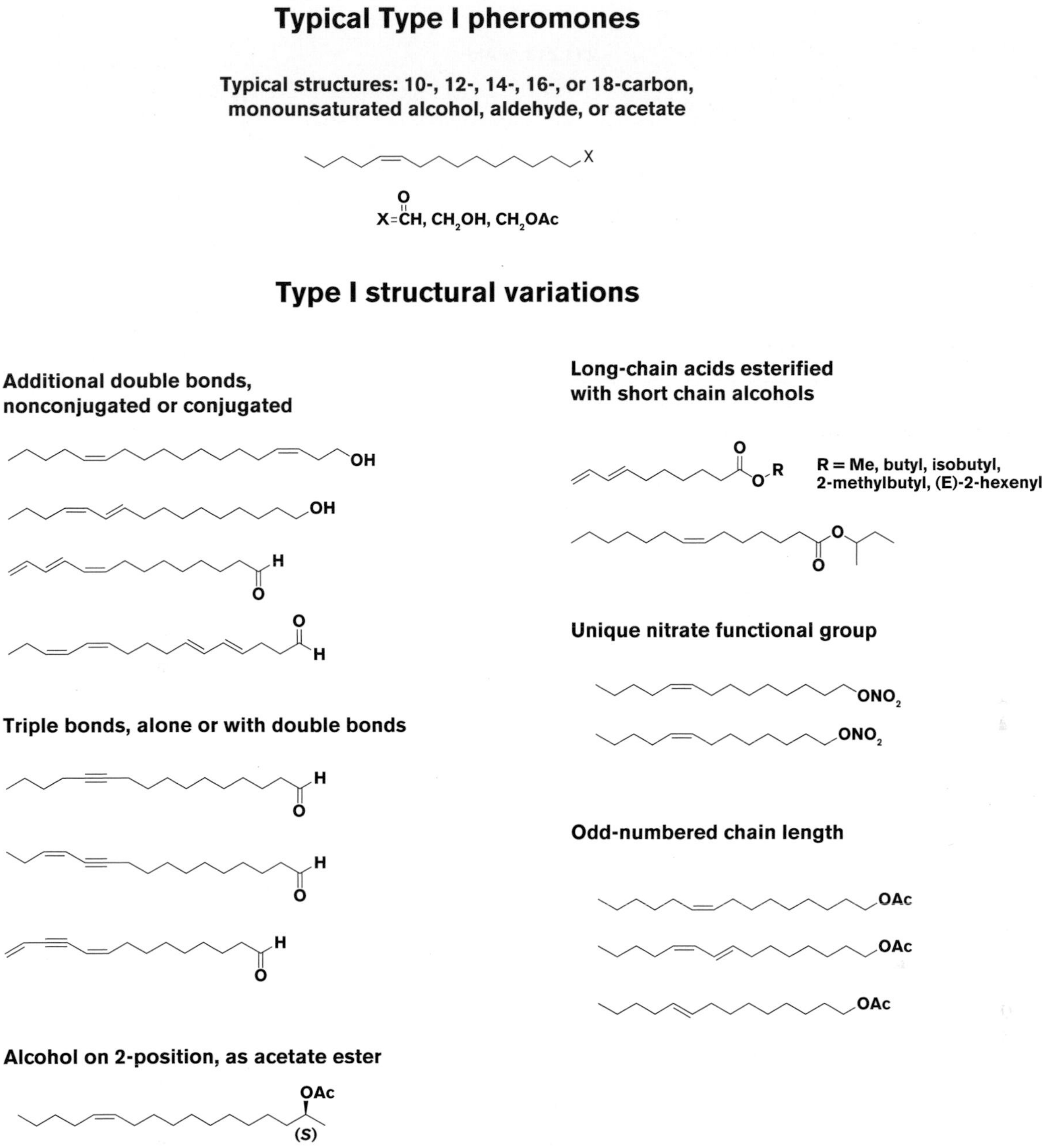

Typical Type I pheromones

Typical structures: 10-, 12-, 14-, 16-, or 18-carbon, monounsaturated alcohol, aldehyde, or acetate

$$X=\overset{\overset{\displaystyle O}{\|}}{C}H,\ CH_2OH,\ CH_2OAc$$

Type I structural variations

Additional double bonds, noconjugated or conjugated

Triple bonds, alone or with double bonds

Alcohol on 2-position, as acetate ester

Long-chain acids esterified with short chain alcohols

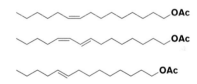

R = Me, butyl, isobutyl, 2-methylbutyl, (E)-2-hexenyl

Unique nitrate functional group

Odd-numbered chain length

FIGURE 4.2 Type I pheromones, showing typical structures composed of even-numbered, 10- to 18-carbon unsaturated aldehydes, acetates, and alcohols, and variations on the typical structures.

mitic (16COOH) and stearic acids (18COOH) are produced in the pheromone gland at the tip of the abdomen of females by typical fatty acid synthesis, although the genes involved in the production of these saturated moth pheromone precursors have not been cloned and functionally assayed (Foster, this volume). Characteristic of biosynthesis of the Type I pheromone compounds is the subsequent modification of the carbon chain by desaturation (introduction of one to four double bonds), and chain shortening by one or more rounds of β-oxidation (leading to chain shortening by two carbon atoms per cycle) or chain elongation by one or more two-carbon units. This combination of reactions explains why most Type I moth pheromone compounds have an even number of carbon atoms (10–18) in the carbon skeleton. After adjustment of the chain length and insertion of the requisite number of double bonds, the terminal carboxyl group is modified in the final step to form a volatile pheromone component, in most cases an alcohol, aldehyde, or acetate (figure 4.3). These general principles have been demonstrated by incorporation of labeled precursors in experiments with a number of moth species which produce typical Type I structures (examples are discussed below). For the more unusual Type I structures shown in figure 4.2, the biosynthetic pathways are in most cases only putative, and need to be confirmed empirically.

The first concerted research on moth pheromone biosynthesis was carried out in the laboratory of Wendell Roelofs in the early 1980s. Roelofs and his coworkers (Lou Bjostad and Walter Wolf) demonstrated that the double bond in many moth pheromone components was introduced by the action of an unusual Δ11-desaturase, different from the ubiquitous Δ9 desaturase that is a key enzyme in the fatty acid metabolism of most organisms. Under the Δ11-desaturase paradigm,

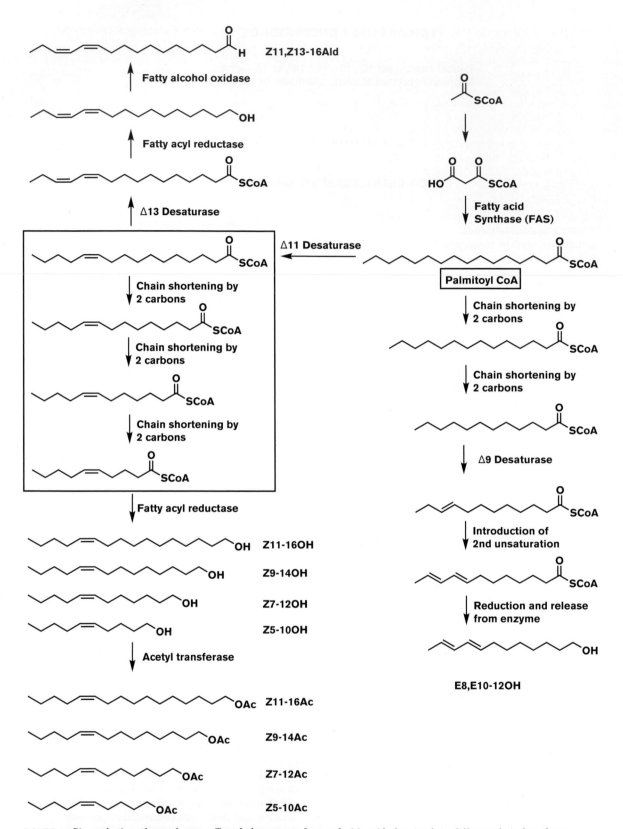

FIGURE 4.3 Biosynthetic pathways for some Type I pheromones from palmitic acid, showing how different chain lengths are constructed by chain elongation or chain shortening, interspersed with desaturation steps to place double bonds in specific, predictable positions. Once the chain is completed, the final steps involve adjustment of the terminal functional group.

Δ11-desaturation and chain shortening in different combinations not only explain the biosynthesis of a large number of monounsaturated pheromone carbon skeletons but may also account for production of doubly unsaturated compounds (see below). As an example, Δ11-desaturation in combination with one or two cycles of chain shortening accounts for production of Z7-12Ac, the major pheromone component of the cabbage looper *Trichoplusia ni* (Noctuidae) as well as many of the minor components such as Z5-12Ac, Δ11-12Ac, Z7-14Ac, and Z9-14Ac that are released from the pheromone gland (Bjostad and Roelofs 1983).

In addition to the characteristic Δ11-desaturases, other desaturases were postulated and proven to account for double bonds in various other positions that would not be accessible via Δ11-desaturation. Thus, tortricid moths were shown to have Δ9- and Δ10-desaturases in addition to Δ11-desaturases (Löfstedt and Roelofs 1985; Roelofs and Wolf 1988; Foster and Roelofs 1988; Löfstedt and Bengtsson 1988), and *Ostrinia furnacalis* (Crambidae) has a Δ14-desaturase (Zhao et al. 1990). Thus, a whole family of desaturases with specific activities in pheromone biosynthesis was postulated and has since been confirmed by means of molecular tools (see, for instance, Roelofs and Rooney 2003; Liénard et al. 2008) (figure 4.4A). Interestingly, two lepidopteran-specific subfamilies of desaturases, different and distinct from the metabolic Δ9-desaturases involved in general fatty acid biosynthesis, were identified, a Δ9 subfamily with preference for C_{18} over C_{16} substrates and the Δ11-desaturase family which contains not only desaturases with Δ11-selectivity but also specialized desaturases that introduce double bonds in other positions, or that have multiple functions in the biosynthesis of doubly unsaturated structures with conjugated double bond systems (see below).

When considering double bond insertions to form doubly- or multiply-unsaturated pheromone components, it is crucial to differentiate between conjugated and nonconjugated double bond systems. Nonconjugated double bonds are typically inserted by different desaturases; for example, in *Antheraea polyphemus* (Saturniidae), the combined actions of a Δ11- and a Δ6-desaturase produce the precursor of its pheromone component E6,Z11-16Ac, and chain shortening by two carbons produces the precursor of the analogous E4,Z9-14Ac and E4,Z9-14Ald, pheromone components of several saturniid species (McElfresh et al. 2001). Similarly, E9,12Z-14Ac, used as a pheromone component by *Cadra cautella* (Pyralidae) and *Spodoptera exigua* (Noctuidae), is formed by action of a Δ11-desaturase on palmitic acid and chain shortening by two carbons to give a Z9-14:Acyl intermediate, which is then desaturated again with a specific Δ12-desaturase (Jurenka 1997).

In contrast, conjugated double bond systems may be biosynthesized by sequential desaturations involving the same desaturase, with or without the substrate undergoing chain shortening in between, or by consecutive action of desaturases with different product specificities. For example, E10,Z12-16OH in *Bombyx mori* is produced from palmitic acid via a Z11-16:Acyl intermediate, which is acted upon a second time by the Δ11-desaturase to create the conjugated diene system (Yamaoka et al. 1984; Moto et al. 2004). In contrast, Z5,E7-12OH from *Dendrolimus punctatus* (Lasiocampidae) is produced by the sequential action of a Z11-desaturase acting on stearic acid and an E11- and/or an E9-desaturase introducing the second double bond into chain-shortened intermediates (Liénard et al. 2010b). Likewise, Z11,Z13-16Ac from *Thaumetopoea processionea* (Thaumetopoeidae) is produced by the

sequential action of a Δ11- and a Δ13-desaturase (Villorbina et al. 2003).

There is also good evidence for analogous multifunctional enzymes in the biosynthesis of more highly unsaturated pheromones, at least in the early steps. For example, the pheromone of *Manduca sexta* (Sphingidae) consists of a complicated blend of as many as eight compounds, including saturated 16Ald, monounsaturated Z9-16Ald, E11-16Ald, and Z11-16Ald, diunsaturated E10,Z12-16Ald and E10,E12-16Ald, and triunsaturated E10,E12,E14-16Ald and E10,E12,Z14-16Ald. Treatment of pheromone glands with labelled 16COOH resulted in incorporation of the label into all of these compounds except Z9-16Ald, whereas treatment with Z11-16COOH resulted in incorporation of label into Z11-16Ald, both of the dienals, and both of the trienals (Fang et al. 1995). Treatment with labelled E11-16COOH resulted in incorporation only into E11-16Ald. A subsequent study cloned a bifunctional desaturase that was shown to have Z11-desaturase and 10,12-desaturase activity, producing the Z11-monounsaturated and 10,12-diunsaturated intermediates (Matoušková et al. 2007). A Δ9-desaturase that catalyzed the transformation of 16COOH to Z9-16:Acyl was also cloned. However, the enzyme(s) involved in the third desaturation to create the triene components have not yet been identified. The pheromone of the carob moth *Ectomyelois ceratoniae* (Pyralidae) is a mixture of the triunsaturated Z9,E11,Δ13-14Ald, Z9,E11-14Ald, and Z9-14Ald (Baker et al. 1989). Based on the components of the pheromone mixture, the biosynthesis of the triene most likely involves three desaturation steps catalyzed by at least two different desaturases, but to date nothing is known about the enzymes involved.

In addition to carbon–carbon double bonds, there are a few examples known of C-C triple bonds (i.e., alkynes or acetylenes) appearing in lepidopteran pheromones. The most well known is probably the pine processionary moth *Thaumetopoea pityocampa* (Thaumetopoeidae), which produces Z13,11yne-16Ac as its main pheromone component. Biosynthetic studies have shown that this compound is synthesized from palmitic acid by Δ11-desaturation, a second Δ11-desaturation to produce the triple bond, and Δ13-desaturation to produce the conjugated enyne. Remarkably, all three desaturations are apparently performed by the same multifunctional enzyme (Arsequell et al. 1990; Serra et al. 2007).

The coding genes and enzymes involved in chain shortening and chain elongation in pheromone biosynthesis have not been characterized, but, as mentioned above, fatty acyl pheromone precursors shorter than 16 carbons are products of one or more cycles of chain shortening of longer chain fatty acyl precursors by two carbons. For example, a shift in pheromone component ratios between 12- and 14-carbon acetates in a mutant strain of the cabbage looper *T. ni* (Noctuidae) appears to be the result of a mutation affecting the chain-shortening reactions (Jurenka et al. 1994). The overall process starts with β-oxidation and involves several steps catalyzed by several enzymes working sequentially, and it can occur either in mitochondria or in peroxisomes. Because β-oxidation in moth pheromone biosynthesis is strictly controlled, it has been suggested to occur in the peroxisomes (Hashimoto 1996; Jurenka 2004). α-Oxidation of fatty acids, resulting in chain shortening by one carbon, is rare compared to the ubiquitous β-oxidation, but it seems likely that the odd-numbered carbon skeletons found in the pheromones of a few moth species are formed by α-oxidation of even-numbered fatty acyl intermediates followed by decarboxylation to remove one carbon. For

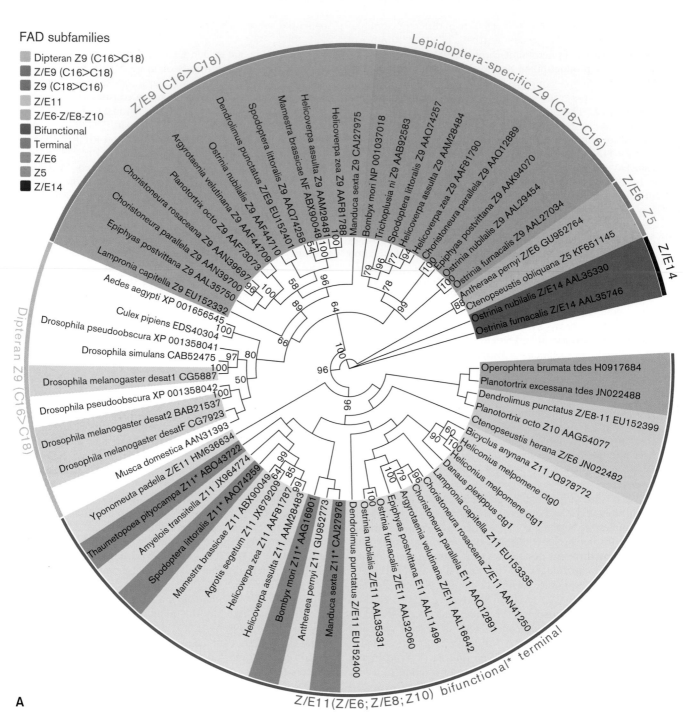

FIGURE 4.4 Phylogenetic trees highlighting fatty-acyl-CoA desaturase (FAD) and fatty-acyl-CoA reductase (FAR) expansions involved in lepidopteran pheromone biosynthesis.

A   FAD phylogenetic tree. A representative set of lepidopteran and dipteran fatty-acyl-CoA desaturases (FADs) were used to reconstruct the maximum-likelihood phylogeny. Accession numbers corresponding to nucleic acid sequences were retrieved from GenBank and the deduced amino acid sequences were used in the phylogeny. Sequence names correspond to the species name followed by a description of the desaturase function when characterized, and the accession number. *Danaus plexippus* and *Heliconius melpomene* sequences were extracted from Liénard et al. (2014). The major functional subfamilies known to be active in pheromone biosynthesis are highlighted as follows: Lepidoptera-specific Δ11-subfamily – Δ11 (light green), bifunctional (dark green), terminal and other desaturation positions (marine green); Δ9 (C16>C18) – fruit fly (gray), moths (blue); lepidopteran Δ9 (C18>C16) (chocolate); moth Δ6 (orange); Δ5 (sand); Δ14 (purple).

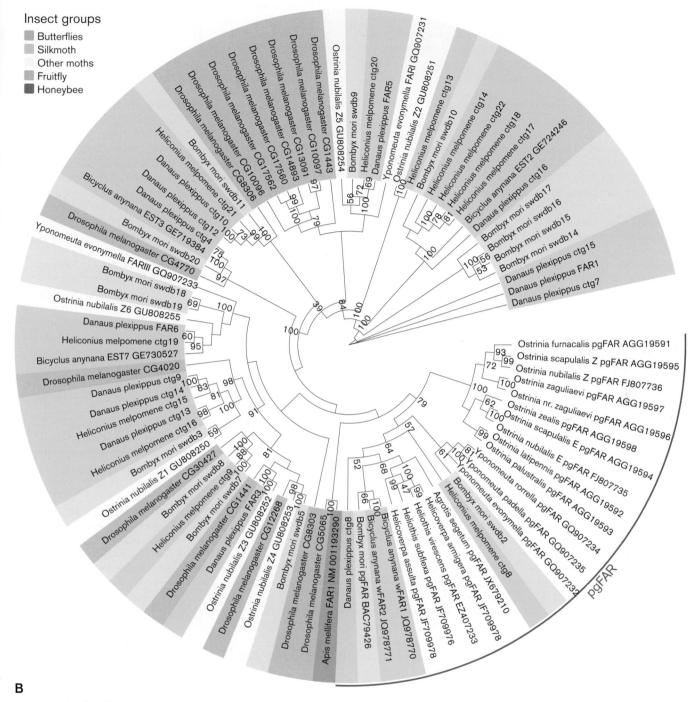

B

FIGURE 4.4 *(continued)*

B   FAR phylogenetic tree. A representative set of lepidopteran, dipteran, and hymenopteran fatty-acyl-CoA reductases (FARs) were used to reconstruct the maximum-likelihood phylogeny. Background colors for each insect group are detailed in the figure legend. Characterized lepidopteran FARs involved in pheromone production belong to the "pgFAR" clade as delineated with a dark green line. Sequence names correspond to the species name followed by the accession or contig number. Accession numbers corresponding to nucleic acid sequences or EST sequences were retrieved from GenBank, or correspond to predicted genes from *Bombyx mori* swdb (Liénard et al. 2010a) or butterfly contigs (Liénard et al. 2014). The deduced amino acid sequences were used to reconstruct the phylogeny.

For both phylogenetic reconstructions, multiple amino acid sequences were generated using MAFFTv7 with the E-INS algorithm and the BLOSUM45 scoring matrix in Geneious® 7.1.7 (Kearse et al. 2012). Maximum-likelihood inference was carried out using the stand-alone version of PhyML68 and the WAG+ I + G model as determined after performing model selection in Topali v2.5 (Milne et al. 2009). Clade support was evaluated using 100 bootstrap replicates. The cladograms were visualized and prepared using the online tool EvolView (Zhang et al. 2012).

example, some *Acrobasis* spp. (Pyralidae) produce 15-carbon pheromones (Tabata et al. 2009) (figure 4.2), and the unusual nitro compound Z8-13ONO$_2$, a minor pheromone component in *Bucculatrix thurberiella* (Bucculatricidae), is probably derived from α-oxidation of Z9-14:Acyl, the likely precursor of the unusually functionalized major pheromone component, Z9-14ONO$_2$, in the same species.

Fatty acids longer than the 16- and 18-carbon products of the cytoplasmic fatty acid synthases (FAS) may be produced by the fatty acid chain elongation system (FACES), enzymes that add two carbon units to the carboxyl end of the fatty acid chain through a four-step sequence (Cook and McMaster 2002; Jurenka 2003). Although the enzyme systems performing chain shortening and chain elongation differ, some of the reactions and intermediates are the same (Jurenka 2003). For example, ketones or the corresponding secondary alcohols with an oxygen at position 2 could result from either chain shortening or chain elongation to produce an intermediate β-ketoacid, which then loses the carboxyl group to yield the corresponding ketone and subsequently the alcohol. Thus, the unusual pheromone of *Kermania pistaciella* (Tineidae), Z12-17:2-OAc, could be formed in this way by chain elongation of Z11-16:Acyl by two carbons, followed by β-oxidation and loss of the carboxyl, analogous to the production of (*Z*)-10-heptadecen-2-one from palmitoleic acid (Z9-16:Acyl) in *Drosophila buzzatii* (Skiba and Jackson, 1994).

The doubly-unsaturated 18-carbon pheromone components in some taxa, like E2,Z13-18Ac, are likely a variation on typical Type I pheromone compounds. They can be plausibly accounted for by Δ11 desaturation of palmitic acid, followed by the formation of the second double bond in connection with the chain elongation, if the chain elongation is interrupted before the final reduction (Roelofs and Bjostad 1984). The suggested steps involve addition of a two-carbon unit, followed by action of a β-ketoacyl reductase to produce the 3-hydroxy intermediate compound, and then dehydration by hydrase enzymes to produce either the E3, Z3, or E2 double bonds typical of pheromones in the Sesiidae and Tineidae.

The tomato pinworm, *Keiferia lycopersicella* (Gelechiidae) uses E4-13Ac as a pheromone. This unusual odd-numbered carbon chain compound could be produced by chain shortening of oleate (Z9-18:Acyl) until the carboxyl group is only one carbon from the double bond (three rounds of β-oxidation and one round of α-oxidation and decarboxylation to give Z2-11:Acyl). The enoyl system isomerizes the double bond to E, and addition of a two-carbon unit would then produce the appropriate precursor fatty acyl compound. A similar sequence of reactions could account for the pheromone components E4,Z7- and E4,Z7,Z10-13Ac of the potato tuberworm *Phthorimaea operculella* (Gelechiidae) from linoleate and linolenate (Persoons et al. 1976). If so, these compounds may be special cases of Type I pheromones. However, this would violate one of the defining characteristics of Type I pheromones, i.e., that they are derived de novo from acetate. This suggests the alternative possibility that they could arise from a Type II pathway (see below), or even a combination of Type I and Type II pathways. In particular, the 1,4-arrangement of the two Z double bonds in E4,Z7,Z10-13Ac, and their position three carbons in from the end of the chain (i.e., the ω-3 position) are strongly suggestive of an origin from linolenate.

Once the penultimate fatty acyl precursors have been formed with the required number of double bonds and the required chain length, the precursors are reduced by a fatty acyl reductase (FAR) that is more or less gland-specific. These FARs belong to an expansion of the FAR tree, which (similar to some of the desaturase subfamilies) appears to be specific to the Lepidoptera (figure 4.4B). The long-chain fatty alcohols formed by FARs may serve directly as pheromone components or, alternatively, they are reoxidized to aldehydes or esterified, typically as acetates. Although both alcohol oxidases and acetyl transferases can be inferred based on experiments with labelled precursors, no such enzymes involved in lepidopteran pheromone biosynthesis have been cloned and characterized. A few examples of esters other than acetates, in some cases in combination with acetates, have been reported, for example, in *Dendrolimus* spp. (Lasiocampidae), or in the sex pheromone of the noctuid *Pseudoplusia includens*, which is a mixture of Z7-12Ac and the corresponding propanoate and butyrate esters (Linn et al. 1987). In another example, (*Z9,E12*)-tetradecadienyl propanoate was reported as a sex attractant for the nolid *Nolathripa* (= *Lamprothripa*) *lactaria* (Ando et al. 1977).

The vast majority of esters used as pheromone components in the Lepidoptera are biosynthesized from a long-chain alcohol and a short-chain acid, but there are examples known, for example, in the Lymacodidae and the Psychodidae, of the inverse, i.e., esters formed between long-chain acids and short-chain alcohols. From their structures, it seems likely that the long-chain acids are biosynthesized using the typical Type I chain elongation and shortening steps, but instead of being reduced, the penultimate acyl precursor is esterified with a short-chain alcohol. These alcohols might in turn be derived from amino acids, based on recent biosynthetic studies of similar short-chain alcohols in butterflies (Wang et al. 2014). However, in the case of the psychids, their unusual pheromones consisting of short-chain acids esterified with branched chain alcohols are not really reminiscent of typical Type I structures (see also the "Compounds That Do Not Fit Any of Types I, II, III, or 0" section below).

In summary, by combining several enzymatic reactions including specific desaturations, elongations by two carbons, chain shortening by one or two carbons, reduction of the terminal functional group, reoxidation, and esterification, an enormous diversity of Type I pheromones are produced in the Lepidoptera. This diversity is further expanded by the use of unusual functional groups, such as a nitro group or esters other than acetates.

Type II Pheromones

(Z3,Z6,Z9)-Heneicosa-3,6,9-triene (Z3,Z6,Z9-21:H) appears to have been the first prototypic Type II pheromone identified, from the arctiine *Utetheisa ornatrix* (Conner et al. 1980), and shortly after Hill and Roelofs (1981) reported that the pheromone blend of the saltmarsh caterpillar moth *Estigmene acrea* (Erebidae: Arctiinae) consisted of a blend of (Z9,Z12)-octadeca-9,12-dienal (Z9,Z12-18Ald), (Z9,Z12,Z15)-octadeca-9,12,15-trienal (Z9,Z12,Z15-18Ald), and (Z3,Z6)-*cis*-9,10-epoxyheneicosa-3,6-diene. Although there were examples of polyunsaturated aldehydes among the Type I pheromones identified up to that point, the 21-carbon diene epoxide clearly did not fit the "ground plan" of Type I pheromones. The aldehydes could obviously be derived from linoleic or linolenic acid, as could the epoxide, and so a new category of lepidopteran pheromones, later referred to as Type II pheromones, was suggested. Now, a substantial number of Type II pheromone components are known from a distinct subset of

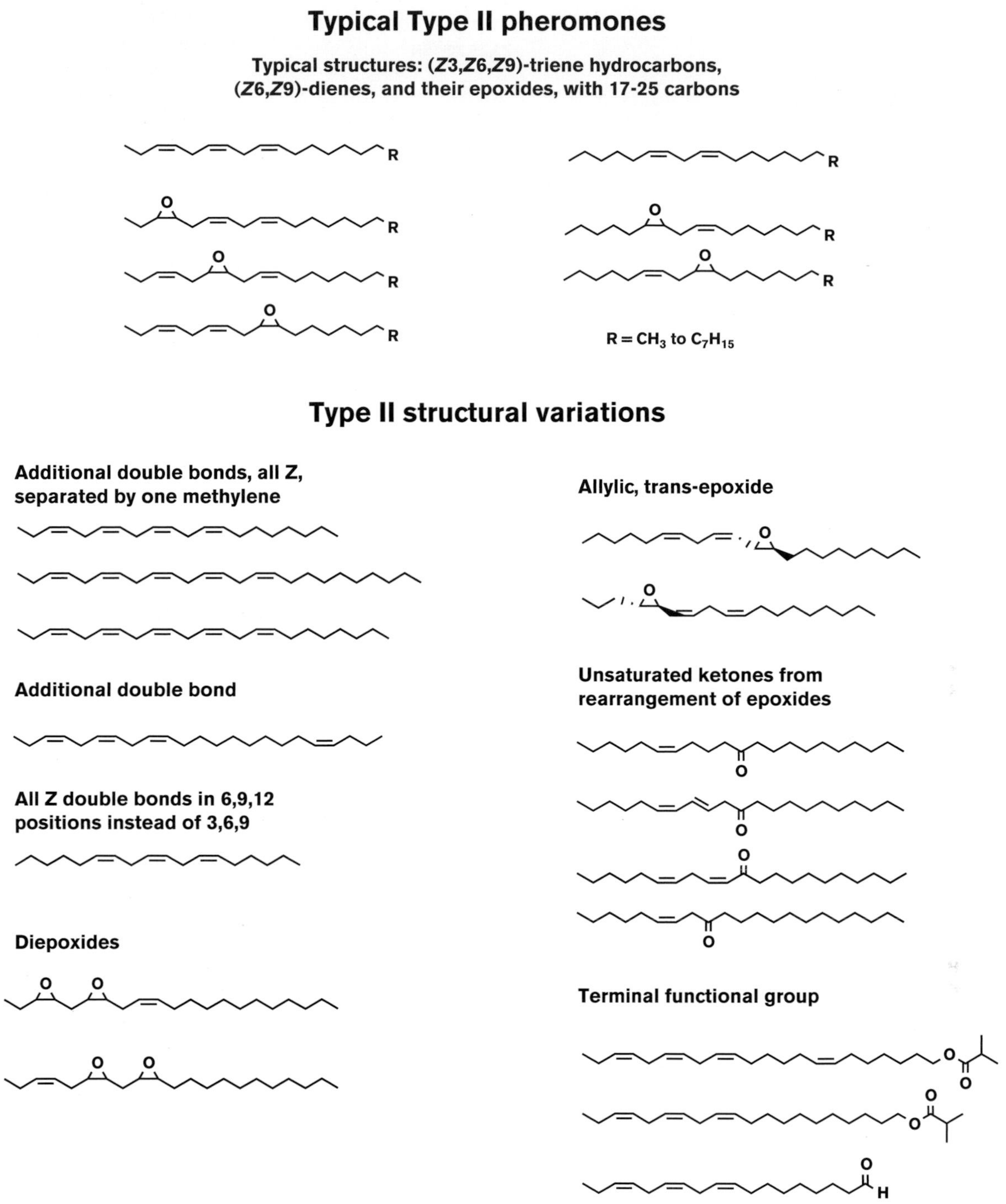

Typical Type II pheromones

**Typical structures: (Z3,Z6,Z9)-triene hydrocarbons,
(Z6,Z9)-dienes, and their epoxides, with 17-25 carbons**

R = CH$_3$ to C$_7$H$_{15}$

Type II structural variations

**Additional double bonds, all Z,
separated by one methylene**

Additional double bond

**All Z double bonds in 6,9,12
positions instead of 3,6,9**

Diepoxides

Allylic, trans-epoxide

**Unsaturated ketones from
rearrangement of epoxides**

Terminal functional group

FIGURE 4.5 Type II pheromones, showing typical structures composed of (Z3,Z6,Z9)- or (Z6,Z9)-unsaturated hydrocarbons with chain lengths of 17–25 carbons, or monoepoxides thereof. Structural variations include additional double bonds, double bonds in (Z6,Z9,Z12) positions, diepoxides, ketones likely derived from rearranged epoxides, and compounds with terminal functional groups.

the lepidopteran families and subfamilies. In 2004, Ando et al. defined Type II pheromones as consisting of polyunsaturated hydrocarbons and their epoxide derivatives with C$_{17}$–C$_{23}$ straight chains, and lacking a terminal functional group (Ando et al. 2004). However, both earlier and more recent data suggest that this definition should be expanded to cover compounds with chain lengths of up to 25 carbons (Grant et al. 2009; Löfstedt et al. 2012), and compounds with a terminal functional group (see examples in figure 4.5).

Several studies of the biosyntheses of Type II pheromone components have elucidated the basic mechanisms by which they are formed (figure 4.6). They are readily distinguished from Type I compounds by three specific characteristics. First, unlike Type I compounds which are biosynthesized de novo from acetate, Type II compounds are derived from linoleic or linolenic acid precursors that must be obtained from the diet because insects are unable to produce these unsaturated fatty acids (Stanley-Samuelson et al. 1988). Furthermore, the pattern of two or three (Z)-double bonds separated by a methylene group that characterizes the structures of linoleic or linolenic acids, respectively, is preserved in the final products, either intact or as more highly derived

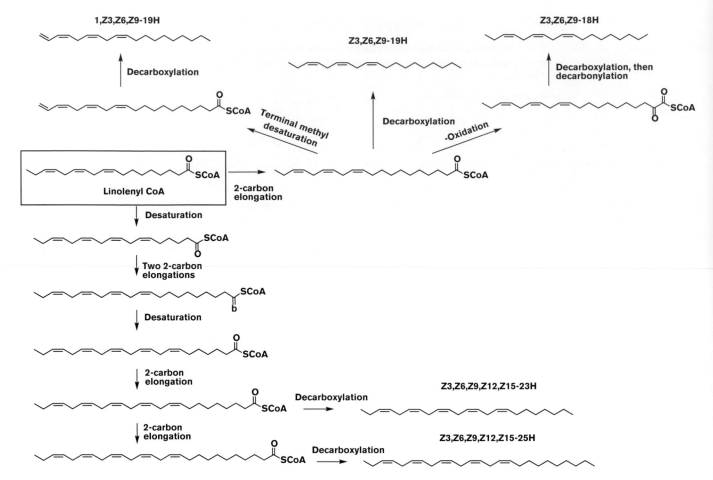

FIGURE 4.6 Biosynthetic pathways for some Type II hydrocarbon pheromones from linolenic acid. In the oenocytes, different chain lengths are constructed by chain elongation or chain shortening, interspersed with desaturation steps to place double bonds in specific, predictable positions. Once the chain is completed, the final steps involve decarboxylation to provide odd-numbered chains, or α-oxidation followed by decarboxylation and decarbonylation to produce even-numbered chains. The completed hydrocarbons are then transported to the pheromone gland for release, before or after epoxidation.

structures, for example, from epoxidation of one or more of the double bonds.

Second, the hydrocarbon skeletons of Type II pheromones are produced in oenocyte cells associated with the integument, rather than being synthesized in the pheromone gland. The hydrocarbons then are transported to the pheromone gland by lipophorin carrier proteins in the hemolymph (Matsuoka et al. 2006), and selectively downloaded to the pheromone gland, from which they are released, either with or without further structural modification. Also, the decarboxylase enzymes that can produce a hydrocarbon from an acyl group are present in the oenocytes, but not present in lepidopteran pheromone glands (Blomquist et al. 1987).

For those Type II pheromones which bear a terminal functional group, the only available piece of evidence indicates that at least the final form of the terminal functional group is generated in the pheromone gland. Specifically, the pheromone blend of the fall webworm *Hyphantrea cunea* (Erebidae: Arctiinae) contains Z9, Z12, Z15-18:Ald, and when deuterium-labeled (*Z9,Z12,Z15*)-octadeca-9,12,15-trien-1-ol (linolenyl alcohol) was applied to the gland, the labelled aldehyde was produced (Kiyota et al. 2011), whereas labelled linolenic acid itself was not incorporated when applied to the gland. However, a previous study with another arctiine species, *E. acrea*, which also produces Z9,Z12,Z15-18:Ald, had shown that

[14]C-labeled linolenic acid was indeed incorporated into the aldehyde when applied to the gland (Rule and Roelofs 1989). This apparent contradiction may be an artifact of the different detection limits of the two types of labelled compounds. However, in neither study did the authors treat the pheromone gland with the corresponding hydrocarbon, (*Z3,Z6,Z9*)-octadeca-3,6,9-triene, and so it remains unknown whether the triene hydrocarbon is produced in the oenocytes and then transported to the gland and terminally functionalized, or whether a terminally functionalized triunsaturated precursor is already present in the gland.

A third characteristic that distinguishes Type II from Type I pheromones is the very different roles of pheromone biosynthesis activating neuropeptides (PBAN), released from the subeosophageal gland, in the two biosynthetic pathways. Thus, biosynthesis of Type I pheromones is under the direct control of PBAN, which upregulates pheromone production in the pheromone gland. In contrast, the available evidence suggests that the synthesis of Type II pheromones is not directly controlled by PBAN. Instead, PBAN in Type II insects may control the uptake of pheromone precursors into the gland from the hemolymph (Wei et al. 2004; Choi et al. 2007; Fujii et al. 2007). Furthermore, a PBAN isolated from the Type II- producing geometrid species *Ascotis selenaria cretacea* had <46% sequence homology with PBANs from 15 Type I pheromone-producing species (Kawai et al. 2007), further accentuating the differences between the two PBAN types.

A fourth possible characteristic that may or may not differentiate Type I and Type II pheromone biosynthesis is that there is evidence to suggest that Type II pheromone biosynthesis begins during the early pupal stage. Specifically, while studying pheromone biosynthesis in *E. acrea*, Rule and Roelofs (1989) discovered that maximum incorporation of label occurred when labelled precursors were injected into 0- to 1-day-old pupae, and from dissection of pupae, pheromone was already present by the time the glands were discernable in 2- to 3-day-old pupae. This early onset of pheromone biosynthesis during the pupal stage seems less likely to occur for Type I pheromones, where pheromone biosynthesis is clearly controlled by PBAN release, unless PBAN is indeed released during the pupal stage. To our knowledge, the developing pheromone glands in Lepidoptera that produce Type I pheromones have not been analyzed to check for the presence of pheromone. A single intriguing report indicates that male codling moths (*Cydia pomonella*, Tortricidae) are attracted to the pupae of females, suggesting that pheromone production and even release may also occur in pupae of moths with Type I pheromones (Duthie et al. 2003). However, these "pupae" may actually be pharate females which have not yet emerged from the pupal case.

A number of studies have elucidated some of the basic details of the biosyntheses of Type II pheromones (figure 4.6). Thus, the scenario of linoleic or linolenic acid precursors being successively chain elongated by two carbon units in the oenocytes to produce even-numbered chains, which then are decarboxylated to hydrocarbons with odd-numbered chains, is supported by several pieces of evidence. Incorporation of labelled linolenic acid into Type II pheromones has been demonstrated in several species from different families (Rule and Roelofs 1989; Choi et al. 2007; Wang et al. 2010a). In contrast, saturated 18COOH was not incorporated, indicating that pheromone synthesis begins with intermediates with the characteristic double bonds already in place (Rule and Roelofs 1989). Furthermore, Z12,Z15,Z18-21COOH was not incorporated, proving that Z13,Z16,Z19-22COOH was decarboxylated to the C_{21} triene, as opposed to the more indirect route of α-oxidation of the Z13,Z15,Z18-22COOH, decarboxylation, and then reduction of the former α-carbon. In addition, some of the postulated chain-extended intermediates en route to the final pheromones, such as Z11,Z14,Z17-20COOH and Z13,Z16,Z19-22COOH, have been found in the oenocytes and their associated fat bodies (Matsuoka et al. 2008). Furthermore, labelled versions of these acids were incorporated into Type II pheromones when they were injected into the hemolymph, but not when they were applied to the pheromone gland (Rule and Roelofs 1989; Wang et al. 2010a), confirming that the hydrocarbon skeletons of the pheromones were not produced in the pheromone gland.

Transport of the hydrocarbon products from the oenocytes to the pheromone gland has also been firmly established, both from finding the hydrocarbons in extracts of the hemolymph (Wei et al. 2003; Matsuoka et al. 2008) and from identification and characterization of the lipophorin carrier proteins (Matsuoka et al. 2007). Interestingly, the lipophorins accepted a series of analogs of the triene pheromone components, indicating that the specificity of the pheromone blend was probably a result of tight control of the production of specific hydrocarbons in the oenocytes rather than specificity of the transport proteins (Wei et al. 2004; Matsuoka et al. 2007).

Finally, several pieces of evidence demonstrate that epoxidation of the unsaturated hydrocarbon precursors occurs in the pheromone gland after the hydrocarbons have been selectively downloaded to the gland from lipophorins. First, epoxides have not been found in the hemolymph (Wei et al. 2003), whereas the unsaturated hydrocarbons have, as mentioned above. Second, when the unsaturated hydrocarbon precursors were applied directly to the pheromone gland, the epoxides were readily produced (Miyamoto et al. 1999; Wei et al. 2004), and further studies localized the monooxygenase activity to a layer of cells in the intersegmental membrane between segments 8 and 9 (Fujii et al. 2007). Furthermore, the monooxygenase enzymes that carried out the epoxidations were specific in terms of the position and geometry of the double bond which they epoxidized, but otherwise relatively nonspecific in terms of accepting analogs and homologs, providing another indication that the pheromone blend ratio is likely a result of tight control of production of the hydrocarbon substrates in the oenocytes or selective transport or downloading to the pheromone gland rather than selectivity of the monooxygenases. Rong et al. (2014) recently characterized one of these monooxygenases as a cytochrome P450, which was specifically expressed in the sex pheromone gland of *H. cunea* and which specifically epoxidized the Z double bond in the ninth position in the pheromone precursor Z3,Z6,Z9-21H.

The biosynthetic pathways described above accommodate the biosynthesis of the prototypic Type II compounds, i.e., the triunsaturated hydrocarbons and epoxides. Plausible mechanisms can be derived for most variations on this base plan. For example, some species produce even-numbered rather than odd-numbered hydrocarbons and epoxides, and these have been shown to be generated by α-oxidation of an even-numbered acyl intermediate, followed by decarboxylation, then loss of the second oxidized carbon as CO or CO_2 (Goller et al. 2007). Other species have pheromone components with additional double bonds, which may arise in several ways, depending on their positions. For example, *U. ornatrix* produces a blend of Z3,Z6,Z9-21H and 1,Z3,Z6,Z9-21H (Jain et al. 1983), and both are derived from linolenic acid. However, 1,Z3,Z6,Z9-21H was not produced when Z3,Z6,Z9-21H was injected into the hemolymph, indicating that the terminal double bond must be inserted prior to the decarboxylation step (Choi et al. 2007). In the geometrid winter moth *Operophtera brumata*, Ding et al. (2011) characterized a terminal fatty acyl desaturase from epidermal tissue that is involved in the formation of the terminal double bond in 1,Z3,Z6,Z9-19H before the final decarboxylation.

The navel orangeworm *Amyelois transitella* (and a number of other pyraloid species) produces the more highly unsaturated pentaenes, Z3,Z6,Z9,Z12,Z15-23H and Z3,Z6,Z9,Z12,Z15-25H, but the only intermediate which has been found in significant amounts in tissue analyses was Z5,Z8,Z11,Z14,Z17-20COOH, suggesting that the additional two unsaturations are introduced at a relatively early stage, and that the subsequent two or three cycles, respectively, of chain elongation by two carbons followed by decarboxylation are tightly coupled, with the intermediates remaining bound to the enzyme complex (Wang et al. 2010a). Chain elongation of linolenic acid and further desaturation to produce Z5,Z8,Z11,Z14,Z17-20COOH is also known to occur in other insects (Stanley-Samuelson et al. 1988).

Other variations in Type II pheromones include variation in the positions (e.g., Z6,Z9,Z12-18H in *Hemithea tritonaria*;

Geometridae) or numbers of double bonds (e.g. Z3,Z6,Z9,Z12-20H in *Thalassodes immissaria intaminata*; Geometridae) (Yamakawa et al. 2009). In analogy with the linoleic and linolenic acid-based starting materials used by other Type II moths, these particular structures could be derived from γ-linolenic acid and Z6,Z9,Z12,Z15-18COOH or Z8,Z11,Z14,Z17-20COOH, respectively, obtained from the insects' diet. In contrast, double bonds in conjugation with one of the existing double bonds (e.g., 1,Z3,Z6,Z9- or Z3,Z6,Z9,Z11-tetraenes) are likely to be generated by a desaturase acting on a triene intermediate, as was indirectly demonstrated with *U. ornatrix* (Choi et al. 2007). Diepoxides likely arise from two successive epoxidations, and although a route to the two allylic epoxides that are known is not immediately obvious, the fact that both structures have Z double bonds in positions 6 and 9 strongly suggests that they are modified Type II structures. Similarly, unsaturated ketones found in some lymantriid moths betray their likely biosynthetic origins from the placement and geometry of their remaining double bonds, and placements of the ketones which could arise from, for example, epoxidation and then rearrangement of the epoxide to a ketone. There are also a few cases of compounds with a Type II structure but terminal functionalization in *Euproctis* spp. (Erebidae: Lymantriinae), where again the placement and geometries of the double bonds strongly suggest a linolenic acid or analogous origin. It remains to be determined whether these compounds are formed from hydrocarbon precursors that are transported to the pheromone gland and then oxidized at the terminal position. Alternatively, because these compounds have even-numbered chains, they could equally well be derived from the corresponding acyl derivatives by reduction to the alcohol and esterification, as occurs during the terminal functionalization of Type I compounds.

Overall, a large diversity of Type II structures can be formed starting from linoleic or linolenic acid, by combinations of chain elongation, decarboxylation/decarbonylation, epoxidation of double bonds, and rearrangement of the epoxides to ketones or alcohols. In this respect, there are strong parallels to the production of Type I compounds, in terms of a relatively small number of enzyme-catalyzed reactions being used in different combinations and different orders to produce a diversity of structures.

Type III Pheromones

Type I and Type II pheromones are all derived from modifications to straight-chain fatty acid biosynthetic pathways, but a number of lepidopteran species produce sex attractant pheromones that contain one or more methyl branches (figure 4.7). These compounds include saturated and unsaturated hydrocarbons, as well as functionalized hydrocarbons. These compounds do not really fit into either the Type I or the Type II category, for the following reasons.

Type I pheromones are synthesized de novo from acetate in the pheromone gland, whereas the Type II pheromones are synthesized from diet-derived unsaturated fatty acid precursors in the oenocytes and then transported to the pheromone gland for release, or final modification and release. In contrast, the available evidence suggests that methyl-branched hydrocarbons are synthesized de novo, but not in the pheromone gland. For example, arctiine moths in the genus *Virbia* (also known by its synonym *Holomelina*) synthesize 2-methylheptadecane de novo as a pheromone component, with the

methyl branch originating from the amino acid leucine, which provides the first four carbons of the chain plus the methyl group, and with the rest of the chain arising from typical fatty acid biosynthesis by incorporation of two-carbon acetate units, terminating with a reductive decarboxylation to produce the terminal hydrocarbon (Charlton and Roelofs 1991; figure 4.8A). Labelled precursors applied directly onto the pheromone gland were not incorporated, nor were any likely precursors to the pheromone found in the gland. In contrast, precursors injected into the hemolymph were readily incorporated, suggesting that the pheromone was biosynthesized in tissues other than the pheromone gland. This was later corroborated by Schal et al. (1998), who demonstrated that the pheromone was synthesized in the oenocytes, and then transported through the hemolymph by lipophorin proteins, and selectively released to the pheromone gland by an as-yet unknown mechanism. In a second example, Jurenka and coworkers showed that the methyl-branched pheromone of gypsy moth (*Lymantria dispar*, Erebidae: Lymantriinae), (7R,8S)-2-methyl-7,8-epoxyoctadecane, was derived from an alkene precursor, (Z)-2-methyloctadec-7-ene, which was synthesized de novo from the methyl-branched amino acid valine (Jurenka and Subchev 2000; Jurenka et al. 2003). Thus, valine was first converted to isobutyrate, which was then chain extended to the saturated C_{19} acid, followed by specific placement of the (Z)-double bond with a Δ12-desaturase, and reductive decarboxylation (figure 4.8B). All of these steps occurred in the oenocytes, with the resulting (Z)-2-methyloctadec-7-ene then being transported by lipophorins to the pheromone gland, where it was selectively taken up and epoxidized before release. Furthermore, the available evidence suggested that PBAN regulated the epoxidation of (Z)-2-methyloctadec-7-ene once it was downloaded to the gland, but PBAN did not appear to directly regulate the production of (Z)-2-methyloctadec-7-ene itself (Jurenka 2004).

The two examples above were both for hydrocarbons with a methyl branch in position 2, which is a special case because the branch is derived from the amino acids valine or leucine, both of which terminate in isopropyl groups. For the more general case of methyl branches in other positions, the evidence from other insects suggests that methyl-branched hydrocarbons and related pheromones may also be derived from fatty acid type biosynthesis, with the branch(es) being derived from incorporation of methylmalonyl CoA in place of malonyl CoA into the elongating chain (Blomquist et al. 2010) (figure 4.8C). Thus, the cycle of additions of two carbon segments into the growing chain is maintained, while also introducing one or more methyl branches. This has the consequence that the positions of the methyl branches should be highly predictable, with methyl branches from the 3 position onwards occurring only on odd-numbered carbons in relation to the end of the chain where biosynthesis started.

Several additional points are worth noting. First, the proposed biosynthetic scheme can accommodate compounds with a single methyl branch, or multiple branch points along the chain. Second, based on what is known from biosynthetic pathways in other organisms, the biosynthesis likely goes through a conjugated enoyl CoA structure, and depending on which face of the alkene is hydrogenated, either enantiomer could be generated. It has recently been shown that 36 methyl-branched cuticular hydrocarbons isolated from nine insect orders all had the (R)-configuration (Bello et al. 2015), but when used as volatile pheromones, Lepidoptera seem to be able to circumvent this general trend and produce either (R)-

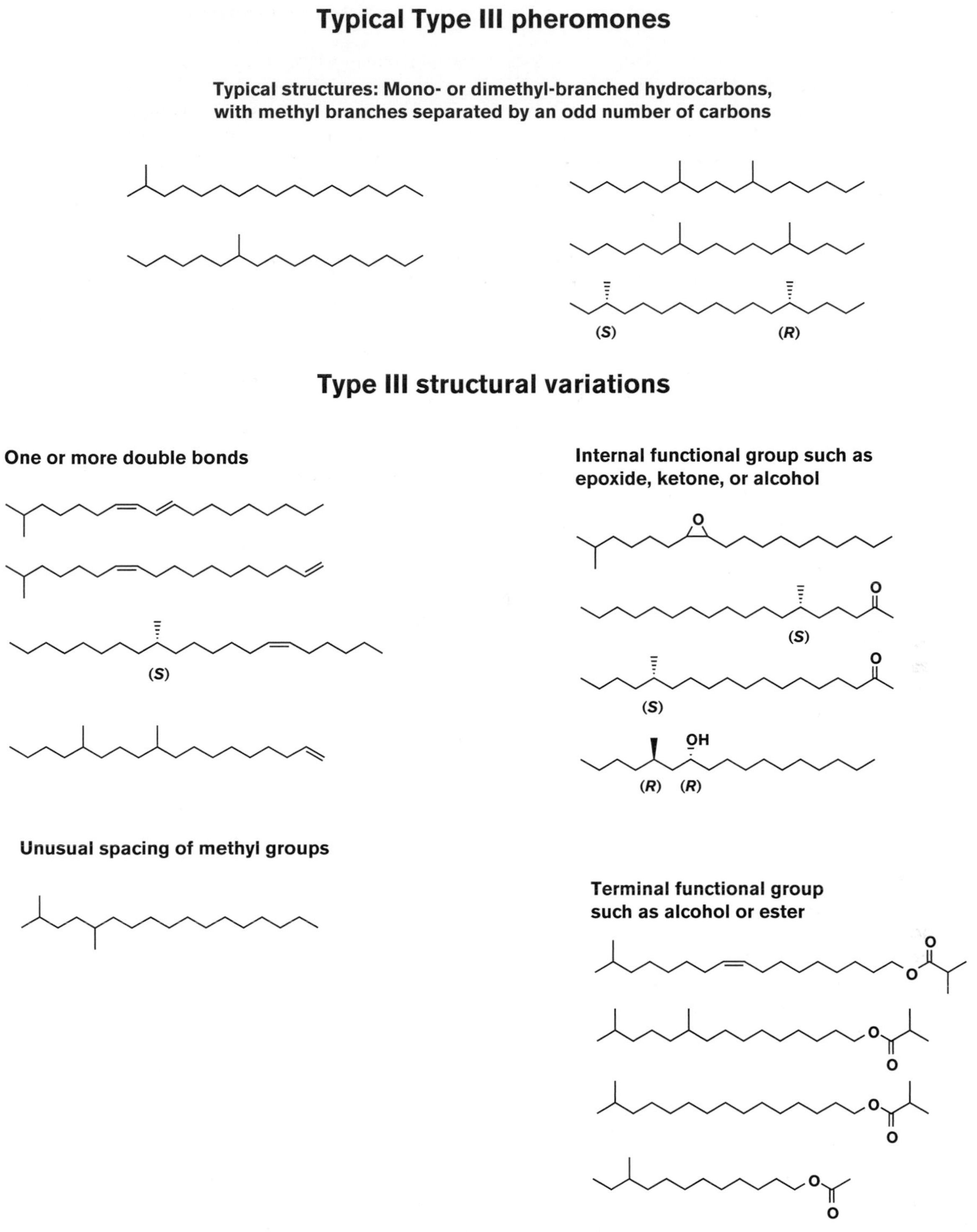

Typical Type III pheromones

Typical structures: Mono- or dimethyl-branched hydrocarbons, with methyl branches separated by an odd number of carbons

Type III structural variations

One or more double bonds

Unusual spacing of methyl groups

Internal functional group such as epoxide, ketone, or alcohol

Terminal functional group such as alcohol or ester

FIGURE 4.7 Methyl-branched Type III pheromones, showing the known structural variations. For compounds where two or more stereoisomeric forms are possible, the pheromones are typically biosynthesized in one stereoisomeric form, and the other(s) vary in activity from antagonistic to benign, to having activity similar to the natural isomer.

or (S)-configurations. For example, the noctuid moths *Anomis texana* and *Alabama argillacea* use (S)-7-methylheptadecane and (S)-9-methylnonadecane, respectively, as pheromones, whereas the geometrid *Nepytia freemani* uses (3S,13R)-3,13-dimethylheptadecane [(3S,13R)-DiMe-17H] as its pheromone (King et al. 1995), and another geometrid, *Lambdina fiscellaria lugubrosa*, uses (5R,11S)-DiMe-17H as its pheromone (Li et al. 1993). However, one must be careful when working with these structures, because the stereochemical notation can be misleading in terms of trying to visualize the three-dimensional structures of these methyl-branched molecules. That is, the nomenclature rules used to assign (R)- or (S)-configurations are based on assigning groups priority according to their sizes, rather than their orientation in space. For example, when drawn out as a straight chain, both of the methyl groups in (3S,13R)-DiMe-17H are on the same face of

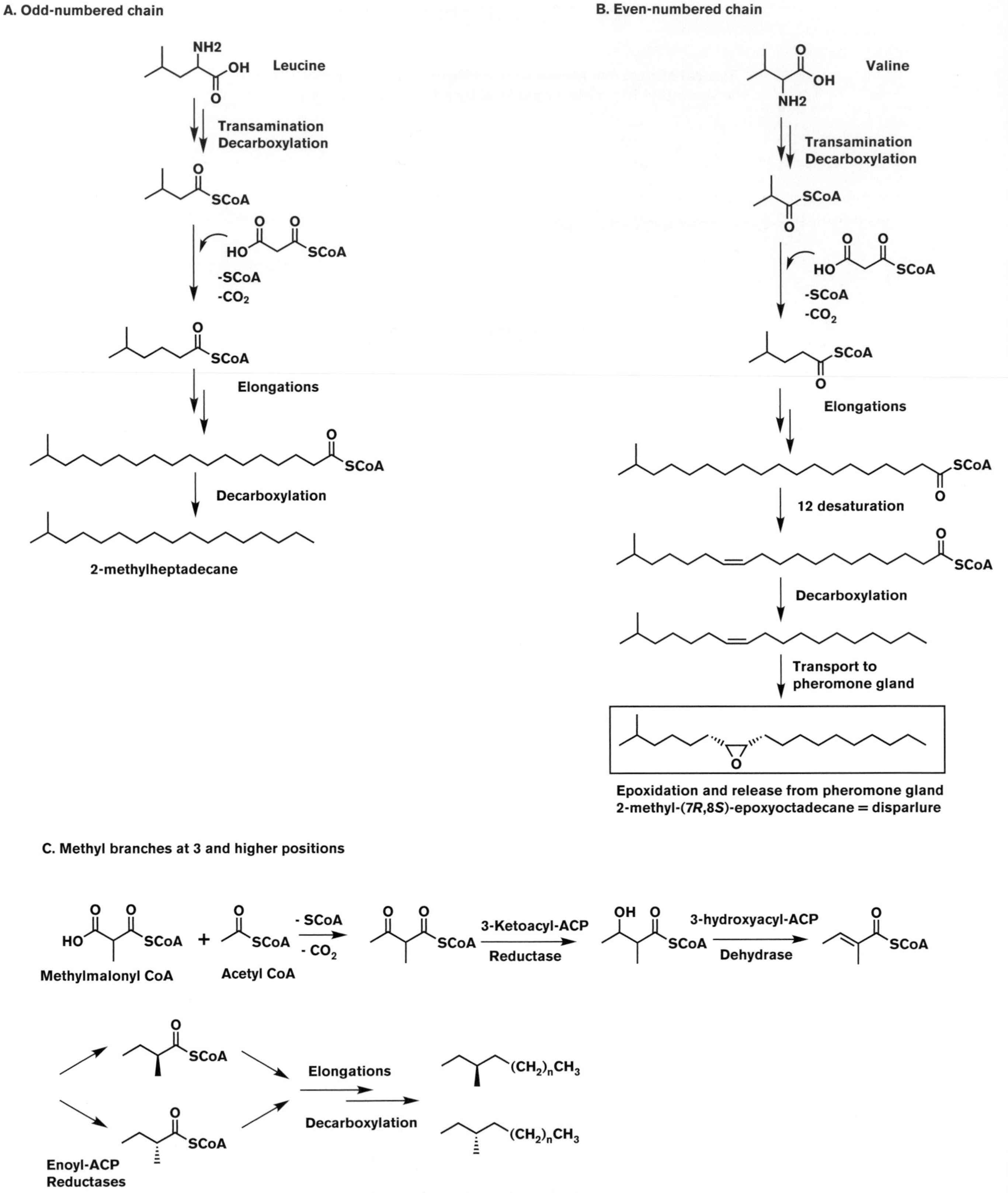

A. Odd-numbered chain

NH2
Leucine

Transamination
Decarboxylation

SCoA

HO SCoA

-SCoA
-CO$_2$

SCoA

Elongations

SCoA

Decarboxylation

2-methylheptadecane

B. Even-numbered chain

O
OH
Valine
NH2

Transamination
Decarboxylation

SCoA
O

HO SCoA

-SCoA
-CO$_2$

SCoA
O

Elongations

SCoA
O

12 desaturation

SCoA
O

Decarboxylation

Transport to
pheromone gland

O

Epoxidation and release from pheromone gland
2-methyl-(7R,8S)-epoxyoctadecane = disparlure

C. Methyl branches at 3 and higher positions

O O
HO SCoA + O SCoA
Methylmalonyl CoA   Acetyl CoA

- SCoA
- CO$_2$

O O
SCoA

3-Ketoacyl-ACP
Reductase

OH O
SCoA

3-hydroxyacyl-ACP
Dehydrase

O
SCoA

O
SCoA

Elongations

(CH$_2$)$_n$CH$_3$

O
SCoA

Decarboxylation

(CH$_2$)$_n$CH$_3$

Enoyl-ACP
Reductases

FIGURE 4.8 Biosynthetic pathways to Type III pheromones. Part A (Charlton and Roelofs 1991) and Part B (Jurenka et al. 2003) have been proven, whereas Part C is speculative, based on what is known from fatty acid biosynthesis.

the molecule, despite the nomenclature rules assigning them (*S*)- and (*R*)- configurations, respectively.

Although there have been no studies to date that have directly addressed the biosynthetic scheme described above for the production of pheromones with single methyl branches in the 3 or higher position, or pheromones with multiple methyl branches, there are several pieces of evidence that provide some indirect support. First, the erebiid moth *Scoliopteryx libatrix* uses (*Z*)-13-methylheneicos-6-ene as its pheromone, and the pheromone has been found in the hemolymph, suggesting that it is made in oenocytes and transported to the pheromone gland for release (Subchev and Jurenka 2001). Second, as mentioned above, the positions of methyl branches should be highly predictable, occurring only on odd-numbered carbons in relation to the beginning of the growing chain, and with one exception (2,5-dimethyl-17:H in *L. f. fiscellaria*, Geometridae), this appears to be the case. However, even this apparent exception can be accommodated by the general pattern of de novo branched chain fatty acid synthesis if the 2-methyl branch comes from leucine-derived isovalerate (as with the *Virbia* spp. discussed above), reacting with methylmalonyl CoA to provide the second, 5-methyl branch, followed by chain elongation with acetate units and final decarboxylation.

A third piece of indirect evidence for de novo synthesis of methyl-branched pheromones is circumstantial, but nevertheless plausible. That is, if the methyl-branched pheromones were derived from methyl-branched fatty acid precursors from the insects' diets, then these specific and highly unusual fatty acids should be present in the diet. Whereas this has not been explicitly checked, it does not seem likely, particularly for polyphagous species.

The methyl-branched structures found in lepidopteran pheromones could theoretically be the products of several biosynthetic pathways, including the fatty acid, acetogenin/polyketide, or terpenoid pathways. The terpenoid pathway can almost certainly be excluded based on the known biosyntheses of the methyl-branched *Virbia* and gypsy moth, *L. dispar*, pheromones, and because the resulting structures would be expected to have more methyl groups and be more highly functionalized. It is also unlikely that the Type III pheromones are produced from the acetogenin/polyketide pathway because these compounds tend to be much more highly functionalized than the structures seen (Khosla et al. 1999), due to the fact that many of the oxygens from the carbonyl groups are retained, or lost through reduction of the carbonyl groups and dehydrations of the resulting alcohols resulting in double bonds being inserted into the growing chain. In contrast, fatty acid biosynthesis proceeds by a repeated cycle of chain elongation by two carbons followed by reduction of the ketone in the position β to the acyl group, dehydration of the resulting alcohol to an α,β-unsaturated carbonyl, and reduction of the C=C double bond so that the growing chain is fully reduced to the saturated alkane before the next elongation step occurs (Khosla et al. 1999). Thus, the fatty acid synthesis pathway generates straight-chain or methyl-branched acyl-CoA products with no other functional groups, which are then acted on by desaturases and epoxidases (as shown with the gypsy moth pheromone; Jurenka et al. 2003), or other enzymes that introduce functional groups such as ketones, at specific places in the chains. The fact that the methyl-branched compounds that we have grouped into Type III are lightly functionalized, with zero, one, or at the most two functional groups or double bonds, would seem to be

most congruent with an origin in the fatty acid biosynthesis pathway, with most if not all of the structure being assembled in the oenocytes (figure 4.8).

It should also be pointed out that when a methyl branch is introduced at a position other than 2, the final molecule has two possible enantiomeric forms. That is, as shown in figure 4.8C, the sequence of steps likely proceeds via a conjugated enoyl-CoA intermediate, which is then reduced. The face of the C=C double bond to which the hydrogen is added determines which enantiomer is produced. Alternatively, the stereochemical outcome could be determined by whether the enoyl-CoA intermediate adopts the *cis*- or *trans*-conformation before being acted on by the enoyl reductase. Either case provides mechanisms whereby either enantiomer of the methyl-branched compound could be produced, and examples of both (*R*)- and (*S*)-configurations are known. Furthermore, male moths can clearly discriminate the enantiomers (e.g. *L. fiscellaria*, Geometridae; Li et al. 1993). It is intriguing to note that this possibly flexible biosynthesis of methyl-branched hydrocarbons used as pheromones is in apparent contrast to the apparently highly conserved biosynthesis of the ubiquitous long-chain methyl-branched hydrocarbons found in insect cuticular lipids, which the available evidence suggests are biosynthesized exclusively as the (*R*)-enantiomers (Bello et al. 2015).

Type 0 Pheromones

Two independent efforts in the 1990s resulted in the identification of a novel type of moth pheromone consisting of short-chain methylcarbinols and methylketones in two primitive lepidopteran lineages. Thus, Tóth and coworkers identified mixtures of (2*S*,6*Z*)-nona-6,8-dien-2-ol and (2*S*,6*E*)-nona-6,8-dien-2-ol as the female-produced sex pheromones of *Stigmella malella* and *S. crataegella* (Nepticulidae) (Toth et al. 1995), whereas a mixture of the (2*R*)-enantiomers attracted a third species. Löfstedt and coworkers identified several related compounds as female-produced sex pheromones in *Eriocrania* species (Eriocraniidae). For example, *E. cicatricella* uses a mixture of (2*R*,4*Z*)-hept-4-en-2-ol and (*R*)-2-heptanol as its sex pheromone (Zhu et al. 1995). Different mixtures of (2*S*)- and (2*R*)-(6*Z*)-non-6-en-2-ol were reported as the sex pheromones of *E. semipurpurella* and *E. sangii* (Kozlov et al. 1996), and several other *Eriocrania* species responded to these and similar compounds (Larsson et al. 2002). Thus, these short-chain secondary alcohols and ketones (C7-C9) appear to represent a distinct category of moth pheromones (figure 4.9). Similar structures have been identified from several species of caddisflies (Trichoptera; Löfstedt et al. 1994, 2008), the sister group of Lepidoptera. In both eriocraniid moths and Trichoptera, these compounds are produced and stored in the sternum V glands, which have been lost in ditrysian moths. Thus, we propose that these short-chain secondary alcohols be categorized as Type 0 pheromones, as befits their status as an ancestral type of moth pheromone distinct from all more highly evolved types.

The biosynthetic pathways that produce Type 0 pheromone components are unknown, but common denominators of these structures suggest that they may be related to Type I pheromones. In addition to the methyl carbinol structures, the corresponding ketones have been found in extracts of pheromone glands from both eriocraniids and caddisflies. The methylcarbinols are easily transformed into the corresponding ketones

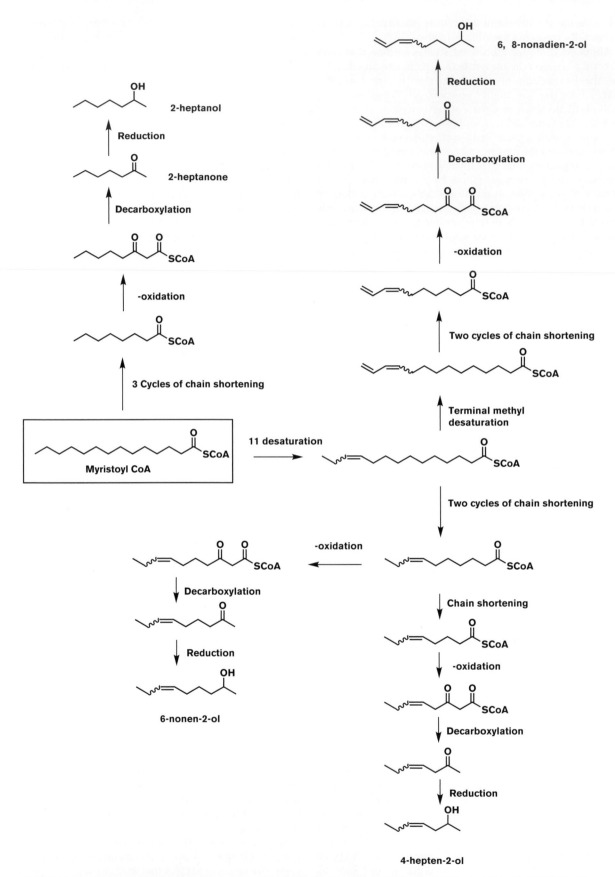

FIGURE 4.9 Typical structures of Type 0 pheromones, and possible biosynthetic routes to them from myristic acid. Wiggly bond indicates either (*Z*)- or (*E*)-stereochemistry.

Structures that do not fit into Types 0-III

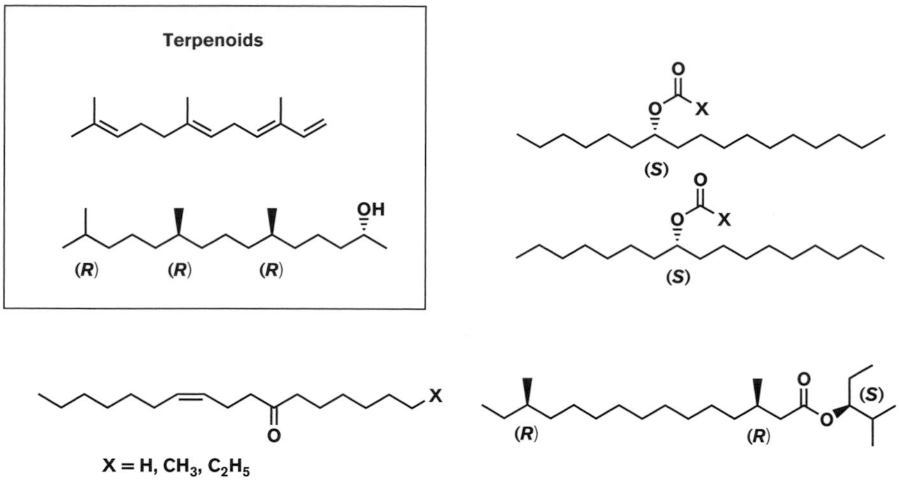

FIGURE 4.10 Lepidopteran pheromones that are not readily classifiable into any of the Type 0 to Type III categories of chemical structures.

and vice versa. (Z)- and (E)-6-nonen-2-one also have been identified from several bark beetles as precursors to the pheromones *exo-* and *endo-*brevicomin. The biosynthetic origin of (Z)- and (E)-6-nonen-2-one in bark beetles is not known, but recent work has ruled out a short-chain desaturase in their biosynthesis (Song et al. 2014a, 2014b), instead suggesting that they are more likely derived from longer-chain unsaturated precursors. Löfstedt and Kozlov (1997) proposed some possible biosynthetic routes to the ketones and alcohols found in primitive moths (see figure 4.9). Thus, both Z6-9:2-OH and Z4-7:2-OH as well as their corresponding *E* isomers could be accounted for by Δ11-desaturation of myristic acid followed by 2 or 3 cycles of β-oxidation and reductive decarboxylation of the β-ketoacyl intermediates produced from the chain shortening. This hypothesis has not yet been tested but in an analogous case, Skiba and Jackson (1994) showed that decarboxylation of a proposed chain-elongated β-ketoacid intermediate resulted in a methylketone in *Drosophila buzzatii*. If this hypothesis were proven correct, then it still remains necessary to explain how the terminal double bond is introduced into the components with conjugated dienes.

The veracity of the proposed biosynthetic pathways leading to the production of Type 0 moth pheromones is not critical for our recognition of these distinct types of compounds as ancestral moth pheromone compounds. The chemical structures and the site of production obviously differ from Type I, Type II, and Type III pheromones. However, it is intriguing that even these primitive moth pheromones might use Δ11-desaturation of saturated fatty acids as a critical step, in analogy to Type I moth pheromones.

Compounds That Do Not Fit Any of Types I, II, III, or 0

There remain a few lepidopteran pheromones that either do not fit into any of the four main types described above or are from obviously different structural classes. Examples of the latter include the geometrid *Hemithea tritonaria*, which uses

(E3,E6)-α-farnesene as one of its pheromone components (Yamakawa et al. 2011) and the pyralid *Corcyra cephalonica* (subfamily Galleriinae), which uses (2R,6R,10R)-6,10,14-trimethylpentadecan-2-ol as a short range attractant (Hall et al. 1987; Mori et al. 1991). Both these structures are clearly terpenoids.

Another group that is not easily categorized comprises the pheromones of several arctiine species in the tribe Lithosiini that include propionate esters of secondary alcohols (Fujii et al. 2013a), a methyl-branched secondary alcohol (Yamakawa et al. 2011), and methyl-branched methylketones (Yamamoto et al. 2007; Do et al. 2009; Adachi et al. 2010) (figure 4.10). None of these structures are clearly or obviously biosynthetically related to any of the Type 0–III classifications, and, to date, too few of them have been identified to get a good idea of the range of variations within these unusual structures that might provide hints as to their biosynthesis. Further complicating matters is the fact that the only other pheromone blend known for a lithosiine moth consists of typical Type II compounds (Fujii et al. 2010).

Another small group of unclassifiable pheromones consisting of straight-chain (Z)-7-alken-11-ones are found in moths in the family Carposinidae. However, their possible biosynthetic origins are hinted at by the placement of the double bond in the 7 position in the 18-, 19-, and 20-carbon alkenones reported as pheromones. This position does not fit any of the common diunsaturated or polyunsaturated acids used as precursors for Type II pheromones, such as linoleic acid, but it could conceivably arise from (Z)-9-hexadecenoic acid (= palmitoleic acid) as a precursor, by chain extensions, decarboxylations, and decarbonylations, combined with a regiospecific oxidation step to place the ketone. The fact that the positions of the double bond and the ketone are fixed, whereas the chain lengths beyond the ketone are variable, would argue for the desaturation and oxidation steps occurring on a common precursor, before elongation (or chain shortening) and the final decarboxylation or decarbonylation steps. Furthermore, these compounds may be made in the oenocytes and transported to the pheromone glands for release because lepidopteran pheromone glands are not known to have decarboxylases that remove a carbon while reducing the residue to a hydrocarbon (Blomquist et al. 1987).

Another group of unusual structures occurs in the few pheromones known from the bagworm moths (Psychidae). Thus, the Paulownia bagworm *Clania variegata* produces (3R,13R,1'S)-1'-ethyl-2'-methylpropyl 3,13-dimethylpentadecanoate as its pheromone (Gries et al. 2006; Mori et al. 2010), i.e., a long-chain, dimethyl-branched acid esterified with a short-chain branched secondary alcohol. Whereas the methyl branching pattern in the acid portion might suggest a Type III structure, the esterification with 2-methypentan-3-ol does not fit with any other suspected Type III structures. Furthermore, the other known psychid pheromones consist of octanoate or decanoate esters of secondary alcohols. Thus, we can only speculate that these pheromones might be formed using elements from two or more of the pathways described above, or entirely new pathways.

Distribution of Pheromone Types among the Major Lineages of Lepidoptera

The Big Picture

Mapping the major pheromone types (Type 0, Type I, Type II, and Type III) onto a phylogenetic hypothesis of Lepidoptera shown in figures 4.11 and 4.12 corroborates preliminary observations mentioned above about the distribution of different pheromone types within the Lepidoptera. The figures also capture the complexity within superfamilies. For example, all noctuoids that have Type II pheromones belong to the newly defined family Erebidae (Zahiri et al. 2011, 2012), which also contains two subfamilies with Type III pheromones. All other noctuoid families appear to have Type I pheromones. In figure 4.11, we have included information on the number of described species in each superfamily (Nieukerken et al. 2011) as well as the number of species for which pheromones have been identified in each taxon based on the information in the comprehensive database of moth pheromones found at http://web.tuat.ac.jp/~antetsu/LepiPheroList.htm (accessed June 22, 2015). Superficially, it may appear that major pheromone types are very stable and unlikely to change evolutionarily but upon closer examination, variations on the major themes show interesting patterns, which mirror critical events in lepidopteran evolutionary history. Deviant structures reported only as sex attractants should be treated with caution, especially when the reports are based on early screening studies. Many of these may include spurious catches in traps baited with compounds that may not be part of the actual pheromone.

Non-Ditrysian Moths and the Outgroup Comparison

Identification of moth pheromones has to a large extent been driven by the economic importance of pest species and the need for pheromone-based tools for their detection, monitoring, and control. Most pest species are found among the higher Lepidoptera (Ditrysia) and this in part explains the very small number of pheromones identified from non-ditrysian moths, even several decades after the identification of the *Bombyx mori* pheromone. The non-ditrysian moths comprise 2617 described species in a total of 14 superfamilies (Nieukerken et al. 2011) as compared to the approximately 10,000 species of caddisflies,

the sister group of Lepidoptera. Pheromone communication, pheromone compounds, and pheromone-producing structures in primitive moths were reviewed previously (Löfstedt and Kozlov 1997). Among these 14 superfamilies, female-produced sex pheromones have been identified in only four, with the Eriocranioidea representing the most primitive lineage for which sex pheromones have been identified. Eriocranioidea produce their methylcarbinol and related methylketone pheromone compounds in glands in sternum V, glands which they share with many of the other basal lineages and with Trichoptera. The outgroup comparison with Trichoptera suggests that both the methylcarbinol pheromones and the sternum V glands represent the ancestral state of lepidopteran pheromone compounds and pheromone-producing structures. This conclusion should be corroborated by studies of other basal superfamilies, including Micropterigoidea (some have and some do not have sternum V glands), Agathiphagoidea, Heterobathmioidea, and Acanthopteroctetoidea, all of which have sternum V glands (Löfstedt and Kozlov 1997, and references therein).

Nepticuloid species also use methylcarbinols as pheromone compounds, but in this superfamily, the pheromone-producing structures are not known. In contrast, within the Adeloidea, both heliozeliid leafminers and the prodoxid *Lampronia capitella* have Type I pheromones. The leafminer *Holocasista capensis* uses Z5-14Ald and Z7-14Ald as pheromone components (Wang et al. 2015a). These structures are likely derived from chain shortening of Z9-16:Acyl and Z9-18:Acyl followed by modification of the functional group. In *L. capitella*, pheromone biosynthesis has been thoroughly investigated and a Δ11-desaturase involved in the biosynthesis of its pheromone, Z9,Z11-14OH, was cloned and characterized (Löfstedt et al. 2004; Liénard et al. 2008). Both heliozeliid moths and *L. capitella* call by extending their terminal abdominal segments, suggesting that the pheromones are released from terminal abdominal glands, as is the case in more advanced Lepidoptera. The occurrence of typical Type I pheromones in these lineages dates the evolution of Type I pheromones and the recruitment of the Δ11-desaturase gene family for pheromone production to approximately 170 Mya (Wahlberg et al. 2013).

Tischeria ekebladella is the only species within the Tischerioidea for which a sex pheromone has been reported. Its sex pheromone contains Z3,Z6,Z9,Z19-23:H, a likely Type II pheromone variant with an unusual Z19 fourth double bond. Whereas the Z3,Z6,Z9 double bonds probably originate from a linolenic acid precursor, the Z19 double bond is likely introduced into a chain-elongated intermediate before the final reductive decarboxylation to the tetraene hydrocarbon.

Tineoid Grade

The tineoid grade contains two families, Psychidae (1350 species) and Tineidae (2393 species), for which pheromones have been reported. These two families do not appear to form a monophyletic group. Rather, Tineidae is the sister group to the rest of Ditrysia and Psychidae is sister to these two (Mutanen et al. 2010; Regier et al. 2013, 2015). Species of Psychidae (bagworms) have an unusual biology, with the females emerging in and calling from their pupal cases, which look like those of caddisflies, and, as described above, they have unusual pheromone structures. Sex pheromones have been identified from four species. The first was identified by Leonhardt et al. (1983) and this and the other known pheromone compounds are "reversed esters," consisting of methyl- or

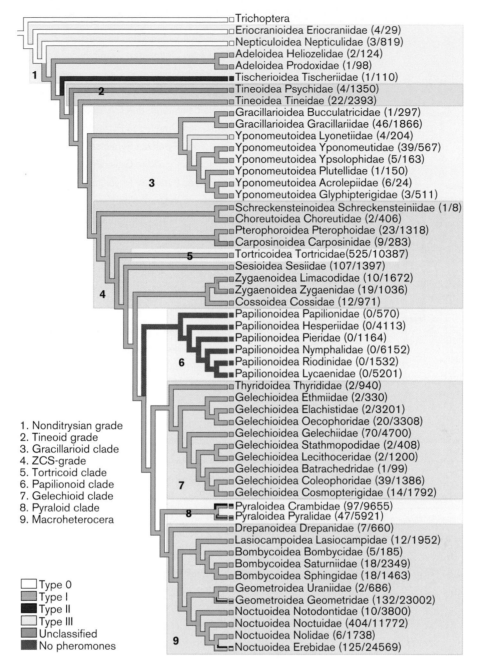

FIGURE 4.11 Pheromone types mapped on to a phylogenetic hypothesis of Lepidoptera using the parsimony criterion. Phylogenetic relationships as in figure 4.1, with unresolved nodes resolved based on Bazinet et al. (2013) for character evolution analysis. Numbers in parentheses after taxa indicate approximate number of pheromones and attractants reported (based on Ando 2015 plus selected additions), followed by the number of species in each taxon according to Nieukerken et al. (2011). Only taxa with reported pheromones or sex attractants are included in the tree.

ethyl-branched, short-chain esters of octanoic, decanoic, or 3,13-dimethylpentadecanoic acids, with most being chiral (e.g., Leonhardt et al. 1983; Rhainds et al. 1994). This type of moth pheromone structure has been found exclusively in the Psychidae. Furthermore, the site of pheromone production in the thorax appears to be unique (Loeb et al. 1989). The sex pheromones are secreted onto detachable scales, which are then shed to attract males. These pheromones are so far unique among lepidopteran pheromones, both because of their chemistry and their unusual site of production.

In contrast to Psychidae, Tineidae are attracted to Type I pheromone compounds, including both typical Type I compounds such as Z11-16OH, Z11-16Ald, and Z9-14Ac, and a less common variant, the $\Delta2,\Delta13$ or $\Delta3,\Delta13$ -18X structures (X = OH, Ac, Ald). Although sex attractants of this type have been found for 19 species in field trials, actual pheromones have been identified from only three species so far. All of these compounds can be arrived at from $\Delta11$-desaturation of palmitate followed by additional steps, including the diunsaturated C_{18} structures (figure 4.13) and the unusual pheromone of *Kermania pistaciella*, Z12-17:2-OAc, as described above. The $\Delta2/\Delta3,\Delta13$-18 theme also occurs in the Zygaenoidea–Cossoidea–Sesioidea (ZCS) group (see below).

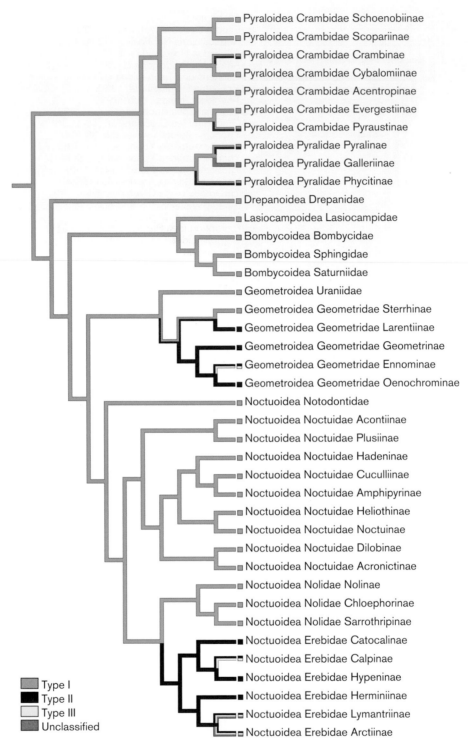

Pyraloidea Crambidae Schoenobiinae
Pyraloidea Crambidae Scopariinae
Pyraloidea Crambidae Crambinae
Pyraloidea Crambidae Cybalomiinae
Pyraloidea Crambidae Acentropinae
Pyraloidea Crambidae Evergestiinae
Pyraloidea Crambidae Pyraustinae
Pyraloidea Pyralidae Pyralinae
Pyraloidea Pyralidae Galleriinae
Pyraloidea Pyralidae Phycitinae
Drepanoidea Drepanidae
Lasiocampoidea Lasiocampidae
Bombycoidea Bombycidae
Bombycoidea Sphingidae
Bombycoidea Saturniidae
Geometroidea Uraniidae
Geometroidea Geometridae Sterrhinae
Geometroidea Geometridae Larentiinae
Geometroidea Geometridae Geometrinae
Geometroidea Geometridae Ennominae
Geometroidea Geometridae Oenochrominae
Noctuoidea Notodontidae
Noctuoidea Noctuidae Acontiinae
Noctuoidea Noctuidae Plusiinae
Noctuoidea Noctuidae Hadeninae
Noctuoidea Noctuidae Cuculliinae
Noctuoidea Noctuidae Amphipyrinae
Noctuoidea Noctuidae Heliothinae
Noctuoidea Noctuidae Noctuinae
Noctuoidea Noctuidae Dilobinae
Noctuoidea Noctuidae Acronictinae
Noctuoidea Nolidae Nolinae
Noctuoidea Nolidae Chloephorinae
Noctuoidea Nolidae Sarrothripinae
Noctuoidea Erebidae Catocalinae
Noctuoidea Erebidae Calpinae
Noctuoidea Erebidae Hypeninae
Noctuoidea Erebidae Herminiinae
Noctuoidea Erebidae Lymantriinae
Noctuoidea Erebidae Arctiinae

Type I
Type II
Type III
Unclassified

FIGURE 4.12 A more detailed look at pheromone type evolution in the pyraloid and macrohete-roceran families and subfamilies. Relationships as in figure 4.1. Only taxa with reported pheromones or sex attractants are included in the tree.

Gracillaroid Clade

The gracillaroid clade consists of two superfamilies, Gracillaroidea (2216 species) and Yponomeutoidea (1756 species). Gracillaroidea comprises three families: Roeslerstammiidae, Bucculatricidae, and Gracillariidae. The pheromone reported from a single species in the Bucculatricidae is a mixture of two unique nitrate compounds, Z9-14ONO$_2$ and Z8-13ONO$_2$, from *Buccula-*

trix thurberiella (Hall et al. 1992), but the simple skeletons are clearly Type I. In Gracillariidae, pheromones have been identified from 17 species and attractants for many more. The known pheromones include 12-, 14-, and 16-carbon alcohols and aldehydes with 1–3 double bonds in both even- and odd-numbered positions. Although some can be derived from Δ11-desaturation, both monounsaturated and diunsaturated structures with double bonds in even-numbered positions are known. Thus, the

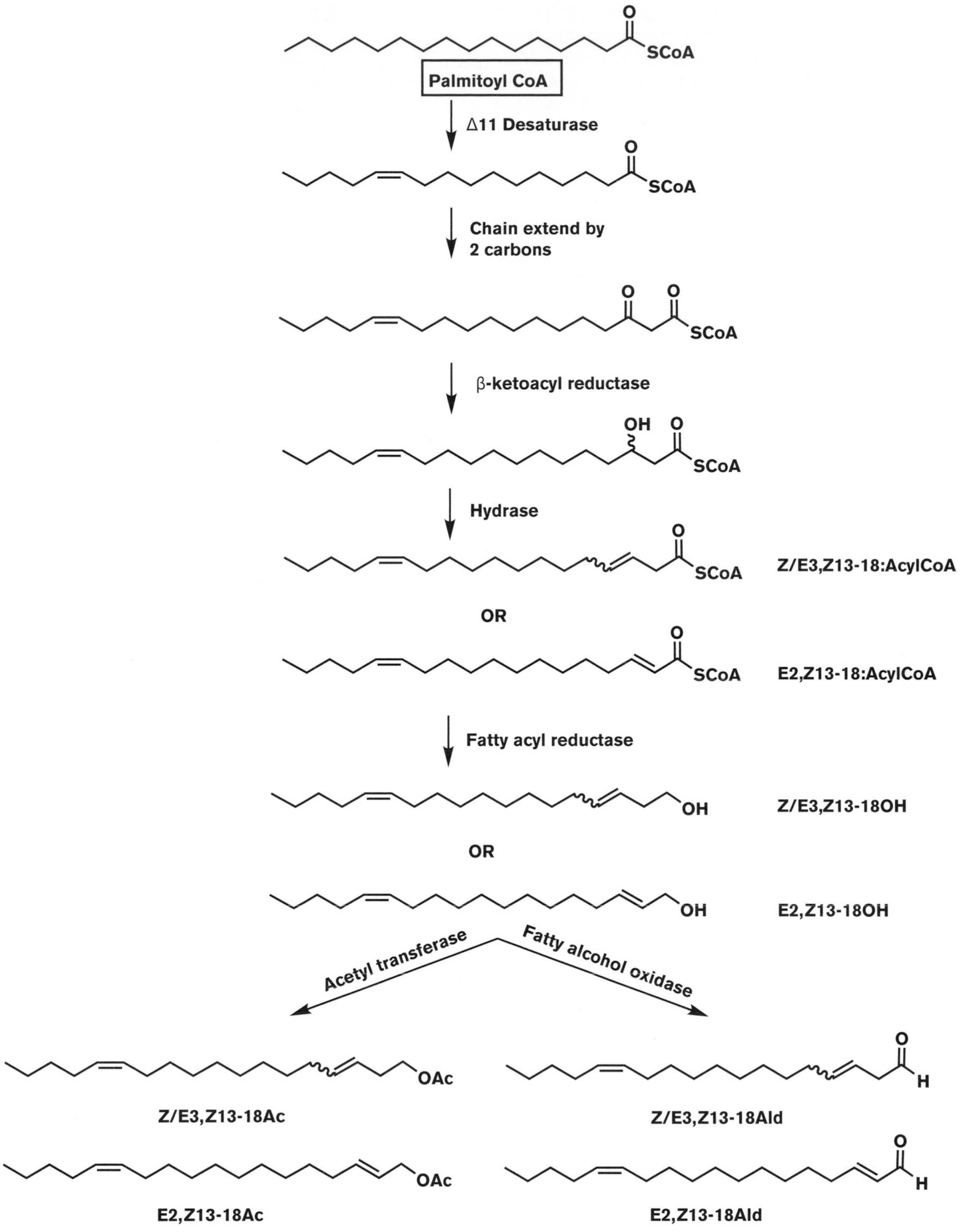

FIGURE 4.13 Putative biosynthetic pathway to the Δ2,Δ13 Type I pheromones found in Tineidae, Sesiidae, and Cossidae.

biosynthesis of a compound such as E4,E10-12Ac, reported from several species, must involve unusual desaturases or biosynthetic mechanisms that are not yet known.

Yponomeutoidea comprises a total of 10 families, of which 5 families have typical Type I pheromones and 1 family, Lyonetiidae, has Type III, methyl-branched pheromones. In phylogenetic analyses based on molecular data (Mutanen et al. 2010; Regier et al. 2013; Sohn et al. 2013), *Lyonetia* and *Leucoptera* do not appear to be closely related, although their phylo-

genetic positions were not well supported in either study. However, both genera have Type III pheromones, suggesting a close relationship and supporting monophyly of Lyonetiidae.

Tortricoidea Superfamily

The leafroller moths (Lepidoptera: Tortricidae) form one of the largest families in the Lepidoptera and their phylogeny is very well supported, although the phylogenetic position of Tortricoidea in the Lepidoptera remains unresolved. Almost all molecular studies suggest that tortricids are somewhere at the base of the ZCS grade, but the exact position changes with each study (Mutanen et al. 2010; Regier et al. 2013; Bazinet et al. 2013; Kawahara and Breinholt 2014).

Many tortricid species are important pests and early on became targets for pheromone studies. Recent molecular studies provide strong support for the monophyly of both Tortricinae and Olethreutinae, the two major subfamilies within Tortricidae, and for grouping of these to the exclusion of the third subfamily Childanotinae (Regier et al. 2012). As mentioned above, Roelofs and Brown (1982) described the general trend that species in the Tortricinae use mostly 14-carbon Type I pheromone components with double bonds in odd-numbered positions (Δ9 or Δ11 isomers), whereas species in the subfamily Olethreutinae use mostly 12-carbon compounds with double bonds in even-numbered positions (Δ8 and Δ10 and doubly unsaturated Δ8,Δ10 isomers). The extensive accumulation of additional data on pheromones from Tortricidae corroborates the pattern of pheromone components in the different subfamilies first suggested by Roelofs and Brown (1982), and molecular studies confirm the paramount role of a lepidopteran-specific Δ11-desaturase gene family in generating pheromone diversity among the vast majority of Lepidoptera (see for instance Roelofs and Rooney 2003; Liénard et al. 2008). Interestingly, in spite of its phylogenetic position within the ZCS grade, known pheromones of Tortricoidea are all typical Type I pheromones with no known examples of the less common, Δ2,Δ13-18C type compounds.

The biosynthetic pathways leading to the pheromone components with double bonds in even-numbered positions, such as the Δ8 and the doubly unsaturated Δ8,Δ10-12-carbon compounds in the Olethreutinae remain unclear. For example, the major pheromone component of the codling moth *Cydia pomonella* is E8,E10-12OH (Roelofs et al. 1971). The oriental fruit moth *Grapholita molesta* is in the same tribe (Grapholitini) (Regier et al. 2012), and its pheromone is a blend of Z8- and E8-12Ac, and Z8-12OH (Cardé et al. 1979 and references therein). Although some endemic New Zealand tortricids in the subfamily Tortricinae use an unusual Δ10-desaturase to produce Δ8- and Δ10-unsaturated compounds (Hao et al. 2002b), precursor analyses and labelling experiments provided no support for a Δ10-desaturase being involved in biosynthesis of the structurally similar E8,E10-12OH in *C. pomonella*. Instead, Löfstedt and Bengtsson (1988) found evidence for a Δ9-desaturase producing E9-12:Acyl which was subsequently transformed into E8,E10-12:Acyl and the corresponding alcohol. The postulated dual function E9-desaturase has not yet been cloned and characterized. The fact that *G. molesta*, which uses Δ8-12Ac and OH pheromone components, belongs to the same tribe as *C. pomonella* makes the elucidation of the pheromone production pathways and the characterization of the desaturases involved in biosynthesis of pheromone components in the Olethreutinae even more interesting. In recent studies using a PCR-based approach and degenerate primers, three desaturases were identified from each of the two species (Ding and Löfstedt 2014), but the postulated dual function E9-desaturase involved in the biosynthesis of the *C. pomonella* pheromone and its ortholog potentially involved in production of the monounsaturated Δ8-dodecenol/acetate components in *G. molesta* have not yet been conclusively cloned and characterized.

Zygaenoidea–Cossoidea–Sesioidea Grade

The ZCS grade is very poorly resolved. In addition to the major superfamilies Zygaenoidea, Cossoidea, and Sesioidea, it comprises several smaller superfamilies with few or no pheromones identified. The relationships among all these superfamilies change with each published study (Mutanen et al. 2010; Regier et al. 2013), including the two recent phylogenomic studies (Bazinet et al. 2013; Kawahara and Breinholt 2014), and usually the superfamily Tortricoidea (see above) is included in the grade. It can be expected that further changes will occur as more detailed transcriptomic studies target this region of the lepidopteran phylogeny (Bazinet et al. 2013). Unfortunately, the poor resolution of this part of the tree complicates interpretations of pheromone evolution.

In Zygaenoidea, with pheromones identified from 10 species and attractants reported from 19 additional species, most pheromones are Type I compounds such as Z5-12Ac, Z7,Z11-16Ac, Z7-12Ald, and Z9,E11-14Ald. These have been found in two families, Zygaenidae and Limacodidae. However, unusual reversed esters are also reported in both families. In Limacodidae, the esters are all decanoates, whereas in Zygaenidae they are tetradecanoates and in one case a dodecanoate. Zygaenidae and Limacodidae are considered close relatives but not sister families (Mutanen et al. 2010; Regier et al. 2013), suggesting that the esters evolved in a common ancestor. This further suggests that unusual esters are likely to be found in other related families, such as Megalopygidae, Lacturidae, and Dalceridae.

Pheromones have been identifed from 26 sesiid species, and sex attractants for another 81 species. In the Cossidae, pheromones are known for eight species, and sex attractants for another four. Sesioidea and Cossoidea include species with pheromones with the Δ2,Δ13-variation on the Type I theme. In the sesiids, all reported pheromones and attractants are of this type.

Of the smaller superfamilies, a single typical Type I pheromone has been reported from a species in Choreutoidea, along with an attractant based on Δ2,Δ13-18C structures. Choreutoidea are occasionally placed on the same major branch as Schreckensteinoidea (Regier et al. 2013) for which no pheromones have been identified. Pterophoroidea has two identified pheromones and 21 species with known attractants and all of the reported structures are typical Type I pheromones. In contrast, the pheromones and sex attractants reported from three and six species, respectively, of Carposinoidea are all unusual ketones (Z7,Ket11-20H, Z7,Ket11-19H, and Z7,Ket11-18H, and similar structures). These structures might all be derived from palmitoleic acid after chain elongation (see the "Compounds That Do Not Fit Any of Types I, II, III, or 0" section above) but at this point, their biosynthesis remains speculative.

Pheromones and sex attractants have not been reported from many of the other superfamilies in the ZCS group, such as Immoidea, Urodoidea, and Galactoidea.

Papilionoid Clade

The ca. 18,000 described species of extant butterflies (Papilionoidea) have diverged extensively from their lepidopteran relatives in a number of life history traits including their diurnal lifestyle, bright colors, body structures, antennal shapes, and mate-finding strategies. Whereas other Lepidoptera rely mostly on chemical signals, visual advertisement is the hallmark of mate finding in butterflies. The use of female-produced long-range sex pheromones appears to have been lost within the whole superfamily. In the context of courtship, however, male chemical signals are widespread, just as in moths (Conner, this volume), and likely have multiple evolutionary origins. Interestingly, in males of the butterfly *Bicyclus anynana*, courtship scents are produced de novo via biosynthetic pathways that show a high degree of conservation with those used by moth species producing Type I pheromones (Liénard et al. 2014). Thus, the Z9-14OH and 16Ald found in male butterfly androconia are produced through the activity of a fatty-acyl Δ11-desaturase and two specialized alcohol-forming fatty-acyl reductases. Males of the closely related *B. martius* produce ethyl, isobutyl, and 2-phenylethyl esters of 16COOH and Z11-16COOH, confirming close structural and biosynthetic relationships between female-produced moth pheromones and the male pheromones in at least some Papilionoidea (Wang et al. 2014). These studies provide evidence of conservation and sharing of ancestral genetic modules for the production of fatty acid derivatives over a long evolutionary timeframe, thereby linking reproductive communication chemistry in moths and butterflies. Although the butterflies lost female-produced sex pheromones during their divergence from other Lepidoptera 110–100 Mya, several genera have retained the biosynthetic machinery needed to produce some of the same compounds that serve as Type I moth sex pheromones, although the compounds are produced by different sexes and used in different contexts between the butterflies and moths.

Gelechioid Clade

Thyridoidea and Calliduloidea appear to form a basal clade within the gelechioid clade and a sister group of Gelechioidea (Kawahara and Breinholt 2014). Sex attractants of Type I have been reported for two thyridoid species, whereas no pheromones or sex attractants have been reported from any Calliduloidea.

Gelechioidea, comprising 16 families (Heikkilä et al. 2014), have Type I pheromones including some unusual "themes." Relationships among the families are very unclear (Heikkilä et al. 2014), and it appears that there was a rapid radiation at the base of the gelechioid clade. The two reported pheromone compounds in Elachistidae are multiply unsaturated 14-carbon aldehydes with double or triple bonds in the 9, 11, and 13 positions. For example, the main component of the *Stenoma catenifer* pheromone has a triple bond in the 11 position, sandwiched between two double bonds. These unusual structures still fit the Type I criteria because of their likely biosynthetic origin from fatty acyl precursors that have been acted on multiple times by specific desaturases. For Ethmidae, no pheromones are known but attraction of two species to E11-14Ac has been reported. The one pheromone reported for an oecophoriid species (*Cheimophila salicella*) is a mixture of E11-14Ac, the corresponding aldehyde and alcohol, and the saturated 14Ac. Sex attractants also have been reported for 19 species in the family, mainly monounsaturated 10-, 12-, 14-, or 16-carbon acetates, plus one diunsaturated 12-carbon compound, a few cases of Z7-12OH, and two instances of attraction to monounsaturated 13-carbon acetates. However, these data should be interpreted with caution until more pheromone structures have been properly identified.

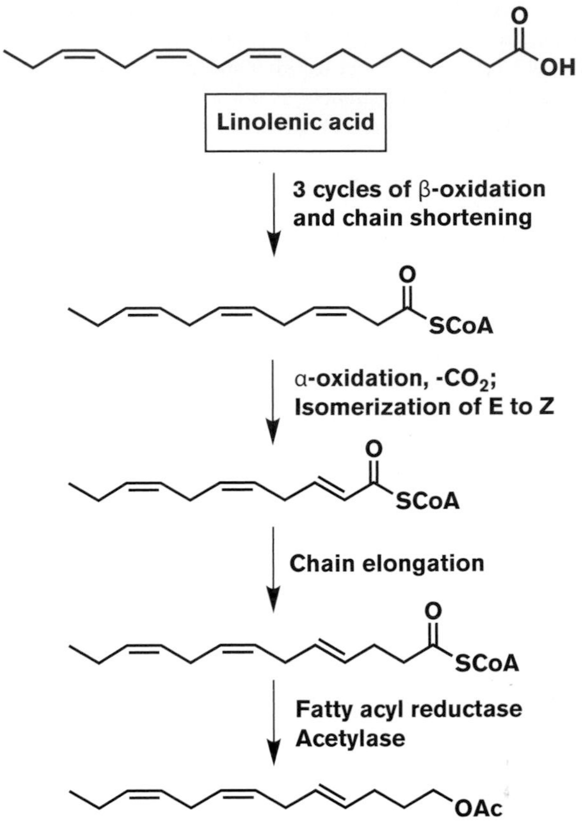

FIGURE 4.14 Putative biosynthetic pathway to E4,Z7,Z10-13Ac, a pheromone component in the potato tuberworm, *Phthorimaea operculella*.

A number of pheromones are known within the family Gelechiidae, including reports of unusual structures like E4,Z7-13Ac and E4,Z7,Z10-13Ac in the potato tuberworm *Pthorimaea operculella*, E3,Z8,Z11-14Ac and E3,Z8-14Ac in the tomato leafminer *Tuta absoluta*, and E3-14Ac, E5-14Ac and E3-12Ac in several species. Although these structures might superficially appear to be unusual Type I pheromone compounds, both the odd-numbered chain lengths and the unusual positions and *E* geometry of some of the double bonds are atypical for Type I structures. Alternatively, the biosynthesis of the *P. operculella* and *T. absoluta* pheromones could start from linoleate and linolenoate (figure 4.14) as suggested by Roelofs and Bjostad (1984) and mentioned above. Their putative origin from linoleate and linolenoate rather than from synthesized palmitic or oleic acid would be a characteristic of Type II rather than Type I pheromones, whereas the acetate functional group is characteristic of many Type I structures. The structures are obviously uncommon, and whether they should be characterized as Type I or Type II or a mixture thereof remains to be determined by a detailed study of their biosynthesis.

Pheromones have been identified from two species of Stathmopodidae. The major pheromone component of *Stathmopoda masinissa* consists of E4,E6-16Ac for Japanese populations (Naka et al. 2003), and a combination of this compound with the corresponding alcohol for Korean populations (Kim et al. 2014). In contrast, the major pheromone component of *S. auriferella* consists of E5-16Ac (Yang et al. 2013). This suggests the possibility of a multifunctional Δ5-desaturase, which can create both the E5 double bond, and the conjugated E4,E6 diene, in analogy to the Δ11-desaturases that create the Δ10,Δ12-16C pheromones in other species.

The pheromone of *Batrachedra amydraula* in Batrachedridae is a mixture of Z5-10Ac, Z5-10OH, and Z4,Z7-10Ac. The first two compounds may be derived from chain shortening of Z11-16:Acyl and thus qualify as classical Type I pheromone compounds, but the biosynthesis of Z4,Z7-10Ac is again open to speculation due to the pattern of methylene-interrupted double bonds that would be more characteristic of a Type II structure. In the Coleophoridae, only one pheromone has been identified (*Coleophora deauratella*, Z7-12Ac + Z7-12OH + Z5-12Ac; Evenden et al. 2010), but sex attractants for 38 species, mostly from the genus *Coleophora*, have been reported. In most cases, the attractants are Z5-10Ac, Z5-10OH, or Z7-12Ac. Z3-10Ac and a mixture of E5-10Ac and the corresponding alcohol also have been reported as attractants. Pheromones of Cosmopterigidae are classical Type I, mainly of 12 or 14 carbons.

The molecular data suggest that the gelechioid clade started diversifying around 100 Mya (Wahlberg et al. 2013). It also seems as if there was rapid divergence of the lineages leading to the families, which may in part explain the variation in pheromone structures among the families.

Pyraloidea Superfamily

The pyraloids form a very well supported clade with two large families containing around 15,000 species in total, and with more than 90 pheromones identified and a further ~50 sex attractants known. The Pyralidae are divided into five subfamilies, Phycitinae, Galleriinae, Epipaschiinae, Chrysauginae, and Pyralinae. Pheromones have been identified from three of these. In Phycitinae, the pheromones are generally classical Type I monounsaturated or diunsaturated 12-, 14-, or 16-carbon acetates, alcohols, and aldehydes. However, relatively recently, some of the first examples of pheromones consisting of blends of Type I and Type II pheromone components were identified from this family. For example, key components of the pheromone of the navel orangeworm, *Amyelois transitella*, were shown to include Z11,Z13-16Ald, Z11,Z13-16OH, and Z3,Z6,Z9,Z12,Z15-23H, with possibly some additional minor components (Leal et al. 2005; Kuenen et al. 2010). This theme of mixed Type I and Type II pheromones has since been found in several pyralids in the genus *Dioryctria*, in which Z3,Z6,Z9,Z12,Z15-25:H was found to be a critical synergist of the main pheromone components, which are otherwise classical Type I compounds (Millar et al. 2005; Miller et al. 2010; Löfstedt et al. 2012). A blend of Type I and Type II compounds was also found in the only pheromone identified to date from the subfamily Pyralinae, for *Pyralis farinalis*, which uses a mix of Z11,Z13-16Ald (Type I) and Z3,Z6,Z9,Z12,Z15-23H and Z3,Z6,Z9,Z12,Z15-25H (Type II) (Landolt and Curtis 1982; Leal et al. 2005; Kuenen et al. 2010).

Remarkably, in the subfamily Galleriinae, the female-produced long-range sex pheromones typical of most other moths appear to have been lost in favor of male-produced sex pheromones. Thus, females of the rice moth *Corcyra cephalonica* produce a short-range courtship attractant, 6,10,14-trimethylpentadecan-2-ol (Hall et al. 1987), but over longer distances, females are attracted by male-produced pheromones consisting of the sesquiterpenes (*E,E*)-farnesal and (*Z,E*)-farnesal (Zagatti et al. 1987). Several other species in the subfamily Galleriinae show the same pattern, including the

greater wax moth *Galleria mellonella*, males of which produce nonanal, undecanal, and 5,11-dimethylpentadecane, with one or more additional compounds remaining to be identified (Leyrer and Monroe 1973; Svensson et al. 2014). Other examples include *Achroia grisella* and *Aphomia sociella* (Svensson et al. 2014). All of these moths are parasites of social bees or wasps, suggesting that there may be a link between their particular lifestyles and the loss of female-produced sex attractant pheromones.

In Crambidae, six subfamilies all appear to have Type I pheromones but similar to some of the pyralids, some species in the subfamilies Crambinae and Pyraustinae also use Type II compounds as important synergists. A number of pheromones are known from species in Crambinae and Pyraustinae, but little is known about pheromones of the other four subfamilies. In the Crambinae, some atypical components are found in *Chilo auricilius*, the pheromone of which consists of a mixture of Z8-13Ac, Z9-14Ac, and Z10-15Ac, whereas in other *Chilo* spp. the pheromones consist of Z11-16 and/or Z13-18 compounds. The *C. auricilius* pheromone appears to be biosynthesized by α-oxidation of Z11-16:Acyl and Z9-14:Acyl, respectively. The pheromone of *Deanolis sublimbalis* is a mixture of Z11-16Ald (Type I) and Z3,Z6,Z9-23H (Type II) (Gibb et al. 2007), whereas *Neoleucinodes elegantalis* uses a pheromone composed of E11-16:OH (Type I) and Z3,Z6,Z9-23H (Type II) (Cabrera et al. 2001). In a curious side note, while field screening a series of compounds as possible aldehyde mimics, Ujvary et al. (1993) discovered that the cranberry girdler, *Chrysoteuchia topiaria*, was attracted to a 97:3 mixture of Z11-16:nitrile and Z9-16:nitrile. The actual pheromone has not yet been identified, but it is indeed possible that the nitrile mimicked an aldehyde function.

Pheromones have been identified from more than 30 species in the subfamily Pyraustinae. A significant fraction of these use diunsaturated Δ10,Δ12-14C or 16C compounds, often in combination with either Δ10 or Δ11 monounsaturated analogs. This suggests that the conjugated dienes could be biosynthesized either by a Δ10-desaturase (or a Δ12-desaturase) inserting the double bonds in even-numbered positions or by a multifunctional Δ11-desaturase inserting both double bonds, as in *Bombyx mori*. The pheromone of *Ostrinia furnacalis* is a mixture of E12- and Z12-14Ac, with the double bond resulting from action of a Δ14-desaturase on palmitic acid, followed by chain shortening (Zhao et al. 1990). In *Conogethes pluto*, the pheromone is another blend of Type I and Type II components, consisting of E10-16Ald, E10,E12-16Ald, and Z3,Z6,Z9-21H.

It also should be noted that among the pyraloid moths, the Z9,E12-14C motif has been found exclusively in the Phycitinae, whereas the Δ10,Δ12-motif occurs only in the Pyraustinae.

Macroheterocera Clade

The Macroheterocera comprise the most advanced Lepidoptera and include approximately half of the described species. Five superfamilies are recognized, and pheromones have been identified from more than 250 species in this clade. In the superfamily Drepanoidea, no pheromones have been identified but sex attractants have been reported for seven species, all with classical Type I structures (E11-14Ac, E11-14OH, E11-14Ald, Z9-14Ald, Z9,E11-14Ald). The next branch, comprising Lasiocampoidea and Bombycoidea, includes some atypical

Type I structures. Pheromones have been identified from 12 species of Lasiocampidae, all of which consist of various $\Delta5,\Delta7$-12C compounds plus, in one case, Z5-12C minor components. Biosynthetic investigations of these structures in *Dendrolimus punctatus* suggested that Z5,E7-12OH is likely produced by interaction of longer fatty acyl moieties with $\Delta11$- and $\Delta9$-desaturases having unusual catalytic properties, followed by chain shortening (Liénard et al. 2010b). It appears that this motif is almost unique within the Lepidoptera (the only exception being pheromones reported in two species of *Thysanoplusia* in Noctuidae), and the lack of structural variation suggests that it is associated with the evolutionary emergence of Lasiocampidae ca. 63 Mya.

Similarly, the superfamily Bombycoidea, divided into 10 families, is typical in having a high frequency of $\Delta10,\Delta12$-16C compounds in the three families for which pheromones have been identified. These three families (Bombycidae, Saturniidae, and Sphingidae) comprise approximately 90% of the species in the superfamily Bombycoidea. For the remaining seven families, no pheromone information is available. $\Delta10,\Delta12$-16C compounds are typical Type I pheromones biosynthesized from palmitic acid by the action of a dual function $\Delta11$ desaturase. In Bombycidae, more variable structures are found in *Andraca bipunctata* that uses E11,E14-18Ald as its major pheromone component, and extracts of the pheromone glands also contained 18Ald, E11-18Ald, and E14-18Ald (Ho 1996).

In Saturniidae, in addition to the $\Delta10,\Delta12$-16C theme, a number of saturniid moth pheromones are comprised of E4,Z9-14 and E6,Z11-16 structures, or more highly unsaturated analogs such as the E4,E6,Z11-structures found in *Samia cynthia ricini* (Bestmann et al. 1989) and *Graellsia* (= *Actias*) *isabellae*, and the tetraene E4,E6,Z11,Z13-16Ald found in *Callosamia promethea* (Gago et al. 2013). Biosynthesis of E6,Z11-16Ald has been elucidated in *Antheraea pernyi* (Wang et al. 2010b). It was found that E6,Z11-16Ald and the corresponding acetate are generated by two biosynthetic routes which implicate a $\Delta6$- and $\Delta11$-desaturase duo but with an inverted reaction order. Here, the two desaturases first catalyze the formation of (*E*)-6-hexadecenoic acid or (*Z*)-11-hexadecenoic acid. Subsequently, each enzyme is able to produce the (E6,Z11)-hexadecadienoic acid intermediate by a site-specific further desaturation of the respective monoene, and the 14-carbon homolog is produced by one round of chain shortening. It is not yet known how and in what order further desaturations to triene and tetraene structures occur. A totally unexpected structure is found in the pheromone of *Nudaurelia cytherea cytherea*, which was reported to consist of (Z)-5-decenyl 3-methylbutanoate (Henderson et al. 1973).

In the Sphingidae, the majority of known pheromone components are $\Delta10,\Delta12$-16C compounds. Exceptions include E11,E13-16Ald in *Agrius convolvuli* (Wakamura et al. 1996), E11-16Ald in *Deilephila elpenor* (Uehara et al. 2012), and E10,E12,E14-16Ald in *Manduca sexta* (Tumlinson et al. 1989). However, all of these compounds can be accommodated by variations on the theme of $\Delta11$-desaturation of palmitic acid, the typical route to Type I pheromone compounds. For example, biosynthetic studies with *M. sexta* showed that isotopically labeled 16COOH and Z11-16COOH were incorporated into the dienals E10,E12-16Ald and E10,Z12-16Ald, and the trienals E10,E12,Z14-16Ald and E10,E12,E14-16Ald (Fang et al. 1995). *Dolbina tancrei*, with E9,Z11-15Ald as its pheromone, appears as another isolated case of chain shortening by α-oxidation before reduction of the fatty acyl pheromone precursor (Uehara et al. 2013).

Thus, the pheromones reported for the Bombycoidea can be generally characterized as variants of Type I pheromones, usually with the known or likely involvement of a $\Delta11$-desaturase with dual function, or by the combined action of a $\Delta11$-desaturase with one or more other desaturases.

Finally, the Geometroidea and Noctuoidea have 24,000 and 42,000 species, respectively, making them the most species-rich lepidopteran superfamilies. The Geometroidea comprises five families (Rajaei et al. 2015). The sex attractants reported for two specis from the Uraniidae are typical C_{14} and C_{16} Type I structures. All identified pheromones are from the Geometridae, to which the vast majority of species (>95%) belong. Within the Geometridae, seven subfamilies are recognized, with the Sterrhinae and Larentiinae being sisters to the other subfamilies (Sihvonen et al. 2011). Pheromone information is available for Sterrhinae, Larentiinae, Ennominae, and Geometrinae. All pheromones and attractants reported for Sterrhinae are Type I (Z7,Z9-12Ac, Z9,Z11-14Ac, Z7-12Ac, etc.), whereas pheromones and attractants from the other subfamilies are Type II, with few exceptions. Five cases of Type III pheromones are known in species of the genera *Lambdina* and *Nepytia* (subfamily Ennominae; summarized in Ando 2015), but the vast majority of known pheromones of Ennominae are Type II. All known pheromones from the Geometrinae and Larentiinae are also Type II.

The Type III pheromones that are restricted to a few taxa in the Ennominae may represent a derived type of pheromone that has evolved within this subfamily. The sisters of Geometridae, and the lineages immediately basal to the Geometroidea, all have Type I pheromones. Thus, it is tempting to suggest that the Sterrhinae with Type I pheromones form the basal lineage within the Geometridae, and that the Type II pheromones evolved once in the early evolution of the Geometroidea. However, molecular data suggest that Sterrhinae and Larentinae are sister groups (Sihvonen et al. 2011; Regier et al. 2013), although this region of the phylogenetic tree is not yet fully resolved and these evolutionary scenarios remain speculative.

The superfamily Noctuoidea are likely to be the sister group of the Geometroidea (Kawahara and Breinholt 2014). Pheromones have been identified from many species in the four most species-rich families within the Noctuoidea. The Notodontidae, with $\Delta11,\Delta13$-16 Type I structures, are sister to the core four families of Noctuoidea (Noctuidae, Euteliidae, Nolidae, and Erebidae), and interestingly, the Noctuoidea contain some of the rare examples of alkynyl pheromones, in the Z13,11yne:16 pheromones of *Thaumetopoea* spp. Relationships among the four core families are still unclear (Zahiri et al. 2011, 2012), but the monophyly of each family is well supported. Among the more than 400 species of Noctuidae for which pheromones or attractants have been reported, all have typical Type I pheromone components. A single possible report of attraction of a noctuid species, *Noctua pronuda*, to a Type II pheromone blend must be interpreted with caution, because this species was caught, along with ~80 other species, in a quarantine detection program for the invasive arctiine moth *Hyphantria cunea* (Ostrauskas 2004). Furthermore, other *Noctua* species for which pheromones have been identified all possess Type I pheromones.

Six species of Nolidae (in the genera *Nolathripa*, *Earis*, *Nola*, and *Uraba*) have been investigated and all have typical Type I pheromones or attractants. In marked contrast, the Erebidae have only Type II and some Type III pheromones (figure 4.12). Specifically, six of the 18 recognized subfamilies within the

Erebidae (Zahiri et al. 2012) have information about phero-mones. The subfamilies Catocalinae, Calpinae, and Hypeni-nae form one clade, whereas Herminiinae, Lymantriinae, and Arctiinae form a sister group. All six of these subfamilies have representatives with typical Type II pheromones. In Catocali-nae, all identified pheromones and attractants are Type II. In Calpinae, the majority of compounds are Type II, but three species have Type III compounds (9Me-19H and 7Me-17:H, Z6,13Me-21H). For Hypeninae, only three attractants have been reported but they are all of Type II. In the sister group, one pheromone and eight attractants of Type II have been reported for the Herminiinae. The reported attraction to 2Me-7,8-epo-18H in *Zanclognatha lunaris* (Minyailo et al. 1977) (Type III) should be confirmed, particularly as a congener was reported to be attracted to Z3,Z9-6,7-epo-19H (Hai et al. 2002), a typical Type II structure.

Finally, the Lymantriinae and Arctiinae are sisters of Her-miniinae. The Lymantriinae and to a lesser extent the Arc-tiinae have a remarkable variety of pheromone structures. In the Lymantriinae, most of the known pheromones are Type II and Type III, but there are also some as yet unclassified struc-tures. Very recently, Wang et al. (2015b) presented a new phy-logenetic classification of Lymantriinae, with seven well-supported tribes, and pheromones/attractants are known from four of these tribes. To date, the Leucomiini have Type II pheromones, whereas the Lymantriini possess Type III struc-tures. The Orgyiini have what appear to be Type II phero-mone compounds, including one "classical" epoxide and many ketone structures, postulated to arise from rearrange-ments of epoxides based on the positions of the ketone func-tions. Finally, the Nygmiini have Type II structures, and some unclassified isobutyrate esters.

The Arctiinae also offer a diversity of structures. Zaspel et al. (2014) recognize four tribes and for three of these, pher-omone information is available. Lithosiini use Type II and unclassified pheromone compounds like the 17:7-OPr in *Bar-sine* sp. (Fujii 2013b). In the Syntomini, pheromone blends composed of Type II compounds are known from three spe-cies, along with a fourth species attracted to a Type II com-pound, and a Type III attractant for a fifth species. A larger number of pheromones and attractants have been reported from Arctiini species, including both Type II (including Z9,Z12,Z15-18Ald) and Type III structures.

The Evolution of Pheromone Types and the Diversity of Pheromone Chemistry

The detailed examination of the distribution of pheromone structures among lepidopteran clades described above sup-ports the linked hypotheses of variation on a limited number of chemical themes, and extensive conservation of phero-mone types within related taxa, in most cases even up to the level of family and superfamily (figure 4.11). Is this picture accurate, or has it possibly been skewed because of the general focus on pest species? Is there a possibility that unusual struc-tures may be shown to be common if "missing" families are examined in more detail? Although the early pheromone identifications indeed targeted important pest insects, the current dataset covers much of the lepidopteran phylogeny, with pheromones or attractants known for 23 out of 53 super-families representing almost 90% of the described species. In addition to pest species, pheromones now have been investi-gated for a substantial number of other species, including rep-

resentatives of groups that are not economically important. It seems unlikely that analyses of species from the missing fam-ilies will reveal entirely novel types of pheromones to any sig-nificant extent, suggesting that we can be fairly confident that four major types of pheromone compounds (Type 0, Type I, Type II, and Type III) will cover most of the chemical diversity to be found in lepidopteran pheromones.

The Evolution of Pheromone Types

There seems little doubt that Type 0 pheromones are the ancestral type, strongly supported by outgroup comparison with the Trichoptera. The shift to Type I pheromones appears as a single, distinct event in the early evolution of the Lepi-doptera. In contrast, the pattern of distribution of Type II and Type III pheromones within the phylogeny indicates that these types of pheromones have evolved independently sev-eral times, with replacement of the basic Type 0 and Type I pheromones by other types of pheromones on separate occa-sions and in well-separated clades. Thus, Type II pheromones appear to have evolved at least four times, one of which includes the combination of Type I and Type II pheromone compounds recently found in a number of species in the superfamily Pyraloidea. The distribution of Type III phero-mones suggests that they have evolved independently on three occasions.

The shift from Type 0 to Type I involved a distinct change in the site of production of the pheromone-producing organ from the fifth sternite to the abdominal tip, and the subse-quent loss of the gland on the fifth sternite in all higher Lepi-doptera. The change in the site of production was accompa-nied by marked differences in the biosynthesis, chemistry, and volatility of Type 0 and Type I pheromones.

In contrast, the shift from Type I to Types II and III may be less dramatic because rather than these being entirely new pheromone systems, several facts instead suggest the recruit-ment of Type II and Type III pheromones from the preexisting cuticular lipid biosynthesis pathways by Type I-producing species. First, the skeletons of Type II and Type III phero-mones are synthesized in the oenocytes and then transported through the hemolymph to the pheromone gland, from which they are released, either unchanged or after addition of one or more functional groups. Thus, the site and mecha-nisms of pheromone release are largely unchanged between Types I, II, and III. Second, for Type II and Type III phero-mones, PBAN does not upregulate the biosynthesis of the car-bon skeletons of the pheromones, but instead seems to con-trol the downloading of the hydrocarbon pheromones from lipophorins to the pheromone gland, or possibly to regulate the epoxidation of the pheromone precursors once they are downloaded to the gland (Jurenka 2004). That is, the mecha-nisms for moving hydrocarbon pheromones or precursors through the hemolymph were already in place, for transport and deposition of cuticular hydrocarbons. All that was needed was to change the function of PBAN from initiating phero-mone biosynthesis to controlling downloading of specific precursors to the pheromone gland, or initiating functional-ization of the precursors in the gland. Third, the recent dem-onstrations of a number of species that use mixtures of Type I and Type II compounds suggests a progressive recruitment of oenocyte products to take advantage of a source of readily available precursors, rather than an abrupt change. We hypothesize that the shift to Type II and/or III pheromones may represent evolutionary streamlining of pheromone production systems, whereby the costs of biosynthesizing Type I pheromones de novo can be largely deleted by replacing them with pheromone compounds or precursors co-opted from another preexisting, essential pathway.

It is also informative to examine the anomalous clades that do not conform to female-produced Type 0 pheromones changing to Type I pheromones, and then the subsequent evolution of Types II and III. As a major event within the Lepidoptera, female-produced sex pheromones have been entirely lost at least once, in the Papilionoidea. However, even this shift may not be an entirely clean break because the biosynthetic machinery used for production of Type I pheromones appears to have been maintained largely intact within at least some butterflies. Specifically, male butterflies of the genus *Bicyclus* produce pheromone structures that are very similar to the female-produced pheromones in other lineages, including retention of the same enzymes/genes. However, retention of the enzymatic pathways was accompanied by a sex reversal in the expression of the necessary genes and production of the compounds. This observation supports the hypothesis that the basic biosynthetic machinery for production of both Type I and Type II pheromones has been present since the early radiation of Lepidoptera, and extends from the basal Tischerioidea through to the more advanced Pyraloidea, Geometroidea, and Noctuoidea. Analogous reasoning could apply to the production of Type III pheromones in Lyonetidae, Geometridae, and Erebidae. It is also intriguing to note the continued production of compounds very similar to female-produced sex pheromones in the hairpencils of male moths of several species. For example, male *Ostrinia nubilalis* (Crambidae) produce courtship pheromones using the same desaturases that are involved in production of sex attractant pheromones by females of the same genus (Lassance and Löfstedt 2009). Thus, the retention of this biosynthetic machinery by male butterflies may be mirrored in males of some moth species as well.

For the superfamilies and families from which pheromones are known, female-produced pheromones appear to have been lost on only one other occasion, for species in the subfamily Gallerinae, including the wax moths. These moths use male-produced chemical (and acoustic) signals to mediate reproductive behaviors. In these cases, the male-produced, long-range sex pheromones have been identified as short-chain aldehydes (possibly derived from cleavage of long-chain unsaturated fatty acids), branched hydrocarbons, or terpenoids. These three types of compounds arise from unrelated biosynthetic pathways, defying any attempt to clearly link them to pheromone biosynthesis in other clades.

The sex role reversals in mate communication within the Gallerinae are in many cases associated with unusual life styles (Greenfield 1981). A tenet of sexual selection theory is that sexual selection is stronger on males than on females. Based on this contention, males should take the reproductive risk and in the case of moth pheromone communication, it is reasonable to assume that it is generally more risky to respond to sexual signals than to produce them. Greenfield (1981) speculated that because male wax mothes call near hives, they are subject to honey bee predation—making male calling (with associated sound production) a risky business. Sex role reversal may also be explained by males providing a nuptial gift or some other resource (like protection) to the responding female, which provides selective pressure for females to respond to males. However, in the Gallerinae, it

remains to be convincingly demonstrated why the role reversal has actually occurred.

Finally, our review of sex pheromones found in different lineages and our classification of the pheromones by their likely biosynthetic origins reveals a number of structures that cannot be assigned to one of the four major types. This is not really remarkable, given the forces of speciation and natural selection that drive the need for unique pheromone channels. What is remarkable is that the majority of the structures can indeed be assigned to only four major types, and that there has not been a massive radiation of pheromone structures. Stated another way, the Lepidoptera are using only a tiny portion of the potential "pheromone space." Thus, it may be as appropriate to ask what forces have constrained the lepidopteran pheromones to essentially only four major types, as it is to ask what forces have driven them to diverge from the single ancestral pheromone type, Type 0, to pheromone Types I–III.

Adaptive Value of Pheromone Chemistry

It is generally accepted that chemical communication was the first form of intraspecific communication to evolve. Organisms are selected to exploit any information that may enhance survival or reproduction. The pheromone signaling systems still present in the most primitive moths and their sister group the Trichoptera suggest that the use of volatile pheromones to bring the sexes together was already well developed in the first ancestral lepidopteran. Many pheromones may have evolved from chemical compounds that originally served other purposes than communication. Outgroup comparison with Trichoptera suggests that the sternum V gland secretions may originally have served as defensive secretions (Ansteeg and Dettner 1991; Löfstedt and Kozlov 1997), a scenario supported by the fact that in Trichoptera these glands occur in both females and males, the amounts of the gland constituents are considerably higher than are typically present in the pheromone glands of most female Lepidoptera, and the toxic effects on other insects have been demonstrated (Ansteeg and Dettner 1991). Sternum V secretions from a large number of Trichoptera were analyzed by Löfstedt and coworkers (Löfstedt et al. 1994, 2008; Bergmann 2001, 2002), and although many elicited antennal responses from conspecifics in electroantennogram studies, sex pheromones (including demonstrated attraction to these compounds in behavioral bioassays) were reported in only a few species (see discussion in Löfstedt et al. 2008). Thus, the primary function of many of these compounds may still be defensive, with the communicative function being secondary and restricted to a subset of the gland components.

In his monumental volume *Sociobiology*, E. O. Wilson argued that pheromone molecules that are transmitted through air can be expected to conform to certain "physical rules" (Wilson 1975), requiring them to have between 5 and 20 carbon atoms and a molecular weight between 80 and 300, based on the following a priori arguments. Below the lower limit, only a relatively small number of different kinds of organic molecules can readily be produced and stored in glands. Above this limit, the possibilities for molecular diversity increase rapidly. "In at least some insects, and for some homologous series of compounds, olfactory efficiency also increases steeply. As the upper limit is approached, molecular diversity becomes astronomical, so that further increase in

are largely speculative at this point. Also, for these and the more well-known pathways for Type I and Type II pheromones, much work remains to be done to isolate and characterize the key enzymes involved, and their coding genes. In particular, characterization of Type 0 pheromone biosynthesis would reveal if there is any relationship between Type 0 and Type I pheromones, or whether the transition from Type 0 to Type I represents a true evolutionary break rather than adaptations to preexisting systems. Biosynthetic studies with specific species would also clarify the origins of the structures which we have labeled as possibly from more than one pathway, as well as elucidating the origins of structures which defied classification into one of the four types.

Finally, characterization of the genes involved in pheromone production may not only elucidate pheromone biosynthetic pathways and deepen our understanding of pheromone evolution, but also provide valuable building blocks in the era of synthetic biology. That is, by assembling the necessary and sufficient genes in different platforms, it may be possible to use genetically modified bacteria or plants to produce moth pheromones (Hagström et al. 2013; Ding et al. 2014).

In conclusion, we would like to express our gratitude for the generous service provided to the chemical ecology community by colleagues who have spent countless hours in compiling and maintaining databases of pheromones. The first, entitled the Pherolist, was compiled in the 1980s by Heinrich Arn in collaboration with Miklos Tóth and Ernst Priesner. For many years, the Pherolist served as the sole comprehensive database on lepidopteran pheromones. The Pherolist has now been replaced by the online database maintained by Tetsu Ando (http://www.tuat.ac.jp/~antetsu/LepiPheroList.htm), and the larger database on insect semiochemicals in general maintained by Ashraf El-Sayed (2016). Without these resources, it would not have been possible to write this chapter.

Acknowledgments

We thank Dr. Marjorie Liénard for generous assistance constructing the phylogenetic trees in figure 4.4. NW acknowledges funding from the Academy of Finland.

References Cited

Adachi, Y., N. D. Do, M. Kinjo, S. Makisako, R. Yamakawa, K. Mori, and T. Ando. 2010. Positions and stereochemistry of methyl branches in the novel sex pheromone component produced by a lichen moth, *Lyclene dharma dharma*. *Journal of Chemical Ecology* 36:814–823.

Allison, J. D., and R. T. Cardé. 2007. Bidirectional selection for novel pheromone blend ratios in the almond moth, *Cadra cautella*. *Journal of Chemical Ecology* 33:2293–2307.

Ando, T. 2015. List of sex pheromones, Ando Laboratory. Available at: http://web.tuat.ac.jp/~antetsu/LepiPheroList.htm

Ando, T., S. Yoshida, and N. Tatsuki. 1977. Sex attractants for male Lepidoptera. *Agricultural and Biological Chemistry* 41:1485–1492.

Ando, T., S. I. Inomata, and M. Yamamoto. 2004. Lepidopteran sex pheromones. Pp. 51–96. In S. Schulz, ed. *The Chemistry of Pheromones and Other Semiochemicals I, Vol. 239: Topics in Current Chemistry*. Berlin: Springer.

Ansteeg, O., and K. Dettner. 1991. Chemistry and possible biological significance of secretions from a gland discharging at the 5th abdominal sternite of adult caddisflies (Trichoptera). *Entomologia Generalis* 156:303–312.

Arsequell, G., G. Fabriàs, and F. Camps. 1990. Sex pheromone biosynthesis in the processionary moth *Thaumetopoea pityocampa* by delta-13 desaturation. *Archives of Insect Biochemistry and Physiology* 14:47–56.

Bacquet, P. M. B., O. Brattström, H.-L. Wang, C. E. Allen, C. Löfstedt, P. M. Brakefield, and C. M. Nieberding. 2015. Selection on male sex pheromone composition contributes to butterfly reproductive isolation. *Proceedings of the Royal Society. B* 282:20142734.

Baker, T. C., W. Francke, C. Löfstedt, B. S. Hansson, J.-W. Du, P. L. Phelan, R. S. Vetter, and R. Youngman. 1989. Isolation, identification and synthesis of sex pheromone components of the carob moth, *Ectomyelois ceratoniae*. *Tetrahedron Letters* 30:2901–2902.

Bazinet, A. L., M. P. Cummings, K. T. Mitter, and C. W. Mitter. 2013. Can RNA-Seq resolve the rapid radiation of advanced moths and butterflies (Hexapoda: Lepidoptera: Apoditrysia)? An exploratory study. *PLOS ONE* 8(12):e82615.

Bello, J. E., J. S. McElfresh, and J. G. Millar. 2015. Isolation and determination of absolute configurations of insect-produced methyl-branched hydrocarbons. *Proceedings of the National Academy of Sciences of the United States of America* 112:1077–1082.

Bergmann, J., C. Löfstedt, V. D. Ivanov, and W. Francke. 2001. Identification and assignment of absolute configuration of methyl-branched ketones from limnephilid caddisflies. *European Journal of Organic Chemistry* 16:3175–3179.

Bergmann, J., C. Löfstedt, V. D. Ivanov, and W. Francke. 2002. Electrophysiologically active volatile compounds from six species of caddisflies. *Nova Supplementa Entomologica* 15:37–46.

Bestmann, H. J., A. B. Attygalle, J. Schwarz, W. Garbe, O. Vostrowsky, and I. Tomida. 1989. Pheromones, 71. Identification and synthesis of female sex pheromone of eri-silkworm, *Samia cynthia ricini* (Lepidoptera: Saturniidae). *Tetrahedron Letters* 30:2911–2914.

Beutel, R. G., F. Friedrich, T. Hörnschemeyer, H. Pohl, F. Hünefeld, F. Beckmann, R. Meier, B. Misof, M. F. Whiting, and L. Vilhelmsen. 2011. Morphological and molecular evidence converge upon a robust phylogeny of the megadiverse Holometabola. *Cladistics* 27:341–355.

Birch, M. C., G. M. Poppy, and T. C. Baker. 1990. Scents and eversible scent structures of male moths. *Annual Review of Entomology* 35:25–54.

Bjostad, L. B., and W. L. Roelofs. 1981. Sex-pheromone biosynthesis from radiolabeled fatty-acids in the redbanded leafroller moth. *Journal of Biological Chemistry* 256:7936–7940.

Bjostad, L. B., and W. L. Roelofs. 1983. Sex pheromone biosynthesis in *Trichoplusia ni*: key steps involve delta-11 desaturation and chain-shortening. *Science* 220:1387–1389.

Blomquist, G. J., D. R. Nelson, and M. de Ronobales. 1987. Chemistry, biochemistry, and physiology of insect cuticular lipids. *Archives of Insect Biochemistry and Physiology* 6:227–265.

Blomquist, G. J., R. Figueroa-Teran, M. Aw, M. Song, A. Gorzalski, N. L. Abbott, E. Chang, and C. Tittiger. 2010. Pheromone production in bark beetles. *Insect Biochemistry and Molecular Biology* 40:699–712.

Cabrera, A., A. E. Eiras, G. Gries, R. Gries, N. Urdaneta, B. Mirás, C. Badji, and K. Jaffe. 2001. Sex pheromone of tomato fruit borer, *Neoleucinodes elegantalis*. *Journal of Chemical Ecology* 27:2097–2107.

Cardé, A. M., T. C. Baker, and R. T. Cardé. 1979. Identification of a 4-component sex-pheromone of the female Oriental fruit moth, *Grapholitha molesta* (Lepidoptera, Tortricidae). *Journal of Chemical Ecology* 5:423–427.

Charlton, R. E., and W. L. Roelofs. 1991. Biosynthesis of a volatile, methyl-branched hydrocarbon sex pheromone from leucine by arctiid moths (*Holomelina* spp.). *Archives of Insect Biochemistry and Physiology* 18:81–97.

Choi, M. Y., H. Lim, K. C. Park, R. Adlo, A. S. Wang, A. Zhang, and R. Jurenka. 2007. Identification and biosynthetic studies of the hydrocarbon sex pheromone in *Utetheisa ornatrix*. *Journal of Chemical Ecology* 33:1336–1345.

Conner, W. E., T. Eisner, R. K. Vander Meer, A. Guerrero, D. Ghiringelli, and J. Meinwald. 1980. Sex attractant of an arctiid moth (*Utetheisa ornatrix*): a pulsed chemical signal. *Behavioral Ecology and Sociobiology* 7:55–63.

Cook, H. W., and C. R. McMaster. 2002. Fatty acid desaturation and chain elongation in eukaryotes. Pp. 181–204. In D. E. Vance and J. E. Vance, eds. *Biochemistry of Lipids, Lipoproteins, and Membranes*. New York: Elsevier.

Ding, B.-J., and C. Löfstedt. 2014. Desaturase orthologues with different specificity, Δ9 and Δ10 respectively, account for differences in sex pheromone biosynthesis between the two tortricid moths *Cydia pomonella* and *Grapholita molesta*. Pp. 59–74. In B.-J. Ding. *On the Way of Making Plants Smell Like Moths*. PhD thesis, Lund University, Sweden.

Ding, B.-J., M. A. Liénard, H.-L. Wang, C.-H. Zhao, and C. Löfstedt. 2011. Terminal fatty-acyl-CoA desaturase involved in sex pheromone biosynthesis in the winter moth (*Operophtera brumata*). *Insect Biochemistry and Molecular Biology* 41:715–722.

Ding, B.-J., P. Hofvander, H.-L. Wang, T.P. Durrett, S. Stymne, and C. Löfstedt. 2014. A plant factory for moth pheromone production. *Nature Communications* 5:e3353.

Do, N. D., M. Kinjo, T. Taguri, Y. Adachi, R. Yamakawa, and T. Ando. 2009. Synthesis and field evaluation of methyl-branched ketones, sex pheromone components produced by Lithosiinae female moths in the family of Arctiidae. *Bioscience Biotechnology and Biochemistry* 73:1618–1622.

Duthie, B., G. Gries, R. Gries, C. Krupke, and S. Derksen. 2003. Does pheromone-based aggregation of codling moth larvae help procure future mates? *Journal of Chemical Ecology* 29: 425–436.

El-Sayed, A.M. 2016. The Pherobase: Database of Pheromones and Semiochemicals. Available at: http://www.pherobase.com.

Evenden, M.L., B.A. Mori, R. Gries, and J. Otani. 2010. Sex pheromone of the red clover casebearer moth, *Coleophora deauratella*, an invasive pest of clover in Canada. *Entomologia Experimentalis et Applicata* 137:255–261.

Fang, N., P.E. Teal, R.E. Doolittle, and J.H. Tumlinson. 1995. Biosynthesis of conjugated olefinic systems in the sex pheromone gland of female tobacco hornworm moths, *Manduca sexta* (L.). *Insect Biochemistry and Molecular Biology* 25:39–48.

Foster, S. P., and W. L. Roelofs. 1988. Sex-pheromone biosynthesis in the leafroller moth Planotortrix excessana by Delta-10 desaturation. *Archives of Insect Biochemistry and Physiology* 8:1–9.

Fujii, T., M. G. Suzuki, T. Kawai, K. Tsuneizumi, A. Ohnishi, M. Kurihara, S. Matsumoto, and T. Ando. 2007. Determination of the pheromone-producing region that has epoxidation activity in the abdominal tip of the Japanese giant looper, *Ascotis selenaria cretacea* (Lepidoptera: Geometridae). *Journal of Insect Physiology* 53:312–318.

Fujii, T., R. Nakano, Y. Takubo, S. G. Qian, R. Yamakawa, T. Ando, and Y. Ishikawa. 2010. Female sex pheromone of a lichen moth *Eilema japonica* (Arctiidae, Lithosiinae): Components and control of production. *Journal of Insect Physiology* 56:1986–1991.

Fujii, T., M. G. Suzuki, S. Katsuma, K. Ito, Y. Rong, S. Matsumoto, T. Ando, and Y. Ishikawa. 2013a. Discovery of a disused desaturase gene from the pheromone gland of the moth *Ascotis selenaria*, which secretes an epoxyalkenyl sex pheromone. *Biochemical and Biophysical Research Communications* 441:849–855.

Fujii, T., R. Yamakawa, Y. Terashima, S. Imura, K. Ishigaki, M. Kinjo, and T. Ando. 2013b. Propionates and acetates of chiral secondary alcohols: novel sex pheromone components produced by a lichen moth *Barsine expressa* (Arctiidae: Lithosiinae). *Journal of Chemical Ecology* 39:28–36.

Gago, R., J.D. Allison, J.S. McElfresh, K.F. Haynes, J. McKenney, A. Guerrero, and J.G. Millar. 2013. A tetraene aldehyde as the major sex pheromone component of the promethea moth (*Callosamia promethea* (Drury)). *Journal of Chemical Ecology* 39:1263–1272.

Gibb, A.R., B. Pinese, D. Tenakanai, A.P. Kawi, B. Bunn, P. Ramankutty, and D.M. Suckling. 2007. (*Z*)-11-Hexadecenal and (3Z,6Z,9Z)-tricosatriene: sex pheromone components of the red banded mango caterpillar *Deanolis sublimbalis*. *Journal of Chemical Ecology* 33:579–589.

Goller, S., G. Szöcs, W. Francke, and S. Schulz. 2007. Biosynthesis of (3Z,6Z,9Z)-octadecatriene: the main component of the pheromone blend of *Erannis bajaria* (Lepidoptera: Geometridae). *Journal of Chemical Ecology* 33:1505–1509.

Grant, G.G., J.G. Millar, and R. Trudel. 2009. Pheromone identification of *Dioryctria abietivorella* (Lepidoptera: Pyralidae) from an eastern North American population: geographic variation in pheromone response. *The Canadian Entomologist* 141:129–135.

Greenfield, M.D. 1981. Moth sex pheromones: an evolutionary perspective. *Florida Entomologist* 64:4–17.

Gries, R., G. Khaskin, Z. X. Tan, B.G. Zhao, G. S. King, A. Miroshnychenko, G. Lin, M. Rhainds, and G. Gries. 2006. (1S)-1-Ethyl-2-methylpropyl 3,13-dimethylpentadecanoate: major sex pheromone component of Paulownia bagworm, *Clania variegata*. *Journal of Chemical Ecology* 32:1673–1685.

Hagström, Å. K., H.-L. Wang, M.A. Liénard, J.M. Lassance, T. Johansson, and C. Löfstedt. 2013. A moth pheromone brewery: production of (*Z*)-11-hexadecenol by heterologous co-expression of two biosynthetic genes from a noctuid moth in a yeast cell factory. *Microbial Cell Factories* 12:125. doi:10.1186/1475-2859-12-125.

Hai, T.V., P.K. Son, S.I. Inomata, and T. Ando. 2002. Sex attractants for moths of Vietnam: field attraction by synthetic lures baited with known lepidopteran pheromones. *Journal of Chemical Ecology* 28:1473–1481.

Hall, D. R., A. Cork, R. Lester, B. F. Nesbitt, and P. Zagatti. 1987. Sex pheromones of rice moth, *Corcyra cephalonica* Stainton. 2. Identification and role of female pheromone. *Journal of Chemical Ecology* 13:1575–1589.

Hall, D. R., P. S. Beevor, D. G. Campion, D. J. Chamberlain, A. Cork, R. D. White, A. Almestar, and T. J. Henneberry. 1992. Nitrate esters: novel sex-pheromone components of the cotton leafperforator, *Bucculatrix thurberiella* Busck (Lepidoptera, Lyonetiidae). *Tetrahedron Letters* 33:4811–4814.

Hao, G., G. Liu, M. O'Connor, and W. L. Roelofs. 2002. Acyl-CoA Z9- and Z10-desaturase genes from a New Zealand leafroller moth species, *Planotortrix octo*. *Insect Biochemistry and Molecular Ecology* 9:961–966.

Hashimoto, T. 1996. Peroxisomal beta-oxidation: enzymology and molecular biology. *Annals of the New York Academy of Sciences* 804:86–98.

Haynes, K.F., and R.E. Hunt. 1990. A mutation in pheromonal communication system of cabbage looper moth, *Trichoplusia ni*. *Journal of Chemical Ecology* 16:1249–1257.

Heikkilä, M., M. Mutanen, M. Kekkonen, and L. Kaila. 2014. Morphology reinforces proposed molecular phylogenetic affinities: a revised classification for Gelechioidea (Lepidoptera). *Cladistics* 30:563–589.

Henderson, H.E., F.L. Warren, O.P.H. Augustyn, B.V. Burger, D.F. Schneider, P.R. Boshoff, H.S.C. Spies, and H. Geertsema. 1973. Isolation and structure of the sex-pheromone of the moth, *Nudaurelia cytherea cytherea*. *Journal of Insect Physiology*, 19:1257–1264.

Hill, A.S., and W.L. Roelofs. 1981. Sex pheromone of the saltmarsh caterpillar moth, *Estigmene acrea*. *Journal of Chemical Ecology* 7:655–668.

Ho, H. Y., Y. T. Tao, R. S. Tsai, Y. L. Wu, H. K. Tseng, and Y. S. Chow. 1996. Isolation, identification, and synthesis of sex pheromone components of female tea cluster caterpillar, *Andraca bipunctata* Walker (Lepidoptera: Bombycidae) in Taiwan. *Journal of Chemical Ecology* 22:271–285.

Jain, S. C., D. E. Dussourd, W. E. Conner, T. Eisner, A. Guerrero, and J. Meinwald. 1983. Polyene pheromone components from an arctiid moth (*Utetheisa ornatrix*): characterization and synthesis. *Journal of Organic Chemistry* 48:2266–2270.

Jones, I. F., and R. S. Berger. 1978. Incorporation of (1-C-14) acetate into cis-7-dodecen-1-ol acetate, a sex pheromone in cabbage looper (*Trichoplusia ni*). *Environmental Entomology* 7:666–669.

Jurenka, R. A. 1997. Biosynthetic pathway for producing the sex pheromone component (*Z, E*)-9,12-tetradecadienyl acetate in moths involves a Δ 12 desaturase. *Cellular and Molecular Life Sciences* 53:501–505.

Jurenka, R. A. 2003. Biochemistry of female moth sex pheromones. Pp. 53–80. In G. Blomquist and R. Vogt, eds. *Insect Pheromone Biochemistry and Molecular Biology*. London, UK/San Diego, CA: Elsevier Academic Press.

Jurenka, R. 2004. Insect Pheromone Biosynthesis. Pp. 97–131. In S. Schulz, ed. *The Chemistry of Pheromones and Other Semiochemicals I, Vol. 239: Topics in Current Chemistry*. Berlin: Springer.

Jurenka, R.A., and M. Subchev. 2000. Identification of cuticular hydrocarbons and the alkene precursor to the pheromone in hemolymph of the female gypsy moth, *Lymantria dispar*. *Archives of Insect Biochemistry and Physiology* 43:108–115.

Jurenka, R. A., K. F. Haynes, R. O. Adlof, M. Bengtsson, and W. L. Roelofs. 1994. Sex-pheromone component ratio in the cabbage looper moth altered by a mutation affecting the fatty-acid chain-shortening reactions in the pheromone biosynthetic pathway. *Insect Biochemistry and Molecular Biology* 24:373–381.

Jurenka, R.A., M. Subchev, J.L. Abad, M.Y. Choi, and G. Fabriàs. 2003. Sex pheromone biosynthetic pathway for disparlure in the gypsy moth, *Lymantria dispar*. *Proceedings of the National Academy of Sciences of the United States of America* 100:809–814.

Kawahara, A. Y., and J. W. Breinholt. 2014. Phylogenomics provides strong evidence for relationships of butterflies and moths. *Proceedings of the Royal Society of London B* 281.

Kawai, T., A. Ohnishi, M. G. Suzuki, T. Fujii, K. Matsuoka, I. Kato, S. Matsumoto, and T. Ando. 2007. Identification of a unique pheromonotropic neuropeptide including double FXPRL motifs from a geometrid species, *Ascotis selenaria cretacea*, which produces an epoxyalkenyl sex pheromone. *Insect Biochemistry and Molecular Biology* 37:330–337.

Kearse, M., R. Moir, A. Wilson, S. Stones-Havas, M. Cheung, S. Sturrock, S. Buxton, A. Cooper, S. Markowitz, C. Duran, T. Thierer, B. Ashton, P. Meintjes, and A. Drummond. 2012. Geneious Basic: an integrated and extendable desktop software platform for the organization and analysis of sequence data. *Bioinformatics* 28:1647–1649.

Khosla, C., R. S. Gokhale, J. R. Jacobsen, and D. E. Cane. 1999. Tolerance and specificity of polyketide synthases. *Annual Review of Biochemistry* 68:219–253.

Kim, J., K. S. Cho, C. Y. Yang, and C. G. Park. 2014. Identification and field evaluation of the female sex pheromone of *Stathmopoda masinissa* in Korea. *Chemoecology* 24:253–259.

King, G. G. S., R. Gries, G. Gries, and K. N. Slessor. 1995. Optical isomers of 3,13-dimethylheptadecane: Sex pheromone components of the western false hemlock looper, *Nepytia freemani* (Lepidoptera: Geometridae). *Journal of Chemical Ecology* 21:2027–2045.

Kiyota, R., M. Arakawa, R. Yamakawa, A. Yasmin, and T. Ando. 2011. Biosynthetic pathways of the sex pheromone components and substrate selectivity of the oxidation enzymes working in pheromone glands of the fall webworm, *Hyphantria cunea*. *Insect Biochemistry and Molecular Biology* 1:362–369.

Kozlov, M. V., J. Zhu, P. Philipp, W. Francke, E. L. Zvereva, B. S. Hansson, and C. Löfstedt. 1996. Pheromone specificity in *Eriocrania semipurpurella* (Stephens) and *E. sangii* (Wood) (Lepidoptera: Eriocraniidae) based on chirality of semiochemicals. *Journal of Chemical Ecology* 22:431–454.

Kristensen, N. P. 1998. *Handbook of Zoology, Vol. 4: Arthropoda: Insecta, Part 35. Lepidoptera, Moths, and Butterflies, Vol. 1: Evolution, Systematics and Biogeography.* Berlin: Walter de Gruyter.

Kristensen, N. P., and A. W. Skalski. 1998. Phylogeny and palaeontology. Pp. 7–25. In N. P. Kristensen, ed. *Handbook of Zoology, Vol. 4: Arthropoda: Insecta, Part 35, Lepidoptera, Moths and Butterflies, Vol. 1: Evolution, Systematics and Biogeography.* Berlin: Walter de Gruyter.

Kuenen, L. P. S., J. S. McElfresh, and J. G. Millar. 2010. Identification of critical secondary components of the sex pheromone of the navel orangeworm (Lepidoptera: Pyralidae). *Journal of Economic Entomology* 103:314–330.

Landolt, P. J., and C. E. Curtis. 1982. Interspecific sexual attraction between *Pyralis farinalis* L. and *Amyelois transitella* (Walker) (Lepidoptera: Pyralidae). *Journal of the Kansas Entomological Society* 55:248–252.

Larsson, M. C., E. Hallberg, M. V. Kozlov, W. Francke, B. S. Hansson, and C. Löfstedt. 2002. Specialized olfactory receptor neurons mediating intra- and interspecific chemical communication in leafminer moths *Eriocrania* spp. (Lepidoptera: Eriocraniidae). *Journal of Experimental Biology* 205:989–998.

Lassance, J.-M., and C. Löfstedt. 2009. Concerted evolution of male and female display traits in the European corn borer, *Ostrinia nubilalis*. *BMC Biology* 7:10.

Leal, W. S., A. L. Parra-Pedrazzoli, K. E. Kaissling, T. I. Morgan, F. G. Zalom, D. J. Pesak, E. A. Dundulis, C. S. Burks, and B. S. Higbee. 2005. Unusual pheromone chemistry in the navel orangeworm: novel sex attractants and a behavioral antagonist. *Naturwissenschaften* 92:139–146.

Leonhardt, B. A., J. W. Neal, J. A. Klun, M. Schwarz, and J. R. Plimmer. 1983. An unusual lepidopteran sex pheromone system in the bagworm moth. *Science* 219:314–316.

Leyrer, R. L., and R. E. Monroe. 1973. Isolation and identification of the scent of the moth, *Galleria melonella*, and a reevaluation of its sex pheromone. *Journal of Insect Physiology* 19:2267–2271.

Li, J., R. Gries, G. Gries, K. N. Slessor, G. G. S. King, W. W. Bowers, and R. J. West. 1993. Chirality of 5,11-dimethylheptadecane, the major sex pheromone component of the hemlock looper, *Lambdina fiscellaria* (Lepidoptera: Geometridae). *Journal of Chemical Ecology* 19:1057–1062.

Liénard, M. A., M. Strandh, E. Hedenström, T. Johansson, and C. Löfstedt. 2008. Key biosynthetic gene subfamily recruited for pheromone production prior to the extensive radiation of Lepidoptera. *BMC Evolutionary Biology* 8:270.

Liénard, M. A., Å. K. Hagström, J.-M. Lassance, and C. Löfstedt. 2010a. Evolution of multi-component pheromone signals in small ermine moths involves a single fatty-acyl reductase gene. *Proceedings of the National Academy of Sciencies of the United States of America* 107:10955–10960.

Liénard, M. A., J.-M. Lassance, H.-L. Wang, C.-H. Zhao, J. Piškur, T. Johansson, and C. Löfstedt. 2010b. Elucidation of the sex-pheromone biosynthesis producing 5,7-dodecadienes in *Dendrolimus punctatus* (Lepidoptera: Lasiocampidae) reveals Δ11- and Δ9-desaturases with unusual catalytic properties. *Insect Biochemistry and Molecular Biology* 40:440–452.

Liénard, M. A., H.-L. Wang, J.-M. Lassance, and C. Löfstedt. 2014. Sex pheromone biosynthetic pathways are conserved between moths and the butterfly *Bicyclus anynana*. *Nature Communications* 5:3957.

Linn, C. E., Jr., J. Du, A. Hammond, and W. L. Roelofs. 1987. Identification of unique pheromone components for soybean looper moth *Pseudoplusia includens*. *Journal of Chemical Ecology* 13:1351–1360.

Loeb, M. J., J. W. Neal, and J. A. Klun. 1989. Modified thoracic epithelium of the bagworm (Lepidoptera: Psychidae): site of pheromone production in adult females. *Annals of the Entomological Society of America* 82:215–219.

Löfstedt, C., and W. L. Roelofs. 1985. Sex pheromone precursors in two primitive New Zealand tortricid moth species. *Insect Biochemistry* 15:729–734.

Löfstedt, C., and M. Bengtsson. 1988. Sex pheromone biosynthesis in the codling moth *Cydia pomonella* involves E9-desaturation. *Journal of Chemical Ecology* 14:903–915.

Löfstedt, C., and M. Kozlov. 1997. A phylogenetic analysis of pheromone communication in primitive moths. Pp. 473–489. In R. T. Cardé and A. K. Minks, eds. *Insect Pheromone Research: New Directions.* New York: Chapman & Hall.

Löfstedt, C., B. S. Hansson, E. Pettersson, P. Valeur, and A. Richards. 1994. Pheromonal secretions from glands on the 5th abdominal sternite of hydropsychid and rhyacophilid caddisflies (Trichoptera). *Journal of Chemical Ecology* 20:153–170.

Löfstedt, C., J. Bergmann, W. Francke, E. Jirle, B. S. Hansson, and V. D. Ivanov. 2008. Identification of a sex pheromone produced by sternal glands in females of the caddisfly *Molanna angustata* Curtis. *Journal of Chemical Ecology* 34:220–228.

Löfstedt, C., G. P. Svensson, E. V. Jirle, O. Rosenberg, A. Roques, and J. G. Millar. 2012. (3Z,6Z,9Z,12Z,15Z)-Pentacosapentaene and (9Z,11E)-tetradecadienyl acetate: sex pheromone of the spruce coneworm *Dioryctria abietella* (Lepidoptera: Pyralidae). *Journal of Applied Entomology* 136:70–78.

Löfstedt, C., J. Zhu, M. V. Kozlov, V. Buda, E. V. Jirle, S. Hellqvist, J. Löfqvist, E. Plass, S. Franke, and W. Francke. 2004. Identification of the sex pheromone of the currant shoot borer, *Lampronia capitella*. *Journal of Chemical Ecology* 30:643–658.

Malm, T., K. A. Johanson, and N. Wahlberg. 2013. The evolutionary history of Trichoptera (Insecta): a case of successful adaptation to life in freshwater. *Systematic Entomology* 38:459–473.

Matsuoka, K., H. Tabunoki, T. Kawai, S. Ishikawa, M. Yamamoto, R. Sato, and T. Ando. 2006. Transport of a hydrophobic biosynthetic precursor by lipophorin in the hemolymph of a geometrid female moth which secretes an epoxyalkenyl sex pheromone. *Insect Biochemistry and Molecular Biology* 36:576–583.

Matsuoka, K., M. Yamamoto, R. Yamakawa, M. Muramatsu, H. Naka, Y. Kondo, and T. Ando. 2008. Identification of novel C_{20} and C_{22} trienoic acids from arctiid and geometrid female moths that produce polyenyl Type II sex pheromone components. *Journal of Chemical Ecology* 34:1437–1445.

Matoušková, P., I. Pichová, and A. Svatoš. 2007. Functional characterization of a desaturase from the tobacco hornworm moth (*Manduca sexta*) with bifunctional Z11- and 10,12-desaturase activity. *Insect Biochemistry and Molecular Biology* 37:601–610.

McElfresh, J. S., J. G. Millar, and D. Rubinoff. 2001. (*E4,Z9*)-Tetradecadienal, a sex pheromone for three North American moth species in the genus Saturnia. *Journal of Chemical Ecology* 27:791–806.

Millar, J. G., G. G. Grant, J. S. McElfresh, W. Strong, C. Rudolph, J. D. Stein, and J. A. Moreira. 2005. (3Z,6Z,9Z,12Z,15Z)-Pentacosapentaene, a key pheromone component of the fir coneworm moth, *Dioryctria abietivorella. Journal of Chemical Ecology* 31:1229–1234.

Miller, D., J.G. Millar, G. Grant, L. MacDonald, and G. DeBarr. 2010. (3Z,6Z,9Z,12Z,15Z)-Pentacosapentane and (9Z,11E)-tetradecadienyl acetate: attractant lure blend for *Dioryctria ebeli* (Lepidopterea: Pyralidae). *Journal of Entomological Science* 45:54–57.

Milne, I., D. Lindner, M. Bayer, D. Husmeier, G. McGuire, D. Marshall, and F. Wright. 2009. TOPALi v2: a rich graphical interface for evolutionary analyses of multiple alignments on HPC clusters and multi-core desktops. *Bioinformatics* 25:126–127.

Minyailo, V.A., B.G. Kovalev, E.I. Kirov, and A.K. Minyailo. 1977. On the attractiveness of disparlure, the sex pheromone of the gypsy moth *Porthetria dispar* (Lepidoptera, Orgyidae) for males of *Zanclognatha lunaris* (Lepidoptera, Noctuidae). *Zoologicheskii Zhurnal* 56:309–310.

Miyamoto, T., M. Yamamoto, A. Ono, K. Ohtani, and T. Ando. 1999. Substrate specificity of the epoxidation reaction in sex pheromone biosynthesis of the Japanese giant looper (Lepidoptera: Geometridae). *Insect Biochemistry and Molecular Biology* 29:63–69.

Mori, K., H. Harada, P. Zagatti, A. Cork, and D. R. Hall. 1991. Pheromone synthesis. 126. Syntesis and biological activity of 4 stereoisomers of 6,10,14-trimethyl-2-pentadecanol, the female-produced sex-pheromone of rice moth (*Corcyra cephalonica*). *Liebigs Annalen der Chemie* 3:259–267.

Mori, K., T. Tashiro, B. Zhao, D.M. Suckling, and A.M. El-Sayed. 2010. Pheromone synthesis. Part 243: synthesis and biological evaluation of (3R,13R,1-S)-1--ethyl-2--methylpropyl 3,13-dimethyl-pentadecanoate, the major component of the sex pheromone of Paulownia bagworm, *Clania variegata*, and its stereoisomers. *Tetrahedron* 14:2642–2653.

Moto, K.I., M.G. Suzuki, J.J. Hull, R. Kurata, S. Takahashi, M. Yamamoto, K. Okana, K. Imai, T. Ando, and S. Matsumoto. 2004. Involvement of a bifunctional fatty-acyl desaturase in the biosynthesis of the silkmoth, *Bombyx mori*, sex pheromone. *Proceedings of the National Academy of Sciences of the United States of America* 101:8631–8636.

Mutanen, M., N. Wahlberg, and L. Kaila. 2010. Comprehensive gene and taxon coverage elucidates radiation patterns in moths and butterflies. *Proceedings of the Royal Society of London B Biological Sciences* 277:2839–2848.

Naka, H., S.I. Inomata, T. Ando, T. Kimura, H. Honda, K. Tsuchida, and H. Sakurai. 2003. Sex pheromone of the persimmon fruit moth, *Stathmopoda masinissa*: identification and laboratory bioassay of (4E, 6Z)-4,6-hexadecadien-1-ol derivatives. *Journal of Chemical Ecology* 29:2447–2459.

Nieukerken, E.J.V., L. Kaila, I.J. Kitching, N.P. Kristensen, D.C. Lees, J. Minet, C. Mitter et al. 2011. Order Lepidoptera. In Z.-Q. Zhang, ed. Animal Biodiversity: an outline of higher-level classification and survey of taxonomic richness. *Zootaxa* 3148:212–221.

Ostrauskas, H. 2004. Moths caught in pheromone traps for American white moth (*Hyphantria cunea* Dr.) (Arctiidae, Lepidoptera) in Lithuania during 2001. *Acta Zoologica Lituanica* 14:66–74.

Persoons, C.J., S. Voerman, P.E.J. Verwiel, F.J. Ritter, W.J. Nooyen, and A.K. Minks. 1976. Sex pheromone of the potato tuberworm moth, *Phthorimaea operculella*: isolation, identification and field evaluation. *Entomologia Experimentalis et Applicata* 20:289–300.

Rajaei, H.S., C. Greve, H. Letsch, D. Stüning, N. Wahlberg, J. Minet, and B. Misof. 2015. Advances in Geometroidea phylogeny, with characterization of a new family based on *Pseudobiston pinratanai* (Lepidoptera, Glossata). *Zoologica Scripta* 44:418–436.

Regier, J. C., A. Zwick, M. P. Cummings, A. Y. Kawahara, S. Cho, S. J. Weller, A. D. Roe et al. 2009. Toward reconstructing the evolution of advanced moths and butterflies (Lepidoptera: Ditrysia): an initial molecular study. *BMC Evolutionary Biology* 9:e280.

Regier, J.C., J.W. Brown, C. Mitter, J. Baixeras, S. Cho, M.P. Cummings, and A. Zwick. 2012. A molecular phylogeny for the leafroller moths (Lepidoptera: Tortricidae) and its implications for classification and life history evolution. *PLOS ONE* 7(4):e35574.

Regier, J.C., C. Mitter, A. Zwick, A.L. Bazinet, M.P. Cummings, A.Y. Kawahara, J.-C. Sohn et al. 2013. A large-scale, higher-level, molecular phylogenetic study of the insect order Lepidoptera (moths and butterflies). *PLOS ONE* 8:e58568.

Regier, J.C., C. Mitter, D.R. Davis, T.L. Harrison, J.-C. Sohn, M.P. Cummings, A. Zwick, and K.T. Mitter. 2015. A molecular phylogeny and revised classification for the oldest ditrysian moth lineages (Lepidoptera: Tineoidea), with implications for ancestral

feeding habits of the mega-diverse Ditrysia. *Systematic Entomology* 40:409–432.

Rhainds, M., G. Gries, J. Li, R. Gries, K.N. Slessor, C.M. Chinchilla, and A.C. Oehlschlager. 1994. Chiral esters: sex pheromone of the bagworm, *Oiketicus kirbyi* (Lepidoptera: Psychidae). *Journal of Chemical Ecology* 20:3083–3096.

Roelofs, W.L., and R.L. Brown. 1982. Pheromones and evolutionary relationships of Tortricidae. *Annual Review of Ecology and Systematics* 13:395–422.

Roelofs, W., and L. Bjostad. 1984. Biosynthesis of lepidopteran pheromones. *Bioorganic Chemistry* 12:279–298.

Roelofs, W.L., and W.A. Wolf. 1988. Pheromone biosynthesis in Lepidoptera. *Journal of Chemical Ecology* 14:2019–2031.

Roelofs, W.L., and A.P. Rooney. 2003. Molecular genetics and evolution of pheromone biosynthesis in Lepidoptera. *Proceedings of the National Academy of Sciencies of the United States of America* 100:9179–9184.

Roelofs, W., A. Comeau, A. Hill, and G. Milicevi. 1971. Sex attractant of codling moth: characterization with electroantennogram technique. *Science* 174:297–299.

Rong, Y., T. Fujii, S. Katsuma, M. Yamamoto, T. Ando, and Y. Ishikawa. 2014. CYP341B14: a cytochrome P450 involved in the specific epoxidation of pheromone precursors in the fall webworm *Hyphantria cunea*. *Insect Biochemistry and Molecular Biology* 54:122–128.

Rule, G.S., and W.L. Roelofs. 1989. Biosynthesis of sex pheromone components from linolenic acid in arctiid moths. *Archives of Insect Biochemistry and Physiology* 12:89–97.

Schal, C., V. Sevala, and R.T. Cardé. 1998. Novel and highly specific transport of a volatile sex pheromone by hemolymph lipophorin in moths. *Naturwissenschaften* 85:339–342.

Serra, M., B. Piña, J.L. Abad, F. Camps, and G. Fabriàs. 2007. A multifunctional desaturase involved in the biosynthesis of the processionary moth sex pheromone. *Proceedings of the National Academy of Sciences of the United States of America* 104:16444–16449.

Sihvonen, P., M. Mutanen, L. Kaila, G. Brehm, A. Hausmann, and H. S. Staude. 2011. Comprehensive molecular sampling yields a robust phylogeny for geometrid moths (Lepidoptera: Geometridae). *PLOS ONE* 6:e20356.

Skiba, P. J., and L. L. Jackson. 1994. Fatty-acid elongation in the biosynthesis of (Z)-10-heptadecen-2-one and 2-tridecanone in ejaculatory bulb microsomes of *Drosophila buzzatii*. *Insect Biochemistry and Molecular Biology* 24:847–853.

Sohn, J.-C., J. C. Regier, C. Mitter, D. Davis, D., J.-F. Landry, A. Zwick, and M.P. Cummings. 2013. A molecular phylogeny for Yponomeutoidea (Insecta, Lepidoptera, Ditrysia) and its implications for classification, biogeography and the evolution of host plant use. *PLOS ONE* 8:e55066.

Song, M., A. Gorzalski, T.T. Nguyen, X. Liu, C. Jeffrey, G.J. Blomquist, and C. Tittiger. 2014a. *exo*-Brevicomin biosynthesis in the fat body of the mountain pine beetle, *Dendroctonus ponderosae*. *Journal of Chemical Ecology* 40:181–189.

Song, M., A. Gorzalski, A., T.T. Nguyen, X. Liu, C. Jeffrey, G.J. Blomquist, and C. Tittiger. 2014b. *exo*-Brevicomin biosynthesis enzymes from the mountain pine beetle, *Dendroctonus ponderosae*. *Insect Biochemistry and Molecular Biology* 53:73–80.

Stanley-Samuelson, D.W., R.A. Jurenka, C. Cripps, G.J. Blomquist, and M. de Renobales. 1988. Fatty acids in insects: composition, metabolism, and biological significance. *Archives of Insect Biochemistry and Physiology* 9:1–33.

Subchev, M., and R.A. Jurenka. 2001. Sex pheromone levels in pheromone glands and identification of the pheromone and hydrocarbons in the hemolymph of the moth *Scoliopteryx libatrix* L.(Lepidoptera: Noctuidae). *Archives of Insect Biochemistry and Physiology* 47:35–43.

Svensson, G.P., E.A. Gündüz, N. Sjöberg, E. Hedenström, J.-M. Lassance, H.-L. Wang, C. Löfstedt, and O. Anderbrant. 2014. Identification, synthesis, and behavioral activity of 5,11-dimethyl-pentacosane, a novel sex pheromone component of the greater wax moth, *Galleria mellonella* (L.). *Journal of Chemical Ecology* 40:387–395.

Tabata, J., M. Minamishima, H. Sugie, T. Fukomoto, F. Mochizuki, and Y. Yoshiyasu. 2009. Sex pheromone components of the pear fruit moth, *Acrobasis pyrivorella* (Matsumura). *Journal of Chemical Ecology* 35:243–249.

Tóth, M., G. Szöcs, E. J. Van Nieukerken, P. Philipp, F. Schmidt, and W. Francke. 1995. Novel type of sex-pheromone structure identified from Stigmella malella (Stainton) (Lepidoptera, Nepticulidae). *Journal of Chemical Ecology* 21:13–27.

Tumlinson, J.H., M.M. Brennan, R.E. Doolittle, E.R. Mitchell, A. Brabham, B.E. Mazomenos, A.H. Baumhover, and D.M. Jackson. 1989. Identification of a pheromone blend attractive to *Manduca sexta* (L.) males in a wind tunnel. *Archives of Insect Biochemistry and Physiology* 10:255–271.

Uehara, T., H. Naka, S. Matsuyama, T. Ando, T., and H. Honda. 2012. Identification and field evaluation of sex pheromones in two hawk moths *Deilephila elpenor lewisii* and *Theretra oldenlandiae oldenlandiae* (Lepidoptera: Sphingidae). *Applied Entomology and Zoology* 47:227–232.

Uehara, T., H. Naka, S. Matsuyama, T. Ando, and H. Honda. 2013. Identification of conjugated pentadecadienals as sex pheromone components of the sphingid moth, *Dolbina tancrei*. *Journal of Chemical Ecology* 39:1441–1447.

Ujvary, I., J.C. Dickens, J.A. Kamm, and L.M. McDonough. 1993. Natural product analogs: stable mimics of aldehyde pheromones. *Archives of Insect Biochemistry and Physiology* 22:393–411.

Villorbina, G., S. Rodríguez, F. Camps, and G. Fabriàs. 2003. Comparative sex pheromone biosynthesis in *Thaumetopoea pityocampa* and *T. processionea*: a rationale for the phenotypic variation in the sex pheromone within the genus *Thaumetopoea*. *Insect Biochemistry and Molecular Biology* 33:155–161.

Wahlberg, N., C.W. Wheat, and C. Peña. 2013. Timing and patterns in the taxonomic diversification of Lepidoptera (butterflies and moths). *PLOS ONE* 8:e80875.

Wakamura, S., T. Yasuda, M. Watanabe, K. Kiguchi, M. Shimoda, and T. Ando. 1996. Sex pheromone of the sweetpotato hornworm, *Agrius convolvuli* (L.) (Lepidoptera: Sphingidae): identification of a major component and its activity in a wind tunnel. *Applied Entomology and Zoology* 31:171–174.

Wang, H.L., C.H. Zhao, J.G. Millar, R.T. Cardé, and C. Löfstedt. 2010a. Biosynthesis of unusual moth pheromone components involves two different pathways in the navel orangeworm, *Amyelois transitella*. *Journal of Chemical Ecology* 36:535–547.

Wang, H.-L., M.A. Liénard, C.-H. Zhao, C.-Z. Wang, and C. Löfstedt. 2010b. Neofunctionalization in an ancestral insect desaturase lineage led to rare Δ^6 pheromone signals in the Chinese tussah silkworm. *Insect Biochemistry and Molecular Biology* 40:742–751.

Wang, H.-L., O. Brattström, P. Brakefield, W. Francke, and C. Löfstedt. 2014. Identification and biosynthesis of novel male-specific esters in the the wings of the tropical butterfly, *Bicyclus martius sanaos*. *Journal of Chemical Ecology* 40:549–559.

Wang, H.-L., H. Geertsema, E.J. van Nieukerken, and C. Löfstedt. 2015a. Identification of the female-produced sex pheromone of the leafminer *Holocacista capensis* infesting grapevine in South Africa. *Journal of Chemical Ecology* 41:724–731.

Wang, H., N. Wahlberg, J.D. Holloway, J. Bergsten, X. Fan, D.H. Janzen, W. Hallwachs, L. Wen, M. Wang, and S. Nylin. 2015b. Molecular phylogeny of Lymantriinae (Lepidoptera, Noctuoidea, Erebidae) inferred from eight gene regions. *Cladistics* 31:579–592.

Wei, W., T. Miyamoto, M. Endo, T. Murakawa, G.Q. Pu, and T. Ando. 2003. Polyunsaturated hydrocarbons in the hemolymph: biosynthetic precursors of epoxy pheromones of geometrid and arctiid moths. *Insect Biochemistry and Molecular Biology* 33:397–405.

Wei, W., M. Yamamoto, T. Asato, T. Fujii, G.Q. Pu, and T. Ando. 2004. Selectivity and neuroendocrine regulation of the precursor uptake by pheromone glands from hemolymph in geometrid female moths, which secrete epoxyalkenyl sex pheromones. *Insect Biochemistry and Molecular Biology* 34:1215–1224.

Wilson, E.O. 1975. *Sociobiology*. Cambridge, MA: Belknap Press of Harvard University Press.

Wunderer, H., K. Hansen, T.W. Bell, D. Schneider, and J. Meinwald. 1985. Sex pheromones of two Asian moths (*Creatonotos transiens*, *C. gangis*; Lepidoptera: Arctiidae): behavior, morphology, chemistry and electrophysiology. *Experimental Biology* 46: 11–27.

Yamakawa, R., N.D. Do, Y. Adachi, M. Kindo, and T. Ando. 2009. (6Z,9Z,12Z)-6,9,12-Octadecatriene and (3Z,6Z,9Z,12Z)-3,6,9,12-icosatetraene, the novel sex pheromone produced by emerald moths. *Tetrahedron Letters* 50:4738–4740.

Yamakawa, R., D.D. Nguyen, M. Kinjo, Y. Terashima, and T. Ando. 2011. Novel components of the sex pheromones produced by emerald moths: Identification, synthesis, and field evaluation. *Journal of Chemical Ecology* 37:105–113.

Yamamoto, M., T. Kamata, N.D. Do, Y. Adachi, M. Kinjo, and T. Ando. 2007. A novel lepidopteran sex pheromone produced by females of a Lithosiinae species, *Lyclene dharma dharma*, in the family of Arctiidae. *Bioscience Biotechnology and Biochemistry* 71:2860–2863.

Yamaoka, R., Y. Taniguchi, and K. Hayashiya. 1984. Bombykol biosynthesis from deuterium-labeled (Z)-11-hexadecenoic acid. *Experientia* 40:80–81.

Yang, C.Y., K.S. Choi, and M.R. Cho. 2013. (E)-5-Hexadecenyl acetate: a novel moth sex pheromone component from *Stathmopoda auriferella*. *Journal of Chemical Ecology* 39:555–558.

Zagatti, P., G. Kunesch, F. Ramiandrasoa, C. Malosse, D.R. Hall, R. Lester, and B.F. Nesbitt. 1987. Sex pheromones of rice moth, *Corcyra cephalonica* Stainton. *Journal of Chemical Ecology* 13:1561–1573.

Zahiri, R., I.J. Kitching, J.D. Lafontaine, M. Mutanen, L. Kaila, J.D. Holloway, and N. Wahlberg. 2011. A new molecular phylogeny offers hope for a stable family-level classification of the Noctuoidea (Lepidoptera). *Zoologica Scripta* 40:158–173.

Zahiri, R., J.D. Holloway, I.J. Kitching, J.D. Lafontaine, M. Mutanen, and N. Wahlberg. 2012. Molecular phylogenetics of Erebidae (Lepidoptera, Noctuoidea). *Systematic Entomology* 37:102–124.

Zaspel, J.M., S.J. Weller, C.T. Wardwell, R. Zahiri, and N. Wahlberg. 2014. Phylogeny and evolution of pharmacophagy in tiger moths (Lepidoptera: Erebidae: Arctiinae). *PLOS ONE* 9:e101975.

Zhang, H., S. Gao, M. Lercher, S. Hu, and W.-H. Cheng. 2012. EvolView, an online tool for visualizing, annotating and managing phylogenetic trees. *Nucleic Acids Research* 40 (W1): W569–W572.

Zhao, C., C. Löfstedt, and X. Wang. 1990. Sex pheromone biosynthesis in the Asian corn borer *Ostrinia furnacalis* (II): biosynthesis of (E)- and (Z)-12-tetradecenyl acetate involves $\Delta 14$ desaturation. *Archives of Insect Biochemistry and Physiology* 15:57–65.

Zhu, J., M.V. Kozlov, P. Philipp, W. Francke, and C. Löfstedt. 1995. Identification of a novel moth sex pheromone in *Eriocrania cicatricella* (Zett.) (Lepidoptera: Eriocraniidae) and its phylogenetic implications. *Journal of Chemical Ecology* 21:29–43.

CHAPTER FIVE

Sexual Selection

MICHAEL D. GREENFIELD

Introduction

One might begin this chapter by asking whether pheromone communication in moths is actually subject to sexual selection. This is not a rhetorical question, as pheromone communication in moth species generally entails a female signaler and a male receiver, an inversion of the roles in classical cases of sexual selection, and it is not immediately clear which signaler and receiver traits might be selected in the context of intra-sexual competition or mate choice. Moreover, a survey of the literature confirms the validity of this opening question, as there have been precious few studies of moth pheromones and their perception in the context of sexual selection (see Svensson 1996; Johansson and Jones 2007 for reviews). Some recent research, however, has yielded several novel findings on the role of sexual selection in shaping pheromone communication in moths. In this chapter, I summarize the typical influences of sexual selection on communication in animal species where males and females assume the roles of signaler and receiver, respectively. I then contrast these "textbook cases" with mating communication in moths, and I discuss how sexual selection might be expected to function when these gender roles are reversed. These expectations are compared with observations and experimental results in various moth species, with emphasis placed on the long-range advertisement pheromones broadcast by females in most species and perception and responses to these advertisements by

males—as well as by females. I will not treat the male pheromones that function over short distances in the context of courtship in many Lepidoptera except in the special case where these signals interface with long-range communication. Male courtship pheromones are treated elsewhere in this volume (Conner and Iyengar, this volume).

Long-Range Advertisement: Sex Roles and Signal Modalities

Long-range advertisement in mating communication implies that an individual, normally a male, broadcasts a signal that announces its species identity, gender, physiological state, and possibly "quality" (Bradbury and Vehrencamp 2011). These signals are emitted whether or not the signaler perceives the presence of potential mates, but they are produced at a time and place when and where the likelihood of receptive mating partners would be high. They are also produced and transmitted in a way that the message, i.e., the presence and characteristics of the signaler, extends over a long distance from the signaler and possibly across a wide transverse arc such that it may reach a great many potential mates. Long-range advertisements may be particularly important in species where mating does not occur at specific points in the landscape, e.g., at host plants, or where these specific points are rather common and greatly outnumber the population of signaling individuals, e.g., host plants that are abundant and widely distributed (Greenfield 1981).

The majority of long-range mating advertisement in animal species is performed by males. Most of these advertisements are transmitted along acoustic and visual channels, with lower percentages using vibrational, electrostatic, and chemical modalities. Sexual selection generates patterns in male acoustic and visual advertisements that are found across a wide range of species. Whereas certain signal characters may provide information on both species identity and individual quality, the characters that are specific for a signaler's quality tend to be those that reflect signal energy or the energy reserves of the signaler (Ryan 1988). For example, among acoustic insects, females tend to choose males whose songs are longer, delivered at a faster pulse repetition rate, continue uninterrupted for a longer time interval, or are broadcast at greater sound amplitude (Greenfield 2002). Among anurans, who have a greater ability to discriminate sound frequency than acoustic insects, females may choose males whose songs are delivered at lower than average carrier frequencies, a character that is normally associated with greater body size (Gerhardt and Huber 2002). The general focus on signal energy may arise because this feature can indicate that the male is capable of transferring or offering more direct benefits (material resources), is in better health, or that he is of "superior" genetic quality and that the offspring he sires will be of similar high quality. A female is also more likely to perceive and localize a high-energy signal, but this "passive" preference would simply select for males capable of broadcasting faster, longer, and more intense signals. In this latter case, females may not necessarily pair with a male of better quality.

Male signal characters that are subject to female choice often exhibit substantial variation along their respective "character axes." This variation may be considered as "signal space" (e.g., Amezquita et al. 2011). From the perspective of the female receiver, a corresponding "recognition space" exists and represents the range of signal character values that would elicit a sexual response. For a given population, recognition space is generally much larger than signal space, as females will often respond to synthetic signals that are faster or more intense than those produced by any male, and they may prefer these exaggerated signals—"supernormal stimuli"—over those actually transmitted by males. This disparity between signal space and recognition space implies that female choice imposes selection, generally directional but sometimes stabilizing, on the male signal. In the case where directional selection is imposed, it is normally assumed that natural selection, e.g., energy limitation or predation, is the factor preventing male signals from matching the features most preferred by females. Here, the female "preference function" tends to be "open-ended" over the range of male signal character values occurring in the population, but this function usually does not rise indefinitely as these character values increase to supernormal levels (Ritchie 1996). Rather, the preference function often attains a maximum value at a certain level of signal character exaggeration, beyond which the function declines, possibly to zero. For example, as the sound intensity of a synthetic male signal is raised, it may eventually inhibit female responses. Nonetheless, the point along the signal character axis at which the preference function reaches its maximum value is normally well beyond the range exhibited by males in the population.

In contrast to male signal characters subject to female choice, characters that serve primarily in species recognition tend to exhibit less within-population variation. The disparity between signal space and recognition space is typically smaller and more symmetrical than that observed where female choice is a factor, and this coincidence of the two spaces has led to the "matched filter" concept in neuroethology. The receiver responds to a range of signals that closely matches the range exhibited by signalers in the population (e.g., Gerhardt et al. 2007; Kostarakos et al. 2008). Selection imposed by the female receivers in this situation is generally stabilizing as opposed to directional.

As for why it is normally the male who transmits long-range signals in species where the two sexes have separate signaler and receiver roles, and why his signals are more often acoustic or visual than chemical, we can speculate based on sexual selection theory. Females are expected to assume the role in pair formation that suffers less exposure to risk because they invest more than males in offspring and their fitness usually does not increase substantially by pairing with multiple mates. Thus, the prevalence of male signaling may imply that this role is generally more dangerous than searching for mates (Alexander and Borgia 1979; Greenfield 1981). The prevalence of sound and light in male signals may reflect the physical properties of these modalities that allow a given signaler to be readily distinguished from his neighbors, a critical feature in the context of female choice and male–male competition. Were pheromones used, a given male's plume could easily be lost in the cloud emanating from a group of neighboring signalers. Whereas signals differing in chemical composition may yet be distinguished (cf. Liu and Haynes 1992; Allison and Cardé 2008), those differing only in concentration might not, and a particularly intense signaler may not enjoy an attractiveness to females commensurate with his intensity. Sound and light also have the advantages of rapid transmission over relatively long distances and some possibility for the signaler to control the direction in which the message is sent. But despite these expectations, male-emitted advertisement pheromones do occur in some insect species (e.g., *Galleria mellonella*; Lepidoptera: Pyralidae: Galleriinae; Röller et al. 1968), and in several insect orders (e.g., Lepidoptera, Trichoptera) and suborders (Hymenoptera: Symphyta) female-emitted advertisement pheromones not only occur but are also the norm. These female advertisement signals tend to be chemical rather than acoustic or visual (Wyatt 2003). Chemical signals may offer females the possibility of remaining inconspicuous to generalist predators, who might eavesdrop on the mating activities of their prey, an option that would normally not be true for acoustic and visual signaling. Whereas a predator may conceivably evolve recognition of the sex pheromone of a specific prey species, recognition of the range of chemical compounds representing the pheromones of a broader group of species and discriminating these compounds from other odors in the environment seems unlikely (cf. Hemmann et al. 2008 on the trade-off between sensitivity and breadth of response).

Pair Formation in Moths: What Is the Opportunity for Sexual Selection?

As noted above, female moths advertise to males via long-range pheromones. Various studies suggest that these female signals function in the context of species recognition and promote pre-mating reproductive isolation between closely related species. For example, pheromones of congeneric species normally differ in one or more chemical characteristics except in those cases where additional mechanisms exist to isolate species (e.g., differences in diel patterns of activity). These studies have traditionally focused on population-level characteristics of signal and receiver traits while ignoring variation among individuals in female pheromones and male preference functions. Beginning in the 1980s, with the advent of improved technology that offers more sensitive measurement of pheromone composition from individual females, it became feasible to quantify these traits with precision, and several studies have now quantified the ratio of the components of the pheromone blend released by a given individual at a given age and time of day (e.g., Collins and Cardé 1985; Barrer et al. 1987; Du et al. 1987; Svensson et al. 1997; Allison and Cardé 2006, 2007). Similarly, tests of the responses of individual males to various blend ratios were conducted allowing measurement of male preference function shape, i.e., the relative responses exhibited along a gradient of blend ratios (e.g., Collins and Cardé 1989). In the almond moth, signal and receiver traits of individual females and males were determined and then used to examine and compare signal and receiver space within a population (Allison and Cardé 2008). As expected, signal space was narrower than receiver space. Moreover, when male preference function shape was determined via choice tests, the unimodal shape and position of the function optima relative to the distribution of female pheromone signals suggested the occurrence of a matched filter and that males impose stabilizing selection on the female pheromone. However, when determined via single-stimulus presentations, the male preference function was flat across a range of blend ratios. Overall, these findings appear to answer the opening question of this chapter. There was no evidence that female advertisement pheromones in almond moths are subject to sexual selection other than that imposed in a very broad sense, which would include species recognition. Thus, to continue our inquiry on sexual selection in moths, we must address the issue from the special perspective of the sex role reversal in pair formation in this group.

When males search and females advertise, male sensitivity to females and their ability to search for and find them may represent traits subject to strong intra-sexual selection. That is, male orientation toward calling females may take the place of the signal competition and physical aggression that normally occur in species where males advertise (Greenfield 1981). In moths the neurophysiological and behavioral sensitivity of male responses to conspecific female pheromone has been well documented, but less attention has been paid to male searching strategies during appetitive flight toward females. That is, do males follow particular orientations with respect to local wind such that they would maximize their rate of encounter with pheromone plumes of females? This question has been difficult to address because most moth species are nocturnal, rendering the monitoring of flight paths difficult in nature. A recent study of a diurnal species (*Virbia lamae*; Erebidae: Arctiinae) circumvented this problem, though, and showed that flight paths were oriented randomly with respect to local wind (Cardé et al. 2012). While this finding might seem to refute the suggestion that searching trajectories maximize interceptions of female pheromone plumes, further analysis by the authors of the study indicates that random orientations could actually increase the encounter rate provided that plumes extend farther downwind than they spread crosswind.

Once males arrive at a signaling female, the possibility exists that several males have been attracted at the same time and that some form of competition ensues. But a survey of the literature on moths indicates little evidence of this general prediction of agonistic behavior. Rather, the basic protocol of racing toward receptive females appears to hold, and male moths may even interrupt their approach to a female if they detect the presence of other males who have already arrived (Hirai et al. 1978; cf. Bijpost et al. 1985; Fitzpatrick et al. 1988; Klun and Graf 1997). This presence may be revealed by the courtship pheromone of the earlier arrivals. One might infer that it would be preferable for a male to search for another female under these circumstances provided that a sufficient number of such mating opportunities exist. Other considerations would include the male's expected survival and the rapidity with which mating ensues following courtship. Should mating proceed quickly, the presence of other males adjacent to a female may generally imply that she will have mated by the time the focal male pinpoints her location and alights. Because female moths typically enter a refractory period after mating or do not re-mate at all, pursuing a courtship under such competition would seldom be worthwhile.

Are Female Sex Pheromones Subject to Sexual Selection?

Are female advertisement signals in role-reversed species also subject to a particular form of sexual selection? Unlike males, females are generally not expected to evolve traits that greatly increase the number of males they attract. Females normally mate relatively few times, and in some species only once. Even in the context of a need to encounter multiple males in order to choose a "superior" individual, an intense advertisement signal would not be expected given the chance that natural enemies may be attracted by the signal or the throng of suitors. Nonetheless, we may ask if there are other ways in which females could signal such that the possibility of attracting superior males is increased. For example, females could select for superior searchers by emitting a low, rather than high, pheromone titer (Lloyd 1979; Greenfield 1981).

It is tempting to propose a low-intensity filtering mechanism as described above for female advertisement pheromones in moths. Pheromone release rates are generally quite low, and male sensitivity and odor tracking ability are impressive, if not legendary. However, further analysis of this purported mechanism encounters several complications. Males who are adept at tracking pheromone plumes will find high-intensity sources as well as low-intensity ones, although on average the males arriving at low-intensity sources should be slightly more sensitive and better trackers. But from the female perspective, various selection pressures in addition to filtering out inferior males may influence pheromone release rate. Whereas females may not be expected to broadcast an exaggerated pheromone signal, they do have to mate at least once, and their generally brief reproductive life span dictates that they have to do so sooner rather than later (Michereff et al. 2004; Stelinski and Gut 2009). This factor may impose an opposing selection pressure that maintains a release rate above some minimum threshold. The low release rates observed may also reflect metabolic limitations (e.g., Foster 2009). Although female advertisement pheromones in moths may not directly reflect specific food substances acquired during larval development (e.g., Miller et al. 1976), a feature that is common in male courtship pheromones, the specific compounds released in the female signal represent the end products of certain biochemical pathways in secondary metabolism (Tillman et al. 1999). Presumably, the specificity of these compounds represents a response to selection pressure for species recognition, but does this specificity then necessitate a reduction in the quantity of pheromone that can be produced and released? Thus, the proposed filtering by low-intensity female signals remains only a hypothesis, and one that is not easily disentangled from other factors influencing pheromone production.

Sexual Selection: How Does Male Choice Arise and Influence the Gambit?

Analysis of sexual selection has traditionally assumed that males compete and females choose. However, more and more studies continue to reveal that variations on this theme exist and that under certain circumstances males may choose among competing females, sometimes in very subtle ways (Bonduriansky 2001). One circumstance under which male choice can arise is when a male's contribution to his mate and offspring extends well beyond the genes in his sperm. Such "material" contribution is likely to be limiting for a male, which would select for some level of discrimination of the "quality" of mating partners. Here, quality may be the number of mature or maturing eggs the female has and that can be fertilized, the potential ability of the female to nourish and care for the offspring, and the likelihood that the female will remate in the near future with another male whose sperm would have precedence in fertilizing her eggs. Other features of quality may involve aspects of genetic compatibility. In insects, the most likely indications of a female's quality are her size and age, and a male might assess these via certain indirect cues and signals. For example, the intensity of a mating signal could represent a reliable proxy for size or general "condition."

Should male contributions to their mating partners and offspring be of sufficient magnitude and importance, some level of competition among females may occur. Such competition may be particularly likely in cases where male choice occurs. Female competition could include active searching for males or, in the event that the female is the sex who advertises, "signal interactions" that feature specialized responses to the advertisements of neighbors. For example, the focal female may increase the intensity, rhythm, or other features of her advertisement (e.g., Jia et al. 2001) or adjust the timing of her signals with respect to her neighbors (Greenfield 2005). And in some cases, overt aggression between females might conceivably arise.

Does male choice occur in moths, and can it generate selection on female advertisement pheromones? To answer this question, I note that several studies on moth species report that males do mate preferentially with larger females, who are expected to be more fecund (e.g., Van Dongen et al. 1998; Jaffe et al. 2007). Moreover, in many Lepidoptera males package compounds in their spermatophore that may influence the survival of the female who receives them, as well as the number and survival of sired offspring (e.g., Svard and McNeil 1994; Cook and Wedell 1996; see South and Lewis 2011 for review). Thus, we may expect that females in some species have evolved mechanisms with which they could compete for males and their material contributions. Such competition could entail enhancing the release of pheromone as well as attention to the pheromone released by neighboring females in the context of signal competition. However, both of these possibilities run counter to the standard paradigm of adver-

tisement pheromones in moths: female signals that are generally released in relatively low concentrations and perceived only by males.

Female–Female Signal Competition

Case Study of *Utetheisa ornatrix*

Standard paradigms may be retained because alternatives are not imagined, or because the alternatives when imagined threaten an established set of principles and explanations with messy complications. In the case of moth pheromones, why would females detect, let alone respond to, conspecific female pheromone? Thus, the vast majority of behavioral and neurophysiological studies of responses to female advertisement pheromones in moths have focused on males and ignored females. Nonetheless, various reports have trickled in since the late 1970s indicating that a certain level of responsiveness may exist in females as well as males. Most of the reports concern antennal electrophysiology (e.g., Cook and Shelton 1978; Ljungberg et al. 1993; Schneider et al. 1998; Kalinová et al. 2001; Groot et al. 2005; Hillier et al. 2006; Stelinski et al. 2006; Gokce et al. 2007; Yang et al. 2009), but some studies have also addressed behavior. The latter include findings on modifications in signaling as well as in spacing among neighboring females. Interestingly, one of the electrophysiological studies (Grant and O'Connell 2000) reported findings in an arctiine moth (Erebidae) where males were known to make a substantial material contribution to the female and her eggs. This arctiine species, *Utetheisa ornatrix*, has been a model for research on the chemical ecology of sexual selection (Eisner and Meinwald 1995; Iyengar and Conner, this volume). Male *U. ornatrix* transfer defensive compounds, obtained from the host plant during larval development, in their spermatophore. These compounds increase the survival of females and their eggs by rendering them less palatable to predators (Dussourd et al. 1988; Iyengar and Eisner 1999). Could a connection exist between the female electrophysiological response to female advertisement pheromone and the male investment?

The above question launched a project that concentrated on the fundamental issue of this chapter: do female advertisement pheromones in moths exhibit traits that have evolved under the pressure of sexual selection? We decided to pursue this issue with *U. ornatrix*, and we organized the investigation in four parts:

1. Do neighboring *U. ornatrix* females exhibit signal interactions while they are releasing pheromone, and do such interactions reflect female–female competition?
2. If signal interactions are observed between neighboring females, what cues and signals are responsible for this behavior?
3. Are observed signal interactions in females associated with specific movement and spacing among neighboring individuals?
4. How do signaling *U. ornatrix* females interact with one another in natural populations in the field?

Our study of potential signal interactions among female *U. ornatrix* was motivated by earlier observations that when multiple females are held collectively with multiple males in the same cage, a given female is more likely to signal and to

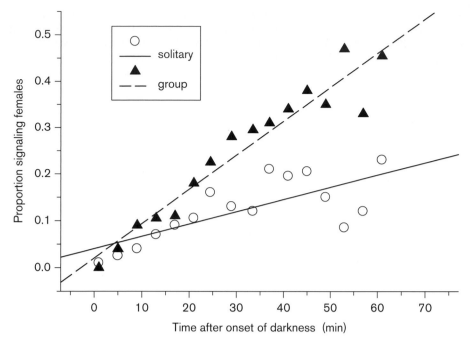

FIGURE 5.1 Proportion of *Utetheisa ornatrix* females signaling when olfactory contact with neighboring females was permitted (group: dashed line and triangles) and when such contact was prevented (solitary: solid line and circles). Observations were taken at 4-min intervals following the beginning of the night in a laboratory population (adapted from Greenfield 2005 and Lim and Greenfield 2007).

mate than a female held individually with a single male. This enhanced mating rate did not result from forced copulation. We thus designed an experiment to test the influence of neighboring females on the signaling activity of a focal female. The study was conducted in the laboratory and relied on cages that could limit the transmission of olfactory and/or visual cues and signals of neighbors from reaching the focal female; acoustic recording did not reveal airborne sound production in *U. ornatrix*, and the cages blocked the transmission of any potential substrate vibration. Olfactory signaling was readily scored because female *U. ornatrix* pump their abdomen rhythmically when releasing pheromone (Conner et al. 1980) and release very little or no pheromone when not pumping. By monitoring this special activity, which serves to extrude the location from which pheromone is released at the tip of the abdomen, we determined that females were more likely to signal when in the presence of olfactory cues and signals of neighboring females as opposed to visual ones or none at all. The heightened signaling effort included (1) an earlier onset of pumping at the beginning of the nightly activity period, (2) a higher proportion of females pumping at any given time during the activity period, and (3) a faster pumping rhythm (figure 5.1). These exaggerations are typical of those observed in males in acoustic species such as Orthoptera and anurans, where the phenomenon is termed chorusing. Thus, we dubbed the observed group phenomenon "female pheromonal chorusing" (Lim and Greenfield 2007).

In acoustic species, it is hypothesized that chorusing interactions often arise because males must equal or exceed the signaling of their neighbors in order to achieve mating success (e.g., Greenfield and Minckley 1993; Gerhardt et al. 2000; Jia et al. 2001). Would the same principle hold in *U. ornatrix* females? An earlier study reported that female *U. ornatrix* choose males who can transfer a greater quantity of defensive compounds, and that female choice is mediated by male

courtship pheromone titer (Iyengar et al. 2001). In our own study, we observed that the operational sex ratio in our laboratory population of *U. ornatrix* was female biased. That is, at any given time the number of sexually receptive females normally exceeded the number of sexually active males, and we inferred that this bias arose because (1) females sought to mate with multiple males in order to obtain a large quantity of defensive compounds and (2) males became refractory for several days following a mating because they could not immediately form another large spermatophore. These findings and inferences suggest that female *U. ornatrix* might have to compete for sexually active males, and pheromonal chorusing might be their means of doing so (Lim and Greenfield 2007).

We did not observe all of the features typical of chorusing in our study of *U. ornatrix*. In acoustic insects and anurans chorusing typically entails precise temporal relationships between the calls of neighboring males, i.e., displays of collective synchrony or alternation (Greenfield 2005). Although individual *U. ornatrix* females display a signaling rhythm in which they pump their abdomen at approximately 100 extrusions·min[-1], no coordinated temporal interactions between female neighbors are observed other than an advancement of the nightly onset of signaling. This advancement occurs because once one individual begins signaling, neighboring females follow suit within several min. The absence of fine-level temporal structure most probably reflects the turbulent diffusion of pheromone in air, which would generally preclude a female from perceiving a neighbor's rhythm: except directly downstream from the signaling female, where a pulsed-pheromone plume might retain its integrity over a moderate distance and time interval, temporal fluctuations in concentration at the pheromone source are likely obscured beyond a very short range such that a continuous concentration would be perceived several centimeter distant (Greenfield 2002). It might also be expected that male *U. ornatrix* would exhibit some level of mate choice, thereby increasing competition among females. However, studies in another laboratory (Iyengar and Eisner 2004) found no evidence of male choice of females in *U. ornatrix*, and suggested that the short nightly activity period constrained males from discriminating among females.

Having established the existence of pheromonal chorusing in *U. ornatrix* females, we then sought to determine the specific olfactory stimuli that elicited this collective behavior. Given that females exhibit electrophysiological responses to female pheromone, we hypothesized that the behavioral responses observed were made to the pheromone or its components, but the possibility remained that chorusing was elicited by responses to some general body odor of neighboring females. Tests with whole pheromone extracted from females and with individual synthetic components of the pheromone showed that the sex pheromone was the responsible stimulus. Individual behavior that generated chorusing in groups of females was specifically elicited by the extracted pheromone or by either one of two specific pheromone components from among the five known ones (Choi et al. 2007; Lim et al. 2007). One feature of the individual behavior of chorusing females is that they pump their abdomen at a higher rate than when signaling in solo; however, we did not find that a faster pumping rate was correlated with a higher amount of released pheromone (Lim et al. 2007). This surprising result may reflect technical limitations in quantifying pheromone concentration.

When males in acoustic animal species chorus, a spatial component of the collective behavior is often present. Signaling males may be passively aggregated because of a scarcity of preferred habitat, but active aggregation in which males show mutual attraction can also occur. The latter may reflect "lekking" behavior that arises when females tend to approach male groups in order to avail themselves of the opportunity to make simultaneous, as opposed to sequential, comparisons among available mates (Hoglund and Alatalo 1995). Thus, males would avoid signaling alone regardless of their signaling prowess. Lekking can also arise because inferior male signalers may be attracted to the vicinity of superior ones (Beehler and Foster 1988). We therefore asked whether *U. ornatrix* females showed some level of active aggregation when they signal, and whether the same female pheromonal stimuli that elicit collective signaling at the beginning of the night also elicit grouping in space. We conducted a series of laboratory olfaction tests with a Y-tube apparatus and, as above, found that whole pheromone extracted from females and several synthetic components of the pheromone influenced females to move, orient toward the odor stimulus, and aggregate nearby. The female pheromonal stimuli had similar effects on males, except that males were somewhat more sensitive than females (Lim and Greenfield 2008).

The litmus test for evaluating the role of behavior is its occurrence in natural populations in the field, and we therefore examined whether *U. ornatrix* females exhibit pheromone-mediated aggregation in the wild. Populations of *U. ornatrix* had been studied for many years in the vicinity of the Archbold Biological Station in Highlands County, Florida, and some reports noted aggregations of adults. However, adults are probably attracted to their host plant (*Crotalaria mucronata*), and these observations per se could not be taken as evidence of active aggregation. When cages containing either six virgin *U. ornatrix* females, most of whom signaled, or no females at all (empty "control" cages) were placed in the field, we found a higher number of mating pairs adjacent to the female cages than the controls. Moreover, the mating pairs at the female cages were invariably situated on the downwind side, whereas those pairs at the control cages were distributed in all directions (Lim 2006). Thus, some level of active female aggregation, probably mediated by female pheromone, was apparent. The per female mating success may be higher in small aggregations than for solitary females, but presumably this advantage peaks and then declines as aggregations increase beyond a certain number or density. But as for why *U. ornatrix* females might gather in aggregations, this behavior remains enigmatic given that males do not appear to choose among local females.

Overview

Are the pheromone-mediated behaviors in *Utetheisa ornatrix* that appear to reflect sexual selection special traits only representative of species with exceptional biology? Several recent reports suggest they may not be. Once biologists recognized that the female signaler/male receiver paradigm for moth communication might not characterize all interchanges that occur among conspecifics, female perception of conspecific female pheromone was tested in various species, and examples of female detection of conspecific pheromone began to accumulate (Table 5.1). In some moth species, females were found to advance or otherwise increase their daily emission of pheromone when exposed to females or their pheromone (Palaniswamy and Seabrook 1978, 1985; Den Otter et al. 1996; Stelinski et al. 2006; Harari et al. 2011; Sadek et al. 2012), as in *U. ornatrix*, but in other species delays in emission were observed (Noguchi and Tamaki 1985; Gokce et al. 2007; Yang et al. 2009). Similarly, exposure to females or their pheromone elicited female aggregation in some species (Den Otter et al. 1996), whereas increased spacing resulted in others (Saad and Scott 1981). These divergent results might reflect the relative importance of clustering in mate attraction versus competition between individual signalers. But without more detailed information on the biology of each species, adaptive interpretations of the results would be overly speculative. A more systematic survey of these behaviors among moth clades and species would be needed to contend with the small number of studies and the possibility that publication biases influenced those studies that appeared in print (Table 5.1).

Unlike the situation in *U. ornatrix*, one of the above studies reporting female responses to female pheromone also noted the occurrence of male choice. Males in the olethreutine tortricid moth *Lobesia botrana* prefer to mate with larger females, who are more fecund, and laboratory trials indicate that they discern female size by evaluating the titer of emitted pheromone (Harari et al. 2011), an ability that will have to be confirmed under more realistic circumstances. Female *L. botrana*, for their part, increase their pheromone emission when in the company of neighboring females, but they appear to pay a cost in terms of reduced oviposition and longevity for their greater signaling effort. Possibly, the selection pressure imposed by male choice favors females who opt for gaining a current mating at the expense of lower fecundity and fewer mating opportunities in the future. *L. botrana* males transfer a sizeable spermatophore, and the potential value of this transfer to females may influence their competitiveness. Importantly, the *L. botrana* study demonstrates that female pheromone signals may be costly under certain circumstances, notably in cases where females are selected to broad-

TABLE 5.I

Female–female signal competition in moths

The entries represent cases in which females are reported to respond neurally or behaviorally to the sexual advertisement pheromone emitted by conspecific females

Family	Species	EAG response[1]	Timing of pheromone emission[2]	Spacing[3]	References[4]
Gelechiidae	*Pectinophora gossypiella*	+			1
Tortricidae	*Adoxophyes* sp.		−		13
	Argyrotaenia velutinana	+	−		3
	Choristoneura fumiferana		+		14, 15
	Choristoneura rosaceana	+	−		3
	Cydia fagiglandana		+	+	2
	Cydia splendana		+	+	2
	Grapholita molesta	+	+		19
	Homona magnanima		−		13
	Lobesia botrana		+		6
Sphingidae	*Manduca sexta*	+			8
Erebidae, Arctiinae	*Panaxia quadripunctaria*	+			18
	Utetheisa ornatrix	+	+	+	4, 9, 10, 11
Noctuidae	*Heliothis armigera*			−	16
	Heliothis subflexa	+			5
	Heliothis virescens	+			7
	Heliothis zea		−		17
	Spodoptera exigua	+	−		20
	Spodoptera littoralis	+	+		12, 17

1. (+) Electroantennogram response to advertisement pheromone of conspecific females confirmed.

2. (+) Female emission of advertisement pheromone is advanced or otherwise increased in the presence of conspecific pheromone or pheromone components. (−) female emission of pheromone is delayed or otherwise diminished in the presence of conspecific pheromone.

3. (+) Conspecific pheromone or pheromone components elicits mutual attraction of females. (−) conspecific pheromone or pheromone components influence females to move apart from one another.

4. (1) Cook and Shelton 1978; (2) Den Otter et al. 1996; (3) Gokce et al. 2007; (4) Grant and O'Connell 2000; (5) Groot et al. 2005; (6) Harari et al. 2011; (7) Hillier et al. 2006; (8) Kalinová et al. 2001; (9) Lim and Greenfield 2007; (10) Lim et al. 2007; (11) Lim and Greenfield 2008; (12) Ljungberg et al. 1993; (13) Noguchi and Tamaki 1985; (14) Palaniswamy and Seabrook 1978; (15) Palaniswamy and Seabrook 1985; (16) Saad and Scott 1981; (17) Sadek et al. 2012; (18) Schneider et al. 1998; (19) Stelinski et al. 2006; (20) Yang et al. 2009.

cast a reliable indication of their reproductive potential to choosy males.

Summary

Most moths reproduce sexually, and thus some level of sexual selection is expected to influence the various aspects of their pair formation, including the broadcast of and response to female sex pheromones. But sexual selection in moths with respect to female pheromones has seldom been examined, and biologists may have assumed that these signals function largely within the realm of species recognition. The general scheme of female signalers and male receivers found in moths does not fit the classical model of female choice and male competition. Moreover, this reversed scheme may obscure those traits of the female signals that have been shaped by sexual selection. Nonetheless, some recent analyses do reveal the subtle ways in which female pheromonal signals are subject to sexual selection, especially in species where males make substantial material contributions to females and offspring, factors that can promote competition among females for males. The most evident competitive trait observed is the behavioral response in which females accelerate the onset and output of their pheromone emission when neighboring females are present and themselves broadcasting sex pheromone.

At a fundamental level, female–female signal competition in moths derives from the ability of females to perceive conspecific sex pheromone. This ability also falls outside the classical model in which only males were assumed to perceive and respond to the female signals. Thus, a full understanding of sexual selection of female pheromones in moths must initially account for this perceptual ability. Because there is no obvious a priori explanation for female response to female pheromones, one could suppose that the perception exists because of relatively low sex-specific expression of a trait that is strongly favored by selection in males. If so, would this rationale imply that female perception of conspecific female pheromone is a widespread, perhaps ancestral, condition among moths, and that female–female competition then evolved in those species where ecological and behavioral circumstances strongly favored it, as where male choice occurs? Possibly, but an alternative scenario is that the sex-specific aspect of expression is weaker, i.e., female perception of pheromone is stronger, in some moth clades and species than others for any number of reasons, and it is in these species that female–female competition, such as pheromonal chorusing, evolved more readily. And, it is also possible that the selection pressure from ecological and behavioral circumstances was sufficiently strong in some species that it favored evolution of both the requisite perceptual machinery in females and female–female competition. At present, we lack the information for discriminating among these possibilities. Our questions on the origins of female–female perception and competition are yet another example of the importance of understanding neurobiology and mechanism in addressing problems in the evolution of behavior. As shown in other chapters of this volume, progress will depend on interdisciplinary studies.

Acknowledgments

I am grateful to Hang-Kyo Lim for having pursued the question of female–female signal competition in the arctiine moth *Utetheisa ornatrix* for his PhD dissertation in my laboratory. This case study using *U. ornatrix* benefited greatly from the general advice of William E. Conner, Mark Deyrup, and Craig LaMunyon, field and laboratory assistance by Nadine Appenbrink, Katrina M. Larson, Clarissa E. Owen, and Anthony M. Swatek, help in rearing host plants by Kathleen Nus, and work by Thomas C. Baker, Man-Yeon Choi, Russell Jurenka, and Kye-Chung Park on pheromone chemistry. I thank Ali Harari for general discussions that led to the synthesis presented in this chapter and Jeremy Allison and Ring Cardé for their expert editorial advice.

References Cited

Alexander, R.D., and G. Borgia. 1979. On the origin and basis of the male–female phenomenon. Pp. 417–440. In M.S. Blum and N.A. Blum, eds. *Sexual Selection and Reproductive Competition in Insects.* New York: Academic Press.

Allison, J.D., and R.T. Cardé. 2006. Heritable variation in the sex pheromone of the almond moth, *Cadra cautella. Journal of Chemical Ecology* 32:621–641.

Allison, J.D., and R.T. Cardé. 2007. Bidirectional selection for novel pheromone blend ratios in the almond moth, *Cadra cautella. Journal of Chemical Ecology* 33:2293–2307.

Allison, J.D., and R.T. Cardé. 2008. Male pheromone blend preference function measured in choice and no-choice wind tunnel trials with almond moths, *Cadra cautella. Animal Behaviour* 75:259–266.

Amezquita, A., S.V. Flechas, A.P. Lima, H. Gasser, and W. Hodl. 2011. Acoustic interference and recognition space within a complex assemblage of dendrobatid frogs. *Proceedings of the National Academy of Sciences of the United States of America* 108:17058–17063.

Barrer, P.M., M.J. Lacey, and A. Shani. 1987. Variation in relative quantities of airborne sex pheromone components from individual female *Ephestia cautella* (Lepidoptera, Pyralidae). *Journal of Chemical Ecology* 13:639–653.

Beehler, B.M., and M.S. Foster. 1988. Hotshots, hotspots, and female preference in the organization of Lek mating system. *The American Naturalist* 131:203–219.

Bijpost, S.C.A., G. Thomas, and J.P. Kruijt. 1985. Olfactory interactions between sexually active males in *Adoxophyes orana* (Fvr) (Lepidoptera, Tortricidae). *Behaviour* 95:121–137.

Bonduriansky, R. 2001. The evolution of male mate choice in insects: a synthesis of ideas and evidence. *Biological Reviews* 76:305–339.

Bradbury, J.W., and S.L. Vehrencamp. 2011. *Principles of Animal Communication.* 2nd ed. Sunderland, MA: Sinauer Associates.

Cardé, R.T., A.M. Cardé, and R.D. Girling. 2012. Observations on the flight paths of the day-flying moth *Virbia lamae* during periods of mate location: do males have a strategy for contacting the pheromone plume? *Journal of Animal Ecology* 81:268–276.

Choi, M.-Y., H.-K. Lim, K.C. Park, R. Adlof, S. Wang, A. Zhang, and R. Jurenka. 2007. Identification and biosynthetic studies of the hydrocarbon sex pheromone in *Utetheisa ornatrix. Journal of Chemical Ecology* 33:1336–1345.

Collins, R.D., and R.T. Cardé. 1985. Variation in and heritability of aspects of pheromone production in the pink bollworm moth, *Pectinophora gossypiella* (Lepidoptera, Gelechiidae). *Annals of the Entomological Society of America* 78:229–234.

Collins, R.D., and R.T. Cardé. 1989. Heritable variation in pheromone response of the pink bollworm, *Pectinophora gossypiella* (Lepidoptera, Gelechiidae). *Journal of Chemical Ecology* 15:2647–2659.

Conner, W.E., T. Eisner, R.K. Vander Meer, A. Guerrero, D. Ghiringelli, and J. Meinwald. 1980. Sex attractant of an arctiid moth (*Utetheisa ornatrix*): a pulsed chemical signal. *Behavioral Ecology and Sociobiology* 7:55–63.

Cook, B.J., and W.D. Shelton. 1978. Antennal responses of the pink bollworm to gossyplure. *The Southwestern Entomologist* 3:141–146.

Cook, P.A., and N. Wedell. 1996. Ejaculate dynamics in butterflies: a strategy for maximizing fertilization success? *Proceedings of the Royal Society of London: B* 263:1047–1051.

Den Otter, C.J., A. De Cristofaro, K.E. Voskamp, and G. Rotundo. 1996. Electrophysiological and behavioural responses of chestnut moths, *Cydia fagiglandana* and *C. splendana* (Lep., Tortricidae), to sex attractants and odours of host plants. *Journal of Applied Entomology* 120:413–421.

Du, J.W., C. Lofstedt, and J. Lofqvist. 1987. Repeatability of pheromone emissions from individual female ermine moths *Yponomeuta padellus* and *Yponomeuta rorellus. Journal of Chemical Ecology* 13:1431–1441.

Dussourd, D.E., K. Ubik, C. Harvis, J. Resch, J. Meinwald, and T. Eisner. 1988. Defense mechanisms of arthropods. 86. Biparental defensive endowment of eggs with acquired plant alkaloid in the moth *Utetheisa ornatrix. Proceedings of the National Academy of Sciences of the United States of America* 85:5992–5996.

Eisner, T., and J. Meinwald. 1995. The chemistry of sexual selection. *Proceedings of the National Academy of Sciences of the United States of America* 92:50–55.

Fitzpatrick, S.M., J.N. McNeil, and S. Dumont. 1988. Does male pheromone effectively inhibit competition among courting true armyworm males (Lepidoptera, Noctuidae). *Animal Behaviour* 36:1831–1835.

Foster, S. 2009. Sugar feeding via trehalose haemolymph concentration affects sex pheromone production in mated *Heliothis virescens* moths. *Journal of Experimental Biology* 212:2789–2794.

Gerhardt, H.C., and F. Huber. 2002. *Acoustic Communication in Insects and Anurans: Common Problems and Diverse Solutions.* Chicago, IL: The University of Chicago Press.

Gerhardt, H.C., J.D. Roberts, M.A. Bee, and J.J. Schwartz. 2000. Call matching in the quacking frog (*Crinia georgiana*). *Behavioral Ecology and Sociobiology* 48:243–251.

Gerharct, H.C., C.C. Martinez-Rivera, J.J. Schwartz, V.T. Marshall, and C.G. Murphy. 2007. Preferences based on spectral differences in acoustic signals in four species of treefrogs (Anura: Hylidae). *Journal of Experimental Biology* 210:2990–2998.

Gokce, A., L.L. Stelinski, L.J. Gut, and M.E. Whalon. 2007. Comparative behavioral and EAG responses of female oblique-banded and redbanded leafroller moths (Lepidoptera: Tortricidae) to their sex pheromone components. *European Journal of Entomology* 104:187–194.

Grant, A.J., and R.J. O'Connell. 2000. Responses of olfactory receptor neurons in *Utetheisa ornatrix* to gender-specific odors. *Journal of Comparative Physiology: A* 186:535–542.

Greenfield, M.D. 1981. Moth sex pheromones: an evolutionary perspective. *Florida Entomologist* 64:4–17.

Greenfield, M.D. 2002. *Signalers and Receivers: Mechanisms and Evolution of Arthropod Communication*. Oxford: Oxford University Press.

Greenfield, M.D. 2005. Mechanisms and evolution of communal sexual displays in arthropods and anurans. *Advances in the Study of Behavior* 35:1–62.

Greenfield, M.D., and R.L. Minckley. 1993. Acoustic dueling in tarbush grasshoppers: settlement of territorial contests via alternation of reliable signals. *Ethology* 95:309–326.

Groot, A., C. Gemeno, C. Brownie, F. Gould, and C. Schal. 2005. Male and female antennal responses in *Heliothis virescens* and *H. subflexa* to conspecific and heterospecific sex pheromone compounds. *Environmental Entomology* 34:256–263.

Harari, A., T. Zahavi, and D. Thiery. 2011. Fitness cost of pheromone production in signaling female moths. *Evolution* 65: 1572–1582.

Hemmann, D.J., J.D. Allison, and K.F. Haynes. 2008. Tradeoff between sensitivity and specificity in the cabbage looper moth response to sex pheromone. *Journal of Chemical Ecology* 34:1476–1486.

Hillier, N.K., C. Kleineidam, and N.J. Vickers. 2006. Physiology and glomerular projections of olfactory receptor neurons on the antenna of female *Heliothis virescens* (Lepidoptera: Noctuidae) responsive to behaviorally relevant odors. *Journal of Comparative Physiology A* 192:199–219.

Hirai, K., H.H. Shorey, and L.K. Gaston. 1978. Competition among courting male moths: male-to-male inhibitory pheromone. *Science* 202:644–645.

Hoglund, J., and R.V. Alatalo. 1995. *Leks*. Princeton, NJ: Princeton University Press.

Iyengar, V.K., and T. Eisner. 1999. Female choice increases offspring fitness in an arctiid moth (*Utetheisa ornatrix*). *Proceedings of the National Academy of Sciences of the United States of America* 96:15013–15016.

Iyengar, V.K., and T. Eisner. 2004. Male indifference to female traits in an arctiid moth (*Utetheisa ornatrix*). *Ecological Entomology* 29:281–284.

Iyengar, V.K., C. Rossini, and T. Eisner. 2001. Precopulatory assessment of male quality in an arctiid moth (*Utetheisa ornatrix*). *Behavioral Ecology and Sociobiology* 49:283–288.

Jaffe, K., B. Miras, and A. Cabrera. 2007. Mate selection in the moth *Neoleucinodes elegantalis*: evidence for a supernormal chemical stimulus in sexual attraction. *Animal Behaviour* 73:727–734.

Jia, F.-Y., M.D. Greenfield, and R.D. Collins. 2001. Ultrasonic signal competition between male wax moths. *Journal of Insect Behavior* 14:19–33.

Johansson, B.G., and T.M. Jones. 2007. The role of chemical communication in mate choice. *Biological Reviews* 82:265–289.

Kalinová, B., M. Hoskovec, I. Liblikas, C.R. Unelius, and B.S. Hansson. 2001. Detection of sex pheromone components in *Manduca sexta* (L.). *Chemical Senses* 26:1175–1186.

Klun, J.A., and J.C. Graf. 1997. Contextual chemical ecology: male to male interactions influence European corn borer (Lepidoptera: Pyralidae) male behavioral response to female sex pheromone in a flight tunnel. *Journal of Entomological Science* 32:472–477.

Kostarakos, K., M. Hartbauer, and H. Römer. 2008. Matched filters, mate choice and the evolution of sexually selected traits. *PLOS ONE* 3(8), Article Number: e3005.

Lim, H.-K. 2006. Female pheromonal chorusing and aggregation in an arctiid moth, *Utetheisa ornatrix*. PhD thesis, University of Kansas, Lawrence.

Lim, H.-K, and M.D. Greenfield. 2007. Female pheromonal chorusing in an arctiid moth, *Utetheisa ornatrix*. *Behavioral Ecology* 18:165–173.

Lim, H.-K., and M.D. Greenfield. 2008. Female arctiid moths, *Utetheisa ornatrix*, orient towards and join pheromonal choruses. *Animal Behaviour* 75:673–680.

Lim, H.-K, K.C. Park, T.C. Baker, and M.D. Greenfield. 2007. Perception of conspecific female pheromone stimulates female calling in an arctiid moth, *Utetheisa ornatrix*. *Journal of Chemical Ecology* 33:1257–1271.

Liu, Y.B., and K.F. Haynes. 1992. Filamentous nature of pheromone plumes protects integrity of signal from background chemical noise in cabbage looper moth, *Trichoplusia ni*. *Journal of Chemical Ecology* 18:299–307.

Ljungberg, H., P. Anderson, and B.S. Hansson. 1993. Physiology and morphology of pheromone-specific sensilla on the antennae of male and female *Spodoptera littoralis* (Lepidoptera: Noctuidae). *Journal of Insect Physiology* 39:253–260.

Lloyd, J.E. 1979. Sexual selection in luminescent beetles. Pp. 293–342. In M.S. Blum and N.A. Blum, eds. *Sexual Selection and Reproductive Competition in Insects*. New York: Academic Press.

Michereff, M.F.F., E.F. Vilela, M. Michereff, D.M.S. Nery, and J.T. Thiebaut. 2004. Effects of delayed mating and male mating history on the reproductive potential of *Leucoptera coffeella* (Lepidoptera: Lyonetiidae). *Agricultural and Forest Entomology* 6:241–247.

Miller, J.R., T.C. Baker, R.T. Cardé, and W.L. Roelofs. 1976. Reinvestigation of oak leaf roller sex pheromone components and hypothesis that they vary with diet. *Science* 192:140–143.

Noguchi, H., and Y. Tamaki. 1985. Conspecific female sex pheromone delays calling behavior of *Adoxophyes* sp. and *Homona magnanima* (Lepidoptera: Tortricidae). *Japanese Journal of Applied Entomology and Zoology* 29:113–118.

Palaniswamy, P., and W.D. Seabrook. 1978. Behavioral responses of the female eastern spruce budworm *Choristoneura fumiferana* (Lepidoptera, Tortricidae) to the sex pheromone of her own species. *Journal of Chemical Ecology* 4:649–655.

Palaniswamy, P., and W.D. Seabrook. 1985. The alteration of calling behaviour by female *Choristoneura fumiferana* when exposed to synthetic sex pheromone. *Entomologia Experimentalis et Applicata* 37:13–16.

Ritchie, M.G. 1996. The shape of female mating preferences. *Proceedings of the National Academy of Sciences of the United States of America* 93:14628–14631.

Röller, H., L. Biemann, J.S. Bjerke, D.W. Norgard, and W.H. McShan. 1968. Sex pheromones of pyralid moths. I. Isolation and identification of sex attractant of *Galleria mellonella* L. (greater waxmoth). *Acta Entomologica Bohemoslovaca* 65: 208–211.

Ryan, M.J. 1988. Energy, calling, and selection. *American Zoologist* 28:885–898.

Saad, A.D., and D.R. Scott. 1981. Repellency of pheromones released by females of *Heliothis armigera* and *H. zea* to females both species. *Entomologia Experimentalis et Applicata* 30:123–127.

Sadek, M.M., G. von Wowern, C. Lofstedt, W.-Q. Rosen, and P. Anderson. 2012. Modulation of the temporal pattern of calling behavior of female *Spodoptera littoralis* by exposure to sex pheromone. *Journal of Insect Physiology* 58:61–66.

Schneider, D., S. Schulz, E. Priesner, J. Ziesmann, and W. Francke. 1998. Autodetection and chemistry of female and male pheromone in both sexes of the tiger moth *Panaxia quadripunctaria*. *Journal of Comparative Physiology A* 182:153–161.

South, A., and S.M. Lewis. 2011. The influence of male ejaculate quantity on female fitness: a meta-analysis. *Biological Reviews* 86:299–309.

Stelinski, L.L., and L.J. Gut. 2009. Delayed mating in tortricid leafroller species: simultaneously aging both sexes prior to mating is more detrimental to female reproductive potential than aging either sex alone. *Bulletin of Entomological Research* 99:245–251.

Stelinski, L.L., A.L. Il'Ichev, and L.J. Gut. 2006. Antennal and behavioral responses of virgin and mated oriental fruit moth (Lepidoptera: Tortricidae) females to their sex pheromone. *Annals of the Entomological Society of America* 99:898–904.

Svard, L., and J.N. McNeil. 1994. Female benefit, male risk: Polyandry in the true armyworm *Pseudaletia unipuncta*. *Behavioral Ecology and Sociobiology* 35:319–326.

Svensson, M. 1996. Sexual selection in moths: the role of chemical communication. *Biological Reviews* 71:113–135.

Svensson, M.G.E., M. Bengtsson, and J. Lofqvist. 1997. Individual variation and repeatability of sex pheromone emission of female turnip moths *Agrotis segetum. Journal of Chemical Ecology* 23:1833–1850.

Tillman, J.A., S.J. Seybold, R.A. Jurenka, and G.J. Blomquist. 1999. Insect pheromones: an overview of biosynthesis and endocrine regulation. *Insect Biochemistry and Molecular Biology* 29:481–514.

Van Dongen, S., E. Matthysen, E. Sprengers, and A.A. Dhondt. 1998. Mate selection by male winter moths *Operophtera brumata* (Lepidoptera, Geometridae): adaptive male choice or female control? *Behaviour* 135:29–42.

Wyatt, T.D. 2003. *Pheromones and Animal Behaviour: Communication by Smell and Taste.* Cambridge: Cambridge University Press.

Yang, M.-W., S.-L. Dong, and L. Chen. 2009. Electrophysiological and behavioral responses of female beet armyworm *Spodoptera exigua* (Hübner) to the conspecific female sex pheromone. *Journal of Insect Behavior* 22:153–164.

CHAPTER SIX

Genetic Control of Moth Sex Pheromone Signal and Response

KENNETH F. HAYNES

Introduction/Overview

The connection between genes and communication phenotypes is readily apparent in species of moths that are amenable to laboratory study. Results from selection experiments and crosses between closely related strains or species, parent–offspring correlation, commonalities between siblings (or half-siblings), and introgression of a specific Quantitative Trait Locus (QTL) into the gene pool of a closely related species have demonstrated that genes with major and additive effects act on both the signal and response phenotype. The relative importance of genes with additive and major effects differs widely among species, even closely related species. When evidence for all species is considered, both autosomal and sex-linked genes have been implicated in both signal and response characteristics. Many of the genes that affect the pheromone blend ratio have been shown or are presumed to act on components of the biosynthetic pathway. The response of male moths could be influenced at many levels, including sensory receptors, binding proteins, enzymatic degradation, central integration, and others. Although pleiotropy (one gene affecting multiple phenotypes) or linkage disequilibrium (failure of genes to assort randomly) of signal and response genes has been implicated with other modalities of communication (such as singing pattern by male crickets and female prefer-

ence for song), the evidence for these is not strong for olfactory communication in moths. An understanding of the genetic bases for signal and response characteristics in moths will contribute to our understanding of biosynthetic pathways, sensory and central processing of olfactory signals, and evolution of communication, including the potential for evolution or resistance to synthetic pheromones.

Inheritance of the Female Produced Signal

Impact of Single Genes

The impact of a gene on a signal phenotype is more clearly illustrated by examples from chemical communication in Lepidoptera than from any other modality or taxonomic group. Typically the pattern of inheritance is determined by crossing closely related species or races (e.g., parental types P1 and P2 in figure 6.1), and examining the pheromone blend phenotypes of the first filial generation (F1) hybrids, the second generation (F2), and backcrosses to the maternal and paternal strains (BC) (figures 6.1 and 6.2). From these crosses one can distinguish between single-gene effects and quantitative characters influenced by more than one gene, as well as between sex-linked (sex and ancestry of partners needs to be tracked in each generation) and autosomal inheritance. Females are the heterogametic sex in Lepidoptera (designated as ZW as opposed to XY), and therefore females only inherit a Z chromosome from their fathers, whereas males inherit a Z chromosome from both mother and father (ZZ). Dominance of one allele over another can also be inferred in the F1. Of course the success or failure of crosses between strains may be influenced by other genes that affect phenotypes that limit copulation itself, or lead to low hybrid fertility. Thus, our understanding of genetic control of chemical communication through crossing experiments is limited to pairs of species, strains, or races in which genetic divergence is not so pronounced that crosses are impossible.

The European corn borer, *Ostrinia nubilalis* (Crambidae), has two pheromone strains that have female phenotypes that are distinctly different. The E-morph produces a pheromone

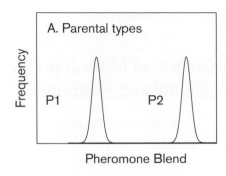

A. Parental types

Frequency

P1   P2

Pheromone Blend

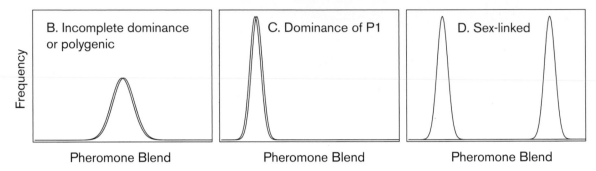

B. Incomplete dominance or polygenic

Frequency

Pheromone Blend

C. Dominance of P1

Pheromone Blend

D. Sex-linked

Pheromone Blend

FIGURE 6.1 Hypothetical patterns of inheritance from crosses between two strains (P1 and P2) that differ in their pheromone blend ratios: (A) Frequency of pheromone blend of P1 and P2; (B) incomplete dominance or polygenic inheritance would result in F1 females with intermediate blend ratios; (C) strict resemblance of one parent, irrespective of the direction of the cross, indicates one allele is dominant (in this case the allele inherited from P1 is dominant); (D) resemblance to females from the father's strain indicates sex-linked inheritance because the female is the heterogametic sex. For example, the curve at the far right indicates that the father was from the P2 strain.

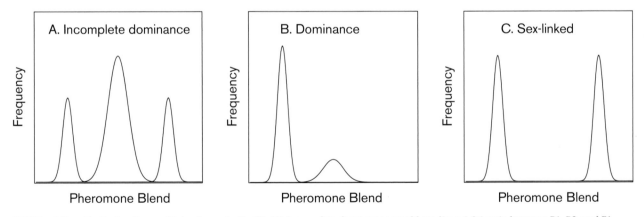

A. Incomplete dominance

Frequency

Pheromone Blend

B. Dominance

Frequency

Pheromone Blend

C. Sex-linked

Frequency

Pheromone Blend

FIGURE 6.2 Hypothetical patterns of inheritance in the F2: (A) Incomplete dominance would predict a 1:2:1 ratio between P1, P2, and F1 blend ratios; (B) if the P1 allele is dominant, then a 3:1 ratio of phenotypes is expected; (C) with sex-linked inheritance on the Z chromosome, both P1 and P2 phenotypes are recovered in the F2. (The F1 father would carry alleles from both of his parents, but he can only pass on one or the other to his female offspring. The genotype of the mother is irrelevant in this case.)

that is predominantly (E)-11-tetradecenyl acetate (E11-14Ac) with ca. 1% (Z)-11-tetradecenyl acetate (Z11-14Ac). In contrast, the Z-morph has a 3:97 blend of E11-14Ac to Z11-14Ac (Klun and Maini 1979; Roelofs et al. 1987). Crosses between the two strains produce female offspring with an intermediate pheromone phenotype centered on 65:35 E11-14Ac to Z11-14Ac. The intermediate phenotype of F1 females excludes the potential of sex-linked inheritance (unless there are minor effects hidden by the major autosomal effect), because under the sex-linked model the female would only inherit a phero-

mone phenotype determining allele from her father, and thus she would resemble females from her father's strain. However, the observed pattern could indicate simple Mendelian inheritance of a trait with incomplete dominance or polygenic differences between the parental strains. Limited variance in the phenotype of the F1 does not indicate whether the trait is influenced by one major gene (with two alleles) or more than one gene because all of the genes would be heterozygous. If one allele or the other was dominant, the F1 offspring (including both reciprocal crosses) should resemble one parent. In

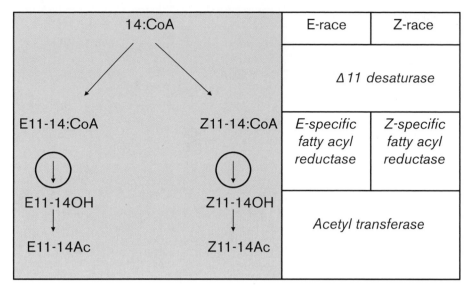

	E-race	Z-race
14:CoA	Δ11 desaturase	
E11-14:CoA → Z11-14:CoA	*E-specific fatty acyl reductase*	*Z-specific fatty acyl reductase*
E11-14OH, Z11-14OH	*Acetyl transferase*	
E11-14Ac, Z11-14Ac		

FIGURE 6.3 The biosynthetic pathway in two strains of *Ostrinia nubilalis* is the same except for the specificity of fatty acyl reductase (circled). In both strains, the pool of E and Z11-14:CoA has the same composition, but the fatty acyl reductase selectively reduces the E-precursor in the E-strain and the Z-precursor in the Z-strain.

addition with dominance, the ratio of phenotypes should be 3:1 in the F2, but instead it actually was 1:2:1 (homozygote: heterozygote: homozygote) (Roelofs et al. 1987). Polygenic inheritance was excluded by results in the F2 crosses, because the pheromone blends fall into discrete classes (1:2:1 ratio of parental EE, hybrid EZ, and parental ZZ phenotypes) and backcrosses (1:1:0 or 0:1:1 ratio of phenotypes). When the F1 hybrid offspring are intermediate between parental types, distinguishing between single gene and polygenic inheritance always requires continuing crosses to the F2.

Difference in substrate specificity of one or more of the enzymes that are involved in pheromone biosynthesis in Lepidoptera could explain pheromone differences between closely related strains, such as the E- and Z-strains of *O. nubilalis*, with the simplest explanation being a change in the gene that codes for the critical enzyme. With *O. nubilalis* and many other moths the biosynthetic pathway includes Δ-11 desaturase, fatty acyl reductase, and acetyl transferase (Tillman et al. 1999). Recent research indicates that the two strains of *O. nubilalis* differ in specificity of their fatty acyl reductases (figure 6.3). Thus, a mutation in this gene would be sufficient to explain differences in the pheromone blends (i.e., desaturases and acetyl transferase could be identical) (Zhu et al. 1996b; Lassance et al. 2010). Lassance et al. (2010) definitively connected the gene coding for substrate specificity of fatty acyl reductase to the phenotypic differences between the E- and Z-strains.

Superimposed on the clear impact of a major gene with two alleles on the pheromone blend of *O. nubilalis* are additional alleles or possibly genes whose impact on the pheromone blend is normally not apparent (Zhu et al. 1996a). A bimodal distribution of pheromone phenotypes in the F1 cross between Z- and E-strains was found with one peak centered about 65% E and the second at about 85% E. Neither additive models nor single gene models would predict a bimodal distribution of intermediate pheromone phenotypes. These authors reasonably suggest that there are two types of Z-alleles in the original Z-strain that they used, only one of which was present in the Z-strain used previously (Roelofs et al. 1987). The results of Zhu et al. (1996a) suggest that some of the genotypic variation that may occur in field populations is normally hidden within the different strains, as a result of canalization.

Other studies with the genus *Ostrinia* have shown additional ways that genotype and phenotype are connected. A contrast between pheromone biosynthesis in *O. nubilalis* and *O. furnacalis* (the Asian corn borer) indicates the involvement of distinct desaturase genes in generating pheromone components (Roelofs and Rooney 2003). *O. nubilalis* uses a Δ-11 desaturase to produce Z and E11-14:fatty acyl pheromone precursors from 14:Acyl. In contrast, *O. furnacalis* uses Δ-14 desaturase to produce Z and E14-16:Acyl, which are subsequently chain shortened, reduced, and acetylated to Z and E12-14Ac. Interestingly, genes for the reciprocal desaturases in the two species are present but not expressed (e.g., Δ-14 desaturase is not found in *O. nubilalis*, but the gene that would encode for it is there). Roelofs and Rooney (2003) suggest that gene duplication and gene loss are likely to be major factors in generating pheromone diversity. Furthermore, this genetic diversity is not necessarily expressed but may facilitate the evolution of new pheromone phenotypes.

Blend ratio differences between *O. zealis*, *O. zaguliaevi*, and *O.* sp. near *zaguliaevi* and interspecific crosses suggested the existence of both autosomal and sex-linked genes (Tabata and Ishikawa 2011). F1 offspring resulting from crosses between *O. zealis* and *O.* sp. near *O. zaguliaevi* resembled *O. zealis* with respect to the proportion of E11-14Ac out of E11-14Ac plus Z11-14Ac, but resembled near *O. zaguliaevi* with respect to the proportion of Z9-14Ac in Z9-14Ac, Z11-14Ac, and E11-14Ac. These researchers interpret the inheritance pattern (including backcross results) as suggesting a single autosomal gene, but with three different alleles in *O. zaguliaevi*, *O. zealis*, and *O.* sp. near *zaguliaevi*. Crosses between *O. zaguliaevi* and *O.* sp. near *zaguliaevi* suggest that genes on the sex chromosomes are also involved.

Crosses between related species in other families of moths showed similarities and differences with the results from *Ostrinia*. The sibling species *Ctenopseustis obliquana* and *C. herana* (Tortricidae) use pheromone blends consisting of an 80:20 blend of (*Z*)-8-tetradecenyl acetate (Z8-14Ac) and (*Z*)-5-

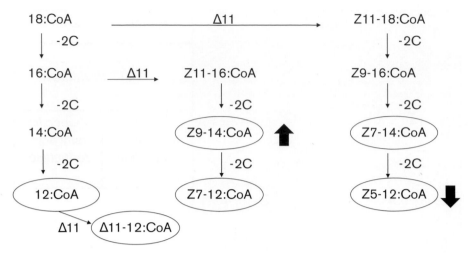

FIGURE 6.4 Precursors enclosed in ellipses are reduced and acetylated to the pheromone components of *Trichoplusia ni* (Bjostad et al. 1984). Bold arrows indicate two critical changes in pheromone gland content that is found in females that carry two mutant alleles. The quantity of Z9-14Ac increases 20-fold over the normal quantity, while the quantity of Z5-12Ac decreases 30-fold. Modification of the chain-shortening step is consistent with the changes seen in homozygous recessive females.

tetradecenyl (Z5-14Ac) or Z5-14Ac only, respectively (Foster et al. 1997). The resemblance between F1 females and their *C. obliquana* parent (for both directions) would indicate dominance, but the F1 from *C. herana* (female) by *C. obliquana* (male) were more variable than expected. Results from the F2 were broadly consistent with a single autosomal gene, but the paternal backcrosses revealed sex linkage. The inheritance of the pheromone blend ratio was not entirely consistent with a single autosomal gene with two alleles or a sex-linked gene, and thus at a minimum would involve two genes.

The sibling species *Spodoptera latifascia* and *S. descoinsi* (Noctuidae) differ in their blend of pheromone components, including the blend (Z,E)-9,12-tetradecadienyl acetate $(Z,E$-9,12:14Ac) and (Z)-9-tetradecenyl acetate (Z9-14Ac) (Monti et al. 1997). F1 hybrids between the two produce an intermediate blend that overlaps somewhat with that of *S. descoinsi*, but not *S. latifascia*. Only one of the two possible directions of hybridization was well represented in the F1 because of frequent infertile copulations between *S. descoinsi* females and *S. latifascia* males. There was a very small, but significant difference between the blend ratios which were biased toward the male parent, suggesting a small modifying impact of a sex-linked gene on this ratio. A major impact of a single autosomal gene (with dominance) was inferred from the F2 and backcross results, but this interpretation may be complicated by lethality of some crosses (e.g., the *S. descoinsi* phenotype was not recovered in the F2).

Helicoverpa armigera and *H. assulta* (Noctuidae) share two pheromone components, (Z)-11:hexadecenal and (Z)-9-hexadecenal (Z11-16Ald and Z9-16Ald), but in different blend ratios (100:2 and 6:100, respectively). The female F1 hybrids (female *H. assulta* × male *H. armigera*) resemble the father, and are clearly not intermediate (Wang et al. 2008). The reciprocal F1 produced no females, so the pattern cannot be distinguished between dominance and sex-linked inheritance. However, while not all backcrosses were possible, all of those that were obtained were consistent with a single autosomal gene. There was more variation in two of the backcrosses than would be expected under this model, so modifying genes should be expected.

Crossing of closely related and naturally occurring strains has provided much information on the nature of genetic control of pheromone blends. Ultimately that genetic diversity comes from mutation, which is filtered by natural (or sexual) selection. Decades of studies of mutant *Drosophila* have provided key information on how phenotypes are regulated by genes (Morgan 1917; Hall 1986; Sokolowski 2001). The discovery of a mutation in the cabbage looper moth, *Trichoplusia ni* (Noctuidae), provided information that complements the crossing studies discussed above. We found a rare mutant pheromone phenotype in our laboratory colony (Haynes and Hunt 1990b). These individuals were characterized by a 30-fold reduction in the quantity of (Z)-5-dodecenyl acetate (Z5-12Ac) and a 20-fold increase in the quantity of (Z)-9-tetradecenyl acetate (Z9-14Ac), among other changes in the six-component pheromone system (e.g., a threefold decrease in the quantity of the major component (Z)-7-dodecenyl acetate (Z7-12Ac)). It is very unlikely that females with this genotype would attract males in the field, but they readily mate in the laboratory. Once a colony was established in which all the females expressed this phenotype, we were able to conduct the appropriate set of crosses (F1, F2, and backcrosses) to determine that the mutant allele was recessive and that the inheritance of this trait followed Mendelian inheritance for an autosomal trait. As in *O. nubilalis*, enzymatic steps involved in pheromone biosynthesis in *T. ni* include fatty acid synthesis, Δ11-desaturation, β-oxidation, reduction, and acetylation. In the case of the mutant *T. ni*, a difference in the chain-shortening or β-oxidation step was implicated as sufficient to explain the divergent composition of mutant and wild-type pheromone blends (figure 6.4) (Jurenka et al. 1994). (Z)-9-Fatty acyl coenzyme A (Z9-14:CoA) is only one cycle of chain-shortening downstream of the fatty acyl product of Δ11-desaturation. In contrast, Z5-12:CoA is three cycles downstream of Z11-18:CoA (figure 6.4). With β-oxidation playing a key role in many steps of pheromone biosynthesis, multiple effects on the pheromone phenotype should be expected and were found. We have not observed pleiotropic effects of this mutation on life history traits. One advantage to this system is that it is likely that this single gene mutation

TABLE 6.1

Genetic correlations (with SE) between pheromone components in a laboratory colony of *Trichoplusia ni*
originally from Gainesville, FL

	12Ac	Z5-12Ac	Z7-12Ac	11-12Ac	Z7-14Ac
Z5-12Ac	−0.61 (0.35)				
Z7-12Ac	0.32 (0.38)	0.02 (0.51)			
11-12Ac	0.67 (0.27)	−0.45 (0.46)	0.53 (0.32)		
Z7-14Ac	0.58 (0.20)	0.68 (0.20)	0.01 (0.28)	−0.06 (0.31)	
Z9-14Ac	0.83 (0.09)	0.23 (0.34)	0.86 (0.07)	0.93 (0.06)	0.35 (0.17)

SOURCE: Gemeno et al. (2001).

is expressed in a genetic background that is identical to normal females, and thus more complicated interpretations of the inheritance patterns are not likely.

Additive Genetic Variation in Quantity and Composition

Genes coding for critical enzymes in the biosynthetic pathway can have diverse and large effects on the pheromone blend. Thus, single gene effects can result in changes that are akin to those observed in the divergence of pheromone blends as part of speciation (e.g., Baker 2002; Roelofs et al. 2002) at least with respect to the female pheromone blend. Subtler changes in the specificity of biosynthetic enzymes, or the genetic milieu in which they act (e.g., epitasis), could potentially result in finer changes in the pheromone blend. These kinds of subtle changes could explain pheromone variation that is often observed in species across a broad geographic range, with the most distant locales having populations that are dramatically different. Clinal differences among populations would be consistent with a gradual evolution of divergent pheromone blends.

There is minor variation between pheromone blends of *Trichoplusia ni* among field populations (Haynes and Hunt 1990a; Hunt et al. 1990; Haynes 1997), although these differences are dwarfed by the impact of the single gene mutation discussed above (Haynes and Hunt 1990b). Crossing experiments between a long-maintained laboratory population and a field population from California indicated that more than one gene differed between these populations, and that both autosomal and sex-linked genes were involved (Gemeno et al. 2001). A half-sib breeding design was used to estimate heritabilities, coefficients of additive genetic variation, and phenotypic and genotypic correlations (Table 6.1) between pheromone components within a normal and mutant strain. Heritabilities of quantities of individual pheromone components were moderate to high in a normal colony that originated from Gainesville, Florida. Genetic correlations between pheromone components ranged from significantly negative to significantly positive. Based on these calculations, if one selected for a higher quantity of Z5-12Ac, the quantity of Z7-12Ac would be expected to be unaffected, whereas the quantity of 11-dodecenyl acetate (11-12Ac) would decrease and the quantities of Z7-14Ac and Z9-14Ac would increase. The direction and amplitude of the genetic correlations between pheromone components suggest that the evolutionary trajectories of pheromone blends may be restricted by their biosynthetic pathways.

The two-component pheromone of the pink bollworm moth, *Pectinophora gossypiella* (Gelechiidae), provides a more straightforward example of how genetically based variation within a population was estimated. Collins and Cardé (1985) used a full-sib design to estimate genetic correlation between two components (which was very high at 0.99) and heritabilities of quantity and blend ratios. Heritability of total quantity and blend ratio (± SE) of the two pheromone components, (*Z,E*)- and (*Z,Z*)-7,11-hexadecadienyl acetate (ZE and ZZ-7,11-16Ac), were 0.41 ± 0.09 and 0.34 ± 0.08, respectively (Collins and Cardé 1985). These characteristics would thus be expected to respond to selection, which was subsequently verified (Collins and Cardé 1989b; Collins et al. 1990). Twelve generations of selection for an increased blend ratio of ZE-7,11-16Ac resulted in a change from 42.9% to 48.2% ZE-7,11-16Ac (Collins and Cardé 1989b), yielding a realized heritability (± SE) of 0.50 ± 0.04. Selection in the opposite direction did not result in change. A concomitant decrease in the quantity of pheromone that accompanied the increase in blend ratio suggests either a negative genetic correlation between quantity and blend ratio or some form of inbreeding depression. When selection was stopped, there was a partial return toward the original mean blend after six generations, which could reflect genetic drift or some sort of trade-off with an unknown selective factor. Six generations of selection for an increased total titer of the two components resulted in a 91% increase (Collins et al. 1990).

The pheromone blend of the almond moth, *Cadra cautella* (Pyralidae), is highly variable, being influenced by the age of females and time of scotophase, as well as environmental factors (Allison and Cardé 2006). A half-sib genetic analysis indicated that there was a high heritability (± SE) of the titer of Z9-14Ac (0.75 ± 0.24), ZE9,12-14Ac (1.20 ± 0.32) and a moderate heritability of the ratio of these components (0.46 ± 0.17). Thus, additive genetic variance, a prerequisite for gradual evolution via selection, was present. Bidirectional selection for change in the pheromone blend ratio resulted in an average increase of 54% in the high lines, and a decrease in 43% in the low lines over five generations (Allison and Cardé 2007). These results gave an estimate of realized heritability that was higher than the narrow-sense heritability established with the half-sib experimental design (0.77 vs. 0.46). It is clear that under selection the pheromone blend of this species would evolve in the absence of countering selection.

In addition to selection experiments and full-sib or half-sib designs for estimating heritabilities, mother–daughter regression has been used. Using this approach, Svensson et al. (2002) estimated the heritability of blend ratio of two

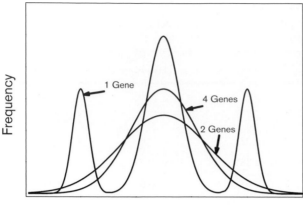

FIGURE 6.5 Hypothetical effect of one or more loci on a pheromone blend in the F2 assuming autosomal inheritance on alleles without dominance. As the number of genes increases, the variance approaches that of the F1 (see figure 6.1). The probability of recovering a parental pheromone genotype in the F2 is inversely related to the number of genes involved.

components (ZE-9,12:14Ac and (Z,E)-9,12-tetradecadienyl (ZE-9,12-14OH)) of the Indian meal moth, *Plodia interpunctella* (Pyralidae) (Svensson et al. 2002) at 0.65. Because the trait is not expressed in males, the heritability is two times the slope of the mother–daughter regression line.

Across a broad geographic range the pheromone blend of the turnip moth, *Agrotis segetum* (Noctuidae), shows substantial variation in the composition and blend ratio of pheromone components (Lofstedt et al. 1986; Löfstedt 1993). Scandinavian and Zimbabwean populations represent extreme pheromone phenotypes within this species (Wu et al. 1999), and thus a crossing experiment was undertaken to gain an understanding of the genetic basis of this difference (LaForest et al. 1997). The genetic basis for pheromonal differences between these geographically distant populations was complex. Incomplete dominance in one or more genes affected the ratio of Z5-12Ac to Z7-12Ac, but the Scandinavian genotype was dominant for the ratio of Z9-14Ac to Z7-12Ac. The variance of a character in the F2 should be greater than that in the F1. If the mean values of the parental character and the variance of the F1 and F2 are known, it is possible to estimate the minimum number of genes involved in a quantitative character (Lande 1981) using the equation:

$$(\mu_{P1} - \mu_{P2})^2/8(\sigma_{F2}^2 - \sigma_{F1}^2)$$

As the number of independently assorting loci increases, the variance in the F2 decreases (figure 6.5). Variance from the backcrosses can also be incorporated into this estimate (see Lande 1981). For the log ratio of Z5-10Ac/Z7-12Ac, the similarity of the variances in the F2 and F1 suggests that many genes are involved. The inheritance of the log ratio of Z9-14Ac/Z7-12Ac is difficult to interpret, because the variance in the F2 is greater than the F1. Autosomal dominance is implicated with F1 phenotypes being similar to Scandinavian parent (irrespective to sex). However, the backcross data are not consistent with this hypothesis.

Selection for high blend ratios of Z11-14Ac to E11-14Ac in the red-banded leafroller, *Argyrotaenia velutinana* (Tortricidae), resulted in small changes in the blend ratio in the expected direction that were not sustained after selection was stopped (Roelofs et al. 1986; Sreng et al. 1989). This blend

ratio appears to be canalized. Sreng et al. (1989) hypothesize that this would make gradual evolutionary change in the blend ratio more difficult. In contrast, the ratio of E9-12Ac to E11-14Ac (two minor components) did respond to selection, with a realized heritability of over 1 in both the line selected for higher ratios (a change from a starting point of about 25–42% after three generations) and lower ratios (a change from 25% to about 14%). Evolution of blend ratio in multicomponent systems is complicated and may be limited by canalization and common biosynthetic pathways, resulting in genetic correlations (both positive and negative).

A chromosome mapping approach was used to explore the genetic basis of pheromone differences between *Heliothis virescens* and *H. subflexa* (Noctuidae) (Groot et al. 2004, 2009a; Sheck et al. 2006). These investigators used an AFLP (Amplified Fragment Length Polymorphism) analysis to identify many potential genetic markers. Those markers that were present in both the *H. virescens* parent and the F1 offspring, but were not found in the *H. subflexa* parent, were useful for this analysis. Because there is no crossing over in female Lepidoptera, a QTL can be localized to a specific chromosome (30 autosomes and one sex chromosome in *Heliothis* [Sheck et al. 2006]). At least 9 of 31 chromosomes had an impact on the blend ratio of one or more compounds in the pheromone gland. Clearly, the differences between these two species are polygenic; however, single genes may have major effects which are superimposed on subtler variation. A more detailed exploration of the impact of one QTL (chromosome 22) indicated that it suppressed biosynthesis of several acetate compounds (Groot et al. 2004), which are likely behavioral antagonists for *H. virescens*. Two chromosomes interact to explain 53% of phenotypic variation in the total quantity of these acetates. Chromosome 24 from *H. subflexa* in a *H. virescens* background increased production of Z9-16Ald (Groot et al. 2009b). This compound is a component of the pheromone blend of *H. subflexa*.

Inheritance of the Male Behavioral Response

Genetic studies of male behavioral response to pheromones are inherently more difficult to conduct and interpret, in part because the key characteristic is not quantitative; either a response was seen or it was not. If a given male does not complete the sequence of behaviors necessary to locate a female, one does not know that this one observation represents his inherent probability of response. Thus, the true phenotype of an individual may be difficult to determine. Presumably his probability of responding to a range of blends would show an underlying optimum. Repeated testing of an individual male can be problematic because prior experience may influence subsequent responses. Even with this limitation various studies have shown both major gene effects and additive genetic variation for male response to sex pheromone.

Here again, the studies of *Ostrinia nubilalis* have been key to our understanding of how major genes can affect the response to the pheromone blend. While the specificity of the behavioral response (flying upwind and contacting a pheromone source) was once thought to be autosomal, it is now clear that it is influenced by a single sex-linked gene (Roelofs et al. 1987). This pattern of inheritance cannot be distinguished from a single autosomal gene with incomplete dominance in the F1, but instead is revealed in the F2 and backcrosses (figure 6.6). Because crosses between the Z- and E-strains result in two genotypes of females in the F1 (Z^ZW and Z^EW where the super-

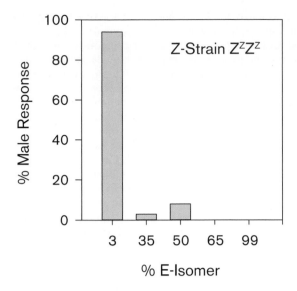

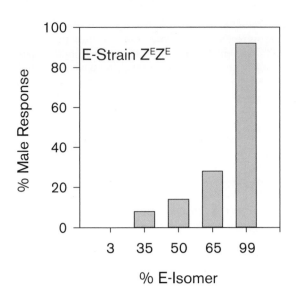

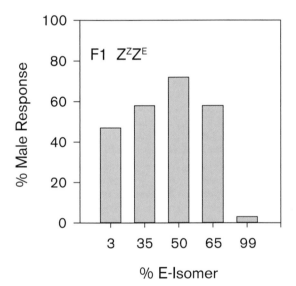

F1 females are Z^ZW or Z^EW carrying, but not expressing, the behavioral response allele of their fathers

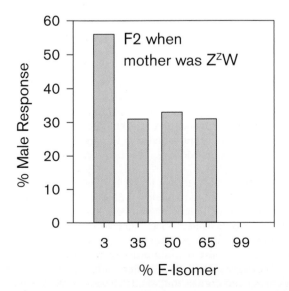

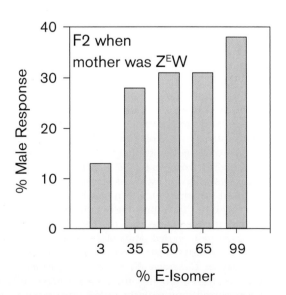

FIGURE 6.6 Behavioral responses of male *Ostrinia nubilalis* resulting from F1 and F2 crosses between E- and Z-strains. The results indicate that a gene influencing the male response is carried on the sex chromosome (Z chromosome). A female's genotype is inherited only from her father. However, Z-linked behavioral response characteristics in moths are not expressed in females, so their impact needs to be examined in subsequent generations (modified from Roelofs et al. 1987).

script represents the "pheromone response allele"), males in an F1 cross receive a behavioral response allele from each parent Z^ZZ^E, but females only carry the allele from their fathers. The result is that the F2 male offspring have genotype Z^ZZ^Z and Z^ZZ^E if the mother had a Z^ZW genotype and Z^EZ^E and Z^EZ^Z if the mother was Z^EW. Common sex-linked characters in humans like color-blindness and some types of hemophilia are exposed in the heterogametic sex (males-XY), but moth pheromone response has no phenotypic expression in females.

The male behavioral phenotype is a product of characteristics of his peripheral nervous system and how this information is integrated within the Central Nervous System (CNS). Roelofs et al. (1987) found that there was a gene that affects spike amplitude of sensory neurons in *O. nubilalis*. The larger spiking neuron within a sensillum paralleled the behavioral type in the parental strain (i.e., E-strain males have larger spikes in the E-neuron). Löfstedt et al. (1989) determined that there was no linkage between the gene that affects the pheromone blend (see above) and this spike amplitude gene. Furthermore, detailed studies of electrophysiological and behavioral characteristics of F2 males indicated that spike amplitude and behavioral specificity segregate independently (Cossé et al. 1995), as must be the case when one character is sex-linked and the other is autosomal. Thus, individuals with E-type neurons (large spike to E-isomer and small spike to Z-isomer) could respond preferentially to the Z-isomer. However, a recent study indicated that amplitude of axon potentials in the periphery is more complex and appears to be influenced by both sex-linked (with E-strain dominance) and autosomal genes (Olsson et al. 2010). The topology of the macroglomerular complex in F1 hybrids is similar to that of the E-parent, despite the fact that these males would prefer an intermediate blend (Kárpáti et al. 2010). However, the volume of the medial glomerulus is normally twice that of the lateral glomerulus (in both E- and Z-strains), but in the F1 hybrids the glomeruli are equally sized. Subsequent crosses showed that this size characteristic was sex-linked. While the pieces of this story are not altogether consistent, overall it is clear that sex-linked and autosomal genes play a role at some level.

Inheritance of pheromone responses in crosses between the sibling species *Ctenopseustis obliquana* and *C. herana* (Tortricidae) showed evidence of sex-linkage in both neurophysiological characteristics at the sensory level (large spiking neurons for the major component) (Hansson et al. 1989) and behavioral response preference to pheromone blends (Foster et al. 1997). Male *C. obliquana* antennae had one type of sensillum that contained a large spiking neuron responsive to Z8-14Ac (major component) and a small spiking neuron responsive to Z5-14Ac and 14Ac (minor components). Sensilla in *C. herana* contain a large spiking neuron responsive to the major component Z5-14Ac (and to a lesser extent to 14Ac) and a small spiking neuron responsive to a minor component, Z8-14Ac. However, the F1 male offspring showed some atypical neurons that could not be related to either parental type, a pattern that carried over to the F2, and three out of four types of backcrosses (only one backcross fit the specific expectations of sex-linked inheritance). Taken together the results suggest that other genes, including one or more that is not sex-linked, influence the type of neurons present in sensilla. The behavioral response to varied ratios of Z8-14Ac and Z5-14Ac more closely resembled the parental *C. herana* in the F1, which suggests that the *C. herana* allele is at least partially dominant (either sex-linked or autosomal). Backcrosses and F2 crosses were most consistent with sex-linked inheritance.

In the cabbage looper, *Trichoplusia ni*, the wild-type upwind flight response is influenced by a six-component pheromone blend, in which some of the minor components are redundant, but the male is still tuned to the blend produced by wild-type females (Linn et al. 1984). Our first evidence that male response could evolve came from a colony in which all females expressed a mutant pheromone phenotype that had multiple changes in quantity and blend ratio of components (see earlier section and Haynes and Hunt 1990b). Initially males from this colony were much more likely to respond to the wild-type phenotype, as one would expect if there was no linkage between signal and response phenotypes. We followed the response profile of this colony (and a second colony) for many generations and observed a broadening of the male response phenotype (Liu and Haynes 1994). After nearly 50 generations males did not discriminate between mutant and wild-type phenotypes. Genetic drift seems an unlikely explanation for the observed change because of the consistency of the direction of change among generations and replicated independent mutant colonies. We hypothesized that selection was occurring in these colonies because males that would not respond to the mutant blend would go unmated in these colonies (this hypothesis is consistent with the Asymmetric Tracking Hypothesis [Phelan 1992]). Because the mutant-type females have 20-fold more Z9-14Ac in their released blend (Haynes and Hunt 1990b), the broadening of the behavioral response could be mediated by an acceptance of higher levels of input to the CNS from the Z9-14Ac sensory neurons, or by a decrease in the peripheral input for this neuron with maintenance of required balance within the CNS. Domingue et al. (2009) found that the sensitivity of the peripheral neuron responsive to Z9-14Ac was attenuated in males from mutant-type colonies (figure 6.7). This attenuation leads to the sensory input from a mutant and wild-type pheromone blend being roughly equivalent in terms of spike ratios for Z9-14Ac and Z7-12Ac in the mutant-type males. At the behavioral level, although mutant-type males responded equally well to the two pheromone blends, they were less sensitive overall, suggesting a trade-off between sensitivity and specificity of response (Hemmann et al. 2008).

The observed evolution of male response phenotype in colonies in which all females have this altered blend suggests that there was heritable variation in our wild-type colony. Using a father–son regression technique for a threshold character, Evenden et al. (2002) found that heritability of male behavioral response to the mutant pheromone blend was 0.25 within the normal colony, but was at or near 0 in the mutant colony. This suggested that normal males would respond to selection in the same way that the response had evolved within isolated mutant colonies. Selection of normal males over three generations resulted in an improved response to the mutant blend (Evenden et al. 2002). The low heritability of the response of mutant males to the mutant pheromone may indicate that additive genetic variance had been exhausted by selection.

Collins and Cardé (1989a) found heritable variation in a wing-fanning response to blends of ZE and ZZ-7,11-16Ac, the pheromone components of the pink bollworm moth, *Pectinophora gossypiella* (Gelechiidae). They were able to generate a quantitative indicator of male specificity by measuring the duration of the wing-fanning response to normal (44% ZE) and off-blends (25% ZE or 65% ZE), with the ratio of these times being the critical statistic. Each male was allowed to recover from the first exposure for one day before the next

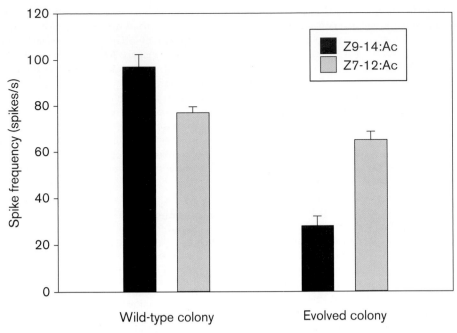

FIGURE 6.7 The spike frequency of neurons sensitive to Z7-12Ac and Z9-14Ac in two colonies of *Trichoplusia ni*. The rate of firing of the Z9-14Ac neuron is reduced in the evolved colony. This reduction allows these males to respond to a pheromone blend that contains very high quantities of Z9-14Ac (a characteristic of females with a single gene mutation affecting chain shortening).

evaluation was made. Using parent–offspring regression, they found that this index had a heritability of 0.2 for the low blend ratio (25% ZE) (significantly different from 0 at $\alpha = 0.05$), but was −0.04 (not different from 0) to a high blend ratio (65% ZE). This result indicates that there should be an asymmetrical response to selection for response to low and high blend ratios.

A QTL approach similar to that used to study divergence in the pheromone blend of *Heliothis virescens* and *H. subflexa* has been applied to species differences in male behavioral response in these species (Gould et al. 2010). For example, chromosome 27 from *H. virescens* has an impact on specificity mediated by Z9-14Ald, which is a species-specific compound from *H. virescens*. Without this introgressed chromosome *H. subflexa* would not respond to a blend containing this compound. Similarly, chromosome 27 from *H. subflexa* was correlated with an improved response in backcrosses to blends containing Z9-16Ald. The same chromosome influences positive (*H. subflexa*) or negative (*H. virescens*) responses to Z11-16Ac. Four odorant receptor genes map to this QTL.

Coupling or Independence of Signal and Response

A gene or linkage between genes that affect signal and response traits involved in mating would facilitate speciation. The coiling pattern of snails is an intriguing example of single gene speciation (Ueshima and Asami 2003). In this case hermaphroditic snails with opposite coiling patterns fail to copulate because of the physical restrictions imposed by the shell. The direction of coiling is controlled by a single gene. In this very specialized case, prezygotic mating isolation from the coiling direction leads to reproductive isolation without the need for ecological or geographical isolation. Obviously, this model of speciation does not apply to moths where male and

female communication characteristics are dependent on very different physiological characteristics. The same should be said for communication systems of most animals, and convincing evidence for linkage has been weak (Butlin and Ritchie 1989). Recent studies of visual communication in two *Heliconius* butterflies (Kronforst et al. 2006), and acoustical communication in two Hawaiian *Laupala* crickets (Wiley et al. 2012), have provided evidence that such linkage sometimes exists in other communication modalities. In *H. cydno* and *H. pachinus* female wing-patch color and male wing-patch color preferences map to the same QTL. In the Hawaiian crickets *L. kohalensis* and *L. paranigra* (Orthoptera: Gryllidae), male-produced song characteristics (pulse rate) and female preference for pulse rate are genetically correlated in second-generation backcrosses. Four generations of backcrossing identified four QTLs for song pulse rate that were physically linked to genes influencing preference (Wiley et al. 2012). In theory chromosomal linkage between signal and response could facilitate coordinated evolution and maintenance of these traits. Crosses of strains of *Ostrinia nubilalis* (Roelofs et al. 1987), *Ctenopseustis* spp. (Foster et al. 1997), *Heliothis* spp. (Gould et al. 2010), and *Trichoplusia ni* have failed to find linkage. Allison et al. (2008) did not find correlated changes in the male response phenotype after selection resulted in changes in the female pheromone blend ratio. In contrast, male response specificity (as measured by a wing-fanning assay) was influenced by directional selection on the female pheromone blend ratio in a direction that would facilitate assortative mating in *Pectinophora gossypiella* (Collins and Cardé 1989b). Similarly, in the red-banded leafroller, *Argyrotaenia velutinana*, there is evidence that selecting male parents for response to a high blend ratio of E11-14Ac to Z11-14Ac contributed to female offspring with higher blend ratios of E to Z11-14Ac (Roelofs et al. 1986). It is possible that in the future stronger evidence of linkage between signal and response will be discovered, but, if it is, the evidence to date suggests that it will be an exception rather than a rule that could broadly explain the evolution of pheromone diversification.

Periodicity of Calling Behavior

The periodicity of signaling and response is another dimension of communication that has proven to be critical to reproductive isolation (Roelofs and Cardé 1971; Greenfield and Karandinos 1979; Hendrikse 1979; Löfstedt et al. 1985; Haynes and Birch 1986). Periodicity of these activities in females and males could be influenced by a common gene, but there is no evidence for this conjecture. If this were the case a mutation that affects that gene would be likely to influence both timing of calling and response. One study that looked at the timing of calling behavior found that it had a genetic basis (Monti et al. 1997). The mean onset time of calling was 4.2 and 7.8 hours after the initiation of scotophase for *Spodoptera descoinsi* and *S. latifascia*, respectively. The F1 hybrid females were intermediate at 5.4 hours after initiation of scotophase. Overall the successful crosses that were possible indicated that this aspect of calling behavior was influenced by more than one gene (the variance in the F2 was higher than the F1, but was still unimodal). A study of *Choristoneura fumiferana* and *C. pinus* (Tortricidae) suggested that the cyclical initiation of calling behavior was influenced by a sex-linked gene(s) (Sanders et al. 1977) because the timing of calling in the F1 resembled the father in both reciprocal crosses. In contrast, F1 offspring of crosses between *Euxoa declarata* and *E. rockburnei* (Noctuidae) indicated that calling periodicity was influenced by both maternal effects and autosomal loci (Teal 1985). The female offspring of crosses between female *E. declarata* and male *E. rockburnei* resembled their mother (i.e., a maternal effect), which would not be the prediction of either Z-chromosome-linked inheritance or autosomal inheritance. The F1 offspring of the reciprocal cross had a broad period of calling that was intermediate between the two species: a finding consistent with incomplete dominance of autosomal alleles. In this crossing direction maternal effects were not apparent. To date, the results of crossing experiments with a limited number of species do not suggest a general pattern of inheritance of genes affecting periodicity.

Summary/Discussion

Most frequently our understanding of the genetic basis of traits is gained through crossing experiments between closely related species, races, or strains. These types of experiments can only be informative if they are followed to the F2 and backcrosses. It is apparent from these studies that single or closely linked genes with major effects on blend ratios of pheromone components mediated most often by direct or indirect effects on the biosynthetic pathway are common. However, polygenic effects have also been observed and at least in part explain pheromone variation in some species. Some of these studies find surprisingly high heritability in characteristics of the pheromone blend (both quantity and blend ratio). The specificity of male response to pheromone blends has been examined from a genetic perspective in a few species. Major gene effects have been found in some species, but additive genetic variation has also been implicated in others. Linkage disequilibrium between signal and response appears to be uncommon. QTL studies have been effective in relating both signal and response characteristics to specific

chromosomes. As molecular, neuroanatomical and neurophysiological tools are linked with the more long-standing crossing, parent–offspring, and selection experiments (Baker et al. 2006; Vickers 2006; Rooney 2008; Groot et al. 2009a; Gould et al. 2010; Kárpáti et al. 2010; Olsson et al. 2010), a more comprehensive picture will emerge, and perhaps some generalities about genetic control of chemical communication. These connections will certainly help us to gain insights into variation in chemical communication systems, reproductive isolation, phylogenetic patterns of pheromone use, and the potential for evolution of resistance to synthetic pheromones that have been used in agriculture and forestry.

References Cited

Allison, J.D., and R.T. Cardé. 2006. Heritable variation in the sex pheromone of the almond moth, *Cadra cautella*. *Journal of Chemical Ecology* 32:621–641.

Allison, J.D., and R.T. Cardé. 2007. Bidirectional selection for novel pheromone blend ratios in the almond moth, *Cadra cautella*. *Journal of Chemical Ecology* 33:2293–2307.

Allison, J.D., D.A. Roff, and R.T. Cardé. 2008. Genetic independence of female signal form and male receiver design in the almond moth, *Cadra cautella*. *Journal of Evolutionary Biology* 21:1666–1672.

Baker, T.C. 2002. Mechanism for saltational shifts in pheromone communication systems. *Proceedings of the National Academy of Sciences of the United States of America* 99:13368–13370.

Baker, T.C., C. Quero, S.A. Ochieng, and N.J. Vickers. 2006. Inheritance of olfactory preferences II. Olfactory receptor neuron responses from *Heliothis subflexa* x *Heliothis virescens* hybrid male moths. *Brain Behavior and Evolution* 68:75–89.

Bjostad, L.B., C.E. Linn, J.W. Du, and W.L. Roelofs. 1984. Identification of new sex pheromone components in *Trichoplusia ni*, predicted from biosynthetic precursors. *Journal of Chemical Ecology* 10:1309–1323.

Butlin, R.K., and M.G. Ritchie. 1989. Genetic coupling in mate recognition systems—what is the evidence. *Biological Journal of the Linnean Society* 37:237–246.

Collins, R.D., and R.T. Cardé. 1985. Variation in and heritability of aspects of pheromone production in the pink bollworm moth, *Pectinophora gossypiella* (Lepidoptera:Gelechiidae). *Annals of the Entomological Society of America* 78:229–234.

Collins, R.D., and R.T. Cardé. 1989a. Heritable variation in pheromone response of the pink bollworm, *Pectinophora gossypiella* (Lepidoptera: Gelechiidae). *Journal of Chemical Ecology* 15:2647–2659.

Collins, R.D., and R.T. Cardé. 1989b. Selection for altered pheromone-component ratios in the pink bollworm moth, *Pectinophora gossypiella* (Lepidoptera: Gelechiidae). *Journal of Insect Behavior* 2:609–621.

Collins, R.D., S.L. Rosenblum, and R.T. Cardé. 1990. Selection for increased pheromone titer in the pink bollworm moth, *Pectinophora gossypiella* (Lepidoptera: Gelechiidae). *Physiological Entomology* 15:141–147.

Cossé, A.A., M.G. Campbell, T.J. Glover, C.E. Linn, J.L. Todd, T.C. Baker, and W.L. Roelofs. 1995. Pheromone behavioral-responses in unusual male European corn borer hybrid progeny not correlated to electrophysiological phenotypes of their pheromone-specific antennal neurons. *Experientia* 51:809–816.

Domingue, M.J., K.F. Haynes, J.L. Todd, and T.C. Baker. 2009. Altered olfactory receptor neuron responsiveness is correlated with a shift in behavioral response in an evolved colony of the cabbage looper moth, *Trichoplusia ni*. *Journal of Chemical Ecology* 35:405–415.

Evenden, M.L., B.G. Spohn, A.J. Moore, R.F. Preziosi, and K.F. Haynes. 2002. Inheritance and evolution of male response to sex pheromone in *Trichoplusia ni* (Lepidoptera: Noctuidae). *Chemoecology* 12:53–59.

Foster, S.P., S.J. Muggleston, C. Löfstedt, and B. Hansson. 1997. A genetic study on pheromonal communication in two *Ctenopseustis* moths. Pp. 514–524. In R.T. Cardé and A.K. Minks, eds. *Insect Pheromone Research: New Directions*. London: Chapman & Hall.

Gemeno, C., A.J. Moore, R.F. Preziosi, and K.F. Haynes. 2001. Quantitative genetics of signal evolution: a comparison of the pheromonal signal in two populations of the cabbage looper, *Trichoplusia ni*. *Behavior Genetics* 31:157–165.

Gould, F., M. Estock, N.K. Hillier, B. Powell, A.T. Groot, C.M. Ward, J.L. Emerson, C. Schal, and N.J. Vickers. 2010. Sexual isolation of male moths explained by a single pheromone response QTL containing four receptor genes. *Proceedings of the National Academy of Sciences of the United States of America* 107:8660–8665.

Greenfield, M.D., and M.G. Karandinos. 1979. Resource partitioning of the sex communication channel in clearwing moths (Lepidoptera: Sesiidae) of Wisconsin. *Ecological Monographs* 49:403–426.

Groot, A.T., M.L. Estock, J.L. Horovitz, J. Hamilton, R.G. Santangelo, C. Schal, and F. Gould. 2009a. QTL analysis of sex pheromone blend differences between two closely related moths: insights into divergence in biosynthetic pathways. *Insect Biochemistry and Molecular Biology* 39:568–577.

Groot, A.T., O. Inglis, S. Bowdridge, R.G. Santangelo, C. Blanco, J.D. Lopez, A.T. Vargas, F. Gould, and C. Schal. 2009b. Geographic and temporal variation in moth chemical communication. *Evolution* 63:1987–2003.

Groot, A.T., C. Ward, J. Wang, A. Pokrzywa, J. O'Brien, J. Bennett, J. Kelly, R.G. Santangelo, C. Schal, and F. Gould. 2004. Introgressing pheromone QTL between species: towards an evolutionary understanding of differentiation in sexual communication. *Journal of Chemical Ecology* 30:2495–2514.

Hall, J.C. 1986. Learning and rhythms in courting, mutant *Drosophila*. *Trends in Neurosciences* 9:414–418.

Hansson, B.S., C. Lofstedt, and S.P. Foster. 1989. Z-linked inheritance of male olfactory response to sex-pheromone components in 2 species of tortricid moths, *Ctenopseustis obliquana* and *Ctenopseustis* sp. *Entomologia Experimentalis et Applicata* 53:137–145.

Haynes, K.F. 1997. Genetics of pheromone communication in the cabbage looper moth, *Trichoplusia ni*. Pp. 525–534. In R.T. Cardé and A.K. Minks, eds. *Insect Pheromone Research: New Directions*. London: Chapman & Hall.

Haynes, K.F., and M.C. Birch. 1986. Temporal reproductive isolation between two species of plume moths (Lepidoptera: Pterophoridae). *Annals of the Entomological Society of America* 79:210–215.

Haynes, K.F., and R.E. Hunt. 1990a. Interpopulational variation in emitted pheromone blend of cabbage looper moth, *Trichoplusia ni*. *Journal of Chemical Ecology* 16:509–519.

Haynes, K.F., and R.E. Hunt. 1990b. A mutation in pheromonal communication system of cabbage looper moth, *Trichoplusia ni*. *Journal of Chemical Ecology* 16:1249–1257.

Hemmann, D.J., J.D. Allison, and K.F. Haynes. 2008. Trade-off between sensitivity and specificity in the cabbage looper moth response to sex pheromone. *Journal of Chemical Ecology* 34:1476–1486.

Hendrikse, A. 1979. Activity patterns and sex-pheromone specificity as isolating mechanisms in eight species of (Lepidoptera: Yponomeutidae). *Entomologia Experimentalis et Applicata* 25:172–180.

Hunt, R.E., B.G. Zhao, and K.F. Haynes. 1990. Genetic aspects of interpopulational differences in pheromone blend of cabbage looper moth, *Trichoplusia ni*. *Journal of Chemical Ecology* 16:2935–2946.

Jurenka, R.A., K.F. Haynes, R.O. Adlof, M. Bengtsson, and W.L. Roelofs. 1994. Sex-pheromone component ratio in the cabbage looper moth altered by a mutation affecting the fatty-acid chain-shortening reactions in the pheromone biosynthetic pathway. *Insect Biochemistry and Molecular Biology* 24:373–381.

Kárpáti, Z., S. Olsson, B.S. Hansson, and T. Dekker. 2010. Inheritance of central neuroanatomy and physiology related to pheromone preference in the male European corn borer. *BMC Evolutionary Biology* 10.

Klun, J.A., and S. Maini. 1979. Genetic basis of an insect chemical communication system: the European corn borer. *Environmental Entomology* 8:423–426.

Kronforst, M.R., L.G. Young, D.D. Kapan, C. McNeely, R.J. O'Neill, and L.E. Gilbert. 2006. Linkage of butterfly mate preference and wing color preference cue at the genomic location of wingless. *Proceedings of the National Academy of Sciences of the United States of America* 103:6575–6580.

LaForest, S., W.Q. Wu, and C. Löfstedt. 1997. A genetic analysis of population differences in pheromone production and response between two populations of the turnip moth, *Agrotis segetum*. *Journal of Chemical Ecology* 23:1487–1503.

Lande, R. 1981. The minimum number of genes contributing to quantitative variation between and within populations. *Genetics* 99:541–553.

Lassance, J.M., A.T. Groot, M.A. Lienard, B. Antony, C. Borgwardt, F. Andersson, E. Hedenstrom, D.G. Heckel, and C. Löfstedt. 2010. Allelic variation in a fatty-acyl reductase gene causes divergence in moth sex pheromones. *Nature* 466:486–489.

Linn, C.E., L.B. Bjostad, J.W. Du, and W.L. Roelofs. 1984. Redundancy in a chemical signal—behavioral-responses of male *Trichoplusia ni* to a 6-component sex-pheromone blend. *Journal of Chemical Ecology* 10:1635–1658.

Liu, Y.B., and K.F. Haynes. 1994. Evolution of behavioral responses to sex pheromone in mutant laboratory colonies of *Trichoplusia ni*. *Journal of Chemical Ecology* 20:231–238.

Löfstedt, C. 1993. Moth pheromone genetics and evolution. *Philosophical Transactions of the Royal Society of London Series B: Biological Sciences* 340:167–177.

Löfstedt, C., B.S. Lanne, J. Lofqvist, M. Appelgren, and G. Bergstrom. 1985. Individual variation in the pheromone of the turnip moth, *Agrotis segetum* Schiff (Lepidoptera: Noctuidae). *Journal of Chemical Ecology* 11:1181–1196.

Löfstedt, C., J. Löfqvist, B.S. Lanne, J.N.C.Van Der Pers, and B.S. Hansson. 1986. Pheromone dialects in European turnip moths *Agrotis segetum*. *Oikos* 46:250–257.

Löfstedt, C., B.S. Hansson, W. Roelofs, and B.O. Bengtsson. 1989. No linkage between genes controlling female pheromone production and male pheromone response in the European corn borer, *Ostrinia nubilalis* Hübner (Lepidoptera: Pyralidae). *Genetics* 123:553–556.

Monti, L., J. Genermont, C. Malosse, and B. Lalanne-Cassou. 1997. A genetic analysis of some components of reproductive isolation between two closely related species, *Spodoptera latifascia* (Walker) and *S. descoinsi* (Lalanne-Cassou and Silvain) (Lepidoptera: Noctuidae). *Journal of Evolutionary Biology* 10:121–134.

Morgan, T.H. 1917. The theory of the gene. *American Naturalist* 51:513–544.

Olsson, S.B., S. Kesevan, A.T. Groot, T. Dekker, D.G. Heckel, and B.S. Hansson. 2010. *Ostrinia* revisited: evidence for sex linkage in European corn borer *Ostrinia nubilalis* (Hübner) pheromone reception. *BMC Evolutionary Biology* 10.

Phelan, P.L. 1992. Evolution of sex pheromones and the role of asymmetric tracking. Pp. 265–314. In B.D. Roitberg and M.B. Isman, eds. *Insect Chemical Ecology: An Evolutionary Approach*. New York: Chapman & Hall.

Roelofs, W.L., and R.T. Cardé. 1971. Hydrocarbon sex pheromone in tiger moths (Arctiidae). *Science* 171:684–686.

Roelofs, W.L., and A.P. Rooney. 2003. Molecular genetics and evolution of pheromone biosynthesis in Lepidoptera. *Proceedings of the National Academy of Sciences of the United States of America* 100:9179–9184.

Roelofs, W.L., J.W. Du, C. Linn, T.J. Glover, and L.B. Bjostad. 1986. The potential for genetic manipulation of the redbanded leafroller moth sex pheromone blend. Pp. 263–272. In M.D. Huettel, ed. *Evolutionary Genetics of Invertebrate Behavior: Progress and Prospects*. New York: Plenum Press.

Roelofs, W., T. Glover, X.H. Tang, I. Sreng, P. Robbins, C. Eckenrode, C. Löfstedt, B.S. Hansson, and B.O. Bengtsson. 1987. Sex-pheromone production and perception in European corn borer moths is determined by both autosomal and sex-linked genes. *Proceedings of the National Academy of Sciences of the United States of America* 84:7585–7589.

Roelofs, W.L., W.T. Liu, G.X. Hao, H.M. Jiao, A.P. Rooney, and C.E. Linn. 2002. Evolution of moth sex pheromones via ancestral genes. *Proceedings of the National Academy of Sciences of the United States of America* 99:13621–13626.

Rooney, A.P. 2008. Molecular evolution of moth sex pheromone desaturases. *Chemical Senses* 33:S7–S8.

Sanders, C.J., G.E. Daterman, and T.J. Ennis. 1977. Sex-pheromone responses of *Choristoneura* spp. and their hybrids (Lepidoptera: Tortricidae). *The Canadian Entomologist* 109:1203–1220.

Sheck, A.L., A.T. Groot, C.M. Ward, C. Gemeno, J. Wang, C. Brownie, C. Schal, and F. Gould. 2006. Genetics of sex pheromone blend differences between *Heliothis virescens* and *Heliothis subflexa*: a chromosome mapping approach. *Journal of Evolutionary Biology* 19:600–617.

Sokolowski, M.B. 2001. *Drosophila*: genetics meets behaviour. *Nature Reviews Genetics* 2:879–890.

Sreng, I., T. Glover, and W. Roelofs. 1989. Canalization of the redbanded leafroller moth sex pheromone blend. *Archives of Insect Biochemistry and Physiology* 10:73–82.

Svensson, G.P., C. Ryne, and C. Löfstedt. 2002. Heritable variation of sex pheromone composition and the potential for evolution of resistance to pheromone-based control of the Indian meal moth, *Plodia interpunctella*. *Journal of Chemical Ecology* 28:1447–1461.

Tabata, J., and Y. Ishikawa. 2011. Genetic basis regulating the sex pheromone blend in *Ostrinia zealis* (Lepidoptera: Crambidae) and its allies inferred from crossing experiments. *Annals of the Entomological Society of America* 104:326–336.

Teal, P.E.A. 1985. The effect of hybridization on the periodicity of sex pheromone release of *Euxoa* species (Lepidoptera: Noctuidae). *Physiological Entomology* 10:463–466.

Tillman, J.A., S.J. Seybold, R.A. Jurenka, and G.J. Blomquist. 1999. Insect pheromones—an overview of biosynthesis and endocrine regulation. *Insect Biochemistry and Molecular Biology* 29:481–514.

Ueshima, R., and T. Asami. 2003. Single-gene speciation by left-right reversal—a land-snail species of polyphyletic origin results from chirality constraints on mating. *Nature* 425:679–679.

Vickers, N.J. 2006. Inheritance of olfactory preferences I. Pheromone-mediated behavioral responses of *Heliothis subflexa* x *Heliothis virescens* hybrid male moths. *Brain Behavior and Evolution* 68:63–74.

Wang, H.L., Q.L. Ming, C.H. Zhao, and C.Z. Wang. 2008. Genetic basis of sex pheromone blend difference between *Helicoverpa armigera* (Hübner) and *Helicoverpa assulta* (Guenée) (Lepidoptera: Noctuidae). *Journal of Insect Physiology* 54:813–817.

Wiley, C., C.K. Ellison, and K.L. Shaw. 2012. Widespread genetic linkage of mating signals and preferences in the Hawaiian cricket *Laupala*. *Proceedings of the Royal Society B—Biological Sciences* 279:1203–1209.

Wu, W.Q., C.B. Cottrell, B.S. Hansson, and C. Löfstedt. 1999. Comparative study of pheromone production and response in Swedish and Zimbabwean populations of turnip moth, *Agrotis segetum*. *Journal of Chemical Ecology* 25:177–196.

Zhu, J.W., C. Löfstedt, and B.O. Bengtsson. 1996a. Genetic variation in the strongly canalized sex pheromone communication system of the European corn borer, *Ostrinia nubilalis* Hübner (Lepidoptera: Pyralidae). *Genetics* 144:757–766.

Zhu, J.W., C.H. Zhao, F. Lu, M. Bengtsson, and C. Löfstedt. 1996b. Reductase specificity and the ratio regulation of E/Z isomers in pheromone biosynthesis of the European corn borer, *Ostrinia nubilalis* (Lepidoptera: Pyralidae). *Insect Biochemistry and Molecular Biology* 26:171–176.

CHAPTER SEVEN

Contextual Modulation of Moth Pheromone Perception by Plant Odors

TEUN DEKKER and ROMINA B. BARROZO

Moth pheromones are often viewed as simple, stand-alone signals in a complex olfactory world. From the first deciphering of a moth pheromone (Butenandt et al. 1959), to studies that detailed the olfactory circuitry underlying their detection (Schneider et al. 1964), pheromones have typically been viewed as a distinct olfactory entity, mirrored by a devoted olfactory sub-circuitry in the male moth (see Galizia and Rössler 2010). This may have supported the concept that pheromones are detected and processed in isolation from other odors. In nature, however, pheromone plumes are superimposed on a miasma of local odors, and moths can be far from indifferent to such complex plumes. Wind-tunnel and field-capture studies with pheromones have demonstrated behavioral interactions with non-pheromonal odors and recent neurophysiological studies have shown that pheromone and plant odor detection and processing are not independent in moths. Thus, although conceptually and evolutionarily pheromones may be best regarded as separate from non-pheromonal odors, from a behavioral, ecological, and applied perspective, boundaries between the two odor classes are often less well defined. The modes and mechanisms of plant odors and pheromone interactions are reviewed here.

Interactions Defined: Labeled Line versus Combinatorial Coding

Insect olfaction is a highly multidimensional sense. Odors are detected by several tens of olfactory receptor (Ors; Hallem and Carlson 2006) and ionotropic receptor (IRs; Silbering and Benton 2010) types, expressed in olfactory sensory neurons (OSNs), which together define the highly multidimensional olfactory space of an insect. OSNs vary in their breadth of response and detect a variety of often-related compounds. A single compound, on the other hand, generally elicits a response across several different OSN types (Hallem and Carlson 2006). Accordingly, general odors and blends are thought to be coded and discriminated through unique activation patterns across an ensemble of OSN types. This type of coding is called "across-fiber" coding (Erickson 1963).

Traditionally, however, moth pheromone perception has been categorized as a separate sub-circuitry, devoted to the detection of and response to intraspecific olfactory messages (Hansson et al. 1992). Pheromone signals are relatively simple, often binary or ternary blends, produced in specialized glands and distinct among species. Similarly, sensory neurons detecting these signals (pheromone-sensitive olfactory sensory neurons [Phe-OSNs]) are narrowly tuned, project to a small subset of glomeruli in the antennal lobes (ALs), and produce highly stereotypic behaviors that can be readily evoked in a wind tunnel. To contrast the pheromone coding system with the previously mentioned general odor circuitry, the term "labeled-line" coding was used early on (O'Connell 1975). Historically, labeled lines were defined as capturing, and coding for, single odor qualities. In terms of pheromones, in a labeled-line system a single pheromone component evokes a response exclusively in one neuron type, which in turn is exclusively tuned to the detection of that particular compound (i.e., exhibits a nonoverlapping response profile). This seemed fitting for pheromone coding in *Bombyx mori* (Bombycidae), which detects bombykol (E,Z)-10,12-hexadecadienol (Butenandt et al. 1959) using a single devoted OSN type (Schneider et al. 1964). In addition to peripheral detection, Phe-OSNs in males faithfully project to a unique set of enlarged glomeruli in the brain, the macroglomerular complex (MGC), in which each glomerulus is devoted to the processing of a single pheromone component (Hansson et al. 1992; Hildebrand 1995).

Later studies called into question the validity of the concept of labeled-line coding in pheromone perception in moths, following the finding that sensory neurons of the redbanded leafroller moths (*Argyrotaenia velutinana*, Tortricidae) cross-responded to two of its pheromone components (O'Connell 1975), although perhaps in part caused by artificially high doses used in stimulation (Akers and O'Connell 1988, 1991). Overlapping activation patterns of different Phe-OSN types were also described for *Yponomeuta* spp. (Yponomeutidae; Van Der Pers and Den Otter 1978) and the moth *Ostrinia furnacalis* (Crambidae; e.g., Takanashi et al. 2006; Domingue et al. 2007). In addition, the fact that at higher levels of organization in the brain, neurons downstream of these labeled input lines extensively "cross-talk" and show integrative, rather than separate lines of information flow (Kanzaki et al. 1989; Hansson et al. 1994; Trona et al. 2010a; Varela et al. 2011; Chaffiol et al. 2012), seems inconsistent with the concept of labeled lines in pheromone coding, in the sense of "all the way through from detection to behavioral output."

In some recent studies, the definition of labeled-line coding in olfaction seems to have shifted to a more OSN-to-behavior focus, rather than an odorant perspective. In those studies, labeled line typifies an OSN type that is, more or less irrespective of its tuning breadth, "necessary and sufficient" to evoke a defined behavioral response, such as repulsion, feeding, or oviposition (e.g., Stensmyr et al. 2012; Dweck et al. 2013). The use of labeled line in this context thus includes the behavioral significance of stimulating a particular neuronal pathway, but does not necessarily require narrow tuning. This is potentially helpful, as under this concept a broader tuning curve does not necessarily negate a neuron type to be labeled line, as long as it is necessary and sufficient to evoke a defined behavior. Also, it is increasingly clear that many so-called broadly tuned general odor OSNs are in fact narrowly tuned (i.e., labeled line *sensu stricto*) at ecologically relevant concentrations (Hallem and Carlson 2006), but do not evoke a particularly defined behavioral response alone, blurring the tuning-breadth concept of labeled-line coding (Galizia and Rössler 2010). The necessary-and-sufficient criterion, however, would potentially disqualify moth pheromone coding as labeled line, as most compounds in a pheromone blend are by themselves not sufficient to evoke the full behavioral repertoire. Galizia and Rössler (2010) advocated a further refinement of labeled-line coding, and distinguished "true" labeled-line systems (exemplified by CO_2 and geosmin detection in moths and *Drosophila*, respectively) and combinatorial-labeled line, the latter being typical for moth pheromones that generally use specific ratios of pheromone blends. In both labeled-line and combinatorial labeled-line coding, the information flow remains largely segregated from other AL glomeruli.

The necessary-and-sufficient criterion also does not preclude interactions with other odors. In this chapter, we highlight how the combinatorial labeled-line systems of pheromone coding is modulated by local odors at several levels of integration. We highlight studies that have dissected the modulatory effects of non-pheromonal odors on this combinatorial labeled-line system. We provide an overview from field to laboratory studies, from behavior to physiology, and from the periphery to the CNS. Finally, we evaluate the current state of understanding of the ecology and evolution of such interactions, as well as the literature on "sites" of interactions, their mechanisms, and their possible application in the management of moth pests.

Context Dependency of Pheromone Signaling in Moths, Its Adaptive Significance, and Behavioral Mechanisms

Although evolutionarily and ecologically plant odors and pheromones fall into different categories, each with a different set of selection pressures, the odor world presents itself to an organism as a whole, and one may expect these odor realms to interact. The fact that male moths, whose primary objective is mate finding, have OSNs and glomeruli dedicated to plant odors (see Kárpáti et al. 2008, and references therein), suggests a significant role of plant odors in male ecology, including pheromone orientation. Indeed, pheromone and plant-odor detection and preference may act in unison, rather than being temporally and spatially segregated. Synergisms between pheromones (both sex and aggregation pheromones) have frequently been observed in various beetle species (for an overview, see Landolt and Phillips 1997; Reddy and Guerrero 2004), but these have less frequently been reported for moths.

Some early observations on moth sexual communication noted that females called preferentially near host plants (Shorey 1974) and that some females perched on their host plants while calling (Kaae and Shorey 1973). Sexual communication in females of the sunflower moth, *Homoeosoma electellum* (Pyralidae), also appeared context sensitive; females initiated calling earlier and longer, and had an increased rate of egg maturation in the presence of odors of sunflower pollen (McNeil and Delisle 1989a). Similarly, in heliothine species host stimuli can regulate calling behavior (Raina 1988; Raina et al. 1992). In the cabbage looper, *Trichoplusia ni* (Noctuidae), exposure to volatiles from cotton plants modulated the calling of females, which resulted in increased male attraction (Landolt et al. 1994). In monophagous Yponomeutidae, females preferred to call on their host, and odors of these plants increased calling rates. This was matched by male pheromone orientation: host plant volatiles (HPVs) synergized male attraction to pheromone, and females perched on host plants caught more males than pheromone alone (Hendrikse and Vos-Buennemeyer 1987). In the codling moth, *Cydia pomonella* (Tortricidae), synergy was also noted for a combination host-emitted pear ester [(E,Z)-2,4-decadienaote] and codlemone [(E,E)-8,10-dodecadienol], depending on the type of orchard (pear, walnut, or apple) in which the tests were performed (Knight and Light 2005; Kutinkova et al. 2005; Light and Knight 2005; Knight et al. 2012). Similar synergy between pheromone and HPVs was observed for the grapevine moth, *Lobesia botrana* (Tortricidae) (von Arx et al. 2012).

Recent studies on host plant preference in the polyphagous species *Spodoptera littoralis* (Noctuidae) capitalized on observations that moth pheromone preference is sensitive to plant odor (Thöming et al. 2013). Males were given a choice between two different host plants, each with a pheromone lure, in a wind tunnel. Moths reliably showed in-flight preference among five plant species. However, this "innate" preference was dramatically modulated by the plant they had been reared on (the experimental design eliminated priming to these plant odors during eclosion; Thöming et al. 2013). In addition, a mating experience on one of these five host plants increased male preference for combinations of this plant plus female pheromone in subsequent dual-choice wind tunnel tests (Proffit et al. 2015). These experiments suggest that, besides the fact that olfactory memory is retained during metamorphosis, preference for host volatiles as background odor can modulate pheromone orientation, even in a polyphagous species.

In addition to direct effects of plant odors on sexual communication, there are also indirect effects. During hairpencil displays while courting females, males of many moth species emit plant-odor-type compounds, either synthesized de novo or sequestered from host plants. Compounds emitted by hairpencils facilitate mate courtship in various species through attracting, arresting, or inducing acceptance by females. They can also be used by other males in orientation (Fitzpatrick and McNeil 1988; Birch et al. 1990, Hillier and Vickers 2004). In many moth species the hairpencils release general odor compounds such as farnesal, linalool, various phenolics, and short-chain fatty acids (Birch et al. 1990; Landolt and Heath 1990). Some nymphalid butterflies and arctiine moths sequester pyrrolizidine alkaloids from their host plants to produce volatile hydropyrrolizines in male hairpencil perfumes. These compounds are important in mating success, and are transferred to females to protect eggs from predation (Eisner and Meinwald 1987; Landolt and Philips 1997; Schultz et al. 2004). The fact that in several lepidopteran species males release "general plant odors" during sexual communication demonstrates that plant-type odors are important in sexual communication.

Interactions between pheromones and plant odors can take place at various levels of integration in the olfactory circuitry (details below). Schneider et al. (1964) noted that the activity of Phe-OSNs was reduced when pheromone stimuli were combined with geraniol and other monoterpenoids in *Antheraea pernyi* (Saturniidae). Similar observations were made on *B. mori* (Kaissling et al. 1989). In an elegant experiment on a locomotor compensator, Kramer (1992) demonstrated how suppressed neuronal activity can indeed have pronounced effects on ensuing behavior. (*E,Z*)-9,11-hexadecadiene, a pheromone analog (not naturally occurring), induced tonic firing (prolonged responses up to 10 minutes post stimulus) in pheromone sensory neurons, and did not induce behavioral orientation. By applying pulsed linalool onto male *B. mori*, whose Phe-OSNs were firing in response to a continuous presentation of pheromone, males were induced to orient upwind. Neither linalool nor the hexadecadiene induced orientation alone. This demonstrates the biological significance of the linalool-induced reduction in the Phe-OSN response, and indicated that observed physiological effects of plant odor on pheromone signaling could be important in ecologically relevant settings. Although these studies used neat linalool on a filter paper, which calls in question the ecological relevance of the findings, later studies have confirmed the interaction between pheromone detection and several plant odor compounds, including monoterpenoids (see below). Rouyar et al. (2011) and Party et al. (2013) found that linalool, which reduces Phe-OSN activity, transiently disrupted walking orientation of wingless *S. littoralis* males on a locomotor compensator.

In wind, a plume emanating from a point source is shredded by turbulent forces into a filamentous structure. Filaments originating from two nearby sources may never fully merge spatiotemporally, and male moths orienting to pheromone are extraordinarily sensitive to these intricacies (Baker et al. 1998). In physiological studies, whereas linalool synergized Phe-OSN response to pheromone, an offset by 300 milliseconds did not (Ochieng et al. 2002). Similarly, stimulation of different parts of the antenna did not synergize pheromone and (*Z*)-3-hexenol responses in Phe-OSN of *B. mori* (Namiki et al. 2008). However, in addition to the necessity for high spatial–temporal coherence of the two stimuli, brief pre-exposure to plant odor may also sensitize moths to pheromone 3–24 hours after exposure (Minoli et al. 2012). Apparently, modulatory effects of plant odors need not be immediate, an aspect that has been overlooked in most studies.

Generally, studies on interactions between pheromone and plant odors have focused on their effects on moth behavior (measured in terms of captures; see below), with little reference to the adaptive significance of these interactions. However, contextual sensitivity serves some potential fitness benefits, including increased sensitivity and specificity in sexual communication, signaling of "good" host-plant patches, signaling quality of mates, and increased ability to track plumes. For each of these scenarios there is some evidence. Den Otter and Thomas (1979) hypothesized that in monophagous and oligophagous species, host-plant odors would strengthen the response to pheromone, whereas non-HPVs would weaken sexual attraction, as these plant odors would signal host quality and survival probability of offspring. Such interactions could lead to the evolution of tightly knit pheromone-HPV signaling, and support sympatric speciation. Indeed, host plants as "rendezvous" sites have been inferred for pheromone communication in some monophagous Yponomeutidae: pheromones alone provide insufficient species separation. Instead, a combination of pheromones and specific HPV seems to account for lack of heterospecific matings in nature (Hendrikse and Vos-Buennemeyer 1987; Liénard and Löfstedt, this volume). Thus, high host specificity may cause more relaxed selection on specificity of pheromone communication channels (in spite of sympatric congeners with similar blends), and an increased dependency on plant odor signaling in sexual communication.

Modulation of male pheromone preference by HPVs would seem also adaptive for nonspecialist moth species. A female calling on a host plant can oviposit directly after mating, without first dispersing. The combination of pheromone + host-plant odor thus signals to a male a ready-to-reproduce mate in a habitat that supports larval survival. As this likely enhances the success of offspring, this forms a reliable signal of reproductive success for orienting males. This would imply that males might have put selection pressure on females to call from suitable host plants. Indeed, in *S. littoralis*, a generalist species, HPVs influence male choice of female pheromone (Thöming et al. 2013; Hatano et al. 2015). Also, HPV caused earlier onset of female calling and prolonged calling in the sunflower moth *H. electellum*, indicating the importance of HPV in pheromone-mediated behaviors (e.g., McNeil 1986; McNeil and Delisle 1989b). Similarly, females of *Helicoverpa zea* (Noctuidae), which delay reproductive behaviors until after finding a suitable host plant, can be induced to produce and release pheromone by odors of their hosts (Raina et al. 1992). The significance of HPVs in signaling mate quality to male pheromone behavior also was shown in Yponomeutidae

lateral protocerebrum (Anton and Homberg 1999; Galizia and Rössler 2010).

In the AL of moths, the sex pheromone- and plant-related glomeruli are spatially segregated. Phe-OSNs project exclusively to a cluster of large, sexually dimorphic glomeruli known as the MGC, whereas the Pl-OSNs send axons to a separate array of ordinary glomeruli (OG) (Hansson and Anton 2000; Christensen and Hildebrand 2002). As shown by calcium imaging, pheromonal signal activity is confined to the MGC in males, whereas activity evoked by plant odors occurred in the OG in both male and female moths (Galizia et al. 2000; Carlsson et al. 2002; Deisig et al. 2012). In both the MGC and OG, olfactory information is encoded by combinatorial activation of glomeruli (Galizia and Rossler 2010).

There is substantial evidence of interaction between the MGC and OG. OG neurons were found to respond upon pheromone stimulation and conversely responses to plant volatiles were found within MGC neurons (Kanzaki et al. 1989; Hansson et al. 1994; Anton and Hansson 1995; Trona et al. 2010a; Varela et al. 2011; Chaffiol et al. 2012). In the moth *Grapholita molesta* (Tortricidae), pheromone processing takes place mainly in OG rather than within the MGC (Varela et al. 2011). Likewise, Trona et al. (2010a) demonstrated that many PNs and LNs in *C. pomonella* responded to stimulation with pheromonal compounds and plant odors, and their projections do not always innervate exclusively the MGC. This illustrates that pheromone encoding may not be restricted to the MGC, or that the MGC is not solely restricted to the processing of pheromones. Although there is a clear subdivision of the olfactory system, some overlap of responses among the subsystems of the AL of moths adds complexity to the system.

Systematic studies of the interconnectivity between the MGC and OG have offered ways to start deciphering pheromone–plant encoding in the brain of moths. For example, intracellular recordings in *Manduca sexta* (Sphingidae) have demonstrated asymmetries in the interaction of the MGC and OG. OG receive significant inhibition originating from the MGC, whereas the MGC receives little inhibition from the OG at high doses of plant odors (Reisenman et al. 2008). A large proportion of OG neurons of male *A. ipsilon* responded with excitation to pheromone (Chaffiol et al. 2014). Numerous MGC neurons respond to the plant odor heptanal, some with excitatory and others with inhibitory effects (Chaffiol et al. 2012). Because of the primarily excitatory nature of the OSNs projecting to the AL, inhibitory and excitatory responses to plant odors in the MGC may originate from different processing within the AL.

Interactions among pheromone blend components or among plant odors have been studied extensively (Lei and Vickers 2008; Riffell et al. 2009; Kuebler et al. 2011; Hatano et al. 2015, and references therein). However, studies on the interaction of sex pheromone and plant odor are less common. Despite this, a variety of experimental methods (e.g., intracellular and extracellular recordings, calcium imaging) have documented interactions occurring in the moth's AL, including synergistic, additive, and suppressive effects. In *B. mori*, the response of MGC PNs to pheromone was synergized by a mulberry leaf volatile, (*Z*)-3-hexenol, whereas pheromone had no influence on the responses of OG PNs to plant odors (Namiki et al. 2008). Similarly, in *C. pomonella*, a synergy was found in the response of males to mixtures of sex pheromone and plant volatiles in MGC neurons, whereas suppression was observed for mixtures in the OG neurons (Trona

et al. 2013). Synergistic and additive effects between pheromone and the plant-odor heptanal on OG neurons were found in the male moth *A. ipsilon*, (Barrozo et al. 2010; Chaffiol et al. 2014). Moreover and in contrast to other moths, in *A. ipsilon* suppression was the main effect of the plant-odor heptanal on the pheromone response within neurons in the MGC (Deisig et al. 2012; Chaffiol et al. 2012). These findings demonstrate that whereas mixture processing is species specific, mixtures are not symmetrically processed by both subsystems. Further, as pointed out by Deisig et al. (2014), external factors (other stimuli modalities, such as gustatory, visual, or auditory signals) and internal factors (experience, age, reproduction) can influence not only the behavioral response but also the response of AL neurons to mixtures.

Mixtures modulate the coding properties of MGC neurons (Chaffiol et al. 2012). In *A. ipsilon*, pheromone stimulation evokes first an increase in the spiking frequency of MGC neurons followed by an inhibitory phase, as in several other moths (Anton and Hansson 1999; Lei and Hansson 1999; Han et al. 2005; Jarriault et al. 2009). The co-occurrence of pheromone and heptanal stimulation, however, not only produces a decrease in the firing rate of MGC neurons but also a reduction of the excitatory phase duration, while the latency and the inhibitory phases are increased. Changes in the response characteristics of neurons may have profound effects on temporal coding of stimuli, a factor known to be highly significant in behavior (Baker et al. 1998; Vickers et al. 2001). Thus, Chaffiol et al. (2012) have shown that MGC neurons resolve pulses of mixtures of pheromone and heptanal better than pulses of pheromone alone. A pronounced inhibitory phase (following a shorter excitatory phase) may improve temporal resolution of reiterative stimulus encounters, such as a moth experiences in nature. The loss of the inhibitory phase of MGC neurons in *M. sexta* resulted in the inability of moths to follow a pheromone plume (Lei et al. 2009). Therefore, an "on–off" behavior of MGC PNs might help to resolve the odor plume discontinuity as proposed by Martinez et al. (2013). Similar to the pheromone blends that enhance the ability of MGC PNs to accurately follow pulses (Christensen and Hildebrand 1997; Heinbockel et al. 2004, 2013; Vickers et al. 2005), plant odors might well increase the ability of a male to track an intermittent pheromone plume through temporal contrast enhancement in the ALs.

The origin of the observed mixture interactions differs depending on neuron type and moth species. Evidence for synergy and suppressive effects of plant compounds on pheromone responses have been found at the periphery in several moths. Therefore, both types of interaction at the AL level may originate solely from input interactions at OSNs. Alternatively, mixture interactions may be a consequence of the action of the global or local inhibitory network in the insect AL. The inhibitory inputs to PNs are mediated by GABA-ergic LNs. The involvement of LNs in glomerular processing will thus likely modify the output of PNs, by either enhancing or inhibiting their spiking activity. Thus, lateral interactions between the MGC and OG could account at least in part for mixture processing and behavioral sensitivity to contextual stimuli in insects (Lei and Vickers 2008; Riffell et al. 2009; Clifford and Riffell 2013; Heinbockel et al. 2013). Additionally, synchrony in the firing among PNs may constitute a key mechanism of the neural representations of odors in the brain (Christensen and Hildebrand 2002).

In summary, a multitude of interactions between sex pheromone and plant odors, from the periphery to the ALs, have been documented. Although the behavioral and ecological significance of each of these interactions is not always clear, factors such as cross-sensitivity, synergism, inhibition and contrast enhancement, and increased temporal resolutions could have behaviorally relevant correlates.

Plant Odors and Pheromones in Application: Laboratory and Field Studies

Interest in the interaction of plant odors and pheromone and the effects of these interactions on orientation stems to a large degree from the potential synergism that could be achieved in trapping and perhaps direct control of pests. The use of multi-component blends in pest management programs can become prohibitively expensive. If inexpensive plant odors could synergize trap capture, reduced amounts of pheromone or incomplete blends could be sufficient in the control of pest moths. Attraction to the vicinity of traps could be mediated by plant odor, while at closer range incomplete blends (e.g., only the major pheromone component) could suffice for capturing males (Dickens et al. 1990, 1993). Although complete pheromone blends may have a larger active space, incomplete high-concentration blends may be sufficient at close range (Linn et al. 1987). Few studies have detailed the behavior of male moths to incomplete or underdosed pheromone blend augmented with plant odors (Yang et al. 2004; Schmidt-Büsser et al. 2009; Schmera and Guerin 2011; Varela et al. 2011). Most studies have scored trap capture in the field as a measure of these interactions.

With the exception of one anecdotal report (Creighton et al. 1973), the first field demonstrations of increased trap capture of male moths by combining pheromone with synthetic plant odor were not published until the 1990s. Following demonstration that trap captures of boll weevils (*Anthonomus grandis*, Curculionidae) were higher to combinations of aggregation pheromone and *(E)*-2-hexenol, a GLV from cotton, than to either alone (Dickens et al. 1990), cotton GLVs were tested for effects on the tobacco budworm *Heliothis virescens*. A series of GLVs were tested, six-carbon alcohols, aldehydes, and acetates, of which *(E)*-2- and *(Z)*-3-hexenyl acetates increased trap capture (Dickens et al. 1993). Similarly, trap capture of *H. zea*, a pest in corn, was enhanced by a combination of pheromone and the GLVs *(Z)*-3-hexenyl acetate, with other GLVs failing to augment trap capture (Light et al. 1993). None of the GLVs alone captured moths of either sex, indicating that the increased capture rates was due to the interaction of pheromone with *(Z)*-3-hexenyl acetate. Also, the presence of young corn plants significantly increased trap capture, indicating that in natural settings HPVs influence pheromone orientation in *H. zea*. Similar results were obtained for *C. pomonella*: a blend of GLVs significantly increased trap capture with *(E,E)*-8,10-dodecadienol, the major pheromone component of *C. pomonella* (Light et al. 1993). In the diamondback moth (*Plutella xylostella*, Plutellidae) blends of GLVs increased the attractiveness of pheromone in the wind tunnel and in the field (Reddy and Guerrero 2000). Deng et al. (2004) found that benzaldehyde, phenylacetaldehyde, *(E)*-2-hexanal, and *(Z)*-3-hexenyl acetate each synergized attraction in the wind tunnel and trap capture in the field of *Spodoptera exigua* (Noctuidae). Similarly, the pear and peach volatiles *(Z)*-3-hexenyl acetate and 1-undecanol synergized attraction of male *G. molesta* to pheromone in field capture studies (Yu et al. 2015). Although in

many of the above cases synergy was noted, in others simple additive responses were found, such as between *H. armigera* (Noctuidae) pheromone and phenylacetaldehyde (Kvedaras and Del Socorro 2007).

There is accumulating evidence that the effects of HPVs on male responses to pheromone are not independent of environmental effects. For instance, in both wind-tunnel and field trials with the apple fruit moth, *Argyresthia conjugella* (Yponomeutidae), the plant odors 2-methyl ethanol and anethole did or did not attract individuals, depending on the HPV background (Knudsen et al. 2008). Similarly, ethyl *(E,Z)*-2,4-decadienoate, a pear-derived attractant of *C. pomonella* (Light et al. 1993) with pheromone-like attractiveness in the United States, was less attractive in other parts of the world (Light et al. 2001; Mitchell et al. 2008; El-Sayed et al. 2013, and references therein). Part of the variation seemed to be due to the orchard type (apple varieties, walnut, pear) and associated background odors (Light et al. 2001; Thwaite et al. 2004; Hilton et al. 2005; Knight and Light 2005; Light and Knight 2005). Perhaps due to similar variation in background odors, combinations of codlemone and pear ester increased trap capture (Knight et al. 2005, 2010; Kutinkova et al. 2005), whereas it decreased orientation in wind-tunnel studies (Trona et al. 2010b; Schmera and Guerin 2011). Apparently, host odors, even those highly attractive on their own, do not always augment the attraction to pheromone when combined. The dependency of HPV-pheromone synergies on local settings greatly increases the difficulty to translate wind-tunnel studies into natural settings in the field, and may represent an impediment for HPV use in pest control.

Summary and Outlook

The available literature suggests that physiological and behavioral interactions between plant odors and pheromones are common and can be inhibitory, additive, or synergistic. Although physiological studies have scored inhibition of sensory responses, most behavioral and field studies have shown increased, often synergistic, attraction and trap capture. Strikingly, the repertoire of HPVs that have been tested in these studies has been rather limited. In behavioral and field studies, GLVs (six-carbon-chain alcohols, aldehydes, and esters) that are commonly released by most plants species have been extensively studied, whereas interactions with a set of monoterpenoids (particularly linalool and geraniol) dominate the physiological studies. The reason is simple: GLVs are ubiquitous and cheap to produce, and if they can synergize (incomplete or underdosed) pheromone blends, this would be of applied significance. Conversely, the interaction of monoterpenoids with peripheral detection of pheromones was already noted in the 1960s and these early observations have channeled much of the attention in physiological studies to this class of compounds.

In spite of the solid behavioral and physiological evidence of pheromone–plant odor interactions, the evolutionary-ecological context of these interactions has been less clear. GLVs or monoterpenoids are alone rarely characteristic of a host plant, and therefore do not fit into the classification of HPVs (Bruce et al. 2005; Bruce and Pickett 2011). In addition, in studies on contextual modulations HPVs are often released uncontrolled and in unnaturally high release rates, which makes inferences about their ecological relevance speculative. However, an increasing number of synthetic host odor blends are available for several moth and host-plant species (Hartlieb and Rembold 1996; Tasin et al. 2005, 2007; Bruce and Pickett 2011) that can readily be used to study the interaction of blends in natural ratios and concentrations. Recently, Hatano et al. (2015) created a synthetic cotton mimic that attracted moths in the wind tunnel. Addition of the HIPV DMNT strongly reduced attraction. The same was true for pheromone orientation: addition of DMNT to the lure strongly reduced attraction. This was correlated with suppression of odor-evoked responses in the antenna and AL, which suggests that DMNT and other HIPVs may constitute a signal of poor-quality and high-risk habitats. Studies such as the above shed light on the behavioral, ecological, and evolutionary relevance of contextual modulation, and may stimulate further investigations on the use of HPVs in applied settings.

Acknowledgments

This work was funded through the Linneaus grant Insect Chemical Ecology, Ethology and Evolution (IC-E³) and the Carl Trygger Foundation (TD), and the Argentine National Agency of Science and Technology (ANPCyT) and the Argentine National Research Council (CONICET).

References Cited

Akers, R.P., and R.J. O'Connell. 1988. The contribution of olfactory receptor neurons to the perception of pheromone component ratios in male redbanded leafroller moths. *Journal of Comparative Physiology A* 163:641–650.

Akers, R.P., and R.J. O'Connell. 1991. Response specificity of male olfactory receptor neurones for the major and minor components of a female pheromone blend. *Physiological Entomology* 16:1–17.

Anton, S., and B.S. Hansson. 1995. Sex pheromone and plant-associated odour processing in antennal lobe interneurons of male *Spodoptera littoralis* (Lepidoptera: Noctuidae). *Journal of Comparative Physiology A* 176:773–789.

Anton, S., and B.S. Hansson. 1999. Physiological mismatching between neurons innervating olfactory glomeruli in a moth. *Proceeding of the Royal Society of London B* 266:1813–1820.

Anton, S., and U. Homberg. 1999. Antennal lobe structure. Pp. 98–125. In B.S. Hansson, ed. *Insect Olfaction Pages*. Berlin: Springer.

Baker, T.C., H.Y. Fadamiro, and A.A. Cossé. 1998. Moth uses fine tuning for odour resolution. *Nature* 393:530–530.

Barrozo, R.B., C. Gadenne, and S. Anton. 2010. Switching attraction to inhibition: mating-induced reversed role of sex pheromone in an insect. *The Journal of Experimental Biology* 213:2933–2939.

Birch, M.C., G.M. Poppy, and T.C. Baker. 1990. Scents and eversible scent structures of male moths. *Annual Review of Entomology* 35:25–58.

Brigaud, I., X. Grosmaitre, M.C. Francois, and E. Jacquin-Joly. 2009. Cloning and expression pattern of a putative octopamine/tyramine receptor in antennae of the noctuid moth *Mamestra brassicae. Cell Tissue Research* 335:455–463.

Bruce, T.J.A. and J.A. Pickett. 2011. Perception of plant volatile blends by herbivorous insects: finding the right mix. *Phytochemistry* 72:1605–1611.

Bruce, T.J.A., L.J. Wadhams, and C.M. Woodcock. 2005. Insect host location: a volatile situation. *Trends in Plant Science* 10:269–274.

Butenandt, A., R. Beckman, and E. Hecker. 1959. Über den Sexuallockstoff des Seidenspinners. 1. Der biologische Test und die Isolierung des reinen Sexuallockstoffes Bombykol. *Hoppe-Seylers Zeitschrift für Physiologische Chemie* 324:71–83. [In German]

Carlsson, M.A., C.G. Galizia, and B.S. Hansson. 2002. Spatial representation of odours in the antennal lobe of the moth *Spodoptera littoralis* (Lepidoptera: Noctuidae). *Chemical Senses* 27:231–244.

Chaffiol, A., F. Dupuy, R.B. Barrozo, J. Kroft, M. Renou, J.P. Rospars, and S. Anton. 2014. Pheromone modulates plant odour responses in the antennal lobe of a moth. *Chemical Senses* 39:451–463.

Chaffiol, A., J. Kropf, R.B. Barrozo, C. Gadenne, R.J. Rospars, and S. Anton. 2012. Plant odour stimuli reshape pheromonal representation in neurons of the antennal lobe macroglomerular complex of a male moth. *The Journal of Experimental Biology* 215:1670–1680.

Christensen, T.A., and J.G. Hildebrand. 1997. Coincident stimulation with pheromone components improves temporal pattern resolution in central olfactory neurons. *Journal of Neurophysiology* 77:775–781.

Christensen, T.A., and J.G. Hildebrand. 2002. Pheromonal and host-odor processing in the insect antennal lobe: how different? *Current Opinion Neurobiology* 12:393–399.

Clifford, M.R., and J.A. Riffel. 2013. Mixture and odorant processing in the olfactory systems of insects: a comparative perspective. *Journal of Comparative Physiology A* 199:911–928.

Creighton, C.S., T.L. McFadden, and E.R. Cuthbert. 1973. Supplementary data on phenylacetaldehyde: an attractant for Lepidoptera. *Journal of Economic Entomology* 66:114–115.

De Moraes, C.M., M.C. Mescher, and J.H. Tumlinson. 2001. Caterpillar induced nocturnal plant volatiles repel conspecific females. *Nature* 410:577–580.

Deisig, N., F, Dupuy, S. Anton, and M. Renou. 2014. Responses to pheromones in a complex odor world: sensory processing and behavior. *Insects* 5:399–422.

Deisig, N., J. Kropf, S. Vitecek, D. Pevergne, A. Rouyar, J.C. Sandoz, P. Luca, C. Gadenne, S. Anton, and R.B. Barrozo. 2012. Differential interactions of sex pheromone and plant odour in the olfactory pathway of a male moth. *PLOS ONE* 7:e33159.

Den Otter, C.J., and G. Thomas. 1979. Olfactory preference in insects: a synthesis of behaviour and electrophysiology. Pp. 171–182. In G.H.A. Kroeze, ed. *Preference Behaviour and Chemoreception.* London: Information Retrieval Limited.

Den Otter, C.J., H.A. Schuil, and A.-V. Oosten. 1978. Reception of host-plant odours and female sex pheromone in *Adoxophyes orana* (Lepidoptera: Tortricidae): electrophysiology and morphology. *Entomologia Experimentalis et Applicata* 24:570–578.

Deng, J.-Y., H.-Y. Wei, Y.-P. Huang, and J.-W. Du. 2004. Enhancement of attraction to sex pheromones of *Spodoptera exigua* by volatile compounds produced by host plants. *Journal of Chemical Ecology* 30:2037–2045.

Dickens, J.C., E.B. Jang, D.M. Light, and A.R. Alford. 1990. Enhancement of insect pheromone responses by green leaf volatiles. *Naturwissenschaften* 77:29–31.

Dickens, J.C., J.W. Smith, and D.M. Light. 1993. Green leaf volatiles enhance sex attractant pheromone of the tobacco budworm, *Heliothis virescens* (Lep.: Noctuidae). *Chemoecology* 4:175–177.

Dolzer, J., S. Krannich, K. Fischer, and M. Stengl. 2001. Oscillations of the transepithelial potential of moth olfactory sensilla are influenced by octopamine and serotonin. *The Journal of Experimental Biology* 204:2781–2794.

Domingue, M.J., C.J. Musto, C.E. Linn, Jr., W.L. Roelofs, and T.C. Baker. 2007. Altered olfactory receptor neuron responsiveness in rare *Ostrinia nubilalis* males attracted to the *O. furnacalis* pheromone blend. *Journal of Insect Physiology* 53:1063–1071.

Duchamp-Viret, P., A. Duchamp, and M.A. Chaput. 2003. Single olfactory sensory neurons simultaneously integrate the components of an odour mixture. *European Journal of Neuroscience* 18:2690–2696.

Dweck, H.K.M., S.A.M. Ebrahim, S. Kromann, D. Bown, Y. Hillbur, S. Sachse, B.S. Hansson, and M.C. Stensmyr. 2013. Olfactory preference for egg laying on citrus substrates in *Drosophila. Current Biology* 23:2472–2480.

Eisner, T., and J. Meinwald. 1987. Alkaloid-derived pheromones and sexual selection in *Lepidoptera*. Pp. 341–368. In G.J. Blomquist and R.C. Vogt, eds. *Pheromone Biochemistry and Molecular Biology.* London: Elsevier Academic Press.

El-Sayed, A.M., L. Cole, J. Revell, L.-A. Manning, A. Twidle, A.L. Knight, V.G.M. Bus, and D.M. Suckling. 2013. Apple volatiles synergize the response of codling moth to pear ester. *Journal of Chemical Ecology* 39:643–652.

Erickson, R.P. (1963) Sensory neural patterns and gustation. Pp. 205–213. In Y. Zotterman, ed. *Olfaction and taste*, Vol. I. Oxford: Pergamon Press.

Fitzpatrick, S.M., and J.N. McNeil. 1988. Male scent in lepidopteran communication; the role of male pheromone in mating behaviour of *Pseudaletia unipuncta* (Haw.) (Lepidoptera: Noctuidae). *Memoirs of the Entomological Society of Canada* 146:131–151.

Flecke, C., and M. Stengl. 2009. Octopamine and tyramine modulate pheromone-sensitive olfactory sensilla of the hawkmoth *Manduca sexta* in a time-dependent manner. *Journal of Comparative Physiology A* 195:529–539.

Flecke, C., A. Nolte, and M. Stengl. 2010. Perfusion with cAMP analogue affects pheromone-sensitive trichoid sensilla of the hawkmoth *Manduca sexta* in a time-dependent manner. *The Journal of Experimental Biology* 213:842–852.

Galizia, C.G., and W. Rössler. 2010. Parallel olfactory systems in insects: anatomy and function. *Annual Review of Entomology* 55:399–420.

Galizia, C.G., S. Sachse, and H. Mustaparta. 2000. Calcium responses to pheromones and plant odours in the antennal lobe of the male and female moth *Heliothis virescens. Journal of Comparative Physiology A* 186:1049–1063.

Grosmaitre, X., F. Marion-Poll, and M. Renou. 2001. Biogenic amines modulate olfactory receptor neurons firing activity in *Mamestra brassicae. Chemical Senses* 26:653–661.

Grosse-Wilde, E., A. Svatos, and J. Krieger. 2006. A pheromone-binding protein mediates the bombykol-induced activation of a pheromone receptor in vitro. *Chemical Senses* 31:547–555.

Hallem, E.A., and J.R. Carlson. 2006. Coding of odors by a receptor repertoire. *Cell* 125:143–160.

Han, Q., B.S. Hansson, and S. Anton. 2005. Interactions of mechanical stimuli and sex pheromone information in antennal lobe neurons of a male moth, *Spodoptera littoralis. Journal of Comparative Physiology A* 191:521–528.

Hansson, B.S., and S. Anton. 2000, Function and morphology of the antennal lobe: new developments. *Annual Review of Entomology* 45:203–231.

Hansson, B.S., S. Anton, and T.A. Christensen. 1994. Structure and function of antennal lobe neurons in the male turnip moth, *Agrotis segetum* (Lepidoptera: Noctuidae). *Journal of Comparative Physiology A* 175:547–562.

Hansson, B.S., H. Ljungberg, E. Hallberg, and C. Löfstedt. 1992. Functional specialization of olfactory glomeruli in a moth. *Science* 256:1313–1315.

Hartlieb, E., and H. Rembold. 1996. Behavioral response of female *Helicoverpa* (*Heliothis*) *armigera* HB. (Lepidoptera: Noctuidae) moths to synthetic pigeonpea (*Cajanus cajan* L.) kairomone. *Journal of Chemical Ecology* 22:821–837.

Hatano, E., A.M. Saveer, F. Borrero-Echeverry, M. Strauch, A. Zakir, M. Bengtsson, R. Ignell et al. 2015. A herbivore-induced plant volatile interferes with host plant and mate location in moths through suppression of olfactory signaling pathways. *BMC Biology* 13:75.

Heinbockel, T., T.A. Christensen, and J.G. Hildebrand. 2004. Representation of binary pheromone blends by glomerulus-specific olfactory projection neurons. *Journal of Comparative Physiology A* 190:1023–1037.

Heinbockel, T., V.D.C. Shields, and C.E. Reisenman, C.E. 2013. Glomerular interactions in olfactory processing channels of the antennal lobes. *Journal of Comparative Physiology A* 199:929–946.

Hendrikse, A., and E. Vos-Buennemeyer. 1987. Role of host-plant stimuli in sexual behaviour of small ermine moths (*Yponomeuta*). *Ecological Entomology* 12:363–371.

Hildebrand, J.G. 1995. Analysis of chemical signals by nervous systems. *Proceedings of the National Academy of Sciences of the United States of America* 92:67–74.

Hillier, N.K. and N.J. Vickers. 2004. The role of heliothine hairpencil compounds in female *Heliothis virescens* (Lepidoptera: Noctuidae) behavior and mate acceptance. *Chemical Senses* 29:499–511.

Hillier, N.K., and N.J. Vickers. 2011. Mixture interactions in moth olfactory physiology: examining the effects of odorant mixture, concentration, distal stimulation, and antennal nerve transection on sensillar responses. *Chemical Senses* 36:93–108.

Hilton, R.J., P. Van Buskirk, R.J. Hilton, B.G. Zoller, and D.M. Light. 2005. Monitoring codling moth in four pear cultivars with the pear ester. *Acta Horticulturae* 671:565–570.

Jarriault, D., C. Gadenne, J.P. Rospars, and S. Anton. 2009 Quantitative analysis of sex-pheromone coding in the antennal lobe of the moth *Agrotis ipsilon*: a tool to study network plasticity. *The Journal of Experimental Biology* 212:1191–1201.

Kaae, R.S., and H.H. Shorey. 1973. Sex pheromones of Lepidoptera: influence of environmental conditions on the location of pheromone communication and mating in *Pectinophora gossypiella*. *Environmental Entomology* 2:1081–1084.

Kaissling, K.E., L.Z. Meng, and H.J. Bestmann. 1989. Responses of bombykol receptor cells to (Z,E)-4,6-hexadecadiene and linalool. *Journal of Comparative Physiology A* 165:147–154.

Kanzaki, R., E.A. Arbas, N.J. Strausfeld, and J.G. Hildebrand. 1989. Physiology and morphology of projection neurons in the antennal lobe of the male moth *Manduca sexta. Journal of Comparative Physiology A* 165:427–453.

Kárpáti, Z., T. Dekker, and B.S. Hansson. 2008. Reversed functional topology in the antennal lobe of the male European corn borer. *The Journal of Experimental Biology* 211:2841–2848.

Knight, A., J. Haworth, B. Lingren, and V. Hebert. 2010. Combining pear ester with codlemone improves management of codling moth. *IOBC/wprs Bulletin* 2010:345–348.

Knight, A.L., and D.M. Light. 2005. Factors affecting the differential capture of male and female codling moth (Lepidoptera: Tortricidae) in traps baited with Ethyl (*E, Z*)-2,4-Decadienoate. *Environmental Entomology* 34:1161–1169.

Knight, A.L., R. Hilton, and D.M. Light. 2005. Monitoring codling moth (Lepidoptera: Tortricidae) in apple with blends of ethyl (*E, Z*)-2,4-decadienoate and codlemone. *Environmental Entomology* 34:598–603.

Knight, A.L., D.M. Light, and V. Chebny. 2012. Monitoring codling moth (Lepidoptera: Tortricidae) in orchards treated with pear ester and sex pheromone combo dispensers. *Journal of Applied Entomology* 137:214–224.

Knudsen, G.K., M. Bengtsson, S. Kobro, G. Jaastad, T. Hofsvang, and P. Witzgall. 2008. Discrepancy in laboratory and field attraction of apple fruit moth *Argyresthia conjugella* to host plant volatiles. *Physiological Entomology* 33:1–6.

Kramer, E. 1992. Attractivity of pheromone surpassed by time-patterned application of 2 nonpheromone compounds. *Journal of Insect Behavior* 5:83–97.

Kuebler, L.S., S.B. Olsson, R. Weniger, and B.S. Hansson. 2011. Neuronal processing of complex mixtures establishes a unique odor representation in the moth antennal lobe. *Frontiers in Neural Circuits* 5:7.

Kutinkova, H., M. Subchev, and D.M. Light. 2005. Interactive effects of ethyl (*E,Z*)-2, 4-decadienoate and sex pheromone lures to codling moth: apple orchard investigations in Bulgaria. *Journal of Plant Protection Research* 45:49–52.

Kvedaras, O.L., and A.P. Del Socorro. 2007. Effects of phenylacetaldehyde and (*Z*)-3-hexenyl acetate on male response to synthetic sex pheromone in *Helicoverpa armigera* (Hübner) (Lepidoptera: Noctuidae). *Australian Journal of Entomology* 46:2240–230.

Landolt, P.J., and R.R. Heath. 1990. Sexual role reversal in mate-finding strategies of the cabbage looper moth. *Science* 249:1026–1028.

Landolt, P.J., and T.W. Phillips. 1997 Host plant influences on sex pheromone behavior of phytophagous insects. *Annual Review of Entomology* 42:371–391.

Landolt, P.J., R.R. Heath, J.G. Millar, K.M. Davis-Hernandez, B.D. Dueben, and K.E. Ward. 1994. Effects of host plant, *Gossypium hirsutum* L., on sexual attraction of cabbage looper moths, *Trichoplusia ni* (Hübner) (Lepidoptera: Noctuidae). *Journal of Chemical Ecology* 20:2959–2974.

Lei, H., and B.S. Hansson. 1999. Central processing of pulsed pheromone signals by antennal lobe neurons in the male moth *Agrotis segetum. Journal of Neurophysiology* 81:1113–1122.

Lei, H., and N. Vickers. 2008. Central processing of natural odor mixtures in insects. *Journal Chemical Ecology* 34:915–927.

Lei, H., J.A. Riffell, S.L. Gage, and J.G. Hildebrand. 2009. Contrast enhancement of stimulus intermittency in a primary olfactory network and its behavioral significance. *Journal of Biology* 8:21.

Light, D.M., and A.L. Knight. 2005. Specificity of codling moth (Lepidoptera: Tortricidae) for the host plant kairomone, ethyl (*E, Z*)-2,4-decadienoate: field bioassays with pome fruit volatiles, analogue, and isomeric compounds. *Journal of Agricultural and Food Chemistry* 53:4046–4053.

Light, D.M., R.A. Flath, R.G. Buttery, F.G. Zalom, R.E. Rice, J.C. Dickens, and E.B. Jang. 1993. Host-plant green-leaf volatiles synergize the synthetic sex pheromones of the corn earworm and codling moth (Lepidoptera). *Chemoecology* 4:145–152.

Light, D.M., A.L. Knight, C.A. Henrick, and D. Rajapaska. 2001. A pear-derived kairomone with pheromonal potency that attracts male and female codling moth, *Cydia pomonella* (L.) *Naturwissenschaften* 88:333–338.

Linn, C.E., Jr., and W.L. Roelofs. 1986. Modulatory effects of octopamine and serotonin on male sensitivity and periodicity of response to sex pheromone in the cabbage looper moth

Trichoplusia ni. Archives of Insect Biochemistry and Physiology 3:161–172.

Linn, C. E., Jr., M. G. Campbell, and W. L. Roelofs. 1987. Pheromone components and active spaces: what do moths smell and where do they smell it? *Science* 237:650–652.

Linn, C. E., Jr., M. G. Campbell, and W. L. Roelofs. 1992. Photoperiod cues and the modulatory action of octopamine and 5-hydroxy-tryptamine on locomotor and pheromone in male gypsy moths, *Lymantria dispar. Archives of Insect Biochemistry and Physiology* 20:265–284.

Mafra-Neto, A., and R. T. Cardé. 1994. Fine-scale structure of pheromone plumes modulates upwind orientation of flying moths. *Nature* 369:142–144.

Martinez, D., A. Chaffiol, N. Voges, Y. Gu, S. Anton, J. P. Rospars, and P. Lucas. 2013. Multiphasic on/off pheromone signalling in moths as neural correlates of a search strategy. *PLOS ONE* 8:e61220.

McNeil, J. N., 1986. Calling behavior: can it be used to identify migratory species of moths? *The Florida Entomologist* 69:78–84.

McNeil, J. N., and J. Delisle. 1989a. Are host plants important in pheromone-mediated mating systems of Lepidoptera? *Cellular and Molecular Life Sciences* 45:236–240.

McNeil, J. N., and J. Delisle. 1989b. Host plant pollen influences calling behavior and ovarian development of the sunflower moth, *Homoeosoma electellum. Oecologia* 80:201–205.

Minoli, S., I. Kauer, V. Colson, V. Party, M. Renou, P. Anderson, C. Gadenne, F. Marion-Poll, and S. Anton. 2012. Brief exposure to sensory cues elicits stimulus-nonspecific general sensitization in an insect. *PLOS ONE* 7:e34141.

Mitchell, V. J., L.-A. Manning, L. Cole, D. M. Suckling, and A. M. El-Sayed. 2008. Efficacy of the pear ester as a monitoring tool for codling moth *Cydia pomonella* (Lepidoptera: Tortricidae) in New Zealand apple orchards. *Pest Management Science* 64:209–214.

Munch, D., B. Schmeichel, A. F. Silbering, and G. C. Galizia. 2013. Weaker ligands can dominate an odor blend due to syntopic interactions. *Chemical Senses* 38:293–304.

Namiki, S., S. Iwabuchi, and R. Kanzaki. 2008. Representation of a mixture of pheromone and host plant odor by antennal lobe projection neurons of the silkmoth *Bombyx mori. Journal of Comparative Physiology A* 194:501–515.

Nikonov, A. A., and W. S. Leal. 2002. Peripheral coding of sex pheromone and a behavioral antagonist in the Japanese beetle, *Popillia japonica. Journal of Chemical Ecology* 28:1075–1089.

Ochieng, S. A., K. C. Park, and T. C. Baker 2002. Host plant volatiles synergize responses of sex pheromone-specific olfactory receptor neurons in male *Helicoverpa zea. Journal of Comparative Physiology A* 188:325–333.

O'Connell, R. J., 1975. Olfactory receptor responses to sex pheromone components in the redbanded leafroller moth. *The Journal of General Physiology* 65:179–205.

O'Connell, R. J. 1985 Responses to pheromone blends in insect olfactory receptor neurons. *Journal of Comparative Physiology A* 156:747–761.

Party, V., C. Hanot, I. Said, D. Rochat, and M. Renou. 2009. Plant terpenes affect intensity and temporal parameters of pheromone detection in a moth. *Chemical Senses* 34:763–774.

Party, V., C. Hanot, D. Schmidt Büsser, D. Rochat, and M. Renou. 2013. Changes in odor background affect the locomotory response to pheromone in moths. *PLOS ONE* 8:e52897.

Pophof, B. 2000. Octopamine modulates the sensitivity of silkmoth pheromone receptor neurons. *Journal of Comparative Physiology A* 186:307–313.

Pophof, B., and W. van der Goes van Naters. 2002. Activation and inhibition of the transduction process in silkmoth olfactory receptor neurons. *Chemical Senses* 27:435–443.

Pregitzer, P., M. Schubert, H. Breer, B. S. Hansson, S. Sachse, and J. Krieger. 2012. Plant odorants interfere with detection of sex pheromone signals by male *Heliothis virescens. Frontiers in Cellular Neuroscience* 6:42.

Proffit, M., M. A. Khallaf, D. Carrasco, M. C. Larsson, and P. Anderson. 2015. 'Do you remember the first time?' Host plant preference in a moth is modulated by experiences during larval feeding and adult mating. *Ecology Letters* 18:365–374.

Raina, A. K. 1988. Selected factors influencing neurohormonal regulation of sex pheromone production in *Heliothis* species. *Journal of Chemical Ecology* 14:2063–2069

Raina, A. K., T. G. Kingan, and A. K. Mattoo. 1992. Chemical signals from host plant and sexual behavior in a moth. *Science* 255:592–594.

Reddy, G. V. P., and A. Guerrero. 2000. Behavioral responses of the diamondback moth, *Plutella xylostella*, to green leaf volatiles of *Brassica oleracea* subsp. *capitata. Journal of Agricultural and Food Chemistry* 48:6025–6029.

Reddy, G. V. P., and A. Guerrero. 2004. Interactions of insect pheromones and plant semiochemicals. *Trends in Plant Science* 9:253–261.

Reisenman, C. E., T. Heinbockel, and J. G. Hildebrand. 2008. Inhibitory interactions among olfactory glomeruli do not necessarily reflect spatial proximity. *Journal of Neurophysiology* 100:554–564.

Riffell, J. A., H. Lei, and J. G. Hildebrand. 2009. Neural correlates of behavior in the moth *Manduca sexta* in response to complex odors. *Proceedings of the National Academy of Sciences of the United States of America* 106:19219–19226.

Rospars, J. P. 1988. Structure and development of the insect antennodeutocerebral system. *International Journal of Insect Morphology and Embryology* 17:243–294.

Rospars J. P., P. Lansky, M. Chaput, and P. Duchamp-Viret P. 2008. Competitive and noncompetitive odorant interactions in the early neural coding of odorant mixtures. *Journal of Neuroscience* 28:2659–2666.

Rouyar, A., V. Party, J. Presern, A. Blejec, and M. Renou. 2011. A general odorant background affects the coding of pheromone stimulus intermittency in specialist olfactory receptor neurones. *PLOS ONE* 6:e26443.

Schmera, D., and P. M. Guerin. 2011. Plant volatile compounds shorten reaction time and enhance attraction of the codling moth (*Cydia pomonella*) to codlemone. *Pest Management Science* 68:454–461.

Schmidt-Büsser, D., M. Arx, and P. M. Guerin. 2009. Host plant volatiles serve to increase the response of male European grape berry moths, *Eupoecilia ambiguella*, to their sex pheromone. *Journal of Comparative Physiology A* 195:853–864.

Schneider, D. 1964. Insect antennae. *Annual Review of Entomology* 9:103–122.

Schneider, D., V. Lacher, and K.-E. Kaissling. 1964. Die Reaktions-weise und das Reaktionsspektrum von Riechzellen bei *Antheraea pernyi* (Lepidoptera, Saturniidae). *Journal of Comparative Physiology A* 48:632–662. [In German]

Schultz, S., G. Beccaloni, K. S. Brown, M. Boppre, A. V. L. Freitas, P. Ockenfels, and J. R. Trigo. 2004. Semiochemicals derived from pyrrolizidine alkaloids in male ithomiine butterflies (Lepidoptera: Nymphalidae: Ithomiinae). *Biochemical Systematics and Ecology* 32:699–713.

Shorey, H. H., 1974. Environmental and physiological control of insect sex pheromone behavior. Pp. 60–82. In M. C. Birch, ed. *Pheromones*. Noord Holland: Elsevier.

Silbering, A. F., and R. Benton. 2010. Ionotropic and metabotropic mechanisms in chemoreception: 'chance or design'? *EMBO Reports* 11:173–179.

Stensmyr, M. C., H. K. M. Dweck, A. Farhan, I. Ibba, A. Strutz, L. Mukunda, J. Linz et al. 2013. A conserved dedicated olfactory circuit for detecting harmful microbes in *Drosophila. Cell* 151:1345–1357.

Su, C. Y., K. Menuz, J. Reisert, and J. R. Carlson. 2012. Non-synaptic inhibition between grouped neurons in an olfactory circuit. *Nature* 492:66–71.

Takanashi, T., Y. Ishikawa, P. Anderson, Y. Huang, C. Löfstedt, S. Tatsuki, and B. S. Hansson. 2006. Unusual response characteristics of pheromone-specific olfactory receptor neurons in the Asian corn borer moth, *Ostrinia furnacalis. The Journal of Experimental Biology* 209:4946–4956.

Tasin, M., G. Anfora, C. Ioriatti, S. Carlin, A. De Cristofaro, S. Schmidt, M. Bengtsson, G. Versini, and P. Witzgall. 2005. Antennal and behavioral responses of grapevine moth *Lobesia botrana* females to volatiles from grapevine. *Journal of Chemical Ecology* 31:77–87.

Tasin, M., A.-C. Bäckman, M. Coracini, D. Casado, C. Ioriatti, and P. Witzgall. 2007. Synergism and redundancy in a plant volatile blend attracting grapevine moth females. *Phytochemistry* 68:203–209.

Thöming, G., M. C. Larsson, B. S Hansson, and P. Anderson. 2013. Comparison of plant preference hierarchies of male and female moths and the impact of larval rearing hosts. *Ecology* 94:1744–1752.

Thwaite, W.G., M.A. Eslick, and R. Van de Ven. 2004. Susceptibility of three apple cultivars to petroleum-derived spray oils applied in late blossom. *Entomological Society of New South Wales* 33:29–55.

Trona, F., G. Anfora, M. Bengtsson, P. Witzgall, and R. Ignell 2010a. Coding and interaction of sex pheromone and plant volatile signals in the antennal lobe of the codling moth *Cydia pomonella*. *The Journal of Experimental Biology* 291:4291–4303.

Trona, F., D. Casado, D., M. Coracini, M. Bengtsson, C. Ioriatti, and P. Witzgall. 2010b. Flight tunnel response of codling moth *Cydia pomonella* to blends of codlemone, codlemone antagonists and pear ester. *Physiological Entomology* 35:249–254.

Trona, F., G. Anfora, A. Balkenius, M. Bengtsson, M. Tasin, A. Knight, N. Janz, P. Witzgall, and R. Ignell. 2013. Neural coding merges sex and habitat chemosensory signals in an insect herbivore. *Proceedings of the Royal Society Biological Sciences* 280:20130267.

Van der Pers, J.N.C., and C.J. Den Otter. 1978. Single cell responses from olfactory receptors of small ermine moths to sex-attractants. *Journal of Insect Physiology* 24:337–343.

Van der Pers, J.N.C., G. Thomas, and C.J. Den Otter. 1980. Interactions between plant odours and pheromone reception in small ermine moths (Lepidoptera: Yponomeutidae). *Chemical Senses* 5:367–371.

Varela, N.J. Avilla, C. Gemeno, and S. Anton. 2011. Ordinary glomeruli in the antennal lobe of male and female tortricid moth *Grapholita molesta* (Busck) (Lepidoptera: Tortricidae) process sex pheromone and host-plant volatiles. *The Journal of Experimental Biology* 214:637–645.

Vermeulen, A., and J.P. Rospars. 2004. Why are insect olfactory receptor neurons grouped into sensilla? The teachings of a model investigating the effects of the electrical interaction between neurons on the transepithelial potential and the neuronal transmembrane potential. *European Biophysics Journal* 33:633–643.

Vickers, N.J., and T.C. Baker, 1994. Reiterative responses to single strands of odor promote sustained upwind flight and odor source location by moths. *Proceedings of the National Academy of Sciences of the United States of America* 91:5756–5760.

Vickers, N.J., T.A. Christensen, T.C. Baker, and J.G. Hildebrand. 2001. Odour-plume dynamics influence the brain's olfactory code. *Nature* 410:466–470.

Vickers, N.J., K. Poole, and C.E. Linn, Jr. 2005. Plasticity in central olfactory processing and pheromone blend discrimination following interspecies antennal imaginal disc transplantation. *Journal of Comparative Neurology* 491:141–156.

Von Arx, M., D. Schmidt-Büsser, and P.M. Guerin. 2012. Plant volatiles enhance behavioral responses of grapevine moth males, *Lobesia botrana* to sex pheromone. *Journal of Chemical Ecology* 38:222–225.

Von Nickisch-Rosenegk, E., J. Krieger, S. Kubick, R. Laage, J. Strobel, J. Strotmann, and H. Breer. 1996. Cloning of biogenic amine receptors from moths (*Bombyx mori* and *Heliothis virescens*). *Insect Biochemistry and Molecular Biology* 26:817–827.

Yang, Z., M., M. Bengtsson, and P. Witzgall. 2004. Host plant volatiles synergize response to sex pheromone in codling moth, *Cydia pomonella*. *Journal of Chemical Ecology* 30:619–629.

Yu, H., J. Feng, Q. Zhang, and H. Xu. 2015. (Z)-3-hexenyl acetate and 1-undecanol increase male attraction to sex pheromone trap in Grapholita molesta (Busck) (Lepidoptera: Tortricidae). *International Journal of Pest Management* 61:30–35.

Zakir, A., M. Bengtsson, M.M. Sadek, B.S. Hansson, P. Witzgall, and P. Anderson. 2013. Specific response to herbivore-induced de novo synthesized plant volatiles provides reliable information for host plant selection in a moth. *Journal of Experimental Biology* 216:3257–3263.

Toward a Quantitative Paradigm for Sex Pheromone Production in Moths

STEPHEN P. FOSTER

> If I cannot overwhelm with my quality, I will overwhelm with my quantity.
>
> EMILE ZOLA

Introduction

The incredible ability of male moths to find conspecific female moths over great distances, mediated primarily by the odor produced and released by a female, has fascinated entomologists and chemists alike since the phenomenon was first described by the French naturalist Jean-Henri Fabre in the late nineteenth century. However, it was not until the first identification of a pheromone compound that the truly minute amounts of chemical that elicited this attraction were appreciated. Butenandt et al. (1959) extracted pheromone from nearly half a million female silkworm moths, *Bombyx mori* (Bombycidae), in order to isolate a mere 6.4 mg (i.e., roughly 13 ng/female) of a compound that they identified as (10*E*, 12*Z*)-hexadeca-dien-1-ol ("bombykol"). The subsequent development of more sensitive analytical techniques greatly facilitated the identification of these small amounts of chemicals such that, today, roughly 400 compounds are known as sex pheromone components from over 600 species of moths, representing nearly 50 families (El-Sayed 2014).

As more sex pheromones of moths were identified, it became apparent that a pheromone of a given species typically consisted of more than one component (i.e., a "blend" of components) and that many of these chemicals were structurally, and therefore biosynthetically, related to common fatty acids. Studies on the biosynthesis of these compounds over the last 30 years have largely been conducted under a "blend" or "quality" paradigm. That is, the pheromone blend (qual-

ity), as represented by the different chemicals and relative ratios of components, is of paramount importance in determining successful intraspecific communication between mates (Roelofs 1979). Consequently, much of the work on pheromone biosynthesis has focused on how the structures and ratios of components are produced by female moths, and how qualitative changes in the pheromone blend combined with parallel changes in male response might lead to the evolution of a new chemical communication system and ultimately speciation (e.g., Phelan 1992; Löfstedt 1993; Roelofs et al. 2002; Symonds and Elgar 2008; Liénard et al. 2010a).

In contrast to the great effort on pheromone quality, there has been little research on quantitative aspects of pheromone biosynthesis (Johansson and Jones 2007), other than recording titers of pheromone components or their fatty acid precursors at specific times or at different physiological states, such as age and mating. While the ability of male moths to locate conspecific females over great distances via minute quantities of pheromone produced by females is often commented on (e.g., Symonds et al. 2012), we actually know little about what controls the quantities of pheromone produced by individual females, outside of the daily cycle of pheromone production/nonproduction stimulated by the pheromone biosynthesis-activating neuropeptide (PBAN; Raina 1993; Rafaeli and Jurenka 2003). Similarly, it is not known what proximate and ultimate mechanisms result in some species producing relatively large amounts (e.g., micrograms in *Virbia lamae* [Erebidae, Arctiinae]; Schal et al. 1987), and others producing relatively small amounts (<1 pg in *Acrobasis nuxvorella* [Pyralidae]; Millar et al. 1996) of pheromone. Moreover, while significant changes in pheromone quality are likely to be relatively rare, because of their potentially profound effects on intraspecific communication (Symonds and Elgar 2008), large quantitative variations, both within (e.g., Haynes et al. 1983) and among (e.g., Miller and Roelofs 1980) individuals, are common (Johansson and Jones 2007) and may strongly influence mating success (e.g., Foster and Johnson 2011), as well as intraspecific competition among females (Johansson and Jones 2007).

But what actually is pheromone quantity and what is a useful measure of this concept? Pheromone quantity has typically been considered the amount of pheromone in the gland at a particular time (titer) or the mass of pheromone released by a female over a given time (release rate). While these measures are useful for comparative purposes (e.g., how does the titer or release rate of a female differ at different times?), neither gives a real measure of the mass of metabolites that a female has actually expended in toto in order to produce a time-related titer or release rate. Therefore, in this review, pheromone quantity is defined as the mass of pheromone produced per unit time. Determination of such a measurement allows the mass of pheromone (and precursor, see later) to be calculated for a given time interval (hours, night, lifetime, etc.). The practical drawback of using this definition is that pheromone quantity cannot be measured statically (e.g., as a titer) but has to be measured kinetically, as the mass produced (synthesized) per unit time. Unfortunately, such measurements of pheromone are rare (but see Foster and Anderson 2011) and cannot be done readily for individual females. For the purpose of this review, I have largely had to infer pheromone quantity from measurements of titer and/or release rate.

In this review, I will briefly outline our knowledge on the biosynthesis of pheromone quality and quantity in moths, before outlining a framework for a more quantitative understanding of pheromone biosynthesis. For in-depth reviews of biosynthesis of pheromone quality, as well as the effect of PBAN on pheromone biosynthesis, the reader is referred to other recent reviews (Jurenka 2003; Rafaeli and Jurenka 2003; Rafaeli 2005; Matsumoto 2010; Blomquist et al. 2011; Jurenka and Rafaeli 2011).

Pheromone Biosynthesis in Moths: The Quality Paradigm

In species of moths utilizing a female-produced, volatile sex pheromone, pheromone biosynthesis and/or release occur in a daily rhythmic cycle, usually in synchrony with the sexually active period (Webster and Cardé 1982; Raina et al. 1991). Moth sex pheromones (i.e., the chemical(s) produced and released by a female *AND* responded to by a male) typically consist of a blend of more than one, usually biochemically related, compounds (El-Sayed 2014). In most species, the sex pheromone is produced and released from a specialized gland located between the eighth and ninth abdominal segments (Ma and Ramaswamy 2003).

Under the blend paradigm, pheromone quality is considered the production and release of a component, or set of components in a specific ratio (e.g., Linn and Roelofs 1989; Symonds and Elgar 2008; Zavada et al. 2011). For a given species, the ratio(s) of components should vary over a limited range such that, in general, effective intraspecific communication still occurs (Linn and Roelofs 1989). Numerous selec-

tion studies have demonstrated that pheromone quality varies within a species, and is relatively intransigent to significant change outside the normal range over multiple generations of selection pressure in the laboratory (e.g., Miller and Roelofs 1980; Collins and Cardé 1985, 1989; Roelofs et al. 1986; Sreng et al. 1989; Zhu et al. 1996a; Allison and Cardé 2006, 2007). In general, ratios of geometric isomers tend to vary less than ratios of other components in pheromone blends (Schlyter and Birgersson 1989).

Although a large number of compounds (and potentially much larger, with pheromones or attractants[1] known for roughly 1% of the known 150,000 described species of moths; Ando et al., 2004; El-Sayed 2014) are known as moth sex pheromone components, they fall largely into only two distinct structural classes (Ando et al. 2004; El-Sayed 2014). The first class of compound, commonly known as "Type 1" (Ando et al. 2004), is typically, but not exclusively, unbranched with an even number of carbons (10–18), unsaturated (1–3 double bonds), and has a terminal oxygenated functional group, usually alcohol, aldehyde, or acetate ester. The second class consists of long-chain (C_{17}–C_{23}) hydrocarbons, which can be further distinguished as polyunsaturated or epoxide derivatives, commonly known as "Type 2" compounds (Millar 2000; Ando et al. 2004), and methyl-branched hydrocarbons. Of the two structural classes, the first type is much more widespread (Ando et al. 2004), being found in most families for which pheromones have been identified. In contrast, Type 2 and other hydrocarbon pheromone components have, to date, largely been found in species in the families Noctuidae, Erebidae, and Geometridae (Ando et al. 2004; El-Sayed 2012). Although most taxa produce either one or the other structural class, both Type 1 and Type 2 compounds are produced by some species of Crambidae and Pyralidae (e.g., Cabrera et al. 2001; Gibb et al. 2007; Miller et al. 2010).

Biosynthesis of Type 1 Compounds

As more Type 1 pheromone components were identified, their structural similarity, and particularly their similarity to common fatty acids, was noted. Early experiments confirmed that Type 1 compounds were biosynthesized de novo from acetate (Jones and Berger 1978; Bjostad and Roelofs 1981). Subsequent studies demonstrated that biosynthesis of Type 1 compounds was a modification of "normal" fatty acid biosynthesis and metabolism (Blomquist et al. 2011), utilizing many similar steps but occurring only within the cells of the pheromone gland (Bjostad et al. 1981; Wolf and Roelofs 1983; Bjostad and Roelofs 1984). Biosynthesis of these compounds involves three general processes (Jurenka 2003): de novo biosynthesis of saturated fatty acids from acetyl CoA, modification of the carbon chain, and modification of the carboxyl functional group (figure 8.1a).

(I) FATTY ACID SYNTHESIS

Studies with labeled acetate precursor have shown that pheromone is biosynthesized de novo in the gland from acetyl CoA, through production of saturated fatty acid(s) (Bjostad and

---

[1] Attractant is used in the sense of a compound found to attract male moths, but yet to be confirmed as produced by female moths.

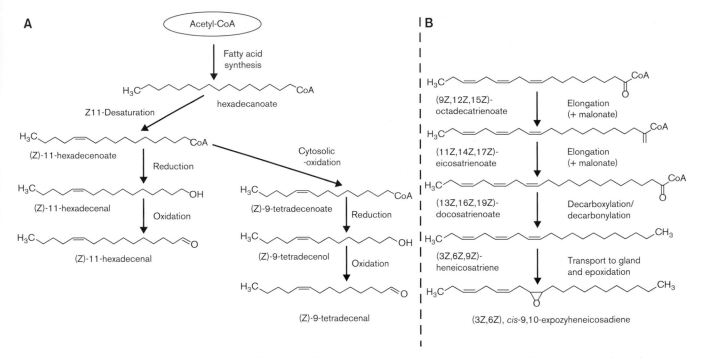

FIGURE 8.1 Examples of the biosynthetic routes for (A) Type 1 and (B) Type 2 sex pheromone components in moths. The Type 1 route shows the example of the biosynthesis of two components, (Z)-11-hexadecenal and (Z)-9-tetradecenal, of the sex pheromone of *Heliothis virescens* from acetyl-CoA (after Choi et al. 2005). The Type 2 route shows the biosynthesis of (3Z,6Z,), *cis*-9,10-epoxyheneicosadiene from linolenic acid ((9Z,12Z,15Z)-octadecatrienoate) in *Estigmene acrea* and *Phragmatobia fuliginosa* (both Erebidae, Arctiinae; after Rule and Roelofs 1989).

Roelofs 1983; Bjostad et al. 1987; Tsfadia et al. 2008). To date, none of the genes that code for the enzymes involved in the synthesis of saturated fatty acids in the pheromone gland have been sequenced or cloned. The synthesis of fatty acids in eukaryotes involves two enzyme systems, acetyl-CoA carboxylase, which catalyzes the formation of malonyl CoA from acetyl CoA, and the multienzyme complex fatty acid synthase, which successively condenses further acetyl-CoA units to malonyl CoA to produce a medium-chain length fatty acid (Salati and Goodridge 1996). Large quantities of both hexadecanoate and octadecanoate are found in the pheromone glands of moths (Bjostad et al. 1987). It has been suggested, based on greater levels of radiolabeled acetate incorporated into octadecanoate than into hexadecanoate in *Argyrotaenia velutinana* (Torticidae), that the former may be the major product of fatty acid synthase in the pheromone gland (Tang et al. 1989). However, in general, the major product of animal fatty acid synthase is hexadecanoate (Salati and Goodridge 1996), with octadecanoate produced by specific elongation of hexadecanoate (Hashimoto et al. 2008). The greater amount of label in octadecanoate in the pheromone gland of *A. veluti-nana* may be due to synthesis of labeled octadecanoate by a relatively fast elongation reaction (i.e., adding a labeled acetate to already synthesized hexadecanoate), whereas labeled hexadecanoate can only be synthesized by a slower de novo synthesis. This interpretation is supported by our own tracer studies on *Heliothis virescens* (Noctuidae) fed U-^{13}C-labeled glucose, in which octadecanoate has a lower apparent precursor enrichment (i.e., the proportion of precursor acetate, following feeding, that originates from U-^{13}C-labeled glucose) than hexadecanoate and pheromone, indicating dilution of the label in the pool of octadecanoate by stores of previously formed (unlabeled) hexadecanoate (Foster and Anderson 2012).

The involvement of acetyl-CoA carboxylase in pheromone biosynthesis has been inferred from in vitro studies using crude pheromone gland preparations, in which malonyl CoA, but not acetyl CoA, is converted to saturated fatty acids and pheromone in the absence of PBAN (Tang et al. 1989; Jurenka et al. 1991; Jacquin et al. 1994), as well as from inhibition of pheromone biosynthesis by the herbicide dicyclofop, which is known to inhibit acetyl-CoA carboxylase in grasses (Eliyahu et al. 2003a).

(II) MODIFICATION OF THE CARBON CHAIN

Studies from the laboratory of Wendell Roelofs elucidated how a diversity of novel pheromone structures could be synthesized by a limited number of biosynthetic steps (Bjostad et al. 1987). Working on various species of Tortricidae and Noctuidae, Roelofs and coworkers identified two key enzyme systems involved in shaping the carbon chain portion of pheromone components: a cytosolic β-oxidase, which shortens the carbon chain by two carbons (Wolf and Roelofs 1983), and a Δ11-desaturase, which inserts a double bond at the 11-position of the chain of saturated fatty acids (Bjostad and Roelofs 1983). Differences in substrate-, regio- or stereo-specificities of these enzymes account for the biosynthesis of a wide range of carbon chain structures, both within and among species (Roelofs and Bjostad 1984; Bjostad et al. 1987). For instance, a Δ11-desaturase acting specifically on hexadec-anoate, in combination with a β-oxidase with specificity for unsaturated acids, could produce Δ11-16, Δ9-14, and Δ7-12 carbon chains, while a Δ11-desaturase with general specificity for saturated acids, combined with a β-oxidase with specificity for saturated acids, could produce Δ11-16, Δ11-14, and Δ11-12 carbon chains (figure 8.2). Genes for Δ11-desaturases that

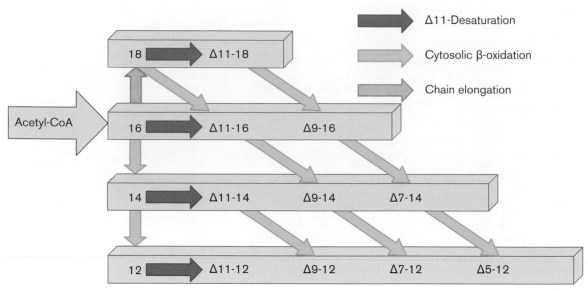

FIGURE 8.2 The variety of straight chain Type 1 C_{12}-C_{18} structures that can be produced from hexadecanoic acid by combinations of Δ11-desaturation, cytosolic β-oxidation (two-carbon chain shortening), and chain elongation (for C_{18} structures). *Cis-* and *trans-*geometries for unsaturation (excluding Δ11-12) further increase the variety of structures possible (after Roelofs and Bjostad 1984).

produce isomerically pure products, either *cis-* (e.g., Knipple et al. 1998) or *trans-* (e.g., Liu et al. 2002a), as well as mixtures of both (e.g., Liu et al. 2002b), have been sequenced, cloned, and functionally characterized (Rooney 2009; Liénard et al. 2010b). In addition to Δ11-desaturases, a number of other functionally important desaturases with different regio-specificities have been identified or implicated in moth pheromone biosynthesis, including Δ14 (Roelofs et al. 2002), Δ10 (Foster and Roelofs 1988; Hao et al. 2002), Δ9 (Rodriguez et al. 2004; Liénard et al. 2010b), Δ6 (Wang et al. 2010a), and Δ5 (Foster and Roelofs 1996). Other multifunctional desaturases involved in the production of dienes or unusual monoenes have also been characterized (Löfstedt and Bengtsson 1988; Moto et al. 2004; Serra et al. 2006, 2007).

(III) MODIFICATION OF THE CARBOXYL FUNCTIONAL GROUP

The first (and in the case of alcohol components, the only) step in converting a fatty acyl precursor to the pheromone component is fatty acyl reduction (Jurenka 2003). In other biological systems, fatty acyl reduction to an alcohol generally involves two enzymes, a fatty acid reductase and an aldehyde reductase (Morse and Meighen 1987b). However, in those species of moths in which the molecular and functional basis of this step in pheromone biosynthesis has been characterized, a single enzyme performs the function (Moto et al. 2003; Antony et al. 2009; Lassance et al. 2010; Liénard et al. 2010a; Hagström et al. 2012). Most of the fatty acid reductases that have been functionally characterized have high substrate specificity for their respective pheromone component precursor acids, although the fatty acid reductases of four *Heliothis* (Noctuidae) species have relatively broad substrate specificity to a range of fatty acids of different chain length and position of unsaturation (Hagström et al. 2012).

To produce aldehyde pheromone components, alcohol precursors must be oxidized. Studies on *Heliothis virescens* and

Helicoverpa zea (Noctuidae; Teal and Tumlinson 1986, 1988), and *Manduca sexta* (Sphingidae; Fang et al. 1995b; Luxová and Svatoš 2006), which produce a number of aldehydes in their respective pheromones, demonstrated cuticular oxidase activity, on or in the pheromone gland, that converted alcohols to aldehydes. These oxidases had rather broad specificity to primary alcohols, with little discrimination of other substrate features, such as chain length and position and geometry of unsaturation.

The esterification of alcohols to acetate esters is carried out by an acetyl-CoA fatty alcohol acetyltransferase (Morse and Meighen 1987a). In the spruce budworm, *Choristoneura fumiferana* (Tortricidae), a species that produces and stores relatively large amounts of (*Z*)- and (*E*)-11-tetradecenyl acetates (Z- and E11-14:Acs) as presumptive precursors to the pheromone aldehydes, this enzyme had broad specificity to medium chain-length (C_{12}–C_{15}) alcohols, and to unsaturated over saturated alcohols (Morse and Meighen 1987a). In a number of species of tortricid moths that produce different blends of Z- and E11-14:Acs as pheromone components, the acetyl transferases exhibit preferential conversion of the Z-isomer, consistent with their production of greater amounts of Z11-14:Ac over E11-14:Ac (Jurenka and Roelofs 1989). In contrast, acetylation in three strains of *Ostrinia nubilalis* (Crambidae), which produce different ratios of Z- and E11-14:Acs, showed no specificity for either geometric isomer (Jurenka and Roelofs 1989).

Biosynthesis of Type 2 and Other Hydrocarbon Compounds

Much less is known about the biosynthesis of Type 2 and other hydrocarbon pheromone components than is known about the biosynthesis of Type 1 compounds. However, it is clear that the biosynthesis of Type 2 compounds differs substantially from that of Type 1 compounds. Perhaps the most significant difference is that hydrocarbon components are not biosynthesized de novo in the gland, but are biosynthesized in oenocytes from essential dietary components, such as polyunsaturated fatty acids or amino acids, and transported to the pheromone gland by lipophorins (Schal et al. 1998), where they may be further modified, before being released (Jurenka 2003).

Production of Type 2 compounds from lineoleic or linolenic acids (Rule and Roelofs 1989; Goller et al. 2007) in oenocytes usually involves one or more cycles of two-carbon chain elongation by malonate prior to decarboxylation (figure 8.1b). In some species, further desaturation can take place (Wang et al. 2010b) before decarboxylation, to produce more highly unsaturated (tetraene) hydrocarbons. Odd-numbered hydrocarbons result from removal of the terminal carboxyl group during decarbonylation or decarboxylation (Rule and Roelofs 1989; Jurenka 2003), while even-numbered hydrocarbons from removal of the terminal carbon by α-oxidation, prior to reduction and decarbonylation or decarboxylation (Jurenka 2003; Goller et al. 2007).

Methyl-branched hydrocarbons are also pheromone components in some moth species (El-Sayed 2014). In several *Virbia* spp., the methyl branch of the hydrocarbon pheromone component, 2-methyl heptadecane, originates from leucine, which is likely converted to isovaleric acid prior to chain biosynthesis (Charlton and Roelofs 1991). No acyl precursors of the pheromone compound were found in the pheromone gland, indicating that it was biosynthesized elsewhere (probably in oenocytes). In the gypsy moth, *Lymantria dispar* (Erebidae, Lymantriinae), the methyl branch of the hydrocarbon pheromone component, $(7R,8S)$-2-methyl-7,8-epoxyoctadecane, originates from valine. Biosynthesis in oenocytes involves chain elongation to C_{19}, followed by $\Delta12$-desaturation to produce 18-methyl-(Z)-12-nonadecenoate, and decarboxylation to 2-methyl-(Z)-7-octadecene. The hydrocarbon is then transported from the oenocytes to the pheromone gland, where epoxidation to the pheromone component takes place (Jurenka et al. 2003).

To date, only one study (Wang et al. 2010b) has investigated biosynthesis of Types 1 and 2 pheromone components in the same species, with results proving consistent with studies on species that produce only one of the compound types. *Amyelois transitella* (Pyralidae) produces a mixture of $(Z11, Z13)$-hexadecadienal and several long-chain pentaenes (Coffelt et al. 1979; Leal et al. 2005; Miller et al. 2010), which are biosynthesized from hexadecanoate (16:Acyl) and linoleate, respectively (Wang et al. (2010b).

In Support of a Quantity Paradigm

Work under the quality paradigm has successfully explained the biosynthesis and diversity of many moth sex pheromone components. However, the paradigm is limited in what it can explain about pheromone biosynthesis in moths, because it fails to account for quantitative differences among pheromone blends, e.g., how does an individual or taxon produce more pheromone than another individual/taxon? The quantity of pheromone produced and released by a female may be critical for ensuring mating success. For example, females producing and releasing greater quantities of pheromone may be more likely to elicit flight responses from more males over greater distances (e.g., Linn et al., 1987; Foster and Johnson 2011), as well as over greater time periods, both within a diel period and over a lifetime. A quantitative advantage in pheromone would be of particular benefit in low-density popula-

tions or in polyandrous species, for which mating at a younger age (Foster et al. 1995) or multiple times (Torres-Vila et al. 2004), respectively, results in increased fecundity. Pheromone quantity may also influence competitive mate selection (Jaffe et al. 2007; Johansson and Jones 2007), with females that produce and release greater quantities of pheromone potentially attracting higher-quality males, e.g., virgin males, in better physical condition than mated males, and therefore capable of flying greater distances, and with greater protein reserves available for transfer to the female in a spermatophore (Torres-Vila and Jennions 2005). In *Heliothis virescens*, sugar-fed females attract greater numbers of males than do starved females (Foster and Johnson 2011), and are also more fecund (Ramaswamy et al. 1997). Thus, a greater quantity of pheromone may be an honest signal of female quality, as indicated by realized fecundity (Foster and Johnson 2011).

Pheromone biosynthetic studies on quality have tended to focus on the production of unusual compounds or on the production of a relatively invariant ratio of two components, particularly geometric isomers (Jurenka 2003). Thus, we have a good understanding of the biosynthetic processes that form the structures of these compounds and ones that contribute to shaping populational average ratios of geometric isomers. For instance, the production of the 92:8 ratio of Z:E-11-14:Acs in *Argyrotaenia velutinana* is shaped by a $\Delta11$-desaturase that produces a 6:1 mixture of the two precursor acids (Liu et al. 2002b), and further shaped by a fatty acid reductase and an acetyl transferase with greater specificity for (Z)-11-(Z11-14:OH) over (E)-11-tetradecenol (E11-14:OH; Jurenka and Roelofs 1989). However, we know little about the production of ratios of components that are not geometric isomers or what causes interindividual variation in ratios, or perhaps even intraindividual variation in ratios over time (Svensson et al. 1997). A more quantitative approach to pheromone biosynthesis would account for differences in amounts of total pheromone produced, among individuals or taxa, as well as for the production of, and variation in specific ratios of components, by explaining production of quantities of individual components and how production of these individual component quantities are interdependent through common precursor use.

Furthermore, pheromone biosynthesis by moths is typically considered as an "isolated" reproductive trait, usually the cyclical production and release of a pheromone blend by the female for eliciting mating responses, rather than as a trait integrated with other reproductive physiologies. Given that the metabolites (carbohydrates and fats) used to produce pheromone are used in other reproductive physiologies (e.g., maturation of oocytes; O'Brien et al. 2000; Foster 2009), it follows that allocation of these metabolites to all these reproductive physiologies may be physiologically coordinated, especially if quantities of a key metabolite in a female are limited. Thus, for example, a singly mated polyandrous moth with access to limited nectar sources might shut down pheromone production, and therefore preclude further mating so as to allocate metabolites totally to egg production in order to maximize fitness, whereas the same female with access to unlimited nectar sources might maintain metabolite allocation to pheromone so as to gain from the direct and indirect benefits of polyandry (Torres-Vila et al. 2004; Slatyer et al. 2012). A quantitative analysis of pheromone biosynthesis could account for the metabolic costs of pheromone production and determine whether and how pheromone production is traded off against other life-history traits (Johansson and Jones 2007).

Factors Affecting Pheromone Quantity

There have been few kinetic studies on moth pheromone biosynthesis that have determined pheromone quantity as an actual mass of pheromone produced per unit time. Most studies have attempted to quantify pheromone production by measuring pheromone titer, or at best changes in titer over time, or by measuring the amount of pheromone released over a given time (release rate). While useful for comparison, these measures do not represent true quantification of the metabolic costs of pheromone production. Titer, for instance, is dependent on both rate of synthesis and rate of loss through release and/or degradation (Foster 2000). A constant titer could mean that pheromone is not being produced or that it is being lost (degraded or released) as quickly as it is being produced. Regardless, no information about the rate of synthesis of pheromone (and hence quantity of precursor used) is available through these measurements. Release rate does quantify the mass of pheromone released, but few studies have even attempted to relate this to titer (Symonds et al. 2012), let alone to the actual mass of pheromone females produce. Due to the paucity of data on quantifying the full amount of pheromone produced by female moths, this review will use data measuring pheromone titers and release rates to infer the effects of various factors on full pheromone quantity.

Numerous studies have demonstrated that pheromone quantity, measured as either titer or release rate, varies both through the day and with female age (e.g., Webster and Cardé 1982; Valeur et al. 1999), as well as among individuals and populations (e.g., Miller and Roelofs 1980; Haynes et al. 1983, 1984; Haynes and Baker 1988; El-Sayed et al. 2003; Allison and Cardé 2006). Both heritable and nonheritable factors explain this variation (Haynes, this volume).

Some studies have demonstrated a heritable basis for pheromone quantity, although none have identified exactly what causes a change in quantity produced. Collins and Cardé (1985) selected for pheromone component ratios in the pink bollworm moth, *Pectinophora gossypiella* (Gelechiidae), and found that quantity (titer) declined over successive selected generations. In a comprehensive study on the almond moth, *Cadra cautella* (Pyralidae), Allison and Cardé (2006) showed that pheromone quantity (titer) was not influenced by the environmental/endogenous factors of access to water, larval rearing temperature, and pupal (or adult) weight, but it did have moderate-to-high heritability values. Differences in pheromone quantity (titer) among populations have been noted for other species (e.g., Ono et al. 1990; El-Sayed et al. 2003), but whether these were because of genetic or environmental causes was not investigated.

The most studied changes in pheromone quantity are those that occur within an individual (usually inferred from analyses of different individuals of the same population) over time. The most dramatic of these is the diel difference in titer between the sexually active (i.e., when large quantities of pheromone are produced and released) and inactive (when low quantities of pheromone are produced) periods (Rafaeli 2005). In moths, this diel variation is typically regulated by the circadian release (Závodská et al. 2009) of PBAN, a member of the pyrokinin/myotrophin peptide family, from the corpora cardiaca (Rafaeli 2005). Released PBAN is transported through the circulatory system to the pheromone gland, where it binds to a G-protein-coupled receptor (Choi et al. 2003), causing release of a second messenger cascade, which likely activates a specific enzyme or enzymes in the pheromone biosynthetic pathway (Rafaeli 2005).

Along with this diel variation, pheromone quantity also varies with female age, generally reaching a peak at sexual maturity, and declining gradually thereafter (e.g., Webster and Cardé 1982; Tang et al. 1992; Foster and Johnson 2010). In *Epiphyas postvittana* (Tortricidae), this age-related decline is due to a decline in production of fatty acids, possibly due to a shortage of pheromone precursor (Foster and Johnson 2010), and to a decline in fatty acid reductase activity (Foster and Greenwood 1997).

Mating by female moths can cause permanent (especially for normally monandrous species) or transient (especially for polyandrous species) decreases in pheromone quantity (Raina et al. 1994), largely through transfer of male substances during copulation (Hanin et al. 2012). Both neural and humoral factors have been linked to this suppression of pheromone biosynthesis in mated females (Rafaeli 2005). In *E. postvittana*, a normally monandrous species, mating usually results in permanent suppression of pheromone production. However, if the ventral nerve cord of a female is severed prior to, or during, an early stage of copulation, pheromone titer of the mated female is similar to that of a virgin (Foster 1993). This neural mechanism has been reported in a number of other species of moths (e.g., Jurenka et al. 1993; Ichikawa et al. 1996; Ramaswamy et al. 1996; Delisle and Simard 2002) and appears to function by suppressing PBAN release from the subesophageal ganglion (Foster 1993; Ando et al. 1996; Delisle et al. 2000).

In *Helicoverpa zea* (Noctuidae), pheromone suppression following mating involves both a neural signal and one or more pheromonostatic peptides, originating from male accessory glands and transferred to the female during copulation (Kingan et al. 1995). Pheromonostatic peptides have also been implicated in suppression of pheromone production in *Heliothis armigera* (Noctuidae; Eliyahu et al. 2003b; Hanin et al. 2012).

Outside of diel and mating effects, the most profound endogenous effect on pheromone quantity so far reported is female hemolymph trehalose concentration (Foster 2009; Foster and Johnson 2010). In *H. virescens*, sugar feeding by adult females results in rapid increases in hemolymph trehalose concentration which, in turn, can cause a rapid increase in titers of fatty acids and sex pheromone (Foster 2009; Foster and Johnson 2010), as well as in sex pheromone release rate (Foster and Johnson 2011). Increased pheromone titers were only apparent in females that had low hemolymph trehalose concentrations (i.e., were sugar stressed by mating or starvation), with a threshold of around 10 mM required for production of "normal" pheromone quantities.

To date, this effect has only been demonstrated in *H. virescens*. However, hemolymph carbohydrate is likely a ubiquitous pool of acetyl-CoA precursor for lipogenesis and pheromone biosynthesis in moths, and it may play a role in pheromone biosynthesis for most, if not all, species that produce Type 1, or even perhaps Type 2, components. In short-lived, nonfeeding species, pheromone is typically produced prior to eclosion, with females mating shortly thereafter (Ramaswamy et al. 1997). Therefore, hemolymph carbohydrate may be utilized in pheromone biosynthesis prior to eclosion. In longer-lived species that feed on nectar, especially polyandrous species in which pheromone production and release take place reiteratively throughout most of the adult life (Ramaswamy et al. 1997; Wäckers et al. 2007), regular nectar feeding will maintain a high hemolymph trehalose concentration for production of high quantities of pheromone throughout the adult life.

Ambient temperature can also affect pheromone titer (Raina 2003), although any effect it has on titer must be tempered by the fact that it affects both release rate (Liu and Haynes 1994) and periodicity of pheromone release (McNeil 1991), both of which may change titer at a certain sampling time, but not necessarily the quantity of pheromone produced over a period. Kinetic studies on synthetic rates would demonstrate whether the effect of temperature on pheromone production is quantitative or just a temporal shift.

Precursor Flux Determines Pheromone Quantity

Type 1 sex pheromone components are polymeric products built from a monomer precursor, acetyl CoA, followed by modification of the polymer along a biosynthetic pathway (see previous sections). The mass or quantity of a pheromone component produced must be a function of the fluxes (mass/ unit time) of the acetate precursor and subsequent precursor acids through a specific biosynthetic pathway. Therefore, factors that limit precursor flux, either of monomeric acetate or of any of the precursor acids, through the various (i.e., a network of) component pathways must determine the total quantity of pheromone produced by a female. This principle also applies to Type 2 components, with regard to the various precursor pools used in their biosynthesis.

The most studied limit on precursor flux has been the control of pheromone biosynthesis by PBAN, which acts by controlling the activity of, and therefore precursor flux through, one or more enzymes in the biosynthetic pathways of pheromone components (Rafaeli and Jurenka 2003; Eltahlawy et al. 2007; Tsfadia et al. 2008). However, even during PBAN stimulation (maximal pheromone biosynthesis), precursor flux must be limited, otherwise pheromone biosynthesis would proceed unabated, with a very large quantity produced quickly, likely resulting in a short "burst" of pheromone produced before acetate precursor resources were drained. The most likely limits on precursor flux for producing pheromone are the size of precursor pools and the flux capacities of enzymes in the biosynthetic pathway (Suarez et al. 1997; Watt and Dean 2000; Eanes et al. 2006).

Precursor Pools

What, potentially, are the precursor pools available for pheromone biosynthesis in moths? For Type 1 pheromone components, the primary precursor for de novo pheromone biosynthesis is acetyl CoA (Jurenka 2003). Although small amounts of cytosolic acetyl CoA can be produced from catabolism of amino acids, most cytosolic acetyl CoA for fatty acid synthesis in eukaryotes is synthesized from citrate, produced from pyruvate or fatty acid oxidation in the mitochondria, transported to the cytosol (Salati and Goodridge 1996). Carbohydrate, in the form of hemolymph trehalose (Thompson 2003), is readily accessible for production of cytosolic acetyl CoA, while both pheromone gland fatty acids and fatty acids in the fat body (after transport to the gland) could be mobilized for β-oxidation to produce cytosolic acetyl CoA. Stored glandular fatty acids in glycerolipids (Matsumoto et al. 2002) could also be used intact for the final steps of pheromone biosynthesis (figure 8.3).

There is evidence to support both carbohydrate and stored glandular fats being utilized for biosynthesis of Type 1 compounds. Hemolymph trehalose is rapidly glycolyzed and converted to acetyl CoA for biosynthesis of the major pheromone component, (Z)-11-hexadecenal (Z11-16:Ald), in *Heliothis virescens* (Foster 2009; Foster and Johnson 2010; Foster and Anderson 2011). Although demonstrated so far only in *H. virescens*, this carbohydrate pool is likely to have widespread use for pheromone production among insects that biosynthesize Type 1 components (see above).

The best evidence for stored glandular fatty acids being mobilized and used for producing Type 1 pheromone components comes from studies on *Bombyx mori*. Large stores of the precursor acids of bombykol are found in triacylglycerols (Matsumoto et al. 2002), contained within cytosolic lipid droplets in pheromone gland cells (Fonagy et al. 2001). Stimulation of the gland with PBAN results in production of larger numbers of droplets of smaller volume, presumably facilitating hydrolysis of precursor fatty acids (Fonagy et al. 2001). Recently, a diacylglycerol transferase 2 gene (Du et al. 2012b), the product of which likely catalyzes the last step in the synthesis of triacylglycerols, and seven lipase genes (Du et al. 2012a), the products of which may catalyze the hydrolysis of glycerolipids, were found to be expressed in the pheromone gland of *B. mori*. RNAi knockdown of these genes showed that reduced expression of the diacylglycerol transferase 2 and four of the lipases coincided with decreased production of bombykol, implicating their involvement in pheromone biosynthesis. Mobilization of fatty acid pheromone precursors from triacylglycerols by PBAN has also been implicated in pheromone biosynthesis in *Manduca sexta* (Fang et al. 1995a,1996).

Given that animals shift between carbohydrate and fat, according to starvation level, as fuels for general metabolism (Bernlohr and Simpson 1996), and the general similarity of moth pheromone biosynthesis to normal intermediary metabolism (Blomquist et al. 2011), it seems likely that most insects producing Type 1 compounds utilize both sugar and fat as precursor pools for pheromone biosynthesis. Evidence exists that suggests both carbohydrate and recycled fatty acids are used as precursors for fatty acid and pheromone biosynthesis in *H. virescens*. Using a tracer–tracee approach, female *H. virescens* were fed U-^{13}C-glucose at the start of the scotophase and the enrichment of precursor in newly synthesized glandular fatty acids was determined over the next 24 hours (Foster and Anderson 2012). The precursor enrichment of newly synthesized fatty acids plateaued 6 hours after feeding, and had declined substantially by 24 hours. Since the proportion of labeled glucose in hemolymph trehalose remained steady throughout this period (the insects did not feed again), the decrease in precursor enrichment implies that synthesis of new fatty acids utilized a more label-diluted acetate precursor pool; most likely from β-oxidation of recycled (unlabeled) fatty acids. It is not known whether this increased use of recycled fats for de novo synthesis during the photophase occurred because of a decrease in hemolymph trehalose concentration (Foster and Johnson 2010) or because of the lower pheromone absolute synthetic rate (Foster and Anderson 2011), during much of the period, due to the absence of PBAN.

Precursor Flux

Other than determining what activities of biosynthetic enzymes are controlled by PBAN (Rafaeli 2005), there has been limited study on determining precursor flux in pheromone biosynthesis in moths. Much of the reason for this is

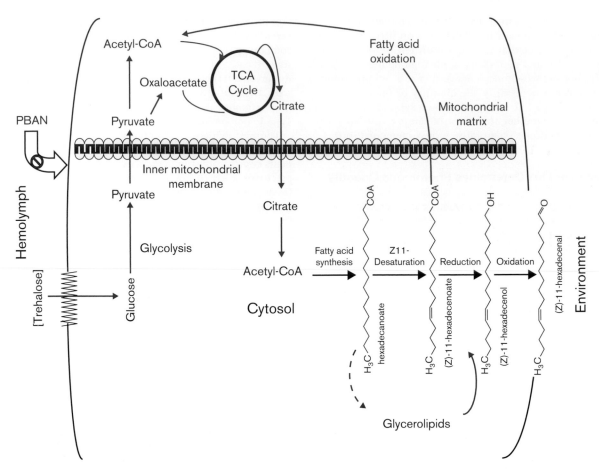

FIGURE 8.3 Potential physiological pools and their routes (shown with blue arrows) for supplying precursors for biosynthesis of Type 1 sex pheromone components in moths. Acetyl-CoA for de novo fatty acid synthesis can be supplied by glycolysis (of glucose from trehalose) or β-oxidation of fats. Fats stored in glycerolipids in gland cells may also be mobilized directly into the pheromone biosynthetic pathway.

that it is laborious to isolate enzymes in pathways for in vitro determination of enzyme flux, while extrapolation of in vitro kinetic parameters of individual enzymes to in vivo fluxes through pathways can be misleading (van Eunen et al. 2012). There are few examples in which a moth sex pheromone biosynthetic enzyme has been characterized kinetically. Although not explicitly measuring precursor flux, Jurenka and Roelofs (1989) used an in vitro assay to calculate kinetic parameters for the acetylation of Z- and E11-14:OH by the fatty alcohol acetyltransferase in *Argyrotaenia velutinana*. All kinetic parameters (V_{max}, K_m) were greater for Z11-14:OH than for E11-14:OH, consistent with greater reaction velocity and flux capacity for the Z11-14:OH pheromone precursor. Using in vitro assays, Morse and Meighen (1987b) partially characterized the kinetics of several pheromone biosynthetic enzymes, including a fatty acid synthase, a fatty alcohol acetyltransferase, and a putative alcohol oxidase, in the pheromone gland of *Choristoneura fumiferana*.

Our work (Foster 2009; Foster and Johnson 2010) on sugar feeding in *Heliothis virescens* demonstrated that a shortage of carbohydrate in hemolymph (caused by starvation or mating) results in reduced fatty acid and pheromone titers, and that feeding on sugar rapidly restores both hemolymph trehalose concentration, and fatty acid and pheromone titers (Foster and Johnson 2010). Therefore, a shortage of cytosolic acetyl CoA in the appropriate precursor pool must limit precursor flux through the pheromone biosynthetic pathway. The pla-

teauing of pheromone titers at hemolymph trehalose concentrations above approximately 10 mM suggests that not only does acetate precursor availability limit precursor flux through the pathway, but also that downstream enzymes in the biosynthetic pathway do so under conditions when acetate precursor availability is not limiting.

The large amounts of pheromone produced by *H. virescens* females after feeding on sugar suggested to us an approach for determining precursor fluxes under in vivo conditions, using a stable isotope-labeled precursor and the technique of mass isotopomer distribution analysis (MIDA; Hellerstein and Neese 1992). MIDA is a combinatorial solution for the patterns of precursor labeling in polymeric products (Hellerstein and Neese 1999) that allows determination of precursor enrichment and product synthetic rates (Chinkes et al. 1996; Hellerstein and Neese 1999; Wolfe and Chinkes 2005). An important advantage of this approach is that it allows both fractional (FSR) and absolute (ASR) synthetic rates of a polymer product to be determined through a pathway at natural levels of precursor. In our case, we fed virgin *H. virescens* females U-^{13}C-glucose (generating ^{13}C$_2$-acetate precursor) and determined the synthetic rates of the major pheromone component, Z11-16:Ald, during the scotophase (i.e., when PBAN is released) and photophase (i.e., when PBAN is not released).

As expected, both ASR and FSR were much higher during the scotophase than during the photophase (Foster and Anderson 2011). Moreover, during the scotophase, phero-

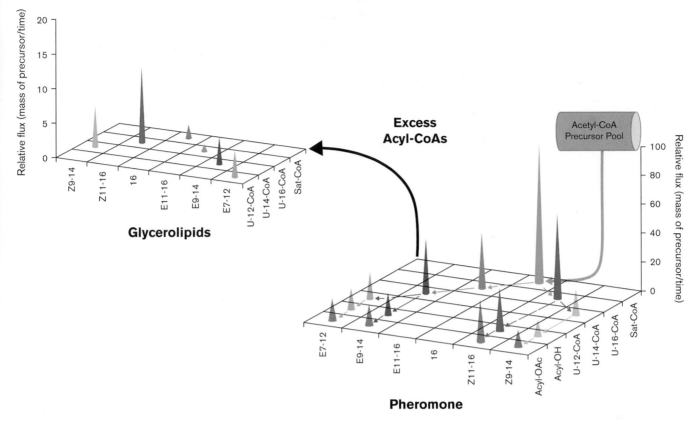

FIGURE 8.4 Schematic showing relative precursor fluxes, originating at the acetyl-CoA pool, that pass through the pheromone biosynthetic network (that produces the whole pheromone blend), comprised of distinct, but linked, pathways, each producing a single component of the blend. Acetate pheromone component fluxes are shown in gray. Acyl-CoAs and transient alcohol precursors (not pheromone components) are individually color-coded. Inherent flux limits through the pathways result in excess acyl-CoAs, which are stored in the glycerolipid pool, but may later be used for biosynthesis of pheromone. The ratio of pheromone components will be representative of precursor fluxes through the pathways of the network. The ratio of stored acids in the glycerolipids will be representative of the flux limits of substrates through the pathways of the network, i.e., the lower the flux of a given fatty acyl substrate that passes through a particular step in a pathway, the more it will be represented in the glycerolipids. For example, there is a big decrease in precursor flux through the pheromone pathways after U-16:CoA; this is complemented by a large amount of U-16:CoA stored in the glycerolipids.

mone was turned over rapidly with an FSR of roughly 97% per hour. Next, we determined synthetic rates of the total glandular content of two key pheromone precursor fatty acids, 16:Acyl and (Z)-11-hexadecenoate (Z11-16:Acyl). Since both acids are pheromone precursors, we reasoned that their ASRs should be at least equal or greater than that of the pheromone component. Surprisingly, both FSRs and ASRs of the two acids were substantially lower than those of the pheromone component. In fact, the ASR of Z11-16:Acyl was only 63% of that of Z11-16:Ald, while that of 16:Acyl was only 3% of that of the pheromone component (Foster and Anderson 2012). This suggested to us that we were really determining the synthetic rate of a "dead end" storage pool of these acids (Bjostad et al. 1987), rather than that of a pool that is rapidly converted to pheromone. This was confirmed by determining the synthetic rates of these acids in the glycerolipid fraction in the gland and finding that the respective ASRs were similar to those determined for total glandular content. This result is consistent with a very rapid turnover of a small, steady-state pool of these fatty acids (as transient acyl CoAs) to either pheromone or glycerolipids, with most of the precursor flux going to synthesis of pheromone, and that which is not converted to pheromone (i.e., surplus acids) being stored in glycerolipids (figure 8.4).

This net synthesis of stored fatty acids in glycerolipids during this period implies that specific biosynthetic steps limit precursor acid flux, thereby controlling the quantity of pheromone produced. In particular, the much higher ASR of Z11-16:Acyl over that of 16:Acyl in glandular glycerolipids (Foster and Anderson 2012) suggests that fatty acid reduction exerts greater control (i.e., is more limiting) over precursor flux through the pathway than does desaturation. Interestingly, the low synthetic rates (especially of 16:Acyl) of these acids in this species, compared to that of the pheromone component, also indicate that once these acids are stored in glycerolipids they are relatively inaccessible for pheromone synthesis, at least over the relatively short period of time studied. Overall, this suggests that the quantities of individual pheromone components produced (i.e., "the pheromone blend") represent the fluxes of precursor through the individual component pathways of the entire blend network, while the individual quantities of precursor acids stored in the glycerolipids represent the flux limits of individual pathway steps for these fatty acid substrates (figure 8.4). This is illustrated by the example of the Z-strain of *Ostrinia nubilalis*, which produces a 97:3 blend of Z11-14:Ac: E11-14:Ac (Kochansky et al. 1975). The respective precursor acids, (Z)-11-tetradecenoate (Z11-14:Acyl) and (E)-11-tetradecenoate (E11-14:Acyl), are produced in a roughly 60:40 ratio by a Δ11-desaturase (Roelofs et al. 2002), while the final 97:3 pheromone blend is determined by the fatty acyl reductase, which has greater specificity (i.e., flux capacity) for the Z-acid, in this strain (Zhu et al. 1996b). This results in more Z11-14:Ac being produced, but a large excess of E11-14:Acyl remaining. Excess acids are stored in the glycerolipids, with the consequence that this pool contains much more E11-14:Acyl than Z11-14:Acyl (Foster 2004).

A consequence of the storage of precursor acids in glycerolipids is that the full acetate precursor cost of producing pheromone may not be represented entirely by the mass of pheromone produced per unit time, but may also need to include acetate precursor used in the net changes in fatty acid stores. This is relatively straightforward to determine for a species such as *H. virescens* over a specific time period, in which there is a net increase in stored acids. However, if these acids are largely recycled (e.g., as for *Bombyx mori*; Matsumoto 2010), calculations of precursor cost will need to account for the net differential in stored acids.

While the above discussion relates to Type 1 pheromone components, similar principles could apply to the biosynthesis of Type 2 or methyl-branched hydrocarbon components, in spite of their differences in site of synthesis and precursors. Although these hydrocarbon compounds are primarily biosynthesized in oenocytes (Blomquist et al. 2011), they could still utilize hemolymph trehalose as a pool for acetyl-CoA precursor for chain elongation prior to transport to the pheromone gland. Species utilizing these compounds as pheromone components must obtain other precursors, such as polyunsaturated acids or the essential amino acids valine and leucine, through larval feeding (Jurenka 2003). In spite of this limitation in acquisition, the large quantities of stored linoleate and linolenate in adult female Lepidoptera (Thompson 1973), as well as the large titers of these types of pheromone components produced by some species (e.g., micrograms of 2-methylheptadecane in *Virbia lamae*; Schal et al. 1987), suggest pools of these essential precursors are not greatly limited.

Final Comments

The blend paradigm of pheromone biosynthesis has emphasized how and why female moths make diverse, but biosynthetically limited, blends of chemicals as sex pheromone components. In recent years, this has led to the identification of a number of genes that are responsible for the structural features of these chemicals, and that help shape pheromone blend quality. However, these studies have, at best, yielded only rudimentary knowledge of the quantitative aspects of sex pheromone biosynthesis in moths, particularly in regard to how and why the quantities of pheromone produced by different individuals or taxa vary so greatly. In this review, I advocate a kinetic, flux-type approach to quantifying pheromone production, rather than the usual static single-time measurement of pheromone quantity through titer, because such an approach yields information on the total mass of pheromone females produce in order to attract males throughout their lifetime. These fluxes can be converted to acetate precursor units, accounting for different pheromone component chain length, and ultimately to energetic costs (Moore and Hopkins 2009) for comparison with energetic costs of other reproductive activities of female moths. Thus, the true metabolic cost of producing pheromone, under a range of endogenous and exogenous conditions, may be determined.

The sex pheromones of most species of moths are multicomponent blends of biosynthetically related components (Jurenka 2003; Byers 2006). Collectively, the individual pathways of biosynthetically related components form a network, with precursor fluxes through the network producing the quantities of individual components, (i.e., the blend). Flux limits through the network result in excess precursor acids being stored in glandular glycerolipids (see figure 8.4). This interlinking of precursor fluxes/stored acids may help provide "blend homeostasis," an emergent property of a network (Prill et al. 2005), which may help stabilize blend ratios through stable acyl-CoA concentrations and fluxes. By defining these fluxes, under a variety of conditions and/or constraints, it should be possible to model precursor fluxes quantitatively through a pheromone network, along the lines used for modeling metabolic fluxes in single cell organisms (Nielsen 2003). Such modeling should be capable of testing the stability of blend production by female moths with respect to changes in fluxes though individual steps, and help predict how pheromone blends produced by females might change with specific gene mutations or in different environments. An improved understanding of the proximate causes underlying the quantities and interdependence of individual pheromone components produced by female moths should help us to understand better how pheromone quality of a species can change.

References Cited

Allison, J., and R. T. Cardé. 2006. Heritable variation in the sex pheromone of the almond moth, *Cadra cautella*. *Journal of Chemical Ecology* 32:621–641.

Allison, J., and R. T. Cardé. 2007. Bidirectional selection for novel pheromone blend ratios in the almond moth, *Cadra cautella*. *Journal of Chemical Ecology* 33:2293–2307.

Ando, T., S. Inomata, and M. Yamamoto. 2004. Lepidopteran sex pheromones. Pp. 51–96. In S. Schulz, ed. *The Chemistry of Pheromones and Other Semiochemicals I*. Berlin/Heidelberg: Springer.

Ando, T., K. Kasuga, Y. Yajima, H. Kataoka, and A. Suzuki. 1996. Termination of sex pheromone production in mated females of the silkworm moth. *Archives of Insect Biochemistry and Physiology* 31:207–218.

Antony, B., T. Fujii, K. Moto, S. Matsumoto, M. Fukuzawa, R. Nakano, S. Tatsuki, and Y. Ishikawa. 2009. Pheromone-gland-specific fatty-acyl reductase in the adzuki bean borer, *Ostrinia scapulalis* (Lepidoptera: Crambidae). *Insect Biochemistry and Molecular Biology* 39:90–95.

Bernlohr, D. A., and M. A. Simpson. 1996. Adipose tissue and lipid metabolism. Pp. 257–281. In D. E. Vance and J. Vance, eds. *Biochemistry of Lipids, Lipoproteins and Membranes*. Amsterdam: Elsevier.

Bjostad, L. B., and W. L. Roelofs. 1981. Sex pheromone biosynthesis from radiolabeled fatty acids in the redbanded leafroller moth. *Journal of Biological Chemistry* 256:7936–7940.

Bjostad, L. B., and W. L. Roelofs. 1983. Sex pheromone biosynthesis in *Trichoplusia ni*: key steps involve Delta-11 desaturation and chain shortening. *Science* 220:1387–1389.

Bjostad, L. B., and W. L. Roelofs. 1984. Biosynthesis of sex pheromone components and glycerolipid precursors from sodium (1-14C) acetate in redbanded leafroller moth. *Journal of Chemical Ecology* 10:681–691.

Bjostad, L. B., W. A. Wolf, and W. L. Roelofs. 1981. Total lipid analysis of the sex pheromone gland of the redbanded leafroller moth, *Argyrotaenia velutinana*, with reference to pheromone biosynthesis. *Insect Biochemistry* 11:73–79.

Bjostad, L. B., W. A. Wolf, and W. L. Roelofs. 1987. Pheromone biosynthesis in lepidopterans: desaturation and chain shortening. Pp. 77–120. In G. D. Prestwich and G. J. Blomquist, eds. *Pheromone Biochemistry*. New York: Academic Press.

Blomquist, G. J., R. Jurenka, C. Schal, and C. Tittiger. 2011. Pheromone production: biochemistry and molecular biology. Pp. 523–567. In L. I. Gilbert, ed. *Insect Endocrinology*. San Diego, CA: Academic Press.

Butenandt, A., R. Beckman, D. Stamm, and E. Hecker. 1959. Über den sexuallockstoff des seidenspinner Bombyx mori, reidarstellung und konstitution. *Zeitschrift für Naturforschung B* 14:283–284. [In German]

Byers, J.A. 2006. Pheromone component patterns of moth evolution revealed by computer analysis of the Pherolist. *Journal of Animal Ecology* 75:399–407.

Cabrera, A., A.E. Eiras, G. Gries, R. Gries, N. Urdaneta, B. Mirás, C. Badji, and K. Jaffe. 2001. Sex pheromone of tomato fruit borer, *Neoleucinodes elegantalis*. *Journal of Chemical Ecology* 27:2097–2107.

Charlton, R.E., and W.L. Roelofs. 1991. Biosynthesis of a volatile, methly-branched hydrocarbon sex pheromone from leucine by arctiid moths (*Holomelina* spp.). *Archives of Insect Biochemistry and Physiology* 18:81–97.

Chinkes, D.L., A. Aarsland, J. Rosenblatt, and R.R. Wolfe. 1996. Comparison of mass isotopomer dilution methods used to compute VLDL production in vivo. *American Journal of Physiology—Endocrinology and Metabolism* 271:E373–E383.

Choi, M.Y., E.J. Fuerst, A. Rafaeli, and R. Jurenka. 2003. Identification of a G protein-coupled receptor for pheromone biosynthesis activating neuropeptide from pheromone glands of the moth *Helicoverpa zea*. *Proceedings of the National Academy of Sciences of the United States of America* 100:9721–9726.

Choi, M.Y., A. Groot, and R.A. Jurenka. 2005. Pheromone biosynthetic pathways in the moths *Heliothis subflexa* and *Heliothis virescens*. *Archives of Insect Biochemistry and Physiology* 59:53–58.

Coffelt, J.A., K.W. Vick, P.E. Sonnet, and R.E. Doolittle. 1979. Isolation, identification, and synthesis of a female sex pheromone of the navel orangeworm, *Amyelois transitella* (Lepidoptera: Pyralidae). *Journal of Chemical Ecology* 5:955–966.

Collins, R.D., and R.T. Cardé. 1985. Variation in and heritability of aspects of pheromone production in the pink bollworm moth, *Pectinophora gossypiella* (Lepidoptera: Gelechiidae). *Annals of the Entomological Society of America* 78:229–234.

Collins, R.D. and R.T. Cardé. 1989. Heritable variation in pheromone response of the pink bollworm, *Pectinophora gossypiella* (Lepidoptera: Gelechiidae). *Journal of Chemical Ecology* 15:2647–2659.

Delisle, J., and J. Simard. 2002. Factors involved in the post-copulatory neural inhibition of pheromone production in *Choristoneura fumiferana* and *C. rosaceana* females. *Journal of Insect Physiology* 48:181–188.

Delisle, J., J.F. Picimbon, and J. Simard. 2000. Regulation of pheromone inhibition in mated females of *Choristoneura fumiferana* and *C. rosaceana*. *Journal of Insect Physiology* 46: 913–921.

Du, M., X. Yin, S. Zhang, B. Zhu, Q. Song, and S. An. 2012a. Identification of lipases involved in PBAN stimulated pheromone production in *Bombyx mori* using the DGE and RNAi approaches. *PLOS ONE* 7:e31045.

Du, M., S. Zhang, B. Zhu, X. Yin, and S. An. 2012b. Identification of a diacylglycerol acyltransferase 2 gene involved in pheromone biosynthesis activating neuropeptide stimulated pheromone production in Bombyx mori. *Journal of Insect Physiology* 58:699–703.

Eanes, W.F., T.J.S. Merritt, J.M. Flowers, S. Kumagai, E. Sezgin, and C.-T. Zhu. 2006. Flux control and excess capacity in the enzymes of glycolysis and their relationship to flight metabolism in *Drosophila melanogaster*. *Proceedings of the National Academy of Sciences of the United States of America* 103:19413–19418.

El-Sayed, A.M. 2014. The Pherobase: Database of Insect Pheromones and Semiochemicals. Available at: http://www.pherobase.com.

El-Sayed, A.M., J. Delisle, N. De Lury, L.J. Gut, G.J.R. Judd, S. Legrand, W.H. Reissig, W.L. Roelofs, C.R. Unelius, and R.M. Trimble. 2003. Geographic variation in pheromone chemistry, antennal electrophysiology, and pheromone-mediated trap catch of North American populations of the obliquebanded leafroller. *Environmental Entomology* 32:470–476.

Eliyahu, D., S. Applebaum, and A. Rafaeli. 2003a. Moth sex-pheromone biosynthesis is inhibited by the herbicide diclofop. *Pesticide Biochemistry and Physiology* 77:75–81.

Eliyahu, D., V. Nagalakshmi, S.W. Applebaum, E. Kubli, Y. Choffat, and A. Rafaeli. 2003b. Inhibition of pheromone biosynthesis in *Helicoverpa armigera* by pheromonostatic peptides. *Journal of Insect Physiology* 49:569–574.

Eltahlawy, H., J.S. Buckner, and S.P. Foster. 2007. Evidence for two-step regulation of pheromone biosynthesis by the pheromone biosynthesis-activating neuropeptide in the moth *Heliothis virescens*. *Archives of Insect Biochemistry and Physiology* 64:120–130.

Fang, N., P.E.A. Teal, and J.H. Tumlinson.1995a. PBAN regulation of pheromone biosynthesis in female tobacco hornworm moths, *Manduca sexta* (L.). *Archives of Insect Biochemistry and Physiology* 29:35–44.

Fang, N., P.E.A. Teal, and J.H. Tumlinson. 1995b. Characterization of oxidase(s) associated with the sex pheromone gland in *Manduca sexta* (L) females. *Archives of Insect Biochemistry and Physiology* 29:243–257.

Fang, N., P.E.A. Teal, J.H. Tumlinson, and N.B. Fang. 1996. Effects of decapitation and PBAN injection on amounts of triacylglycerols in the sex pheromone gland of *Manduca sexta* (L.). *Archives of Insect Biochemistry and Physiology* 32:249–260.

Fonagy, A., N. Yokoyama, and S. Matsumoto. 2001. Physiological status and change of cytoplasmic lipid droplets in the pheromone-producing cells of the silkmoth, *Bombyx mori* (Lepidoptera, Bombycidae). *Arthropod Structure and Development* 30:113–123.

Foster, S.P. 1993. Neural inactivation of sex pheromone production in mated lightbrown apple moths, *Epiphyas postvittana* (Walker). *Journal of Insect Physiology* 39:267–273.

Foster, S.P. 2000. The periodicity of sex pheromone biosynthesis, release and degradation in the lightbrown apple moth, *Epiphyas postvittana* (Walker). *Archives of Insect Biochemistry and Physiology* 43:125–136.

Foster, S.P. 2004. Fatty acid and sex pheromone changes and the role of glandular lipids in the Z-strain of the European corn borer, *Ostrinia nubilalis* (Hübner). *Archives of Insect Biochemistry and Physiology* 56:73–83.

Foster, S.P. 2009. Sugar feeding via trehalose haemolymph concentration affects sex pheromone production in mated *Heliothis virescens* moths. *Journal of Experimental Biology* 212:2789–2794.

Foster, S.P., and K.G. Anderson. 2011. The use of mass isotopomer distribution analysis to quantify synthetic rates of sex pheromone in the moth *Heliothis virescens*. *Journal of Chemical Ecology* 37:1208–1210.

Foster, S.P., and K.G. Anderson. 2012. Synthetic rates of key stored fatty acids in the biosynthesis of sex pheromone in the moth *Heliothis virescens*. *Insect Biochemistry and Molecular Biology* 42:865–872.

Foster, S.P., and D.R. Greenwood. 1997. Change in reductase activity is responsible for senescent decline in sex pheromone titre in the lightbrown apple moth, *Epiphyas postvittana* (Walker). *Journal of Insect Physiology* 43:1093–1100.

Foster, S.P., A.J. Howard, and R.H. Ayers. 1995. Age-related changes in reproductive characters of four species of tortricid moths. *New Zealand Journal of Zoology* 22:271–280.

Foster, S.P., and C.P. Johnson. 2010. Feeding and hemolymph trehalose concentration influence sex pheromone production in virgin *Heliothis virescens* moths. *Journal of Insect Physiology* 56:1617–1623.

Foster, S.P., and C.P. Johnson. 2011. Signal honesty through differential quantity in the female-produced sex pheromone of the moth *Heliothis virescens*. *Journal of Chemical Ecology* 37: 717–723.

Foster, S.P., and W.L. Roelofs. 1988. Sex pheromone biosynthesis in the leafroller moth *Planotortrix excessana* by Δ10 desaturation. *Archives of Insect Biochemistry and Physiology* 8:1–9.

Foster, S.P., and W.L. Roelofs. 1996. Sex pheromone biosynthesis in the tortricid moth, *Ctenopseustis herana* (Felder & Rogenhofer). *Archives of Insect Biochemistry and Physiology* 33:135–147.

Gibb, A., B. Pinese, D. Tenakanai, A. Kawi, B. Bunn, P. Ramankutty, and D. Suckling. 2007. (*Z*)-11-Hexadecenal and (*3Z, 6Z, 9Z*)-trico-satriene: sex pheromone components of the red banded mango caterpillar *Deanolis sublimbalis*. *Journal of Chemical Ecology* 33:579–589.

Goller, S., G. Szöcs, W. Francke, and S. Schulz. 2007. Biosynthesis of (3Z,6Z,9Z)-3,6,9-Octadecatriene: the main component of the pheromone blend of *Erannis bajaria*. *Journal of Chemical Ecology* 33:1505–1509.

Hagström, Å. K., M.A. Liénard, A.T. Groot, E. Hedenström, and C. Löfstedt. 2012. Semi-selective fatty acyl reductases from four heliothine moths influence the specific pheromone composition. *PLoS ONE* 7:e37230.

Hanin, O., A. Azrielli, S.W. Applebaum, and A. Rafaeli. 2012. Functional impact of silencing the *Helicoverpa armigera* sex-peptide receptor on female reproductive behaviour. *Insect Molecular Biology* 21:161–167.

Hao, G., W. Liu, O. C. M, and W. L. Roelofs. 2002. Acyl-CoA Z9- and Z10-desaturase genes from a New Zealand leafroller moth species, *Planotortrix octo. Insect Biochemistry and Molecular Biology* 32:961–966.

Hashimoto, K., A. C. Yoshizawa, S. Okuda, K. Kuma, S. Goto, and M. Kanehisa. 2008. The repertoire of desaturases and elongases reveals fatty acid variations in 56 eukaryotic genomes. *Journal of Lipid Research* 49:183–191.

Haynes, K. F., and T. C. Baker. 1988. Potential for evolution of resistance to pheromones: worldwide and local variation in chemical communication system of pink bollworm moth, *Pectinophora gossypiella. Journal of Chemical Ecology* 14:1547–1560.

Haynes, K. F., L. K. Gaston, M. M. Pope, and T. C. Baker. 1983. Rate and periodicity of pheromone release from individual female artichoke plume moths, *Platyptilia carduidactyla* (Lepidoptera: Pterophoridae). *Environmental Entomology* 12:1597–1600.

Haynes, K. F., L. K. Gaston, M. M. Pope, and T. C. Baker. 1984. Potential for evolution of resistance to pheromones: interindividual and interpopulational variation in chemical communication system of pink bollworm moth. *Journal of Chemical Ecology* 10: 1551–1565.

Hellerstein, M. K., and R. A. Neese. 1992. Mass isotopomer distribution analysis: a technique for measuring biosynthesis and turnover of polymers. *American Journal of Physiology— Endocrinology and Metabolism* 263:E988–E1001.

Hellerstein, M. K., and R. A. Neese. 1999. Mass isotopomer distribution analysis at eight years: theoretical, analytic, and experimental considerations. *American Journal of Physiology - Endocrinology And Metabolism* 276:E1146–E1170.

Ichikawa, T., T. Shiota, and H. Kuniyoshi. 1996. Neural inactivation of sex pheromone production in mated females of the silkworm moth, *Bombyx mori. Zoological Science* 13:27–33.

Jacquin, E., R. A. Jurenka, H. Ljungberg, P. Nagnan, C. Löfstedt, C. Descoins, and W. L. Roelofs. 1994. Control of sex pheromone biosynthesis in the moth *Mamestra brassicae* by the pheromone biosynthesis activating neuropeptide. *Insect Biochemistry and Molecular Biology* 24:203–211.

Jaffe, K., B. Miras, and A. Cabrera. 2007. Mate selection in the moth *Neoleucinodes elegantalis*: evidence for a supernormal chemical stimulus in sexual attraction. *Animal Behaviour* 73:727–734.

Johansson, B., and T. Jones. 2007. The role of chemical communication in mate choice. *Biological Reviews of the Cambridge Philosophical Society* 82:265–289.

Jones, I. F., and R. S. Berger. 1978. Incorporation of (1-14C)acetate into *cis*-7-dodecen-1-ol acetate, a sex pheromone in the cabbage looper (*Trichoplusia ni*). *Environmental Entomology* 7:666–669.

Jurenka, R. A. 2003. Biochemistry of female moth sex pheromones. Pp. 54–80. In G. J. Blomquist and R. C. Vogt, eds. *Insect Pheromone Biochemistry and Molecular Biology*. Amsterdam: Elsevier.

Jurenka, R. A., and A. Rafaeli. 2011. Regulatory role of PBAN in sex pheromone biosynthesis of heliothine moths. *Frontiers in Endocrinology* 2:46.

Jurenka, R. A., and W. L. Roelofs. 1989. Characterization of the acetyltransferase used in pheromone biosynthesis in moths: specificity for the *Z* isomer in Tortricidae. *Insect Biochemistry* 19:639–644.

Jurenka, R. A., G. Fabrias, S. Ramaswamy, and W. L. Roelofs. 1993. Control of pheromone biosynthesis in mated redbanded leafroller moths. *Archives of Insect Biochemistry and Physiology* 24:129–137.

Jurenka, R. A., E. Jacquin, and W. L. Roelofs. 1991. Control of the pheromone biosynthetic pathway in *Helicoverpa zea* by the pheromone biosynthesis activating neuropeptide. *Archives of Insect Biochemistry and Physiology* 17:81–91.

Jurenka, R. A., M. Subchev, J. L. Abad, M. Y. Choi, and G. Fabrias. 2003. Sex pheromone biosynthetic pathway for disparlure in the gypsy moth, *Lymantria dispar. Proceedings of the National Academy of Sciences of the United States of America* 100:809–814.

Kingan, T. G., W. M. Bodnar, A. K. Raina, J. Shabanowitz, and D. F. Hunt. 1995. The loss of female sex pheromone after mating in the corn earworm moth *Helicoverpa zea*: identification of a male pheromonostatic peptide. *Proceedings of the National Academy of Sciences of the United States of America* 92:5082–5086.

Knipple, D. C., C. L. Rosenfield, S. J. Miller, W. T. Liu, J. Tang, P. W. K. Ma, and W. L. Roelofs. 1998. Cloning and functional expression of a cDNA encoding a pheromone gland-specific acyl-CoA Delta(11)-desaturase of the cabbage looper moth, *Trichoplusia ni. Proceedings of the National Academy of Sciences of the United States of America* 95:15287–15292.

Kochansky, J., R. T. Cardé, J. Liebherr, and W. L. Roelofs. 1975. Sex pheromone of the European corn borer, *Ostrinia nubilalis* (Lepidoptera: Pyralidae), in New York. *Journal of Chemical Ecology* 1:225–231.

Lassance, J.-M., A. T. Groot, M. A. Lienard, B. Antony, C. Borgwardt, F. Andersson, E. Hedenstrom, D. G. Heckel, and C. Löfstedt. 2010. Allelic variation in a fatty-acyl reductase gene causes divergence in moth sex pheromones. *Nature* 466:486–489.

Leal, W. S., A. L. Parra-Pedrazzoli, K. E. Kaissling, T. I. Morgan, F. G. Zalom, D. J. Pesak, E. A. Dundulis, C. S. Burks, and B. S. Higbee. 2005. Unusual pheromone chemistry in the navel orangeworm: novel sex attractants and a behavioral antagonist. *Naturwissenschaften* 92:139–146.

Liénard, M. A., Å. K. Hagström, J.-M. Lassance, and C. Löfstedt. 2010a. Evolution of multicomponent pheromone signals in small ermine moths involves a single fatty-acyl reductase gene. *Proceedings of the National Academy of Sciences of the United States of America* 107:10955–10960.

Liénard, M. A., J.-M. Lassance, H.-L. Wang, C.-H. Zhao, J. Piškur, T. Johansson, and C. Löfstedt. 2010b. Elucidation of the sex-pheromone biosynthesis producing 5,7-dodecadienes in *Dendrolimus punctatus* (Lepidoptera: Lasiocampidae) reveals Δ11- and Δ9-desaturases with unusual catalytic properties. *Insect Biochemistry and Molecular Biology* 40:440–452.

Linn, C. E., and W. L. Roelofs. 1989. Response specificity of male moths to multicomponent pheromones. *Chemical Senses* 14:421–437.

Linn, C. E., M. G. Campbell, and W. L. Roelofs. 1987. Pheromone components and active spaces: what do moths smell and where do they smell it? *Science* 237:650–652.

Liu, Y.-B., and K. F. Haynes. 1994. Temporal and temperature-induced changes in emission rates and blend ratios of sex pheromone components in *Trichoplusia ni. Journal of Insect Physiology* 40:341–346.

Liu, W., H. M. Jiao, N. C. Murray, M. O'Connor, and W. L. Roelofs. 2002a. Gene characterized for membrane desaturase that produces (*E*)-11 isomers of mono- and diunsaturated fatty acids. *Proceedings of the National Academy of Sciences of the United States of America* 99:620–624.

Liu, W. T., H. M. Jiao, M. O'Connor, and W. L. Roelofs. 2002b. Moth desaturase characterized that produces both Z and E isomers of Delta 11-tetradecenoic acids. *Insect Biochemistry and Molecular Biology* 32:1489–1495.

Löfstedt, C. 1993. Moth pheromone genetics and evolution. *Philosophical Transactions of the Royal Society of London B* 340:167–177.

Löfstedt, C., and M. Bengtsson. 1988. Sex pheromone biosynthesis of (*E,E*)-8,10-dodecadienol in codling moth *Cydia pomonella* involves E9 desaturation. *Journal of Chemical Ecology* 14:903–915.

Luxová, A., and A. Svatoš. 2006. Substrate specificity of membrane-bound alcohol oxidase from the tobacco hornworm moth (*Manduca sexta*) female pheromone glands. *Journal of Molecular Catalysis B: Enzymatic* 38:37–42.

Ma, P. W. K., and S. B. Ramaswamy. 2003. Biology and ultrastructure of sex pheromone-producing tissue. Pp. 19–51. In G. J. Blomquist and R. C. Vogt, eds. *Insect Pheromone Biochemistry and Molecular Biology*. London: Elsevier Academic Press.

Matsumoto, S. 2010. Molecular mechanisms underlying sex pheromone production in moths. *Bioscience, Biotechnology, and Biochemistry* 74:223–231.

Matsumoto, S., A. Fonagy, M. Yamamoto, F. Wang, N. Yokoyama, Y. Esumi, and Y. Suzuki. 2002. Chemical characterization of cytoplasmic lipid droplets in the pheromone-producing cells of the silkmoth, *Bombyx mori. Insect Biochemistry and Molecular Biology* 32:1447–1455.

McNeil, J. N. 1991. Behavioral ecology of pheromone-mediated communication in moths and its importance in the use of pheromone traps. *Annual Review of Entomology* 36:407–430.

Millar, J. G. 2000. Polyene hydrocarbons and epoxides: a second major class of lepidopteran sex attractant pheromones. *Annual Review of Entomology* 45:575–604.

Millar, J. G., A. E. Knudson, J. S. McElfresh, R. Gries, G. Gries, and J. H. Davis. 1996. Sex attractant pheromone of the pecan nut casebearer (Lepidoptera: Pyralidae). *Bioorganic & Medicinal Chemistry* 4:331–339.

Miller, D.R., J.G. Millar, A. Mangini, C.M. Crowe, and G.G. Grant. 2010. (3Z,6Z,9Z,12Z,15Z)-Pentacosapentaene and (Z)-11-hexadecenyl acetate: sex attractant blend for *Dioryctria amatella* (Lepidoptera: Pyralidae). *Journal of Economic Entomology* 103:1216–1221.

Miller, J.R., and W.L. Roelofs. 1980. Individual variation in sex pheromone component ratios in two populations of the red-banded leafroller moth, *Argyrotaenia velutinana*. *Environmental Entomology* 9:359–363.

Moore, I.T., and W.A. Hopkins. 2009. Interactions and trade-offs among physiological determinants of performance and reproductive success. *Integrative and Comparative Biology* 49:441–451.

Morse, D., and E. Meighen. 1987a. Biosynthesis of the acetate ester precursor of the spruce budworm sex pheromone by an acetyl CoA: fatty alcohol acetyltransferase. *Insect Biochemistry* 17:53–59.

Morse, D., and E. Meighen. 1987b. Pheromone biosynthesis: enzymatic studies in Lepidoptera. Pp. 121–158. In G.D. Prestwich and G.J. Blomquist, eds. *Pheromone Biochemistry*. New York: Academic Press.

Moto, K., M.G. Suzuki, J.J. Hull, R. Kurata, S. Takahashi, M. Yamamoto, K. Okano, K. Imai, T. Ando, and S. Matsumoto. 2004. Involvement of a bifunctional fatty-acyl desaturase in the biosynthesis of the silkmoth, *Bombyx mori*, sex pheromone. *Proceedings of the National Academy of Sciences of the United States of America* 101:8631–8636.

Moto, K., T. Yoshiga, M. Yamamoto, S. Takahashi, K. Okano, T. Ando, T. Nakata, and S. Matsumoto. 2003. Pheromone gland-specific fatty-acyl reductase of the silkmoth, *Bombyx mori*. *Proceedings of the National Academy of Sciences of the United States of America* 100:9156–9161.

Nielsen, J. 2003. It is all about metabolic fluxes. *Journal of Bacteriology* 185:7031–7035.

O'Brien, D.M., D.P. Schrag, and C. Martinez del Rio. 2000. Allocation to reproduction in a hawkmoth: a quantitative analysis using stable carbon isotopes. *Ecology* 81:2822–2831.

Ono, T., R.E. Charlton, and R.T. Cardé. 1990. Variability in pheromone composition and periodicity of pheromone titer in potato tuberworm moth, *Phthorimaea operculella* (Lepidoptera: Gelechiidae). *Journal of Chemical Ecology* 16:531–542.

Phelan, P.L. 1992. Evolution of sex pheromones and the role of asymmetric tracking. Pp. 265–314. In B.D. Roitberg and M.B. Isman, eds. *Insect Chemical Ecology: An Evolutionary Approach*. New York: Chapman & Hall.

Prill, R.J., P.A. Iglesias, and A. Levchenko. 2005. Dynamic properties of network motifs contribute to biological network organization. *PLoS Biology* 3:e343.

Rafaeli, A. 2005. Mechanisms involved in the control of pheromone production in female moths: recent developments. *Entomologia Experimentalis et Applicata* 115:7–15.

Rafaeli, A., and R. Jurenka. 2003. PBAN regulation of pheromone biosynthesis in female moths. Pp. 107–136. In G.J. Blomquist and R.C. Vogt, eds. *Insect Pheromone Biochemistry and Molecular Biology*. Amsterdam: Elsevier.

Raina, A.K. 1993. Neuroendocrine control of sex pheromone biosynthesis in Lepidoptera. *Annual Review of Entomology* 38:329–349.

Raina, A.K. 2003. Pheromone production in corn earworm: effect of temperature and humidity. *Southwestern Entomologist* 28:115–120.

Raina, A.K., J.C. Davis, and E.A. Stadelbacher. 1991. Sex pheromone production and calling in *Helicoverpa zea* (Lepidoptera: Noctuidae): effect of temperature and light. *Environmental Entomology* 20:1451–1456.

Raina, A.K., T.G. Kingan, and J.M. Giebultowicz. 1994. Mating-induced loss of sex pheromone and sexual receptivity in insects with emphasis on *Helicoverpa zea* and *Lymantria dispar*. *Archives of Insect Biochemistry and Physiology* 25:317–327.

Ramaswamy, S.B., Y. Qiu, Y. Park, Y. Qiu, and Y.I. Park. 1996. Neuronal control of post-coital pheromone production in the moth *Heliothis virescens*. *Journal of Experimental Biology* 274:255–263.

Ramaswamy, S.B., S. Shu, S. Shu, Y.I. Park, and F. Zeng. 1997. Dynamics of juvenile hormone-mediated gonadotropism in the Lepidoptera. *Archives of Insect Biochemistry and Physiology* 35:539–558.

Rodriguez, S., G.X. Hao, W.T. Liu, B. Pina, A.P. Rooney, F. Camps, W.L. Roelofs, and G. Fabrias. 2004. Expression and evolution of Delta(9) and Delta(11) desaturase genes in the moth *Spodoptera littoralis*. *Insect Biochemistry and Molecular Biology* 34:1315–1328.

Roelofs, W.L. 1979. Production and perception of lepidopterous pheromone blends. Pp. 159–168. In F.J. Ritter, ed. *Chemical Ecology: Odour Communication in Animals*. New York: Elsevier.

Roelofs, W.L., and L.B. Bjostad. 1984. Biosynthesis of lepidopteran pheromones. *Bioorganic Chemistry* 12:279–298.

Roelofs, W.L., J.-W. Du, C.E. Linn, T.J. Glover, and L.B. Bjostad. 1986. The potential for genetic manipulation of the redbanded leafroller moth sex pheromone blend. Pp. 263–272. In M.D. Huettel, ed. *Evolutionary Genetics of Invertebrate Behavior: Progress and Prospects*. New York: Plenum Press.

Roelofs, W.L., W.T. Liu, G.X. Hao, H.M. Jiao, A.P. Rooney, and C.E. Linn. 2002. Evolution of moth sex pheromones via ancestral genes. *Proceedings of the National Academy of Sciences of the United States of America* 99:13621–13626.

Rooney, A.P. 2009. Evolution of moth sex pheromone desaturases. *Annals of the New York Academy of Sciences* 1170:506–510.

Rule, G.S., and W.L. Roelofs. 1989. Biosynthesis of sex pheromone components from linolenic acid in arctiid moths. *Archives of Insect Biochemistry and Physiology* 12:89–97.

Salati, L.M., and A.G. Goodridge. 1996. Fatty acid synthesis in eukaryotes. Pp. 101–128. In D.E. Vance and J. Vance, eds. *Biochemistry of Lipids, Lipoproteins and Membranes*. Amsterdam: Elsevier.

Schal, C., R.E. Charlton, and R.T. Cardé. 1987. Temporal patterns of sex pheromone titers and release rates in *Holomelina lamae* (Lepidoptera: Arctiidae). *Journal of Chemical Ecology* 13:1115–1129.

Schal, C., V. Sevala, and R.T. Cardé. 1998. Novel and highly specific transport of a volatile sex pheromone by hemolymph lipophorin in moths. *Naturwissenschaften* 85:339–342.

Schlyter, F., and G. Birgersson. 1989. Individual variation in bark beetle and moth pheromones: a comparison and an evolutionary background. *Holarctic Ecology* 12:457–465.

Serra, M., B. Piña, J.L. Abad, F. Camps, and G. Fabriàs. 2007. A multifunctional desaturase involved in the biosynthesis of the processionary moth sex pheromone. *Proceedings of the National Academy of Sciences of the United States of America* 104:16444–16449.

Serra, M., B. Piña, J. Bujons, F. Camps, and G. Fabriàs. 2006. Biosynthesis of 10,12-dienoic fatty acids by a bifunctional Δ11desaturase in *Spodoptera littoralis*. *Insect Biochemistry and Molecular Biology* 36:634–641.

Slatyer, R.A., B.S. Mautz, P.R.Y. Backwell, and M.D. Jennions. 2012. Estimating genetic benefits of polyandry from experimental studies: a meta-analysis. *Biological Reviews* 87:1–33.

Sreng, I., T. Glover, and W. Roelofs. 1989. Canalization of the red-banded leafroller moth sex pheromone. *Archives of Insect Biochemistry and Physiology* 10:73–82.

Suarez, R.K., J.F. Staples, J.R.B. Lighton, and T.G. West. 1997. Relationships between enzymatic flux capacities and metabolic flux rates: nonequilibrium reactions in muscle glycolysis. *Proceedings of the National Academy of Sciences of the United States of America* 94:7065–7069.

Svensson, M.E., M. Bengtsson, and J. Löfqvist. 1997. Individual variation and repeatability of sex pheromone emission of female turnip moths *Agrotis segetum*. *Journal of Chemical Ecology* 23:1833–1850.

Symonds, M.R.E., and M.A. Elgar. 2008. The evolution of pheromone diversity. *Trends in Ecology and Evolution* 23:220–228.

Symonds, M.R.E., T.L. Johnson, and M.A. Elgar. 2012. Pheromone production, male abundance, body size, and the evolution of elaborate antennae in moths. *Ecology and Evolution* 2:227–246.

Tang, J.D., R.E. Charlton, R.T. Cardé, and C.M. Yin. 1992. Diel periodicity and influence of age and mating on sex pheromone titer in gypsy moth, *Lymantria dispar* (L.). *Journal of Chemical Ecology* 18:749–760.

Tang, J.D., R.E. Charlton, R.A. Jurenka, W.A. Wolf, P.L. Phelan, L. Sreng, and W.L. Roelofs. 1989. Regulation of pheromone biosynthesis by a brain hormone in two moth species. *Proceedings of the National Academy of Sciences of the United States of America* 86:1806–1810.

Teal, P.E.A., and J.H. Tumlinson. 1986. Terminal steps in pheromone biosynthesis by *Heliothis virescens* and *H. zea*. *Journal of Chemical Ecology* 12:353–366.

Teal, P.E.A., and J.H. Tumlinson. 1988. Properties of cuticular oxidases used for sex pheromone biosynthesis by *Heliothis zea*. *Journal of Chemical Ecology* 14:2131–2145.

Thompson, S. N. 1973. A review and comparative characterization of the fatty acid compositions of seven insect orders. *Comparative Biochemistry and Physiology Part B: Comparative Biochemistry* 45:467–482.

Thompson, S. N. 2003. Trehalose—the insect "blood" sugar. *Advances in Insect Physiology* 31:205–285.

Torres-Vila, L. M., and M. D. Jennions. 2005. Male mating history and female fecundity in the Lepidoptera: do male virgins make better partners? *Behavioral Ecology and Sociobiology* 57:318–326.

Torres-Vila, L. M., M. C. Rodríguez-Molina, and M. D. Jennions. 2004. Polyandry and fecundity in the Lepidoptera: can methodological and conceptual approaches bias outcomes? *Behavioral Ecology and Sociobiology* 55:314–324.

Tsfadia, O., A. Azrielli, L. Falach, A. Zada, W. Roelofs, and A. Rafaeli. 2008. Pheromone biosynthetic pathways: PBAN-regulated rate-limiting steps and differential expression of desaturase genes in moth species. *Insect Biochemistry and Molecular Biology* 38:552–567.

Valeur, P. G., B. S. Hansson, and C. Löfstedt. 1999. Real-time measurement of pheromone release from individual female moths and synthetic dispensers in a wind tunnel by recording of single receptor-neurone responses. *Physiological Entomology* 24:240–250.

van Eunen, K., J. A. L. Kiewiet, H. V. Westerhoff, and B. M. Bakker. 2012. Testing biochemistry revisited: how in vivo metabolism can be understood from in vitro enzyme kinetics. *PLOS Computational Biology* 8:e1002483.

Wäckers, F. L., J. Romeis, and P. van Rijn. 2007. Nectar and pollen feeding by insect herbivores and implications for multitrophic interactions. *Annual Review of Entomology* 52:301–323.

Wang, H.-L., M. A. Liénard, C.-H. Zhao, C.-Z. Wang, and C. Löfstedt. 2010a. Neofunctionalization in an ancestral insect desaturase lineage led to rare Δ6 pheromone signals in the Chinese tussah silkworm. *Insect Biochemistry and Molecular Biology* 40:742–751.

Wang, H.-L., C.-H. Zhao, J. G. Millar, R. T. Cardé, and C. Löfstedt. 2010b. Biosynthesis of unusual moth pheromone components involves two different pathways in the navel orangeworm, *Amyelois transitella*. *Journal of Chemical Ecology* 36:535–547.

Watt, W. B., and A. M. Dean. 2000. Molecular-functional studies of adaptive genetic variation in prokaryotes and eukaryotes. *Annual Review of Genetics* 34:593–622.

Webster, R. P., and R. T. Cardé. 1982. Relationship among pheromone titre, calling and age in the omnivorous leafroller moth (*Platynota stultana*). *Journal of Insect Physiology* 28:925–933.

Wolf, W. A., and W. L. Roelofs. 1983. A chain-shortening reaction in orange tortrix moth sex pheromone biosynthesis. *Insect Biochemistry* 13:375–379.

Wolfe, R. R., and D. L. Chinkes. 2005. *Isotope Tracers in Metabolic Research*. 2nd ed. Hoboken, NJ: John Wiley & Sons, Inc.

Zavada, A., C. L. Buckley, D. Martinez, J.-P. Rospars, and T. Nowotny. 2011. Competition-based model of pheromone component ratio detection in the moth. *PLOS ONE* 6:e16308.

Závodská, R., G. von Wowern, C. Löfstedt, W. Rosén and I. Sauman. 2009. The release of a pheromonotropic neuropeptide, PBAN, in the turnip moth *Agrotis segetum*, exhibits a circadian rhythm. *Journal of Insect Physiology* 55:435–440.

Zhu, J. W., C. Löfstedt, and B. O. Bengtsson. 1996a. Genetic variation in the strongly canalized sex pheromone communication system of the European corn borer, *Ostrinia nubilalis* Hübner (Lepidoptera, Pyralidae). *Genetics* 144:757–766.

Zhu, J. W., C.-H. Zhao, F. Lu, M. Bengtsson, and C. Löfstedt. 1996b. Reductase specificity and the ratio regulation of E/Z isomers in pheromone biosynthesis of the European corn borer, *Ostrinia nubilalis* (Lepidoptera: Pyralidae). *Insect Biochemistry and Molecular Biology* 26:171–176.

CHAPTER NINE

Molecular Biology of Reception

WALTER S. LEAL

Overview

Olfactory processing in insects starts with the reception of semiochemicals by specialized organs at the periphery. In moths, sex pheromones are detected by olfactory receptor neurons (ORNs) housed in thousands of antennal sensilla, whereas other appendages (e.g., maxillary palps) may have sensors for other odorants. After being transduced, pheromone signals are relayed to the macroglomerular complex in the antennal lobe. Here, signals from constituents of a sex pheromone blend are integrated, and sometimes suppressed by behavioral antagonist(s). Next, the outcome of this integration is relayed by the antennal lobe, integrated with other sensory modalities in the protocerebrum, and finally translated into behavioral response by sending electrical signals to the muscular system for locomotion via walking and/or flight. The word perception has a connotation of a "mental image" and thus is better suited for neural processes. Thus, perception of pheromones involves neural processing, whereas reception or detection of pheromones is a biochemical process that takes place at the periphery. To separate it from perception, the early olfactory processing is sometimes called perireceptor events (Getchell et al. 1984), which involves uptake, binding, transport, and inactivation of odorants, as well as receptor activation and signal transduction. To

put it simply, reception (or detection) is the biochemical processing of odorants, whereas perception is the neural processes dealing with electrical signals (derived from those odorants). Pheromones are, therefore, odorants (chemical molecules), not odors. Thus, flight toward a pheromone source is "odorant-induced navigation," but to sustain flight activity male moths rely on reception and perception of pheromones. Moths must detect pheromone molecules hitting the antennae and inactivate them so rapidly that the brain can perceive their next encounter with pheromone. Perception and signal integration are covered elsewhere (Baker and Hansson, this volume). Here, I will focus on reception of pheromones, and discuss the biochemical and molecular biology details that lead to the highly selective and extremely sensitive olfactory system of moths.

Pheromone-Binding Proteins: Pheromone Uptake and Release at the Right Place and the Right Time

Pheromones, and particularly moth sex pheromones, are hydrophobic compounds with a long hydrophobic carbon chain and usually a polar functional group. These compounds are "greasy," given that the contribution for the hydrophobicity by the nonpolar moiety is larger than the hydrophilicity of the functional groups; in some cases, the pheromone molecule has no polar moiety (e.g., polyunsaturated or saturated hydrocarbons). The receptors for sex pheromones, the odorant receptors (ORs), are embedded in dendritic membranes of ORNs, which are in turn housed inside pheromone-detecting sensilla. ORs are, therefore, protected from harmful chemicals from the environment by an aqueous solution—the sensillar lymph that surrounds dendrites. Pheromone molecules hitting and entering sensilla through pore tubules must travel through the aqueous solution before reaching ORs. Odorant-binding proteins (OBPs), also referred to as pheromone-binding proteins (PBPs) when the natural ligand is a pheromone, solubilize pheromones by holding the hydrophobic molecule with van der Waals interactions in a binding cavity. In the case of the silkworm moth, *Bombyx mori* (Bombycidae), PBP, BmorPBP1, negatively and positively charged amino acid residues on the surface of the protein make the BmorPBP1–bombykol complex highly water-soluble (Sandler et al. 2000). How is pheromone transported from the uptake location (pore tubules) to the ORs? Given the high concentration of PBPs in the sensillar lymph (e.g., 3 mM BmorPBP1 [Syed et al. 2006]), one early hypothesis was "pheromone hopping." In short, pheromone would hop in and out of PBP molecules on the way to receptors. This would be somewhat similar to what protons do in water. Here, however, one proton hops in and another proton hops out from a water molecule, so that protons (from an acid) pipetted into water do not have to travel to the pH meter; they can be sensed remotely. By contrast, receptors must be activated by the same molecules (that enter pore tubules), not by a "surrogate." Furthermore, for PBPs, this notion of "hopping" is not even consistent with the release rate (or the off rate) of binding of pheromones to PBPs. Pheromones bind rapidly to PBPs and their release by mass action is very slow, i.e., in the timescale of minutes compared to the millisecond timescale of olfaction (Leal et al. 2005). Additionally, stray molecules like a pheromone in transit from one PBP molecule to the next would be quickly degraded by aggressive pheromone-degrading enzymes (PDE) in the sensillar lymph (see below). The simplest explanation for pheromone transport is that concentration gradients lead to diffusion away from pore tubules.

There is now consensus that PBPs solubilize pheromones and ferry them to ORs. How they do this is still under debate (see below). PBPs also contribute to the sensitivity of the olfactory system. As demonstrated with two species of mosquito, *Culex quinquefasciatus* (Pelletier et al. 2010) and *Anopheles gambiae* (Biessmann et al. 2010), knockdown of OBPs dramatically reduced the sensitivity of the insect's olfactory systems. Sensitivity is of paramount importance in moth's chemical communication as females release minute amounts of sex pheromone. Therefore, males have evolved the ability to detect these low-amplitude signals concealed in a noisy odorant environment. First, antennal structure affects the capture of odorants from the air and subsequent detection by thousands of pheromone-detecting sensilla (e.g., 17,000 in a male *B. mori* antenna [Schneider and Kaissling 1957; Steinbrecht 1970]). Plumose antennae rapidly and nearly completely remove pheromones from the air surrounding the sensory hairs, and therefore an increase in airflow through the antennae (e.g., pulled by wing fanning) increases the rate of odorant interception, by at least two orders of magnitude in *B. mori* (Loudon and Koehl 2000). For a review of theoretical methods for calculating rates of odorant interception by sensory hairs, see Loudon (2003); for a review on pheromone capture, see Kaissling (2009). Secondly, and consistent with their high concentrations in the sensillar lymph, PBPs act as a "sponge" to capture odorants reaching pore tubules. Lastly, ORs must respond to pheromones with a very low threshold. All these factors contribute to the extraordinary sensitivity of the moth's olfactory system as shown with male *B. mori*, which displays a behavioral reaction when about 1% of the bombykol receptor cells fire one nerve impulse per second above the spontaneous firing rate (natural background) (Kaissling 1996).

There is consensus that OBPs solubilize and transport odorants, as initially postulated (Vogt and Riddiford 1981). There is considerable debate, however, whether ORs are activated by a pheromone–PBP complex or a pheromone molecule per se

(Kaissling 2013). In addition, there are dichotomous hypotheses regarding specificity of PBPs, with one school favoring nonselective transport and another suggesting that PBPs contribute to the selectivity of the olfactory system.

As will be discussed later in depth, ORs can be studied in heterologous systems with surrogate OBPs or in systems devoid of OBPs altogether. Both types of studies have been frequently misinterpreted as an indication that OBPs are not essential for odorant reception. One of the most elegant systems for functional analysis of ORs is the "empty neuron" system from the fruit fly, *Drosophila melanogaster* (Dobritsa et al. 2003). This system utilizes a Δhalo mutant that has all the biochemical machinery, including OBPs, for odorant reception, except that neuron-A in the ab3 sensillum lacks OR22a/b—hence the term "empty neuron." Strictly speaking, ab3A neuron is not empty; it is just missing ORs, and the sensillum has an intact biochemical milieu. A test OR can be expressed in neuron ab3A by a series of genetic manipulations and then the neuron expressing the test receptor can be interrogated electrophysiologically by single-sensillum recording. With this system, expression of a test OR is specific to the ab3A neuron as it is driven by the promoter of *OR22a/b*—the promoter of the receptors deleted in Δhalo mutant. Flies obtained by P-element-mediated transformation and carrying the target gene downstream of UAS sequence are crossed with Δhalo mutants to generate UAS lines with the Δhalo background. Then, these *UAS-target OR* flies are crossed with *OR22a/promoter-GAL4* flies with the Δhalo background. GAL4 will bind to the UAS site and trigger expression of the target OR gene. This system works very well for *Drosophila* (Hallem and Carlson 2006) and other dipteran insects like the malaria mosquito (Carey et al. 2010). A simple explanation is that OBPs expressed in ab3 sensilla may function well as surrogates for OBPs co-expressed with these test receptors in their natural environment. Although raising questions about specificity of OBPs, these results are not inconsistent with the hypothesis that OBPs solubilize and transport odorants, and enhance the sensitivity of the olfactory system. This empty neuron system, however, performs poorly with a silkworm moth pheromone OR, BmorOR1 (Syed et al. 2006), which was fully functional only when expressed in trichoid sensilla using a different expression system (Syed et al. 2010). One possible explanation is that OBPs from the ab3 sensilla are not suitable substitutes for moth PBPs. Indeed, moth PBPs and OBPs from the fruit fly are very distinct (Leal 2013).

Interpretation of the physiological role(s) of OBPs can be even more misleading when ORs are tested using the *Xenopus* oocyte recording system. The *Xenopus* system was employed in insect olfaction soon after OR genes were identified from *D. melanogaster* (Clyne et al. 1999; Gao and Chess 1999; Vosshall et al. 1999) to de-orphanize for the first time an insect OR, DmelOR43a (Wetzel et al. 2001). It was later realized that the odorant receptor co-receptor (Orco) (Larsson et al. 2004) is necessary for the proper formation of ion-channels (Nakagawa et al. 2005). Thereafter, many ORs have been de-orphanized by co-expression with their respective Orcos in *Xenopus* oocytes. This heterologous expression system, albeit convenient, is "incomplete" compared to the molecular and biochemical machinery in an insect's sensillum. A common misinterpretation is that, given that there are no OBPs in the *Xenopus* oocyte system, OBPs are not essential for the function of the olfactory system. However, it should be recognized that these "naked" receptor experiments are performed with hydrophobic odorants solubilized with dimethyl sulfoxide

(DMSO) and the environments of ORs expressed on oocytes and in neurons are remarkably different. The oocytes are bathed with a buffer solution completely devoid of other olfactory proteins, whereas the sensillar lymph is rich in olfactory proteins, including aggressive odorant-degrading enzymes (ODEs, see below).

In the naked system, DMSO performs one of the postulated functions of OBPs—solubilization of odorants. However, in the absence of ODEs, odorants do not need to be protected from degradation. Using another type of heterologous expression, modified HEK293 cells, it has been demonstrated that "naked" ORs respond with 2–3 orders of magnitude lower threshold (higher sensitivity) when pheromones were dissolved with PBPs as compared to DMSO (Grosse-Wilde et al. 2006, 2007, 2009; Forstner et al. 2009). Attempts to enhance sensitivity of BmorOR1•BmorOrco-expressing oocytes showed the opposite effect: receptor activity was dramatically reduced (by 66–87%) when the ligands were solubilized with BmorPBP1 as compared to DMSO (Xu et al. 2012b). Two possible explanations are that pheromones remained trapped in BmorPBP1 because there were no low pH environments to trigger their release from the BmorPBP1•pheromone complex and/or vitelline membrane surrounding the oocytes prevented transport of ligands close to Ors. More recently, it has been reported that two ORs from the diamondback moth, *Plutella xylostella* (Plutellidae), Pxyl0R1 and PxylOR4, gave 1–2 orders of magnitude higher responses to Z11-16Ald and Z9-14Ac, respectively, when they were dissolved with PxylPBPs as compared to DMSO (Sun et al. 2013). By contrast, PxylOR4•PxylOrco-expressing oocytes gave lower responses to Z9E12-14Ac when it was dissolved with PBPs as compared to DMSO. The results of Xu et al. (2012b) and Sun et al. (2013) are not consistent. Although both studies used the same *Xenopus* oocyte recording system, they tested olfactory proteins from two different moth species and differed in the protocols for expression of PBPs. Xu et al. (2012b) used a pET-22b(+) vector system that generated intact, functional Bmor-PBP1, identical to the native protein (Wojtasek and Leal 1999). By contrast, Sun et al. (2013) expressed PxylPBPs with a pET-30a(+) vector that generates His-tagged proteins for purification and, subsequent, removal of the His-tag. In the absence of additional information, it is impossible to conclude if PBP expression and function were significant factors.

The above-described experiments with the *Xenopus* oocyte recording system are inconsistent with one of the earliest modes of action proposed for OBPs (Pelosi 1994)—the OBP•odorant complex hypothesis in that OBPs trigger olfactory receptors when bound to odorant molecules. Later, this model gained substantial support when an OBP from the fruit fly, LUSH (=DmelOBP76a) was demonstrated to be functional in vivo (Xu et al. 2005). While OR67d in T1 trichoid sensilla in wild type flies responded to (*Z*)-11-octadecen-1-yl acetate (Z11-18Ac, *cis*-vaccenyl acetate), T1 sensilla in a *lush* mutant defective for the expression of LUSH gave no response. Activity in *lush* mutants was restored when LUSH expression was genetically rescued or added by injection of recombinant LUSH into the T1 sensilla. By contrast, expression of ApolPBP1, then named APOL3, in *lush* mutant failed to restore activity (Xu et al. 2005), implying specificity of LUSH. This was the first and most elegant demonstration of the functional role of an OBP in vivo (Xu et al. 2005), and is consistent with more recent findings by gene knockdown of mosquito (Biessmann et al. 2010; Pelletier et al. 2010) and fruit fly (Swarup et al. 2011) OBPs. It is, therefore, clear that OBPs are essential for a functional olfactory system, but there is no consensus on the hypothesis that OBP•odorant complexes activate ORs.

The structure of LUSH bound to Z11-18Ac showed the ligand completely encapsulated by the protein, with the polar acetate moiety interacting with Thr-57 and Ser-52 at one end of the binding pocket (Laughlin et al. 2008). The positioning of the acetate group was reinforced by interactions of Z11-18Ac with the aromatic residues Phe-64, Phe-113, and Trp-123. It was then suggested that interactions of the pheromone molecule with Phe-121 mediated specific conformational shifts of amino acid residues in the C-terminus, thus disrupting a salt bridge between Asp-118 and Lys-87, which was present in the apo-LUSH structure. These structural observations led to the hypothesis that LUSH itself acts as the ligand for the Z11-18Ac receptor, OR67d. The proposed mode of action (sometimes referred to as conformational activation model) assumed that the sole function of Z11-18Ac was to trigger the above-described conformational change. Laughlin et al. (2008) reasoned that this conformational change could be mimicked by mutations in LUSH residues. First, they surmised that substitutions of Phe-121 by smaller residues might reduce this conformational change, whereas large residues would enhance it. These hypotheses were tested by infusion of recombinant proteins through the recording electrode into the T1 sensilla of *lush* mutants. When Phe-121-Ala-LUSH was infused, T1 neuronal activities were ca. 50-fold less sensitive to Z11-18Ac compared to the wild type protein at the same doses. By contrast, T1 sensilla infused with Phe-121-Trp-LUSH showed fivefold enhanced responses to Z11-18Ac compared to wild type LUSH. Next, they postulated that disruption of the salt bridge between residues Asp-118 and Lys-87 by mutagenesis might produce an active conformation, which would activate the receptor in the absence of ligand. Indeed, infusion of Asp-118-Ala-LUSH into T1 sensilla of lush null mutants induced high levels of neuronal activity in the absence of Z11-18Ac. Therefore, they concluded that LUSH is the ligand for the pheromone-sensitive neurons, and the role of Z11-18Ac is to trigger its conversion into the activating conformation. Despite the unambiguous evidence by multiple laboratories showing that "naked" ORs (reviewed in Leal 2013), expressed in *Xenopus* oocytes or other heterologous systems, can be activated by odorants in the complete absence of OBPs, the paradigm of LUSH remained for the supporters of the OBP•odorant complex model. Recently, the OBP•ligand hypothesis was refuted (Gomez-Diaz et al. 2013). Instead of infusion of recombinant LUSH into the sensillar lymph of T1 (through recording electrode), LUSH was provided via a transgenic genetic rescue construct that appears to contain all necessary transcriptional and translational regulatory sequences to recapitulate endogenous LUSH expression. Neuronal activity elicited by Z11-18Ac in T1 sensilla of wild type, Phe-121-Ala-LUSH and Phe-121-Trp-LUSH mutants, was very similar across a 10,000-fold range of Z11-18Ac doses. Furthermore, Asp-118-Ala-LUSH did not exhibit the elevated level of spontaneous activity reported for infused recombinant Asp-118-Ala-LUSH. In sum, the data generated with transgenic flies do not support the conformational activation model proposed on the basis of infusion of recombinant proteins. Although the reduction in Z11-18Ac sensitivity in T1 neurons in *lush* null mutants remains the most elegant demonstration of the importance of OBPs for signal transduction in insect olfaction (see above), it is highly unlikely in light of recent evidence that LUSH itself, after a conformational change, activates OR67d neurons.

In the sunset of the last century, my laboratory serendipitously discovered that the silkworm moth PBP, BmorPBP (renamed BmorPBP1 when multiple PBPs were discovered later), undergoes a pH-dependent conformational change (Wojtasek and Leal 1999). Although their structures were unknown at that time, we postulated that there were two forms of PBPs that were important for pheromone binding and release. We prepared pure samples of recombinant, labeled BmorPBP1, which we shipped to Kurt Wüthrich's laboratory for NMR analysis. Although the prepared samples were pure as indicated by multiple chromatographic techniques (native and SDS gel electrophosresis as well as liquid chromatography coupled with electronspray ionization-mass spectrometry), NMR analysis suggested two species. We hypothesized that the protein sample had undergone degradation during shipment to Switzerland; however, reanalysis of the sample returned from Switzerland using the same techniques did not detect an impurity. A thorough analysis of BmorPBP1 by circular dichroism (CD) and fluorescence demonstrated that the protein was also very stable. It could be heated to near the boiling point and retained the CD spectrum when returned to room temperature: a seemingly "rock-solid" protein. We discovered that this remarkably stable protein was very sensitive to pH changes. As indicated by near-UV CD, the tertiary structure of the protein changed dramatically when the pH was reduced, with a clear transition between pH 6 and pH 5, but the secondary structure (as suggested by far-UV CD) remained nearly unchanged. These observations substantiated our hypothesis of the two forms of BmorPBP1 (Wojtasek and Leal 1999).

After attempts that consumed several months, the Jon Clardy group, then at Cornell University, was able to crystallize BmorPBP1, but as the structure had no suitable models, the diffraction data could not be analyzed by molecular replacement with a related protein. Preparation of a sample of selenomethionyl BmorPBP1 allowed Sandler et al. (2000) to solve the phase problem and determine the structure after years of work. The structure of BmorPBP1 (Sandler et al. 2000) was a surprise: six α-helices, four of them converging to form a large flask-shaped cavity that encapsulated bombykol. Thus, BmorPBP1 solubilizes bombykol by holding the pheromone (mostly with van der Waals interactions) in a hydrophobic pocket in the core of the protein. Twenty-one glutamate or aspartate residues and 14 lysine or arginine residues on the outer surface of the protein make BmorPBP1 as well as BmorPBP1-bombykol highly water-soluble. To crystallize BmorPBP1, it was necessary to use concentrations at least twice what one would have expected for a protein of its size.

Almost simultaneously Horst et al. (2001) elucidated the structure of BmorPBP1 at low pH (4.5). The NMR structure provided unambiguous evidence for a pH-dependent conformational change. At low pH, the C-terminus forms a seventh α-helix, which replaced bombykol in the binding site. Thus, the two forms of BmorPBP1 we hypothesized earlier on the basis of CD and fluorescence data are indeed conformations of the protein, i.e., BmorPBP1[A] (the acid form) and BmorPBP1[B] (the basic form). Their 3D structures differed markedly, particularly with respect to the C-terminus, which is at the surface of BmorPBP1[B] and located in the binding pocket of BmorPBP1[A]. This is physiologically relevant, given that the surface of dendrites where the receptors are located is negatively charged (Keil 1984) and thus in the vicinity of the membrane the pH is reduced due to accumulation of protons. Measuring the pH at the surface membrane in vivo remains a challenge despite several attempts by Yuko Ishida in my laboratory who used trans-

genic flies to express a pHluorin GFP engineered by structure-directed combinatorial mutagenesis into a pH-sensitive mutant of green fluorescent protein (Miesenbock et al. 1998). In order to measure bulk and localized pH of sensillar lymph in the sensillum cavity, a bacterial expression vector, pET-22b(+)-mDme-lOBP28a (former name, pheromone-binding protein-related protein 2, PBPRP2) ratiometric pHluorin GFP was constructed, the protein expressed, purified, and its changes in fluorescence at various pH conditions measured. The purified mature Dme-lOBP28a-ratiometric pHluorin GFP showed a reversible excitation ratio change between pH 7 and 5.5 as an original pHluorin GFP, suggesting that our recombinant protein could be used to generate a calibration curve. The pH-reporter OBP was expressed in the sensillum cavity of the Drosophila antenna by GAL4-UAS system using DmelOBP28a-GAL4 fly as a driver strain (Brand and Perrimon 1993; Park et al. 2000). Although GFP-fluorescence was emitted in the sensilla of the established flies, differences in the pattern of fluorescence were not observed in the expected localization-dependent manner by image analysis, most likely because autofluorescence of the cuticle of the sensilla masked the signal of the pH sensor. The measurement of pH at the surface of dendrites must await development of novel pH sensors with excitation frequencies that do not overlap with innate cuticular fluorescence signals. Although the exact pH is yet to be measured, we have demonstrated with stopped flow fluorescence analysis that BmorPBP1[B] is rapidly converted into BmorPBP1[A] (half lifetime ≈9 ms) (Leal et al. 2005) when pH is lowered. Thus, the pH-mediated conformational change is within the timescale of pheromone reception (on average 500 ms) (Kaissling 2013).

BmorPBP1 seems promiscuous in binding to many ligands when tested in vitro (Grater et al. 2006; Lautenschlager et al. 2007), thus suggesting that PBPs are nonselective transporters of odorants. If so, why would insects invest so much effort in making dozens (or in some cases even hundreds) of OBPs? It turned out that in fact OBPs contribute to the selectivity of insect's olfactory system, but measuring affinity under specific conditions may be misleading. This is because many compounds can bind tightly at pH 7 and be released at pH 5, but they may not be delivered to the right place at the right time (Damberger et al. 2013).

Recently, NMR studies demonstrated that in solution at the pH of the sensillar lymph (pH 6.5), unliganded BmorPBP1 exists predominantly in the conformation BmorPBP1[A], but the equilibrium is shifted toward the B conformation in the presence of ligands which effectively compete with the C-terminal helix for the binding site (Damberger et al. 2013). pH titration of BmorPBP1 showed that the B → A transition midpoint was at pH 7.3 for the apo-protein, 5.4 for BmorPBP1-bombykol and 5.8 when the protein was bound to hexadecane-1-ol or hexadecanal. The pH at the surface of the dendrites (or pore tubules) is lower than the bulk pH (pH of the sensillar lymph) increasing with the distance from negatively charged surfaces. At the lower pH in the environs of the pore tubules, the stability of the A-conformation of BmorPBP1 is enhanced and weak binding ligands are unable to displace the C-terminal helix, whereas the affinity of bombykol is sufficient to displace the C-terminal α-helix from the binding pocket. Pheromone is then taken up into BmorPBP1[B]-bombykol, and diffuses toward the region of higher pH (away from the membrane). BmorPBP therefore neatly performs the trick of filtering against weaker ligands and yet enhancing the stability of the complex during its journey toward the receptors. Bombykol is thus made "soluble" and, in contrast to weaker binding ligands, it is protected from aggressive enzymes in the sensillar lymph. When reaching dendritic surfaces the complex undergoes a conformational change with the C-terminus occupying the binding pocket and ejecting bombykol to activate ORs. Ligands with higher affinity than bombykol would exhibit a lower midpoint in the pH transition and the C-terminal helix would be unable to compete for the binding cavity (figure 9.1). Therefore, such a ligand would remain trapped and would be unable to access the receptors. Although, a high-affinity ligand might be released to a certain extent by mass action, the off rate or ligand-release rate is very slow without the help of the C-terminal dodecapeptide. Indeed, the BmorPBP1 conformational change expedites release of pheromone by 10,000-fold (Leal et al. 2005). These slow-release kinetics would make the strongly binding ligand ineffective for olfaction. On the other hand, a ligand with lower affinity than bombykol would exhibit a conformational transition at a higher pH midpoint and therefore both uptake at the pore tubules and release to the receptor would occur farther from the pore or the receptor (figure 9.1), respectively, making weaker binding ligands more vulnerable to degradation. In summary, many compounds may bind to BmorPBP1, but some will remain trapped and others will be dropped out of reach of receptors. Bombykol, and perhaps just a few other closely related compounds, are released at "the right time and right place." The second layer of filtering is then achieved by selectivity of the receptors (see below).

Odorant-Degrading Enzymes: Inactivation by Degradation

Gerhard Kasang postulated that molecules in the sensillar lymph, be they a ligand prematurely dropped by an OBP or a pheromone released from its receptor, are prone to degradation by ODEs (see Kaissling 2013 for a review). This was demonstrated empirically by Vogt and Riddiford (1981). For more in-depth kinetics studies, PDEs from moths and beetles were isolated to purity, and their internal amino acid sequences determined, cDNA clones and proteins expressed (Ishida and Leal 2005, 2008). These studies showed that the kinetic properties of both recombinant and native PDEs at the pH of the sensillar lymph are consistent with rapid inactivation of pheromones. Recently, it has been demonstrated that an esterase from *Drosophila melanogaster*, Est-6, expressed 6.5-fold more in male than in female antennae, is essential for the temporal dynamics of antennal responses to Z11-18Ac and male–male courtship behavior (Chertemps et al. 2012). The support for enzymatic rapid inactivation of pheromones is thus far limited to compounds with an ester moiety (E6Z11-16Ac [Ishida and Leal 2005]; (5R)-5-[(Z)-dec-1-enyl]oxolan-2-one = japonilure [Ishida and Leal 2008]; and Z11-18Ac [Chertemps et al. 2012), which are degraded by sensillar esterases, but pheromone molecules are chemically diverse. As reviewed by Vogt (2005) and (Leal 2013), there is experimental evidence that various types of pheromones are degraded in vitro by antennal aldehyde oxidases, aldehyde dehydrogenases, epoxide hydrolases, glutathione-S-transferases, and cytochrome P450s. However, unambiguous evidence that most of these enzymes are indeed involved in signal termination requires the following: (1) demonstration that these enzymes are part of the sensillar lymph, particularly given that most of them are well-known cytosolic enzymes, and (2) isolation and cDNA cloning of these enzymes, coupled with evidence that

recombinant enzymes rapidly degrade pheromones provided. Earlier biochemical studies gave convincing evidence that an aldehyde oxidase from *Manduca sexta* (Sphingidae) is antennal specific and rapidly degrades a pheromone constituent, bombykal (Rybczynski et al. 1989). On the basis of the extraction procedure, it was also suggested that the bombykal-degrading aldehyde oxidases are components of the sensillar lymph (Rybczynski et al. 1989, 1990). Lastly, it was demonstrated that these antennal aldehyde oxidases do not require a soluble cofactor to oxidize aldehydes (Rybczynski et al. 1989), implying they degrade pheromone molecules in the sensillar lymph. Antennae-specific aldehyde oxidases have been cloned from the silkworm moth and *Mamestra brassicae* (Noctuidae) (Merlin et al. 2005; Pelletier et al. 2007). As expected for this type of enzyme, they do not have a signal peptide and are likely cytosolic. Recently, an antennae-specific aldehyde oxidase, AtraAOX2, was isolated from the navel orangeworm, *Amyelois transitella* (Pyralidae), and expressed (Choo et al. 2013). Transcripts for *AtraAOX2* were detected in male and female antennae, but not in other tissues: legs, wings, thorax, and abdomen. AtraAOX2 has common features with AOXs from the silkworm moth, BmorAOX1 and BmorAOX2 (Pelletier et al. 2007): two putative iron–sulfur redox centers, flavin-containing region, molybdenum cofactor-binding site, and eight Cys residues likely to be linked to iron ions, as expected for aldehyde oxidases (Garattini et al. 2003). Since all three AOXs lack signal peptide, it is unlikely that they are secreted in the sensillar lymph, unless they are translocated by an unknown mechanism. Activity of recombinant AtraAOX2 on various substrates in vitro suggests that AtraAOX2 might be involved in degradation of host-plant volatile compounds, pheromone, and plant-derived xenobiotics (Choo et al. 2013). For degradation by cytosolic enzymes in vivo, the substrates would have to pass through the dendritic membrane to reach these enzymes in the cytosolic compartment. Given the amphipathic nature of lepidopteran sex pheromones, it is not inconceivable that pheromone molecules could diffuse across the dendritic membrane. This process is expected to be slow for signal inactivation, and could be better suited for catabolism of pheromones. There could be yet unknown molecular mechanism(s) to expedite transport through the membrane and make the overall process of degradation occur in the timescale of rapid inactivation. It is worth mentioning that sensory neuron membrane proteins (SNMPs) (Rogers et al. 1997) are embedded in the dendritic membrane in the vicinity of receptors, and their relatives in vertebrate animals internalize multiple ligands, including fatty acids, but as of now there is no evidence that SNMPs internalize pheromones.

An alternative hypothesis to enzymatic degradation yet to be rigorously tested is that pheromones are trapped and inactivated by OBPs like those with a high transition midpoint pH (figure 9.1).

Odorant Receptors: The Ultimate Gatekeepers of Selectivity

It was not until the turn of the century that ORs were discovered (Clyne et al. 1999; Gao and Chess 1999; Vosshall et al. 1999). Ironically, initial attempts to discover *OR* genes in the genome of the fruit fly, *Drosophila melanogaster*, were based mostly on a bioinformatics approach designed to locate G-protein-coupled receptor (GPCR)-like olfactory genes

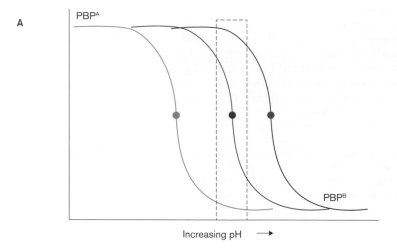

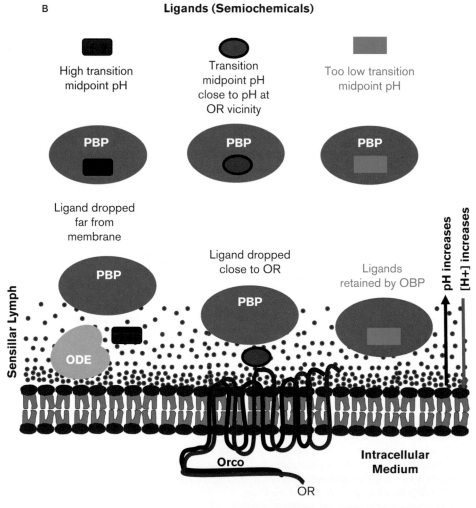

FIGURE 9.1 Schematic view of the proposed mechanism of pheromone release at the vicinity of membrane-standing pheromone receptors.

A pH-dependent conformational transition for three hypothetical ligands, with small, medium, and high pH midpoint transitions. Dashed box represents an "ideal" pH range for a midpoint transition in the vicinity of the membrane.

B Release or retention of ligands as predicted by their midpoint transitions. Ligands with a high midpoint transition (blue rounded rectangle) have a midpoint pH that can be reached far from the membrane, thus dropping ligand prematurely. By contrast, those with too low midpoint pH (depicted as a green rectangle) are retained, because the membrane is not acidic enough to trigger release of ligand. Only ligands for which the transition midpoint is close enough to the pH are released in the immediate vicinity of receptors (depicted as a red oval). pH gradient is depicted by small red circles representing protons (H+) with their concentration decreasing as the distance from the membrane increases.

(Clyne et al. 1999). We now know that insect ORs are not GPCRs; they are seven transmembrane ion channels, with "inverse" topology (Benton et al. 2006) as compared to GPCRs, i.e., the N and C termini are intracellular and extracellular, respectively. The discovery of the first insect *OR* genes thus underscores the importance of a working hypothesis; the hypothesis is a driver, important for the advancement of science. Shortly after ORs were discovered, an obligatory co-receptor, initially named Or83b, was discovered (Larsson et al. 2004). Thereafter, orthologs of the noncanonical OR83b have been identified in moths, mosquitoes, and many other insect species (reviewed in Leal 2013). Because they all apparently perform the same function, OR83b and its orthologs were renamed Orco. Orco is a seven-transmembrane protein with the same topology as insect ORs. The groundbreaking discoveries of ORs and Orco provided the foundation for the spectacular explosion in our understanding of the molecular basis of odorant reception. Together, OR and Orco form ligand-gated nonselective cation channels (OR–Orco complexes) (Sato et al. 2008; Smart et al. 2008; Wicher et al. 2008), with odorants activating only the canonical unit; the OR thus is sometimes referred to as the receptor "binding unit." As insect genomes have been sequenced, putative ORs from multiple insect species, including but not limited to moths, mosquitoes, honeybee, and ants, have been identified at a pace faster than anyone could have imagined even a decade ago. The entire repertoire of *OR* genes in the fruit fly has been elegantly de-orphanized and mapped to ORNs and sensilla type (Hallem and Carlson 2006) as well as ORN projections in the antennal lobe (Couto et al. 2005). The above-described functional assay, the "empty neuron" system, played a significant part is this endeavor. This system has also been used to painstakingly de-orphanize ORs from the malaria mosquito, *Anopheles gambiae* (Carey et al. 2010). Additionally, heterologous expression of ORs, particularly in the *Xenopus* oocyte recording system, helped de-orphanize ORs from moths (see examples below), mosquitoes (Hughes et al. 2010; Pelletier et al. 2010; Wang et al. 2010), and the European honeybee (Wanner et al. 2007). Despite these and other new approaches (Ueira-Vieira et al. 2014), de-orphanization is still the rate-limiting step in full characterization of insect ORs. With the advent of next-generation sequencing, the number of putative OR genes will grow exponentially. The massive genomic data that will be generated in the next 5–10 years is certain to be 100-fold, or perhaps 1000-fold, larger than the data generated in the last 50+ years after the discovery of bombykol. To catalyze this fast-growing field, state-of-the-art high-throughput approaches to de-orphanize ORs are required. It is now possible to investigate the effect of any environmental and/or physiological conditions on the olfactory system by studying the entire repertoire of genes affected. We can expose moths over extended periods to sex pheromones, inflict honeybees with environmental contaminants, blood feed mosquitoes, let them starve, expose them to repellents, infect them with pathogens, and then investigate the effects of these treatments on the expression of all olfactory genes. Sequencing of the *D. melanogaster* genome was a large-scale project, which among many benefits, led to the discovery of OR genes. Very soon, sequencing the genome of a new species will be a master's or even an undergraduate thesis project. The future is bright, but handling the data will be challenging. It was just about a decade ago that our laboratory isolated and identified the first olfactory protein from a mosquito species—the *Culex*

quinquefasciatus odorant-binding protein CquiOBP1—by extracting proteins from antennae and legs, running gel electrophoresis, sequencing bands, and cloning (Ishida et al. 2002). We have just investigated the entire repertoire of olfactory genes in the same species by transcriptome (RNA-Seq analysis). The difference being that we have now identified hundreds of genes, including the complete set of OBPs and 36 novel OR genes (Leal et al. 2013). Despite all these advancements and the state-of-the-art technologies to study olfaction, functional analysis of ORs is still the rate-limiting step in these studies.

A picture regarding selectivity of moth ORs is already emerging with the pheromone receptors identified up to the time of this writing (Table 9.1). There are pheromone ORs narrowly tuned to a constituent of the sex pheromone, ORs responding mainly to one of the constituents but still being sensitive to other components at reasonable albeit higher doses, and ORs broadly activated by multiple constituents of the sex pheromone system. For all pheromone ORs from moth species reported to date, there is at least one OR per species, which is highly selective, e.g., BmorOR1, HvirOR13, MsepOR1, OscaOR1, OnubOR6, SlitOR6, and AsegOR9 (Table 9.1). For ORs that have been investigated by multiple methods (e.g., BmorOR1), there are discrepancies in the literature with reports of very high and moderate specificity. Also, it is common to find one or more OR per species with low or very low specificity (Table 9.1). In summary, apparently there is at least one OR per species, which is an ultimate gate of specificity. For those with moderate or low specificity, PBPs might augment the overall specificity of the system.

Although selective, even the sex pheromone system in moths is not completely "bullet-proof." For example, early work discovered that a formate two-carbons shorter than the natural ligand can substitute behaviorally for the unstable aldehyde pheromones (MacLaughlin et al. 1972; Mitchell et al. 1975) A few decades later, it was demonstrated that these analogs indeed interact with the PBPs and ORs involved in the reception of aldehydes (Xu et al. 2012a).

Many ORs are found silent, or give minimal response to physiological doses of sex pheromones (for the most recent example, see AsegOR10 in Zhang and Lofstedt 2013) when tested in the *Xenopus* oocyte recording system. The just-above-background response levels suggest that important constituent(s) of the sensory machinery is missing in AsegOR10●AsegOrco-expressing oocytes and/or the oocytes were not challenged with the best ligand(s). An SNMP is essential for the olfactory ensemble of pheromone-detecting ORNs in T1 trichoid sensilla, but it is not required for functioning of basiconic sensilla (Benton et al. 2007). When the bombykol receptor BmorOR1 was expressed in the empty neuron (ab3A), thus devoid of an SNMP, the ab3A neurons responded to bombykol with low sensitivity (Syed et al. 2006). Ectopic expression in SNMP-endowed T1 neurons led to a sensitive detection of bombykol comparable to that of the native sensilla in the silkworm moth (Syed et al. 2010). Reception of long-chain pheromones in the fruit fly requires SNMP even for ORs transplanted from moths. The lack of an SNMP is manifested in lower sensitivity as clearly demonstrated by ectopic expression of a moth OR in T1 neurons with or without SNMP. HvirOR13 expressed in DmelOR67-neurons conferred responsiveness to Z11-16Ald. The response was almost abolished in *snmp* mutants and restored by genetic rescue of SNMP (Benton et al. 2007). At this time, it is not possible to rule out the possibility that all experiments reported to date

TABLE 9.1

Specificity of "naked" moth pheromone receptors

Species	De-orphanized OR	Specificity	Ligand	Reference
Bombyx mori	BmorOR1	Very high	E10Z12-16OH	Nakagawa et al. (2005)
	BmorOR1	Low/moderate	E10Z12-16OH	Grosse-Wilde et al. (2006)
			E10Z12-16Ald	
	BmorOR1	High	E10Z12-16OH	Xu et al. (2012b)
			E10Z12-16Ald	
			(E10Z12-14OH)[1]	
Bombyx mori	BmorOR3	Very high	E10Z12-16Ald	Nakagawa et al. (2005)
				Grosse-Wilde et al. (2006)
Heliothis virescens	HvirOR13 (=HR13)	Low	Z11-16Ald	Grosse-Wilde et al. (2007)
			Z11-16Ac	
			Z9-14Ald	
			Z9-16Ald	
Heliothis virescens	HvirOR14 (=HR14)	Very low	Z11-16Ac	Grosse-Wilde et al. (2007)
			Z11-16Ald	
			Z9-16Ald	
			Z9-14Ald	
Heliothis virescens	HvirOR13	High	Z11-16Ald	Wang et al. (2011)
			Z9-14Ald	
Heliothis virescens	HvirOR14	High	Z11-16Ac	Wang et al. (2011)
			Z9-14Ald	
Heliothis virescens	HvirOR6	Very high	Z9-14Ald	Wang et al. (2011)
			Z9-16Ald	
			Z11-16OH	
Heliothis virescens	HvirOR16	Very high	Z11-16OH	Wang et al. (2011)
			Z9-14Ald	Forstner et al. (2009)
			Z11-16Ac	
Antheraea polyphemus	ApolOR1	Very low	E6Z11-16Ac	
			E6Z11-16Ald	
			E4Z9-14Ac	
Plutella xylostella	PxylOR1	High	Z11-16Ald	Mitsuno et al. (2008)
			E11-16Ald	
Mythimna separata	MsepOR1	Very high	Z11-16Ac	Mitsuno et al. (2008)
			E11-16Ac, not tested	
Diaphania indica	DindOR1	High	E11-16Ald	Mitsuno et al. (2008)
			Z11-16Ald	
Ostrinia scapulalis	OscaOR1	Very high	(E11-14OH)[2]	Miura et al. (2009)
			Z11-16Ac	
Ostrinia nubilalis	OnubOR1	Moderate	E12-14Ac	Wanner et al. (2010)
			Z9-14Ac	
			Z11-14Ac	
Ostrinia nubilalis	OnubOR3	Low	E12-14Ac	
			Z12-14Ac	
			E11-A4Ac	
Ostrinia nubilalis	OnubOR5	Low	E12-14Ac	
			E11-14Ac	
			Z12-14Ac	
			E12-14Ac	

Ostrinia nubilalis	OnubOR6	Very high	Z11-16Ac	
Ostrinia scapulalis	OscaOR3	Very low	Z9-14Ac E11-14Ac Z11-14Ac E11-14Ac E12-14Ac	Miura et al. (2010)
Ostrinia scapulalis	OscaOR4	High	E11-14Ac Z11-14Ac	
Amyelois transitella	AtraOR1[3]	High	Z11Z13-16Ald Z11Z13-16OH (Z9Z11-14For)[4]	Xu et al. (2012a)
Amyelois transitella	AtraOR3	High	Z11-16Ald Z11Z13-16Ald Z13-16Ald	
Ostrinia furnacalis	OfurOR3	Low	E12-14Ac Z12-14Ac E11-14Ac Z11-14Ac	Leary et al. (2012)
Ostrinia nubilalis	OnubOR3	Low	(Z11-14Ac)[5]	Leary et al. (2012)
Spodoptera littoralis	SlitOR6	Very high	Z9E12-14Ac	Montagne et al. (2012)
Agrotis segetum	AsegOR9 AsegOR4 AsegOR5 AsegOR1 AsegOR6	Very high High Moderate Low Low	Z5-10Ac Z7-12Ac Z9-14Ac Z8-12Ac Z7-12Ac Z5-10OH Z5-10Ac	Zhang and Lofstedt (2013)

1. This pheromone analog with two omega carbons truncated had apparent slightly higher affinity for BmorOR1 than bombykol.
2. Sex pheromone of a congeneric species.
3. Female-biased OR.
4. Formate analog of the major constituent of the sex pheromone, which has been previously used for mating disruption.
5. The best ligand identified in dose-dependent studies differed from early screening that identified E12-14Ac as the main ligand (Wanner et al. 2010).

were performed with "missing element(s)." It could be that the high currents recorded from pheromone receptors in the *Xenopus* oocyte systems (e.g., BmorOR1 response to 1 µM bombykol, 800 nA, [Xu and Leal 2013]) could be masking the absence of important olfactory components. If so, the lack of a "missing element" is exacerbated in systems with low currents. Thus, it remains an interesting area for future research to determine, for example, if incorporation of SNMPs into the vicinity of receptors would enhance the responses of OR•Orco-expressing oocytes to pheromones.

Solving 3D structures of odorant (or pheromone) receptors in complex with the activating signal molecules is the Holy Grail of insect olfaction. For now, it has been established that (i) the binding pocket is located on the extracellular halves of its transmembrane (TM) domains (Guo and Kim 2010), (ii) a residue located at the predicted interface between TM-3 and extracellular loop (ECL) 2 plays a role in activation in *Drosophila* (Nichols and Luetje 2010), (iii) membrane-embedded binding pockets might be covered by ECL-2 in ORs from moths and mosquitoes (Xu and Leal 2013), and (iv) TM-3 is part of the binding pocket in a moth OR and one of its residues is essential for pheromone specificity (Leary et al. 2012). Experiments with the oocyte interrogation system have demonstrated that the bombykol receptor from the silkworm moth, BmorOR1, is more sensitive to bombykol, but responds to the related aldehyde, bombykal, with about one order of magni-

tude higher threshold (Xu et al. 2012b). Interestingly, the bombykol ORN does not respond to bombykal, except when it is challenged at >10,000-fold higher doses (Kaissling et al. 1978). The discrepancies between the responses recorded from the intact neuronal system and the "naked" receptor thus support the notion that BmorPBP1 plays an essential role in augmenting selectivity. By contrast, BmorPBP1 seems to indiscriminately bind to stereoisomers of bombykol (Hooper et al. 2009), but the receptor is very selective. BmorOR1•BmorOrco-expressing oocytes responded to the natural isomer (10*E*,12*Z*)-bombykol, but not to the other three isomers (Xu et al. 2012b). Taken together, these findings suggest that PBPs and ORs work in a two-step filtering system, with both contributing to the extraordinary selectivity of moth's olfactory system.

Acknowledgments

Research in my laboratory is supported by the National Institute of Allergy and Infectious Diseases of the National Institutes of Health under award R01AI095514, the USDA National Institute of Food and Agriculture under award 2010-65105-20582), and gifts from various donors. We thank Dr. James E. Rothman at Memorial Sloan-Kettering Cancer Center for sharing plasmid vectors of pHluorin GFP and Dr. Claudio Pikielny at Geisel School of Medicine at Dartmouth for sharing *DmelOBP28a-GAL4* fly, respectively, which we mentioned in this review. I am grateful to Drs. Yuko Ishida, Pingxi Xu, Fred Damberger, Karl-Ernst Kaissling, and Julien Pelletier for their helpful comments on earlier versions of the manuscript.

References Cited

Benton, R., Sachse, S., Michnick, S.W., and Vosshall, L.B. 2006. Atypical membrane topology and heteromeric function of *Drosophila* odorant receptors in vivo. *PLOS Biology* 4(2):e20. doi:10.1371/journal.pbio.0040020.

Benton, R., Vannice, K.S., and Vosshall, L.B. 2007. An essential role for a CD36-related receptor in pheromone detection in *Drosophila*. *Nature* 450(7167):289–293. doi:10.1038/nature06328.

Biessmann, H., Andronopoulou, E., Biessmann, M.R., Douris, V., Dimitratos, S.D., Eliopoulos, E., Guerin, P.M. et al. 2010. The *Anopheles gambiae* odorant binding protein 1 (AgamOBP1) mediates indole recognition in the antennae of female mosquitoes. *PLOS ONE* 5(3):e9471. doi:10.1371/journal.pone.0009471.

Brand, A.H. and Perrimon, N. 1993. Targeted gene expression as a means of altering cell fates and generating dominant phenotypes. *Development* 118:401–415.

Carey, A.F., Wang, G., Su, C.Y., Zwiebel, L.J., and Carlson, J.R. 2010. Odorant reception in the malaria mosquito *Anopheles gambiae*. *Nature* 464(7285):66–71. doi:10.1038/nature08834.

Chertemps, T., Francois, A., Durand, N., Rosell, G., Dekker, T., Lucas, P. and Maibeche-Coisne, M. 2012. A carboxylesterase, Esterase-6, modulates sensory physiological and behavioral response dynamics to pheromone in *Drosophila*. *BMC Biology* 10:56. doi:10.1186/1741-7007-10-56.

Choo, Y.M., Pelletier, J., Atungulu, E., and Leal, W.S. 2013. Identification and characterization of an antennae-specific aldehyde oxidase from the navel orangeworm. *PLOS ONE* 8(6):e67794. doi:10.1371/journal.pone.0067794.

Clyne, P.J., Warr, C.G., Freeman, M.R., Lessing, D., Kim, J., and Carlson, J.R. 1999. A novel family of divergent seven-transmembrane proteins: candidate odorant receptors in *Drosophila*. *Neuron* 22(2):327–338.

Couto, A., Alenius, M., and Dickson, B.J. 2005. Molecular, anatomical, and functional organization of the *Drosophila* olfactory system. *Current Biology* 15(17):1535–1547. doi:10.1016/j.cub.2005.07.034.

Damberger, F.F., Michel, E., Ishida, Y., Leal, W.S., and Wuthrich, K. 2013. Pheromone discrimination by a pH-tuned polymorphism of the *Bombyx mori* pheromone-binding protein. *Proceedings of the National Academy of Sciences of the United States of America* 110(46):18680–18685. doi:10.1073/pnas.1317706110.

Dobritsa, A.A., van der Goes van Naters, W., Warr, C.G., Steinbrecht, R.A., and Carlson, J.R. 2003. Integrating the molecular and cellular basis of odor coding in the *Drosophila* antenna. *Neuron* 37(5):827–841.

Forstner, M., Breer, H., and Krieger, J. 2009. A receptor and binding protein interplay in the detection of a distinct pheromone component in the silkmoth Antheraea polyphemus. *International Journal of Biological Sciences* 5(7):745–757.

Gao, Q. and Chess, A. 1999. Identification of candidate *Drosophila* olfactory receptors from genomic DNA sequence. *Genomics* 60(1):31–39. doi:10.1006/geno.1999.5894.

Garattini, E., Mendel, R., Romao, M.J., Wright, R., and Terao, M. 2003. Mammalian molybdo-flavoenzymes, an expanding family of proteins: structure, genetics, regulation, function and pathophysiology. *Biochemical Jouirnal* 372(Pt 1):15–32. doi:10.1042/BJ20030121.

Getchell, T.V., Margolis, F.L., and Getchell, M.L. 1984. Perireceptor and receptor events in vertebrate olfaction. *Progress in Neurobiology* 23(4):317–345.

Gomez-Diaz, C., Reina, J.H., Cambillau, C., and Benton, R. 2013. Ligands for pheromone-sensing neurons are not conformationally activated odorant binding proteins. *PLOS Biology* 11(4):e1001546. doi:10.1371/journal.pbio.1001546.

Grater, F., Xu, W., Leal, W., and Grubmuller, H. 2006. Pheromone discrimination by the pheromone-binding protein of *Bombyx mori*. *Structure* 14(10):1577–1586. doi:10.1016/j.str.2006.08.013.

Grosse-Wilde, E., Svatos, A., and Krieger, J. 2006. A pheromone-binding protein mediates the bombykol-induced activation of a pheromone receptor in vitro. *Chemical Senses* 31(6):547–555. doi:10.1093/chemse/bjj059.

Grosse-Wilde, E., Gohl, T., Bouche, E., Breer, H., and Krieger, J. 2007. Candidate pheromone receptors provide the basis for the response of distinct antennal neurons to pheromonal compounds. *European Journal of Neuroscience* 25(8):2364–2373. doi:10.1111/j.1460-9568.2007.05512.x.

Guo, S. and Kim, J. 2010. Dissecting the molecular mechanism of drosophila odorant receptors through activity modeling and comparative analysis. *Proteins* 78(2):381–399. doi:10.1002/prot.22556.

Hallem, E.A. and Carlson, J.R. 2006. Coding of odors by a receptor repertoire. *Cell* 125(1):143–160. doi:10.1016/j.cell.2006.01.050.

Hooper, A.M., Dufour, S., He, X., Muck, A., Zhou, J.J., Almeida, R., Field, L.M., Svatos, A., and Pickett, J.A. 2009. High-throughput ESI-MS analysis of binding between the Bombyx mori pheromone-binding protein BmorPBP1, its pheromone components and some analogues. *Chemical Communications* (38):5725–5727. doi:10.1039/b914294k.

Horst, R., F. Damberger, P. Luginbuhl, P. Guntert, G. Peng, L. Nikonova, W.S. Leal, and K. Wuthrich. 2001. NMR structure reveals intramolecular regulation mechanism for pheromone binding and release. *Proceedings of the National Academy of Science of the United States of America* 98(25):14374–14379.

Hughes, D.T., Pelletier, J., Leutje, C.W., and Leal, W.S. 2010. Odorant receptor from the southern house mosquito narrowly tuned to the oviposition attractant skatole. *Journal of Chemical Ecology* 36(8):797–800.

Ishida, Y. and Leal, W.S. 2005. Rapid inactivation of a moth pheromone. *Proceedings of the National Academy of Sciences of the United States of America* 102(39):14075–14079. doi:10.1073/pnas.0505340102.

Ishida, Y. and Leal, W.S. 2008. Chiral discrimination of the Japanese beetle sex pheromone and a behavioral antagonist by a pheromone-degrading enzyme. *Proceedings of the National Academy of Sciences of the United States of America* 105(26):9076–9080. doi:10.1073/pnas.0802610105.

Ishida, Y., Cornel, A.J., and Leal, W.S. 2002. Identification and cloning of a female antenna-specific odorant-binding protein in the mosquito *Culex quinquefasciatus*. *Journal of Chemical Ecology* 28(4):867–871.

Kaissling, K.E. 1996. Peripheral mechanisms of pheromone reception in moths. *Chemical Senses* 21:257–268.

Kaissling, K.E. 2009. Olfactory perireceptor and receptor events in moths: a kinetic model revised. *Journal of Comparative Physiology A: Neuroethology, Sensory, Neural, and Behavioral Physiology* 195(10):895–922.

Kaissling, K.E. 2013. Kinetics of olfactory responses might largely depend on the odorant-receptor interaction and the odorant deactivation postulated for flux detectors. *Journal of Comparative Physiology A: Neuroethology, Sensory, Neural, and Behavioral Physiology* 199(11):879–896. doi:10.1007/s00359-013-0812-z.

Kaissling, K.E., Kasang, G., Bestmann, H.J., Stransky, W., and Vostrowsky, O. 1978. A new pheromone of the silkworm moth *Bombyx mori*. Sensory pathways and behavioral effect. *Naturwissenschaften* 65:382–384.

Keil, T.A. 1984. Surface coats of pore tubules and olfactory sensory dendrites of a silkmoth revealed by cationic markers. *Tissue and Cell* 16(5):705–717.

Larsson, M.C., Domingos, A.I., Jones, W.D., Chiappe, M.E., Amrein, H., and Vosshall, L.B. 2004. Or83b encodes a broadly expressed odorant receptor essential for *Drosophila* olfaction. *Neuron* 43(5):703–714. doi:10.1016/j.neuron.2004.08.019.

Laughlin, J.D., Ha, T.S., Jones, D.N., and Smith, D.P. 2008. Activation of pheromone-sensitive neurons is mediated by conformational activation of pheromone-binding protein. *Cell* 133(7):1255–1265. doi:10.1016/j.cell.2008.04.046.

Lautenschlager, C., Leal, W.S., and Clardy, J. 2007. *Bombyx mori* pheromone-binding protein binding nonpheromone ligands: implications for pheromone recognition. *Structure* 15(9):1148–1154. doi:10.1016/j.str.2007.07.013.

Leal, W.S. 2013. Odorant reception in insects: roles of receptors, binding proteins, and degrading enzymes. *Annual Review of Entomology* 58:373–391. doi:10.1146/annurev-ento-120811-153635.

Leal, W.S., Chen, A.M., Ishida, Y., Chiang, V.P., Erickson, M.L., Morgan, T.I., and Tsuruda, J.M. 2005. Kinetics and molecular properties of pheromone binding and release. *Proceedings of the National Academy of Sciences of the United States of America* 102(15):5386–5391. doi:10.1073/pnas.0501447102.

Leal, W.S., Choo, Y.-M., Xu, P., da Silva, C.S.B., and Ueira-Vieira, C. 2013. Differential expression of olfactory genes in the southern house mosquito and insights into unique odorant receptor gene isoforms. *Proceedings of the National Academy of Sciences of the United States of America* 110(46):18704–18709. doi:10.1073/pnas.1316059110.

Leary, G.P., Allen, J.E., Bunger, P.L., Luginbill, J.B., Linn, C.E., Jr., Macallister, I.E., Kavanaugh, M.P., and Wanner, K.W. 2012. Single mutation to a sex pheromone receptor provides adaptive specificity between closely related moth species. *Proceedings of the National Academy of Sciences of the United States of America* 109(35):14081–14086. doi:10.1073/pnas.1204661109.

Loudon, C. 2003. The biomechanical design of an insect antenna as an odor capture device. Pp. 609–630. In G.J. Blomquist and R.G. Vogt., eds. *Insect Pheromone Biochemistry and Molecular Biology: The Biosynthesis and Detection of Pheromones and Plant Volatiles*. New York: Elsevier Academic Press.

Loudon, C. and Koehl, M.A.R. 2000. Sniffing by a silkworm moth: wing fanning enhances air penetration through and pheromone interception by antennae. *Journal of Experimental Biology* 203:2977–2990.

McLaughlin, J.R., H.H. Shorey, L.K. Gaston, R.S. Kaae, F.D., and Stewart. 1972. Sex pheromone of Lepidoptera. 31. Disruption of sex pheromone communication in *Pectinophora gossypiella* with hexalure. *Environmental Entomology* 67:347–350.

Merlin, C., Francois, M.-C., Bozzolan, F., Pelletier, J., Jacquin-Joly, E., and Maibeche-Coisne, M. 2005. A new aldehyde oxidase selectively expressed in chemosensory organs of insects. *Biochemical and Biophysical Research Communications* 332(1):4-10. doi:10.1016/j.bbrc.2005.04.084.

Miesenbock, G., Angelis, D.A.D., and Rothman, J.E. 1998. Visualizing secretion and synaptic transmission with pH-sensitive green fluorescent proteins. *Nature* 394:192–195.

Mitchell, E.R., M. Jacobson, and A.H. Baumhover. 1975. *Heliothis* spp: disruption of pheromone communication with (Z)-9-tetradecen-1-ol formate. *Environmental Entomology* 4:577–579.

Mitsuno, H., Sakurai, T., Murai, M., Yasuda, T., Kugimiya, S., Ozawa, R., Toyohara, H., Takabayashi, J., Miyoshi, H., and Nishioka, T. 2008. Identification of receptors of main sex-pheromone components of three Lepidopteran species. *European Journal of Neuroscience* 28(5):893–902. doi:10.1111/j.1460-9568.2008.06429.x.

Miura, N., Nakagawa, T., Tatsuki, S., Touhara, K., and Ishikawa, Y. 2009. A male-specific odorant receptor conserved through the evolution of sex pheromones in Ostrinia moth species. *International Journal of Biological Sciences* 5(4):319–330.

Miura, N., Nakagawa, T., Touhara, K., and Ishikawa, Y. 2010. Broadly and narrowly tuned odorant receptors are involved in female sex pheromone reception in Ostrinia moths. *Insect Biochemistry and Molecular Biology* 40(1):64–73. doi:10.1016/j.ibmb.2009.12.011.

Montagne, N., Chertemps, T., Brigaud, I., Francois, A., Francois, M.C., de Fouchier, A., Lucas, P., Larsson, M.C., and Jacquin-Joly, E. 2012. Functional characterization of a sex pheromone receptor in the pest moth Spodoptera littoralis by heterologous expression in Drosophila. *European Journal of Neuroscience* 36(5):2588–2596. doi:10.1111/j.1460-9568.2012.08183.x.

Nakagawa, T., Sakurai, T., Nishioka, T., and Touhara, K. 2005. Insect sex-pheromone signals mediated by specific combinations of olfactory receptors. *Science* 307(5715):1638–1642. doi:10.1126/science.1106267.

Nichols, A.S. and Luetje, C.W. 2010. Transmembrane segment 3 of Drosophila melanogaster odorant receptor subunit 85b contributes to ligand-receptor interactions. *Journal of Biological Chemistry* 285(16):11854–11862. doi:10.1074/jbc.M109.058321.

Park, S.-K., Shanbhag, S.R., Wang, Q., Hasan, G., Steinbrecht, R.A., and Pikielny, C.W. 2000. Expression patterns of two putative odorant-binding proteins in the olfactory organs of *Drosophila melanogaster* have different implications for their functions. *Cell & Tissue Research* 300(1):181–192.

Pelletier, J., Bozzolan, F., Solvar, M., Francois, M.C., Jacquin-Joly, E., and Maibeche-Coisne, M. 2007. Identification of candidate aldehyde oxidases from the silkworm *Bombyx mori* potentially involved in antennal pheromone degradation. *Gene* 404(1–2): 31–40. doi:10.1016/j.gene.2007.08.022.

Pelletier, J., Guidolin, A., Syed, Z., Cornel, A.J., and Leal, W.S. 2010. Knockdown of a mosquito odorant-binding protein involved in the sensitive detection of oviposition attractants. *Journal of Chemical Ecology* 36(3):245–248. doi:10.1007/s10886-010-9762-x.

Pelosi, P. 1994. Odorant-binding proteins. *Critical Reviews in Biochemistry and Molecular Biology* 29(3):199–228. doi:10.3109/10409239409086801.

Rogers, M.E., Sun, M., Lerner, M.R., and Vogt, R.G. 1997. Snmp-1, a novel membrane protein of olfactory neurons of the silk moth *Antheraea polyphemus* with homology to the CD36 family of membrane proteins. *Journal of Biological Chemistry* 272(23):14792–14799.

Rybczynski, R., Reagan, J., and Lerner, M.R. 1989. A pheromone-degrading aldehyde oxidase in the antennae of the moth *Manduca sexta*. *Journal of Neuroscience* 9(4):1341–1353.

Rybczynski, R., Vogt, R.G., and Lerner, M.R. 1990. Antennal-specific pheromone-degrading aldehyde oxidases from the moths *Antheraea polyphemus* and *Bombyx mori*. *Journal of Biological Chemistry* 265(32):19712–19715.

Sandler, B.H., Nikonova, L., Leal, W.S., and Clardy, J. 2000. Sexual attraction in the silkworm moth: structure of the pheromone-binding-protein-bombykol complex. *Chemistry and Biology* 7(2):143–151.

Sato, K., Pellegrino, M., Nakagawa, T., Nakagawa, T., Vosshall, L.B., and Touhara, K. 2008. Insect olfactory receptors are heteromeric ligand-gated ion channels. *Nature* 452(7190):1002–1006. doi:10.1038/nature06850.

Schneider, D. and Kaissling, K.-E. 1957. Der Bau der Antenne des Seidenspinners *Bombyx mori* L. II. Sensillen, cuticulare Bildungen und innerer Bau. *Zoologische Jahrbücher Abteilung für Anatomie und Ontogenie der Tiere* 76:224–250. (In German)

Smart, R., Kiely, A., Beale, M., Vargas, E., Carraher, C., Kralicek, A.V., Christie, D.L., Chen, C., Newcomb, R.D., and Warr, C.G. 2008. *Drosophila* odorant receptors are novel seven transmembrane domain proteins that can signal independently of heterotrimeric G proteins. *Insect Biochemistry and Molecular Biology* 38(8):770–780. doi:10.1016/j.ibmb.2008.05.002.

Steinbrecht, R.A. 1970. Zur Morphometrie der Antenne des Seidenspinners, *Bombyx mori* L.: Zahl und Verteilung der Riechsensillen (Insecta: Lepidoptera). *Zeitschrift fur Morphologie Der Tiere* 68:93–126. (In German)

Sun, M., Liu, Y., Walker, W.B., Liu, C., Lin, K., Gu, S., Zhang, Y., Zhou, J., and Wang, G. 2013. Identification and characterization of pheromone receptors and interplay between receptors and pheromone binding proteins in the diamondback moth, *Plutella xyllostella*. *PLOS ONE* 8(4):e62098. doi:10.1371/journal.pone.0062098.

Swarup, S., Williams, T.I., and Anholt, R.R. 2011. Functional dissection of Odorant binding protein genes in *Drosophila melanogaster*. *Genes, Brain, and Behavior* 10(6):648–657. doi:10.1111/j.1601-183X.2011.00704.x.

Syed, Z., Ishida, Y., Taylor, K., Kimbrell, D.A., and Leal, W.S. 2006. Pheromone reception in fruit flies expressing a moth's odorant receptor. *Proceedings of the National Academy of Sciences of the United States of America* 103(44):16538–16543. doi:10.1073/pnas.0607874103.

Syed, Z., Kopp, A., Kimbrell, D.A., and Leal, W.S. 2010. Bombykol receptors in the silkworm moth and the fruit fly. *Proceedings of the National Academy of Sciences of the United States of America* 107(20):9436–9439. doi:10.1073/pnas.1003881107.

Ueira-Vieira, C., Kimbrell, D.A., de Carvalho, W.J., and Leal, W.S. 2014. Facile functional analysis of insect odorant receptors expressed in the fruit fly: validation with receptors from taxonomically distant and closely related species. *Cellular and Molecular Life Sciences* 71(23):4675–4680. doi:10.1007/s00018-014-1639-7.

Vogt, R.G. 2005. Molecular basis of pheromone detection in insects. Pp. 753–804. In L.I. Gilbert, and K. Iatro, S. Gill, eds. *Comprehensive Insect Physiology, Biochemistry, Pharmacology, and Molecular Biology*. London: Elsevier.

Vogt, R.G. and Riddiford, L.M. 1981. Pheromone binding and inactivation by moth antennae. *Nature* 293:161–163.

Vosshall, L.B., Amrein, H., Morozov, P.S., Rzhetsky, A., and Axel, R. 1999. A spatial map of olfactory receptor expression in the *Drosophila* antenna. *Cell* 96(5):725–736.

Wang, G., A.F. Carey, J.R. Carlson, and L.J. Zwiebel. 2010. Molecular basis of odor coding in the malaria vector mosquito *Anopheles gambiae*. *Proceedings of the National Academy of Sciences of the United States of America* 107(9):4418–4423. doi:10.1073/pnas.0913392107.

Wang, G., Vasquez, G.M., Schal, C., Zwiebel, L.J., and Gould, F. 2011. Functional characterization of pheromone receptors in the tobacco budworm *Heliothis virescens*. *Insect Molecular Biology* 20(1):125–133. doi:10.1111/j.1365-2583.2010.01045.x.

Wanner, K.W., Nichols, A.S., Walden, K.K., Brockmann, A., Luetje, C.W., and Robertson, H.M. 2007. A honey bee odorant receptor for the queen substance 9-oxo-2-decenoic acid. *Proceedings of the National Academy of Sciences of the United States of America* 104(36):14383–14388. doi:10.1073/pnas.0705459104.

Wanner, K.W., Nichols, A.S., Allen, J.E., Bunger, P.L., Garczynski, S.F., Linn, C.E., Robertson, H.M., and Luetje, C.W. 2010. Sex pheromone receptor specificity in the European corn borer moth, *Ostrinia nubilalis*. *PLOS ONE* 5(1):e8685. doi:10.1371/journal.pone.0008685.

Wetzel, C.H., Behrendt, H.J., Gisselmann, G., Stortkuhl, K.F., Hovemann, B., and Hatt, H. 2001. Functional expression and characterization of a Drosophila odorant receptor in a heterologous cell system. *Proceedings of the National Academy of Sciences of the United States of America* 98(16):9377–9380. doi:10.1073/pnas.151103998.

Wicher, D., Schafer, R., Bauernfeind, R., Stensmyr, M.C., Heller, R., Heinemann, S.H., and Hansson, B.S. 2008. *Drosophila* odorant receptors are both ligand-gated and cyclic-nucleotide-activated cation channels. *Nature* 452(7190):1007–1011. doi:10.1038/nature06861.

Wojtasek, H. and Leal, W.S. 1999. Conformational change in the pheromone-binding protein form *Bombyx mori* induced by pH and by interaction with membrane. *The Journal of Biological Chemistry* 274(43):30950–30956.

Xu, P. and Leal, W.S. 2013. Probing insect odorant receptors with their cognate ligands: insights into structural features. *Biochemical and Biophysical Research Communications* 435(3):477–482. doi:10.1016/j.bbrc.2013.05.015.

Xu, P., Atkinson, R., Jones, D.N., and Smith, D.P. 2005. *Drosophila* OBP LUSH is required for activity of pheromone-sensitive neurons. *Neuron* 45(2):193–200. doi:10.1016/j.neuron.2004.12.031.

Xu, P., Garczynski, S.F., Atungulu, E., Syed, Z., Choo, Y.M., Vidal, D.M., Zitelli, C.H., and Leal, W.S. 2012a. Moth sex pheromone receptors and deceitful parapheromones. *PLOS ONE* 7(7):e41653. doi:10.1371/journal.pone.0041653.

Xu, P., Hooper, A.M., Pickett, J.A., and Leal, W.S. 2012b. Specificity determinants of the silkworm moth sex pheromone. *PLOS ONE* 7(9):e44190. doi:10.1371/journal.pone.0044190.

Zhang, D.D. and Lofstedt, C. 2013. Functional evolution of a multigene family: Orthologous and paralogous pheromone receptor genes in the turnip moth. *PLOS ONE* 8(10):e77345. doi:10.1371/journal.pone.0077345.

CHAPTER TEN

Moth Sex Pheromone Olfaction

Flux and Flexibility in the Coordinated Confluences of Visual and Olfactory Pathways

THOMAS C. BAKER and BILL S. HANSSON

> The moth doesn't really follow his nose: he follows the wind when his nose tells him to.
>
> GONICK (1995)

INTRODUCTION

SEEING THE WIND VIA OLFACTORY AND VISUAL CONFLUENCE IN SPACE

Spectral space: colors and odors
Structural space: flux and flicker

SEX PHEROMONE OLFACTION PATHWAYS OF MOTHS

VISUALLY MEDIATED BEHAVIORAL RESPONSES TO PHEROMONE

No chemotaxis: moths use two indirect responses to pheromone strands and clean air
Pheromone-mediated optomotor anemotaxis involves longitudinal and transverse image flow; a turn-reversal involves rotatory image flow
Three behavioral responses to pheromone strands coordinated by three protocerebral neuropils known to receive both visual and olfactory inputs
Three observed behavioral outcomes in natural point-source pheromone plumes as a result of different frequencies of plume-strand contact

THE LEPIDOPTERAN VISUAL SYSTEM

Projections of Edge-Motion Information from the Optic Lobes to the Protocerebrum
 Inputs from the optic lobes to mushroom body calyxes
 Inputs from the lobula to optic glomeruli in the lateral protocerebrum
 Key players: local neurons interconnecting optic glomeruli
 Inputs from the lobula plate to optic glomeruli

EVIDENCE FOR PHEROMONE-STIMULATED IMAGE-FLOW ENHANCEMENT

Response to pheromone strands requires high-speed temporal integration of pheromone spectral odor space with visual inputs

RESPONSE PROPERTIES OF NEURONS AND NEUROPILS ALONG MOTH SEX PHEROMONE OLFACTION PATHWAYS

Trichoid sensilla
Pheromone-component-responsive olfactory sensory neurons
 Optimal resolution of pheromone in both odor space and odor time by co-localization of olfactory sensory neurons within single trichoid sensilla
Antennal lobe local neurons
 Local neuron morphology
 Local neuron physiology: GABA-ergic lateral inhibition
 Local neuron physiology is malleable and influenced by age, mating status
Antennal lobe projection neurons
 Reconciling results of various studies on moth projection neuron morphologies and physiologies
 Projection neuron morphologies: three axonal tracts to protocerebral neuropils
 Projection neuron morphologies: odorant-specific synaptic regions in the MBCs
 Projection neuron morphologies: odorant-specific synaptic regions in the ILPt
 General odorant-tuned projection neuron morphologies
 Projection neuron physiologies: antennoprotocerebral tracts are related to conveyance of odor time vs. odor space information
 Projection neuron physiologies: poor temporal plume-strand resolution by chromatic projection neurons?
 Projection neuron physiologies: GABA-ergic projection neurons using the ml-APT

THE LATERAL ACCESSORY LOBE: SITE FOR COUNTERTURN GENERATION, CONVERGENCE OF MULTIMODAL OLFACTORY-VISUAL INPUTS, AND DESCENDING PREMOTOR NEURONS

Introduction

The above-mentioned quote by scientific cartoonist Larry Gonick (1995) from *Discover* magazine conveys the essence of this chapter's message. To truly appreciate how the moth sex pheromone olfactory system is constructed, we must understand that the orientational behavioral response to sex pheromone by a flying male moth is entirely a *visual* response to *wind*, i.e., its direction and speed. Olfaction does not steer the male. It merely drives a visual response to vertical, horizontal, and turn-generated rotational image-motion feedback from the edges of objects in the environment that the male uses to steer either upwind or crosswind and gauge its progress toward a pheromone-emitting female, on a sub-second basis. The olfactory and visual processing systems must be intimately associated in the moth's brain to accomplish this behavioral feat. The abstract, *odor space sensation* of "pheromone" in the brain must be placed into a real-world, four-dimensional *visual-moving-edge* context for steering while the moth is flying and suspended in a moving medium, the wind, via motion-vision.

For a detailed discussion of the maneuverings of flying male moths in response to wind-carried pheromone plumes in field and laboratory studies, see Cardé (this volume). In our chapter, we juxtapose aspects of the flying male moth's brain representation of the visual world, especially its *motion*, with the brain's reconstruction of the pheromone-odor world. We hope this confluence will put the moth sex pheromone olfactory system in context with the stereotyped flight-related behavioral responses that result in the location of a female. We discuss how the feature-extracting neuronal pathways of both the olfactory and the visual systems converge and are integrated in the higher centers of the moth brain. Apparent image-motion and molecular flux representations of pheromone odor are shown to drive sub-second, rapidly reversing turns and visually mediated "surges" to pheromone plume-strand contact. Possible neuronal networks involved in long-lasting, slower reversing turns in pockets of clean air that result in visually mediated crosswind "casting flight" are also described (see Cardé, this volume).

Seeing the Wind via Olfactory and Visual Confluence in Space

Spectral Space: Colors and Odors

The representation of the pheromone-odor world in a moth's brain involves a high-resolution integration of the pheromone's molecular component composition, its spectral sensation of "odor-blend quality," and its temporal fluctuations in intensity—*flux*—that are experienced by the pheromone-olfactory system during encounters with the variable fine-plume-strand structure of the pheromone plume itself. In the stark physical world of chemicals and electromagnetic emanations, spectral sensations of color and odor do not exist; they are only formed *within* the brain. They do not even exist in the excitations of single types of sensory neurons. They only exist in the brain in distinct and identifiable regions that we call, for the purposes of this chapter, *spectral space*.

All sensory neurons are *achromatic* and unable by themselves to discriminate stimulus quality. For example, a visual sensory neuron cannot discriminate, by itself, which wavelength of electromagnetic energy is stimulating it (e.g., discriminate color), due to a confounding of luminance intensity with spectral frequency. Because a stronger stimulus from a suboptimal part of the spectrum can excite a sensory neuron as much as a lower intensity stimulus from an optimal portion of the spectrum, action potential frequencies from an achromatic neuron do not by themselves allow spectral classification, e.g., discrimination. Similarly, each type of differentially tuned insect antennal olfactory sensory neuron (OSN) (figure 10.1A) responds in a graded manner to a panel of different odorants, and even though each OSN has an optimum at some point along that spectrum of odorants, it cannot by itself discriminate chemical quality due to a confounding of a higher concentration of a suboptimal odorant molecule with a lower concentration of an optimal molecule. Thus, insect OSNs are achromatic. The quality of an electromagnetic energy emanation or of a blend of volatile chemicals cannot be discriminated until the sensory inputs from a variety of differentially tuned achromatic sensory neurons are reported to, and cross-compared by, integrative neurons residing in higher neuronal networks. These integrative neurons can then begin to resolve the position of an odor blend, such as a sex pheromone, in spectral *odor space* (de Bruyne et al. 1999; Dobritsa et al. 2003; Hallem and Carlson 2004).

Structural Space: Flux and Flicker

Light has properties that can be sensed and used by insects to reconstruct representations of the *structural space* of the physical world. Reflected light from physical objects in the environment has different intensities of photon flux that define edge-related luminance discontinuities of objects positioned in three-dimensional (3D-structural) space. When time as a fourth dimension is added to 3D-structural space, the visual system can then detect and monitor the ON–OFF flickerings of luminance edges and their relative apparent motions as the insect moves in straight lines, turns, or is displaced by wind.

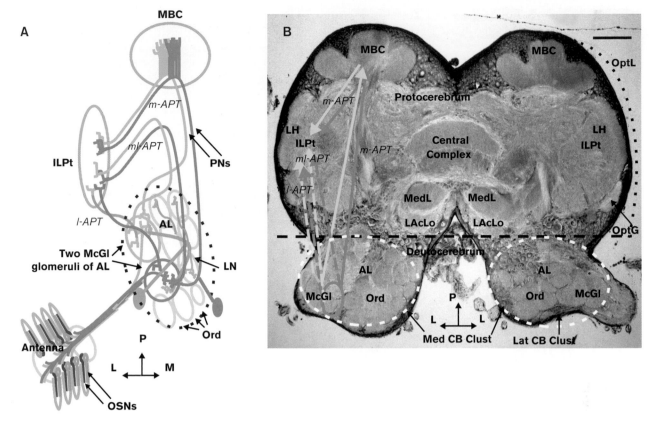

FIGURE 10.1 Neuronal architecture of lepidopteran olfactory system.

A Schematic horizontal (top-down) view of the neurons and neuropils involved in moth sex pheromone olfactory pathways. Two types of olfactory sensory neurons (OSNs, orange and blue) are depicted as being co-located in each of many antennal sensilla. In a two-component sex pheromone blend, these OSNs report the relative molecular abundances of the sex pheromone components to which they are tuned. Each type of OSN arborizes in its own pheromone-component-specific glomerulus in the macroglomerular complex (McGl) of the antennal lobe (AL). There, they each arborize with myriad local neurons (LNs, green) that shape the reports that go to pro-tocerebral neuropils via the axons of projection neurons (PNs). Solid green oval depicts an LN's cell body residing in the lateral cell body cluster. The solid orange oval depicts a PN's cell body residing in the medial cell body cluster. Sexually isomorphic ordinary glomeruli (Ord, gray ovals) receive axons from antennal neurons tuned to general odorants (not shown). The dashed oval depicts the boundaries of the AL. PNs leaving the McGl and projecting their axons via the medial antennoprotocerebral tract (*m-APT*; arrows, blue and orange lines) send collaterals to the calyces of the mushroom body (MBCs) before continuing on to arborize in the inferior lateral protocerebrum (ILPt). PNs projecting along the mediolateral antennoprotocerebral tract (*ml-APT*) and lateral antennoprotocerebral tract (*l-APT*) send terminal arbors directly to the ILPt without visiting the MBCs. Scale bar is 100 μm.

ABBREVIATIONS: P, posterior; L, lateral; M, medial.

B Horizontal section through the brain of a male *Helicoverpa zea* showing positions of neuropils in the brain (adapted from Lee et al. 2006a). Cell bodies of neurons are stained blue; neuropils are stained pink.

ABBREVIATIONS: LAcLo, lateral accessory lobe; LH, lateral horn; MedL, medial lobe of the mushroom body; OptG, one of many optic glomeruli (others not visible) in the lateral protocerebrum receiving inputs from the optic lobe (OptL), whose approximate margin is depicted by curved dashed line. Med CB Clust, small portion of the medial cell body cluster of AL neurons; Lat CB Clust, small portion of the Lat CB Clust of AL neurons. Other abbreviations are as in (A).

NOTE: Some abbreviations (labels) are different from the abbreviations in our text for the same structures; we have retained the original abbreviations from the cited work.

Strands of odor that have sheared off from a point-source emitter, such as a female moth's pheromone gland (see Cardé, this volume), are highly variable in their molecular concen-trations and also have large pockets of relatively pheromone-free (clean) air in between the pheromone strands. Because these molecules move through the environment on the mov-ing air masses that comprise wind, a time dimension is added that involves molecular *flux*, i.e., molecules per second that contact the insect antenna. The speed of motion of molecules on wind through the environment and by the moth's own movement through the air (its airspeed) creates further varia-tions in molecular flux intensities contacting the antenna.

Odorant flux analysis by the insect's olfactory system involves defining the pheromone strand's temporal edges, i.e., a strand's onset upon contact with a sensillum and its offset upon departure during the moth's encounter with a clean-air pocket between strands. These odor-strand ONs and OFFs can then be integrated simultaneously with, and placed in the context of, the brain's visual representation of four-dimensional visual space-motion. In the visual system, when a change in lumi-nance occurs sequentially across an array of photoreceptors, it provides the basis for motion detection that includes integra-tion of direction and speed. Most of the neuronal pathways in the insect visual system are dedicated to motion detection, beginning in the lamina and then through the medulla, lobula, and lobula plate of the optic lobe (figure 10.2) to resolve the speeds and directions of image flows (Strausfeld and Campos-Ortega 1977; Strausfeld 2003, 2012). Visual motion detectors

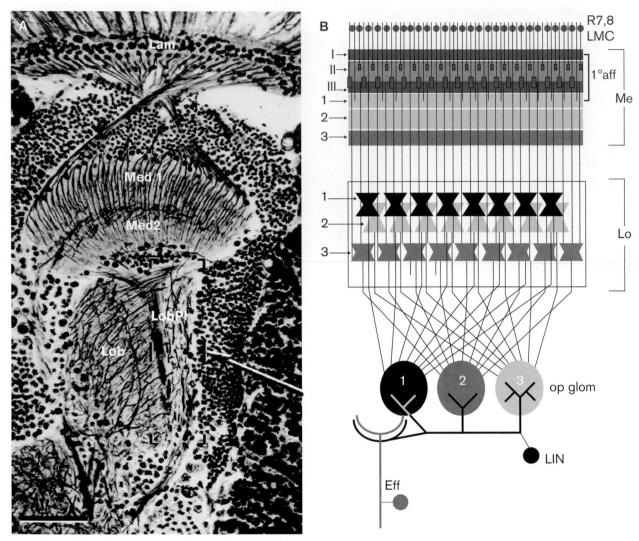

FIGURE 10.2

A Silver/cobalt preparation showing organization of the compound eye and optic lobe of a calliphorid fly (adapted from Strausfeld 2012),
   illustrating the retinotopic cascade of vertically ascending axons through the neuropils of the lamina (Lam), the outer and inner medulla
   (Med 1 and Med 2, respectively), lobula (Lob), and lobula plate (LobPl; in black brackets). Black nodules around all the neuropils are the
   cell bodies supplying neurites to those neuropils (black arrow from right points to cell bodies supplying the Lob Pl).

B Schematic diagram of how visual information, including directionally specific and directionally selective image-motion from different
   retinotopic layers via columnar connections in the optic lobe, ascends to arborize in optic glomeruli (op glom) in the lateral protocerebrum. The op glom are connected via local interneurons (LIN) that have been shown to respond to various directional image-motions
   (Okamura and Strausfeld 2007; Strausfeld and Okamura 2007; Strausfeld et al. 2007). They perform integrations of activities among the
   op glom before efferent neurons (Eff) carry this information to other protocerebral neuropils or perhaps to thoracic ganglia for motor
   outputs. Various horizontal, cross-integrative tangential layers in the medulla (Me) and lobula (Lo) are depicted here in various narrow-
   or wide-field shapes and in different shades of gray. Each optic cartridge in a retinotopic column begins at the periphery (top) with a set
   of primary visual afferents (1° aff) from an ommatidium, consisting of three flicker-sensitive achromatic large monopolar cells (LMC)
   carrying pooled synaptic inputs from receptor neurons R1–R6, plus the long visual fibers of receptor neurons 7 and 8 (R7, 8). Three op
   glom (1, 2, and 3) are depicted that each receive a vision-specific input from one tangential layer in the lobula. The arbors from one local
   interneuron (LIN) are shown that arborize with different architectures in the op glom as well as with the efferent neuron that projects
   its axon to other neuropils. I, II, III and 1, 2, 3 denote different narrow- and wide-field tangential neuropil layers in the lamina, medulla,
   and lobula that integrate at each of those layers different motion (and other) visual information across the vertical retinotopic columns
   transecting those layers.

SOURCE: From Strausfeld and Okamura (2007).

NOTE: Please note that some abbreviations (labels) are different from the abbreviations in our text for the same structures; we have retained the
original abbreviations from the cited work.

and odorant plume-strand flux-change detectors work together to guide the male moth in its progress through wind.

Sex Pheromone Olfaction Pathways of Moths

We define pheromone olfaction pathways as neuronal pathways that are involved with transducing, transmitting, and integrating pheromone-component stimuli to resolve the pheromone's positions in *pheromone-odor space* and *pheromone-odor time*. In the general architecture for sex pheromone olfaction (figure 10.1), pheromone-component-tuned OSNs project from the antenna to the macroglomerular complex (McGl) of the antennal lobe (AL), where they synapse with multitudes of local neurons (LNs) and projection neurons (PNs) (figure 10.1A). PN outputs from individual McGl glomeruli or combinations of glomeruli follow three different tracts to project processed pheromone-component information from the McGl to the inferior lateral protocerebrum (ILPt) of the lateral protocerebrum (LP) (figures 10.1A and 10.1B). One pathway takes a medial route through the brain and first sends collaterals to visit the mushroom body calyces (MBCs) before terminating in areas of the ILPt. The other two PN tracts are more lateral, traveling directly to the ILPt with terminal arbors there, bypassing the MBCs completely (figures 10.1A and 10.1B). OSNs responsive to general (e.g., plant-related) odorants send their axons to ordinary glomeruli in the AL, and interact synaptically with LNs there, with PNs then projecting from these ordinary glomeruli along the same three routes to the LP, except that the axons of these general odorant PNs arborize in a slightly different LP neuropil, the lateral horn (LH) (figure 10.1B).

The post-OSN, toward- or within-protocerebral pathways of sex pheromone-component-sensitive neurons thus involve three highly synaptic integrative neuropils that perform complex integration of sex pheromone-component inputs, all involving intrinsic LNs within these neuropils as well as associated efferents that project out of them. The first integrative neuropil is in the AL of the brain's deutocerebrum; it consists of networks of LNs interconnecting glomeruli within the McGl and also interconnecting McGl glomeruli with those of the entire AL, with the PNs then projecting to protocerebral neuropils. The second integrative neuropil is the mushroom body (MB) of the protocerebrum that receives inputs from the AL via PNs that arborize in the MBCs (figure 10.1B). Kenyon cells (KCs) of the MB are its intrinsic LNs and give it its shape. Their long parallel fibers extend along the length of the MB from the MBCs down through the various mushroom body lobes (MBLs). There are neurons extrinsic to the MBLs, many of which are efferent neurons that integrate information from the arrays of KCs across the MBLs and that project outgoing MB-processed information to other protocerebral neuropils. The third major pheromone-component integrative center includes the ILPt of the LP (figure 10.1B). Within the LP, any olfaction-integrating LNs of the ILPt for pheromone information or in the LH for general odorants have not yet been characterized in moths. Such LNs might be the same LNs used by insect visual systems that interconnect the tens of newly described insect "optic glomeruli" residing in the LP (Okamura and Strausfeld 2007; Strausfeld and Okamura 2007; Strausfeld et al. 2007).

To the above-mentioned three pheromone olfaction integration centers, a fourth center might be added, the lateral accessory lobe (LAcLo), because the LAcLo has been repeatedly implicated in producing turn-reversals in flying or walking males. The paired LAcLos are located medially in the protocerebrum, ventral and slightly laterally to either side of the central complex (figure 10.1B). Inhibitory protocerebral neurons (PrtCNs) connect the two LAcLos and produce alternating "flip-flopping" of excitation related to turn-reversals during pheromone stimulation. With one exception, the LAcLo receives inputs only from purely within-PrtCNs that project there from both the ILPt and the MBLs. These PrtCNs will likely be carrying already processed pheromone-visual information to the LAcLo. However, the one exception involving the moth *Agrotis segetum* (Noctuidae) did show two pheromone-sensitive PNs projecting from the McGls *directly* to the LAcLos. These two PNs also sent collateral arbors to the ILPts on both sides of the male's protocerebrum (Wu et al. 1996). Intrinsic LNs residing within the LAcLos interconnect each LAcLo with a companion neuropil immediately ventral to it called the ventral protocerebrum (VPrtC). These LAcLo/VPrtC LNs respond to pheromone-component stimulation of the antennae by producing long-lasting excitation (LLE) in response to a single puff of pheromone. The LAcLo thus is a neuropil critical to producing premotor neuronal outputs that drive strand-initiated single turns as well as long-lasting, oscillating, clockwise–counterclockwise turn-reversals.

Visually Mediated Behavioral Responses to Pheromone

No Chemotaxis: Moths Use Two Indirect Responses to Pheromone Strands and Clean Air

For a detailed treatment of pheromone-mediated behaviors of male moths in response to pheromone plumes in the field and laboratory, see Cardé (this volume). Here, we summarize aspects of pheromone-mediated behavior for the purpose of placing these behaviors in a neuroethological context related to integrated olfactory and visual pathways.

Decades of research have resulted in the understanding that there are two major systems involved in upwind in-flight progress to a source of pheromone odor, and neither of them involves chemotaxis; i.e., there is no *direct* steering response to chemical concentration gradients. Rather, two *indirect* responses (Kennedy 1983) are now understood to be performed after encounters with the individual strands of pheromone in a plume and with the pockets of clean air between the strands: (1) pheromone-induced optomotor anemotaxis (Kennedy 1940; Kennedy and Marsh 1974), a wind-steering response to global wide-field translatory (linear) flow-field motion across or along the eyes, and (2) a pheromone-triggered, self-steered turn generator. The pheromone-triggered turns are reversed in direction (e.g., clockwise to counterclockwise) with each contact with a pheromone strand (Kanzaki et al. 1992; Mafra-Neto and Cardé 1994; Vickers and Baker 1994; Mishima and Kanzaki 1998, 1999; Iwano et al. 2010). The turn generator also drives a long-lasting program of turn-reversals in clean air that free-runs after the last contact with a pheromone strand, both with wind (Kennedy and Marsh 1974; Kanzaki et al. 1992) and without wind (Baker and Kuenen 1982). That the clockwise–counterclockwise reversals are produced by males in zero wind shows that they are "self-steered" according only to their self-generated optical image flows or to other, non-optomotor-anemotactic or nonvisual feedback (Baker and Kuenen 1982).

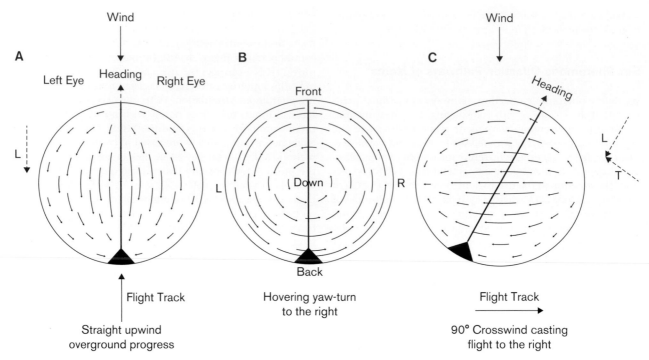

A   B   C

Straight upwind
overground progress

Hovering yaw-turn
to the right

90° Crosswind casting
flight to the right

FIGURE 10.3 Visual image-flow fields experienced by the left and right eyes of a male moth flying just above a plant canopy over open ground. The lengths and directions of the vector arrows indicate the speeds and directions of the flows, with the speeds being fastest directly below the moth. Wind direction is indicated by arrows.

A Male making straight upwind progress; only longitudinal (L) front-to-back image-motion is experienced.

B Male making a clockwise turn using X-Y yaw rotation alone while keeping stationary over the ground. If the turn were to be executed by adding a clockwise roll component to the yaw-turn by reducing lift on the right side and increasing lift on the left, the flow would be more complex and involve down-to-up flow plus left-to-right flow on the right eye and up-to-down flow plus left-to-right flow on the left eye, unless the moth moved its head to minimize such flow.

C Male during a 90° crosswind casting flight-leg; the two components of image flow now include both transverse (T) cross-body and longitudinal (L) front-to-back image motion.

SOURCE: After David (1986).

Pheromone-Mediated Optomotor Anemotaxis Involves Longitudinal and Transverse Image Flow; a Turn-Reversal Involves Rotatory Image Flow

Translational image motion occurs during linear, straight-line flight tracks and is characterized by flow fields in the same direction across all the ommatidia of one eye and therefore across all the ommatidia of both eyes (figures 10.3A and 10.3C). Two types of translatory image flows are experienced by a pheromone strand-stimulated moth flying in wind. Flying straight upwind immediately after contacting a strand produces front-to-back translatory image flow along the long axis of the moth's body, called *longitudinal image flow* (figure 10.3A) (David 1986). During off-windline flight, such as during the straight-across-wind flight track legs occurring between each counterturn during clean-air casting flight, *transverse image flow* across the long axis of the moth's body also will occur (figure 10.3C) (David 1986). During these casting-flight straight legs, longitudinal flow is also present and is integrated with the transverse flow in approximately a 1:1 ratio to steer the moth's resultant flight track straight across the windline (David 1986).

The execution of a turn or loop causes *rotatory image flow* in both walking and flying moths. For example, *Bombyx mori* (Bombycidae) first zigzag and then loop for long periods after emerging into clean air after a pheromone pulse, and flying moths behave the same when they execute turns at the termini of each of their left–right straight-across-wind "casts." Turning-induced rotatory image-flow fields are characterized by opposing-direction flow fields along different portions of the eye (or, e.g., along the left vs. the right eye). In rotatory flow fields, front-to-back motion will occur along one eye, for instance, and back-to-front motion along the other eye (figure 10.3B). These opposing left-side, right-side flows across the long axis of the moth's body will be experienced by a walking moth that circles or turns from left to right or vice versa, and a flying moth that executes a turn via a yaw response (around its X-Y plane). Walking moths are not known to use any kind of image flow to steer in response to wind; they can sense wind from pressure differences across their antennae or bodies because they are fixed to the ground via tarsal contact. If a flying moth were to help execute a clockwise or counterclockwise yaw-turn by also rolling its body to the right or left, respectively (in its Y-Z plane), the flow would be more complex than illustrated in figure 10.3B, with additional rotational image flow involving up-and-down motion in that plane. If a moth were to keep its head steady while rolling or yawing its thorax and abdomen to execute a turn, rotational image flow might be minimized. Compensatory head rotation would not reduce translational flow caused by wind-induced course-track differentials, however.

Three Behavioral Responses to Pheromone Strands Coordinated by Three Protocerebral Neuropils Known to Receive Both Visual and Olfactory Inputs

Contact with a single pheromone strand involves a "pheromone ON" input during strand contact and a "pheromone OFF" input after flight into the clean air after the strand has passed. The pheromone ON stimulation produces a *resultant* flight track called an "upwind surge." The pheromone OFF (clean-air) stimulation eventually produces a resultant flight track called "casting flight" (see Cardé, this volume). However, during the transition to casting flight in long periods of clean air or during intermediate-frequency contact with plume strands in a natural point-source plume, there is an intermediate, upwind-zigzagging type of flight track. Thus, we can identify three behavioral outcomes that are mediated by the two indirect pheromone response systems due to this ON–OFF stimulation. The actual performances and integration of these two indirect response systems occur in the following temporal sequence when observed in response to a single strand of pheromone. We propose that the following protocerebral (central nervous system [CNS]) neuropils are involved in these behaviors that integrate olfaction and vision.

1. *Behavior: A turn-reversal is generated in response to the pheromone ON stimulation from the strand.* The turn-reversal is executed either clockwise to counterclockwise or vice versa in a flying or walking moth and needs no optomotor anemotactic feedback (Baker and Kuenen 1982). However, optomotor anemotaxis will polarize the turn in an upwind direction (Baker et al. 1984).
   CNS: The LAcLo/VPrtC complex generates the turn-reversal in response to a strand. A turn-reversal generator has been shown to reside in the "flip-flop" circuitry connecting the LAcLo/VPrtCs on each side of the protocerebrum of both flying (Kanzaki et al. 1991a, 1991b) and walking moths (Kanzaki and Shibuya 1986, 1992; Kanzaki et al. 1994; Kanzaki and Mishima 1996; Mishima and Kanzaki 1998, 1999; Iwano et al. 2010). The LAcLos receive inputs from the MBLs (Lei et al. 2001), and LH/ILPt (Kanzaki et al. 1991a), both of which could be sources of fast initiation of LAcLo activity after contact with a pheromone strand. In addition, a direct connection between the ALs and the LAcLos has been found in *Agrotis segetum* (Wu et al. 1996) that could also be a fast initiator of a turn-reversal. The LAcLos also receive other inputs from local interneurons that visit them as well as the optic glomeruli of the LP plus its olfactory regions (Kanzaki et al. 1991a, 1991b). A turn-reversal would generate rotatory image-motion optical feedback (figure 10.3B) from the LP to the LAcLos during the execution of a turn that reverses the direction of rotatory image flow from what had been experienced during the execution of the prior reversal.
2. *Behavior: The turn-reversal produced by contact with a strand is followed immediately by approximately direct-upwind flight.* The upwind turn-surge always involves a significant sub-second period of upwind progress (Mafra-Neto and Cardé 1994; Vickers and Baker 1994). Upwind progress (it is always an upwind-oriented turn that occurs) cannot be performed without optomotor anemotaxis that involves predominantly longitudinal, along-body-axis image flow (figure 10.3A).
   CNS: The ILPt, having LN connections with optic glomeruli in the LP, should be involved in the upwind optomotor anemotactic steering of an upwind surge after a turn-reversal. The ILPt, which receives sex pheromone-component inputs, is located in proximity to, and in potential synaptic contact with, LNs that arborize among the abundant optic glomeruli that reside in the LP (Strausfeld and Okamura 2007; Strausfeld et al. 2007). Vertical and horizontal motion detectors in the optic lobe that are known to be involved in optomotor anemotaxis feed into this system of optic glomeruli located in the LP.
3. *Behavior: Increasingly long-duration contact with clean air after a strand contact allows for the complete performance of a long-lasting motor program to play out in which spontaneous (self-steered) oscillatory turn-reversals are executed.* The reversal program involves ever-increasing periods between reversals (slower turn-reversal tempo) and increasingly greater crosswind steering angles in flying moths during casting flight. In walking moths, this relaxation of the tempo is expressed as increasingly continuous one-directional clockwise or counterclockwise looping without the faster tempo, clockwise–counterclockwise looping shifts that immediately followed contact with clean air.
   CNS: The LAcLo/VPrtC complexes on each side generate the LLE after pheromone loss that drives long-term casting flight response in flying males in clean air or long-term looping by walking males after pheromone loss. The mutually inhibitory bilateral LAcLo-to-LAcLo connections during this time drive the alternating left–right, counterclockwise–clockwise spontaneous turn-reversals in clean air. The neurons of the LAcLo-LAcLo system exhibit an oscillatory, spontaneous flip-flopping of action potential activity manifested by increasingly longer duration flip-flopping tempos with time after loss (Kanzaki and Shibuya 1992; Kanzaki et al. 1992, 1994; Mishima and Kanzaki 1998, 1999; Wada and Kanzaki 2005; Iwano et al. 2010). Such flip-flopping will thus be the source of increasingly longer intervals between in-flight crosswind turn-reversals during casting flight by flying males. Flip-flopping tempo in clean air in walking males is manifested by the retardation of the intervals between clockwise–counterclockwise looping reversals. In flying males, the ILPt and its intimate synaptic capabilities with optic glomeruli in the LP should together be involved in a relaxation of the anemotactic vertical–horizontal image flow from mostly upwind (along the body axis) to now crosswind (blend of T and L [figure 10.3C], across-body-axis- plus along-body-axis flows).

Three Observed Behavioral Outcomes in Natural Point-Source Pheromone Plumes as a Result of Different Frequencies of Plume-Strand Contact

In a natural point-source pheromone plume, the male will receive irregular sub-second exposures to pheromone strands and clean-air pockets (Vickers et al. 2001), resulting in three characteristically observed behavioral flight-track outcomes.

1. *Straight-upwind flight: predominantly LAcLo-turn-reversal mediation.* This outcome occurs when there is such

rapid exposure to pheromone plume strands that the turn-reversals plus upwind anemotaxis triggered by pheromone strands occur so frequently that the moth can never go far left or right from the windline as it moves upwind. The rapid turn-reversals keep it aligned with the windline and flying straight up it (Mafra-Neto and Cardé 1994; Vickers and Baker 1994).

2. *Zigzagging upwind flight: mixture of LAcLo-turn-reversal mediation plus ILPt-LP-mediated resetting of the optomotor anemotactic angle to allow more transverse image motion.* This behavior will occur when the frequency of strand contact is at an intermediate level, allowing longer contacts with clean-air pockets and a transition toward casting flight during these longer clean-air pockets. This behavior is the most commonly observed resultant behavior for a male moth flying upwind in a plume and should occur because more of the turn-reversals are endogenously initiated from the flip-flop circuitry of the LAcLo and do not stem from a contact with a new pheromone strand. The increasing relaxation of the anemotactic angle to have more of a transverse component in the optomotor anemotactic circuitry of the ILPt-LP will allow the resultant flight-track angles to be angled more toward crosswind.

3. *Crosswind casting flight in long-duration clean air: a combination of the LAcLo turn-reversal generator's increasingly slower activity plus the ILPt-LP's optomotor anemotactic angle set to ca. a 1:1 ratio of transverse-to-longitudinal image flow.* With no more pheromone strands to be contacted, the turn-reversals from the endogenously operating flip-flop circuitry of the LAcLo become increasingly slower in tempo, and the anemotactic steering angle of each straight flight-track leg between reversals becomes increasingly greater (off-windline).

The Lepidopteran Visual System

Insect visual systems are constructed to render optimally a depiction of edges (luminance discontinuities) occurring in physical (structural) space and the apparent motion of these edges along, across, and around the insect's body. As such, the architecture of the insect optic lobe is organized into highly regular arrays of vertically arranged *optic cartridges* having their afferent axons ascending in parallel toward the protocerebrum from each ommatidium of the compound eye (figure 10.2) (Strausfeld 1970; Strausfeld and Blest 1970). In moths, as in flies, optic cartridges consist of columnar arrays of parallel axonal fibers synapsing multitudes of times with neurons residing in four integrative layers of visual neuropil in the optic lobe: the lamina, medulla, lobula, and lobula plate.

Research on moths' visual motion resolution within these layers (Collett and Blest 1966; Collett 1970, 1972; Strausfeld and Blest 1970; Ibbotson et al. 1991; Maddess et al. 1991; Shimohigashi and Tominaga 1991, 1999; Milde 1993; Cutler et al. 1995; Hämmerle and Kolb 1997; Wicklein and Varju 1999; Wicklein and Strausfeld 2000; Briscoe et al. 2003; Kelber et al. 2003; Stavenga and Arikawa 2006) shows that image-motion integration through these levels does not appear to differ in significant ways from that of flies, upon which most of the classic work in image-motion integrative pathways has been performed (Strausfeld and Campos-Ortega 1977; Douglass

and Strausfeld 1996, 2003, 2007). As in flies, six receptor neurons (R1–R6) of the eight in each ommatidia of moths and butterflies are tuned to the same single wavelength optimum, similar to the six R1–R6 receptor neurons in fly ommatidia that synapse into several flicker-sensitive optic cartridge channels of the large monopolar cells (LMCs) in the lamina (figure 10.2B). Most of the information extracted from the environment by these achromatic neurons is devoted to delineating luminance discontinuities and their apparent motion. Thus, the lepidopteran visual system is organized similarly to the system of flies in being chiefly devoted to "achromatic" (colorless) edge detection and edge image motion (Shimohigashi and Tominaga 1991, 1999; Cutler et al. 1995; Hämmerle and Kolb 1997; Briscoe et al. 2003; Kelber et al. 2003; Stavenga and Arikawa 2006). In Lepidoptera, the other two receptor types, R7 and R8, in each ommatidium, as in flies, are tuned to two different wavelengths different also from those of R1–R6, and their inputs are integrated in the medulla and beyond with some of the R1–R6 (LMC) inputs for wavelength (color) discrimination.

Huge numbers of retinotopic reports from individual visual sensory neurons in ommatidia that view and respond to one small section of visual, structural space in the environment are integrated first in the lamina. Here, edge discontinuities in luminance and their apparent motion are resolved by elementary motion detector lamina neurons that report motion, but with no directional component.

Resolution of edges and their motions in the lamina is aided by an abundance of amacrine cells that release gamma-aminobutyric acid (GABA) and whose delays regarding changes in luminance across the retinotopic arrays of ommatidial receptor neurons provide the mechanism for elementary motion detection (Strausfeld and Campos-Ortega 1977; Strausfeld 2012). Edge-motion-detection is projected via the axons of laminar integrative neurons to the medulla.

In Lepidoptera, the medulla is where the first level of directional edge-motion information resides (Collett and Blest 1966; Collett 1970, 1972; Ibbotson et al. 1991; Maddess et al. 1991; Milde 1993). Here, the outputs of vertically arranged neurons representing small or large groups of retinotopic columns project further toward the protocerebrum, synapsing in the multitudinous horizontal layers of the next integrative regions, the lobula, the lobula plate, or both.

In the lobula, directional motion in flies is further refined to include direction-specific edge-motion as well as object shapes, sizes, and orientations (Strausfeld 2012). In moths, the lobula has similar direction-specific edge-motion abilities (Wicklein and Strausfeld 2000). These integrations are aided by inhibitory and other modulatory activities of synapses in these lobula layers. Many of these layers also include the arbors of wide-spreading centrifugal neurons that provide feedback of processed information from the protocerebrum back out to the lobula and medulla to further modulate and refine the activities along the incoming visual retinotopic pathways (Collett 1970; Milde 1993; Wicklein and Strausfeld 2000).

In sphinx moths (Sphingidae), neuroanatomical studies have shown that specific types of motion information such as "looming" and "anti-looming" (expanding or contracting image-motion parallax) are reported by neurons that arborize in the lobula, the lobula plus lobula plate, and sometimes also in the medulla (Wicklein and Strausfeld 2000). In the lobula plate of *Manduca sexta*, types of wide-spreading tangential neurons that detect and respond to vertical and horizontal image motion also have been neuroanatomically character-

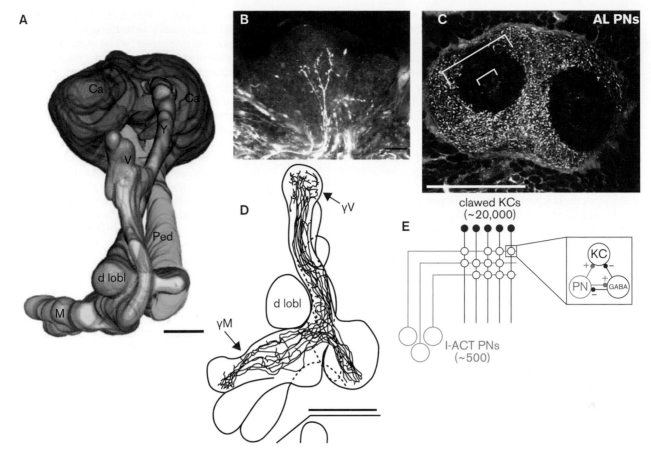

FIGURE 10.4

A Frontal view of a three-dimensional reconstruction of the mushroom body of *Spodoptera littoralis* with the calyces and lobes colored to denote the main sections. Ca, calyx; d lobl, dorsal lobulet; M, medial lobe; Ped, pedunculus; V, vertical lobe; Y, Y-lobe. Scale bar, 100 µm.

B Confocal microscope image of a dextran-stained neuron from the ipsilateral optic lobe of a *Spodoptera littoralis* that terminates with three arbors with profuse presynaptic varicosities in the extreme inner margin of the mushroom body calyx. Such neurons in *S. littoralis* and other moths may provide direct image-motion inputs to the mushroom bodies to be integrated with pheromone-strand contact excitation from antennal lobe projection neurons. Scale bar, 10 µm (from Sjöholm et al. 2005).

C Confocal microscope image of the arborizations of multitudes of antennal lobe projection neurons (AL PNs) in the outer zones of the two conjoined mushroom body calyxes of *Spodoptera littoralis*. PNs had been injected with rhodamine-dextran to aid visualization. This image highlights the lack of olfactory input to the inner zone of the mushroom body calyxes (large bracket), a zone known to receive direct input from at least some neurons from the optic lobe (see B), as well as a lack of olfactory input to the central zone (small bracket with asterisk). Scale bar, 100 µm (from Sinakevitch et al. 2008).

D Posterior view of mushroom body lobes (calyx not shown) with just a few of the multitudes of Kenyon cells that have been Golgi stained for visualization. Kenyon cells are the intrinsic (local) neurons that comprise most of the mushroom body's shape. Note the multitudes of neurites of this type of "clawed" Kenyon cell run in parallel along the lengths of the mushroom body lobes, here shown in just the vertical and medial lobes. Of many laminar sheets that comprise and run along the lengths of the mushroom body lobes, only the Kenyon cells in the "gamma" layer are depicted (γM and γV). The parallel and tightly packed arrangement of Kenyon cells, their vast numbers and arborizations with sensory projection neurons in the calyx, and their synaptic multimodal inputs and outputs with protocerebral neurons along their entire lengths in the mushroom body lobes all facilitate information processing and dispensation in the mushroom body that is essential to the life and reproduction of an insect (from Sjöholm et al. 2005).

E Depiction of how the axons of three different olfactory projection neurons (PNs) arborize in synaptic boutons with different Kenyon cells (KC) to produce odor-specific across-fiber (across-KC) patterns of activity. I-ACT PNs are the PNs of the "Inner Antenno-cerebral Tract" named by Homberg et al. (1988) that is now named "medial antennoprotocerebral tract" (*m*-APT; Galizia and Rössler 2010) (from Szyszka et al. 2005).

INSET: The details of one synaptic bouton are magnified to show how it consists of both GABA-ergic and excitatory interactions between the PN and the KCs at the synaptic junction (from Szyszka et al. 2005).

ized (Wicklein and Varju 1999). Such lobula plate neurons, called "vertically sensitive" (VS) and "horizontally sensitive" (HS), are a common feature of many dipteran and lepidopteran lobula plates. These neurons have a well-established function involving general optomotor stabilization of flight altitude and the orientation of the insect in wind with regard to the apparent horizontal motion of edges along and across the body axis.

Projections of Edge-Motion Information from the Optic Lobes to the Protocerebrum

INPUTS FROM THE OPTIC LOBES TO MUSHROOM BODY CALYXES

There is evidence in moths that direct projections from neurons in the optic lobes are sent to the MBCs (figure 10.4A) to arborize in an "inner rim" region adjoining the arborization points of pheromone-component-sensitive PNs (figures 10.4B and 10.4C) (Sjöholm et al. 2005, 2006). Dye-filled neurons from the optic lobes were shown to terminate in this inner rim region of the MBCs in *Spodoptera littoralis* (Noctuidae) (figure 10.4B). Further work demonstrated that these inner rim areas of the MBCs in *S. littoralis* are devoid of bouton synapse inputs from olfactory PNs arriving from the AL (figure 10.4C). The inner rim was thus demonstrated to be non-olfactory, with the AL PN collaterals only arborizing in the typical lepidopteran outer rim zones of the MBCs (figure 10.4C) (Sjöholm et al. 2006; Sinakevitch et al. 2008). It was not clear from which level of the optic lobe the neurons projected (Sjöholm et al. 2005), but it likely would have been from either the medulla or the lobula. That these visual fibers arborized in the MBCs with synaptic boutons would indicate that they should be visual afferents supplying optic inputs to the MBCs rather than outgoing feedback (centrifugal) neurons modulating activity in the optic lobe. An inner rim zone of the MBCs in the cockroach *Periplaneta americana* (Blattidae) likewise receives direct inputs from the optic lobes, with an outermost layer of the calyces receiving pheromone-component-tuned PNs from the McGl and from other olfactory AL PNs (Nishino et al. 2012b).

INPUTS FROM THE LOBULA TO OPTIC GLOMERULI IN THE LATERAL PROTOCEREBRUM

Vision-related, densely grouped neuropils in the lateral and posterior LP that over the past century had been given names such as "optic foci" and "optic tubercles" have recently been recognized as comprising "optic glomeruli" (figures 10.5A–10.5C) (Okamura and Strausfeld 2007; Strausfeld and Okamura 2007; Strausfeld et al. 2007). In flies, visual information, most certainly including many types of image motion, from each of the many successive layers of the lobula (figures 10.2B and 10.6B), is now understood to be sent to individual optic glomeruli representing specific types of visual outputs from those layers (figures 10.2B and 10.6B). The collective output from within each different lobula layer is suggested to represent a pool of a certain type of visual sensation, and each glomerulus would thus represent a different pool that was gathered from a different depth of neuropil within the lobula (figure 10.6B) (Okamura and Strausfeld 2007; Strausfeld and Okamura 2007; Strausfeld et al. 2007). The 27 or more optic glomeruli known to be present in some species of flies are now recognized to populate huge volumes of each LP (figure 10.5C) (Okamura and Strausfeld 2007). It is not known whether the lepidopteran LP has the same abundance of optic glomeruli as the dipteran LP, but they indeed have been identified and diagrammed there (cf. Heinz and Reppert 2012). As in flies, the optic glomeruli of moths are in proximity to the integrative pheromone and general odor olfactory zones of the ILPt and LH (figures 10.1B and 10.5A).

KEY PLAYERS: LOCAL NEURONS INTERCONNECTING OPTIC GLOMERULI

Of further interest is that optic glomeruli have been shown to be interconnected by LNs, many of which are directionally motion selective, that are suggested to perform the same function as LNs in the AL, e.g., integrating the different activities occurring among groups of glomeruli (Strausfeld et al. 2007). The strong similarity between the architectures of AL olfactory

and lateral protocerebral optic glomeruli has been recognized previously (Strausfeld et al. 2007), with the key common feature being the integration performed by LNs of the pure pools of specific olfactory (figure 10.6A) or visual information (figure 10.6B) residing in different glomeruli. Thus, an optic glomerular LN arborizing in many different glomeruli would seemingly be able to integrate the excitations occurring within those various glomerular neuropils to achieve some sort of overall, moment-to-moment visual blend sensation. For image-flow stimuli, such integrations might be an achromatic vector integration of all the different types of narrow- and wide-field image flows from the environment at any instant. These integrations might not necessarily be based upon a retinotopic representation of any specific features of the environment, but possibly upon combinations of, e.g., motion sensations related to rolling, pitching, and yawing, among other movements.

In flies, directional motion information shown to be integrated by LNs arborizing in optic glomeruli (Strausfeld et al. 2007) has been implicated as extending into other lateral protocerebral regions such as the LH, where visual information from these LNs may be further integrated with odor-related information (Strausfeld et al. 2007). Strausfeld and Okamura (2007) and Strausfeld et al. (2007) have strongly suggested that such LNs are likely to be able to integrate incoming odor-related inputs to the LH with visual information from optic glomeruli.

It is clear that the LP is emerging as an important set of neuropils in which multimodal olfactory-visual information may be occurring via the activities of LNs that arborize across optic glomeruli as well as across olfactory PN arborization locations. Thus, two of the protocerebral projection destinations of neurons carrying visual edge-motion information in moths have been identified as being the same two protocerebral regions to which sex pheromone olfactory information is processed: the MBCs and the LP.

INPUTS FROM THE LOBULA PLATE TO OPTIC GLOMERULI

The main task of the arrangement of wide-field VS and HS large tangential neurons in the lobula plate of flies and other flying insects such as moths is for general flight stabilization via optomotor feedback. It is now evident (Strausfeld et al. 2007; Strausfeld 2012) that a major portion of VS and HS lobula plate axons synapse first with LNs in the LP before these synapse with descending neurons. In *Macroglossum stellatarum* (Sphingidae), some VS and HS neurons from the lobula plate were implicated as arborizing in knots of neuropil, e.g., "optic foci," indicative of what are now known to be optic glomeruli (Wicklein and Varju 1999). These VS and HS neuronal inputs to LH neuropils were shown to be responsive to horizontal, vertical, or diagonal image motion and could be important to pheromone-mediated optomotor anemotaxis via pheromonal ILPt-LN-related integration.

Evidence for Pheromone-Stimulated Image-Flow Enhancement

There is behavioral and neuroanatomical evidence in male moths that attention to optical flow is poor unless pheromone stimulation is present. Preiss and Kramer (1983) and Preiss and Futschek (1985) demonstrated that tethered male *Lymantria dispar* (Erebidae) ignored different optical flow-

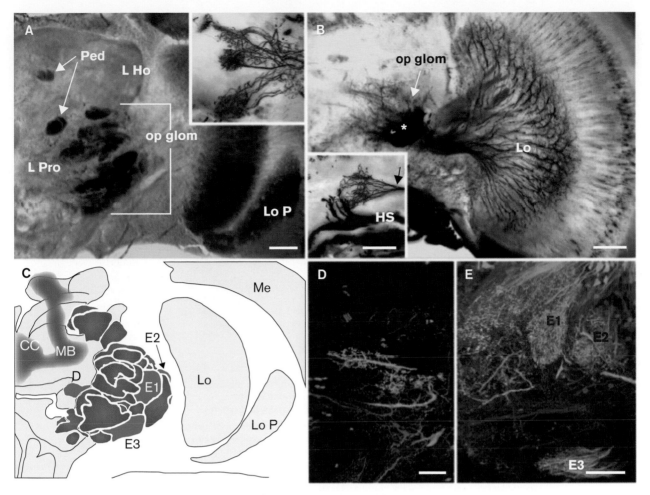

FIGURE 10.5 Optic glomeruli in the lateral protocerebrum (LPro) of the calliphorid fly *Phaenicia sericata*.

A The axons of many optic glomeruli (op glom; bracketed) from the lobula and lobula plate (Lo P) converge and terminate in knots of dense neuropils (inset) in the LPro. Op glom lie adjacent to the lateral horn (L Ho).

B When an optic glomerulus is stained with silver as is shown here, the convergence of inputs from specific retinotopic layers in the Lo is revealed. Thus, each optic glomerulus represents a specific type of visual input, perhaps image-motion specific. The inset illustrates the difference between fine retinotopic fibers ascending to a glomerulus from the Lo (arrow) versus stout fibers of the large tangential horizontally sensitive (HS) neurons from the Lo P.

C Reconstruction in top-down (horizontal) view of the entire suite of op glom that were characterized in the lateral protocerebrum of *Phaenicia sericata*, illustrating their position with regard to the central complex (CC), the ipsilateral mushroom body (MB), and Lo, Lo P, and medulla (Me).

D Double-stained confocal image of fibers in an optic glomerulus (green) amidst fibers from descending neurons (magenta) that reveal no direct connections between the two neuronal types. The optic fibers in this image are located in the green optic glomerulus labeled *D* in (C).

E This image shows as well that the fibers of descending neurons (magenta) do not synapse with the fibers (green) of three op glom labeled E1, E2, and E3, whose locations in the LP are indicated in green in (C). Thus, it appears that it may be the local neurons not visualized in these images that interconnect op glom and also provide the synapses that descending neurons articulate with.

SOURCE: From Strausfeld et al. (2007).

NOTE: Scale bars, 50 μm. Some abbreviations (labels) are different from the abbreviations in our text for the same structures; we have retained the original abbreviations from the cited work.

related sensations indicative of the males either descending or ascending until pheromone was simultaneously presented.

But the most definitive and intriguing demonstrations of the integration of pheromone and visual flow stimuli come from studies of descending neurons in *L. dispar* by Olberg and Willis (1990) and in *Manduca sexta* by Kanzaki et al. (1991b). Descending neurons are premotor command neurons that receive inputs from the LAcLos and other protocerebral neuropils. Their axons project caudally through the subesophageal ganglion and neck toward their termination points in

thoracic motor neurons in one or more of the three thoracic ganglia. The key areas in which descending neurons arborize in the protocerebrum are in the ventral medial and ventral LP (Olberg and Willis 1990; Kanzaki et al. 1991b, 1994; Wada and Kanzaki 2005; Iwano et al. 2010). These areas include both the LAcLo/VPrtC and the ILPt/LH, both of which seem to receive pheromonal, visual, and multimodal inputs.

Olberg and Willis (1990) performed recordings on descending neurons from the severed cervical (neck) connectives of male *L. dispar* ventral nerve chords (figure 10.7A). Seventeen

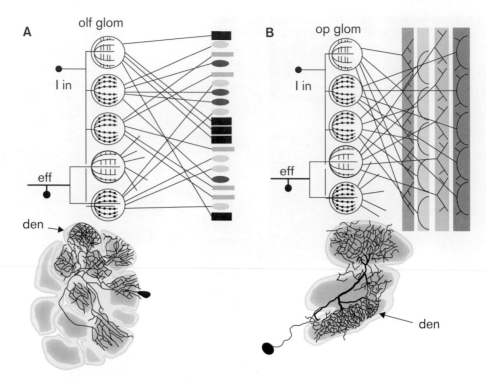

FIGURE 10.6 Schematic diagrams illustrating the similarities between olfactory and optic glomeruli that were recognized by Strausfeld et al. (2007) and the importance of local neurons in integrating the activities occurring among glomeruli in both olfactory and visual systems.

A Each olfactory glomerulus (olf glom) receives inputs from a specific type of identically odorant-tuned olfactory sensory neuron (differently shaped and shaded geometric figures at right). A local interneuron (l in) is illustrated as arborizing in different ways in different glomeruli to modulate and integrate the activities among the glomeruli. Olfactory system efferent neurons (eff) as depicted here are the projection neurons that arborize in different olfactory glomeruli and convey their integrated excitations to the mushroom body and inferior lateral protocerbrum/lateral horn in the protocerebrum. The arborization architecture of an olfactory local neuron in the glomeruli of an antennal lobe is illustrated in the lower left (den, dendrites of the local interneuron).

B Each optic glomerulus (op glom) is illustrated here as receiving vision-specific inputs, including image-motion-specific inputs, from different layers of the lobula, illustrated in different shades of gray on the right. A local interneuron (l in) is shown arborizing in the optic glomeruli, with an efferent neuron (eff) conveying the integrated outputs to other protocerebral neuropils or to thoracic ganglia. Arborization architecture of a visual local interneuron distributing its dendrites (den) among three optic glomeruli is illustrated in the lower right.

SOURCE: From Strausfeld et al. (2007).

NOTE: Please note that some abbreviations (labels) are different from the abbreviations in our text for the same structures; we have retained the original abbreviations from the cited work.

of 70 pheromone-responding descending neurons responded to visual left–right motion in front of the male, and nine of these neurons were also directionally motion sensitive (responding only to either leftward or rightward movement of a visual grid pattern). Five of these neurons exhibited significantly heightened response to the visual grid motion in their preferred direction when 100 ng of *L. dispar* pheromone had been simultaneously presented to the male (figure 10.7B). The initial arborization locations of the pheromone-enhanced directional motion-sensitive neurons were in the posterior LP before they projected their axons caudally to the pterothoracic ganglia (figure 10.7A).

Kanzaki et al. (1991b) found descending neurons in *M. sexta* that responded with LLE of the greatest intensity in response to the optimal two-component pheromone-blend ratio compared to a single component. Several of the neurons were multimodal (visual-olfactory), being not only responsive to the pheromone blend but also to lights-on and lights-off, as well as to hand movement from the researchers. A few of

these multimodal neurons changed firing status *only* when pheromone and visual stimuli were presented closely together in time. Thus, many of the descending neurons were visual-olfactory multimodal neurons, exactly the type of neuron that should be important to pheromone source location by a flying male moth. Kanzaki et al. (1991b), and others (Li and Strausfeld 1999; Strausfeld and Li 1999; Strausfeld 2003), have pointed out that this multimodal integration will most likely have already occurred in protocerebral neuropils before being synaptically transmitted to descending neurons and then to pterothoracic ganglia motor neurons.

Response to Pheromone Strands Requires High-Speed Temporal Integration of Pheromone Spectral Odor Space with Visual Inputs

High-temporal-resolution pathways that represent changes in molecular flux at the OSN level up through higher centers in

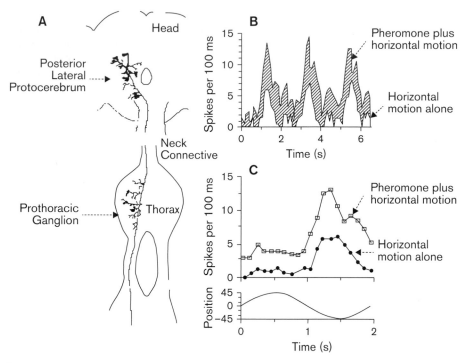

FIGURE 10.7 Descending neuron in *Lymantria dispar* that exhibited heightened responsiveness to the horizontal motion of a bar pattern across its eyes when sex pheromone was presented simultaneously with the motion.

A Top view of the stained descending neuron whose postsynaptic arbors in the posterior lateral protocerebrum projected and arborized with presynaptic arbors in the thoracic ganglia on the ipsilateral side.

B Response of this neuron to 1/sec rightward bar motion alternating with 1/sec leftward motion. Lower (white) curve is the action potential frequency of the neuron in clean air with peaks every 2 sec during rightward bar motion. Upper (darkened) curve is the response of the neuron to the same bar motion in the presence of pheromone.

C A single cycle of bar motion (lower curve) with bar position through time indicated at 45° to the left or right in front of the moth. Middle curve shows action potential frequency in response to the bar in clean air and upper curve is the response of the same neuron to the same bar motion in the presence of pheromone.

SOURCE: Adapted from Olberg and Willis (1990).

the brain should be a crucial part of successful, sustained upwind flight to the source. Upwind flight behavior in response to a species' pheromone blend is known to be elicited on a strand-by-strand, clean-air-pocket-by-clean-air-pocket basis (Mafra-Neto and Cardé 1994; Vickers and Baker 1994). Optimal upwind progress toward a pheromone source also includes a high-temporal-resolution assessment of the positions of pheromone strands in spectral odor space. Contact by a flying male with a single conspecific pheromone strand tainted with a heterospecific behavioral antagonist results in a truncated upwind turn-reversal compared to a longer duration reversal after contact with a strand of the pure pheromone alone (Vickers and Baker 1997; Quero et al. 2001). Similarly, upwind flight is compromised even if the components of the correct blend are emitted separately in a staggered manner such that they arrive in separate, interleaved strands but at what should be a behaviorally optimal sub-second frequency (Vickers and Baker 1992). Thus, temporal resolution of strand ONs and OFFs with apparent image motion is not the only aspect involved in response to pheromone. The visual reaction to wind, the latency to a turn-reversal, or both depend also on the position of the single pheromone strand in pheromone-odor space.

Response Properties of Neurons and Neuropils along Moth Sex Pheromone Olfaction Pathways

The sex pheromone blends of moths are usually comprised of two, sometimes three (or more), chemical components that evoke upwind flight and source location when they are emitted at precise ratios (see Allison and Cardé, this volume). Each of the pheromone OSNs that is tuned to respond optimally to each component is an achromatic neuron, and nearly all of the neurons that conduct pheromone-component information further up into higher levels of the brain are also achromatic (i.e., they have not integrated pheromone-component blend-ratio information even at these deeper projection zones of the protocerebrum). When the inputs of two or three types of pheromone-component-tuned achromatic neurons are integrated eventually by higher order neurons, these neurons are producing *chromatic* pheromone-odor sensations that can be pinpointed by the moth at some location in spectral odor space. It has become increasingly clear, however, that very few chromatic pheromone-blend-sensitive neurons have been shown to exist in the male moth brain. The majority of the

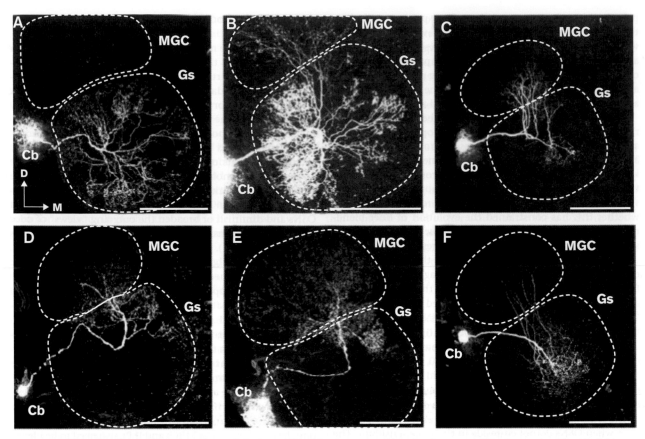

FIGURE 10.8 Examples of the wide variety of morphologies exhibited by the antennal lobe local neurons of *Bombyx mori*, with wide-field global as well as asymmetric and sparse local neuron arborizations occurring among ordinary (Gs) and macroglomerular complex (MGC) glomeruli. (A) Some local neurons arborize only among ordinary glomeruli and do not visit the MGC. (B–F) Various combinations of wide-field versus narrow-field, symmetric versus asymmetric local neuron arborizations. Cb, cell body. Arborizations may be sparse or profuse (C and D vs. A and B). Local neurons may exhibit extreme differences in the intensities of the varicosities of their arbors both from one antennal lobe zone to another, e.g., ordinary glomerulus zone to MGC as in (B) and (E) and within only one zone (A). Hundreds of local neurons are estimated to be in the antennal lobes of *Manduca sexta* and *B. mori*, so it is noteworthy that so much variation in local neuron morphology can be exhibited by just this handful of stained local neurons.

SOURCE: From Seki and Kanzaki (2008).

NOTE: All scale bars, 100 μm. Please note that some abbreviations (labels) are different from the abbreviations in our text for the same structures; we have retained the original abbreviations from the cited work.

connections and either excitatory or inhibitory synapses across varied ensembles of glomeruli. Some wide-field LNs interconnect McGl glomeruli with ordinary glomeruli across the entire AL (figures 10.8A and 10.8B) and may be involved in the upregulation or downregulation of overall "gain" (global sensitivity-activity levels) of the entire ensemble of glomeruli (Martin et al. 2011). Other LNs, denoted as "relay LNs" (Strausfeld 2012), have more localized interconnections with particular glomerular ensembles (figures 10.8C–10.8F) (Matsumoto and Hildebrand 1981; Christensen et al. 1993; Anton and Hansson 1995; Seki and Kanzaki 2008; Reisenman et al. 2011). These more restricted interconnections may serve to sharpen interglomerular ratios of activity via inhibitory GABA transmission or to heighten activity in one direction, e.g., from plant odor-plus-pheromone blends to heighten McGl glomerular excitation while either suppressing or not affecting ordinary glomerular excitation (Namiki et al. 2008; Trona et al. 2010; Chaffiol et al. 2012). More global LN networks in the AL may serve to broaden or narrow the ranges of PN output activities exiting ensembles of AL glomeruli (Martin et al. 2011) via their up- or down-modulation by biogenic amines or neuropeptides from centrifugal neurons (Anton and Homberg 1999) that visit the AL from areas deeper within

the protocerebrum (Homberg et al. 1988). The arborization patterns of LNs *within* the volume of a glomerulus and the varied types of fine vesiculations of their dendrites are also highly variable. Some LNs arborize only at the perimeter of a glomerulus, some only in the core, and some in only one hemisphere of the glomerulus or the other, whereas other LNs invest the entire glomerular volume (Seki and Kanzaki 2008; Reisenman et al. 2011).

LOCAL NEURON PHYSIOLOGY: GABA-ERGIC LATERAL INHIBITION

The majority of AL LNs in moths are inhibitory GABA-ergic (GABA-releasing) neurons and impose lateral inhibition on neighboring glomeruli and on the neurites within individual glomeruli (Hoskins et al. 1986; Waldrop et al. 1987; Homberg et al. 1988, 1990; Seki and Kanzaki 2008; Reisenman et al. 2011; Heinbockel et al. 2013). Combinations of crossglomerular LN synaptic inhibition via their impositions on various combinations of other McGl AL neurons are able to produce PN outputs to protocerebral centers that convey a more precise temporal representation of pheromone odor-strand flux (Waldrop et al. 1987; Christensen and Hildebrand 1988, 1997; Christensen et al. 1993, 1998, 2000, 2003; Heinbockel et al. 1999, 2004, 2013). However, precise "tracking" of major pheromone-component strands (pulses) has been demonstrated to occur in PNs of other species without any extra temporal sharpening occurring from the addition of secondary pheromone components via LN-related lateral inhibition (Vickers et al. 1998; Lei and Hansson 1999; Vickers et al. 2001, 2005; Vickers 2006).

Lateral inhibition by LNs that improve the ability of AL PNs to follow pulsatile stimulation with pheromone components has been shown to occur in LNs arborizing only in ordinary glomeruli, with no arborizations in the McGl (Heinbockel et al. 2013). These ordinary glomeruli-spanning LNs must thus be being inhibited by other LNs that arborize in the McGl and synapse with them in some way (Heinbockel et al. 2013). Using multielectrode recording techniques, Christensen et al. (1993) showed that LNs have the ability to excite otherwise silent PNs in specified glomeruli via GABA-ergic LN–LN synaptic disinhibition. Lateral inhibition imposed on another inhibitory LN can thus disinhibit an output PN and create a different output pattern using the same hardwiring but with different GABA-related outcomes according to which odorant blend is presented to the antenna.

Although GABA-ergic inhibitory LNs have been most commonly described thus far, there are indications that there are other types of LNs that use other neurotransmitters that may not be inhibitory, even releasing different types of neurotransmitters and neuropeptides that can shift subsequent PN activities within the AL in different directions (Homberg et al. 1990; Berg et al. 2007; Utz et al. 2008). Berg et al. (2007) found several neuropeptides present in the AL LNs of *Heliothis virescens*, including A-type allatostatins, *Manduca sexta* allatotropin, FMRFamide-related peptides, and tachykinin-related peptides.

LOCAL NEURON PHYSIOLOGY IS MALLEABLE AND INFLUENCED BY AGE, MATING STATUS

The PN outputs of male *Agrotis ipsilon* (Noctuidae) in response to pheromone or plant odorants were shown to change significantly depending on the physiological state (e.g., mated vs. unmated) or age (young vs. old) of the moth. These differences in AL LN modulations of PN outputs were proven to be due at least in part to alterations in juvenile hormone (JH) levels (Anton and Gadenne 1999; Anton et al. 2007).

In further studies with *A. ipsilon*, during a refractory period after copulation, males were shown to lose interest in mating (Barozzo et al. 2011). During the 24 h after mating, the pheromone sensitivity of AL PNs decreased, but OSNs responsive to pheromone components remained at the same level of sensitivity, as did AL neurons processing plant-related odors (Barrozo et al. 2010, 2011). The mating-induced modulation of sensitivity thus was occurring at the AL level and was restricted to PN outputs that resulted from AL LN–PN processing. After mating, JH concentration drops to a level comparable to that of immature males, so here again JH seems to be playing an important role.

McGl PN sensitivity to pheromone can also increase within minutes due to preexposure to pheromone, which again seems to orchestrate LN modulation of AL activity along with sensitization of OSNs due to preexposure (Anderson et al. 2003; Guerrieri et al. 2012). In experiments involving male *Spodoptera littoralis* preexposed to pheromone, Guerrieri et al. (2012) found that the major pheromone-component-specific McGl glomerulus became significantly enlarged after behaviorally relevant preexposures; extra synaptic interactions there could explain an observed increased behavioral response by preexposed males. Together, these findings indicated that modulation of AL LNs can increase the overall gain of PNs and help the male become more sensitive to calling females in a matter of minutes after first receiving a brief pheromone exposure.

A final consideration regarding the output malleabilities of the LN networks in the AL is the knowledge that centrifugal neurons (Homberg et al. 1988) originating in regions of the protocerebrum, including the pars intercerebralis and subesophageal ganglia, project back out to the AL. These neurons are known to release various neuromodulators, such as histamine, dopamine, serotonin, and neuropeptides, that can change the input–output gain of AL LNs, or shift the balance of neurotransmission activity between and among asymmetric LNs of the AL (Homberg et al. 1990; Homberg and Hildebrand 1991). Thus, the AL is not just a relay station for more refined blend-ratio reporting and precise temporal renderings of pheromone-strand onsets and offsets. Rather, the interactions among LNs can be adjusted and affect the outputs of PNs that conduct action potentials to higher protocerebral centers, significantly altering the olfactory information that is transmitted to these integrative deep-brain locations.

Antennal Lobe Projection Neurons

After all the highly complex processing that goes on in the AL from various interactions between OSN inputs, LN–LN interactions, and LN–PN interactions, the outputs from the McGl glomeruli are carried by PNs from the deutocerebral AL level of the brain to two higher olfactory centers in the protocerebrum: the MB, arborizing in its calyces, and the ILPt.

RECONCILING RESULTS OF VARIOUS STUDIES ON MOTH PROJECTION NEURON MORPHOLOGIES AND PHYSIOLOGIES

The study of moth protocerebral neuroanatomy is an extremely difficult and slow-going endeavor. It is rare that a researcher can impale a PN and record from it for a long enough period of time to run through a full palette of stimuli, and then inject a visible dye with sufficient success to trace its anatomy from the McGl to either the MBC or the ILPt. In addition, only a few research groups, with a few moth species, have incorporated rapidly pulsed pheromone-component stimulation into their multi-pheromone-component-and-blends stimulus regimes, with the majority using just a single stimulus pulse. Thus, it is difficult to compare complete neuroanatomical results across species because attention to temporal aspects of the stimulus has not routinely been addressed by all research groups. It is common for staining of neurons to seem to have been accomplished only to find later that it failed or was incomplete.

Thus, it is not surprising that tens of thousands of PNs of various species have been impaled over the decades, but very few have had their physiological response profiles to pheromone components and pulsatile stimulations fully characterized and in addition had their full anatomies revealed through complete staining to protocerebral destinations along their entire lengths. Inability to accurately visualize the structural details of the neuropil destinations targeted even by completely stained PNs is another factor that has limited more precise descriptions of the targeted neuropils and perhaps the functions of many of these PNs' transmissions.

Most of the earlier studies, such as on *Manduca sexta* (Christensen and Hildebrand 1988; Kanzaki et al. 1989), traditionally impaled pheromone-sensitive PNs from the medial cell cluster of the AL, an area that has been found throughout the decades to be highly populated with the cell bodies of PNs that use the medial antennoprotocerebral tract (*m*-APT) route to the MBCs and then to the ILPt. However, the departure routes of a few of these medial cell cluster PNs in which complete staining was accomplished has shown that a small percentage of these PNs bypass the MBCs and use the mediolateral antennoprotocerebral tract (*ml*-APT) route (cf. Seki et al. 2005; Namiki et al. 2013). Thus, assessing a stained neuron as having a cell body in the medial cell cluster and exiting on what appears to be the *m*-APT (to first visit the MBCs) might not always be correct.

Results from many moth species have shown that when recording from PNs whose cell bodies reside in the lateral cell cluster of the AL, pheromone-sensitive PNs nearly always will project via the *ml*-APT or lateral antennoprotocerebral tract (*l*-APT) directly to the ILPt and terminate there (Vickers et al. 1998; Vickers and Christensen 2003; Vickers 2006; Kárpáti et al. 2008, 2010). However, only more recently has more attention been paid to recording from lateral cell cluster PNs, and researchers are thus starting to gather more information about the physiologies of these *l*-APT PNs, which seem to have a greater tendency to be blend synergist (chromatic) PNs.

PROJECTION NEURON MORPHOLOGIES: THREE AXONAL TRACTS TO PROTOCEREBRAL NEUROPILS

The major PN tract leading to the protocerebrum is the *m*-APT (Galizia and Rössler 2010) (figure 10.1), formerly known as the inner antenno-cerebral tract (Homberg et al. 1988). This is also the route on which most of the neurophysiological recordings and stainings of pheromone PNs have been performed. These PNs, whose cell bodies characteristically reside in the medial cell cluster of the AL (toward the midline of the brain), send axons toward the MB, with collaterals visiting and arborizing in the MBCs before their main axons continue on to terminate in the ILPt.

The MB is the dominant, most distinctive neuropil in the protocerebrum of insects (Strausfeld 2003, 2012; Farris 2005; Strausfeld et al. 2009). In the Lepidoptera, the MB on either side of the protocerebrum has a complex structure with prominent, paired, conjoined calyces occurring dorsally on the pedunculus (stalk) (figures 10.4A and 10.4C) (Pearson 1971; Sjöholm et al. 2005, 2006; Rø et al. 2007; Sinakevitch et al. 2008; Fukushima and Kanzaki 2009). The pedunculus furcates vertically and medially into vertical (alpha), medial (beta), and gamma lobes of the MBLs. In the Lepidoptera, there is another distinct MBL, the Y-lobe (figure 10.4A) (Pearson 1971; Sjöholm et al. 2005, 2006; Rø et al. 2007; Sinakevitch et al. 2008; Fukushima and Kanzaki 2009).

Tens of thousands of intrinsic KCs run in parallel along the lengths of the MBs, from the calyces down through the ends of the lobes (figure 10.4D). The MB is somewhat similar to the AL in that it consists primarily of intrinsic LNs, but in the MB these intrinsic cells appear to us as highly organized arrays of KCs that, because of their number and density, give the MB its characteristically stalked, multilobed shape (figure 10.4A). The swollen region of KCs comprising the MB calyces has repeatedly been shown to be the major MB reception area for sensory afferents in insects (Strausfeld 2003, 2012; Farris 2005; Strausfeld et al. 2009). However, the MBLs have more recently been identified as also being significantly involved in receiving many types of sensory and multimodal inputs (Strausfeld 2003, 2012; Farris 2005; Strausfeld et al. 2009).

Pheromone-sensitive PNs that arborize in the outer zone of the MBCs do so using tiny synaptic connections that are considered to comprise "microglomeruli" and are called boutons (figures 10.4C and 10.4E), each with excitatory–inhibitory PN-GABA-ergic integrative capabilities (Szyszka et al. 2005). The synaptic pattern of PN/KC boutons on the array of KC parallel fibers (figures 10.4D and 10.4E) has the potential to produce great numbers of different cross-fiber patterns on KCs from the inputs of many PNs, each with their own bouton distributions and different physiological response profiles. This vast array of achromatic temporally precise pheromone-component PN input patterns can be integrated across KCs to resolve pheromone-odor time and odor space as well as be integrated with any visual image-motion inputs that might arrive at the MB directly from the optic lobe (figures 10.4B and 10.4C).

A large proportion of the KCs in moth MBs are GABA-ergic along their lengths of the MBLs, as well as in the MBCs, and should impose sharpening and "sparsening" of the integrated olfactory pattern in the MBCs to be transmitted through the lobes (Szyszka et al. 2005). There are many other neuromodulatory and neurotransmitter substances present in KCs; so, highly complex integration of odor, visual, and other modality inputs should be expected throughout the MB. Recurrent neurons also are known to provide feedback between the lobes and calyces and can thus modulate the activities of the KCs throughout the MBCs (Sjöholm et al. 2005; Rø et al. 2007; Sinakevitch et al. 2008; Fukushima and Kanzaki 2009). All of these activities add to the ability of the MB to be a highly refined integrator of temporally precise achromatic pheromone-component information into many types of outputs, possibly even strand-specific chromatic odor space resolution that can be transmitted to other protocerebral neuropils.

A second McGl PN route is out from the McGl glomeruli through the *ml*-APT (Galizia and Rössler 2010) (figure 10.1), formerly known as the middle antenno-cerebral tract (Homberg et al. 1988). These PN axons directly visit the ILPt and terminate with arborizations there, bypassing the MBs. They seem to have their cell bodies in either the lateral or the medial cell clusters (Vickers et al. 1998; Vickers and Christensen 2003; Vickers 2006; Kárpáti et al. 2008, 2010; Namiki et al. 2013).

The third route taken by PN axons that leave McGl glomeruli is called the *l*-APT (Galizia and Rössler 2010; figure 10.1), formerly called the outer antenno-cerebral tract (Homberg et al. 1988). The cell bodies of these PNs characteristically lie in the lateral cell cluster of the AL and their axons project directly to the ILPt, terminating with arborizations there and again bypassing the MBs (Vickers et al. 1998; Vickers and Christensen 2003; Vickers 2006; Kárpáti et al. 2008, 2010; Namiki et al. 2013).

Because all pheromone-component-specific information arrives at the ILPt from the three main APT pathways and the PNs of only one pathway visit the MBCs, it is becoming increasingly clear that the ILPt of the protocerebrum is important for pheromone olfaction (Anton et al. 1997; Kanzaki et al. 2003; Kárpáti et al. 2008, 2010) and deserves an increasingly intense research effort to unravel more knowledge about male moth pheromone olfaction and behavior. In *Ostrinia nubilalis* (Crambidae), Anton et al. (1997) found that the arborizations of PNs in the ILPt were more extensive and profusely arborizing than in other moth species, and they remarked that this could mean that *O. nubilalis* PNs terminating in the LP might have a greater opportunity to be involved in "multimodal" integration there. Strausfeld (cf. 2003, 2012; Strausfeld et al. 2007) has emphasized that the LP, including the LH and ILPt, should be a place where olfactory and visual image-motion information is likely to be integrated because of the multitudes of optic glomeruli connected by networks of LNs residing there. Because the ILPt and LH, unlike the MB and its calyces, are comprised of ill-defined neuropil features, it has been difficult to delineate different subzones within them and thus to decipher anatomically how pheromone and general odors might be integrated there.

Two "pheromone-component-specific" achromatic PNs were identified in *Agrotis segetum* and found to be bilateral in their projections to both the contralateral and the ipsilateral LAcLos, also arborizing in corresponding ILPts on both sides (Wu et al. 1996). Thus far, these PNs are the only PNs in any moth species that have been found to directly connect the AL with the LAcLo and provide a potentially "reflexive" pathway (Wu et al. 1996) to behavioral response.

PROJECTION NEURON MORPHOLOGIES: ODORANT-SPECIFIC SYNAPTIC REGIONS IN THE MBCs

In moths, achromatic pheromone-component-sensitive PNs leaving the McGl along the *m*-APT route send collaterals to first arborize in the MBCs before continuing on to terminate with arborizations in the ILPt. In the MBCs, the PN collaterals synapse with multitudes of microglomeruli (Szyszka et al. 2005) on the intrinsic neurons of the MB, the KCs at characteristic locations around the outer rim zone of the MBCs (figures 10.4C and 10.9A–10.9C).

In *Bombyx mori*, these PN arborizations can be profuse (figure 10.9A) or sparse (figure 10.9B), depending on whether a bombykol pheromone component or bombykal behavioral-antagonist McGl glomerulus was the arborization origin of that PN (figure 10.9) (Seki et al. 2005; Namiki et al. 2013). Bombykol is the only definitive pheromone component known for *B. mori* (Butenandt et al. 1959). Bombykal was later identified from female *B. mori* glands, but curiously, it was shown to be behaviorally antagonistic when added to bombykol (Kaissling et al. 1978). Bombykal has distinct OSN pathways to the "cumulus" of the McGl that differ from the bombykol-tuned OSNs that arborize in the "toroid" of the McGl. The McGl glomerular integration of bombykal inputs with those of bombykol may have more to do with bombykal being a heterospecific antagonist (Daimon et al. 2012) than a *B. mori* pheromone component. Regardless, it is now also known that PNs of *B. mori* traveling along the *m*-APT from either the bombykal-sensitive cumulus or the bombykal–bombykol-sensitive "horseshoe" glomeruli of the McGl have synaptic boutons on the KCs of the MBC that exhibit broad overlap with boutons of PNs carrying general odorant information from AL ordinary glomeruli (figures 10.9B, 10.9C, and 10.10). In contrast, bombykol (pheromone)-responding PNs projecting to the

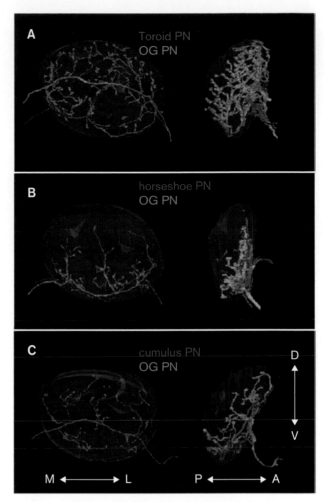

FIGURE 10.9 Confocal microscope images (A–C) of *Bombyx mori* projection neuron (PN) arborization locations in the mushroom body calyxes showing different arborization destinations of PNs carrying different pheromone-component-related information. Pheromone-component-related PNs projecting from glomeruli of the macroglomerular complex are labeled in green to compare with the arborization locations of plant-volatile-sensitive PNs (OG PN) projecting from ordinary glomeruli in the antennal lobe that are labeled in magenta. Top-down (horizontal) views of the mushroom body calyxes are on the left and side views (vertical) of the same preparations are on the right. Note the highly restricted varicose arbors of the PN from the toroid glomerulus of the macroglomerular complex (Toroid PN, green neurites in A) that reports bombykol-related activity, which do not overlap to any large degree with the arbors of neurons projecting from ordinary glomeruli in the antennal lobe (magenta neurites). Compare this arborization pattern to that of the PN from the cumulus of the macroglomerular complex (Cumulus PN, green arbors in B) that reports bombykal activity. The PN from the cumulus has more profuse varicose MBC arbors than the bombykol-reporting PN from the toroid glomerulus and these overlap with the arbors from an ordinary glomerulus PN (magenta in B). The mushroom body calyx arbors of the PNs from the horseshoe glomerulus of the macroglomerular complex (green neurites in C) also overlap to a much greater degree with the mushroom body calyx arbors of PNs projecting from ordinary glomeruli than do the mushroom body calyx arbors of the toroid PNs (A).

ABBREVIATIONS: M, medial; L, lateral; P, posterior; A, anterior; D, dorsal; V, ventral.

SOURCE: From Namiki et al. (2013).

NOTE: Some abbreviations (labels) are different from the abbreviations in our text for the same structures; we have retained the original abbreviations from the cited work.

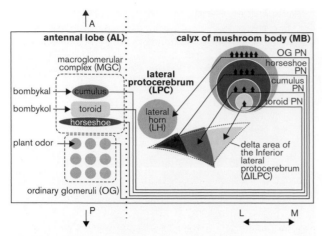

FIGURE 10.10 Illustration of the different semiochemical-related projection destinations that are now known for the medial antennoprotocerebral tract projection neurons (PNs) of *Bombyx mori* that are responsive to bombykol pheromone (toroid PN); bombykal (cumulus PN); and to either bombykol or bombykal or both (horseshoe PN) as well as to plant odorants (OG PN). PNs that arborize in the toroid glomerulus of the macroglomerular complex (MGC) and are responsive to bombykol project collaterals to a restricted area of the mushroom body calyx that arborize there before the main axons terminate with arbors in a more medial area of a pheromone-component-specific region ("delta area") of the inferior lateral protocerebrum, here called the ΔILPC. PNs arborizing in the cumulus of the MGC that are responsive to bombykal, a pheromone-related behavioral antagonist, send axon collaterals that arborize in a broader area of the mushroom body calyces before projecting their main axons to terminate with arbors in a more lateral region of the ΔILPC. A third type of PN that arborizes in the MGC does so in the horseshoe glomerulus, but usually it also has other arbors in the cumulus. This third type is responsive to combinations of bombykol and/or bombykal and sends collaterals to arborize in a still broader region of the mushroom body calyces before terminating with arbors in a lateral region of the ΔILPC. PNs responsive to plant-related volatiles that arborize in one, or in large combinations of various OG project collaterals to an even broader area of the mushroom body calyxes. These plant-volatile-sensitive PNs do not send terminal arbors to the ILPC, but rather terminate in another distinct lateral protocerebral neuropil called the lateral horn (LH).

ABBREVIATIONS: L, lateral; M, medial; A, anterior; P, posterior.

SOURCE: From Namiki et al. (2013).

NOTE: Some abbreviations (labels) are different from the abbreviations in our text for the same structures; we have retained the original abbreviations from the cited work.

MBCs from the toroid glomerulus of the McGl have boutons arborizing within a narrow zone of the MBCs and have little overlap with MBC boutons from cumulus, horseshoe, or ordinary glomerulus PNs (figures 10.9A and 10.10). Similarly, in *Periplaneta americana*, two distinctive McGl PN arborization destinations in two subregions of the MBCs were observed, with one of the PN types responding to the major and the other responding to the minor pheromone component of this species (Nishino et al. 2012a).

PROJECTION NEURON MORPHOLOGIES: ODORANT-SPECIFIC SYNAPTIC REGIONS IN THE ILPt

In *Bombyx mori*, the same *m*-APT PNs carrying achromatic information about different pheromone components to distinctly different zones of the MBCs continue on to terminate with arborizations in different and slightly distinctive

subregions of the ILPt (figure 10.10) (Seki et al. 2005; Namiki et al. 2013). This discovery informs us that in moth pheromone-olfactory systems, there must be further integration needed at the ILPt protocerebral level for the separate incoming pheromone-component-specific inputs. Local integrative neurons (LNs) may process this information to produce higher precision information within the ILPt about the chromatic quality (blend ratios) of the pheromone odor, the temporal aspects of its intermittency, or both. The ratios of excitation coming from the achromatic PNs differentially tuned to pheromone components reaching the ILPt might be integrated to produce chromatic pheromone-blend outputs via the activities of ILPt/LH LNs, or perhaps via optic-olfactory multimodal LNs arborizing in the ILPt/LH and optic glomeruli of the LP (Okamura and Strausfeld 2007; Strausfeld et al. 2007; see "The Lepidopteran Visual System" above). Chromatic pheromone-blend information would then be sent from the ILPt out to other protocerebral areas, e.g., the MBLs or the LAcLo via trans-protocerebral PNs, or to thoracic ganglia for motor output via descending neurons synapsing in with ILPt/LH LNs in the posterior LP.

GENERAL ODORANT-TUNED PROJECTION NEURON MORPHOLOGIES

General odorant-tuned PNs of all moth species send their axons out of the AL from ordinary glomeruli to the protocerebrum via the *m*-APT, *ml*-APT, and the *l*-APT. Unlike pheromone-sensitive PNs, however, the arborization destination of general odorant-tuned PNs in the LP is the LH, an area of neuropil adjacent and dorsal to the ILPt (cf. Anton and Hansson 1994; Seki et al. 2005; Namiki et al. 2013). In all moth species, PNs exiting ordinary glomeruli via the *m*-APT send collaterals that first arborize in the MBCs and then continue on to terminate in the LH (Namiki and Kanzaki 2011; Namiki et al. 2013) (figure 10.10).

PROJECTION NEURON PHYSIOLOGIES: ANTENNOPROTOCEREBRAL TRACTS ARE RELATED TO CONVEYANCE OF ODOR TIME vs. ODOR SPACE INFORMATION

Here, we explain the use of different PN tracts exiting the McGl of male moths based on their functional physiologies, regardless of the numbers of AL glomeruli in which they arborize. Our approach contrasts with overviews that have attempted to link PN glomerular arborization anatomies in the AL to being "uniglomerular" or "multiglomerular." Such overviews have characterized the axons of uniglomerular PNs arborizing in a single AL glomerulus as using predominantly *m*-APT, whereas multiglomerular PNs arborizing in two or more glomeruli have been characterized as using *l*- or *ml*-APT (Galizia and Rössler 2010; Martin et al. 2011).

One reason for functionally characterizing a PN before linking it to one of the APT routes that it takes is that the molecular range of responsiveness of even uniglomerular achromatic PNs can be increased or decreased by LN activities in the AL to include greater collective inputs from differentially odorant-tuned, interconnected glomeruli (see figure 3 of Martin et al. 2011). Even though these PNs may be uniglomerular and seemingly should respond to just one odorant, they now respond to many more odorants due to LNs increasing the gathering of OSN activities from within greater numbers of glomeruli (see figure 3 of Martin et al. 2011). Also, despite their increased molecular range of responsiveness, such PNs remain achromatic, although reporting now across a broader molecular range along a graded, two-dimensional odorant spectrum. Such adjustments in the molecular receptive range of a PN will not change its MBC arborization location in the protocerebrum; they will only change how narrow or broad its view of pheromone-odor space is in making its report to those same locations.

It is becoming increasingly clear that there is usually little correspondence between the number of McGl glomeruli in which a PN arborizes, the APT route on which it travels or its responsiveness to one versus more than one pheromone component. For example, in heliothine moth species, PNs have been found that exit the McGl along the *m*-APT, respond only to the major pheromone component, and yet have multiglomerular McGl arborizations (Christensen et al. 1991; Vickers et al. 1998; Vickers 2006). Hansson et al. (1994), working with *Agrotis segetum*, likewise found multiglomerular McGl PNs using the *m*-APT, yet these PNs responded only to single components. They also found multiple-component-responding *m*-APT-projecting PNs that had only a uniglomerular arborization in the McGl. The lack of correspondence between multiglomerular or uniglomerular McGl arborizations and subsequent PN response profiles was further confirmed in this species by Wu et al. (1996).

In *Trichoplusia ni* (Noctuidae) (Anton and Hansson 1999), *Spodoptera littoralis* (Anton and Hansson 1995), *Ostrinia nubilalis* (Anton et al. 1997), *Bombyx mori* (Kanzaki et al. 2003), *Cydia pomonella* (Trona et al. 2010), and *Grapholita molesta* (Varela et al. 2011), a proliferation of non-correspondences between the number of McGl glomerular arborization locations and PN response characteristics has been similarly noted. In most of these studies, responses to pheromone components were even observed coming from PNs that arborized in ordinary glomeruli, with many instances of PN responses to plant volatiles that arborized in McGl glomeruli.

Thus, we think that it may be more instructive to examine how well PNs respond temporally to rapid pheromone pulses of single pheromone components versus blends, or how much of a dynamic range they have in response to various pheromone components and plant volatiles, rather than whether they are uniglomerular or multiglomerular in AL glomeruli, as outlined via two points below regarding odor time and odor space.

1. Odor time: the m-APT carries achromatic, high-fidelity plume-strand- and clean-air-pocket temporal information to the MBC, then to the ILPt

The *m*-APT, the most commonly traveled of the three APT routes and which first includes a visit to the MBCs, should be involved in the generation of turn-reversal upwind surges in response to plume-strand contact. The PNs that project their axons along the *m*-APT have been the ones most commonly shown to be able to "follow" rapid sub-second pulses of pheromone components or blends of components. Depending on the species, they may originate in only one, or in several of the many McGl glomeruli, or may even have additional arborizations in ordinary glomeruli as described above. Like the columnar visual neurons in the medulla and lobula of the optic lobe, most of these PNs carry high-resolution achromatic olfactory edge-motion (temporal strand-edge) information to higher brain centers.

In the *m*-APT PNs of *Manduca sexta*, *Heliothis virescens*, *H. subflexa* (Noctuidae), *Helicoverpa zea*, *A. segetum*, and *A. ipsilon*,

pulsed pheromone-component stimulation was used to examine the temporal acuity of pheromone-sensitive PNs leaving McGl glomeruli. All but two of the fully anatomically characterized *m*-APT PNs that were examined were achromatic, responding to only one pheromone component or several, but with no greater-than-additive action potential frequencies exhibited when two or more pheromone components were presented. In all of these studies, the PNs could "follow" and lock-on to at least 5-Hz short pulses (*H. virescens*: Vickers et al. 1998, 2001; Vickers and Christensen 2003; Vickers 2006; *H. zea*: Vickers et al. 1998; *H. subflexa*: Vickers and Christensen 2003; Vickers 2006; *M. sexta*: Christensen and Hildebrand 1988, 1997; Heinbockel et al. 1999, 2004; *A. ipsilon*: Chaffiol et al. 2012; and *A. segetum*: Lei and Hansson 1999). Several studies showed that the phase-locking to the pulses was sharpened, i.e., exhibited greater temporal acuity, when a blend of two or more components was puffed (Christensen and Hildebrand 1997; Vickers et al. 1998). Enhanced temporal resolution to 5-Hz pulses was found in *A. ipsilon m*-APT PNs when a blend of pheromone plus a plant odorant, heptanal, was puffed compared to the pheromone alone (Chaffiol et al. 2012).

Double impalements of *m*-APT PNs of *M. sexta* innervating the different glomeruli of the McGl showed that two PNs branching in the same McGl glomerulus display a more synchronized firing pattern than those branching in different glomeruli (Christensen and Hildebrand 1988, 1997; Christensen et al. 1998; Heinbockel et al. 1999, 2004, 2013; Martin et al. 2013). PNs in different glomeruli also tend to inhibit each other's activities. Such LN-related lateral inhibition was shown to produce an increased temporally precise tuning of responses to pheromone pulses. The temporal patterns observed are formed in the AL, and they seem to depend on interactions between *m*-APT PNs and LNs. Experiments suppressing the action of inhibitory LNs showed that pulse-following acuity was abolished in *m*-APT PNs (Christensen and Hildebrand 1988, 1996; Christensen et al. 1998; Lei et al. 2009) and that upwind flight in a pheromone plume to its source was hampered (Lei et al. 2009).

The crisp, phasic-bursting response pattern observed in *m*-APT PNs when all pheromone components are present thus seems to be a prerequisite for optimal orientation in a pheromone plume. In further experiments with *M. sexta*, when the natural pheromone blend was mimicked, using a 1:1 ratio of bombykal (*E,Z*-10,12-hexadecadienal) to a behaviorally effective synthetic hydrocarbon analog of the natural aldehyde second component (*E,E,Z*-10,12,14-hexadecatrienal), the pulse-tracking bursts of action potentials from the PNs were sharper and more accurate in response to the sub-second odor pulses than when either component was pulsed alone (Heinbockel et al. 2004). Further studies showed that a 2:1 blend of the two actual pheromone components (using the triene aldehyde this time and not the hydrocarbon analog, again mimicking the natural blend) created a synchronous firing of the *m*-APT PNs arborizing in the same McGl glomerulus compared to when single components or off-blends were used (Martin et al. 2013). A blend-enhanced synchronous firing of PNs exiting the same glomerulus, rather than an elevation of action potential frequency by such PNs was suggested to be a mechanism for pheromone-blend-quality encoding by such neurons (Martin et al. 2013).

The relationship between some observed highly regular oscillatory patterns called local field potentials (LFPs) resonating between the protocerebrum and the AL due to their circuitries, and the possible synchronization of PN responses to plume strands, was studied previously (Heinbockel et al. 1999; Christensen et al. 2000, 2003; Vickers et al. 2001). Vickers et al. (2001) found that the male moth pheromone-processing system relies not on LFP oscillations, but rather on non-LFP-related excitations of PNs that are directly caused by the arrival of each temporally irregular sub-second plume strand. The precise reactions to plume strands were behaviorally assessed and temporally monitored via electroantennograms, and the *m*-APT PN responses were recorded, all within natural point-source plumes (Vickers et al. 2001).

The temporal sharpening that occurs in the AL in response to blends and its output in the McGl-PNs of the *m*-APT is similar to improved edge-resolution of the visual system due to lateral inhibition (Strausfeld and Campos-Ortega 1977; Strausfeld 2003). Multiple layers of AL LNs shorten the temporal differences between the rise and fall times of the firing of PNs to each pheromone puff and thereby sculpt the temporal edges of the firings. Thus, the predominant type of information that travels along the *m*-APT PNs up to the MBCs involves numerous achromatic reports from differentially tuned PNs related to variations in plume-strand flux.

We should remember that these temporally sharp reports by *m*-APT PNs to the MBCs go on to terminate in the ILPt. In the MBs, integration into chromatic blend outputs on either a strand-by-strand or a time-averaged basis may be occurring. However, these temporally acute *m*-APT reports also may be processed in the ILPt for integrating plume-strand encounters with perhaps any achromatic or chromatic blend-quality reports that are sent to the ILPt from the other two APT pathways.

2. Odor space: the very small amount of chromatic pheromone-blend information projecting out of the AL is carried more by ml- and l-APT PNs than by m-APT PNs

The type of PNs described by Vickers et al. (1998) as "blend synergist" PNs is the type we are calling chromatic PNs. Chromatic, pheromone-blend synergist PNs exhibit a level of action potential frequency in response to a blend of pheromone components that is significantly greater than the sum of the excitations elicited by the same components presented individually. These PNs constitute a tiny proportion of the thousands of PNs that have been recorded from thus far. The few pheromone-blend chromatic PNs that have been described are thus capable of sending to protocerebral neuropils at least a primitive rendering of the blend's position in odor space as a result of only AL-level integration. Chromatic blend synergist PNs seem to be absent in some species. Out of hundreds of recordings and complete stainings by the Kanzaki and Hildebrand groups, no chromatic pheromone-blend PNs have been found in either *M. sexta* or *B. mori*, respectively. However, because *B. mori* seems to use bombykol as a one-component pheromone communication system, with bombykal only serving as a heterospecific behavioral antagonist, it is understandable why efforts to understand the effects of bombykol plus bombykal blends have been unfruitful.

PROJECTION NEURON PHYSIOLOGIES: POOR TEMPORAL PLUME-STRAND RESOLUTION BY CHROMATIC PROJECTION NEURONS?

When chromatic pheromone PNs have been fully neuroanatomically characterized, a slight majority (three of five) have been shown to use the *ml*- or *l*-APT and project to arborize in the ILPt. These PNs were found in *Heliothis virescens* (Vickers et al. 1998). When challenged with 5-Hz pulsed blend stimulation, these chromatic PNs were very poor at following pheromone-blend pulses, whereas nearly all the achromatic PNs following the *m*-APT in this same species did so extremely well (Vickers et al. 1998). Interestingly, there was also a lack of ability to resolve 5-Hz pulses by two chromatic blend synergist PNs of *H. subflexa* and *H. virescens* hybrids (these PNs have not been anatomically fully characterized; Vickers and Christensen 2003).

In both *H. virescens* (Vickers et al. 1998) and *Agrotis segetum* (Hansson et al. 1994), one completely stained and physiologically characterized chromatic pheromone-blend PN was found that used the *m*-APT. The chromatic PN in *H. virescens* following the *m*-APT was also unable to resolve 5-Hz pheromone pulses. The PN in *A. segetum* was not challenged with pulsed stimulation (Hansson et al. 1994). Both of these PNs were shown to send collaterals that arborized in the MBCs and axons that continued on to terminate with their arbors in the ILPt (Hansson et al. 1994; Vickers et al 1998).

Although the projection routes have not been characterized for some PNs, it is possible to hypothesize the projection routes on the basis of whether a PN's cell body is located in the medial or the lateral cell cluster of the AL. Lateral cell cluster PNs have been found to nearly always use either the *ml*-APT or *l*-APT routes (Anton et al. 1997; Kárpáti et al. 2008, 2010) and thus only arborize in the ILPt, not the MBCs. Thus far, no PN has been found having a lateral cell body cluster that projects along the *m*-APT. For PNs shown to have their cell bodies in the medial cluster, there is a bit more uncertainty. In some cases (e.g., *Bombyx mori*), a small percentage of medial cluster cell body PNs leaving via what appears to be the *m*-APT divert their paths and instead use the *ml*-APT, bypassing the MBCs with terminal arbors directly in the ILPt (Kanzaki et al. 2003; Namiki et al. 2013).

Out of many hundreds of recordings from moth PNs with cell bodies in the medial cell cluster, two chromatic pheromone PNs were characterized in *Helicoverpa zea* (Christensen et al. 1991), two in *H. subflexa* (Vickers and Christensen 2003), one in *H. virescens* (Christensen et al. 1995), one in *Spodoptera littoralis* (Anton and Hansson 1995), and seven in *A. segetum* (Hansson et al. 1994; Wu et al. 1996; Hartlieb et al. 1997; Lei and Hansson 1999). It is unknown how many of these chromatic PNs actually followed the *m*-APT.

Most of the PNs from other species having cell bodies in the lateral cluster and whose axons can be inferred to use the *l*- or *ml*-APT have been characterized as chromatic blend synergist PNs. Much of this latter data comes from studies on the PNs of *Ostrinia nubilalis* (Anton et al. 1997; Kárpáti et al. 2008, 2010). These PNs were not challenged with pulsed pheromone stimuli, so their temporal acuity in response to blends is unknown.

A few unstained or incompletely stained chromatic pheromone PNs have been shown to be able to discriminate blends differing from each other only by 5–10% in composition ratios of their pheromone components (Wu et al. 1996; Anton et al. 1997). This precision of blend-ratio discrimination was exhibited by chromatic PNs in both *A. segetum* (Wu et al. 1996) and *O. nubilalis* (Anton et al. 1997). These results suggest that there has been some selection pressure for odor space discrimination to take place at the AL level before the information is transmitted to the protocerebrum. It also highlights once again the subtleties of the complex integrative networks of LNs, OSNs, and PNs at this level.

PROJECTION NEURON PHYSIOLOGIES: GABA-ERGIC PROJECTION NEURONS USING THE ml-APT

The neurotransmitter used by most PNs is understood to be acetylcholine (Hoskins et al. 1986; Homberg et al. 1988, 1990; Seki and Kanzaki 2008; Seki et al. 2005; Reisenman et al. 2011); acetylcholine will transmit excitatory inputs to the MBCs as well as to the ILPt. Two immunocytochemical studies from *Manduca sexta* and *Heliothis virescens*, however, have indicated that many, and possibly most, of the PNs traveling over the *ml*-APT to the ILPt exhibit GABA-like immunoreactivity (GLIR) (Hoskins et al. 1986; Berg et al. 2009). These PNs terminate in the ILPt without continuing on to the MBCs. However, Hoskins et al. (1986) note that this is a tract that also showed the majority of PNs being cholinergic, and they suggested that it is possible that the *ml*-APT PNs might be both cholinergic and GABA-ergic. If *ml*-APT PNs are able to be GABA-ergic, then enhanced discrimination of pheromone-blend ratios and a sharpening of temporal synchrony in the ILPt may be aided by the integration of these inhibitory PN inputs arriving at the ILPt and excitatory inputs arriving there via the *ml*- and *l*-APTs.

The Lateral Accessory Lobe: Site for Counterturn Generation, Convergence of Multimodal Olfactory-Visual Inputs, and Descending Premotor Neurons

The LAcLos are paired neuropils in the midline of the protocerebrum slightly anteroventral to the CC (figures 10.1, 10.11A, and 10.11B). The LAcLos have a large degree of neuronal connectivity with the CC (Strausfeld and Hirth 2013), and they also receive inputs from multimodal neurons projecting to them from other protocerebral neuropils. The LAcLos are key premotor centers in the insect brain that promote the execution of many behaviors, including pheromone-mediated counterturning (Iwano et al. 2010). Pheromone-sensitive neurons of the LAcLo of several moth species arborize with premotor descending neurons in the ventrolateral and inferior lateral protocerebral neuropils (Olberg 1983; Olberg and Willis 1990; Kanzaki et al. 1991b, 1994; Kanzaki and Shibuya 1992; Wada and Kanzaki 2005; Iwano et al. 2010). These neuropils are also key sites where pheromone-olfactory and visual image-motion information is likely to become integrated, either from direct optic inputs to the LAcLo or via multimodal PrtCNs (Strausfeld et al. 2007).

The LAcLo on each side of the protocerebrum is connected via intrinsic LNs to a corresponding VPrtC and forms a LAcLo/VPrtC complex (figure 10.11C) (Iwano et al. 2010). The LAcLo/VPrtC LN network on either side of the protocerebrum is indicated as being the source of the neuronal circuit's LLE. This LLE produces prolonged behavioral excitation by setting up fast-recurrent oscillation between the LAcLo and the VPrtC to drive a continuous LLE output in efferent bilateral LAcLo-to-LAcLo neurons in response to a brief pheromone pulse (Iwano et al. 2010). For both *Manduca sexta* and *Bombyx mori*, the LLE is optimally evoked when the species' pheromone is presented, compared to an off-ratio of pheromone components or of antagonists and plant odor (Kanzaki et al., 1991b, 1994; Kanzaki and Shibuya 1992; Wada and Kanzaki 2005; Iwano et al. 2010).

The bilateral LAcLo/VPrtC neurons connecting each LAcLo/VPrtC unit on either side of the protocerebrum are

GABA-ergic and inhibitory (Iwano et al. 2010); they have been shown to impose an inhibitory LLE to suppress the LLE of the corresponding LAcLo/VPrtC unit on the other side. Thus, a currently dominating LLE produced by one side becomes overtaken and "flipped" to the off-position, by the new dominance of the other side's LLE from its LAcLo/VPrtC unit. The flip-flop occurs either spontaneously after a long time of firing, such as during a sustained period of postpheromone clean air, or else by the arrival of a new pheromone strand (Iwano et al. 2010).

Thus, the brief excitation in one LAcLo/VPrtC from a pheromone strand contacting either or both of the antennae appears to be able to trigger an immediate turn-reversal-related flip-flop of firing activity in the pair of bilateral LAcLo-to-LAcLo neurons. The pheromone strand will also simultaneously initiate the LLE that will drive a many-seconds-long oscillating series of flip-flops during exposure to clean air after the strand. In *B. mori*, an intermediate time frame, high-frequency, endogenous flip-flopping is also expressed in the activities of bilateral LAcLo neurons (Iwano et al. 2010) that is correlated with short-term zigzagging upwind walking in the 200–300 ms between straight-upwind walking and long-term looping in clean air. In flying moths, such a several-hundred-millisecond period would be related to the period of quick zigzag turn-reversals in clean air during the transition to long-term casting flight after a single upwind turn-surge in response to a single strand (Mafra-Neto and Cardé 1994; Vickers and Baker 1994). In most moth species, the counterturning during casting flight occurs with a more or less regular, oscillatory tempo, but in a few species, such as *M. sexta* (Willis and Arbas 1991) and *Lymantria dispar* (Kuenen and Cardé 1994), the counterturn tempos during clean-air casting are much less regular.

The sub-second reaction of a counterturn in response to a pheromone plume strand would seem to imply a very fast connection between the contact of peripheral OSNs with a strand and the LAcLo. There is one direct pheromone-olfactory input that has been found in *Agrotis segetum* by Wu et al. (1996). They reported two "pheromone-component-specific" PNs from the McGl that projected to the contralateral as well as the ipsilateral LAcLos and ILPts on both sides of the protocerebrum. We do not know whether such AL PNs exist in other moth species, but their presence in *A. segetum* is informative in perhaps showing in at least one species the presence of a direct communication channel with the LAcLo-turn generator in response to contact with individual plume strands. Such a direct connection between the AL and the LAcLo has never been found in *M. sexta* or *B. mori*, despite thousands of AL PNs having been recorded from and stained.

Protocerebral Neurons Connecting the Mushroom Body, Inferior Lateral Protocerebrum, and Lateral Accessory Lobe

Inferior Lateral Protocerebrum-to-Mushroom Body and Lateral Horn-to-Mushroom Body Protocerebral Neurons

Some pheromone-olfactory PrtCNs connecting the ILPt with the MBC have been found in *Bombyx mori* (Namiki et al. 2013). One type of neuron, called "ΔILPC" neurons (figures 10.12A and 10.12B), has smooth postsynaptic connections within the ILPt and blebby presynaptic outputs to the same outer collar region of the MBC that corresponds with the

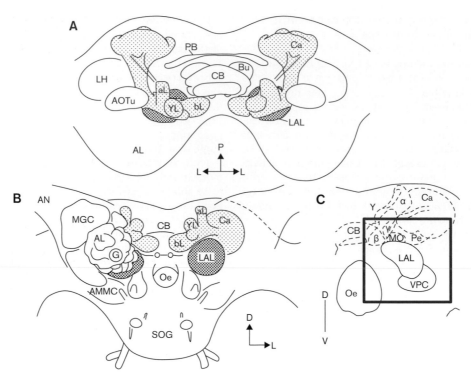

FIGURE 10.11 Position of the lateral accessory lobes of *Bombyx mori* with respect to other protocerebral neuropils, including the ventral protocerebral neurons that synapse with the lateral accessory lobes and promote long-lasting excitation that drives long-lasting, clean-air looping behavior in walking *B. mori* males.

A Vertical (top-down) view of the protocerebrum, which is the same perspective as in figure 10.1B. The positions of the lateral accessory lobes (LAL) on each side of the central body (CB) and posterior to the medial (bL), vertical (aL), and Y-lobes (YL) of the mushroom body can be clearly seen. The anterior positioning of the antennal lobe (AL), the dorsal-anterior positioning of one prominent optic glomerulus (the anterior optic tubercle [AOTu]), and the ventro-posterior positioning of the lateral horn (LH) are depicted as well.

ABBREVIATIONS: Ca, mushroom body calyx; PB, protocerebral bridge; Bu, buttress; L, lateral; P, posterior.

B Frontal (anterior) view of the *Bombyx mori* brain showing one of the anteriorly situated antennal lobes (ALs) with one of its glomeruli noted (G) and its magroclomerular complex (MGC), and the positions of the ALs and MGC with respect to the more posterior LAL, which is not labeled but is the dark gray neuropil oval on that side. The more ventral position of the LAL with respect to the medial (bL), vertical (aL), and Y-lobes (YL) of the mushroom body and also its calyces (Ca) also can be seen.

ABBREVIATIONS: CB, central body; Oe, esophagous; SOG, subesophageal ganglion; AMMC, antennal mechanosensory and motor center. D, dorsal; L, lateral.

C Frontal view (same as in B) of the *Bombyx mori* protocerebrum showing the position of the neuropil of the ventral protocerebrum (VPC) whose neurons synapse with the LAL to produce LLE on that side.

ABBREVIATIONS: α, vertical lobe; β, medial lobe; γ, gamma lobe; Y, Y-lobe; Pe, pedunculus; Ca, calyx of the mushroom body; MO, median olive.

SOURCE: (A) and (B) are from Kanzaki and Shibuya (1992); (C) is from Iwano et al. (2010).

NOTE: Some abbreviations (labels) are different from the abbreviations in our text for the same structures; we have retained the original abbreviations from the cited work.

MBC arborization locations of *m*-APT PNs that have projected there from the McGl. Thus, the ΔILPC neurons appear to be pheromone-related feedback or feed-forward neurons from the ILPt to the MBC.

Another type of PrtCN, named LH neurons (figures 10.12C and 10.12D), connects the plant volatile-related LH region of the LP with the MBCs (Namiki et al. 2013). These LH-MBC neurons have smooth postsynaptic connections in the LH and arborize with presynaptic blebs in the same regions of the MBCs that receive inputs from plant volatile-related PNs leaving ordinary glomeruli from the AL (figures 10.12C and 10.12D)

(Namiki et al. 2013). GABA immunostaining of many more of the LH neurons showed that the neurons connecting the LH with the MBCs did not exhibit GLIR and would thus not be inhibitory. They could possibly be feedback or feed-forward neurons communicating LH-integrated plant volatile-related information to the MBCs' plant volatile reception regions. Although these ΔILPC and LH PrtCNs have been shown thus far to be only olfactory, they have not been challenged with visual or mechanosensory stimuli; so, it is possible that these PrtCNs may be capable of sending multimodal visual-olfactory information to the MBCs.

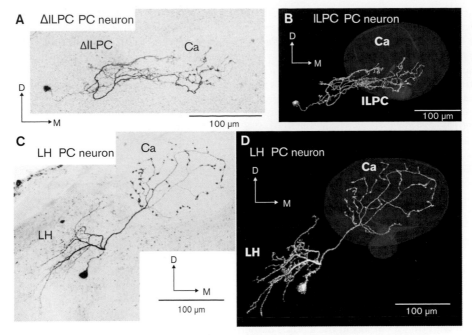

FIGURE 10.12 Examples of protocerebral neurons (PC neurons) that connect either the inferior lateral protocerebrum (ILPC) or the lateral horn (LH) with mushroom body calyxes.

A Confocal image of a "ΔILPC" PC neuron that has smooth arbors in the pheromone-component-targeted region (delta region) of the inferior lateral protocerebrum (ΔILPC) and then has varicose arbors in the mushroom body calyxes. Smooth arbors are considered postsynaptic, and varicose arbors are considered as being presynaptic. Therefore, this type of PC neuron should be transmitting pheromone-related or even multimodal pheromone-plus-visual information from the ΔILPC to the mushroom body calyxes.

B Three-dimensional reconstruction of this neuron with its targeted mushroom body calyx visualized for better perspective of its arbors in the mushroom body calyxes.

C Confocal image of a "LH" PC neuron. It has smooth arbors in the LH of the lateral protocerebrum and varicose arbors in the mushroom body calyxes. Because the LH is the target of projection neurons from ordinary antennal lobe glomeruli that transmit plant-volatile-related information, this "LH" PC neuron is likely to be transmitting plant odor information, or even plant odor-plus-visual information to the mushroom body calyxes from the LH.

D Three-dimensional reconstruction of this neuron with its targeted mushroom body calyx visualized for better perspective of its arbors.

ABBREVIATIONS: Ca, calyx of the mushroom body; D, dorsal; M, medial.

SOURCE: From Namiki et al. (2013).

NOTE: Some abbreviations (labels) are different from the abbreviations in our text for the same structures; we have retained the original abbreviations from the cited work.

Protocerebral Neurons Connecting the Mushroom Body or the Inferior Lateral Protocerebrum/LH with the Lateral Accessory Lobe Are Likely to Be Multimodal Visual-Olfactory Inputs

Only a single pheromone-olfactory-only type of direct connection between McGl AL PNs and the LAcLo has been found thus far, suggesting that the majority of inputs to the LAcLo in moths are likely to come from olfactory-visual multimodal PrtCNs from within the protocerebrum itself. Comprehensive neuroanatomical studies of PrtCNs of *Periplaneta americana* have elucidated how multimodal moth PrtCNs might integrate visual and pheromone-olfactory inputs (Li and Strausfeld 1997, 1999; Strausfeld and Li 1999). These studies demonstrated that there are many multimodal neurons extrinsic to the MB (e.g., non-KC protocerebral interneurons synapsing with the MB and carrying information directly to or away from the MB) that are responsive to the *P. americana* pheromone and to visual stimuli. These PrtCNs were shown to arborize in discrete tufts in the vertical and medial MB lobes as well as in other neuropils, such as the LH/ILPt neuropils (Li and Strausfeld 1997).

Many of these visual-, olfactory-, and sound-responsive extrinsic MB neurons were efferents that projected from the MB and terminated in the LP (Li and Strausfeld 1999). Li and Strausfeld (1999) strongly suggested that such neurons relay information from the MB to the LP about the "sensory context" of olfactory, visual, and other stimuli where the LP can process and integrate it with, for instance, olfaction-only inputs from AL olfactory PNs (Li and Strausfeld 1999). Several different response types of these PrtCNs were found to have stereotypical arborization regions on the MB lobes. For instance, visual-olfactory multimodal MB efferents from the alpha lobes (= vertical lobes) of different individuals were found to send axons to medial protocerebral neuropil (Li and Strausfeld 1997). Conversely, multimodal MB efferents from the beta lobes (= medial lobes) sent axons to the ILPt and superior LP (Li and Strausfeld 1999). Multimodal efferent neurons recurrently connecting the vertical with the medial lobes were found that projected out to the ILPt (Li and Strausfeld 1999).

Instead of the MBCs just receiving unimodal olfactory or visual sensory pathway inputs, it is now clear that they can receive multimodal inputs from PrtCNs as well. In *P. americana*, many multimodal visual-olfactory afferent PrtCNs project from visual-olfactory neuropils to arborize with inputs to the MB pedunculus and calyces (Strausfeld and Li 1999). Nishino et al. (2012a, 2012b) recorded from and stained a PrtCN of *P. americana* that responded to both odor and visual stimuli whose neurites originated in both the medulla of the optic lobe and the LH and then projected to arborize in both the inner- and outer-rim zones of the MBCs.

Inferior Lateral Protocerebrum/Lateral Horn-to-Lateral Accessory Lobe Protocerebral Neurons

In moths, there are several examples of PrtCN synaptic connections between the pheromone olfaction-related ILPt/LH and the LAcLo, and also between the MB and the LAcLo. Kanzaki et al. (1991a) found a pheromone-responsive PrtCN in *Manduca sexta* that arborized with smooth arbors in the ILPt/LH of the ipsilateral side of the brain from where antennal pheromone stimulation came, and projected with varicose arbors to the LAcLo/VPrtC neuropils on both the ipsilateral and contralateral sides (figure 10.13A). This PrtCN responded with brief excitation to the pheromone, and considering the analysis and model of Iwano et al. (2010) for *Bombyx mori*, this type of ILPt-LAcLo PrtCN could be implicated as providing pheromone-olfactory input from the ILPt to trigger LLE in the LAcLo/VPrtC neurons on one side as well as a turn-reversal-related flip-flop by bilateral LAcLo neurons upon contact with a pheromone strand.

Local Pheromone-Sensitive Protocerebral Neurons Connecting the Lateral Protocerebrum, Lateral Accessory Lobe, and Optic Glomeruli

Another LAcLo-visiting PrtCN in *Manduca sexta* that was excited by pheromone was determined to be an "LN" (figure 10.13B) that invested smooth arbors in the LP on one side only (Kanzaki et al. 1991a). This local PrtCN that produced a brief burst of action potentials in response to pheromone projected with varicose arbors to the LAcLo, the ventral LP, optic glomeruli (which had not yet been recognized as such) called "posterior optic foci," and the ventral medial protocerebrum. A few other LN PrtCNs also arborized in the LAcLo of one side and visited other optic glomeruli, intermingling either along the "anterior optic tubercle" or with "posterior optic foci" on that side (figure 10.13C) (Kanzaki et al. 1991a).

These types of ProC LNs may be involved with integrating pheromone-olfactory inputs to the ILPt and optical image flow from optic glomeruli, to trigger LLE/flip-flop-related, turn-reversals. Because no optical stimuli were used during this study (Kanzaki et al. 1991a), it cannot be determined for certain whether these PrtCNs would be multimodal pheromone-visual neurons (Kanzaki et al. 1991a). However, because numerous descending neurons in a companion study of *M. sexta* (Kanzaki et al. 1991b) were found to be multimodal, responding to hand movement, lights-on and lights-off, and pheromone stimulation, it may be inferred that the source of this multimodality was in PrtCNs, such as those that were found by Kanzaki et al. (1991a).

In *M. sexta* PrtCNs, Lei et al. (2013) found a highly pheromone-responsive PrtCN that arborized with smooth processes in the VPrtC and sent a neurite to the lobula of the contralateral optic lobe. They suggested that this PrtCN might be involved in multimodal integration of visual and pheromonal stimuli, although visual stimuli were not tested.

Mushroom Body-to-Lateral Accessory Lobe Protocerebral Neurons

In *Agrotis segetum*, amidst the large number of pheromone-responsive PrtCNs that were recorded from and stained, one PrtCN was found that visited both the LAcLo and the MBLs (figure 10.13D) (Lei et al. 2001). This neuron responded to all pheromone components, could follow 3-Hz pulses of pheromone stimulation, but displayed LLE in response to plant odors or the pheromone behavioral antagonist. It exhibited dense, smooth (postsynaptic) arbors in discrete bulbous areas in two portions of the MB lobes, bifurcated also with arbors in the MB medial lobes, and projected to the LAcLo where it exhibited varicose (presynaptic) blebby synapses (figure 10.13D) (Lei et al. 2001). Such a neuron could be transmitting plume-strand-related pheromone flux information integrated with optic flow-field information to descending neurons found by Lei et al (2001) in their *A. segetum* preparations that exhibited smooth, spiny postsynaptic arbors in the LAcLo.

Similarly, in *Bombyx mori*, quite a few of the LAcLo/VPrtC bilateral neurons found to exhibit only brief excitation or inhibition were shown to receive inputs from extrinsic, efferent MB lobe neurons (Iwano et al. 2010). These MB inputs to the LAcLo/VPrtC bilateral neurons came from either the alpha or beta lobes of the MB (Iwano et al. 2010).

In *Manduca sexta*, Kanzaki et al. (1991a) found some PrtCNs out of the many that were recorded from that had arbors in the MB lobes, but none of these neurons visited the LAcLo. One such PrtCN exhibited LLE and invaded not only two distinctly different zones at the junction of all the MB lobes and the pedunculus but also arborized in the LP, which would include olfaction-vision-related areas of the LH and ILPt. The other few PrtCNs arborizing in the MB lobes all exhibited brief excitation to pheromone and appeared to be wide-ranging protocerebral LNs. Some of these exhibited characteristic tuft-like arbors, called "glomeruli" by Kanzaki et al. (1991a), in a particular MB zone of the lobes. Some tufts such as these have been found in extrinsic MB neurons of *Periplaneta americana* and *Apis mellifera* (Apidae) and can be either outgoing efferent PrtCNs or incoming afferents depending upon the types of spiny or varicose synaptic connections these extrinsic neurons have with the KCs of the MBLs (Farris 2005). Regardless, for *M. sexta* there thus far appears to be a greater direct set of connections with the ILPt and LAcLos than between the MB and LAcLos.

Thus, there is some evidence for *B. mori* and *A. segetum* that temporally sharpened and perhaps quickly integrated visual-olfactory outputs from the MBLs might be received directly by the LAcLo/VPrtC to help trigger counterturns and LLE to visually stabilize crosswind casting behavior. These connections also could explain the attainment of an upwind flight orientation due to optomotor longitudinal image motion during a single surge. In heliothine moths, a single upwind surge is known to be directed quickly upwind and to persist for longer durations when the correct pheromone blend is used rather than a blend of pheromone tainted with a trace amount

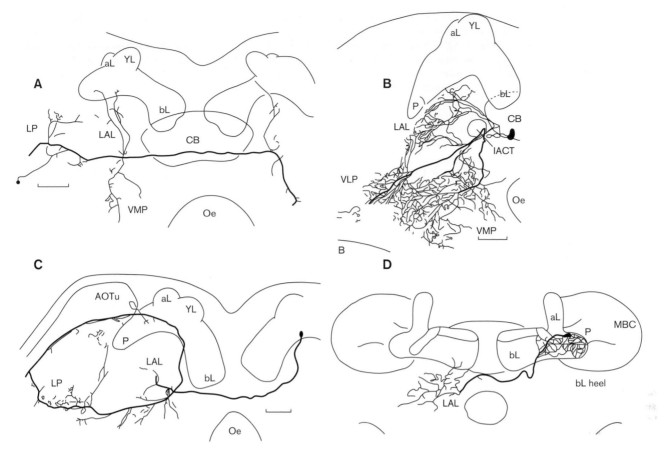

FIGURE 10.13 Reconstructions of stained protocerebral neurons of *Manduca sexta* (A–C) (Kanzaki et al. 1991b) and *Agrotis segetum* (D) (Lei et al. 2001) that respond to their conspecific sex pheromones and innervate the lateral accessory lobe/ventral protocerebrum with varicose arbors. These neurons receive inputs with their smooth arbors that originate either in the lateral protocerebrum (A–C) or next to the mushroom body lobes (D). Reconstructions are all illustrated in frontal view.

A This protocerebral neuron has smooth arbors in the lateral protocerebram (LP) ipsilateral to the antenna receiving sex pheromone stimulation and projects with varicose arbors to the ipsilateral and contralateral lateral accessory lobe (LAL)/ventral protocerebrum.

ABBREVIATIONS: aL, mushroom body vertical lobe; bL, mushroom body medial lobe; YL, mushroom body Y-lobe; CB, central complex; Oe, esophagus; VMP, ventromedial protocerebrum.

B A protocerebral neuron considered by Kanzaki et al. (1991a) to be a "local neuron" (not bilateral) with smooth arbors in the ipsilateral LP and projecting to the ipsilateral LAL, ventro-lateral protocerebrum (VLP), and optic glomeruli (the "posterior optic foci").

ABBREVIATIONS: P, mushroom body pedunculus; IACT, inner antennocerebral tract (nomenclature of Homberg et al. 1988, for the *m*-APT).

C This protocerebral neuron arborized in the LP and responded to pheromone stimulation on the antenna ipsilateral to this LP arborization but contralateral to its cell body (black oval at right). It arborized also in the LAL and visited the length of the optic glomerulus known as the anterior optic tubercle (AOTu).

D This pheromone-responding protocerebral neuron in *Agrotis segetum* arborized with varicose arbors in the contralateral LAL, having come from the mushroom body, where it had smooth arbors in two dense tufts in the "heel" near the base of the medial lobe and pedunculus (bL heel). It also bifurcated to visit the medial lobe in two places before projecting to arborize in the contralateral LAL (Lei et al. 2001).

NOTE: Scale bars (B, C), 100 μm. Some abbreviations (labels) are different from the abbreviations in our text for the same structures; we have retained the original abbreviations from the cited work.

of behavioral antagonist (Vickers and Baker 1997; Quero et al. 2001); so, it is possible that chromatic blend information is integrated by the MBs before at least a small proportion of such MB-PrtCN efferents project to the LAcLo.

Lateral Accessory Lobe-to-Mushroom Body and Other Protocerebral Neuropils

Iwano et al. (2010) found that LLE, pheromone-responding bilateral LAcLo neurons seemed to be involved in sending outputs to other protocerebral regions, including one bilateral LAcLo neuron that sent its output to the beta lobes of the MB. Other such LLE-exhibiting LAcLo bilateral neurons sent

arbors to dorsal protocerebral areas, including one such neuron that also sent arbors to the "posterior slope," an area in which optic glomeruli LNs can intermingle with the axons of lobula–lobula plate visual neurons involved with optomotor regulation of flight and other visual image–motion inputs (Wicklein and Strausfeld 2000).

Protocerebral Chromatic Pheromone-Blend Neurons

In *Manduca sexta* and *Agrotis segetum*, a surprisingly small proportion of PrtCNs have been found that respond preferentially to their pheromone blends, considering that these are two species for which blends are significantly better at eliciting upwind flight of males than individual components (Kanzaki et al. 1991a; Lei et al. 2001, 2013). Thus, at this higher integrative level of the brain, the majority of pheromone information being transmitted involves individual pheromone components and not the blend itself; only a small percentage of PrtCNs convey chromatic information about pheromone-blend quality. With regard to temporal plume-strand information, one study (Lei et al. 2001) on *A. segetum* PrtCNs showed that only a few of these neurons could respond to pulsed pheromone-component stimulation up to 3 Hz; the majority were only able to follow pulses at 1 Hz.

Conclusions

The sex pheromone olfactory system of flying male moths is designed for sub-second pheromone-plume-strand and clean-air processing so that quick steering reactions to the wind can be made via visual image-flow inputs from the optic lobes. The sex pheromone olfactory pathways of male moths converge on many of the same protocerebral neuropils as edge-motion-sensitive optic pathways, such that the two non-chemotactic, indirect behavioral responses to pheromone strands and clean air can be precisely executed: (1) a turn-reversal plus optomotor anemotaxis in response to an individual pheromone strand and (2) an endogenous program of repetitive turn-reversals plus optomotor anemotaxis that plays out in pockets of clean air between strands.

The neuropils receiving pheromone-olfactory-only and edge-motion-vision-only information from primary sensory neuropils are olfactory glomeruli in the AL and optic glomeruli in the LP, respectively. Although these two neuropils appear to be highly separated from each other, residing in two different brain regions, the AL's first-order-processed pheromone information in fact projects to the LP along three different PN axonal routes, so is not separated even early on from optic flow-field information. Pheromone information thus should be able to be integrated with visual edge-motion information via LNs that populate the LP and synapse among optic glomeruli as well as with pheromone-sensitive AL PNs terminating in the ILPt.

The MBCs are also a site of convergence of visual and pheromone-component olfactory information, and may play a role in mediating in-flight behavioral responses to pheromone plume strands and clean air. The MBCs appear to be the preferred neuropil for receiving highly phasic, strand-related fast-tempo inputs from AL PNs projecting along the medial-antennoprotocerebral tract. As such, the MB may be able to provide outputs to other protocerebral neuropils, such as the LAcLos, that can drive strand-triggered turn-reversals related to the upwind surges made by males when contacting an individual pheromone strand.

The ILPt receives the same highly phasic strand-related pheromone-component PN inputs as the MBCs, but it also preferentially receives inputs from two other PN pathways (the mediolateral- and lateral-antennoprotocerebral tracts) that seem to be more involved with carrying integrated, less temporally precise pheromone odor-blend information with less emphasis on phasic, flux-related pheromone-component information. As such, the ILPt and its potential optic glomeruli LN interconnections make it appear to be a site more strongly associated with long-lasting, clean-air casting flight behavior driven by the correct pheromone blend.

Casting flight is a highly optomotor anemotactically guided behavior that depends upon a precise maintenance of a balanced ratio of transverse and longitudinal image flows that reverses in polarity between each endogenously generated turn-reversal.

Thus, at present, we would propose that the protocerebral neuropils that are more medially located, e.g., the mushroom bodies and the LAcLos, are more likely to be involved with promoting straight-upwind flight via the rapid phasic generation of turn-reversals and optomotor anemotactic longitudinal flows upon plume-strand contacts. The more laterally located neuropils of the LP connected with the LAcLo's self-generated flip-flop oscillator appear more likely to be involved with long-term, crosswind casting flight in clean air. Prolonged casting flight requires an LLE-related after effect of the correct pheromone blend to optimally drive the LAcLo flip-flop circuit's endogenous turn-reversal-related program. It also requires a precise integration of longitudinal plus transverse image flows, and such flows are likely to be integrated by optic glomerular LNs that also can synapse with the pheromone-blend-sensitive areas in the ILPt.

Protocerebral neuropils are interconnected by neurons that are most likely all multimodal-responding neurons (Strausfeld 2003, 2012). Thus, the LAcLos' turn-generating outputs may be orchestrated in concert with integrated pheromone-plus-visual-edge-motion inputs from both the mushroom bodies and the LP. Some pheromone-sensitive PrtCNs have been described in moths that have their synaptic origins along the MBLs and terminal arbors in the LAcLos. Other PrtCNs have been described that originate in the LP and have their terminal arbors in the LAcLos. It remains to be seen what such neurons would do if they were to also be challenged with visual motion stimuli.

It is not surprising that pheromone-sensitive PrtCNs such as these were not also presented with visual stimuli because researchers in insect olfaction in the past have not been aware that the response to odor by a flying insect is nearly entirely a *visual* response. Strausfeld et al. (2007) emphasized that an organized regime of multimodal stimuli should be used on the LNs connecting optic glomeruli in the LH and LP to wholly understand their full complements of stimulus-related behavior. They also predicted that these lateral protocerebral LNs known to invade optic glomeruli should prove to be odor- as well as visual-responsive.

We hope that an awareness of this type of multimodal stimulus regime can be applied more often in future neuroethological work, such as when undertaking difficult and time-limited recordings from AL PNs, or even OSNs during antennal single-cell recordings, to at least rule out definitively any types of centrifugal neuron motion-vision or olfactory feedback influencing these more peripheral neurons. Surprises may be in store that will inform us more completely about the degree to which motion-vision and pheromone olfaction may be integrated to modulate all levels of the male moth sex pheromone olfactory system. Attention also should be paid to what we now know to be the significant amount of malleability of integrative neuropils, such as those comprising the AL. Neuronal responses to pheromone components from these neuropils can be influenced by the male's age, mating status, or recent preexposure to stimuli, in addition to the admixing of host plant odorants with pheromone components. These factors all have the potential to change these neurons' reports that characterize their current views of odor time and odor space.

References Cited

Almaas, T.J., and H. Mustaparta. 1991. *Heliothis virescens*: response characteristics of receptor neurons in *sensilla trichodea* type 1 and type 2. *Journal of Chemical Ecology* 5:953–972.

Almaas, T.J., Christensen, T.A., and H. Mustaparta. 1991. Chemical communication in heliothine moths. I. Antennal receptor neurons encode several features of intra- and interspecific odorants in the male corn earworm moth *Helicoverpa zea. Journal of Comparative Physiology A* 169:249–258.

Anderson, P. Sadek, M.M., and B.S. Hansson. 2003. Pre-exposure modulates attraction to sex pheromone in a moth. *Chemical Senses* 28:285–291.

Anton, S., and C. Gadenne. 1999. Effect of juvenile hormone on the central nervous processing of sex pheromone in an insect. *Proceedings of the National Academy of Sciences of the United States of America* 96:5764–5767.

Anton, S., and B.S. Hansson. 1994. Central processing of sex pheromone, host odour and oviposition deterrent information by interneurons in the antennal lobe of female *Spodoptera littoralis* (Lepidoptera: Noctuidae). *Journal of Comparative Neurology* 350:199–214.

Anton, S., and B.S. Hansson. 1995. Sex pheromone and plant-associated odour processing in antennal lobe interneurons of male *Spodoptera littoralis* (Lepidoptera: Noctuidae). *Journal of Comparative Physiology A* 176:773–789.

Anton, S., and B.S. Hansson. 1999. Physiological mismatching between neurons innervating olfactory glomeruli in a moth. *Proceedings of the Royal Society of London B* 266:1813–1820.

Anton, S., and U. Homberg 1999. Antennal lobe structure. Pp. 98–124. In B.S. Hansson, ed. *Insect Olfaction*. Berlin: Springer.

Anton, S., C. Löfstedt, and B.S. Hansson. 1997. Central nervous processing of sex pheromones in two strains of the European corn borer *Ostrinia nubilalis* (Lepidoptera: Pyralidae). *Journal of Experimental Biology* 200:1073–1087.

Anton, S., M.-C. Dufour, and C. Gadenne. 2007. Plasticity of olfactory-guided behavior and its neurobiological basis: lessons from moths and locusts. *Entomologia Experimentalis et Applicata* 123:1–11.

Baker, T.C. 1990. Upwind flight and casting flight: complimentary phasic and tonic systems used for location of sex pheromone sources by male moths. Pp. 18–25. In K. Døving, ed. *ISOT X: Proceedings of the 10th International Symposium on Olfaction and Taste*. Oslo: Graphic Communication System.

Baker, T.C., and L.P.S. Kuenen. 1982. Pheromone source location by flying moths: a supplementary non-anemotactic mechanism. *Science* 16:424–427.

Baker, T.C., and K.F. Haynes. 1987. Manoeuvres used by flying male oriental fruit moths to relocate a sex pheromone plume in an experimentally shifted wind-field. *Physiological Entomology* 12:263–279.

Baker, T.C., M.A. Willis, and P.L. Phelan. 1984. Optomotor anemotaxis polarizes self-steered zigzagging in flying moths. *Physiological Entomology* 9:365–376.

Baker, T.C., H.Y. Fadamiro, and A.A. Cossé. 1998. Moth uses fine tuning for odour resolution. *Nature* 393:530.

Baker, T.C., S.A. Ochieng', A.A. Cossé, S.G. Lee, J.T. Todd, C. Quero, and N.J. Vickers. 2004. A comparison of responses from olfactory receptor neurons of *Heliothis subflexa* and *Heliothis virescens* to components of their sex pheromone. *Journal of Comparative Physiology A* 190:155–165.

Baker, T.C., M.J. Domingue, and A.J. Myrick. 2012. Working range of stimulus flux transduction determines dendrite size and relative number of pheromone component receptor neurons in moths. *Chemical Senses* 37:299–313.

Barrozo, R.B., C. Gadenne, and S. Anton. 2010. Switching attraction to inhibition: mating-induced reversed role of sex pheromone in an insect. *Journal of Experimental Biology* 213:2933–2939.

Barrozo, R.B., D. Jarriault, N. Deisig, C. Gemeno, C. Monsempes, P. Lucas, C. Gadenne, and S. Anton. 2011. Mating-induced differential coding of plant odour and sex pheromone in a male moth. *European Journal of Neuroscience* 33:1841–1850.

Berg, B.G., T.J. Almaas, J.G. Bjaalie, and H. Mustaparta. 1998. The macroglomerular complex of the antennal lobe in the tobacco budworm moth *Heliothis virescens*: specified subdivision in four compartments according to information about biologically significant compounds. *Journal of Comparative Physiology A* 183:669–682.

Berg, B.G., T.J. Almaas, J.G. Bjaalie, and H. Mustaparta. 2005. Projections of male-specific receptor neurons in the antennal lobe of the oriental tobacco budworm moth, *Helicoverpa assulta*: a unique glomerular organization among related species. *Journal of Comparative Neurology* 486:209–220.

Berg, B.G., J. Schachtner, S. Utz, and U. Homberg. 2007. Distribution of neuropeptides in the primary olfactory center of the heliothine moth, *Heliothis virescens. Cell & Tissue Research* 327:385–398.

Berg, B.G., J. Schachtner, and U. Homberg. 2009. Gamma-aminobutyric acid immunostaining in the antennal lobe of the moth *Heliothis virescens* and its colocalization with neuropeptides. *Cell & Tissue Research* 335:593–605.

Briscoe, A.D., G.D. Bernard, A.S. Szeto, L.M. Nagy, and R.H. White. 2003. Not all butterfly eyes are created equal: rhodopsin absorption spectra, molecular identification, and localization of ultraviolet-, blue-, and green-sensitive rhodopsin-encoded mRNAs in the retina of *Vanessa cardui. Journal of Comparative Neurology* 458:334–349.

Butenandt, A., R. Beckmann, D. Stamm, and E. Hecker. 1959. Über den Sexual-Lockstoff des Seidenspinners *Bombyx mori*—Reindarstellung und Constitution. *Zeitschrift für Naturforschung B* 14:283–284. [In German]

Chaffiol, A., J. Kropf, R.B. Barrozo, C. Gadenne, J.-P. Rospars, and S. Anton. 2012. Plant odour stimuli reshape pheromonal representation in neurons of the antennal lobe macroglomerular complex of a male moth. *Journal of Experimental Biology* 215:1670–1680.

Christensen, T.A., and J.G. Hildebrand. 1988. Frequency coding by central olfactory neurons in the sphinx moth *Manduca sexta. Chemical Senses* 13:123–130.

Christensen, T.A., and J.G. Hildebrand. 1996. Olfactory information processing in the brain: encoding chemical and temporal features of odors. *Journal of Neurobiology* 30:82–91.

Christensen, T.A., and J.G. Hildebrand. 1997. Coincident stimulation with pheromone components improves temporal pattern resolution in central olfactory neurons. *Journal of Neurophysiology* 77:775–781.

Christensen, T.A., H. Mustaparta, and J.G. Hildebrand. 1991. Chemical communication in heliothine moths. II. Central processing of intra- and interspecific olfactory messages in the male corn earworm moth *Helicoverpa zea. Journal of Comparative Physiology A* 169:259–274.

Christensen, T.A., B.R. Waldrop, E.D. Harrow, and J.G. Hildebrand. 1993. Local interneurons and information processing in the olfactory glomeruli of the moth *Manduca sexta. Journal of Comparative physiology A* 173:385–399.

Christensen, T.A., H. Mustaparta, and J.G. Hildebrand. 1995. Chemical communication in heliothine moths VI. Parallel pathways for information processing in the macroglomerular complex of the male tobacco budworm moth *Heliothis virescens. Journal of Comparative Physiology A* 177:545–557.

Christensen, T.A., B.R. Waldrop, and J.G. Hildebrand. 1998. Multitasking in the olfactory system: context-dependent responses to odors reveal dual GABA-regulated coding mechanisms in single olfactory projection neurons. *Journal of Neuroscience* 18:5999–6008.

Christensen, T.A., V.M. Pawlowski, H. Lei, and J.G. Hildebrand. 2000. Multi-unit recordings reveal context-dependent modulation of synchrony in odor-specific neural ensembles. *Nature Neuroscience* 3:927–931.

Christensen, T.A., H. Lei, and J.G. Hildebrand. 2003. Coordination of central odor representations through transient, non-oscillatory synchronization of glomerular output neurons. *Proceedings of the National Academy of Sciences of the United States of America* 100:11076–11081.

Collett, T. 1970. Centripetal and centrifugal visual cells in the medulla of the insect optic lobe. *Journal of Neurophysiology* 33:239–256.

Collett, T. 1972. Visual neurons in the anterior optic tract of the privet hawk moth. *Journal of Comparative Physiology* 78:396–433.

Collett, T., and A.D. Blest. 1966. Binocular directionally selective neurons, possibly involved in the optomotor response of insects. *Nature* 212:1330–1333.

Cossé, A.A., J.L. Todd, and T.C. Baker. 1998. Neurons discovered on male *Helicoverpa zea* antennae that correlate with pheromone-mediated attraction and interspecific antagonism. *Journal of Comparative Physiology A* 182:585–594.

Cutler, D.E., R.R. Bennet, R.D. Stevenson, and R.H. White. 1995. Feeding behavior in the nocturnal moth *Manduca sexta* is mediated mainly by blue receptors, but where are they located in the retina? *Journal of Experimental Biology* 198:1909–1917.

Daimon, T., T. Fujii, M. Yago, Y.-F. Hsu, Y. Nakajima, T. Fujii, S. Katsuma, Y. Ishikawa, and T. Shimada. 2012. Female sex pheromone and male behavioral responses of the bombycid moth *Trilocha varians*: comparison with those of the domesticated silkmoth *Bombyx mori*. *Naturwissenschaften* 99:207–215.

David, C.T. 1986. Mechanisms of directional flight in wind. Pp. 49–57. In T.L. Payne, M.C. Birch, and C. Kennedy, eds. *Mechanisms in Insect Olfaction*. Oxford: Oxford University Press.

De Bruyne, M., and T.C. Baker. 2008. Odor detection in insects: volatile codes. *Journal of Chemical Ecology* 34:882–897.

De Bruyne, M., P.J. Clyne, and J.R. Carlson. 1999. Odor coding in a model olfactory organ: the *Drosophila* maxillary palp. *Journal of Neuroscience* 19:4520–4532.

Deisig, N., J. Kropf, S. Vitecek, D. Pevergne, A. Rouyar, J.-C. Sandoz, P. Lucas, C. Gadenne, S. Anton, and R. Barrozo. 2012. Differential interactions of sex pheromone and plant odour in the olfactory pathway of a male moth. *PLOS ONE* 7:e33159.

Dobritsa, A.A., W. Van der Goes van Naters, C.G. Warr, R.A. Steinbrecht, and J.R. Carlson. 2003. Integrating the molecular and cellular basis of odor coding in the *Drosophila* antenna. *Neuron* 37:827–841.

Douglass, J.K., and N.J. Strausfeld. 1996. Visual motion-detection circuits in flies: parallel direction- and non-direction-sensitive pathways between the medulla and lobula plate. *Journal of Neuroscience* 16:4551–4562.

Douglass, J.K., and N.J. Strausfeld. 2003. Retinotopic pathways providing motion-selective information to the lobula from peripheral elementary motion-detecting circuits. *Journal of Comparative Neurology* 457:326–344.

Douglass, J.K., and N.J. Strausfeld. 2007. Diverse speed response properties of motion-sensitive neurons in the fly's optic lobe. *Journal of Comparative Physiology A* 193:233–247.

Fadamiro, H.Y., A.A. Cossé, and T.C. Baker. 1999. Fine-scale resolution of closely spaced pheromone and antagonist filaments by flying male *Helicoverpa zea*. *Journal of Comparative Physiology A* 185:131–141.

Farris, S.M. 2005. Evolution of insect mushroom bodies: old clues, new insights. *Arthropod Structure & Development* 34:211–234.

Fukushima, R., and R. Kanzaki. 2009. Modular subdivision of mushroom bodies by Kenyon cells in the silkmoth. *Journal of Comparative Neurology* 513:315–330.

Galizia, C.G., and W. Rössler. 2010. Parallel olfactory systems in insects: anatomy and function. *Annual Review of Entomology* 55:399–420.

Gonick, L. 1995. Phero-flying. *Discover*. Waukesha, WI: Kalmbach Publishing. Available at: http://discovermagazine.com.

Grant, A.J., R.J. O'Connell, and A.M. Hammond. 1988. A comparative study of pheromone perception in two species of noctuid moths. *Journal of Insect Behavior* 1:75–95.

Guerrieri, F., C. Gemeno, C. Monsempes, S. Anton, E. Jacquin-Joly, P. Lucas, and J.-M. Devaud. 2012. Experience-dependent modulation of antennal sensitivity and input to antennal lobes in male moths (*Spodoptera littoralis*) pre-exposed to sex pheromone. *Journal of Experimental Biology* 215:2334–2341.

Hallberg, E., B.S. Hansson, and R.A. Steinbrecht. 1994. Morphological characteristics of antennal sensilla in the European cornborer *Ostrinia nubilalis* (Lepidoptera: Pyralidae). *Tissue and Cell* 26:489–502.

Hallem, E.A., and J.R. Carlson. 2004. The odor coding system of Drosophila. *Trends in Genetics* 20:453–459.

Hämmerle, B., and G. Kolb. 1997. Organization of the lamina ganglionaris of the optic lobe of the butterfly *Pararge aegeria* (Linné) (Lepidoptera: Satyridae). *International Journal of Insect Morphology and Embryology* 26:139–147.

Hansson, B.S., M. Tóth, C. Löfstedt, G. Szócs, M. Subcheve, and J. Löfqvist. 1990. Pheromone variation among eastern European and a western Asian population of the turnip moth *Agrotis segetum*. *Journal of Chemical Ecology* 16:1611–1622.

Hansson, B.S., T.A. Christensen, and J.G. Hildebrand. 1991. Functionally distinct subdivisions of the macroglomerular complex in the antennal lobe of the male sphinx moth *Manduca sexta*. *Journal of Comparative Physiology* 312:264–278.

Hansson, B.S., S. Anton, and T.A. Christensen. 1994. Structure and function of antennal lobe neurons in the male turnip moth, *Agrotis segetum* (Lepidoptera: Noctuidae). *Journal of Comparative Physiology A* 175:547–562.

Hansson, B.S., T.J. Almaas, and S. Anton. 1995. Chemical communication in heliothine moths V: antennal lobe projection patterns of pheromone-detecting olfactory receptor neurons in the male *Heliothis virescens* (Lepidoptera: Noctuidae). *Journal of Comparative Physiology A* 177:535–543.

Hartlieb, E., S. Anton, and B.S. Hansson. 1997. Dose-dependent response characteristics of antennal lobe neurons in the male moth *Agrotis segetum* (Lepidoptera: Noctuidae). *Journal of Comparative Physiology A* 181:469–476.

Heinbockel, T., T.A. Christensen, and J.G. Hildebrand. 1999. Temporal tuning of odor responses in pheromone-responsive projection neurons in the brain of the sphinx moth *Manduca sexta*. *Journal of Comparative Neurology* 409:1–12.

Heinbockel, T., T.A. Christensen, and J.G. Hildebrand. 2004. Representation of binary pheromone blends by glomerulus-specific olfactory projection neurons. *Journal of Comparative Physiology A* 190:1023–1037.

Heinbockel, T., V.D.C. Shields, and C.E. Reisenman. 2013. Glomerular interactions in olfactory processing channels of the antennal lobes. *Journal of Comparative Physiology A* 199:929–946.

Heinz, S., and S.M. Reppert. 2012. Anatomical basis of sun compass navigation I: the general layout of the monarch butterfly brain. *Journal of Comparative Neurology* 520:1599–1628.

Homberg, U., and J.G. Hildebrand. 1991. Histamine-immunoreactive neurons in the midbrain and suboesophageal ganglion of the sphinx moth *Manduca sexta*. *Journal of Comparative Neurology* 307:647–657.

Homberg, U., R.A. Montague, and J.G. Hildebrand. 1988. Anatomy of antenno-cerebral pathways in the brain of the sphinx moth *Manduca sexta*. *Cell & Tissue Research* 254:255–281.

Homberg, U., T.G. Kingan, and J.G. Hildebrand. 1990. Distribution of FMRFamide-like immunoreactivity in the brain and suboesophageal ganglion of the sphinx moth *Manduca sexta* and colocalization with SCP$_B$-, BPP-, and GABA-like immunoreactivity. *Cell & Tissue Research* 259:401–419.

Hoskins, S.G., U. Homberg, T.G. Kingan, T.A. Christensen, and J.G. Hildebrand. 1986. Immunocytochemistry of GABA in the antennal lobes of the sphinx moth *Manduca sexta*. *Cell & Tissue Research* 244:243–252.

Ibbotson, M.R., T. Maddes, and R. DuBois. 1991. A system of insect neurons sensitive to horizontal and vertical image motion connects the medulla and midbrain. *Journal of Comparative Physiology A* 169:355–367.

Iwano, M., E.S. Hill, A. Mori, T. Mishima, K. Ito, and R. Kanzaki. 2010. Neurons associated with the flip-flop activity in the lateral accessory lobe and ventral protocerebrum of the silkworm moth brain. *Journal of Comparative Neurology* 518:366–388.

Kaissling, K.-E. 1990. Sensory basis of pheromone-mediated orientation in moths. *Verhandlungen der deutschen Zoologischen Gesellschaft* 83:109–131.

Kaissling, K.-E. 1998. Flux detectors versus concentration detectors: two types of chemoreceptors. *Chemical Senses* 23:99–111.

Kaissling, K.-E., G. Kasang, H.J. Bestmann, W. Stransky, and O. Vostrowsky. 1978. A new pheromone of the silkworm moth *Bombyx mori*. *Naturwissenschaften* 65:382–384.

Kanzaki, R., and T. Shibuya. 1986. Descending protocerebral neurons related to the mating dance of the male silkworm moth. *Brain Research* 377:378–382.

Kanzaki, R., and T. Shibuya. 1992. Long-lasting excitation of protocerebral bilateral neurons in the pheromone-processing pathways of the male moth *Bombyx mori*. *Brain Research* 587:211–215.

Kanzaki, R., and T. Mishima. 1996. Pheromone-triggered "flipflopping" neural signals correlated with activities of neck motor neurons of a male moth, *Bombyx mori*. *Zoological Science* 13:79–87.

Kanzaki, R., E.A. Arbas, N.J. Strausfeld, and J.G. Hildebrand. 1989. Physiology and morphology of projection neurons in the antennal lobe of the male moth *Manduca sexta*. *Journal of Comparative Physiology A* 165:427–453.

Kanzaki, R., E. A. Arbas, and J. G. Hildebrand. 1991a. Physiology and morphology of protocerebral olfactory neurons in the male moth *Manduca sexta*. *Journal of Comparative Physiology A* 168:281–298.

Kanzaki, R., E. A. Arbas, and J. G. Hildebrand. 1991b. Physiology and morphology of descending neurons in pheromone-processing olfactory pathways in the male moth *Manduca sexta*. *Journal of Comparative Physiology A* 169:1–14.

Kanzaki, R., N. Sugin, and T. Shibuya. 1992. Self-generated zigzag turning of *Bombyx mori* males during pheromone-mediated upwind walking. *Zoological Science* 9:515–527.

Kanzaki, R., N. Sugin, and T. Shibuya. 1994. Morphology and physiology of pheromone-triggered flipflopping descending interneurons of the male silkworm moth, *Bombyx mori*. *Journal of Comparative Physiology A* 175:1–14.

Kanzaki, R., K. Soo, Y. Seki, and S. Wada. 2003. Projections to higher olfactory centers from subdivisions of the antennal lobe macroglomerular complex of the male silkmoth. *Chemical Senses* 28:113–130.

Kárpáti, Z., T. Dekker, and B. S. Hansson. 2008. Reversed functional topology in the antennal lobe of the mal European corn borer. *Journal of Experimental Biology* 211:2841–2848.

Kárpáti, Z., S. Olsson, B. S. Hansson, and T. Dekker. 2010. Inheritance of central neuroanatomy and physiology related to pheromone preference in the male European corn borer. *BMC Evolutionary Biology* 10:286–297.

Kelber, A., A. Balkenius, and E. J. Warrant. 2003. Colour vision in diurnal and nocturnal hawkmoths. *Integrative and Comparative Biology* 43:571–579.

Kennedy, J. S. 1940. The visual responses of flying mosquitoes. *Proceedings of the Zoological Society of London A* 109:221–242.

Kennedy, J. S. 1983. Zigzagging and casting as a response to windborne odor: a review. *Physiological Entomology* 8:109–120.

Kennedy, J. S., and D. Marsh. 1974. Pheromone-regulated anemotaxis in flying moths. *Science* 184:999–1001.

Koutroumpa, F. A., Z. Kárpáti, C. Monsempes, S. R. Hill, B. S. Hansson, J. Krieger, and T. Dekker. 2014. Shifts in sensory neuron identity parallel differences in pheromone preference in the European corn borer. *Frontiers in Ecology and Evolution* 2:65. doi:10.3389/fevo.2014.00065.

Kuenen, L. P. S., and R. T. Cardé. 1994. Strategies for reconciling a lost pheromone plume: casting and upwind flight in the male gypsy moth. *Physiological Entomology* 19:15–29.

Lawrence, P. A. 1966. Development and determination of hairs and bristles in the milkweed bug *Oncopeltus fasciatus* (Lygaeidae, Hemiptera). *Journal of Cell Science* 1:475–498.

Lee, S.-G., and T. C. Baker. 2008. Incomplete electrical isolation of sex-pheromone responsive olfactory receptor neurons from neighboring sensilla. *Journal of Insect Physiology* 54:663–671.

Lee, S.-G., M. A. Carlsson, B. S. Hansson, J. L. Todd, and T. C. Baker. 2006a. Antennal lobe projection destinations of *Helicoverpa zea* male olfactory receptor neurons responsive to heliothine sex pheromone components. *Journal of Comparative Physiology A* 192:351–363.

Lee, S.-G., N. J. Vickers, and T. C. Baker 2006b. Glomerular targets of *Heliothis subflexa* male olfactory receptor neurons housed within long trichoid sensilla. *Chemical Senses* 9:821–834.

Lei, H., and B. S. Hansson. 1999. Central processing of pulsed pheromone signals by antennal lobe neurons in the male moth *Agrotis segetum*. *Journal of Neurophysiology* 81:1113–1122.

Lei, H., S. Anton, and B. S. Hansson. 2001. Olfactory protocerebral pathways processing sex pheromone and plant odor information in the male moth *Agrotis segetum*. *Journal of Comparative Neurology* 432:356–370.

Lei, H., J. A. Riffell, S. L. Gage, and J. G. Hildebrand. 2009. Contrast enhancement of stimulus intermittency in a primary olfactory network and its behavioral significance. *Journal of Biology* 8:21.

Lei, H., H.-Y. Chiu, and J. G. Hildebrand. 2013. Responses of protocerebral neurons in *Manduca sexta* to sex-pheromone mixtures. *Journal of Comparative Physiology A* 199:997–1014.

Li, Y., and N. J. Strausfeld. 1997. Morphology and sensory modality of mushroom body extrinsic neurons in the brain of the cockroach, *Periplaneta americana*. *Journal of Comparative Neurology* 387:631–650.

Li, Y., and N. J. Strausfeld. 1999. Multimodal efferent and recurrent neurons in the medial lobe of the cockroach mushroom body. *Journal of Comparative Neurology* 409:647–663.

Linn, C. E., Jr., C. J. Musto, M. J. Domingue, T. C. Baker, and W. L. Roelofs. 2007. Support for (*Z*)-11-hexadecenal as a pheromone antagonist in *Ostrinia nubilalis*: flight tunnel and single sensillum studies with a New York population. *Journal of Chemical Ecology* 33:909–921.

Ljungberg, H., P. Anderson, and B. S. Hansson. 1993. Physiology and morphology of pheromone specific sensilla on the antennae of male and female *Spodoptera littoralis* (Lepidoptera: Noctuidae). *Journal of Insect Physiology* 39:253–260.

Maddess, T., R. A. Dubois, and M. R. Ibbotsen. 1991. Response properties and adaptation of neurons sensitive to image motion in the butterfly, *Papilio aegeus*. *Journal of Experimental Biology* 161:171–199.

Mafra-Neto, A., and R. T. Cardé. 1994. Fine-scale structure of pheromone plumes modulates upwind orientation of flying moths. *Nature* 369:142–144.

Martin, J. P., A. Beyerlein, A. M. Dacks, C. E. Reisenman, J. A. Riffell, H. Lei, and J. G. Hildebrand. 2011. The neurobiology of insect olfaction: sensory processing in a comparative context. *Progress in Neurobiology* 95:427–447.

Martin, J. P., H. Lei, J. A. Riffell, and J. G. Hildebrand. 2013. Synchronous firing of antennal-lobe projection neurons encodes the behaviorally effective ratio of sex pheromone components in male *Manduca sexta*. *Journal of Comparative Physiology A* 199:963–979.

Matsumoto, S. G., and J. G. Hildebrand. 1981. Olfactory mechanisms in the moth *Manduca sexta*: response characteristics and morphology of the central neurons in the antennal lobes. *Proceedings of the Royal Society of London B* 213:249–277.

Milde, J. J. 1993. Tangential medulla neurons in the moth *Manduca sexta*. Structure and responses to optomotor stimuli. *Journal of Comparative Physiology A* 173:783–799.

Mishima, T., and R. Kanzaki. 1998. Coordination of flipflopping neural signals and head turning during pheromone-mediated walking in a male silkworm moth, *Bombyx mori*. *Journal of Comparative Physiology A* 183:273–282.

Mishima, T., and R. Kanzaki. 1999. Physiological and morphological characterization of olfactory descending interneurons of the male silkworm moth, *Bombyx mori*. *Journal of Comparative Physiology A* 184:143–160.

Namiki, S., and R. Kanzaki. 2011. Heterogeneity in dendritic morphology of moth antennal lobe projection neurons. *Journal of Comparative Neurology* 519:3367–3386.

Namiki, S., S. Iwabuchi, and R. Kanzaki. 2008. Representation of a mixture of pheromone and host plant odor by antennal lobe projection neurons of the silkmoth *Bombyx mori*. *Journal of Comparative Physiology A* 194:501–515.

Namiki, S., T. Mitsuko, Y. Seki, T. Kazawa, R. Fukushima, C. Iwatsuki, and R. Kanzaki. 2013. Concentric zones for pheromone components in the mushroom body calyx of the moth brain. *Journal of Comparative Neurology* 521:1073–1092.

Nishino, H., M. Iwasaki, I. Kamimura, and M. Mizunami. 2012a. Divergent and convergent projections to the two parallel olfactory centers from two neighboring, pheromone-receptive glomeruli in the male American cockroach. *Journal of Comparative Neurology* 520:3428–3445.

Nishino, H., M. Iwasaki, K. Yasuyama, H. Hongo, H. Watanabe, and M. Mizunami. 2012b. Visual and olfactory input segregation in the mushroom body calyces in a basal neopteran, the American cockroach. *Arthropod Structure & Development* 41:3–16.

Ochieng', S. A., P. Anderson, and B. S. Hansson. 1995. Antennal lobe projection patterns of olfactory receptor neurons involved in sex pheromone detection in *Spodoptera littoralis* (Lepidoptera: Noctuidae). *Tissue & Cell* 27:221–232.

Okamura, J.-Y., and N. J. Strausfeld. 2007. Visual system of calliphorid flies: motion- and orientation-sensitive visual interneurons supplying dorsal optic glomeruli. *Journal of Comparative Neurology* 500:189–208.

Olberg, R. M. 1983. Pheromone-triggered flip-flopping interneurons in the ventral nerve cord of the silkworm moth, *Bombyx mori*. *Journal of Comparative Physiology* 152:297–307.

Olberg, R. M., and M. A. Willis. 1990. Pheromone-modulated optomotor response in male gypsy moths, *Lymantria dispar* L.:

directionally selective visual interneurons in the ventral nerve cord. *Journal of Comparative Physiology A* 167:707–714.

Pearson, L. 1971. The corpora pedunculata of *Sphinx ligustri* L. and other Lepidoptera: an anatomical study. *Philosophical Transactions of the Royal Society of London B Biological Sciences* 259:477–516.

Preiss, R., and L. Futschek. 1985. Flight stabilization by pheromone-enhanced optomotor responses. *Naturwissenschaften* 72:435–436.

Preiss, R., and E. Kramer. 1983. Stabilization of altitude and speed in tethered flying gypsy moth males: influence of (+) and (–) disparlure. *Physiological Entomology* 8:55–68.

Quero, C., H.Y. Fadamiro, and T.C. Baker. 2001. Responses of male *Helicoverpa zea* to single pulses of sex pheromone and behavioural antagonist. *Physiological Entomology* 26:106–115.

Reisenman, C.E., T. Heinbockel, and J.G. Hildebrand. 2008. Inhibitory interactions among olfactory glomeruli do not necessarily reflect spatial proximity. *Journal of Neurophysiology* 100:554–564.

Reisenman, C.E., A.M. Dacks, and J.G. Hildebrand. 2011. Local interneuron diversity in the primary olfactory center of the moth *Manduca sexta*. *Journal of Comparative Physiology A* 197:653–665.

Rø, H., D. Müller, and H. Mustaparta. 2007. Anatomical organization of antennal lobe projection neurons in the moth *Heliothis virescens*. *Journal of Comparative Neurology* 500:658–675.

Seki, Y., and R. Kanzaki. 2008. Comprehensive morphological identification and GABA immunocytochemistry of antennal lobe local interneurons in *Bombyx mori*. *Journal of Comparative Neurology* 506:93–107.

Seki, Y., H. Aonuma, and R. Kanzaki. 2005. Pheromone processing center in the protocerebrum of *Bombyx mori* revealed by nitric oxide-induced anti-cGMP immunocytochemistry. *Journal of Comparative Neurology* 481:340–351.

Shimohigashi, M., and Y. Tominaga. 1991. Identification of UV, green and red receptors, and their projection to lamina in the cabbage butterfly, *Pieris rapae*. *Cell & Tissue Research* 263:49–59.

Shimohigashi, M., and Y. Tominaga. 1999. Synaptic organization in the lamina of the superposition eye of a skipper butterfly, *Parnara guttata*. *Journal of Comparative Neurology* 408:107–124.

Sinakevitch, I., M. Sjöholm, B.S. Hansson, and N.J. Strausfeld. 2008. Global and local modulatory supply to the mushroom bodies of the moth *Spodoptera littoralis*. *Arthropod Structure & Development* 37:260–272.

Sjöholm, M., I. Sinakevitch, R. Ignell, N.J. Strausfeld, and B.S. Hansson. 2005. Organization of Kenyon cells in subdivisions of the mushroom bodies of a lepidopteran insect. *Journal of Comparative Neurology* 491:290–304.

Sjöholm, M., I. Sinakevitch, N.J. Strausfeld, R. Ignell, and B.S. Hansson. 2006. Functional division of intrinsic neurons in the mushroom bodies of male *Spodoptera littoralis* revealed by antibodies against aspartate, taurine, FMRF-amide, Mas-allatotropin and DC0. *Arthropod Structure & Development* 35:153–168.

Stavenga, D.G., and K. Arikawa. 2006. Evolution of color and vision of butterflies. *Arthropod Structure & Development* 35:307–318.

Strausfeld, N.J. 1970. Golgi studies on insects part II: the optic lobes of Diptera. *Philosophical Transactions of the Royal Society of London B Biological Sciences* 258:135–223.

Strausfeld, N.J. 2003. Brain and optic lobes. Pp. 121–130. In V.R. Resh and R.T. Cardé, eds. *Encyclopedia of Insects*. Cambridge, MA: Academic Press.

Strausfeld, N.J. 2012. Beneath the faceted eye. Pp. 125–184. In N.J. Strausfeld, ed. *Arthropod Brains: Evolution, Functional Elegance, and Historical Significance*. Cambridge, MA: Belknap Press of Harvard University Press.

Strausfeld, N.J., and A.D. Blest. 1970. Golgi studies on insects. Part I: the optic lobes of Lepidoptera. *Philosophical Transactions of the Royal Society of London Series B Biological Sciences* 258:81–134.

Strausfeld, N.J., and J.A. Campos-Ortega. 1977. Vision in insects: pathways possibly underlying neural adaptation and lateral inhibition. *Science* 195:894–897.

Strausfeld, N.J., and Y. Li. 1999. Organization of olfactory and multimodal afferent neurons supplying the calyx and pedunculus of the cockroach mushroom bodies. *Journal of Comparative Neurology* 409:603–625.

Strausfeld, N.J., and J.-Y. Okamura. 2007. Visual system of calliphorid flies: organization of optic glomeruli and their lobula complex efferents. *Journal of Comparative Neurology* 500:166–188.

Strausfeld, N.J., and F. Hirth. 2013. Deep homology of arthropod central complex and vertebrate basal ganglia. *Science* 340:157–161.

Strausfeld, N.J., I. Sinakevitch, and J.-Y. Okamura. 2007. Organization of local interneurons in optic glomeruli of the dipterous visual system and comparisons with the antennal lobes. *Developmental Neurobiology* 67:1267–1288.

Strausfeld, N.J., I. Sinakevitch, S.M. Brown, and S.M. Farris. 2009. Ground plan of the insect mushroom body: functional and evolutionary implications. *Journal of Comparative Neurology* 513:265–291.

Szyszka, P., M. Ditzen, A. Galkin, C.G. Galizia, and R. Menzel. 2005. Sparsening and temporal sharpening of olfactory representations in the honeybee mushroom bodies. *Journal of Neurophysiology* 94:3303–3313.

Todd, J.L., and T.C. Baker. 1999. Function of peripheral olfactory organs. Pp. 67–96. In B.S. Hansson, ed. *Insect Olfaction*. Berlin: Springer-Verlag.

Todd, J.L., K.F. Haynes, and T.C. Baker. 1992. Antennal neurones specific for redundant pheromone components in normal and mutant *Trichoplusia ni* males. *Physiological Entomology* 17:183–192.

Todd, J.L., S. Anton, B.S. Hansson, and T.C. Baker. 1995. Functional organization of the macroglomerular complex related to behaviourally expressed olfactory redundancy in male cabbage looper moths. *Physiological Entomology* 20:349–361.

Trona, F., G. Anfora, M. Bengtsson, P. Witzgall, and R. Ignell. 2010. Coding and interaction of sex pheromone and plant volatile signals in the antennal lobe of the codling moth *Cydia pomonella*. *Journal of Experimental Biology* 213:4291–4303.

Trona, F., G. Anfora, A. Blkenius, M. Bengtsson, M. Tasin, A. Knight, N. Janz, P. Witzgall, and R. Ignell. 2013. Neural coding merges sex and habitat chemosensory signals in an insect herbivore. *Proceedings of the Royal Society of London B* 280:20130267.

Utz, S., Huetteroth, W., Vömel, M., and J. Schachtner. 2008. Mas-allatotropin in the developing antennal lobe of the sphinx moth *Manduca sexta*: distribution, time course, developmental regulation, and colocalization with other neuropeptides. *Developmental Neurobiology* 68:123–142.

Varela, N., J. Avilla, C. Gemeno, and S. Anton. 2011. Ordinary glomeruli in the antennal lobe of male and female tortricid moth *Grapholita molesta* (Busck) (Lepidoptera: Tortricidae) process sex pheromone and host-plant volatiles. *Journal of Experimental Biology* 214:637–645.

Vickers, N.J. 2006. Inheritance of olfactory preferences. III. processing of pheromonal signals in the antennal lobe of *Heliothis subflexa × Heliothis virescens* hybrid male moths. *Brain Behavior & Evolution* 68:90–108.

Vickers, N.J., and T.C. Baker. 1992. Male *Heliothis virescens* sustain upwind flight in response to experimentally pulsed filaments of their sex-pheromone. *Journal of Insect Behavior* 5:669–687.

Vickers, N.J., and T.C. Baker. 1994. Reiterative responses to single strands of odor promote sustained upwind flight and odor source location by moths. *Proceedings of the National Academy of Sciences of the United States of America* 91:5756–5760.

Vickers, N.J., and T.C. Baker. 1997. Chemical communication in heliothine moths. VII. Correlation between diminished responses to point-source plumes and single filaments similarly tainted with a behavioral antagonist. *Journal of Comparative Physiology A* 180:523–536.

Vickers, N.J., and T.A. Christensen. 2003. Functional divergence of spatially conserved olfactory glomeruli in two related moth species. *Chemical Senses* 28:325–338.

Vickers, N.J., T.A. Christensen, and J.G. Hildebrand. 1998. Combinatorial odor discrimination in the brain: attractive and antagonist odor blends are represented in distinct combinations of uniquely identifiable glomeruli. *Journal of Comparative Neurology* 400:35–56.

Vickers, N.J., T.A. Christensen, T.C. Baker, and J.G. Hildebrand. 2001. Odour-plume dynamics influence the brain's olfactory code. *Nature* 410:466–470.

Vickers, N.J., K. Poole, and C.E. Linn, Jr. 2005. Plasticity in central olfactory processing and pheromone blend discrimination following interspecies antennal imaginal disc transplantation. *Journal of Comparative Neurology* 491:141–156.

Wada, S., and R. Kanzaki. 2005. Neural control mechanisms of the pheromone-triggered programmed behavior in male silk-moths revealed by double-labeling of descending interneurons and a motor neuron. *Journal of Comparative Neurology* 484:168–182.

Waldrop, B.R., T.A. Christensen, and J.G. Hildebrand. 1987. GABA-mediated synaptic inhibition of projection neurons in the antennal lobes of the sphinx moth, *Manduca sexta*. *Journal of Comparative Physiology A* 161:23–32.

Wicklein, M., and N.J. Strausfeld. 2000. Organization and significance of neurons that detect change of visual depth in the hawk moth *Manduca sexta*. *Journal of Comparative Neurology* 424:356–376.

Wicklein, M., and D. Varju. 1999. Visual system of the European hummingbird hawkmoth, *Macroglossum stellatarum* (Sphingidae, Lepidoptera): motion-sensitive interneurons of the lobula plate. *Journal of Comparative Neurology* 408:272–282.

Willis, M.A., and E.A. Arbas. 1991. Odor-modulated upwind flight of the sphinx moth, *Manduca sexta* L. *Journal of Comparative Physiology A* 169:427–440.

Wu, W., S. Anton, C. Löfstedt, and B.S. Hansson. 1996. Discrimination among pheromone component blends by interneurons in male antennal lobes of two populations of the turnip moth, *Agrotis segetum*. *Proceedings of the National Academy of Sciences of the United States of America* 93:8022–8027.

Yang, Z., M. Bengtsson, and P. Witzgall. 2004. Host plant volatiles synergize response to sex pheromone in codling moth, *Cydia pomonella*. *Journal of Chemical Ecology* 30:619–629.

CHAPTER ELEVEN

Moth Navigation along Pheromone Plumes

RING T. CARDÉ

Introduction

One of the most remarkable adaptations of moths is the ability of a male to detect the scent of a conspecific female over considerable distances—from tens to perhaps hundreds of meters upwind—and then navigate a course along the pheromone plume to her side. This mate-finding system likely enables moths to persist at densities far lower than insects chiefly reliant on either visual or acoustic signals. The great distances over which moths can communicate by pheromones and the notable species specificity of the pheromone message in concert likely have fostered the profusion of moth species, currently over 140,000. The navigational mechanisms that moths use in tracking plumes have been the subject of many experimental studies, mainly conducted in wind-tunnel settings. This artificial environment allows systematic manipulations of the interplay of visual cues, the structural and chemical composition of the pheromone plume, and the airflow regime. Because the vast majority of moths are nocturnal, field observations and recordings of orientation have been limited almost exclusively to a handful of diurnal species. This review summarizes our current understanding of the mechanisms used by moths to find sources of female pheromone, in part, focusing on how differences in habitat, light levels, and structural features of the plume and the wind that propels it modulate the cues available for navigation. Recent reviews of finding and navigating along plumes by flying insects include those of Kaissling (1997), Vickers (2000, 2006), Cardé and Willis (2008), and Cardé and Gibson (2010). Elkinton and Cardé (1984) and Murlis et al. (1992) have reviewed the structure of wind-borne odor plumes. The odor-guided maneuvers that males use to locate a female are also very likely used by some female moths in their upwind flight to host-plant odors (e.g., Mechaber et al. 2002; Riffell et al. 2009). Figure 11.1 depicts a generalized model of a moth finding and orienting along a plume, starting with ranging flight to contact the plume, zigzag upwind flight while in plume contact, casting if contact with the plume is lost, and an upwind surge when filaments of pheromone are contacted at rates above 5 Hz. These elements of mate location and their sensory inputs will be considered in subsequent sections.

That the ancestral state in Lepidoptera is female attraction of males has been attributed to a disparity in parental investment (Cardé and Baker 1984; see also Trivers 1972; Thornhill 1979). The number of a male's offspring is dependent on how many matings can be secured, whereas a female's production is limited by her egg complement. Males produce more gametes than females, but females furnish each egg with nutrients, thereby investing more per gamete. Males therefore

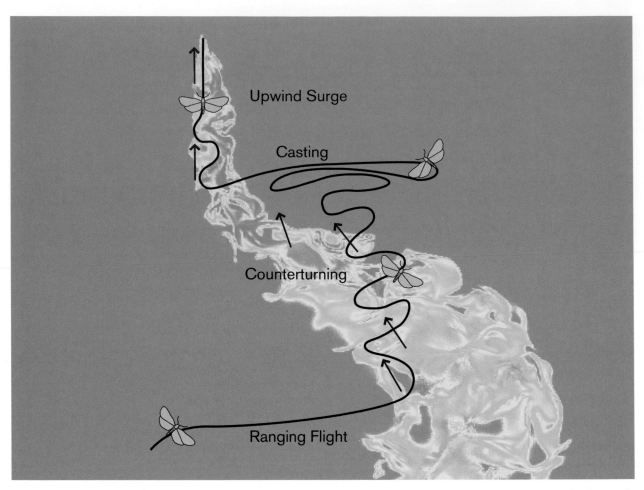

FIGURE 11.1 Template of moth maneuvers as governed by sequential interactions with filaments of pheromone, encounters with "clean air," and wind flow. White areas depict fluxes of pheromone above threshold. In this representation, the wind changes direction but not velocity, so that the wind vectors always point at the source of pheromone as in figure 11.3B. (See figure 11.3C for the effects of changes of wind direction *and* velocity.) Initially a moth in ranging flight encounters the plume and turns upwind. The flight has a zigzag form because the moth encounters filaments at a fairly low rate (<5 Hz). Note that the heading of the moth differs from its course (see figure 11.4). Where the plume snakes left because of a shift in wind direction, the moth continues due upwind, exiting the plume. Casting allows the moth to recontact the plume. A straightened-out surge upwind to the source is caused by high (>10 Hz) rates of encounter with filaments. Very near the source, velocity slows, possibly because of an increase in overall concentration and a narrowing of the plume's outer envelope. The flight track depicted is diagrammatic and not to a consistent scale. Small zigzags occurring several times a second would not be visible.

should compete in finding the limiting resource of females, with those males being able to detect and navigate to females from afar being at a reproductive advantage. A male's detection of volatile chemicals emitted by females likely was an initial stage in the origin of this mate-finding system.

There are, however, many secondarily evolved exceptions to the usual convention of females luring males by a pheromone message. Male wax moths (Pyralidae, Gallerinae) attract females with a pheromone from wing glands that is coupled with an acoustic signal produced by wing fanning (e.g., Greenfield and Coffelt 1983; Kindl et al. 2011). These males generally call from a source of larval food and therefore "offer" a resource beyond sperm. Males of the ghost moth *Hepialus humuli* (Hepialidae) emit pheromone from brush-like organs on their metathoracic tibae and hover in leks. When a female enters the lek, usually one male will follow her until she settles on the ground and copulation then ensues (Mallet 1984). Hepialids generally have a very high egg laying capacity (29,000 recorded from a single *Trictena* female by Tindale [1932]), so that they are not the limited sex in terms of parental investment; many species of hepialid, however, follow the usual convention of male attraction to females. Males of

Estigmene acrea (Erebidae, Arctiinae) attract females in early evening with pheromone released from everted coremata (sometimes in leks), whereas later on in the evening sexual roles are reversed, with the usual convention of females luring males (Willis and Birch 1982; Jordon et al. 2005). Dependent on their host plant, larvae often ingest a pyrrolizidine alkaloid that is converted into the "courtship" pheromone hydrodanaidal (Jordon et al. 2005). Perhaps hydrodanaidal-rich males offer a nuptial gift of significance (such as a contribution of alkaloids to her defensive complement and possibly to her eggs). In the Castinidae, long-distance communication by pheromones has been abandoned in favor of a butterfly-like strategy for mate seeking (Sarto i Monteys et al. 2012). In these diurnal moths, patrolling males locate females by visual signals and females lack the sex pheromone-producing gland that typifies the ancestral state of ditrysian moths.

Structure of Wind-Borne Pheromone Plumes

The active space—the region where pheromone concentration is above behavioral threshold—of wind-formed odor

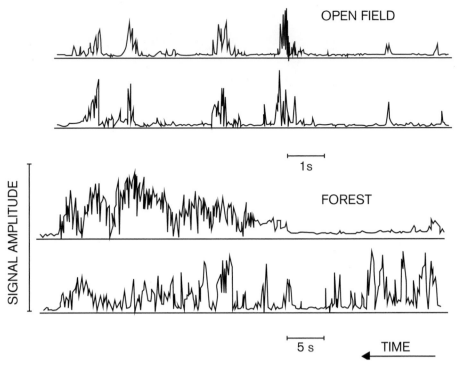

OPEN FIELD

SIGNAL AMPLITUDE

1 s

FOREST

5 s   TIME

FIGURE 11.2 The fine-scale features of odor plumes dispersed in open forest and open field environ-ments. Surrogate plumes were generated from a point source of negatively charged ions and their intensities measured downwind with Langmuir probes. Amplifier gains are adjusted for each distance. Note that the instantaneous strength of the signal is highly variable in both habitats and that when the signal is present it persists in the forest for longer intervals than in the open field (from Murlis et al. 2000).

plume is often depicted as a smooth surfaced, semi-ellipsoid that is drawn out along the wind's mean direction (Bossert and Wilson 1963). The envelope of the active space is set at behavioral threshold. The Bossert and Wilson depiction of the active space was a time average of concentration (3 min in the case of the Sutton equation used by Bossert and Wilson), rather than the instantaneous concentration that dictates behavioral response. Although the relevance to plume model-ing of instantaneous concentration to behavior was noted by Aylor et al. (1976) and Miksad and Kittredge (1979), Murlis and Jones (1981) were the first to measure the instantaneous intensity of a surrogate odor plume at a time and spatial scale relevant to moth perception of pheromone. They used a small source of negatively charged ions and measured signal inten-sity with detectors positioned at distances of up to 15 m downwind in an open field. The odor plume crossed a fixed downwind sampling spot in discrete bursts, typically ≈0.1 s long and ≈0.5 s apart. The signal when present is highly inter-mittent, being comprised of discrete bursts that vary in inter-nal structure and intensity (figure 11.2).

The orientation maneuvers that moths use in plume track-ing thus must contend with a plume that the forces of turbu-lent diffusion have fragmented into wispy filaments while simultaneously creating gaps free of pheromone or at least having concentrations of pheromone that fall below the threshold of detection (Murlis and Jones 1981; Murlis et al. 1992, 2000). As the plume is transported downwind, rela-tively high concentrations of pheromone can persist within some filaments for many meters, thereby enabling detection over considerable distances. Plume fragmentation and dilu-tion can be accentuated by passing through foliage; if the

calling female is situated within the canopy, the plume is immediately shredded as it passes out through the surround-ing foliage. Such gaps in the pheromone plume obviously can impede tracking, particularly when contact with pheromone is lost for several seconds. Moreover, shifts in wind direction mean that the direction upwind and along the plume's long axis can be misaligned (David et al. 1982; Elkinton et al. 1987; Brady et al. 1989), so that heading upwind within the plume's overall envelope often can lead a moth out of the plume (figure 11.3). Objects in the plume's path also can degrade the plume's filamentous structure or cause the plume to flow around them. When the pheromone signal is present, turbu-lent forces also cause its intensity to fluctuate markedly many times within a second, a feature of the signal that can be cru-cial to orientated response.

The structure of the signal also is influenced by habitat and time of day. In open fields, the signal persists at a given down-wind sampling point for shorter intervals than in forests (Murlis et al. 2000). Measurements by Thistle et al. (2004) of plume dispersal using a surrogate odor in three kinds of forest habitats suggest that plumes meander more widely from late morning to late afternoon when the boundary layer is unsta-ble. Conversely, when atmospheric conditions are relatively stable, as typically occurs near dusk and dawn, then the plume has more directional consistency and contains higher odor concentrations. In daytime, plumes may be subject to buoyancy caused by air rising in pockets warmed by the sun, such as occurs in forest openings (Fares et al. 1983). Upward vertical movement of surrogate odor plumes also has been documented at nighttime in tropical rain forests (Schal 1982) and in an orchard (Girling et al. 2013).

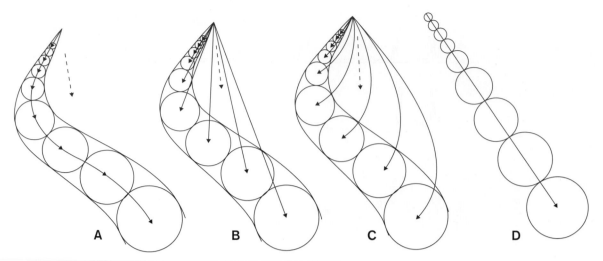

FIGURE 11.3 Diagrammatic view from above of pheromone dispersal under steady and shifting wind directions. The instantaneous wind direction at given points in the plume is indicated by the arrows and the gradual expansion of a simple, time-averaged plume of pheromone is represented as circular "puffs."

A Represents what many assumed (incorrectly) was the relationship between a "snaking" plume and the instantaneous wind direction, prior to the studies of David et al. (1982), Elkinton et al. (1987), and Brady et al. (1989).

B Illustrates the effects of a shifting wind direction when the wind velocity holds steady, following David et al. (1982).

C Gives a shifting wind direction when the wind velocity varies.

D Depicts dispersion with an unvarying wind direction. When the long axis of the plume and the upwind direction are so aligned, an insect can make rapid progress toward the odor source (from Elkinton et al. 1987).

In contrast to turbulent diffusion, molecular diffusion plays essentially no role in either plume structure or the ratio of its components. A 16-carbon, straight-chain alcohol (hexadecanol), for example, has a diffusion coefficient in air of only 2.5×10^{-3} mm s^{-1} (Loudon 2003), so that the distribution of molecules within the plume is governed almost entirely by turbulent forces (but see the hypothesis of Symonds et al. [2011] that ties elaborate antennal structure in male moths to lower molecular weight pheromones with increased diffusivity). The ratio of components within the plume thus should remain essentially constant over time. The diffusivity of a compound is the inverse of the square root of the compound's molecular weight (Loudon 2003), and so, even when components of a pheromone blend differ by two or even four carbons, as is the case with *Agrotis segetum* (Noctuidae), their diffusivities will be very similar.

In *Heliothis virescens* (Noctuidae) and likely in other moths, the complete pheromone blend needs to be present in each filament of pheromone to sustain upwind flight (Vickers and Baker 1992). The fine-scale structure of pheromone plumes is crucial to both moment-to-moment in-flight maneuvers and the overall flight path, as will be considered shortly.

A final issue in plume description is how to characterize stimulus intensity. Concentration is the most oft-used metric and this measure is appropriate in still air for a stationary receiver, but when the air moves across the sensor because of airflow or movement of the insect (or both), then flux provides an accurate measure of molecules impinging on the antennae. Given a source with a constant rate of emission and neglecting the effect of wind velocity on plume structure, however, flux at a given downwind sampling point will remain fairly constant regardless of wind speed, so that the effect of dilution is effectively cancelled (Elkinton et al. 1984).

In the case of the silkworm moth *Bombyx mori* (Bombycidae), the concentration threshold evoking activation within 1 s is about 1000 molecules cm^{-3} in air stream of 60 cm s^{-1} (Kaissling 1996). (This entirely domesticated species is no longer capable of flight; pheromone response in this species is measured by wing fanning, antennal movement, and walking.) Each antenna possesses about 17,000 sensory hairs. Given the area of the antennae impinging on the plume, this suggests a per second flux of about 4 molecules per 100 sensory hairs, in turn evoking nerve impulses of 1 per 100 sensory hairs per second. Kaissling (1996) calculated that if 1% of the receptor cells fire one impulse per second in response to pheromone, then the Central Nervous System (CNS) has detected this elevation over the spontaneous firing rate of 0.1 per cell per second. Thus, the CNS is detecting the 170 pheromone-induced firings over the 1700 spontaneous impulses produced by the 17,000 receptor cells of one antenna. How this extraordinary sensitivity translates into distance of communication in the field in its pre-domesticated flying ancestor has not been measured. In this review usually concentration is used as the metric for stimulus intensity, but it is recognized that flux generally provides the most accurate description of stimulus intensity.

Finding Pheromone Plumes

Although a pheromone plume might engulf a stationary (quiescent) male, generally it is assumed that males encounter a pheromone plume while engaged in ranging or as it is occasionally called "appetitive" flight. (Kennedy [1986] cautioned against the use of the term appetitive as it is a teleological concept.) As the plume elongates in shape as it is carried downwind (assuming a relatively invariant wind direction), males that favor a crosswind trajectory should increase their probability of initial contact with the plume while decreasing the time to and energetic cost of its location (Cardé 1981). As characterized by Greenfield (1981), males are in "a race to locate females." The optimal strategies for rapid plume contact and minimizing energetic cost have been explored as theoretical models (Sabelis and Schippers 1984; Dusenbery 1989, 1990; Cardé et al. 2012; Bau and Cardé 2015). Depending on a given model's assumptions of energetic costs of flight, the speed and directional variability of the winds, the resulting pattern of plume dispersion, and a moth's dubious ability to keep track of its and the wind direction's recent cardinal positions over more than several seconds, the models can predict trajectories for rapid contact that could favor either upwind, downwind, or crosswind headings. Keeping track of current and past headings, so as to head upwind when the wind direction varies over more than 60° and therefore the plume becomes very wide, as proposed in the Sabelis and Schippers (1984) model, presumably exceeds the computational and positional memory ability of any moth.

We have, however, relatively few field observations that bear on this issue. In apparent absence of nearby calling females, diurnal male gypsy moths, *Lymantria dispar* (Erebidae, Lymantriine) (Elkinton and Cardé 1983), and diurnal male bog tiger moths *Virbia lamae* (Erebidae, Arctiinae) (Cardé et al. 2012) flew random directions in the horizontal plane with respect to current wind flow. In the case of *V. lamae*, further analysis showed that random headings of flight trajectories with respect to contemporaneous wind flow was in effect a de facto crosswind strategy, as the two crosswind sectors combined encompassed twice the total degree bearings of either the upwind or downwind sectors. Another experimental approach used harmonic radar tracking of individual *Agrotis segetum* (Noctuidae) males. A general downwind displacement of moths prior to upwind plume following was detected (Reynolds et al. 2007), but at coarser spatial and temporal scales than in the *Lymantria* and *Virbia* studies. Over time, some downwind drift is to be expected, if a moth's heading with respect to current wind direction is random. Given, however, that we have few field observations on the heading preference of male moths in steady or shifty winds, and those that are available all differ in their temporal precision and the spatial scale of measuring moth and simultaneous wind movement, the question of whether males employ optimal strategies for contacting a pheromone plume remains unresolved. An alternative strategy would be to assess *current* variability (e.g., over the last several seconds) in wind direction and then direct ranging flight along the wind line when direction is relatively stable and crosswind when it is not, as seems to be the strategy adopted by *Drosophila* flies using odor to search for food based on wind-tunnel manipulations (Zanen et al. 1994).

Distance of Communication

In the eighteen century, London Aurelians used the principle of "sembling" to collect day-flying moths. Females were taken

> "in a box with a gauze lid into the vicinity of the woods, where, if the weather be favourable, she never fails to attract a numerous train of males, whose only business appears to be an incessant, rapid undulating flight in search of the females. One of these is no sooner descried, than they become so enamored of their fair kinswoman, as absolutely to lose all fear for their own personal safety." (HAWORTH 1832)

It was recognized then that males used "the sense of smell" in their discovery of the females' location. The French naturalist

Jean-Henri Fabre (1916) confirmed the power of the female's scent by caging females of the great peacock moth (*Saturnia pyri*) (Saturniidae), luring as many as 150 males to his house in an evening. Fabre presumed that males traveled "from afar, very far, within a radius of a mile and a half or more." Fabre equivocated, however, about whether males were lured by female odors or their creation of a mysterious electrical field.

Just how far males actually perceive an upwind female and then successfully navigate a course to her side has been much debated (reviewed by Wall and Perry 1987). The distance or range of communication seems to be greatest in moth groups such as the Saturniidae and Lymantriinae, clades with males that possess plumose antennae fitted with tens of thousands of olfactory sensilla and therefore, it is presumed, correspondingly low threshold for pheromone perception (see also Symonds et al. 2011 for a listing of other moth groups with elaborate antennae). Before considering navigational mechanisms that enable such orientation, it will be constructive to consider the available evidence on just how far females actually lure males.

The most common approach for establishing this distance has been to release males at some distance from a caged female (or a source of synthetic pheromone, ideally set to release at a female's rate) and then determine what proportion of males locate the source of pheromone and how quickly they do so. If location is fairly rapid, meaning that the time from release to arrival at the source (transit time) suggests that the flight was direct given the moths potential flight speed, this is solid evidence that pheromone was detected at very near to the time of release and plume following was continuous. Ideally, these observations would be accompanied by enough information on the direction of airflow to verify that the wind was generally from the pheromone source toward the release site. When the transit time is long enough to suggest that flight was not continuous, however, then there are two other possibilities: either males may have flown some proportion of the flight "randomly," that is not under the guidance of pheromone, and then entered the plume; alternatively, males may have lost contact with the plume along the way (possibly many times) and then recontacted the plume, and continued upwind orientation.

The difficulty of choosing among these possibilities is illustrated by Mell's (1922) experiments in southern China with the saturniid *Actias selene*. Marked males were released 4 and 11.6 km from his house where three caged females were held. Remarkably, 8 of 20 and 4 of 15 of the males, respectively, located the females by 23:00 that evening. Of course the time of release, wind speed, and wind direction would also provide important information. The substantial proportion of males arriving is incomplete evidence of potential communication over these distances.

Rau and Rau (1929) conducted extensive experiments in St. Louis, Missouri, with several saturniid species, releasing males at various distances from their house where they caged virgin females on the roof, providing an omnidirectional dispersal of pheromone. As expected, the farther away the site of release from the females, generally the lower the proportion of males finding their way to the females. Rau and Rau were fully aware of the difficulties in interpreting communication distance. Transit times, in so far as they could be calculated, generally suggested that the flight might not be continuous and direct. For example, with *Hyalophora cercopia* (Saturniidae) 16 of 140 released males (11%) found their way to the caged females over 4.8 km. At these longer distances, however, males sometimes reached the female only on the second, third, or even fourth night after release. As captivating as these experiments are, they do not enable us to distinguish between a random flight that fortuitously enters a detectable pheromone plume close to the source versus detection and successful orientation from the point of release.

Similar experiments with the gypsy moth also were suggestive of communication over hundreds of meters. Males were marked and time of capture at female-baited stations recorded (Fernald 1896). In all but one trial, each trap housed 5–12 females, but in most trials none of the males released in the putative downwind direction reached the traps. In one experiment, one of eight released males reached two cages each holding six females over 402 m distant in 40 min. If flight were continuous and direct, it would have a velocity of 0.17 m s^{-1}. Using a single female as a lure, 1 of 6 released males reached the female over 390 m in 80 min (0.08 m s^{-1}) and 1 of 50 released males reached the female over 652 m in 105 min (0.10 m s^{-1}). These minimum flight velocities are not consistent with continuous and direct flight to females; males orienting along a plume in the field have a flight velocity around 0.5 m s^{-1} (Willis et al. 1994).

Collins and Potts (1932) conducted similar field experiments, but these too are difficult to interpret. Some tests used caged females on offshore New England islands free of gypsy moth infestations; arriving males were assumed to have been attracted over open water from the nearest known infestations. Other trials released marked males and assumed attraction over the distance of transit. Perhaps the best evidence comes from one experiment in which 4 of 75 males released 1.6 km away reached the caged females within 55–105 min. The fastest male traversed this distance at approximately 0.5 m s^{-1}, a velocity (Willis et al. 1994) that is consistent with a fairly direct course and continuous flight. There remain two problems with this interpretation. Initially, some males may have headed in the direction of the caged females without being influenced by pheromone, at some unknown point sensed pheromone, and then oriented along the plume. Furthermore, Collins and Potts used varying numbers of females per cage (when stated, from 15 to over 50!), so that how far an individual female gypsy moth communicates cannot be inferred in these trials.

Elkinton et al. (1984) used arrays of individually caged gypsy moth males in a forest with an open understory to establish both distance of detection (monitored by the initiation of wing fanning) and, in further tests (Elkinton et al. 1987), subsequent orientation of activated males to a source of synthetic pheromone located up to 120 m downwind. At this range, 65% of males reacted to pheromone by departure from their release cages. Of these, 8% reached the source in an average minimum transit speed (per replicate) of about 0.2 m s^{-1} and one of the 390 activated males traveled to the source within 3.5 min at a rate of 0.6 m s^{-1}. This male traversed the 120 m during an interval when the wind direction held very steady from the pheromone source toward the males. The comparatively long transit times for most other males indicate that they did not track plumes continuously, presumably either because they encountered gaps within the plume or because a varying wind direction would at times lead them out of the plume, or both. The 100 μg pheromone source used here would have exceeded the release rate from calling females (Elkinton et al. 1987). Together these findings coupled with the early experiments of Fernald (1896) and Collins and Potts (1932) suggest that the range of attraction of

a female gypsy moth in terms of continuous detection and orientation by males would surpass 100 m infrequently.

A variant on this method has been to cage males individually at set distances away from the pheromone source and measure perception by the proportion of males that are activated; in some trials, males are then permitted to orient to the pheromone source. Wall and Perry (1987) found that day-active male pea moths (*Cydia nigricana*) (Tortricidae) were activated from quiescence by a 100 μg lure situated as distant as 500 m upwind in an open field, and, if released, seemed to orient upwind from 200 m downwind. How closely a 100 μg lure mimics a female's rate of emission is not known, but it could be well above the female's natural rate.

Baker and Roelofs (1981) observed activation and orientation of males of the closed related *Grapholita molesta* (Tortricidae) at various distances downwind of synthetic sources charged with 1, 10, 100, and 1000 μg of pheromone. Although there was some daily variation principally due to temperature effects on male threshold, males were activated a mean of up to 77 m downwind with the 1000 μg stimulus, but at this dose males terminated flight approximately ≈1.5 m from the source; similarly a 100 μg source evoked orientation up to a mean of 29 m downwind, but again the high concentration of pheromone in the plume caused cessation of orientation ≈20 cm from the source. A 10 μg stimulus evoked the closest flight to the source (≈2 cm) with an average maximum distance of 12 m for initiation of upwind flight. These tests were conducted in an open field that would have allowed an accurate estimate of the distance of communication uninfluenced by foliage interfering with the plume's dispersion. A 10 μg source emits pheromone at 1.2 ng hr^{-1} and 100 μg source at 12 ng hr^{-1} (Baker et al. 1980), whereas females release at a mean of 8.5 ng hr^{-1}, ranging up to 25.3 ng hr^{-1} (Lacey and Saunders 1992). Together these observations place the upper limit of distance of communication of females as somewhat beyond 12 m. This tortricid may be somewhat atypical in that its upwind flight is terminated by a sharply defined upper boundary of pheromone concentration.

Taken together, these field experiments do not provide a definitive answer to how far female moths communicate their presence. It is likely that active species vary among species from a few meters to perhaps a kilometer or more, with saturniid moths most likely to represent the upper end. It is also possible to define distance of communication to be the distance of initial detection, even if flight to the source is not continuous and the proportion of responders eventually locating the source is low. There remains, however, some tendency to conflate distance of male flight with distance of attraction, leading, for example, Morton (2009) to interpret flight of a marked *Callosamia promethea* (Saturniidae) male over 33 km over 3 days before capture in a female-baited trap as attraction over this distance.

Clearly females could expand their range of communication by emitting more pheromone and males might respond to females from greater distances by possessing a lower pheromone threshold. Directional selection for increased distance of communication, however, may be counterbalanced by added biosynthetic costs for the female (see Foster, this volume) coupled with a diminishing probability of males being able to navigate a course successfully over these extended ranges. Greenfield (1981) advanced the counterintuitive hypothesis that it could be advantageous for females to emit *less* pheromone, because then females might select for the most sensitive males and therefore the most competent mate-finders. In so doing, however, females might diminish their opportunity to secure a mate. Indeed, at low population levels not all females lure a mate (e.g., Sharov et al. 1995), suggesting a countervailing selective pressure to increase pheromone emission.

Orientation along Wind-Borne Plumes

Role of Optomotor Anemotaxis

The first experimental demonstration of how flying insects orient upwind when stimulated by an odor was performed by Kennedy (1940) using a small wind tunnel and the yellow fever mosquito, *Aedes aegypti*. Using his breath (and therefore an elevation in carbon dioxide and possibly other odors) as a stimulus and a projected floor pattern that could be moved with or opposite the direction of the airflow, it was possible to simulate the visual sensation of displacement upwind or downwind. Flight directly upwind generates a front-to-rear image flow below, whereas an upwind heading off of the wind line adds transverse image flow. By moving the floor pattern with and against the airflow, Kennedy verified that female yellow fever mosquitoes gauged their movement in airflow, not by mechanoreception, but by a visual feedback system. Indeed, it is not possible to gauge the direction of flow of air by simple mechanoreception, unless an organism has contact with a substrate. Another widely accepted mechanism assumes that moths follow an increasing concentration gradient, but at distances of decimeters and beyond from the source, the concentration gradient is too shallow and variable to be used as a reliable orientation cue (Murlis et al. 1992). Using flow of the visual surround to gauge wind direction is an optomotor reaction and this navigation system is termed optomotor anemotaxis. Farkas and Shorey (1972) challenged this mechanism, suggesting instead that male pink boll-worms, *Pectinophora gossypiella* (Gelechiidae), oriented along wind-borne pheromone plumes simply by sensing the plume's boundaries as they zigzagged upwind (termed transverse klinokinesis) rather than by using airflow that is detected by the optomotor response. This contention seemed supported by the ability of some males to continue tracking along a wind-generated plume after airflow in the tunnel was suddenly cut off and, in their experiments, the failure of males to respond to movement of the wind tunnel's floor pattern.

This explanation was refuted by Kennedy and Marsh's (1974) demonstration that that male *Plodia interpunctella* (Pyralidae), like the yellow fever mosquito, responded to a moving pattern on the tunnel's floor. Movement of the floor pattern with the direction of the airflow caused males to speed up (because they do not seem to be making progress upwind), whereas moving the floor pattern against the direction of airflow (thereby creating the illusion of upwind progress) caused males to slow their ground speed or, if the pattern was moved rapidly enough, even drift downwind. All moths are now assumed to orient along a wind-borne pheromone plume by heading upwind using the same visual feedback system. Both upwind and downwind trajectories, however, present a fore-to-aft image flow. During flight polarization of heading toward upwind is presumably accomplished entirely by visual detection of drift (e.g., Baker et al. 1984). A possible role for mechanosensory input in optomotor anemotaxis remains speculative.

Sane et al. (2007) found that flight stability of flying *Manduca sexta* moths (Sphingidae) requires input from the John-ston's organs at the base of the antennae. The importance of some mechanoreceptor input to flight tracks also has been verified with by direct stimulation of the antennal muscles of flying *M. sexta* (Hinterwith et al. 2012). Perhaps antennal deflection and motor output are employed to detect whether a moth is orienting into the wind. Mechanoreceptor input also likely governs detection of upwind direction of a perched moth before takeoff; in wind-tunnel assays, moths usually turn toward upwind before launching.

In wind-tunnel studies, visual cues typically are presented from below (on the tunnel's floor) and they are uniform in size, so that these subtend the same angle when moths fly, as they tend to do, at a relatively constant altitude above the floor. In nature the visual surround is heterogeneous, and likely requires males as they move upwind to adjust to cues that vary markedly in the angle they subtend. At night the most prominent visual cues in forest environments may be sky openings in the foliage. Sanders (1985) found with spruce budworm males *Christoneura fumiferana* (Tortricidae), a forest denizen, that the optomotor reaction (regulation of apparent "ground" speed) could be evoked by moving a pattern in the ceiling of a wind tunnel.

A comprehensive study of the influence of the position of visual cues on the success of orientation was undertaken by Vickers and Baker (1994b). They used a uniformly white, translucent tube placed within a wind tunnel as a visually featureless environment (save the upwind end of the tunnel) and added visual cues as rows of dots placed below, above, or on both sides of the tube. Without any supplementary visual cues and therefore with essentially no optomotor feedback, very few *Heliothis virescens* males were able to navigate along a pheromone plume; males appeared disoriented and frequently collided with the sides of the tube. When rows of dots were placed below, above, or on each side, however, many males successfully navigated the plume. The greatest success occurred when the pattern of dot placement enabled both the perception of longitudinal and transverse image motion.

Counterturning across the Wind Line

The path taken by moths as they orient along a plume (figure 11.1) usually seems to assume a zigzag form that, upon close examination, involves far greater lateral excursions than vertical movement. It would appear that moths are "feeling" the plume's edge, but studies in wind tunnels have shown that males can zigzag along one edge of a corridor of pheromone (equivalent to an edge of a large plume), with counterturns on one side within the column of pheromone where there is no boundary of clean air (Kennedy et at. 1981; Willis and Baker 1984). Right and left counterturns along the plume thus seem to be controlled by a CNS pattern generator (Kennedy 1983), rather than by reiterative contact with plume's boundaries. In *Lymantria dispar* at the apex of each turn, the moth rolls, a maneuver that enables the moth to aim its trajectory sufficiently toward upwind to align the crosswind leg with a preferred course angle (Zanen and Cardé 1999). The degree of roll thus must vary with wind speed.

An alternative explanation for counterturning is that in-flight orientation toward due upwind is inherently imprecise and that reiterative course corrections yield a zigzag path (Preiss and Kramer 1986). This mechanism was challenged by David and Kennedy (1987) who accepted their findings as applicable to moths that were tethered, the assay used in the

Preiss and Kramer study, but not to moths in free flight. That the timing of the turns in *L. dispar* have a narrow range near 3.5–4 turns s^{-1} when many other features of the zigzag such as track velocity and width vary with concentration and temperature (Charlton et al. 1993), plume structure and light level (Cardé and Knols 2000), and field conditions (Willis et al. 1994), all argue for an endogenous source dictating the timing of turns. Another example of metronome-like timing of counterturns was demonstrated in *Epiphyas postvittana* (Tortricidae). Foster and Howard (1999) found that its turning frequency held near 5–6 turns s^{-1} across treatments varying in pheromone concentration, wind speeds, and a variety of visual conditions. It may be, however, that when the moth is headed off of the wind line, a visually perceived change in the direction of transverse image flow induces the turn and the narrow band of the timing of turns simply reflects the minimum time to execute each turn, rather than output of an endogenous turn generator.

Counterturning and the zigzag path it generates also have been proposed to aid in the determination of wind direction (Cardé 1984). When moths are headed slightly off the wind line (e.g., 5° from due upwind) they experience less transverse image flow than they would in a typical zigzag course (e.g., heading 30–40° from due upwind on each leg of the zigzag). Thus, zigzagging should enhance a moth's ability to detect the direction of wind flow and this may be crucial to orientating upwind when wind velocity or light levels are near the optomotor threshold. Lateral (and vertical) movement also may aid in keeping the moth in contact with the plume.

Regulation of Flight Speed

Kennedy (1951) proposed that migrating locust swarms maintain a relatively constant ground speed while in-flight by holding the rate of image flow of the ground features below near a set rate, the so-called preferred retinal velocity. One consequence of this mechanism is that the ground speed should hold relatively constant through a range of airspeeds (i.e., those within an insect's flight capabilities). *Lymantria dispar* males navigating along a pheromone in a wind tunnel had nearly identical ground speeds at airspeeds of 0.7 and 2.5 m^{-1} (Cardé and Hagaman 1979). Support for such an orthokinetic response mediated by visual feedback cues also comes from the wind-tunnel studies of Kuenen and Baker (1982a), Foster and Howard (1999), and Kuenen (2013). Males flying at an increased height over a given floor pattern adjust their airspeed to approximate the same rate of flow of ground speed and therefore maintain similar rates of retinal velocity, in accord with Kennedy's hypothesis. Rigorous tests of this mechanism will require varying the size of the visual cues (and therefore the angle they subtend) and image velocity (by having the plume and therefore the moth fly at differing heights over the floor patterns). Other factors that can modulate the speed of flight include pheromone concentration (e.g., Cardé and Hagaman 1979; Kuenen and Baker 1982b; Foster and Howard 1999) and the plume's structure (e.g., Mafra-Neto and Cardé 1994, 1995; Vickers and Baker 1994a).

Casting after Loss of Plume Contact

Temporally regular counterturning also occurs after a moth loses contact with the plume (David et al. 1983; Kennedy 1983;

Kuenen and Cardé 1994), except that there is (in wind-tunnel trials) little or no upwind progress. In such "casting," typically the width of the lateral excursions increases progressively until the moth "gives up" and departs the area where the plume was last sensed. Casting seems to be an effective strategy for regaining contact with a plume that has been lost either because of its fragmentation by turbulence or because of shifts in wind direction. This was verified by Baker and Haynes (1987) in wind-tunnel experiments. They shifted the wind field 35° over intervals of 1, 2, or 4 s, mimicking sudden changes in wind direction in the field. Many plume-following *Grapholita molesta* males lost contact with the plume; these males promptly changed to casting at ca. 90° to the shifting wind line, usually resulting in recontacting the plume, in turn followed by an upwind surge.

A convincing demonstration of the value of casting to successful navigation along a plume comes from the field study of David et al. (1983) with *Lymantria dispar*. Moth flights were video-recorded in a 400 m^2 area from a 27 m high tower; the position of the pheromone plume was estimated from a train of soap-film bubbles released from just below the source of pheromone. Males navigating along the plume and then loosing contact with a plume because of a shift in wind direction flew as far as 4 m crosswind in an initial lateral sweep. This casting maneuver often enabled recontact with the plume, provided the direction of the cast and the wind shift coincided. Furthermore, David et al. (1983) contended that if casting enabled recontact, this strategy would have brought the moth closer to the pheromone's source than before the cast and so this strategy would be more valuable than drifting downwind. The crosswind distance travelled during casting in the David et al. (1983) study was considerably greater than that observed in the wind-tunnel experiments of Kuenen and Cardé (1994), also with the gypsy moth. The divergent outcomes in these two studies could be attributable to differences to milieus in which each was conducted. Visual cues in the wind tunnel and field differ, with those presented by the wind tunnel possibly inhibiting lateral movement. As well, the plumes' size, internal structure, and concentration would differ between the field and wind tunnel.

Analysis of Maneuvers in Three-Dimensions

Most of the foregoing wind-tunnel studies have recorded and analyzed moth flight tracks in 2-D planar view. It has been generally assumed that moths stabilize their flight height above a wind-tunnel floor by maintaining roughly the same angle subtended of the visual cues below and that lateral displacement in zigzagging flight greatly exceeds vertical movement. This assumption is reflected in the usual method of flight track analysis (figure 11.4), the 2-D triangle of velocities (Marsh et al. 1978), which decomposes flight maneuvers in the horizontal plane into those described by the actual track (e.g., heading with respect to upwind, speed along its flight track, net upwind speed, turning across the wind line), and those described by the insect's heading along its track, given that its heading must compensate for wind-induced drift. Clearly there is *some* vertical movement and the question is whether analyses in 2-D distort our picture of how moths orient, with the question of where moths execute turns and the extent of vertical movement being the most salient issues.

Witzgall (1997) reviewing studies with *Lobesia botrana* (Tortricidae) and *Grapholita molesta* (see also Witzgall and Arn 1990, 1991) characterized their 3-D flight paths as

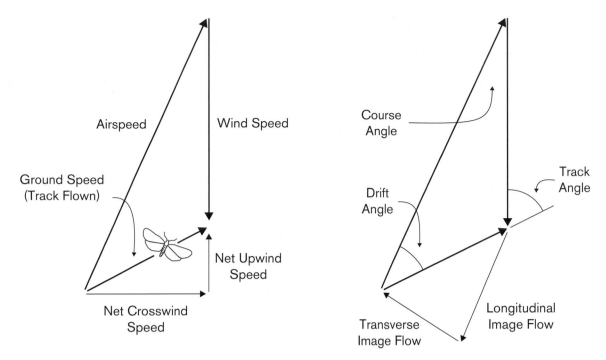

FIGURE 11.4 The "triangle of velocities" method for analysis of flight tracks in 2-D, after Marsh et al. (1978). To fly a course along the track shown, the moth's heading would be aligned with the airspeed vector. The image flow that the moth would see below is broken into longitudinal and transverse components. A moth heading directly upwind would experience only a front-to-rear (or longitudinal) flow.

continuous loops *without* switches across the wind line, evidently in conflict with the definition of "counterturning" or a "cross-wind track reversal" by Marsh et al. (1978). Such spirals around the plume's long axis nonetheless should generate the same reversals in the direction of transverse image flow as in counterturning. The two-video camera setup that was used to record the flight trajectories in these studies had background reflective surfaces to highlight the positions of moths and enable their position to be sensed accurately and their position automatically digitized (Witzgall and Arn 1991). The recording area of the flight arena was essentially devoid of visual cues. This setup may have provided suboptimal visual feedback for stabilization of height, and thus contributed to an increase in vertical displacement. Regardless, Witzgall (1997) documented considerable vertical movement in *L. botrana* and *G. molesta* and further suggested that the 2-D view from above does not provide unambiguous evidence of a counterturn oscillator. Another intriguing aspect of their studies is that an extract of female pheromone induced a much straighter upwind path than the three synthetic pheromone components in both species. Moreover, partial blends (one component for *L. botrana* and two for *G. molesta*) produced even more convoluted tracks. Witzgall (1997) suggested that the bimodal distribution of course and track angles around due upwind (reflecting a zigzag course) in these two tortricids were artifacts caused by an inadequate chemical stimulus, calling into question all studies that do not have a calling female or extract as a control treatment.

In female host-seeking mosquitoes, vertical movement routinely exceeds lateral movement and 3-D analyses would seem mandatory (e.g., Cooperband and Cardé 2006; Dekker and Cardé 2011). A field study by Murlis et al. (1982) with ciné recordings from the side documented considerable vertical movement of male *Spodoptera littoralis* (Noctuidae) orienting to a point source. Several studies have recorded moth flight along plumes in 3-D, including wind-tunnel studies by von Keyserlingk (1984) and Witzgall and Arn (1990) and verified vertical movement. In one analyses of *Cadra cautella* flight, however, there were no important differences in flight maneuvers revealed by expanding to the third dimension, mainly because the preponderance of movement was planar (Wisniewska and Cardé 2012). Vickers and Baker (1996) found that as long as *Heliothis virescens* maintained contact with the plume, vertical deviations were suppressed; however, once moths entered pheromone-free air, the ensuing casting had lateral and vertical components of comparable magnitude. Similarly, in field recordings Baker and Haynes (1996) found that in zigzagging along a plume most displacement of *G. molesta* was in the horizontal plane.

Rutkowski et al. (2009) concluded from a 3-D study of plume following in *Manduca sexta* (a moth considerably larger that most others studied and noted for its maneuverability in flower feeding) that instead of zigzagging or counterturning, cross-plume maneuvers could be most accurately described as "cutting through the plume from all directions with loops of different radii" and that vertical and horizontal displacements were "roughly equivalent." Certainly a 3-D analysis provides the most accurate representation of maneuvering. What remains unclear is how much maneuvers of moths from differing lineages and sizes differ and the value to be gained from analysis in 3-D. *M. sexta* males, for example, are noted for their in-flight maneuverability and have forewing lengths of 51–56 mm, whereas in *C. cautella* the forewing varies between 5 and 9 mm. The approximate antennal expanses of these two moths are approximately 40 and 4 mm, respectively, so that the swaths of the plume sampled and the likelihood of encountering a filament of pheromone differs. The recent availability of relatively automated systems (e.g., Lacey and Cardé 2011) for rapidly converting records of two video cameras into *x, y, z* coordinates may render 2-D recordings obsolete.

Influence of a Plume's Fine-Scale Structure

Kennedy et al. (1981) provided the first evidence of the importance of fluctuating signal intensity to orientation. Males of *Adoxophyes orana* (Tortricidae) were unable to progress upwind in a homogeneous cloud of pheromone. Willis and Baker (1984) extended these trials with *Grapholita molesta*, confirming this tortricid also does not orient upwind in a pheromone cloud of unvarying concentration, whereas males readily flew upwind when the pheromone was presented as a pulsed cloud (Baker et al. 1985). An extension of these findings was Baker's (1990) proposal that successive contacts with filaments of pheromone above a particular frequency induced sustained upwind movement. Definitive proof of this hypothesis came from work with *Cadra cautella* (Pyralidae) (Mafra-Neto and Cardé 1994, 1995, 1996) and *Heliothis virescens* (Vickers and Baker 1994a, 1996). In both species, reiterative contact with experimentally generated puffs of pheromone at rates above about 5 Hz straightened out their paths toward upwind and increased velocity. This form of orientation has been termed a "Surge-Cast" model. These pulse regimes were intended to simulate rates of encountering odor filaments within a turbulent plume, but it is the rate of filament encounter as the moth heads upwind in the plume and not the rate of puff generation at the source that determines the shape and the direction of the flight track (Baker and Vickers 1994; Mafra-Neto and Cardé 1998). At the antennal level, pheromone filaments of 20 ms duration can be resolved at fluxes of 25 Hz (Bau et al. 2002, 2005). When the plume's dynamics were monitored in the antennal lobe of *H. virescens*, the spike patterns correlated at a millisecond timescale with the plume's temporal dynamics and intensity (Vickers et al. 2001).

In *C. cautella*, rates near 10 Hz and above can generate a nearly straight upwind course (Mafra-Neto and Cardé 1994). An unexpected finding with *C. cautella* was the ability of males to orient upwind in homogeneous clouds (Justus and Cardé 2002), although the same study verified that *Pectinophora gossypiella*, like *A. orana* and *G. molesta*, also would not orient upwind in homogeneous clouds. Why *C. cautella* should have this capability in contrast to several other moth species remains unclear, although perhaps its current residence in still-air habitats such as granaries is facilitated by this trait.

Do Males Compare Inputs from Both Antennae?

The possibility that males use differences in concentration or possibly differential time of arrival as perceived by the right and left antennae while orientating along the plume was tested by Vickers and Baker (1991). Flight tracks of intact *Heliothis virescens* males and those with one antenna removed were overall indistinguishable in their course angles, track angles, airspeed, and ground speed. Given that pheromone plumes are highly filamentous in structure, there should be no informative value in comparing either concentration or arrival time at each antenna, because neither of these would not be reliable indicators of a moth's position within the plume (is the moth near the plume's centerline or is straying to the plume's edge?). If moth exits a plume at an angle to due upwind (as can happen in a zigzag maneuver), however, it will experience a sequential loss of contact with pheromone (either beginning from right then proceeding to the left or vice versa) that signifies the direction back toward the plume;

in other words, to regain plume contact, the moth then should turn back toward the direction where pheromone was last sensed (Kennedy 1977); this sort of comparison does not necessitate a differential inputs from each antenna. As turns in zigzag flight are executed several times per second, these comparisons would need to be made on a timescale of less than 100 ms.

Walking orientation very near the female, however, can rely on differential antennal input (tropotaxis). In *Bombyx mori*, for example, walking males turn toward the more pheromone-stimulated side or when both antennae are stimulated toward the last stimulated side which is presumed to be in the direction of the plume's center; both reactions are presumed to facilitate location of the female (Takasaki et al. 2012). This maneuver becomes feasible because very near to a point source of pheromone the plume is relatively small and discrete in structure (figure 11.5) and a walking approach to the source does not require a series of rapidly occurring counterturns, as are typified in zigzagging flight.

The possibility of differential right–left input in mediating upwind plume navigation via odor gradients has been suggested for flying *Drosophila* (Duistermars et al. 2009; Gaudry et al. 2012). Tethered flies turned toward the side with the higher concentration or, after removal of one antenna, toward the intact side. For the reasons mentioned above, orientation within the chaotic plume structure, however, should not be improved by right–left comparisons; in *Drosophila* this strategy would be especially difficult to implement because the antennae are separated by only 1 mm. While it may be that tethered flies are detecting and responding to such differences, it also may be that this is simply a representation of their known ability while walking to use tropotaxis to turn toward a higher concentration of odorant (Borst and Heisenberg 1982) and it may not be indicative of their behavior in free flight. Recent studies with *Drosophila* by Gaudry et al. (2013) have demonstrated that a mere 5% asymmetry in odorant input to right and left antennal lobes was sufficient to direct a walking fly toward the more stimulated antenna within less time than is required to complete a stride. Very close to an odorant source, such changes in odorant concentration should be sufficiently steep to enable tropotaxis.

Observations of Orientation in the Field

Nearly all of the studies considered so far have been conducted in wind tunnels. Recording flight tracks in the field poses considerable technical challenges (especially at night), and it can be unclear what the wind direction and particularly the structure of the plume are in a moth's immediate vicinity. Nonetheless, such observations provide clear records (in 2-D from above) of the range of maneuvers that occur under relatively natural circumstances. Studies with *Lymantria dispar* (David et al. 1983; Willis et al. 1991, 1994), *Grapholita molesta* (Baker and Haynes 1996), and *Heliothis virescens* (Vickers and Baker 1997) verify casting when contact with the plume is lost and a generally elevated ground speed compared to wind-tunnel trials. The latter effect seems due to moths flying at higher altitudes and therefore perceiving a reduced rate of flow of their visual field below, compared to visual feedback in wind-tunnel experiments. A second factor, however, may be that in the field plumes are more dilute, which promotes an increase in ground speed.

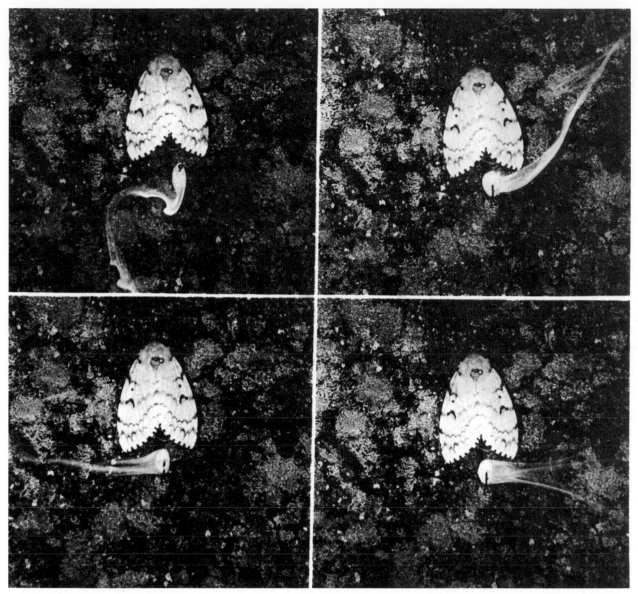

FIGURE 11.5 Visualization of plume structure nearby a dead female gypsy moth, *Lymantria dispar*, pinned in a calling stance onto a tree trunk. "Smoke" from a small source of titanium tetrachloride serves as a visual surrogate for pheromone issuing from female's abdominal tip. Changes in wind flow cause the plume to shift rapidly, but note its relatively cohesive structure, which could be selectively engulfed by a single antenna of a male walking nearby (photos taken over 60 s in a deciduous forest; from Charlton 1988).

Stereotypy versus Flexibility

Although there is some predictability in a given species of how the pattern of encounters with pheromone and clean air dictates in-flight maneuvers, there also can be considerable variability in the flight tracks of individual moths along plumes that should be identical in structure and concentration. Why this should be so is not understood. In the field, the structure of the pheromone plume and wind conditions are highly variable and it may be that variability in response provides a range of orientation "strategies." Alternatively, it may simply be due to the stochastic effects of encounter with filaments of the plume. Speed of flight along the plume, for example, can vary widely. In wind-tunnel trials with 100 ng pheromone sources (an intermediate dose for these trials), males of *Lymantria dispar* flew upwind at mean ground speeds of 4–5 m min^{-1}, but occasional males flew as rapidly as 10–12 m min^{-1}, whereas others flew very slowly, some at less than

1 m min^{-1}. Cardé and Hagaman (1979) proposed that the faster males might be most apt to locate a calling female when wind produced a well-defined plume aligned with the wind, but also most likely to fly upwind out of the plume when the wind shifted course (figure 11.3). When shifting winds and turbulence caused the plume to writhe and twist, perhaps males flying more slowly would be more likely to maintain contact with the plume.

In another approach to understanding the selective value of variability, Belanger and Willis (1996) used existing information on plume structure and moth orientation (particularly of *Manduca sexta*) to create simulation models that tested if flexibility in response to varying plume structure improved the success of orientation. In simulations counterturning (zigzagging) improved a moth's ability to recontact a turbulent plume (one with gaps), but this maneuver diminished ability to track a continuous ("laminar") plume. They pointed out that it may be simplistic to consider responses as strictly reflexive, given that a prior encounter with, for example, a particular pattern of bursts of pheromone could set the next maneuver to increase the probability of continued contact with the plume. Belanger and Willis (1996) have emphasized that the value of flexibility in maneuvers is dependent on environmental context, rather than variation in individual reactions to the same stimuli. The kind of track variability evident in *L. dispar* (Cardé and Hagaman 1979) and *Cadra cautella* (Justus et al. 2002b) could be, however, mediated simply by differences in threshold of response, which varies among individuals (e.g., Charlton et al. 1993). In this explanation, the slow flight of *L. dispar* males would be presumed to have relatively low thresholds of response, and therefore react as if the plume was emanating from a high-dose dispenser.

Orientation along Plumes in Still Air

Farkas and Shorey's (1972) wind-tunnel study was important in demonstrating that male *Pectinophora gossypiella* moths could track a wind-formed pheromone plume after wind flow was stopped. Subsequent work with moths in three additional families *Grapholita molesta* (Baker and Kuenen (1982), *Lymantria dispar* (Willis and Cardé 1990), and *Cadra cautella* (Wisniewska and Cardé 2012) have substantiated that these moths, also having set course in wind along the plume, can continue plume tracking following the sudden cessation of wind flow.

There are several possible explanations for moth flights along these plumes. In these studies, males had already set course along the plume when wind flow was abruptly stopped. Males could have continued along the plume by maintaining the same flow of the visual field as they experienced in airflow, or they could have sensed the plume's position laterally or by changes in concentration along its core (forms of klinotaxis). There is some evidence that once males have set course in airflow, in still air the pattern of visual cues can collimate a moth's trajectory along the plume. Males of *C. cautella* orienting within a very wide plume tended to fly above a "path" of visual cues—a line of solid red dots—even when this linear trail abruptly veered midpoint in the wind-tunnel 15° toward one side. This maneuver occurred in conditions of constant wind or after wind flow was interrupted (Wisniewska and Cardé 2012). The question of how males might orient when they take off in still air and contact a stationary plume was tested with *G. molesta* by Baker et al. (1984). In these circumstances, 21% of the males eventually located the plumes' source with counterturning (zigzag) flight tracks within and near the stationary plume. This study established that the counterturning program does not require the presence of wind and therefore it is independent of anemotaxis. In wind, however, the counterturning program is integrated with anemotaxis and it is an essential component of in-flight plume tracking (Baker et al. 1984).

Although plume tracking over many meters almost certainly requires wind as an orienting cue, there are many environments where wind flow can fall below a detectable level, particularly at night within the forest canopy where low light levels may render optomotor feedback more difficult or perhaps impossible to sense. Even under conditions well above the detection threshold for the optomotor response, light levels can modulate the form and velocity of flight. The amplitude of the zigzag and the flight velocity taken by male gypsy moths in wind-tunnel trials varied with light intensities of 4 and 450 lux (Cardé and Knols 2000). At 4 lux, males flew narrower width zigzags more slowly than at 450 lux. This indicates that light level influences optomotor feedback.

Other evidence of the importance of good night vision for plume tracking was the finding that few night-blind mutants of *Trichoplusia ni* (Noctuidae) were able to fly to a pheromone source in a wind tunnel, although they retained their ability to walk upwind to the same source (Rucker and Haynes 2004). When the eyes of *Manduca sexta* moths were occluded, males were unable to track a pheromone source in a wind tunnel while in flight and also, perhaps unexpectedly, while walking (Willis et al. 2011). Another correlate that suggests that acuity of visual feedback is important to orientation along plumes at night is that eye size is reduced substantially in some diurnal moths, such as those inhabiting very high latitudes where there is no "true" night (*Arctia* spp. Erebidae, Arctiinae; Dubatolov and Philip 2013), other high latitude moths (*Mesothea incertata*, Geometridae; Ferguson 1969), and *Hemileuca* spp. (Saturniidae; Michener 1952).

Thus, the ability of males to locate a distant calling female once its pheromone plume has been detected is contingent on two mechanisms: plume tracking using optomotor anemotaxis and casting if the plume contact is lost temporally. Orientation under zero wind conditions is the least understood mechanisms. If orientation along the plume is based on sensing changes in concentration, it could be based on changes perceived laterally (transverse klinokinesis) or as the moth progresses toward the source (longitudinal kinesis). Successful orientation under suddenly windless conditions could also rely on maintaining the same flow of the visual field as males used when in wind as long as contact with the plume continued (Wisniewska and Cardé 2012). Such orientation has only been shown to occur over distances of approximately a meter, all in wind-tunnel settings.

Another orientation strategy has been documented in the potato tuberworm moth (*Phthorimaea operculella* (Gelechiidae). Males employ short (<1 m), intermittent flights ("hops") to advance to toward a pheromone source. Evidently males use mechanosensory input to detect wind direction while on the ground and then head upwind using an "aim-then-shoot" mechanism. If males lose contact with the plume, their subsequent crosswind casting is by successive hops (Tejima et al. 2013). This moth seems to have lost its ability to navigate by in-flight optomotor anemotaxis.

Orientation at Close Range

Once a male has arrived at the vicinity of a calling female, landing ensues. This maneuver generally requires a visual cue indicating a suitable landing site and also requires the moth to change the flow of its visual field. In upwind orientation, the visual surround is experienced generally as a front-to-rear flow of the visual surround, with avoidance of objects looming directly ahead. A prerequisite to landing, however, is to allow visual expansion of the object looming directly ahead (Maimon et al. 2008; Cardé and Gibson 2010). The specific characteristics of the plume that induce landing are not well understood. Very close to the source (≈1–20 cm) and prior to landing, males typically slow their upwind velocity and decrease the width of lateral excursions, although they also may encounter the plume's edges more frequently, because close to the source the plume also narrows. Many other structural and concentration features of the plume also differ and are candidates for triggering landing, but absolute concentration and burst length (duration of an intermittent pheromone signal) have been suggested as the potential keys to changes in flight behavior close to the source (Justus et al. 2002a). Specific pheromone components that induce the final stages of orientation but do not potentiate long-range attraction have long been the subject of speculation (e.g., Bradshaw et al. 1983, but see Linn et al. 1987). Only recently has a first case been substantiated. In the yellow peach moth *Conogethes punctiferalis* (Crambidae) two long-chain hydrocarbons, (Z9)-heptacosene and (Z3, Z6, Z9)-tricosatriene, added to the two-component attractant, increase orientation to the source but only at close range (Xiao et al. 2012). This finding implies that by focusing only on long-range attractants, a class of orientation responses and the pheromones that induce them may have been overlooked in many other moths.

If the male lands in the plume downwind of the female, orientation can continue by walking upwind until contact with the female. In *Grapholita molesta* a landed male walks a straight upwind course without any of the zigzags that characterize flight along the plume (Willis and Baker 1987). Should the male land outside the plume, the male may engage in local "area restricted" search, a walking path that typically involves repetitive turns until plume contact, allowing plume tracking, or the male's walking path simply may bring the male into contact with the female (e.g., Charlton and Cardé 1990a).

In some species tactile cues from either the females' wing scales (Ono 1977, 1980) or the females' abdominal scales (Charlton and Cardé 1990b), perceived via the males' tarsi, induce arrestment of walking and copulatory attempts. In the case of the tussock moth *Orgyia leucostigma* (Erebidae, Lymantriinae), Grant et al. (1987) found that copulatory behavior was governed by the tactile cues from the abdominal scales augmented by two *n*-alkanes (C-24 and C-25) released from these scales.

In several day-mating species, visual cues presented by the female can have very dissimilar influences. In the courtship sequence of *G. molesta*, visual cues presented by the female direct the male's courtship display and movement until the female contacts the male (Baker and Cardé 1979). Similarly in the wasp mimic *Amata fortunei* (Erebidae, Syntominae), KonDo et al. (2012) found that in-flight close-range orientation and eventual copulatory attempts were augmented by visual models that closely resembled the female abdomen's yellow–black banding. In *Lymantria dispar*, conversely, the female's obvious (to us) visual presence on a tree trunk seems of no importance to the precise landing spot selected, the walking path taken during "area-restricted" search, or his eventual probability of contacting a female and then recognizing her by tactile cues (Charlton and Cardé 1990a, 1990b). In the Arctiinae (Conner 1999) the ultrasonic pulses produced by tymbals and used by arctiines to ward off bat attacks with ultrasonic cries have in some species been co-opted for close-range communication used after males have reached the female's vicinity by conventional plume tracking. Males and females of *Cycnia tenera* (Erebidae, Arctiinae), for example, engage in duets as the male approaches the calling female, with signals from the female presumably aiding a male's ability to localize the female's precise location (Conner 1997).

Summary

The distance over which male moths typically detect and track a pheromone plume to a calling female remains unresolved, although distances of tens of meters likely apply to many species. Whether some saturniid moths do so over a range of a kilometer or more remains to be firmly established. The strategies used to locate a plume also are poorly documented. The principal navigational mechanism used by male moths to find a distant female is to track a wind-borne plume of female pheromone using optomotor anemotaxis. In two moths, *Cadra cautella* and *Heliothis virescens*, sustained upwind progress has been shown to require reiterative contact with filaments of pheromone at rates above ≈5 Hz. Loss of contact with the plume, due to either turbulence-induced gaps or exiting the plume, causes moths to cast, a maneuver that can enable the moth to re-enter the plume. The ability of moths to continue orientation along a wind-formed plume when wind ceases or falls below the optomotor-detectable threshold has been verified in wind-tunnel settings with moths in several lineages and such forms of orientation likely are important to mate-finding in the field when there are similar wind or light conditions. This maneuver may rely on sensing the distribution of pheromone within the plume (forms of klinotaxis), but males also may be collimating their trajectory along the wind-formed plume by holding the flow of the visual field constant, as long as contact with the plume is maintained.

We know a reasonable amount about how a few moth species orient in the well-defined milieu of wind tunnels. This artificial environment does not allow us to mimic a freely flying moth entering a pheromone plume—usually the moth is caged and tested from quiescence—and so its physiological state and likely its olfactory threshold are not comparable to those of freely flying moths reacting in the field. We also generate plumes that fail to recreate the complex structure and concentration of the very dilute and fragmented plumes that exists many meters downwind of a calling female. Odor plumes have many features (e.g., intermittency, burst length, peak concentration, peak-to-mean ratio, mean concentration, overall dimensions of the outer envelope) that may influence maneuvers and their importance to orientation remains speculative. As moths reach the vicinity of the pheromone source, typically velocity slows and the track narrows (in concert with a decreasing plume diameter and increasing overall concentration). Which features of the plume are salient and govern close approach and, in concert with visual cues, induce landing remain to be defined. The visual surrounds that are typically used in wind tunnels to enable expression of the optomotor reaction provide an abundance of cues of consistent size and spacing, but this milieu is hardly comparable to the visually cluttered environment found in nature.

Male moths have an extraordinary capability for finding distant, calling females in the heterogeneous environment of nature, but our understanding of the mechanisms employed are derived mainly from laboratory studies of a few exemplar species over short distances in wind tunnels. Future explorations of this communication phenomenon should expand to species in other lineages, although the general principle of optomotor anemotaxis is expected to be upheld as the universal mechanism for flying upwind during plume tracking. The precise nature of the maneuvers used, however, also may reflect phylogeny, body size and flight capability, variability in plume dispersion as influenced by habitat, the effect of light levels on the optomotor reaction, and a predictable population density. Of particular value will be the use of use experimental setups that attempt to mimic the complex visual backgrounds and spatial structures of plumes encountered in the field.

Acknowledgment

I thank Kenneth Haynes for providing a very helpful review.

References Cited

Aylor, D. E., J.-Y. Parlange, and J. Grannett. 1976. Turbulent dispersion of disparlure in the forest and gypsy moth response. *Environmental Entomology* 5:1026–1032.

Baker, T. C. 1990. Upwind flight and casting flight: complementary and tonic systems used for location of sex pheromone sources by male moths. Pp. 18–25. In K. B. Døving, ed. *Proceedings of the Tenth International Symposium on Olfaction and Taste.* Oslo, Norway: GCS A/S.

Baker, T. C., and R. T. Cardé. 1979. Courtship behavior of the oriental fruit moth (*Grapholitha molesta*): experimental analysis and consideration of the role of sexual selection in the evolution of courtship pheromones in the Lepidoptera. *Annals of the Entomological Society of America* 72:173–188.

Baker, T. C., and K. F. Haynes. 1987. Manoeuvres used by flying male oriental fruit moths to relocate a sex pheromone plume in an experimentally shifted wind-field. *Physiological Entomology* 12: 263–279.

Baker, T. C., and K. F. Haynes. 1996. Pheromone-mediated optomotor anemotaxis and altitude control exhibited by male oriental fruit moths in the field. *Physiological Entomology* 21:20–32.

Baker, T. C., and L. P. S. Kuenen. 1982. Pheromone source location by flying moths: a supplementary non-anemotactic mechanism. *Science* 216:424–427.

Baker, T. C., and W. L. Roelofs. 1981. Initiation and termination of oriental fruit moth male response to pheromone concentrations in the field. *Environmental Entomology* 10:211–218.

Baker, T. C., and N. J. Vickers. 1994. Behavioral reaction times of male moths to pheromone filaments and visual stimuli: determinants of flight track shape and direction. Pp. 838–841. In K. Kurihura, N. Suzuki, and H. Ogawa, eds. *Olfaction and Taste IX.* Tokyo: Springer-Verlag.

Baker, T. C., R. T. Cardé, and J. R. Miller. 1980. Oriental fruit moth pheromone component emission rates measured after collection by glass-surface adsorption. *Journal of Chemical Ecology* 6:749–758.

Baker, T. C., M. A. Willis, and P. L. Phelan. 1984. Optomotor anemotaxis polarizes self-steered zigzagging in flying moths. *Physiological Entomology* 9:365–376.

Baker, T. C., M. A. Willis, K. F. Haynes, and P. L. Phelan. 1985. A pulsed cloud of sex pheromone elicits upwind flight in male moths. *Physiological Entomology* 10:257–265.

Bau, J., and R. T. Cardé. 2015. Modeling optimal strategies for finding a resource-linked, windborne odor plume: theories, robotics and biomimetic lessons from flying insects. *Integrative and Comparative Biology* 55:461–477.

Bau, J., K. A. Justus, and R. T. Cardé. 2002. Antennal resolution of pulsed pheromones plumes in three moth species. *Journal of Insect Physiology* 48:433–442.

Bau, J., K. A. Justus, C. Loudon, and R. T. Cardé. 2005. Electroantennographic resolution of pulsed pheromone plumes in two species of moths with bipectinate antennae. *Chemical Senses* 30:771–780.

Belanger, J. H., and M. A. Willis. 1996. Adaptive control of odor-guided locomotion: behavioral flexibility as an antidote to environmental unpredictability. *Adaptive Behavior* 4:217–253.

Borst, A., and M. Heisenberg. 1982. Osmotropotaxis in *Drosophila melanogaster*. *Journal of Comparative Physiology A* 147:479–484.

Bossert, W. H., and E. O. Wilson. 1963. The analysis of olfactory communication among animals. *Journal of Theoretical Biology* 48:443–469.

Bradshaw, J. W. S., R. Baker, and J. C. Lisk. 1983. Separate orientation and releaser components in a sex pheromone. *Nature* 204:265–267.

Brady, J., G. Gibson, and M. J. Packer. 1989. Odour movement, wind direction, and the problem of host-finding by tsetse flies. *Physiological Entomology* 14:369–380.

Cardé, R. T. 1981. Precopulatory behavior of the adult gypsy moth. Pp. 572–587. In C. C. Doane and M. L. McManus, eds. *The Gypsy Moth: Research toward Integrated Pest Management,* Technical Bulletin 1584. Washington, DC: United States Department of Agriculture.

Cardé, R. T. 1984. Chemo-orientation in flying insects. Pp. 111–124. In W. J. Bell and R. T. Cardé, eds. *Chemical Ecology of Insects.* London: Chapman & Hall Ltd.

Cardé, R. T., and T. C. Baker. 1984. Sexual communication with pheromones. Pp. 335–383. In W. J. Bell and R. T. Cardé, eds. *Chemical Ecology of Insects.* London: Chapman & Hall Ltd.

Cardé, R. T., and G. Gibson. 2010. Long-distance orientation of mosquitoes to host odours and other host-related cues. Pp. 115–141. In W. Takken and B. G. F. Knols, eds. *Ecology of Vector-Borne Diseases. Vol. 2. Olfaction in Vector-Host Interactions.* Wageningen, The Netherlands: Wageningen Academic Publishers.

Cardé, R. T., and T. E. Hagaman. 1979. Behavioral responses of the gypsy moth in a wind tunnel to air-borne enantiomers of disparlure. *Environmental Entomology* 8:475–484.

Cardé, R. T., and B. G. J. Knols. 2000. Effects of light levels and plume structure on the orientation manoeuvres of male gypsy moths flying along pheromone plumes. *Physiological Entomology* 25:141–150.

Cardé, R. T., and M. A. Willis. 2008. Navigational strategies used by flying insects to find distant, wind-borne sources of odor. *Journal of Chemical Ecology* 43:854–866.

Cardé, R. T., A. M. Cardé, and R. D. Girling. 2012. Observations on the flight paths of the day-flying moth *Virbia lamae* during periods of mate location: do males have a strategy for contacting the pheromone plume? *Journal of Animal Ecology* 81:268–276.

Charlton, R. E. 1988. Pheromone-mediated flying and walking orientation and factors promoting mate recognition in the gypsy moth, *Lymantria dispar* (L.). PhD thesis, University of Massachusetts, Amherst.

Charlton, R. E., and R. T. Cardé. 1990a. Orientation of male gypsy moths, *Lymantria dispar* (L.), to pheromone sources: the role of olfactory and visual cues. *Journal of Insect Behavior* 3:443–469.

Charlton, R. E., and R. T. Cardé. 1990b. Factors mediating copulatory behavior and close-range mate recognition in the male gypsy moth, *Lymantria dispar* (L.). *Canadian Journal of Zoology* 68:1995–2004.

Charlton, R. E., H. Kanno, R. D. Collins, and R. T. Cardé. 1993. Influence of pheromone concentration and ambient temperature on flight of the gypsy moth, *Lymantria dispar* (L.), in a sustained-flight wind tunnel. *Physiological Entomology* 18:349–362.

Collins, C. W., and S. F. Potts. 1932. *Attractants for the Flying Gipsy Moths as an Aid in Locating New Infestations,* Technical Bulletin no. 336. Washington, DC: United States Department of Agriculture, 43pp.

Conner, W. E. 1997. Ultrasound: its role in the courtship of the arctiid moth, *Cycnia tenera. Experientia* 43:1029–1031.

Conner, W. E. 1999. "Un chant d'appel amoreuax": acoustic communication in moths. *Journal of Experimental Biology* 202:1711–1723.

Cooperband, M. F., and R. T. Cardé. 2006. Orientation of *Culex* mosquitoes to carbon dioxide-baited traps: flight manoeuvres and trapping efficiency. *Medical and Veterinary Entomology* 26:11–26.

David, C. T., and J. S. Kennedy. 1987. The steering of zigzagging flight by male gypsy moths. *Naturwissenschaften* 74:194–196.

David, C. T., J. S. Kennedy, A. R. Ludlow, J. N. Perry, and C. Wall. 1982. A reappraisal of insect flight towards a distant point source of wind-borne odor. *Journal of Chemical Ecology* 8:1207–1215.

David, C. T., J. S. Kennedy, and A. R. Ludlow. 1983. Finding of a sex pheromone source by gypsy moths released in the field. *Nature* 303:804–806.

Dekker, T., and R. T. Cardé. 2011. Moment-to-moment flight manoeuvres of the female yellow fever mosquito (*Aedes aegypti* L.) in response to plumes of carbon dioxide and human skin odour. *Journal of Experimental Biology* 214:3480–3494.

Dubatolov, V. V., and K. W. Philip. 2013. Review of the northern Holarctic *Arctia caja* complex (Lepidoptera: Noctuidae, Arctiinae). *The Canadian Entomologist* 145:147–154.

Duistermars, B. J., D. M. Chow, and M. A. Frye. 2009. Flies require bilateral sensory inputs to track odor gradients in flight. *Current Biology* 19:1301–1307.

Dusenbery, D. B. 1989. Optimal search direction for an animal flying or swimming in a wind or current. *Journal of Chemical Ecology* 15:2511–2519.

Dusenbery, D. B. 1990. Upwind searching for an odor plume is sometimes optimal. *Journal of Chemical Ecology* 16:1971–1976.

Elkinton, J.S., and R.T. Cardé. 1983. Appetitive flight behavior of male gypsy moths (Lepidoptera: Lymantriidae). *Environmental Entomology* 12:1702–1707.

Elkinton, J.S., and R.T. Cardé. 1984. Odor dispersion, Pp. 73–91. In W.J. Bell and R.T. Cardé, eds. *Chemical Ecology of Insects*. London: Chapman & Hall Ltd.

Elkinton, J.S., R.T. Cardé, and C.J. Mason. 1984. Evaluation of time-average dispersion models for estimating pheromone concentration in a deciduous forest. *Journal of Chemical Ecology* 10:1081–1108.

Elkinton, J.S., C. Schal, T. Ono, and R.T. Cardé. 1987. Pheromone puff trajectory and upwind flight of male gypsy moths in a forest. *Physiological Entomology* 12:399–406.

Fabre, J.H. 1916. *The Life of the Caterpillar*. Translated by A. Teixeira de Mattos. New York: Dodd, Mead and Co.

Fares, Y., P.J.H. Sharpe, and C.E. Magnuson. 1983. Pheromone dispersion in forests. *Journal of Theoretical Biology* 84:355–359.

Farkas, S.R., and H.H. Shorey. 1972. Chemical trail-following by flying insects: a mechanism for orientation to a distant odor source. *Science* 178:67–68.

Ferguson, D.C. 1969. *A Revision of the Moths of the Subfamily Geometrinae of America North of Mexico (Insecta, Lepidoptera)*, Bulletin 29. New Haven, CT: Peabody Museum of Natural History, Yale University.

Fernald, C.H. 1896. The gypsy moth: *Porthetria dispar* (L.). Pp. 255–495. In E.H. Forbush and C.H. Fernald, eds. *The Gypsy Moth: Porthetria dispar (Linn.)*. Boston, MA: Wright & Potter Printing Co.

Foster, S.P., and A.J. Howard. 1999. The effects of source dosage, flight altitude, wind speed and ground pattern on sex pheromone-mediated flight manoeuvres of male lightbrown apple moth, *Epiphyas postvittana* (Walker). *New Zealand Journal of Zoology* 26:97–104.

Gaudry, Q., K.I. Nagel, and R.I. Wilson. 2012. Smelling on the fly: sensory cues and strategies for olfactory navigation in *Drosophila*. *Current Opinion in Neurobiology* 22:216–222.

Gaudry, Q., E.J. Hong, J. Kain, B.J. de Bivort, and R.I. Wilson. 2013. Asymmetric neurotransmitter release enables rapid lateralization in *Drosophila*. *Nature* 493:424–428.

Girling, R.D., B.S. Higbee, and R.T. Cardé. 2013. The plume also rises: trajectories of pheromone plumes issuing from point sources in an orchard canopy at night. *Journal of Chemical Ecology* 39:1150–1160.

Grant, G.G., D. Frech, L. MacDonald, K.N. Slessor, and G.G.S. King. 1987. Scales of the female whitemarked tussock moth, *Orgyia leucostigma* (Lepidoptera: Lymantriidae): identification and behavioral role. *Journal of Chemical Ecology* 13:345–356.

Greenfield, M.D. 1981. Moth sex pheromones: an evolutionary perspective. *Florida Entomologist* 64:4–17.

Greenfield, M.D., and J.D. Coffelt. 1983. Reproductive behaviours of the lesser wax moth, *Achroia grisella* (Pyralidae: Galleriinae): signaling, pair formation, male interactions, and mate guarding. *Behaviour* 84:287–315.

Haworth, A. 1832. *Insect Miscellanies*. Boston, MA: Lily and Wait/Carter and Hendee, pp. 215–216.

Hinterwith, A.J., B. Medina, J. Lockey, D. Otten, J. Voldman, J.H. Lang, J.G. Hildebrand, and T.L. Daniel. 2012. Wireless stimulation of antennal muscles in freely flying hawkmoths leads to flight path changes. *PLOS ONE* 7(12):e52725.

Jordon, A.T., T.H. Jones, and W.E. Conner. 2005. If you've got it, flaunt it: ingested alkaloids affect coremal display in the salt marsh moth, *Estigmene acrea*. *Journal of Insect Science* 5:1, 6pp.

Justus, K.A., and R.T. Cardé. 2002. Flight behaviour of males of two moths, *Cadra cautella* and *Pectinophora gossypiella*, in homogenous clouds of pheromone. *Physiological Entomology* 27:67–75.

Justus, K.A., J. Murlis, C. Jones, and R.T. Cardé. 2002a. Measurement of odor-plume structure in a wind tunnel using a photoionization detector and a tracer gas. *Environmental Fluid Mechanics* 2:115–142.

Justus, K.A., S.W. Schofield, J. Murlis, and R.T. Cardé. 2002b. Flight behaviour of *Cadra cautella* males in rapidly pulsed pheromone plumes. *Physiological Entomology* 27:58–66.

Kaissling, K.-E. 1996. Peripheral mechanism of pheromone reception in moths. *Chemical Senses* 21:257–268.

Kaissling, K.-E. 1997. Pheromone-controlled anemotaxis in moths. Pp. 343–374. In M. Leher, ed. *Orientation and Communication in Arthropods*. Basel: Birkhäuser Verlag.

Kennedy, J.S. 1940. The visual responses of flying mosquitoes. *Proceedings of the Zoological Society of London* 109:221–242.

Kennedy, J.S. 1951. The migration of the desert locust (*Schistocerca gregaria* Forsk.) *Philosophical Transactions of the Royal Society of London B, Biological Sciences* 235:163–290.

Kennedy, J.S. 1977. Olfactory responses to distant plants and other odor sources. Pp. 67–91. In H.H. Shorey and J.J. McKelvey, Jr., eds. *Chemical Control of Insect Behavior: Theory and Application*. New York: John Wiley & Sons.

Kennedy, J.S. 1983. Zigzagging and casting as a programmed response to wind-borne odour: a review. *Physiological Entomology* 8:109–120.

Kennedy, J.S. 1986. Some current issues in orientation to odour sources. Pp. 11–25. In T.L. Payne, M.C. Birch, and C.J.E. Kennedy, eds. *Mechanisms in Insect Olfaction*. Oxford: Oxford University Press.

Kennedy, J.S., and D. Marsh. 1974. Pheromone-regulated anemotaxis in flying moths. *Science* 184:999–1001.

Kennedy, J.S., A.R. Ludlow, and C.J. Saunders. 1981. Guidance of flying male moths by wind-borne sex pheromone. *Physiological Entomology* 6:395–412.

Kindl, J., B. Kalinová, M. Červenka, M. Jílek, and I. Valterová. 2011. Male moth songs tempt females to accept mating: the role of acoustic and pheromonal communication in the reproductive behaviour of *Aphomia sociella*. *PLoS One* 6(10): e26476. doi:10.1371/journal.pone.0026476.

KonDo, Y., H. Naka, and K. Tsuchida. 2012. Pheromones and body coloration affect mate recognition in the Japanese nine-spotted moth, *Amata fortunei* (Lepidoptera: Arctiidae). *Journal of Ethology* 30:301–308.

Kuenen, L.P.S. 2013. Flying faster: flight height affects orthokinetic responses during moth flight to sex pheromone. *Journal of Insect Behavior* 26:57–68.

Kuenen, L.P.S., and T.C. Baker. 1982a. Optomotor regulation of ground velocity in moths during flight to sex pheromone at different heights. *Physiological Entomology* 7:57–68.

Kuenen, L.P.S., and T.C. Baker. 1982b. The effects of pheromone concentration on the flight behaviour of the oriental fruit moth. *Physiological Entomology* 7:423–434.

Kuenen, L.P.S., and R.T. Cardé. 1994. Strategies for recontacting a lost pheromone plume: casting and upwind flight in the male gypsy moth. *Physiological Entomology* 19:15–29.

Lacey, E.S., and R.T. Cardé. 2011. Activation, orientation, and landing of female *Culex quinquefasciatus* in response to carbon dioxide and odour from human feet: 3-D flight analysis in a wind tunnel. *Medical and Veterinary Entomology* 25:94–103.

Lacey, M.J., and C.J. Saunders. 1992. Chemical composition of sex pheromone of oriental fruit moth and rates of release by individual female moths. *Journal of Chemical Ecology* 18:1421–1435.

Linn, C.E., Jr., M.G. Campbell, and W.L. Roelofs. 1987. Pheromone components and active spaces: what do moths smell and where do they smell it? *Science* 237:650–652.

Loudon, C. 2003. The biomechanical design of an insect antenna as an odor capture device. Pp. 609–630. In G.J. Blomquist and R.G. Vogt, eds. *Insect Pheromone Biochemistry and Molecular Biology*. Amsterdam: Elsevier Academic Press.

Mallet, J. 1984. Sex roles in the ghost moth *Hepialus humuli* (L.) and a review of mating in the Hepialidae (Lepidoptera). *Zoological Journal of the Linnaean Society* 79:67–82.

Mafra-Neto, A., and R.T. Cardé. 1994. Fine-scale structure of pheromone plumes modulates upwind orientation of flying moths. *Nature* 369:142–144.

Mafra-Neto, A., and R.T. Cardé. 1995. Influence of plume structure and pheromone concentration on upwind flight of *Cadra cautella* males. *Physiological Entomology* 20:117–133.

Mafra-Neto, A., and R.T. Cardé. 1996. Dissection of the pheromone-modulated flight of moths using single-pulse response as a template. *Experientia* 52:373–379.

Mafra-Neto, A. and R.T. Cardé. 1998. Rate of realized interception of pheromone pulses in different wind speeds modulates almond moth orientation. *Journal of Comparative Physiology A* 182:563–572.

Maimon, G., A.D. Straw, and M.H. Dickinson. 2008. A simple vision-based algorithm for decision making in flying *Drosophila*. *Current Biology* 18:463–470.

Marsh, D., J.S. Kennedy, and A.R. Ludlow. 1978. An analysis of anemotactic zigzagging flight in male moths stimulated by pheromone. *Physiological Entomology* 3:221–240.

Mechaber, W.L., C.T. Capaldo, and J.G. Hildebrand. 2002. Behavioral responses of adult female tobacco hornworms, *Manduca sexta*, to hostplant volatiles change with age and mating status. *Journal of Insect Science* 2(5):1–8. Available online at: insectscience.org/2.5.

Mell, R. 1922. *Biologie und Systematik der südchinesischen Sphingiden*. Berlin: R. Freidlander & Sohn. [In German]

Michener, C.D. 1952. The Saturniidae (Lepidoptera) of the western hemisphere. *Bulletin of the American Museum of Natural History* 98:335–502.

Miksad, R.W., and J. Kittredge. 1979. Pheromone aerial dispersal: a filament model. 14th Conference on Agricultural and Forest Meteorology. American Meteorological Society, pp. 236–243.

Morton, E.S. 2009. The function of multiple mating by female promethea moths, *Callosamia promethea* (Drury) (Lepidoptera: Saturniidae). *The American Midland Naturalist* 162:7–18.

Murlis, J., and C.D. Jones. 1981. Fine-scale structure of odour plumes in relation to distant pheromone and other attractant sources. *Physiological Entomology* 6:71–86.

Murlis, J., B.W. Bettany, J. Kelley, and L. Martin. 1982. The analysis of flight paths of male Egyptian cotton leafworm moths, *Spodoptora littoralis*, to a sex pheromone source in the field. *Physiological Entomology* 7:435–441.

Murlis, J., J.S. Elkinton, and R.T. Cardé. 1992. Odor plumes and how insects use them. *Annual Review of Entomology* 37:505–532.

Murlis, J., M.A. Willis, and R.T. Cardé. 2000. Spatial and temporal structures of pheromone plumes in fields and forests. *Physiological Entomology* 25:211–222.

Ono, T. 1977. The scales as a releaser of the copulation attempt in Lepidoptera. *Naturwissenschaften* 64:386–387.

Ono, T. 1980. Role of scales as a releaser of copulation attempt in the silkworm, *Bombyx mori* (Lepidoptera: Bombycidae). *Kontyu* 44:540–544.

Preiss, R., and E. Kramer. 1986. Mechanism of pheromone orientation in flying moths. *Naturwissenschaften* 73:555–557.

Rau, P., and N.L. Rau. 1929. The sex attraction and rhythmic periodicity in giant saturniid moths. *Transactions of the Academy of Science of St. Louis* 26:83–221.

Reynolds, A.M., D.R. Reynolds, A.D. Smith, G.P. Svensson, and C. Löfstedt. 2007. Appetitive flight patterns of male *Agrotis segetum* moths over landscape scales. *Journal of Theoretical Biology* 245:141–149.

Riffell, J.A., H. Lei, J.A. Christensen, and J.G. Hildebrand. 2009. Characterization and coding of behaviorally significant odor mixtures. *Current Biology* 19:335–340.

Rucker, C.N., and K.F. Haynes. 2004. Impairment of optomotor anemotaxis in yellow-eyed mutant cabbage looper moths, *Trichoplusia ni*. *Journal of Insect Behavior* 17:437–442.

Rutkowski, A.J., R.D. Quinn, and M.A. Willis. 2009. Three-dimensional characterization of the wind-borne pheromone tacking behaviour of male hawkmoths, *Manduca sexta*. *Journal of Comparative Physiology A* 195:39–54.

Sabelis, M.W., and P. Schippers. 1984. Variable wind directions and anemotactic strategies of searching for an odour plume. *Oecologia* 63:225–228.

Sanders, C.J. 1985. Flight speed of male spruce budworm moths in a wind tunnel at different wind speeds and at different distances from a pheromone source. *Physiological Entomology* 10:83–88.

Sane, S.P., A. Dieudonné, M.A. Willis, and T.L. Daniel. 2007. Antennal mechanosensors mediate flight control in moths. *Science* 315:863–866.

Sarto i Monteys, V., P. Acin, G. Rosell, C. Quero, M.A. Jiménez, and A. Guerrero. 2012. Moths behaving like butterflies. Evolutionary loss of long range attractant pheromones in Castniid moths: a *Paysandisia archon* model. *PLOS ONE* 7(1):e29282.

Schal, C. 1982. Intraspecific vertical stratification as a mate finding mechanism in tropical cockroaches. *Science* 215:1405–1407.

Sharov, A.A., A.M. Liebhold, and F.W. Ravlin. 1995. Prediction of gypsy moth (Lepidoptera: Lymantriidae) mating success from pheromone trap counts. *Environmental Entomology* 24:1239–1244.

Symonds, M.R.E., T.L. Johnson, and M.A. Elgar. 2011. Pheromone production, male abundance, body size, and the evolution of elaborate antennae in moths. *Ecology and Evolution* 2:227–246.

Takasaki, T., S. Namiki, and R. Kanzaki. 2012. Use of bilateral information to determine walking direction during orientation to a pheromone source in the silkmoth *Bomby mori*. *Journal of Comparative Physiology A* 198:295–307.

Tejima, S., T. Ono, and M. Sakuma. 2013. Aim-then-shoot anemo-taxis in the hopping approach of the potato tuberworm moth, *Phthorimaea operculella* toward a sex pheromone source. *Physiological Entomology* 38:292–301.

Thistle, H.W., H. Peterson, G. Allwine, B. Lamb, T. Strand, E.H. Holstein, and P. Shea. 2004. Surrogate pheromone plumes in three forest trunk spaces: composite statistics and case studies. *Forest Science* 50:610–625.

Thornhill, R. 1979. Male and female selection and the evolution of mating strategies in insects. Pp. 81–121. In M.S. Blum and N.A. Blum, eds. *Sexual Selection and Reproductive Competition in Insects*. New York: Academic Press.

Tindale, N.B. 1932. Revision of the Australian ghost moths (Lepidoptera Homoneura, family Hepialidae): part 1. *Records of the South Australian Museum* 4:97–536.

Trivers, R.L. 1972. Parental investment and sexual selection. Pp. 136–179. In B. Campbell, ed. *Sexual Selection and the Descent of Man 1871–1971*. Chicago, IL: Aldine.

Vickers, N.J. 2000. Mechanisms of animal navigation in odor plumes. *Biological Bulletin* 198:203–212.

Vickers, N.J. 2006. Winging it: moth flight behavior and responses of olfactory neurons are shaped by pheromone plume dynamics. *Chemical Senses* 31:155–166.

Vickers, N.J., and T.C. Baker. 1991. The effects of unilateral antennectomy on the flight behaviour of male *Heliothis virescens* in a pheromone plume. *Physiological Entomology* 16:497–506.

Vickers, N.J., and T.C. Baker. 1992. Male *Heliothis virescens* maintain upwind flight in response to experimentally pulsed filaments of their sex pheromone (Lepidoptera: Noctuidae). *Journal of Insect Behavior* 5:669–687.

Vickers, N.J., and T.C. Baker. 1994a. Reiterative responses to single strands of odor promote sustained upwind flight and odor source location by moths. *Proceedings of the National Academy of Sciences of the United States of America* 91:5756–5760.

Vickers, N.J., and T.C. Baker. 1994b. Visual feedback in the control of pheromone-mediated flight of *Heliothis virescens* males (Lepidoptera: Noctuidae). *Journal of Insect Behavior* 7:605–632.

Vickers, N.J., and T.C. Baker. 1996. Latencies of behavioral response to interception of filaments of sex pheromone and clean air influence flight track shape in *Heliothis virescens* (F.) males. *Journal of Comparative Physiology A* 178:831–847.

Vickers, N.J., and T.C. Baker. 1997. Flight of *Heliothis virescens* males in the field in response to sex pheromone. *Physiological Entomology* 22:277–285.

Vickers, N.J., T.A. Christensen, T.C. Baker, and J.G. Hildebrand. 2001. Odour-plume dynamics influence the brain's olfactory code. *Nature* 410:466–470.

von Keyserlingk, H. 1984. Close range orientation of flying Lepidoptera to pheromone sources in a laboratory wind tunnel and the field. *Mededelingen van de Faculteit Landbouwwetenschappen, Rijksuniversiteit Gent* 49:683–689.

Wall, C., and J.N. Perry. 1987. Range of attraction of moth sex-attractant sources. *Entomologia Experimentalis et Applicata* 44: 5–14.

Willis, M.A., and T.C. Baker. 1984. Effects of intermittent and continuous pheromone stimulation on the flight behaviour of the oriental fruit moth, *Grapholita molesta*. *Physiological Entomology* 9:341–358.

Willis, M.A., and T.C. Baker. 1987. Comparison of maneuvers used by walking versus flying *Grapholita molesta* males during pheromone-mediated upwind movement. *Journal of Insect Physiology* 33:875–883.

Willis, M.A., and M.C. Birch. 1982. Male lek formation and female calling in a population of the arctiid moth, *Estigmene acrea*. *Science* 218:168–170.

Willis, M.A., and R.T. Cardé. 1990. Pheromone-modulated optomotor response in male gypsy moths, *Lymantria dispar* L.: upwind flight in a pheromone plume in different wind speeds. *Journal of Comparative Physiology A* 167:699–706.

Willis, M.A., J. Murlis, and R.T. Cardé. 1991. Pheromone-mediated upwind flight of male gypsy moths, *Lymantria dispar*, in a forest. *Physiological Entomology* 16:507–521.

Willis, M.A., C.T. David, J. Murlis, and R.T. Cardé. 1994. Effects of pheromone plume structure and visual stimuli on the pheromone-modulated upwind flight of male gypsy moths (*Lymantria dispar*) in a forest (Lepidoptera: Lymantriidae). *Journal of Insect Behavior* 7:385–409.

Willis, M.A., J.L. Avondet, and E. Zheng. 2011. The role of vision in odor-plume tracking by walking and flying insects. *Journal of Experimental Biology* 214:4121–4132.

Wisniewska, J., and R.T. Cardé. 2012. Visual cues collimate the trajectories of almond moth, *Cadra cautella*, males flying in wind and still air within a wind-formed plume of pheromone. *Physiological Entomology* 37:42–52.

Witzgall, P. 1997. Modulation of pheromone-mediated flight in male moths. Pp. 265–274. In R.T. Cardé and A.K. Minks, eds. *Insect Pheromone Research: New Directions*. New York: Chapman & Hall.

Witzgall, P., and H. Arn. 1990. Direct measurement of the flight behaviour of male moths to calling females and synthetic sex pheromones. *Zeitschrift für Naturforschung* 45c:1067–1069.

Witzgall, P., and H. Arn. 1991. Recording flight tracks of *Lobesia botrana* in the wind tunnel. Pp. 187–193. In I. Hirdý, ed. *Insect Chemical Ecology*. The Hague: SPB Academic Publishing.

Xiao, W., S. Matsuyama, T. Ando, J.G. Millar, and H. Honda. 2012. Unsaturated cuticular hydrocarbon synergizes responses to sex attractant pheromone in the yellow peach moth, *Conogethes punctiferalis*. *Journal of Chemical Ecology* 38:1143–1150.

Zanen, P.O., and R.T. Cardé. 1999. Directional control by male gypsy moths of upwind flight along a pheromone plume in three wind speeds. *Journal of Comparative Physiology A* 184: 21–35.

Zanen, P.O., M.W. Sabelis, J.P. Buonaccorsi, and R.T. Cardé. 1994. Search strategies of fruit flies in steady and shifting winds in the absence of food odours. *Physiological Entomology* 19:335–341.

CHAPTER TWELVE

Male Pheromones in Moths

Reproductive Isolation, Sexy Sons, and Good Genes

WILLIAM E. CONNER and VIKRAM K. IYENGAR

Introduction

The amount of information published on the pheromones of moths is voluminous, and the great majority of it focuses on the female attractants of pest insects because of their economic importance (Cardé and Minks 1997; Blomquist and Vogt 2003; Cardé and Millar 2004; El-Sayed 2011; Ando 2012). From the perspectives of an animal behaviorist and an evolutionary biologist, however, male pheromones are of equal if not greater interest. Male courtship pheromones and the structures and behaviors associated with their release are the grist of sexual selection. This chapter is devoted to male moth pheromones, their biology, chemistry, and evolution. It will begin with a discussion of the genetic architecture of sex determination in Lepidoptera and how it predisposes males to extravagance. We will describe the benefits accrued by females who exercise mate choice based on male courtship pheromones, and we will emphasize empirical tests of the models for sexual

selection through mate choice using the case study of *Utetheisa ornatrix* (Erebidae). We will also describe the diversity of male pheromone releasing structures, the propensity for diet dependence in male pheromones, and the unusual behaviors and physiology associated with pheromone precursor collection. Our primary objective is to demonstrate that male pheromones are worth studying and their stories are some of the most interesting in chemical ecology. Previous reviews of male courtship pheromones in moths include those of Birch (1970a, 1970b, 1972, 1974), Weatherston and Percy (1977), Haynes and Birch (1985), Tamaki (1985), Birch and Hefetz (1987), Baker (1989), Birch et al. (1990), and Krasnoff (1997).

Genetic Architecture in Lepidoptera

Lepidopterans and birds share a "reversed" genetic architecture that determines sex. In both birds and Lepidoptera, and

191

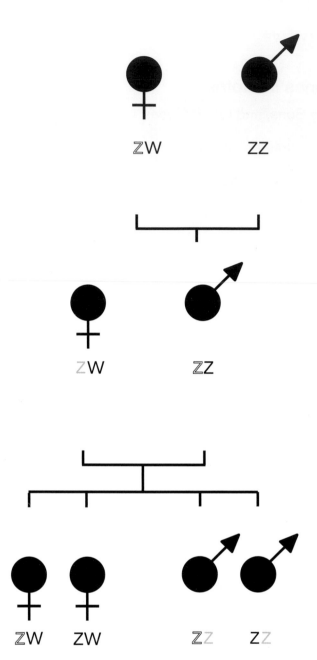

FIGURE 12.1 Inheritance of a Z-linked mating preference. The white Z represents the sex chromosome bearing the preference gene(s), shown here to be introduced into the lineage by the paternal grandmother. The gray Z is that possessed by the mother but not transmitted to daughters. Therefore, grand-daughters can only acquire Z-linked gene(s) by way of their father (after Iyengar et al. 2002 and reprinted with permission from *Nature*).

traits through sexual selection. The protected invasion theory also predicts that the female preference alleles will be correlated with the preferred male trait alleles, a prerequisite for both Fisherian and good genes models of sexual selection (see below). Experimental evidence for the paternal inheritance of a female moth's mating preference and a correlation of female mating preference and male target traits have been found in the mating system of *Utetheisa ornatrix* (see case study).

Mate Choice

Given the diversity of scents and disseminating structures in male moths (see figure 12.2 and Table 12.1), there can be little doubt that male courtship pheromones have arisen through sexual selection (Birch et al. 1990). The roles played by males and females in moth courtship are consistent with those predicted by disparities in parental investment between the two sexes (Trivers 1972; Thornhill 1979; Cardé and Baker 1984). In keeping with their large contribution to offspring in the form of nutrient-rich eggs, females play a less costly role in the communication system (Hammerstein and Parker 1987). Females generally remain stationary and emit vanishingly small amounts of pheromones that draw males upwind to them. Females are said to be the limiting, choosy sex because of the time and energy necessary to produce and replenish eggs (Trivers 1972; Andersson 1994) and they maximize their fitness through the choice of high-quality mates. Because of their smaller per capita contribution to offspring and ability to replenish sperm quickly, males, the non-limiting sex, bear the cost of searching for and competing for females. Male courtship pheromones with their associated elaborate disseminating structures frequently mediate female mate choice (Baker and Cardé 1979; Greenfield 2002) and males maximize their fitness through multiple matings. Phelan (1992) nicely summarized and extended the logic behind the disparate roles of the sexes in moth chemical communication in his asymmetric tracking hypothesis.

Beyond these general patterns the critical questions are those concerned with the specific benefits of female mate choice. In this sense studies of moth courtship and the role of male pheromones have mirrored the trends of the sexual selection literature over the last three decades (Bradbury and Andersson 1987).

Species or Mate Recognition

One of the key ideas of modern biology is the concept of reproductive isolation. It is the foundation of the biological species concept and fundamental to the process by which species form and are maintained (Coyne and Orr 2004). It has been suggested that one important function of male scent disseminating structures in moths and the pheromones that they release is that they promote species recognition, maintain reproductive isolation, and prevent hybridization and the loss of fitness associated with it. Evidence for these functions for male pheromones, however, is scant (Birch et al. 1990). An exception is an insightful meta-analysis by Phelan and Baker (1987) in which they measured whether the presence of scent disseminating structures in members of five families of moths (Phycitinae of North America and Europe, Yponomeutidae of Japan, Tortricidae and Noctuidae of Great

unlike humans, the female is the heterogametic sex. The implications of this are profound and may have preadapted males to evolve exaggerated traits in these prominent taxa (Reeve and Pfennig 2003; Iyengar and Reeve 2010). Female moths are ZW and males are ZZ (figure 12.1). The protected invasion theory of Reeve and Shellman-Reeve (1997) predicts that female mating preference alleles should be inherited from their fathers and will be transmitted to all of a female's sons that have the father's attractive trait (Iyengar et al. 2002). Such alleles will be less vulnerable to chance loss when rare and more likely to fuel the evolution of exaggerated male

Britain, and Ethmiidae from the western United States) was correlated with sharing the same host plant. They reasoned that if two species shared a host plant they would be more likely to make mating mistakes increasing the relative strength of selection for species recognition. They found a significant correlation between sharing a host plant and the presence of male scent disseminating structures across these groups and concluded that male pheromones promote reproductive isolation. Further, they suggest that the mechanism for the evolution of reproductive isolation was runaway sexual selection, a mechanism consistent with the pattern of male pheromones arising repeatedly and independently across taxa. Although correctly criticized for analyzing species as independent units and for not analyzing independent evolutionary events phylogenetically (Krasnoff 1997; Greenfield 2002), the study stands alone as the most comprehensive cross-taxa test of the species-recognition hypothesis in moths. Additional meta-analyses with proper controls are needed (Krasnoff 1997). For a model phylogenetically controlled study of the evolution of male moth pheromones see Wagner and Rosovsky (1991) who studied the evolution of male scent disseminating structures in the Hepialidae, a primitive family of Lepidoptera.

Additional evidence for a species-recognition function has been gathered by focusing on small clades of insects like *Ephestia elutella*, *Cadra figulilella*, and related phycitinae Pyralidae (Phelan and Baker 1990a, 1990b, and references therein) which have complex courtship sequences, the choreography of which involve the delivery of male pheromones. Similarly the multicomponent pheromone blend of *Grapholita molesta* and a complex courtship ritual point to a species-recognition function (Baker and Cardé 1979; Baker et al. 1981). The role of male courtship pheromones in reproductive isolation in members of the genus *Heliothis* (Noctuidae) has received renewed interest (Hillier and Vickers 2004, 2011; Hillier et al. 2007) and the recent studies of Lassance and Löfstedt (2009) suggest a similar function for the hairpencil pheromones of *Ostrinia* species (Crambidae) (for additional examples see Table 12.1).

Yet in some moth groups like the Arctiinae of the Erebidae (formerly the Arctiidae) there is a profound lack of species specificity in male courtship pheromones. This is true despite efforts to identify minor pheromone components (Schulz 2009) and pheromone components, when they do exist, can vary with diet (Krasnoff and Roelofs 1989). This is difficult to reconcile with a species-recognition function and requires a different focus, that of mate assessment.

Mate Assessment

Biologists have also looked at mating systems from the perspective mate assessment, usually exercised through female preferences for male traits (Andersson 1994; Johannson and Jones 2007). Traditionally the benefits of mate assessment have been divided into discrete categories: direct phenotypic or material benefits and indirect genetic benefits including good genes and sexy sons (Fisherian runaway sexual selection). It is important to note that these are not mutually exclusive and indeed may coexist in the same population. We briefly summarize the evidence for mate assessment based on male courtship pheromones in moths and benefits that females may accrue through mate choice.

Benefits of Mate Assessment

Direct (Phenotypic) Benefits

In some species, females have a propensity to choose mates based on tangible resources provided to the female during courtship. Direct (phenotypic) or material benefits are the result of nongenetic quantities that have a positive impact on the survivorship and offspring production of the female such as enhanced paternal care (Davies 1992), the transmission of antipredator defensive chemicals (González et al. 1999), and the acquisition of nutrient-rich nuptial gifts (Thornhill and Alcock 1983; Gwynne 1984).

In Lepidoptera, sperm, nutrients, and defensive chemicals are transferred from the male to the female in the form of a spermatophore during prolonged copulation (Thornhill and Alcock 1983; Dussourd et al. 1988, LaMunyon and Eisner 1994; Vahed 1998). The nongametic constituents have been referred to as nuptial gifts (Thornhill and Alcock 1983) and they can represent a significant fraction of a male's body mass (LaMunyon and Eisner 1994). Much of the information on direct phenotypic benefits comes from tiger moths (Erebidae, Arctiinae) and/or the convergent evolutionary system in the Ithomiine and Danaine butterflies (not covered here; for an excellent review see Boppré 1990). The spermatophores of some arctiines contain nutrients plus pyrrolizidine alkaloids—PAs (Dussourd et al. 1988), defensive chemicals sequestered during larval life (Hartmann 2009) and/or collected from plant sources in adulthood—pharmacophagy (Conner and Jordan 2009). Male pheromones in this group are dihydropyrrolizines derived from the necine base of the PAs with the main pheromonal components being danaidal, hydroxydanaidal, and rarely danaidone (Schulz 2009). The PA-derived pheromones have the potential to act as indicators of male quality in the form of alkaloid load, potential to transfer alkaloids to the female, and additional correlated phenotypic characters.

In the erebid *Cosmosoma myrodora* (figure 12.2) the nuptial gift is bestowed in spectacular fashion (Conner et al. 2000). The male possesses a ventral abdominal pouch filled with deciduous flocculent scales laden with the pyrrolizidine alkaloids intermedine and lycopsamine obtained through adult pharmacophagy. During courtship the male showers the female with up to four doses of flocculent. The alkaloidal cloak bestowed by the male at least temporarily protects the female from predation (as evidenced by the experiments showing that *Nephila clavipes*, the golden orb-weaving spider, cut females cloaked in flocculent scales from their webs). Males also transfer additional alkaloids to the female in a more traditional spermatophore. Males that do not utilize the flocculent in courtship are discriminated against by virgin females.

In erebids the pyrrolizine alkaloids transferred during courtship are quickly transferred to the eggs (Dussourd et al. 1988). In species like *Cosmosoma* this is particularly critical, because the female does not normally forage for alkaloids on her own. As predicted by the parental investment theory described above, it is the male that takes on the costly task of collecting alkaloids and females "forages" only by mating with alkaloid-laden males. The critical question in species with nuptial gifts is whether direct material benefits are the prime benefits for females or whether indirect genetic benefits are more important.

TABLE 12.1

Moth species with known male courtship pheromones (restricted to those with chemical characterization and measured female response)

Taxon	Behavioral role[a]	Disseminating structure	Chemical components	Female response	Male sound production	References
Tortricoidea						
Tortricidae						
Grapholita molesta	RI MA SS	Abdominal hair pencils	Methyl 2-epijasmonate trans-ethyl cinnamate	Short range attraction of female		Baker and Cardé (1979) Baker et al. (1981) Nishida et al. (1982) Löfstedt et al. (1989, 1990)
Pyraloidea						
Pyralidae						
Corcyra cephalonica	RI	Wing glands	(E,E)-farnesal (Z,E)-farnesal	Short range attraction of female	+	Spangler (1987)[s] Zagatti et al. (1987)
Eldana saccharina	RI	Wing gland	(E)-3,7dimethyl-6-octen-4olide (many additional components)	Short range attraction of female	+	Kunesch et al. (1981) Bennett et al. (1991)[s] Burger et al. (1993)
	MA	Abdominal hair pencils	Vanillin 4-hydroxybenzaldehyde (many additional components)	Female acceptance		Zagatti (1981) Zagatti et al. (1981)
Ephestia elutella	RI MA	Costal fold wing gland	(E)-phytol γ-decalactone γ-undecalactone	Ventral flexion of abdomen	+	Krasnoff and Vick (1984) Phelan et al. (1986) Trematerra and Pavan (1994)[s]
Galleria mellonella	RI	Wing glands	*n*-undecanal *n*-nonalal	Attracts female (minor components including aldehydes, alcohols, and ketones)	+	Roller et al. (1968) Spangler (1986)[s] Lebedeva et al. (2002)
Achroia grisella	SS	Wing glands	*n*-undecanal *n*-11-octadecenal	Excites female	+	Dahm et al. (1971, but see Greenfield and Coffelt 1983 Spangler (1984)[s] Collins et al. (1999)[s] Jang and Greenfield (1996)[s] Jia and Greenfield (1997)[s] Jang et al. (1997)[s] Jang and Greenfield (2000)[s]

Crambidae

Taxon	Code	Abdominal and genital hair pencils	Pheromone	Female acceptance behavior	References
Ostrinia nubilalis	RI MA		Hexadecanyl acetate (Z)-9 hexadecenyl acetate (Z)-11 hexadecenyl acetate (Z)-14 hexadecenyl acetate		Lassance and Löfsted (2009) Nakano et al. (2008)ˢ; sound in O. furnacalis

Noctuiodea

Erebidae
Arctiinae
Arctiini

Taxon	Code	Hair pencils	Pheromone	Female acceptance behavior	References
Creatonotus gangis	MA	Abdominal coremata	R-(–)-hydroxydanaidal	Lek formation	Schneider et al. (1982) Bell and Meinwald (1986) Wunderer et al. (1986)
Creatonotus transiens	MA	Abdominal coremata	R-(–)-hydroxydanaidal	Lek formation	Schneider et al. (1982) Bell et al. (1984) Bell and Meinwald (1986) Wink et al. (1988) Egelhaaf et al. (1990) Wink and Schneider (1990) von Nickisch-Rosenegk and Wink (1994) Wunderer et al. (1986)
Estigmene acrea	MA	Abdominal coremata	Hydroxydanaidal	Lek formation	Willis and Birch (1982) Krasnoff and Roelofs (1989) Davenport and Conner (2003) Jordan et al. (2005, 2007) Jordan and Conner (2007)
Phragmatobia fuliginosa (L.)	lost	Abdominal coremata	Danaidal Hydroxydanaidal	Tymbalar sound	– Krasnoff et al. (1987) Krasnoff and Roelofs (1989, 1990) von Nickisch-Rosenegk and Wink (1993)
Pyrrharctia isabella (JE Smith)	lost	Abdominal coremata	Hydroxydanaidal	Tymbalar sounds	Krasnoff et al. (1987) Krasnoff and Yager (1988) Krasnoff and Roelofs (1989, 1990)

Callimorphini (s.l.)

Taxon	Code	Hair pencils	Pheromone	Female acceptance behavior	References
Euplagia quadripunctataria	MA	Abdominal coremata	Hydroxydanaidal Danaidal Ethyl esters	EAG response	Schneider et al. (1998)

(continued)

TABLE 12.1 (continued)

Taxon	Behavioral role[a]	Disseminating structure	Chemical components	Female response	Male sound production	References
Haploa clymene	MA	Abdominal coremata	Hydroxydanaidal	Female acceptance		Davidson et al. (1997)
Pareuchaetes pseudoinsulata (Rego Barros)	MA	Abdominal	Hydroxydanaidal	EAG response		Schneider et al. (1992)
Utetheisa ornatrix	MA MB GG SS SSP	Genitalic coremata	R-(-)-hydroxydanaidal	Female acceptance	–	Conner et al. (1981) Dussourd et al. (1988) Dussourd et al. (1991) Lamunyon and Eisner (1993) Lamunyon and Eisner (1994) Iyengar and Eisner (1999a) Iyengar and Eisner (1999b) Iyengar et al. (2001) Iyengar et al. (2002) Bezzerides et al. (2005) Iyengar and Reeve (2010)
Noctuidae						
Conogethes punctiferalis	RI MA	Abdominal hair pencils	Tiglic acid	Female acceptance		Takayoshi and Hiroshi (1999)
Heliothis virescens	RI MA	Abdominal hair pencils	Tetradecanoic acid Hexadecanol Hexadecanoic acid Hexadecanyl acetate Octadecanol Octadecanoic acid Octadecanyl acetate	Abdominal extension in female; inhibition of upwind flight by male		Teal and Tumlinson (1989) Teal et al. (1989) Hillier and Vickers (2004) Hillier and Vickers (2011) Hillier et al. (2007)
Helicoverpa armigera	RI	Abdominal hair pencils	(Z)-11-hexadecen-1-ol	Inhibition of male response to female		Huang et al. (1996)
Mamestra brassica	RI MA	Abdominal scent brushes	2-phenyl ethanol 2-methyl butanoic acid benzaldehyde 2-methyl propanoic acid Benzyl alcohol Phenol	Female acceptance		Jacquin et al. (1991)

Species		Gland/structure	Chemicals	Function		References
Phlogophora meticulosa	RI MA	Abdominal scent brushes	6-methyl-5-heptene-2-one 6-methyl-5-heptene-2-ol 2-methyl butyric acid	Female acceptance		Aplin and Birch (1970) Birch (1970a)
Plodia interpunctella	RI	Wing gland	Palmitic acid Oleic acid Linoleic acid Ethyl esters	Female exposes abdominal tip between wings	+	Grant (1974) Phelan (1992) Tramaterra and Pavan (1994)[s]
Pseudaletia unipunctata	RI MA	Abdominal scent brushes	Benzyl alcohol Benzaldehyde Acetic acid	Female acceptance		Fitzpatrick and McNeil (1988)
Trichoplusia ni	RI MA	Genitalic hair pencils	(S)-(+) linalool p-cresol m-cresol	Long range attraction of female		Landolt and Heath (1989) Landolt and Heath (1990) Heath et al. (1992) Landolt et al. (1994) Landolt (1995)

a. RI = reproductive isolation; MA = mate assessment; MB = material benefit; GG = good genes; SS = sexy sons; SSP = sexy sperm.
s. Sound production reference.

FIGURE 12.2 (A) Courtship of *Cosmosoma myrodora* (Erebidae). Explosive release of flocculent by the male (right) festooning the female (left) with pyrrolizidine alkaloid-laden scales. (B) Courtship of *Utetheisa ornatrix* (Erebidae). Male (bottom) flexing abdomen and airing genitalic coremata near the antennae of the female (top). (C) Male *Creatonotus transiens* (Erebidae) with abdominal coremata fully inflated (courtesy of Michael Boppré). (D) Abdominal hairpencils of *Ostrinia nubilalis* (Crambidae) (photo provided by Christer Löfsted and reprinted with permission of *BMC Biology*). (E) Abdominal scent brushes of *Heliothis virescens* (Noctuidae) (photo provided by Kirk Hillier and reprinted with permission of *Chemical Senses*).

SOURCE: Photos A through C reprinted with permission of Oxford University Press.

Indirect Genetic Benefits

Indirect (genetic) benefits are those that are experienced in the next generation in the form of increased offspring survivorship ("good genes") and mating success of sons ("sexy sons") (Andersson 1994). There are numerous models (verbal, mathematical, and genetic) that explain the evolution of mate assessment (Bradbury and Andersson 1987), but here we emphasize empirical methods for testing the predictions of the models.

GOOD GENES

To demonstrate a good gene's benefit, four criteria must be met: (1) males in a given population must vary genetically with respect to their survivorship; (2) male behavior and ornamentation (scent disseminating structures, courtship pheromones, and display behaviors for the purposes of this chapter) should provide accurate information on the survival value of their possessors; (3) females must base their choice of mates on male behavior and ornamentation; and (4) offspring should benefit from their mother's choice by increased survival (Heisler et al. 1987). In only one moth species have these criteria been met. In the rattlebox moth, *Utetheisa ornatrix*, the quantity of male courtship pheromone hydroxydanaidal is correlated with male body size and is heritable. Females that choose males with greater quantities of hydroxydanaidal benefit through increased fecundity (Iyengar and Eisner 1999b) and their offspring have an enhanced ability to compete for PA-rich food sources (Kelly et al. 2012).

A subcategory of the good genes argument is the healthy mate hypothesis (Hamilton and Zuk 1982). This idea suggests that females may choose males with strong resistance to diseases and parasites and that courtship cues will sometimes be honest indicators of good health. This idea has never been vetted in the male courtship pheromone arena. Given the recent finding that generalist tiger moth caterpillars preferentially feed on PA-containing host plants when infested with parasitoids (Bernays and Singer 2005; Singer et al. 2009), this hypothesis should be investigated. In this context the benefit of choosing a male with the genetic competence to find and utilize PAs could be the reduced incidence of disease/parasitism in progeny.

SEXY SONS

The idea that the only benefit to a choosy female might be that her male offspring will be chosen in the next generation (Lande 1981) dominated the literature on sexual selection for several decades (Andersson 1994). It is an idea that finds its roots in Darwin's (1871) *The Descent of Man and Selection in Relation to Sex* and was formalized first by Fisher (1930) when he realized that female preference for male traits could theoretically give rise to an evolutionary positive feedback loop. The result would be the exaggeration of male sexually selected traits in a runaway process that continues until the traits become sufficiently deleterious to survival that natural selection brings the process to a halt. Theoretical models are numerous and demonstrate that runaway sexual selection can occur (O'Donald 1967, 1980; Lande 1981; Kirkpatrick 1982), but the underlying assumptions of the models are numerous and critical to outcomes (Andersson 1987). The diversity of male courtship pheromones and the behaviors associated with their release are consistent with Fisherian runaway sexual selection (Baker and Cardé 1979; Baker 1989; Birch et al. 1990). Empirical tests of the Fisherian process are, however, rare, as four criteria must be met. The first is that the signals, courtship pheromones, should arise frequently and randomly across taxa. As a result of their multiple independent origins, courtship pheromones should thus vary in their chemistry and in their mode of release. Studies should have adequate phylogenetic controls as emphasized by Krasnoff (1997). The second requirement is that the male trait be shown to be heritable or correlated with a heritable trait (Iyengar and Eisner 1991a). The third requirement is that the female preference be shown to be heritable (Iyengar et al. 2002). And last it must be shown that the female preference and the male trait are genetically linked (Iyengar et al. 2002). This linkage completes the positive feedback loop that results in runaway sexual selection. This sets the bar for support of Fisherian runaway sexual selection very high.

SEXY SPERM

As described above it has been argued that females will always be the choosy sex and maximize their fitness through the quality of their choices and not their quantity (Williams 1966; Trivers 1972; Parker 1979). Many recent studies have challenged these assumptions and have found that females may be more promiscuous than was predicted (Birkhead and Møller 1998; Arnqvist and Nilsson 2000). Multiple mating could benefit females in many contexts. They could provide the female with serial nuptial gifts and be an unusual form of nutrient and chemical defense "foraging." They could increase the odds of finding a mate of superior genetic quality and they could increase female fecundity (LaMunyon 1997). Keller and Reeve (1995) hypothesized that female promiscuity and male sperm competitiveness could coevolve in a runaway sexual selection process that would ultimately promote sperm competition within the body of the female. This has been dubbed the "sexy sperm" hypothesis. Iyengar and Reeve (2010) have gathered evidence in support of this hypothesis by showing that female promiscuity genes in *Utetheisa ornatrix*, which mates up to 22 times, are Z-linked, a pattern consistent with the sexy sperm hypothesis (see case study on *U. ornatrix*).

It is not possible for all of the potential benefits of mate assessment to be measured in all species; however, they should be addressed for a number of "model" species. Only then will we begin to fully understand the intricacies of moth courtship and the mechanisms that underlie their evolution.

Scent Structures in Male Lepidoptera

The scent disseminating structures of male Lepidoptera are impressive in several respects. They are exceedingly common across taxa, and their varied location, shape, and chemistry suggest repeated evolutionary origins. This pattern is consistent with the genetic bias toward elaboration that stems from the genetic architecture of Lepidoptera described above and can be construed as circumstantial evidence for runaway sexual selection. Depending on their morphology, the disseminating structures are referred to as androconial scales, scent brushes, scent fans, hairpencils, and coremata (Birch et al.

1990). They are also frequently large relative to body size and reminiscent of the feathered displays of male birds (Andersson 1994). The most notable examples are the magnificent abdominal coremata of *Creatonotos* species (figure 12.2). The diversity of male scent disseminating structures is highest in the Gelechiidae, Tortricidae, Pyralidae, Notodontidae, Erebidae, and Noctuidae (Brown et al. 2011).

The newest addition to the list of potential scent disseminating structures involved in courtship are the deciduous flocculent scales associated with pouches located on the venter of the second and third segments of the abdomen (Weller et al. 2000). Although originally described in the context of defense (Blest 1964), deciduous scales that pack these pockets are shed in clouds in the vicinity of the female during courtship as described for *Cosmosoma* above (figure 12.2). They are found in Euchromina arctiines (Erebidae) and have evolved multiple times as evidenced by their association with different segments and fine structure (Weller et al. 2000). Thus far they are restricted to clades associated with PA-containing plants, again strengthening the relationship between defensive chemicals and sex in the Erebidae (Conner and Jordan 2009).

Scent dissemination is usually linked to behaviors associated with normal sexual excitement in males such as wing fanning for wing glands, genital extrusion for scent brushes and coremata associated with male genitalia, and copulatory flexion of the abdomen for the coremata on the intersegmental membranes of the abdomen (figure 12.2). The latter mechanisms take advantage of biomechanical changes during the flexion of the abdomen of the male during copulatory attempts. Abdominal flexion associated with copulatory attempts pressurizes the air-filled tracheal system and hemolymph spaces within the insect. The intersegmental membranes respond to the increased pressure and bulge outward. Pressurization is likely the first evolutionary step in scent dissemination in these species. Elaboration of the intersegmental and genital membranes into coremata and related structures follows. Baker and Cardé (1979) referred to the courtship behavior of *Grapholita molesta* as ritualized copulation referring to the link between copulatory attempts, the scent dissemination mechanism, and sexual selection.

The scent disseminating structures themselves appear to be optimized for the release of small volatile molecules at high flux rates. They are composed of highly modified scales with large surface areas and associated secretory cells (Birch 1970a, 1970b). They are deployed rapidly in time frames frequently measured in milliseconds. Our understanding of the evolution of male courtship pheromone disseminating structures is hampered by a dearth of genetic and developmental studies. The so-called evo-devo studies have illuminated the origins of insect epidermal outgrowths such a sexually selected beetle horns (Moczek et al. 2007). Such studies frequently discover fundamental bauplan genes that can be redeployed to produce new structures and functions. For moth coremata, particularly because they are sexually dimorphic and can be diet dependent, these are tractable questions that would clarify how scent structures arise and diversify. Also as pointed out by Krasnoff (1997), Krasnoff and Roelofs (1990), and Phelan (1992), further gaps in our understanding of the evolution of male courtship pheromones are the result of a lack of comparative phylogenetic studies (but see Phelan and Baker 1990a; Weller et al. 1999). This is being rectified but at a frustratingly slow rate for behaviorists.

Courtship Characteristics

The courtship of moths varies in complexity from the unadorned and simple to the more complicated choreographic exchange of both female and male signals. The primitive condition is exemplified by a female releasing her sex attractant blend (Löfstedt and Kozlov 1997; Greenfield 2002). The male flies upwind and mates with the female without foreplay. We have attempted to arrange the following vignettes in three sections: female responses to male pheromones, male responses to male pheromones, and lekking. The first two sections are generally arranged in order of increasing complexity.

Female Responses to Male Pheromones

INCREASED FEMALE ACCEPTANCE

Removal of scent disseminating structures in male moths frequently diminishes their probability of mating (Birch, 1970a, 1974; Clearwater 1972; Grant 1974, 1976; Hendrikse 1986; Phelan and Baker 1986; Fitzpatrick and McNeil 1988; Cibrian-Tovar and Mitchell 1991; Royer and McNeil 1992; Kimura and Honda 1999). In some cases the proximate cause is not immediately obvious. This prompted the labeling of the male courtship pheromones "aphrodisiac" pheromones—pheromones that increase the likelihood of copulation (Birch 1974).

FEMALE QUIESCENCE

In some cases the only observable behavior is a decrease in the probability of female taking flight or in other ways moving away from the male (Conner 1981; Hillier and Vickers 2004).

SPECIFIC ACCEPTANCE BEHAVIORS

Fine-grained behavioral analysis, often through video recording, has allowed the description of discrete acceptance behaviors. In *Heliothis virescens* abdominal extension, which may be associated with increased female pheromone release, is elicited by natural hairpencil eversion, hairpencil extracts, and blends of synthetic chemicals combined in the proportions of the natural courtship pheromones (Hillier and Vickers 2004). Female acceptance behavior sometimes includes curling of the abdomen toward the substrate, a behavior that has been interpreted as scent marking (Teal et al. 1981; Thibout et al. 1994; Hillier and Vickers 2004). In phycitinae pyralid species female acceptance behavior often involves curling the abdomen upwards between the wings, an adjustment that is necessary for coupling in these species (Krasnoff and Vick 1984; Phelan and Baker 1990a, 1990b; and references therein). Abdominal curling can also be directed toward the male, bringing the genitalia of the female into apposition with the male genitalia and thereby facilitating copulation. An alternative method for exposing the genitalia is to raise the wings allowing a male to approach from the side (Conner 1981; Krasnoff and Roelofs 1990).

FEMALE MOVES AWAY

When exposed to extracts of male courtship pheromones of heterospecific males (*Helicoverpa zea* or *Heliothis subflexa*),

H. virescens females move away suggesting a role of male courtship pheromones in reproductive isolation (Hillier and Vickers 2004).

FEMALE ATTRACTED (SHORT RANGE)

In *Grapholita molesta* males wing fan and extrude their abdominal hairpencils in a pulsatile fashion in the vicinity (<2 cm) of a calling female. Females orient to the wing-generated air currents and move toward the male (Baker and Cardé 1979). Through orientation toward the male, females express a clear "preference" for males with courtship pheromones. Males of the lesser wax moth, *Achroia grisella*, attract females from close range with a combination of courtship pheromones and ultrasound (Dahm et al. 1971; but see Greenfield 2002). In *Galleria mellonella* wing-gland pheromones also attract females from close range (Röller et al. 1968; Leyrer and Monroe 1973; Finn and Payne 1977).

FEMALE ATTRACTED (LONG RANGE)

In *Trichoplusia ni* exposure of genitalic hairpencils attracts females from long (distances greater that those typically associated with moth courtship, i.e., greater than 25 cm) range (Landolt and Heath 1990). Combinations of male pheromones with plant odors or female sex attractants are synergistic in triggering female attraction (Landolt et al. 1994). Female attraction has also been described for *Heliothis virescens* (Heath et al. 1992).

FEMALE DECREASES CALLING

Male hairpencil odors have been reported to trigger a decrease in female calling in tobacco budworm moths, *Heliothis virescens*, and cotton bollworm moths, *Helicoverpa zea* (Hendricks and Shaver 1975; Huang et al. 1996). It is possible that males have exploited a mechanism for the control of female pheromone release that allows them to decrease their competitive interactions with other males. Such a mechanism would operate just prior to copulation, an event that will also decrease female pheromone release.

FEMALE SOUND PRODUCTION

Females of *Phragmatobia fuliginosa* and *Pyrrharctia isabella* (Erebidae) produce tymbalar clicks in response to the eversion of the coremata of males and their dihydropyrrolizine pheromone hydroxydanaidal (Krasnoff and Yager 1988). This behavior can be used as a remarkably sensitive bioassay for hydroxydanaidal. While intriguing, Krasnoff and Roelofs (1989) were unable to demonstrate a function for either eversion of the coremata or the "pheromone" hydroxydanaidal and tentatively suggested that the acoustic response is a vestige of a no longer functional communication system (Krasnoff and Roelofs 1990). Perhaps there is an erebid species for which the pheromone/acoustic exchange remains salient. Cryptic (at least to us) ultrasonic signals are being found to play important roles in the courtship in many moth species previously thought to be silent (Nakano et al. 2008).

Male Responses to Male Pheromones

ATTRACTION OF CONSPECIFIC MALES

Male *Grapholita molesta* are attracted to the hairpencil secretions of conspecifics (Baker 1983). This attraction appears to represent an alternative mating strategy in which males may "sneak" matings in the presence of conspecifics. A similar phenomenon was demonstrated for the male courtship pheromone of *Anticarsia gemmatalis* (Heath et al. 1988), suggesting that exploitation of conspecific male courtship signals by male moths foraging for mates may be common (Birch et al. 1990).

REPULSION OF CONSPECIFIC MALE

The inhibition of upwind orientation of male moths by conspecific male courtship pheromone (Hirai et al. 1978; Huang et al. 1996; LeComte et al. 1998; Hillier and Vickers 2004) is a controversial putative male response to courtship pheromone. While early studies (Hirai et al. 1978) were criticized for technical flaws (Fitzpatrick and McNeil 1988), group selection arguments for why a male would break off an approach, and for the necessity to use excessive amounts of the pheromone to show an effect (Phelan 1992), more recent studies (Huang et al. 1996; Hillier and Vickers 2004) support early reports. A male that detects male courtship pheromone components in the pheromone plume of a conspecific female may conserve energy and search time by breaking off an approach to an already taken female (Greenfield 2002; but see Baker and Cardé 1979). It is also possible that a male may gain a competitive advantage by inserting a "false pheromone component" into the pheromone plume of a conspecific female (Hillier et al. 2007). Downwind males would interpret the pheromone blend as one of a heterospecific female and break off pursuit toward what appears to be an inappropriate target. Also see Davie et al. (2010) for additional competitive interactions among males.

Lekking (Sex Role Reversal)

There are now several examples of sex role reversal in moths (i.e., males attract females from long distance) (Willis and Birch 1982; Wunderer et al. 1986; Wagner and Rosovsky 1991). Parental investment theory predicts that sex role reversal should occur when the male becomes the limiting sex (Phelan 1992, 1997). The way in which this could happen is most clear in the erebid *Creatonotos gangis* with its enormous, four-branched, inflatable coremata deployed for a few hours each night in "lek-like" behavior (Schneider et al. 1982; Wunderer et al. 1986). Their corematal scents attract additional males to the lek and also females. After use the coremata are deflated and fold neatly (without the aid of retractor muscles) into the abdomen leaving no trace of their existence. Bifid versions of the coremata can be seen in the common salt marsh moth *Estigmene acrea* where they play a similar behavioral role (Willis and Birch 1982; Jordan et al. 2005). In *C. gangis*, *C. transiens*, and *E. acrea* male pheromone precursors are potent morphogens—specific chemicals that stimulate growth and development. Males deprived of pheromone precursors (PAs) in their larval food do not produce male pheromone and do not produce fully formed coremata for their release (Boppré and Schneider

1985; Schmitz et al. 1989; Jordan et al. 2005, 2007; Jordan and Conner 2007). This prevents the energy wasting process of construction of a large scent disseminating structure with no signal to convey.

In both *C. gangis* and *E. acrea*, the primitive condition of the mating system is still evident. The females of both species release pheromone and attract males late in the evening. The derived strategy, lekking, occurs earlier in the evening when the males inflate their coremata. One possible explanation is that the males' contribution to the female—a nuptial gift in the form of spermatophore—has become large relative to the contribution of the female to her offspring. At this tipping point a sex role reversal would take place and females would be attracted to males (Gwynne 1984). It is also possible that *Creatonotos* and *Estigmene* are practicing alternative mating strategies depending on larval access to PAs (Jordan and Conner 2007). Males that find PA-containing plants as caterpillars and have an ample supply of PAs develop robust coremata (see morphogenetic effect above), join aggregations, and display. Mating with such males would be highly advantageous to females, especially those that did not find PAs as caterpillars. Males without PAs could revert to the normal mating system in which they detect the female sex attractant and fly upward. The occurrence of lekking would thus depend on the local availability of PAs in the habitat. From the female's perspective alternative mating strategies are also available. If she is PA-deprived because of an inferior larval diet she can attend the lek to obtain PAs from males. Alternatively, if she acquired sufficient PAs during larval feeding she can avoid the costs of searching for a lek by remaining stationary and attracting males. Additional examples of sex role reversal, such as the pheromonal/acoustic leks of the lesser wax moth (Greenfield and Coffelt 1983), are resource related (Emlen and Oring 1977; Phelan 1992).

Source of Male Pheromones

Past reviews have drawn attention to the fact that many male moth pheromones are minimally altered host plant volatiles (figure 12.3). These include floral and fruit odors and pyrrolizidine alkaloid derivatives. The most likely reason is that males have exploited existing chemoreceptors that were present in females for locating their host plants (Baker 1989; Phelan 1992). Preexisting receptors remove one step in the evolutionary pathway to a courtship pheromone and all that remains is changing (sometimes minimally) the behavioral response to the pheromone.

More recent studies indicate that there are some notable exceptions to the plant-derived pheromone trend (Heath et al. 1988; Hillier and Vickers 2004; Lassance and Löfstedt 2009). In these cases the male courtship pheromones are long-chained hydrocarbons with various degrees of unsaturation and/or functional groups and they are chemically related to female pheromone components. They are processed with receptors (Baker et al. 2004) and neural centers (olfactory lobe glomeruli) analogous to those used by males to receive and recognize female sex attractants (Hillier et al. 2006, 2007). Indeed, the production of the male pheromone components is regulated by neuropeptides that they share with females (Bober and Rafaeli 2010). Again this may be an illustration of the expedient use of preexisting features. In this case the evolution of the male signal may be facilitated by preexisting biochemical pathways and the female response by preexisting

chemoreceptors in the female (e.g., Lim and Greenfield 2006, 2008; Lim et al. 2007).

It seems clear the sources of male courtship pheromones are more diverse than originally thought (Birch 1970a, 1970b; Birch et al. 1990) and reflect the different selective advantages of their use. Given that the number of species investigated remains less than 0.01% of species with male scent disseminating structures, it seems likely that additional classes of male pheromones will be discovered in the future.

Quest for Pheromone Precursors

The diet dependence of some male pheromones is no better illustrated than by the extraordinary behavior termed pharmacophagy. As redefined by Boppré (1984) pharmacophagous (literally "drug eating") insects seek out non-host plant species and collect specific chemicals for sexual or defensive (non-nutritional) purposes. Our knowledge of this behavior can be traced to the early collectors of Lepidoptera in central and South America, including William Beebe in Trinidad (Beebe 1955; Beebe and Kenedy 1957). They found that the dead shoots of *Heliotropium indicum* attracted swarms of erebids by night and ithomiine and danaine butterflies by day. Researchers later found that virtually any plant with high concentrations of pyrrolizine alkaloids served as excellent bait (Pliske 1975; Goss 1979; Boppré 1986; DeVries and Stiles 1990; Häuser and Boppré 1997; Brehm et al. 2007; Conner and Jordan 2009). The visitors are most often (but not always) male-biased, in keeping with parental investment theory. Upon arrival at the bait they unfurl their proboscis regurgitate on the surface of the plant and reimbibe an extract of surface alkaloids. These serve as chemical defenses, male courtship pheromones, or their precursors (Krasnoff and Dussourd 1989; Conner et al. 2000; figure 12.4). The tip of the proboscis of pharmacophagous moths possesses clusters of sensilla stylconica that appear to be gustatory chemoreceptors for PAs and their derivatives (Zaspel et al. 2013; figure 12.4D) similar to those found on the maxillary palps of pharmacophagous larvae (Bernays and Singer 2005).

The attraction of male moths to rotting fruit, mud puddles, feces, blood, and tears provides opportunities to collect exogenous chemicals that may be incorporated into courtship and all merit further study (Baker et al. 1989; Smedley and Eisner 1995; Zaspel et al. 2011). It is notable that plants containing cardenolides, a major class of defensive chemicals found in some Erebidae, are not known to be attractive to males of these groups (Baker 1989; Birch et al. 1990). Why plants containing PAs attract erebids and plants containing cardenolides do not remains a mystery.

Role of Sound in Courtship

Although this review has focused exclusively on the chemical modality, it is becoming increasingly apparent that moth courtship involves multimodal cues. Many of the species that use male courtship pheromones combine them with acoustic and usually ultrasonic signals (Nakano et al. 2008; Table 12.1). The acoustic signals can potentially play roles in both reproductive isolation and mate choice. Their frequent use in courtship is likely a result of exploitation of preexisting structures that evolved for bat detection (Conner 1999). The use of multimodal courtship signals raises the bar for studying moth

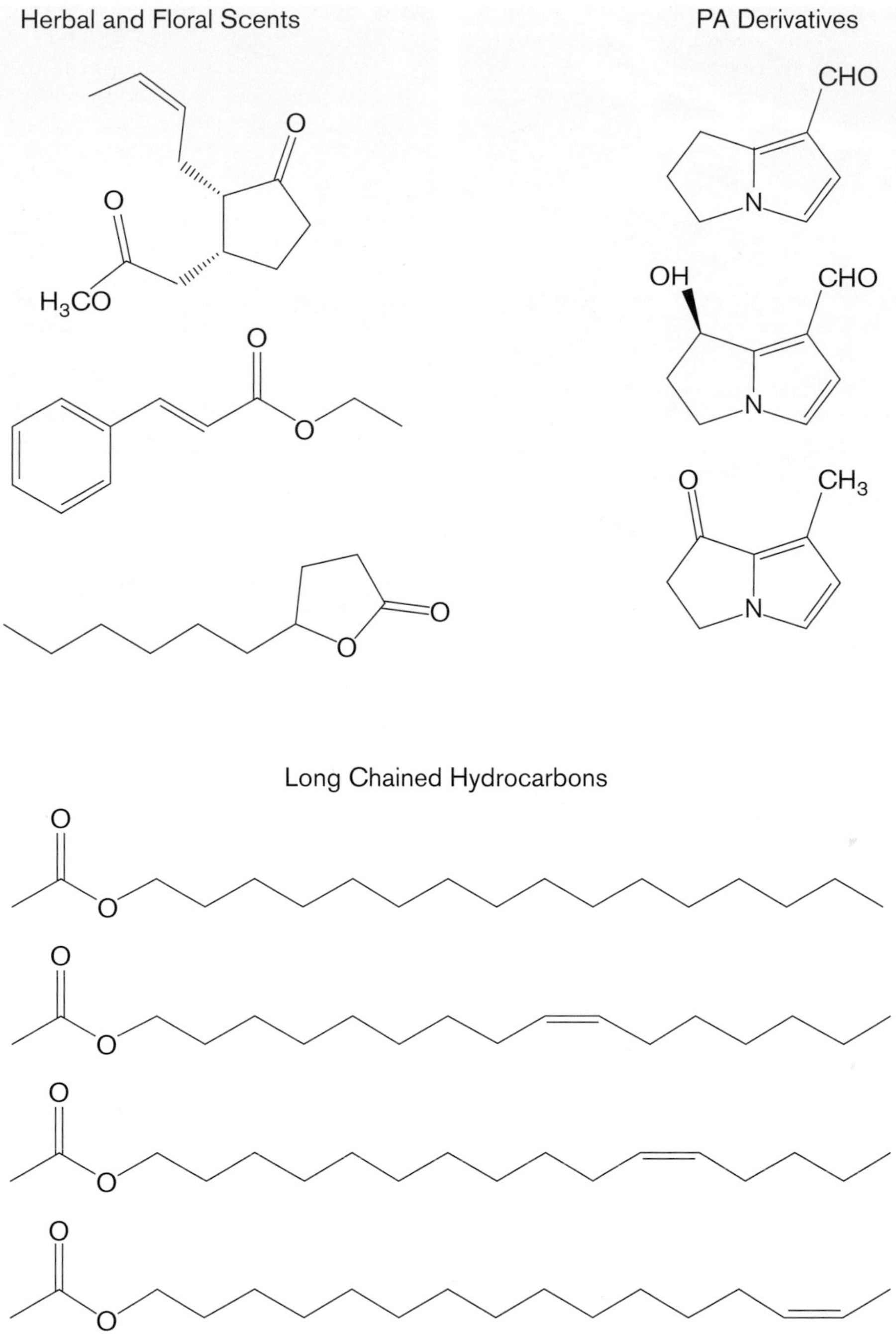

Herbal and Floral Scents

PA Derivatives

Long Chained Hydrocarbons

FIGURE 12.3 Examples of male courtship pheromones. Herbal and floral scents: (1) methyl 2-epijasmonate; (2) ethyl *trans*-cinnamate; (3) γ-decalactone. PA derivatives: (4) danaidal; (5) *R*-hydroxydanaidal; (6) danaidone. Long-chained hydrocarbons: (7) hexadecanyl acetate; (8) (Z)-9 hexadecenyl acetate; (9) (Z)-11 hexadecenyl acetate; (10) (Z)-14 hexadecenyl acetate.

courtship. Researchers will need to use sophisticated technologies in both the fields of chemical ecology and bioacoustics. Experimental protocols will need to incorporate controls for both sound and chemistry and early studies will need to be reexamined in light of the possibility of redundant chemical and acoustic signals.

Future Directions

Detailed studies of male courtship pheromones have been largely restricted to the Erebidae and select noctuids and pyral-

ids. Understudied groups include the Gelechiidae, Tortricidae, and the Notodontidae, all of which have a high diversity of androconial scales. The latter groups deserve special attention.

With the advent of new technologies it is becoming clear that moths are combining chemical and acoustic modalities in their courtship repertoires. Acoustic signals provide another opportunity to study the evolutionary forces influencing moth courtship. Evolutionary studies that map both chemical and acoustic cues on phylogenies will be particularly instructive.

Studies of the evolution and development of male courtship pheromone disseminating structures are needed. They

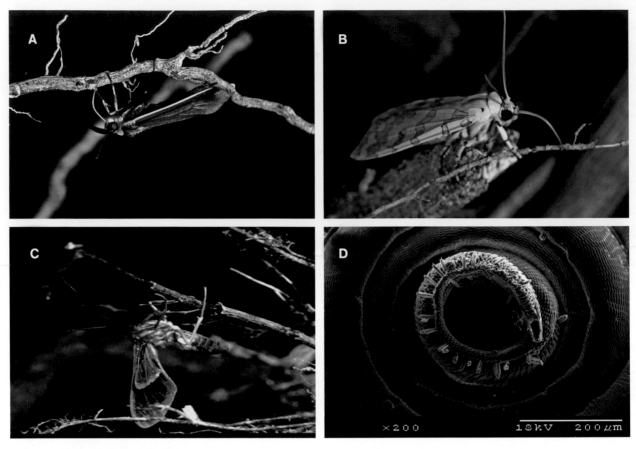

FIGURE 12.4 Pharmacophagy in Erebidae.

A Male *Cisseps fulvicollis* imbibing pyrrolizidine alkaloids from the roots of *Eupatorium capillifolium*.

B Adult *Halysidota tessellaris* collecting alkaloids.

C Male *Cosmosoma myrodora* visiting roots.

D Scanning electron micrograph of presumed alkaloid receptors on the tip of the proboscis of a male *Cosmosoma auge* (photograph provided by Jennifer Zaspel).

SOURCE: Photos A through C reprinted with permission of Cambridge University Press.

will provide clues about how scent structure arise and diversify and would also ultimately explain how specific dietary requirements can control scent structure development. In addition understanding the neural processing of male courtship pheromones would help clarify the early steps in the evolution of male courtship pheromones.

Before conclusions can be made about the selective pressures that gave rise to male courtship pheromones, the assumptions and predictions of the various models of sexual selection must be tested empirically in a variety of species. It is likely that there will be no single answer, because females and males have a number of options for increasing their fitness through mate choice and in some cases they may be using them all. Detailed genetic analyses, physiological studies, quantitative measurements of behavior, and, especially, phylogenetic treatments will play important roles in our understanding of what must be considered one of the pinnacles of chemical communication.

Acknowledgments

We thank Christer Löfstedt, Kirk Hillier, Michael Boppré, and Jennifer Zaspel for providing photographic material. We are grateful to the Archbold Biological Station for providing a congenial atmosphere in which to do research and write, and Wake Forest University which provided a Z. Smith Reynolds leave to WEC during the preparation of this chapter.

References Cited

Andersson, M.B. 1987. Genetic models of sexual selection: some aims, assumptions, and tests. Pp. 41–53. In J.W. Bradbury and M.B. Andersson, eds. *Sexual Selection: Testing the Alternatives.* Chichester: John Wiley and Sons.

Andersson, M.B. 1994. *Sexual Selection.* Princeton, NJ: Princeton University Press.

Ando, T. 2012. Sex pheromones of moths. Available online at: http://www.tuat.ac.jp/~antetsu/LepiPheroList.htm.

Aplin, R.T., and M.C. Birch. 1970. Identification of odorous compounds from male Lepidoptera. *Experientia* 26:1193–1194.

Arnqvist, G., and T. Nilsson. 2000. The evolution of polyandry: multiple mating and female fitness in insects. *Animal Behaviour* 60:145–164.

Baker, T.C. 1983. Variations in male oriental fruit moth courtship patterns due to male competition. *Experientia* 39:112–114.

Baker, T.C. 1989. Origin of courtship and sex pheromones or the oriental fruit moth and a discussion of the role of phytochemicals in the evolution of lepidopteran male scents. Pp. 401–418. In C.H. Chou and G.R. Waller, eds. *Phytochemical Ecology: Allelochemicals,*

Mycotoxins and Insect Pheromones and Allomones. Institute of Botany Sinica Monograph Series, No. 9. Taipei, China.

Baker, T.C., and R.T. Cardé. 1979. Courtship behavior of the oriental fruit moth (*Grapholitha molesta*): experimental analysis and consideration of the role of sexual selection in the evolution of courtship pheromones in the Lepidoptera. *Annals of the Entomological Society of America* 72:173–188.

Baker, T.C., R. Nishida, and W.L. Roelofs. 1981. Close-range attraction of female oriental fruit moths to herbal scent of male hairpencils. *Science* 214:1359–1361.

Baker, T.C., S.A. Ochieng, A.A. Cossé, S.G. Lee, J.L. Todd, C. Quero, and N.J. Vickers. 2004. A comparison of responses from olfactory receptor neurons of *Heliothis subflexa* and *Heliothis virescens* to components of their sex pheromone. *Journal of Comparative Physiology A* 190:155–165.

Beebe, W. 1955. Two little-known selective insect attractants. *Zoologica: Scientific Contributions of the New York Zoological Society.* 40:27–36.

Beebe, W., and R. Kenedy. 1957. Habits, palatability, and mimicry in thirteen ctenuchid moth species from Trinidad, B.W.I. *Zoologica: Scientific Contributions of the New York Zoological Society* 42:147–158.

Bell, T.W., and J. Meinwald. 1986. Pheromones of two arctiid moths (*Creatonotus transiens* and *C. gangis*): chiral components from both sexes and achiral female components. *Journal of Chemical Ecology* 12:385–409.

Bell, T.W., M. Boppré, D. Schneider, and J. Meinwald. 1984. Stereochemical course of pheromone biosynthesis in the arctiid moth, *Creatonotos transiens*. *Experientia* 40:713–714.

Bennett, A.L., P.R. Atkinson, and N.J.S. La Croix. 1991. On communication in the African sugarcane borer, *Eldana saccharina* Walker (Lepidoptera: Pyralidae). *Journal of the Entomological Society of South Africa* 54:243–259.

Bernays, E.A., and M.S. Singer. 2005. Taste alteration and endoparasites. *Nature* 436:476.

Bezzerides, A., V.K. Iyengar, and T. Eisner. 2005. Corematal function in *Utetheisa ornatrix*: interpretation in light of data from field-collected males. *Chemoecology* 15:187–192.

Birch, M.C. 1970a. Structure and function of the pheromone-producing brush organs in males of *Phlogophora meticulosa* (L.) (Lepidoptera: Noctuidae) *Transactions of the Royal Entomological Society of London* 122:277–292.

Birch, M.C. 1970b. Pre-courtship use of abdominal brushes by the nocturnal moth *Phlogophora meticulosa* (L.) (Lepidoptera: Noctuidae). *Animal Behaviour* 18:310–316.

Birch, M.C. 1972. Male abdominal brush-organs in British noctuid moths and their value as a taxonomic character. *The Entomologist* 105:185–205; 233–244.

Birch, M.C. 1974. Aphrodisiac pheromones in insects. Pp. 115–134. In M.C. Birch, ed. *Pheromones*. Amsterdam: North-Holland Publishing.

Birch, M.C., and A. Hefetz. 1987. Extrusible organs in male moths and their role in courtship behavior. *Bulletin of the Entomological Society of America* 33: 222–229.

Birch, M.C., G.M. Poppy, and T.C Baker. 1990. Scents and eversible scent structures in male moths. *Annual Review of Entomology* 35:25–58.

Birkhead, T.R., and A.P. Møller. 1998. *Sperm Competition and Sexual Selection*. San Diego, CA: Academic Press.

Blest, A.D. 1964. Protective display and sound production in some New World arctiid and ctenuchid moths. *Zoologica: Scientific Contributions of the New York Zoological Society* 49:161–181.

Blomquist, G.J., and R.G. Vogt. 2003. *Insect Pheromone Biochemistry and Molecular Biology: The Biosynthesis and Detection of Pheromones and Plant Volatiles*. London: Elsevier Press.

Bober, R., and A. Rafaeli. 2010. Gene-silencing reveals the functional significance of pheromone biosynthesis activating neuropeptide receptor (PBAN-R) in a male moth. *Proceedings of the National Academy of Sciences of the United States of America* 107:16858–16862.

Boppré, M. 1984. Redefining "pharmacophagy". *Journal of Chemical Ecology* 10:1151–1154.

Boppré, M. 1986. Insects pharmacophageously utilizing defensive plant chemicals (pyrrolizidine alkaloids). *Naturwissenschaften* 73:17–26.

Boppré, M. 1990. Lepidoptera and pyrrolizidine alkaloids: exemplification of complexity in chemical ecology. *Journal of Chemical Ecology* 16:165–185.

Boppré, M., and D. Schneider. 1985. Pyrrolizidine alkaloids quantitatively regulate both the scent organ morphogenesis and pheromone biosynthesis in male *Creatonotos* moths (Lepidoptera: Arctiidae). *Journal of Comparative Physiology A* 157:569–577.

Bradbury, J.W., and M.B. Andersson. 1987. *Sexual Selection: Testing the Alternatives*. Chichester: John Wiley and Sons.

Brehm, G., T. Hartmann, and K. Willmott. 2007. Pyrrolizidine alkaloids and pharmacophagous Lepidoptera visitors of *Prestonia amabilis* (Apocynaceae) in a montane forest in Ecuador. *Annals of the Missouri Botanical Garden* 94:463–473.

Brown, R.L., J. Baixeras, and S. Lee. 2011. Male sex scales of moths. Available online at: http://lepcourse.wikispaces.com/file/view /sex+scales.AZ.+aug.12.pdf.

Burger, B.V., A.E. Nell, D. Smit, H.S.C. Spies, W.M. Mackenroth, D. Groche, and P.R. Atkinson. 1993. Constituents of wing gland and abdominal hair pencil secretions of male African sugarcane borer, *Eldana saccharina* Walker (Lepidoptera: Pyralidae). *Journal of Chemical Ecology* 19:2255–2277.

Cardé, R.T., and T.C. Baker. 1984. Sexual communication with pheromones. Pp. 355–383. In W.J. Bell and R.T. Cardé, eds. *Chemical Ecology of Insects*. London: Chapman & Hall Press.

Cardé, R.T., and J.G. Millar. 2004. *Advances in Insect Chemical Ecology*. Cambridge: Cambridge University Press.

Cardé, R.T., and A.K. Minks. 1997. *Insect Pheromone Research: New Directions*. New York: Chapman & Hall.

Cibrian-Tovar, J., and E.R. Mitchell. 1991. Courtship behavior of *Heliothis subflexa* Gn. Lepidoptera, Noctuidae and associated backcross insects obtained from hydridization with *Heliothis virescens* F. *Environmental Entomology* 20:419–426.

Clearwater, J.R. 1972. Chemistry and function of a pheromone produced by male of the southern armyworm *Pseudaletia separata*. *Journal of Insect Physiology* 18:781–789.

Collins, R.D., Y. Jang, K. Rheinhold, and M.D. Greenfield. 1999. Quantitative genetics of ultrasonic advertisement signaling in the lesser waxmoth *Achroia grisella* (Lepidoptera: Pyralidae). *Heredity* 83:644–651.

Conner, W.E. 1999. "Un chant d'appel amoureux": acoustic communication in moths. *Journal of Experimental Biology* 202:1711–1723.

Conner, W.E., and A.T. Jordan. 2009. From armaments to ornaments: the relationship between chemical defense and sex in tiger moths. Pp. 155–172. In W.E. Conner, ed. *Tiger Moths and Woolly Bears: Behavior, Ecology, and Evolution of the Arctiidae*. Oxford: Oxford University Press.

Conner, W.E., T. Eisner, R.K. VanderMeer, A. Guerrero, and J. Meinwald. 1981. Precopulatory sexual interactions in an arctiid moth (*Utetheisa ornatrix*): role of pheromone derived from alkaloids. *Behavioral Ecology and Sociobiology* 9:227–235.

Conner, W.E., R. Boada, F. Schroeder, and T. Eisner. 2000. Chemical defense: bestowal of a nuptial alkaloidal garment by a male moth on its mate. *Proceedings of the National Academy of Sciences of the United States of America* 97:14406–14411.

Coyne, J.A., and H.A. Orr. 2004. *Speciation*. Sunderland, MA: Sinauer.

Dahm, K.H., D. Meyer, W.E. Finn, V. Reinhold, and H. Röller. 1971. The olfactory and auditory mediated sex attraction in *Achroia grisella* (Fabr.). *Naturwissenschaften* 5:265–266.

Darwin, C. 1871. *The Descent of Man, and Selection in Relation to Sex*. London: John Murray.

Davenport, J.W., and W.E. Conner. 2003. Dietary alkaloids and the development of androconial organs in *Estigmene acrea*. *Journal of Insect Science* 3:3. PMCID: PMC524643.

Davidson, R.B., C. Baker, M. McElveen, and W.E. Conner. 1997. Hydroxydanaidal and the courtship of *Haploa* (Arctiidae). *Journal of the Lepidopterists' Society* 51:288–294.

Davie, L.C., T.M. Jones, and M.A. Elgar. 2010. The role of chemical communication in sexual selection: hair pencil displays in the diamondback moth, *Plutella xylostella*. *Animal Behaviour* 79:391–399.

Davies, N.B. 1992. *Dunnock Behaviour and Social Evolution*. Oxford: Oxford University Press.

DeVries, P.J., and F.G. Stiles. 1990. Attraction of pyrrolizidine alkaloid-seeking Lepidoptera to *Epidendrum paniculatum* orchids. *Biotropica* 22:290–297.

Dussourd, D.E., K. Ubik, C. Harvis, J. Resch, J. Meinwald, and Eisner, T. 1988. Biparental defensive endowment of eggs with acquired plant alkaloid in the moth *Utetheisa ornatrix*. *Proceedings of the National Academy of Sciences of the United States of America* 85:5992–5996.

Dussourd, D. E., C. A. Harvis, J. Meinwald, and T. Eisner. 1991. Pheromonal advertisement of a nuptial gift by a male moth (*Utetheisa ornatrix*). *Proceedings of the National Academy of Sciences of the United States of America* 88:9224–9227.

Egelhaaf, A., C. Coelln, B. Schmitz, M. Buck, M. Wink, and D. Schneider. 1990. Organ specific storage of dietary pyrrolizidine alkaloids in the arctiid moth *Creatonotos transiens*. *Zeitschrift für Naturforschung C* 45:115–120.

El-Sayed, A. M. 2011. The pherobase: database of pheromones and semiochemicals. Available online at: http://www.pherobase.com.

Emlen, S. T., and L. W. Oring. 1977. Ecology, sexual selection, and the evolution of mating systems. *Science* 197:215–223.

Finn, W. E., and T. L. Payne. 1977. Attraction of greater wax moths females to male-produced pheromones. *Southwestern Entomologist* 2:62–64.

Fisher, R. A. 1930. *The Genetical Theory of Natural Selection*. Oxford: Clarendon Press.

Fitzpatrick, S. M., and J. N. McNeil. 1988. Male scent in lepidopteran communication: the role of male pheromone in mating behaviour of *Pseudaletia unipuncta* (Haw.) (Lepidoptera: Noctuidae). *Memoirs of the Entomological Society of Canada* 146:131–151.

González, A., C. Rossini, M. Eisner, and T. Eisner. 1999. Sexually transmitted chemical defense in a moth (*Utetheisa ornatrix*). *Proceedings of the National Academy of Sciences of the United States of America* 96:5570–5574.

Goss, G. J. 1979. The interaction between moths and plants containing pyrrolizidine alkaloids. *Environmental Entomology* 8:487–493.

Grant, G. G. 1974. Male sex pheromone from the wing glands of the Indian meal moth, *Plodia interpunctella* (Hbn.) (Lepidoptera: Phycitidae). *Experientia* 30:917–918.

Grant, G. G. 1976. Courtship behavior of a phycitid, *Vitula edmandsae*. *Annals of the Entomological Society of America* 69:445–449.

Greenfield, M. 2002. *Signalers and Receivers: Mechanisms and Evolution of Arthropod Communication*. Oxford: Oxford University Press.

Greenfield, M. D., and J. A. Coffelt. 1983. Reproductive behavior of the lesser waxmoth, *Achroia grisella* (Pyralidae: Galleriinae): signaling, pair formation, male interactions, and mate guarding. *Behaviour* 84:287–315.

Gwynne, D. T. 1984. Courtship feeding increases female reproductive success in bushcrickets. *Nature* 307:361–363.

Hamilton, W. D., and M. Zuk. 1982. Heritable true fitness and bright birds: a role for parasites? *Science* 218:384–387.

Hammerstein, P., and G. A. Parker. 1987. Sexual selection: games between the sexes. Pp. 119–142. In J. W. Bradbury and M. Andersson, eds. *Sexual Selection: Testing the Alternatives*. New York: John Wiley Press.

Hartmann, T. 2009. Pyrrolizidine alkaloids: the successful adoption of a plant chemical defense. Pp. 55–81. In W. E. Conner, ed. *Tiger Moths and Wooly Bears: Behavior, Ecology, and Evolution of the Arctiidae*. Oxford: Oxford University Press.

Häuser, C. L., and M. Boppré. 1997. A revision of the Afrotropical taxa of the genus *Amerila* Walker (Lepidoptera: Arctiidae). *Systematic Entomology* 22:1–44.

Haynes, K. F., and M. C. Birch. 1985. The role of other pheromones, allomones, and kairomones in the behavioral responses of insects. Pp. 225–255. In G. A. Kerkut and L. I. Gilbert, eds. *Comprehensive Insect Physiology, Biochemistry, and Pharmacology, Vol. 9*. London: Pergamon Press.

Heath R. R., P. J. Landolt, N. C. Leppla, and B. D. Dueben. 1988. Identification of a male-produced pheromone of *Anticarsia gemmatalis* (Hübner) (Lepidoptera: Noctuidae) attractive to conspecific males. *Journal of Chemical Ecology* 14:1121–1130.

Heath, R. R., P. J. Landolt, B. D. Dueben, R. E. Murphy, and R. E. Schneider. 1992. Identification of male cabbage looper pheromone attractive to females. *Journal of Chemical Ecology* 18:441–453.

Heisler, L., M. Andersson, S. J. Arnold, C. R. Boake, G. Borgia, G. Hausfater, M. Kirkpatrick et al. 1987. Evolution of mating preferences and sexually selected traits. Pp. 97–118. In J. W. Bradbury and M. Andersson, eds. *Sexual Selection: Testing the Alternatives*. New York: John Wiley and Sons.

Hendricks, D. E., and T. N. Shaver. 1975. Tobacco budworm: male pheromone suppressed emission of sex pheromone by the female. *Environmental Entomology* 3:555–558.

Hendrikse, A. 1986. The courtship behavior of *Yponomeuta padellus*. *Entomologia Experimentalis et Applicata* 42:45–55.

Hillier, N. K., and N. J. Vickers. 2004. The role of heliothine hairpencil compounds in female *Heliothis virescens* (Lepidoptera: Noctuidae) behavior and mate acceptance. *Chemical Senses* 29:499–511.

Hillier, N. K., and N. J. Vickers. 2011. Hairpencil volatiles influence interspecific courtship and mating between two related moth species. *Journal of Chemical Ecology* 37:1127–1136.

Hillier, N. K., C. K. Kleineidam, and N. J. Vickers. 2006. Physiology and glomerular projections of olfactory receptor neurons on the antenna of female *Heliothis virescens* (Lepidoptera: Noctuidae) responsive to behaviorally relevant odors. *Journal of Comparative Physiology A* 192:199–219.

Hillier, N. K., D. Kelly, and N. J. Vickers. 2007. A specific male olfactory sensillum detects behaviorally antagonistic hairpencil odorants. *Journal of Insect Science* 7:1–12.

Hirai, K., H. H. Shorey, and L. K. Gaston. 1978. Competition among courting male moths: male-to-male inhibitory pheromone. *Science* 202:644–645.

Huang, Y., S. Xu, X. Tang, Z. Zhao, and J. Du. 1996. Male orientation inhibitor of cotton bollworm: identification of compounds produced by male hairpencil glands. *Insect Science* 4:173–181.

Iyengar, V. K., and T. Eisner. 1999a. Heritability of body mass, a sexually selected trait, in an arctiid moth (*Utetheisa ornatrix*). *Proceedings of the National Academy of Sciences of the United States of America* 96:9169–9171.

Iyengar, V. K., and T. Eisner. 1999b. Female choice increases offspring fitness in an arctiid moth (*Utetheisa ornatrix*). *Proceedings of the National Academy of Sciences of the United States of America* 96:15013–15016.

Iyengar, V. K., and H. K. Reeve. 2010. Z linkage of the female promiscuity genes in the moth *Utetheisa ornatrix*: support for the sexy-sperm hypothesis? *Evolution* 64:1267–1272.

Iyengar, V. K., C. Rossiniand, and T. Eisner. 2001. Precopulatory assessment of male quality in an arctiid moth (*Utetheisa ornatrix*): hydroxydanaidal is the only criterion of choice. *Behavioral Ecology and Sociobiology* 49:283–288.

Iyengar, V. K., H. K. Reeve, and T. Eisner. 2002. Paternal inheritance of a female moth's mating preference. *Nature* 419:830–832.

Jacquin, E., P. Nagnan, and B. Frerot. 1991. Identification of hairpencil secretion from male *Mamestra brassicae* (L.) (Lepidoptera: Noctuidae) and electroantennogram studies. *Journal of Chemical Ecology* 17:239–247.

Jang, Y., and M. D. Greenfield. 1996. Ultrasonic communication and sexual selection in wax moths: female choice based on energy and asynchrony of male signals. *Animal Behavior* 51:1095–1106.

Jang, Y., and M. D. Greenfield. 2000. Quantitative genetics of female choice in an ultrasonic pyralid moth, *Achroia grisella*: variation and evolvability of preference along multiple dimensions of the male advertisement signal. *Heredity* 84:73–80.

Jang, Y., R. D. Collins, and M. D. Greenfield. 1997. Variation and repeatability of ultrasonic sexual advertisement signals in *Achroia grisella* (Lepidoptera: Pyralidae). *Journal of Insect Behavior* 10:87–98.

Jia, F.-Y., and M. D. Greenfield. 1997. When are good genes good? Variable outcomes of female choice in wax moths. *Proceedings of the Royal Society of London B* 264:1057–1063.

Johannson, B. G., and T. M. Jones. 2007. The role of chemical communication in mate choice. *Biological Reviews* 82:265–289.

Jordan, A. T., and W. E. Conner. 2007. Dietary basis for developmental plasticity of an androconial structure in the salt marsh moth *Estigmene acrea* (Drury). *Journal of the Lepidopterists' Society* 61:32–37.

Jordan, A. T., T. H. Jones, and W. E. Conner. 2005. If you've got, flaunt it: ingested alkaloids affect corematal display behavior in the salt marsh moth, *Estigmene acrea*. *Journal of Insect Science* 5:1. PMCID: PMC1283882.

Jordan, A. T., T. H. Jones, and W. E. Conner. 2007. Morphogenetic effects of alkaloidal metabolites on the development of the coremata in the salt marsh moth, *Estigmene acrea* (Dru.) (Lepidoptera: Arctiidae). *Archives of Insect Biochemistry and Physiology* 66:183–189.

Keller, L., and H. K. Reeve. 1995. Why do females mate with multiple males? The sexually-selected sperm hypothesis. *Advances in the Study of Behavior* 24:291–315.

Kelly, C. A., A. J. Norbutus, A. F. Lagalante, and V. K. Iyengar. 2012. Male courtship pheromones as indicators of genetic quality in an arctiid moth (*Utetheisa ornatrix*). *Behavioral Ecology* 23:1009–1014.

Kimura, T., and H. Honda. 1999. Identification and possible functions of the hairpencil scent of the yellow peach moth, *Conogethes punctiferalis* (Guenée) (Lepidoptera: Pyralidae). *Applied Entomology and Zoology* 34:147–153.

Kirkpatrick, M. 1982. Sexual selection and the evolution of female choice. *Evolution* 36:1–12.

Krasnoff, S. B. 1997. Evolution of male lepidopteran pheromones: a phylogenetic perspective. Pp. 490–504. In R. T. Cardé and A. K. Minks, eds. *Insect Pheromone Research: New Directions.* New York: Chapman & Hall.

Krasnoff, S. B., and D. D. Dussourd. 1989. Dihydropyrrolizidine attractants for arctiid moths that visit plants containing pyrrolizidine alkaloids. *Journal of Chemical Ecology* 15:47–60.

Krasnoff, S. B., and W. L. Roelofs. 1989. Quantitative and qualitative effects of larval diet on male scent secretions of *Estigmene acrea, Phragmatobia fuliginosa* and *Pyrrharctia isabella* (Lepidoptera: Arctiidae). *Journal of Chemical Ecology* 15:1077–1093.

Krasnoff, S. B., and W. L. Roelofs. 1990. Evolutionary trends in the male pheromone systems of arctiid moths: evidence from studies of courtship in *Phragmatobia fuliginosa* and *Pyrrharctia isabella* (Lepidoptera: Arctiidae). *Zoological Journal of the Linnean Society* 99:319–338.

Krasnoff, S. B., and K. W. Vick. 1984. Male wing gland pheromone of *Ephestia elutella. Journal of Chemical Ecology* 10:667–679.

Krasnoff, S. B., and D. D. Yager. 1988. Acoustic response to a pheromonal cue in the arctiid moth *Pyrrharctia isabella. Physiological Entomology* 13:433–440.

Krasnoff, S. B., L. B. Bjostad, and W. L. Roelofs. 1987. Quantitative and qualitative variation in male pheromones of *Phragmatobia fuliginosa* and *Pyrrharctia isabella* (Lepidoptera: Arctiidae). *Journal of Chemical Ecology* 13:807–822.

Kunesch, G., P. Zagatti, J. Y. Lallemand, A. Debal, and J. P. Vigneron. 1981. Structure and synthesis of the wing gland pheromone of the male African sugar-cane borer: *Eldana saccharina* (Wlk.) (Lepidoptera: Pyralidae). *Tetrahedron Letters* 22:5271–5274.

LaMunyon, C. W. 1997. Increased fecundity, as a function of multiple mating, in an arctiid moth, *Utetheisa ornatrix. Ecological Entomology* 22:69–73.

LaMunyon, C. W., and T. Eisner. 1993. Postcopulatory sexual selection in an arctiid moth (*Utetheisa ornatrix*). *Proceedings of the National Academy of Sciences of the United States of America* 90:4689–4692.

LaMunyon, C. W., and T. Eisner. 1994. Spermatophore size as determinant of paternity in an arctiid moth (*Utetheisa ornatrix*). *Proceedings of the National Academy of Sciences of the United States of America* 91:7081–7084.

Lande, R. 1981. Models of speciation by sexual selection on polygenic traits. *Proceedings of the National Academy of Sciences of the United States of America* 78:3721–3725.

Landolt, P. J. 1995. Attraction of female cabbage looper moths (Lepidoptera: Noctuidae) to males in the field. *Florida Entomologist* 78:96–100.

Landolt, P. J., and R. R. Heath. 1989. Attraction of female cabbage looper moth (Lepidoptera: Noctuidae) to male-produced sex pheromone. *Annals of the Entomological Society of America* 82:520–525.

Landolt, P. J., and R. R. Heath. 1990. Sex role reversal in mate-finding strategies of the cabbage looper moth. *Science* 249:1026–1028.

Landolt, P. J., R. R. Heath, J. G. Millar, K. M. Davis-Hernandez, B. D. Dueben, and K. E. Ward. 1994. Effects of host plant, *Gossypium hirsutum* L., on sexual attraction of cabbage looper moths, *Trichoplusia ni* (Hübner) (Lepidoptera: Noctuidae). *Journal of Chemical Ecology* 20:2959–2974.

Lassance, J.-M., and C. Löfstedt. 2009. Concerted evolution of male and female display traits in the European corn borer *Ostrinia nubilalis. BMC Biology* 7:10.

Lebedeva, K. V., N. V. Vendilo, V. L. Ponomarev, V. A. Pletnev, and D. B. Mitroshin. 2002. Identification of pheromone of the greater wax moth *Galleria mellonella* from the different region of Russia. *International Organization of Biological and Integrated Control of Noxious Animals and Plants: West Palaearctic Regional Section Bulletin* 25:1–5.

LeComte, C., E. Thibout, D. Pierre, and J. Auger. 1998. Transfer, perception, and activity of male pheromone of *Acrolepiopsis assectella* with special reference to conspecific male sexual inhibition. *Journal of Chemical Ecology* 24:655–671.

Leyrer, R. L., and R. E. Monroe. 1973. Isolation and identification of the scent of the moth *Galleria mellonella* and a reevaluation of its sex pheromone. *Journal of Insect Physiology* 19:2267–2271.

Lim, H., and M. D. Greenfield. 2006. Female pheromonal chorusing in an arctiid moth, *Utetheisa ornatrix. Behavioral Ecology* 18:165–173.

Lim, H., and M. D. Greenfield. 2008. Female arctiid moths, *Utetheisa ornatrix*, orient towards and join pheromonal choruses. *Animal Behaviour* 75:673–680.

Lim, H., K. C. Park, T. C. Baker, and M. D. Greenfield. 2007. Perception of conspecific female pheromone stimulates female calling in an arctiid moth, *Utetheisa ornatrix. Journal of Chemical Ecology* 33:1257–1271.

Löfstedt, C., and M. Kozlov. 1997. A phylogenetic analysis of pheromone communication in primitive moths. Pp. 473–489. In R. T. Cardé and A. K. Minks, eds. *Insect Pheromone Research: New Directions.* New York: Chapman & Hall.

Löfstedt, C., N. J. Vickers, W. L. Roelofs, and T. C. Baker. 1989. Diet related courtship success in the oriental fruit moth, *Grapholita molesta* (Tortricidae). *Oikos* 55:402–408.

Löfstedt, C., N. J. Vickers, and T. C. Baker. 1990. Courtship, pheromone titre and determination of male mating success in the oriental fruit moth *Grapholita molesta* (Lepidoptera: Tortricidae). *Entomologia Generalis* 15:121–125.

Moczek, A. P., J. Andrews, T. Kijimoto, Y. Yerushalmi, and D. J. Rose. 2007. Emerging model systems in evo-devo: horned beetles and the origins of diversity. *Evolution and Development* 9: 323–328.

Nakano, R., N. Skals, T. Takanashi, A. Surlykke, T. Koike, K. Yoshida, H. Maruyama, S. Tatsuki, and I. Ishikawa. 2008. Moths produce extremely quiet ultrasonic courtship songs by rubbing specialized scales. *Proceedings of the National Academy of Sciences of the United States of America* 105:11812–11817.

Nishida, R., T. C. Baker, and W. L. Roelofs. 1982. Hairpencil pheromone components of male oriental fruit moths *Grapholitha molesta. Journal of Chemical Ecology* 8:947–959.

O'Donald, P. 1967. A general model of sexual and natural selection. *Heredity* 22:499–518.

O'Donald, P. 1980. *Genetic Models of Sexual Selection.* Cambridge: Cambridge University Press.

Parker, G. A. 1979. Sexual selection and sexual conflict. Pp. 123–166. In M. S. Blum and N. A. Blum, eds. *Sexual Selection and Reproductive Competition in Insects.* New York: Academic Press.

Phelan, P. L. 1992. Evolution of sex pheromones and the role of asymmetric tracking. Pp. 265–314. In B. D. Roitberg and M. B. Isman, eds. *Insect Chemical Ecology: An Evolutionary Approach.* New York: Chapman & Hall.

Phelan, P. L. 1997. Evolution of mate-signaling in moths: phylogenetic considerations and predictions from the asymmetric tracking hypothesis. Pp. 240–256. In J. C. Choe and B. J. Crespi, eds. *Mating Systems in Insects and Arachnids.* Cambridge: Cambridge University Press.

Phelan, P. L., and T. C. Baker. 1986. Male-size-related courtship success and intersexual selection in the tobacco moth, *Ephestia elutella. Experientia* 42:1291–1293.

Phelan, P. L., and T. C. Baker. 1987. Evolution of male pheromones in moths: reproductive isolation through sexual selection. *Science* 235:205–207.

Phelan, P. L., and T. C. Baker. 1990a. Comparative study of courtship in twelve phyticine moths (Lepidoptera: Pyralidae). *Journal of Insect Behavior* 3:303–326.

Phelan, P. L., and T. C. Baker. 1990b. Information transmission during intra- and interspecific courtship in *Ephestia elutella* and *Cadra figulilella. Journal of Insect Behavior* 3:589–602.

Phelan, P. L., P. J. Silk, C. J. Northcott, S. H. Tan, and T. C. Baker. 1986. Chemical identification and behavioral characterization of male wing pheromone *Ephestia elutella* (Pyralidae). *Journal of Chemical Ecology* 12:135–146.

Pliske, T. E. 1975. Attraction of Lepidoptera to plants containing pyrrolizidine alkaloids. *Environmental Entomology* 4:455–473.

Reeve, H. K., and D. W. Pfennig. 2003. Genetic biases for showy males: are some genetic systems especially conducive to sexual selection? *Proceedings of the National Academy of Sciences of the United States of America* 100:1089–1094.

Reeve, H. K., and J. S. Shellman-Reeve. 1997. The general protected invasion theory: sex biases in parental and alloparental care. *Evolutionary Ecology* 11:357–370.

Röller, H., K. Biermann, J. S. Bjerke, D. W. Norgard, and W. H. McShan. 1968. Sex pheromones of pyralid moths. I. Isolation and identification of sex attractant of *Galleria mellonella* L. (greater waxmoth). *Acta Entomologica Bohemoslovaca* 65:208–211.

Royer, L., and J.N. McNeil. 1992. Evidence of a male pheromone in the European corn borer, *Ostrinia nubilalis* (Hübner) (Lepidoptera: Pyralidae). *The Canadian Entomologist* 124:113–116.

Schmitz, B., M. Buck, A. Egelhaaf, and D. Schneider. 1989. Ecdysone and a dietary alkaloid interact in the development of the pheromone gland of a male moth (*Creatonotos*, Lepidoptera: Arctiidae). *Developmental Biology* 198:1–7.

Schneider, D., M. Boppré, J. Zweig, S.B. Horsely, T.W. Bell, J. Meinwald, K. Hansen, and E.W. Diehl. 1982. Scent organ development in *Creatonotos* moths: regulation by pyrrolizidine alkaloids. *Science* 215:1264–1265.

Schneider, D., S. Shulz, R. Kittmann, and P. Kanagaratnam. 1992. Pheromones and glandular structures of both sexes of the weed defoliator moth, *Pareuchaetes pseudoinsulata* Rego Barros (Lep., Arctiidae). *Journal of Applied Entomology* 113:280–294.

Schneider D., S. Schulz, E. Priesner, J. Ziesmann, and W. Francke.1998. Autodetection and chemistry of female and male pheromone in both sexes of the tiger moth, *Panaxia quadripunctaria. Journal of Comparative Physiology A* 182:153–161.

Schulz, S. 2009. Alkaloid-derived male courtship pheromones. Pp. 145–153. In W.E. Conner, ed. *Tiger Moths and Wooly Bears: Behavior, Ecology, and Evolution of the Arctiidae.* Oxford: Oxford University Press.

Singer, M.S., K.C. Mace, and E.A. Bernays. 2009. Self-medication as adaptive plasticity: increased ingestion of plant toxins by parasitized caterpillars. *Public Library of Science ONE* 4: e4796.

Smedley, S., and T. Eisner. 1995. Sodium uptake by puddling in a moth. *Science* 270:1816–1818.

Spangler, H.G. 1986. Functional and temporal analysis of sound production in Galleria mellonella L. (Lepidoptera: Pyralidae). *Journal of Comparative Physiology A* 159:751–756.

Spangler, H.G. 1987. Ultrasonic communication is *Corcyra cephalonica* (Stainton) (Lepidoptera: Pyralidae). *Journal of Stored Product Research* 23:203–211.

Takayoshi, K., and H. Hiroshi. 1999. Identification and possible functions of the hairpencil scent of the yellow peach moth, *Conogethes punctiferalis* Guenée (Lepidoptera: Pyralidae). *Applied Entomology and Zoology* 34:147–153.

Tamaki, Y. 1985. Sex pheromones. Pp. 145–191. In G.A. Kerkut and L.I. Gilbert, eds. *Comprehensive Insect Physiology, Biochemistry, and Pharmacology,* Vol. 9. London: Pergamon Press.

Teal, P.E.A., and J.H. Tumlinson. 1989. Isolation, identification and biosynthesis of compounds produced by male hairpencil glands for *Heliothis virescens* (F.) (Lepidoptera: Noctuidae). *Journal of Chemical Ecology* 15:413–427.

Teal, P.E.A., J.R. McLaughlin, and J.H. Tumlinson. 1981. Analysis of the reproductive behavior of *Heliothis virescens* (F.) (Lepidoptera: Noctuidae) under laboratory conditions. *Annals of the Entomological Society of America* 74:324–330.

Teal, P.E.A., J.R. McLaughlin, and J.H. Tumlinson. 1989. Analysis of the reproductive behavior of *Heliothis virescens. Annals of the Entomological Society of America* 74:324–330.

Thibout, E., S. Ferary, and J. Auger. 1994. Nature and role of the sexual pheromones emitted by males of *Acrolepiopsis assectella. Journal of Chemical Ecology* 20:1571–1581.

Thornhill, R. 1979. Male and female sexual selection and the evolution of mating strategies in insects. Pp. 81–121. In M.M. Blum and N.A. Blum, eds. *Sexual Selection and Reproductive Competition in Insects.* New York: Academic Press.

Thornhill, R., and J. Alcock. 1983. *The Evolution of Insect Mating Systems.* Cambridge: Harvard University Press.

Trematerra, P. and G. Pavan. 1994. Role of ultrasound production and chemical signals in the courtship behavior of *Ephestia cautella* (Walker), *Ephestia kuehniella* Zeller and *Plodia interpunctella*

(Hubner) (Lepidoptera: Pyralidae). *Proceeding of the 6th International Working Conference on Stored-Product Protection* 1:591–594.

Trivers, R.L. 1972. Parental investment and sexual selection. Pp. 136–179. In B. Campbell, ed. *Sexual Selection and the Descent of Man.* London: Aldine Publishing Company.

Vahed, K. 1998. The function of nuptial feeding in insects: a review of empirical studies. *Biological Reviews of the Cambridge Philosophical Society* 73:43–78.

von Nickisch-Rosenegk, E., and M. Wink. 1993. Sequestration of pyrrolizidine alkaloids in several arctiid moths (Lepidoptera: Arctiidae). *Journal of Chemical Ecology* 19:1889–1903.

von Nickisch-Rosenegk, E., and M. Wink. 1994. Influence of previous feeding regimes and ambient temperatures on degradation and storage of pyrrolizidine alkaloids in the moth species *Creatonotos transiens* (Lepidoptera: Arctiidae). *Entomologia Generalis* 19:159–170.

Wagner, D.L., and J. Rosovsky. 1991. Mating systems in primitive Lepidoptera, with emphasis on the reproductive behavior of *Korscheltellus gracilis* (Hepialidae). *Zoological Journal of the Linnean Society* 102:277–303.

Weatherston, J., and J.E. Percy. 1977. Pheromones of male Lepidoptera. Pp. 295–307. In K.G. Adiyodi and R.G. Adiyodi, eds. *Advances in Invertebrate Reproduction, Vol. 1.* Kerala: Peralam-Kenoth.

Weller, S., J., N.L. Jacoben, and W.E. Conner. 1999. The evolution of chemical defenses and mating systems in tiger moths (Lepidoptera: Arctiidae). *Biological Journal of the Linnean Society* 68:557–578.

Weller, S.J., R.B. Simmons, R. Boada, and W.E. Conner. 2000. Abdominal modifications occurring in wasp mimics of the ctenuchine-euchromiine clade (Lepidoptera: Arctiidae). *Annals of the Entomological Society of America* 93:920–928.

Williams, G.C. 1966. *Adaptation and Natural Selection.* Princeton, NJ: Princeton University Press.

Willis, M.A., and M.C. Birch. 1982. Male lek formation and female calling in a population of the arctiid moth, *Estigmene acrea. Science* 218:168–170.

Wink, M., and D. Schneider. 1990. Fate of plant-derived secondary metabolites in 3 moth species (*Syntomis mogadorensis, Syntomeida epilais,* and *Creatonotos transiens*). *Journal of Comparative Physiology B* 160:389–400.

Wink, M., D. Schneider, and L. Witte. 1988. Biosynthesis of pyrrolizidine alkaloid-derived pheromones in the arctiid moth, *Creatonotos transiens:* stereochemical conversion of heliotrine. *Zeitschrift für Naturforschung* 43c:737–741.

Wunderer, H., K. Hansen, T.W. Bell, D. Schneider, and J. Meinwald. 1986. Sex pheromones of two Asian moths (*Creatonotos transiens, C. gangis;* Lepidoptera—Arctiidae): behavior, morphology, chemistry, and electrophysiology. *Experimental Biology* 46:11–27.

Zagatti, P. 1981. Comportement sexuel de la pyrale de la canne à sucre *Eldana saccharina* (Wlk.) lie a deux phéromones émises par le male. *Behaviour* 78:81–89. [In French]

Zagatti, P., G. Kunesch, and N. Morin.1981. La vanilline, constituant majoritaire de la sécrétion aphrodisiaque émise par les androconies du mâle de la Pyrale de la Canne àsucre: *Eldana saccharina* (Wlk.) (Lépidoptères, Pyralidae, Gallerinae). *Comptes Rendus de l' Académie des Sciences, Série III* 292:633–635. [In French]

Zagatti, P., G. Kunesch, F. Ramiandrasoa, C. Malosse, D.R. Hall, R. Lester, and B.F. Nesbit. 1987. Sex pheromones of rice moth, *Corcyra cephalonica* Stainton. I. Identification of male pheromones. *Journal of Chemical Ecology* 13:1561–1573.

Zaspel, J.M., S.J. Weller, and M.A. Branham. 2011. A comparative survey of proboscis morphology and associated structures is fruit-piercing, tear-feeding, and blood-feeding moths in the subfamily Calpinae. *Zoomorphology* 130:203–225.

Zaspel, J.M., S. Coy, K. Habanek, and S.J. Weller. 2013. Presence and distribution of sensory structures on the mouthparts of self-medicating moths. *Zoologicher Anzeiger* 253:6–10.

PART TWO

Small Ermine Moths

Role of Pheromones in Reproductive Isolation and Speciation

MARJORIE A. LIÉNARD and CHRISTER LÖFSTEDT

Introduction

Small ermine moths belong to the genus *Yponomeuta* (Yponomeutidae) that comprises about 75 species distributed globally but mainly in the Palearctic region (Gershenson and
Ulenberg 1998). These moths are a useful model to decipher
the process of speciation, in particular the importance of ecological adaptation driven by host-plant shifts and the utilization of species-specific pheromone mating-signals as prezygotic reproductive isolating mechanisms.

Historically the small ermine moths have presented great
challenges to taxonomists due to the difficulty of identifying
to species the adults of many of the *Yponomeuta*. The earliest
identifications were based on larval food plants and the morphology of larvae and pupae, but this caused problems beginning with Linnaeus when he named the first *Yponomeuta* species. *Y. padella* (Linnaeus 1758) is oligophagous and feeds on
Crataegus spp., *Prunus spinosa, P. domestica,* and *Sorbus aucuparia* but is not associated with the bird cherry *P. padus,* from
which nevertheless it derives its species name *padella*. It may
have been mistaken for *Y. evonymella* (Linnaeus 1758), which
in spite of its name is the species feeding on *P. padus,* and not
on *Euonymus* (figure 13.1 and Table 13.1). After Linnaeus, tax-

onomic investigations were based on examination of adult
morphological characters (e.g., wing-spot size and color, genitalia) (Martouret 1966), which did not allow conclusive discrimination of all species, leading to recognition of the so-
called *padellus*-species complex (Friese 1960) which later
proved to include five species (Wiegand 1962; Herrebout et al.
1975; Povel 1984).

In the 1970s, "the small ermine moth project" was initiated
to include research on many aspects of the small ermine
moth biology. A major aim of the project was to uncover
whether *Y. malinellus,* the apple orchard ermine moth, which
was causing considerable damage in apple orchards in the
Netherlands, was identical to ermine moths found on shrubs
and other plants in the area, a question of considerable practical consequence for pest monitoring (Herrebout et al. 1975).
Coordinated by Wim Herrebout and Jacobus Wiebes, at
Leiden University in the Netherlands (van Bruggen and van
Achterberg 2000), the research was carried out by several
independent groups over two decades, including researchers
from Groningen, Amsterdam, and Wageningen in the Netherlands and the University of Lund in Sweden. It resulted in
12 PhD theses and many more associated research publications (reviewed in Menken et al. 1992). The multidisciplinary

FIGURE 13.1 The small ermine moth project and some European species of small ermine moths.

A Drs. Wim Herrebout (coordinator) and Jan van der Pers (electrophysiologist and founder of Syntech), two of the scientists who took part in the initial "Small ermine moth project" launched in the 1970s.

B-C *Yponomeuta padella* nest and larvae on *Prunus spinosa* (Dalby, Skåne, Sweden).

D-E Pupae and adult *Yponomeuta padella* on *Crataegus* sp.

F Spectacular covering tree webs of larvae and pupae of *Yponomeuta evonymella* (Lund, Skåne, Sweden).

G Adult *Yponomeuta evonymella*.

PICTURE CREDITS: Christer Löfstedt (A), Marjorie Liénard (B–F), and Steph Menken (G).

TABLE 13.1

Ecological and temporal dimensions of reproductive isolation in small ermine moths

| Species | Ecological | | Temporal |
	Host plant (family)	Common name	Female calling period
Yponomeuta cagnagella	*Euonymus europaeus (Celastraceae)*	European spindle tree	Dawn
Yponomeuta irrorella	*Euonymus europaeus (Celastraceae)*	European spindle tree	Dawn
Yponomeuta plumbella	*Euonymus europaeus (Celastraceae)*	European spindle tree	Night
Yponomeuta rorrella	*Salix* species (Salicaceae)	Willow tree	Dawn
Yponomeuta gigas	*Salix canariensis,*§ *Populus alba* (Salicaceae)	Willow tree, silver poplar	Not known
Yponomeuta evonymella	*Prunus padus* (Rosaceae)	Bird cherry tree	Dawn
Yponomeuta mahalebella	*Prunus mahaleb* (Rosaceae)	Mahaleb cherry tree	Dawn
Yponomeuta malinellus	*Malus* sp. (Rosaceae)	Apple tree	Dawn
*Yponomeuta sedella***	*Sedum telephium* (Crassulaceae)	Stonecrops plants	Night
Yponomeuta padella	*Crataegus sp., Prunus spinosa*§ *Prunus domestica, Sorbus aucuparia* (Rosaceae)	Hawthorn, blackthorn, plum tree, rowan	Dawn

SOURCE: Adapted following Friese (1960); Wiegand (1962); Gerrits-Heybroek et al. (1978); Löfstedt et al. (1991).
** *Previously Y. vigintipunctatus.*
§ Primary host.

approach facilitated significant advances in our understanding of phylogenetic relationships of the nine European *Yponomeuta* species, their sex pheromones and host plants, and ultimately the evolution of the genus and factors driving its diversification.

Early efforts were directed to provide rigorous phylogenies by using allozyme techniques (Menken 1980). Subsequently, nine European species were recognized by multivariate analysis of both morphological and biological characters in larvae, pupae, and adults (Povel 1984). In addition to the phylogenetic and taxonomic work, larval food choice (Kooi 1990), larval taste receptors (van Drongelen 1980), larval parasitoids (Dijkerman 1990), larval trail marking (Roessingh 1989), the chemical composition of host plants (Fung 1989), and host race formation (Raijmann, 1996) were investigated in order to obtain insights into the evolution of host-plant relationships and the processes that have led to present-day associations. Finally, the chemical composition of sex pheromones (Löfstedt and van der Pers 1985; Löfstedt et al. 1986; Löfstedt and Herrebout 1988; Löfstedt et al. 1991), odor perception (van der Pers 1978, 1981, 1982; van der Pers and den Otter 1978), and the role of pheromones and host-plant volatiles in reproductive isolation (Hendrikse 1990) were also investigated, providing a comprehensive framework and establishing ermine moths as an important model for study of the evolution of sex pheromones and their role in mate recognition and speciation.

In this chapter, we summarize ecological factors of relevance to mate-finding and pheromone communication in the small ermine moths, and we review the sex pheromone studies that were mainly carried out in the 1980s. Some earlier screening studies have been left out. We also touch upon how application of molecular techniques that have become available during the last decades have more recently started to refine our understanding of the evolution and role of sex pheromone communication in this fascinating group of closely related moth species.

Some of the outstanding questions are the relative importance of host plants versus pheromones and other ecological factors in promoting the early stages of population divergence and reproductive isolation and whether the predominant factors differ between species with different ecological niches. It also remains to be determined whether present-day host-plant associations that evolved from a common ancestor associated with Celastraceae (*Euonymus*) occurred through speciation in allopatry or in sympatry through selection for pheromone divergence and host-plant shifts (e.g., driven by host race enemy-free space mechanisms or competition for resources) (Menken and Roessingh 1998).

The Evolution towards Specialized Host-Plant Associations

The genus *Yponomeuta* radiated early through sequential adaptation, i.e., following the evolution of host plants (Jermy 1984). Currently, there are 76 known species of *Yponomeuta* found almost all around the world, including Australia, New Zealand, Eurasia, Africa, Asia, and North America, except for arctic and desert regions (South America and Antarctica were not investigated) (Gershenson and Ulenberg 1998; Turner et al. 2010). The nine species found in Western Europe (see Table 13.1) occur sympatrically mainly in Palaearctic regions. A closely related 10th species, *Y. gigas*, is endemic to the

Canary Islands and is not considered to be Western European and following biogeographic regions was placed among African species (Cox 2001).

Ermine moth species have strong host associations and are typically monophagous on one host or, if oligophagous, restricted to plants of the same family (Table 13.1). Consequently, host-plant selection has long been hypothesized to be an important element in the speciation process through sympatric divergence via host-race formation (Wiebes 1976; Gerrits-Heybroek et al. 1978).

Phylogenetic analyses of the genus (Menken 1996; Ulenberg 2009; Turner et al. 2010) have increased our understanding of the evolution of ermine moth host-plant associations. The Western European species are hypothesized to have evolved from a common ancestor on Celastraceae, most likely associated with the European spindle tree (*Euonymus europaeus*), a host plant which remains today in association with *Y. cagnagella*, *Y. irrorella*, and *Y. plumbella* (Menken et al. 1992; Turner et al. 2010). A sister taxon of the Yponomeutinae, the Saridoscelinae, also feed exclusively on Celastraceae, which supports the view that the association evolved very early in the Yponomeutidae and in the case of *Yponomeuta* spp. represents the ancestral character state (Gerrits-Heybroek et al. 1978; Ulenberg 2009). To date, host-plant family associations are known for 39 species of Yponomeutidae, among which 32 species are monophagous or oligophagous on one plant genus. At least 27 of these (more than one-third of all species in the genus) still feed on Celastraceae plants, of which 22 species are still associated with the ancestral host genus, *Euonymus* (Ulenberg 2009).

The colonization of new hosts from the Rosaceae, Crassulaceae, and Salicaceae families took place via sequential evolution (reviewed in Menken et al. 1992; Menken 1996). Mapping host plants onto the phylogeny of European *Yponomeuta* species indicates that the genus most likely dispersed from East Asia to the western Palearctic (figure 13.2). Concomitantly a single shift occurred from the ancestral Celastraceae to Rosaceae, involving an ancestor of *Y. mahalebella* and subsequently formed the clade leading to *Y. evonymella*, *Y. padella*, and *Y. malinellus* (Menken 1996; Turner et al. 2010). In contrast, *Y. plumbella*, *Y. irrorella*, and *Y. cagnagella* remained associated with Celastraceae although in the case of *Y. cagnagella*, it is very likely a secondary association following a reversal from Rosaceae (Ulenberg 2009; Turner et al. 2010) as evidenced by its relict sensitivity to benzaldehyde, a chemical abundant in Rosaceae plants but absent from its host *E. europaeus* (Roessingh et al. 2007). Finally, a shift from Rosaceae to Salicaceae is observed for *Y. rorrella* and its close relative *Y. gigas*, whereas *Y. sedella* further specialized to *Sedum* sp. (Crassulaceae) (figure 13.2) (Menken et al. 1992; Turner et al. 2010). *Y. padella* is the sole species exhibiting an oligophagous feeding pattern including several genera of Rosaceae (Table 13.1), while all other European species remain strictly monophagous.

The exact nature of the changes leading to adaptation to new hosts is unknown but could have involved enemy-free space mechanisms, the evolution of new adult host preference, and adaptations of larval host acceptance behavior including new feeding preferences, changes at the sensory level, and changes in digestive or detoxification systems (Kooi 1990; Roessingh et al. 2000). Larval gustatory sensitivity towards various host-plant chemicals was initially studied in the European species by electrophysiological recordings of larval chemoreceptor sensilla (van Drongelen 1979, 1980).

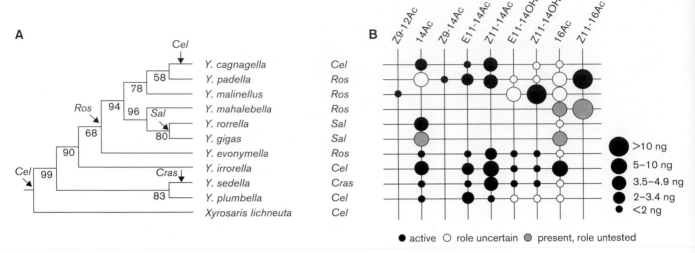

FIGURE 13.2 Phylogeny and pheromone composition.

A Phylogeny of the western European clade of small ermine moths. The MP tree was reconstructed using consensus based on the COXII, 16S, and ITS phylogenies after Turner et al. (2010), and *Xyrosaris lichneuta* as an outgroup. Host plant shifts are indicated by arrows and current affiliations are indicated after the species names. Cel, Celastraceae, ancestral host; Ros, Rosaceae; Sal, Salicaceae; Cras, Crassulaceae. Please note that *Yponomeuta sedella* was sometimes previously named *Y. vigintipunctatus*.

B Pheromone composition and titres per individual small ermine moth female, adapted from Löfstedt et al. (1991) and Löfstedt and Herrebout (1988). Filled circles represent female pheromone components with confirmed behavioral activity in conspecific males, empty circles are compounds present in the gland for which the role in mate attraction is uncertain or not essential. The sex-pheromone of *Yponomeuta gigas*, a species endemic to the Canary Islands, is similar to *Y. rorrella*. Only saturated acetates were found in pheromone gland extracts (C. Löfstedt, unpublished).

Celastraceae-feeders, i.e., *Y. plumbella*, *Y. irrorrella*, and *Y. cagnagella* are sensitive to dulcitol, a feeding stimulant sugar alcohol (van Drongelen 1979; Peterson et al. 1990), whereas Rosaceae-feeders such as *Y. padella* and *Y. evonymella* respond to various degrees to sorbitol, a stereoisomer of dulcitol present at high levels in this plant family (van Drongelen 1979; Roessingh et al. 1999), and to the Rosaceae-specific compound benzaldehyde (Kooi 1988; Roessingh et al. 2007). Although adult females mainly determine the host plant by their choice of oviposition sites in *Yponomeuta*, it has been proposed that the shift from Celastraceae to Rosaceae could also have been facilitated by larval preferences toward the presence of low amounts of dulcitol in Prunoidea, a suborder of Rosaceae (Menken and Roessingh, 1998). Since Prunoidea also contain sorbitol, and both dulcitol and sorbitol are perceived in homologous peripheral sensory cells, a simple sensory shift might have favored the acquisition of a new feeding preference allowing the utilization of Rosaceae as new host plants (Menken and Roessingh, 1998; Roessingh et al. 1999). Interestingly, *Y. mahalabella*, *Y. padella*, and *Y. evonymella* are still sensitive to dulcitol (van Drongelen 1979) despite feeding on Rosaceae hosts containing no or very limited amounts of this sugar, which supports the idea that ancestral larval preferences are maintained among *Yponomeuta* (van Drongelen 1979). As mentioned above, although feeding on Celastraceae, sensory cells in larval maxillary palps of *Y. cagnagella* respond strongly to benzaldehyde, supporting a recent reversal event from a Rosaceae host-plant ancestor common with *Y. padella* (Roessingh et al. 2007; Turner et al. 2010).

To corroborate the current correlations between the phylogenetic pattern and the role of sensory stimuli adaptation in ecological speciation, future studies should further investigate adult oviposition preferences and larval sensory responses to host specific plant compounds.

Sex Pheromones and Other Ecological Factors Involved in Reproductive Isolation

Overview of Sex-Pheromone Composition

Female sex pheromones were demonstrated in the early 1980s to play an essential role in efficient mate finding in small ermine moths (reviewed in Löfstedt et al. 1991). Females bend their abdomen in a characteristic position that leads to the extrusion of the last abdominal segments (figure 13.3A) (Hendrikse 1978), actively displaying the pheromone gland and thereby releasing the pheromone into the surrounding air.

Like many other closely related moth species, *Yponomeuta* sex pheromones are blends of structurally related C_{14} and C_{16} acetates and alcohols (figure 13.2) (Löfstedt et al. 1991). The (Z)-11-tetradecenyl acetate (Z11-14Ac) likely represents an ancestral pheromone component based on its occurrence across the *Yponomeuta* genus (Löfstedt and van der Pers 1985; Löfstedt and Herrebout 1988; Löfstedt et al. 1991). Six of the nine European species, including the four basal species in the clade (*Y. plumbella*, *Y. sedella*, *Y. evonymella*, and *Y. irrorella*) as well as *Y. padella* and *Y. cagnagella* use a mixture of Z11-14Ac together with varying ratios of (E)-11-tetradecenyl acetate (E11-14Ac) (figures 13.2 and 13.4) as primary pheromone components, i.e., pheromone components that cannot be subtracted from the synthetic pheromone without resulting in a significant loss of activity. Additional gland constituents (and likely pheromone components) include combinations of tetradecyl acetate (14Ac), (Z)-9-tetradecenyl acetate (Z9-14Ac), and (Z)-11-hexadecenyl acetate (Z11-16Ac) as well as the corresponding unsaturated C_{14} alcohol compounds, with or without confirmed behavioral activity depending on the species (Löfstedt et al. 1991). Some of the compounds display a more repeatable average ratio than others, which may indi-

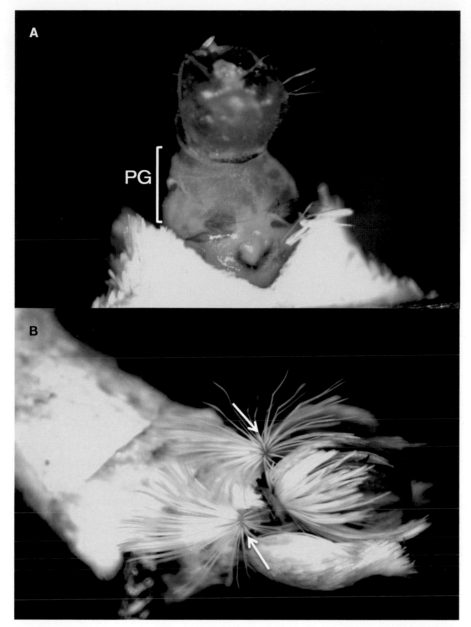

FIGURE 13.3 *Yponomeuta rorrella* female pheromone gland and male hairpencil structures.

A Ventral view of an *Yponomeuta rorrella* female extruding her 8th and 9th to 10th terminal abdominal segments; the sex pheromone gland (PG) is located along the intersegmental integument between the 8th and 9th abdominal segments.

B Lateral view of an *Yponomeuta rorrella* male with extruded hairpencil-like structures. Whether these structures are vestigial or play a role in courtship behavior, as suggested by Hendrikse et al. (1984), remains to be confirmed.

PICTURE CREDITS: Jean-Marc Lassance.

cate evolutionary constraints due to their importance for reproductive isolation (Löfstedt and van der Pers 1985; Du et al. 1987; Löfstedt and Herrebout 1988) or simply reflect the nature of the biosynthetic pathways involved in their production. For instance, low variation was found in the production of E11/Z11-14Ac components in *Y. padella*, in contrast to higher variation found among saturated acetates and the unsaturated (Z)-9-tetradecenyl acetate (Z9-14Ac) in repetitive sampling of individual females (Du et al. 1987).

The species *Y. cagnagella*, *Y. irrorella*, and *Y. plumbella* share *Euonymus* as a host plant and like all European *Yponomeuta*

they are sympatric and in this case also synchronic. The pheromones of these three species are blends of several compounds in addition to Z and E11-14Ac. The pheromone blend of *Y. plumbella* also contains saturated acetates. Female *Y. plumbella* and *Y. irrorella* emit a mixture of 16Ac, 14Ac, E-, and Z11-14Ac. The former emit a 25:46:148:100 ratio, whereas the latter emit a 17:68:56:100 ratio (figure 13.2). The *Y. cagnagella* pheromone blend consists of three components, 14Ac, E11-14Ac, and Z11-14Ac, in a 30:3:100 ratio (Löfstedt and Herrebout 1988). The saturated tetradecyl acetate acts as a synergist in all three species, whereas 16Ac is produced in all species

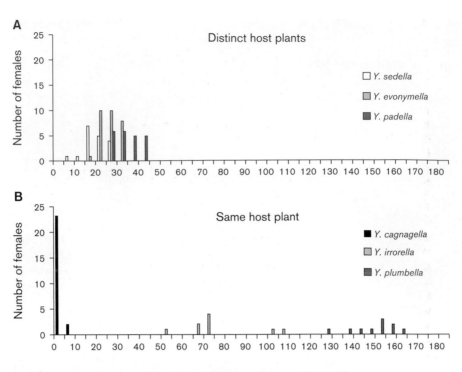

FIGURE 13.4 Frequency distribution of pheromone production in individual females in relation to cross-attraction and host plant. The graphs represent the E11- to Z11-14Ac isomer ratios in individual female pheromone gland extracts analyzed by gas chromatography, from six different Yponomeuta species.

A In species that do not share a host plant, cross-attraction has been observed in flight tunnel experiments as a consequence of overlapping relative amounts of E11-14Ac.

B The three species sharing the spindle tree as host plant show a pattern of reinforcement through adaptive selection for pheromone differences; non-overlapping ranges of E11- to Z11-14Ac actively prevent cross-attraction in flight tunnel experiments and in the field.

SOURCE: Adapted from Löfstedt et al. (1991).

but has confirmed behavioral activity only in *Y. irrorella*. Female glands in all three spindle tree species also contain tetradecanol and the corresponding monounsaturated alcohols (figure 13.2), but the alcohols only appear to be active minor components in *Y. irrorella* (Löfstedt and Herrebout 1988). The range of Z/E11-14Ac ratios produced by each species is essential for conspecific male attraction by females of the three species sharing the spindle tree as host plant. Moreover, subtraction of E11-14Ac in either of the synthetic pheromones of *Y. irrorella* or *Y. plumbella* leads to high catches of *Y. cagnagella* in field-trapping experiments (Löfstedt and Herrebout 1988). The likely explanation is that a majority of *Y. cagnagella* females produce a very low amount of E11-14Ac that averages 3% relative to the Z11-14Ac. Thus, even a trace amount of E11-14Ac occurring as an impurity in the synthetic Z11-14Ac may still be enough to cause significant attraction of *Y. cagnagella*.

Whereas the *Y. cagnagella* and *Y. plumbella* pheromones are highly specific due to very distinct E/Z isomer ratios, the *Y. irrorella* synthetic blends consistently attracted low numbers of *Y. cagnagella* and *Y. padella* in the field, demonstrating that mate discrimination and reproductive isolation are not entirely mediated by pheromones but require additional isolating mechanisms (i.e., different hosts for *irrorella* and *padella*, possibly different time and height of flight [Herrebout and van de Water 1983]) and possibly male courtship phero-

mones, although the latter has not been conclusively demonstrated (Löfstedt and Herrebout 1988).

The two allochronic species, *Y. evonymella* and *Y. sedella*, are nearly identical with respect to their female-produced pheromones. They produce a 100:20 mixture of Z11-14Ac and E11-14Ac, and in addition minor amounts of Z/E11-14OH are reported to synergize trap catches (Löfstedt and van der Pers 1985). Although the E11-14Ac/Z11-14Ac ratio of individual *Y. padella* females overlaps with those of *Y. evonymella* and *Y. sedella* (figure 13.4), *Y. padella* females produce a 34:100:400 mixture of E11-14Ac, Z11-14Ac, and Z11-16Ac (Löfstedt and van der Pers 1985) and the presence of Z11-16Ac efficiently reduces trap catches of *Y. evonymella* and *Y. sedella* (Löfstedt and van der Pers 1985). *Y. padella* is also unique in having small amounts of Z9-14Ac as an important fourth active component (Löfstedt et al. 1991).

Among the three remaining species, *Y. malinellus* uses an unusual combination of (Z)-9-dodecenyl acetate (Z9-12Ac) and Z11-14OH (McDonough et al. 1990), whereas *Y. mahalebella* (only female glands analyzed, no behavioral experiments) and *Y. rorrella* differ by an overall reduced number of acetates and a decrease in pheromone complexity that likely played a role in the adaptation to new communication niches (Löfstedt et al. 1991). *Y. mahalebella* females produce Z11-16Ac plus saturated acetates, whereas *Y. rorrella* females produce a mixture of saturated acetates, of which only 14Ac has been demonstrated to be attractive to males. Among all European small ermine moth species, the *Y. rorrella* pheromone is unique due to the absence of any unsaturated compounds (figure 13.2) and any unsaturated fatty acyl precursors (Löfstedt et al. 1991), pointing to a mechanism disrupting Δ11-desaturation in this species (Löfstedt et al. 1986).

Temporal and Behavioral Niches Contributing to Species Separation

Interspecific encounters are common among European small ermine moths in their natural habitats. Cross-attraction occurs between several species under field (see above) and laboratory conditions (figures 13.4 and 13.5). This reflects similarities in female pheromone production (figure 13.2), mirrored by *Yponomeuta evonymella* males showing strong responses to synthetic blends mimicking *Y. padella*, *Y. irrorella*, and *Y. sedella* female pheromones in the wind tunnel, and males of *Y. padella* responding to *Y. irrorella* synthetic pheromones in the field (figure 13.5) (Löfstedt and Herrebout 1988; Löfstedt et al. 1991). However, hybridization under natural conditions has rarely been reported; among trapping studies that captured tens of thousands of males, discriminant allozyme analyses never identified a single hybrid (Hendrikse 1979). Depending on species, additional prezygotic (e.g., flight height, temporal differences in diurnal and seasonal activity, male courtship behavior) and postzygotic isolating mechanisms may contribute additional dimensions to the pre-mating reproductive isolation provided by sex pheromones.

As mentioned earlier, *Y. evonymella* and *Y. sedella* have a more or less identical pheromone composition (figure 13.2) (Löfstedt et al. 1991), likely accountable for by the absence of selection for pheromone divergence due to differences in geographic distribution and seasonal activity. *Y. evonymella* is active at dawn, whereas *Y. sedella* is active in the middle of the night (Hendrikse 1979). *Y. evonymella* is also univoltine, flying from late June to mid-July in south Sweden, at approximately the same time as most of the other small ermine moths, whereas *Y. sedella* is bivoltine with its first flight occurring well before the flight of *Y. evonymella*, and the second one late in the summer (Löfstedt and van der Pers 1985; Löfstedt et al. 1986; Löfstedt and Herrebout 1988).

Temporal differences can additionally reinforce niche separation. For instance, *Y. plumbella* and *Y. sedella* females call early in the night compared to females of all other species, which preferentially call near the end of the scotophase and at dawn (Hendrikse 1979). Female maturation (i.e., peak in calling activity) varies depending on species from 1 day (*Y. plumbella*) to 10 days (*Y. malinellus* and *Y. cagnagella*) and averaging 2–5 days after emergence for other species (Hendrikse 1979), which may also contribute to temporal niche distinctness.

Regarding the potential role of male pheromones, all *Yponomeuta* males possess two sets of abdominal, eversible hairpencils (see figure 13.3B) that they extrude during courtship while wing fanning at close range toward calling females (Hendrikse 1979). This behavior suggested that male-species-specific signals might help females to discriminate among mates (Hendrikse et al. 1984) and function to enhance reproductive barriers, particularly for those species for which cross-attraction has been observed. In flight tunnel experiments, Hendrikse et al. (1984) demonstrated that *Y. padella* and *Y. evonymella* females rejected more than 90% of heterospecific

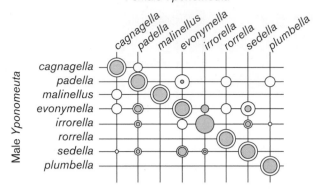

FIGURE 13.5 Male behavioral response to calling females (figure modified after Hendrikse 1986 and Löfstedt et al. 1991). Empty and full circles represent males taking flight and landing, respectively. Response is proportional to the diameter of circles (intraspecific empty circles correspond to 100%).

males. Furthermore, results with *Y. padella* suggested that the male hairpencil display might mediate female mate recognition and acceptance. An inhibitory effect on other males was also noted in *Y. cagnagella*, *Y. evonymella*, and *Y. padella*. When males actively displaying their hairpencils were placed upwind of a calling female, wing-fanning movements in conspecific and heterospecific males placed downwind were decreased (Hendrikse et al. 1984). Although this moth family is not known to possess ears (Scoble 1992), it was not possible to discriminate among the competing hypotheses that acoustic or olfactory cues were involved. Despite behavioral indications, a male pheromone in *Y. padella* acting as a second pre-mating isolation barrier especially against interspecific hybridization with *Y. irrorella* and *Y. evonymella* females (Löfstedt and Herrebout 1988) has not yet been conclusively demonstrated.

Most European ermine moth species have species-specific differences in their female pheromone blends. Whenever pheromones remain similar, temporal and behavioral factors play a more important role. With all ecological factors combined, all nine European species occupy a virtually unique communication channel that ensures reproductive isolation.

Pheromone Biosynthesis and Modulation of Blend Ratios

The biosynthetic machinery of mate signaling in moths is controlled by several multigene families (see Roelofs and Rooney 2003; Blomquist et al. 2005). Fatty-acyl-CoA desaturases and fatty-acyl-CoA reductases have been characterized and confirmed to encode specific enzymatic functions that, in addition to uncharacterized β-oxidases, acetyl-transferases, and oxidases, contribute to defining the final structures and ratio of each component in the species-specific pheromone mixtures (Roelofs and Bjostad 1984; Bjostad et al. 1987). Similar to most moth species for which pheromone biosynthetic pathways have been elucidated, the production of unsaturated components in the *Yponomeuta* spp. pheromone blends was postulated to start from the ubiquitous palmitic acid (hexadecanoate or 16:Acyl) and involve combinations of chain shortening and desaturation by a Δ11-desaturase (Löfstedt et al. 1991). The proposed biosynthetic pathway (figure 13.6) is based on analyses of fatty-acyl precursors in pheromone gland extracts (Löfstedt et al. 1991). In addition to

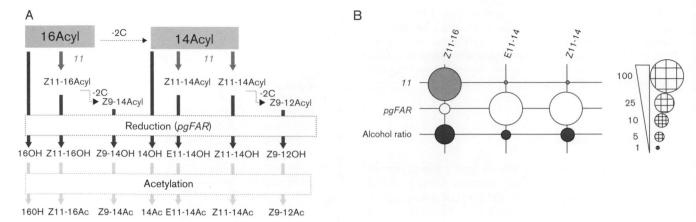

FIGURE 13.6 Pheromone biosynthesis and modulation of pheromone ratios in small ermine moths.

A Pheromone biosynthesis pathway including characterized biosynthetic genes (Δ11, delta11 acyl-CoA-desaturase; pgFAR, pheromone gland specific fatty-acyl-CoA-reductase). Pheromone production starts from palmitic acid (16Acyl) and all unsaturated acyl precursors are synthesized by combination of Δ11-desaturation (Δ11) and β-oxidation (-2C). Subsequently, a single pgFAR reduces the species-specific precursors present in the gland of the respective species to produce intermediate alcohol products as investigated in *Yponomeuta evonymella*, *Y. padella*, and *Y. rorrella* (Liénard et al. 2010) that are acetylated to give the active pheromone compounds. Note that the only European species producing Z9-12Ac as pheromone compound is *Y. malinellus*, and the pathway represented here leading to Z9-12Ac has not yet been confirmed in vivo.

B The reverse chain-length preference of the Δ11-desaturase and pgFAR for acyl substrates with 14 or 16 carbon atoms in *Yponomeuta padella* allows adjusting the alcohol blend ratio, which matches the active blend ratio especially for Δ11-unsaturated compounds. The circled areas are proportional (%) to the *Y. padella* Δ11-desaturase (upper line) and pgFAR substrate preferences (middle line), and to the final ratio between components (bottom line) (adapted from Liénard and Löfstedt 2010).

16:Acyl, precursor analyses revealed the presence of myristic acid (tetradecanoate or 14:Acyl), (*Z*)-9-hexadecenoate (Z9-16:Acyl), and (*Z*)-9-tetradecenoate (Z9-14:Acyl) in all species. (*Z*)- and (*E*)-11-tetradecenoates are found in all species but *Y. rorrella* and *Y. mahalebella*. (*Z*)-11-hexadecenoate (Z11-16:Acyl) is present in *Y. cagnagella*, *Y. padella*, and *Y. mahalebella*. Finally, (*Z*)-9-dodecanoate (Z9-12:Acyl) is found only in *Y. malinellus* (Löfstedt et al. 1991). The species-specific pools of saturated and monounsaturated fatty-acyl precursors are subsequently reduced and acetylated.

Although no experiments with labeled precursors have been carried out in vivo to support the postulated pathways, some of the biosynthetic enzymes have been isolated and functionally characterized, including a pheromone gland-specific Δ11-fatty-acyl-CoA desaturase and a broad-range fatty-acyl-CoA reductase (pgFAR) acting on C_{14} and C_{16} acyl precursors (Liénard and Löfstedt 2010; Liénard et al. 2010). The Δ11-desaturase characterized from *Y. padella* produces large amounts of (*Z*)-11-hexadecenoic acid but also catalyzes the dehydrogenation of tetradecanoic acid to produce minor amounts (around 5% of the produced unsaturated FAs) of (*E*)- and (*Z*)-11-tetradecenoic acids, altogether accounting for the production of all potential intermediate Δ11 fatty-acyl-precursors (Liénard and Löfstedt 2010). The broad range, exquisitely nonselective pgFAR has been functionally characterized in *Y. padella*, *Y. evonymella*, and *Y. rorrella* and accounts for the reduction of a wide range of saturated precursors (from C_{12} to C_{16}) and their corresponding Δ9 and Δ11 unsaturated acyl precursors, thus including all potential saturated and monounsaturated fatty-acyl-precursors found across the *Yponomeuta* genus (Liénard et al. 2010).

Pheromone glands of *Y. rorrella* females differ markedly from other ermine moths by their absence of any Δ11 unsaturated fatty-acyl precursors, suggesting that the Δ11-desaturase is inactive in this species (Löfstedt et al. 1986), yet the exact molecular mechanism remains unknown. Nevertheless, in agreement with the scenario under which the simple *Y. rorrella* pheromone blend derives from an ancestral more complex multicomponent pheromone (Löfstedt et al. 1986), its downstream pgFAR enzyme has retained the ability to reduce all unsaturated compounds. This also suggests that FARs might have a conserved function, if not throughout the genus at least in several of the ermine moth species (Liénard et al. 2010). In *Y. padella*, the pgFAR substrate specificity ($C_{14} > C_{16}$) further counterbalances the inherent chain-length preference ($C_{16} > C_{14}$) of the upstream Δ11-desaturase to modulate the production of the unsaturated alcohols. In vivo, the alcohol outcome depends on the species-specific fatty-acyl pools (Liénard et al. 2010), but essentially the Δ11-alcohol ratios correlated with the final acetate ratios previously reported (Löfstedt et al. 1991).

Among all nine species only female *Y. malinellus* use Z9-12Ac as a sex pheromone compound together with Z11-14OH (McDonough et al. 1990). However, female pheromone glands in that species also contain E11-14OH, 14OH, 16OH, and 16Ac but neither saturated nor unsaturated C_{14} acetates (McDonough et al. 1990). The presence of these compounds in female glands supports the hypothesis that pgFAR has broad range activity similar to that of *Y. padella*, *Y. evonymella*, and *Y. rorrella* and a selective acetyl transferase that does not acetylate the C_{14} alcohols. In addition, this species likely possesses a β-oxidase able to chain-shorten the Z11-14:Acyl to produce the Z9-12:Acyl otherwise absent in other species.

Pheromone biosynthesis in small ermine moths involves several key biosynthetic enzyme-encoding genes, including a single Δ11-desaturase and a single pgFAR. Characterization of these enzymes has shed light on some of the mechanisms involved in determination of the final blend composition in several of the species (Liénard and Löfstedt 2010). Candidate genes involved in chain shortening and acetylation remain to be identified and may reveal the mechanisms involved in adjusting the final ratios between unsaturated and saturated components, and between alcohols and acetates.

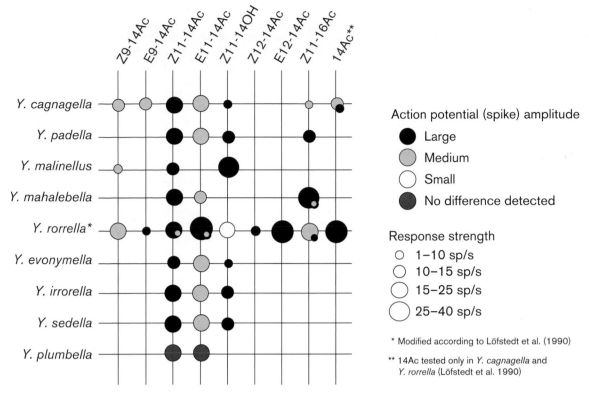

FIGURE 13.7 Electrophysiological responses of male sensilla trichoidea to pheromone compounds. Responses of olfactory sensory neurons to pheromone compounds in nine species of small ermine moths. Activities elicited in receptor cells are presented for the most abundant sensillum type (type I) that harbors one or two cells with different spike amplitudes. The size of a circle reflects the response to 1 μg (van der Pers 1982) or 10 μg (Löfstedt et al. 1990) of doses of the stimulus. The bulk of data is extracted from van der Pers (1982), but the 14Ac and *Yponomeuta rorrella* data as indicated by asterisks (except E9-14Ac and Z11-14OH) are modifications based on Löfstedt et al. (1990).

Male Physiological and Behavioral Response

Detection of Pheromone and Plant Compounds

Like many other moth species, male small ermine moths are capable of detecting host-plant odors and conspecific pheromone components, the latter through specialized pheromone receptor cells localized in sensilla trichodea (Cuperus 1985). Interactions between pheromone compounds and plant odors were studied by electroantennogram (EAG) and single cell recordings, demonstrating that responses of pheromone compound receptors in males were stronger in the presence of host-plant odors (van der Pers et al. 1980). This suggested that a synergism between the two types of odors might occur in the field, although this has not been demonstrated. The observation is nevertheless interesting because of the possible interplay between pheromones and host plants in the process of speciation.

Antennal response profiling was carried out using EAG and single sensillum tip-recording techniques. The results were based on a variable number (8–36) of randomly selected sensilla per species. The first pheromone identifications had not been reported yet and each recorded sensillum was assessed for a broad range of potential pheromone compounds including monounsaturated and doubly unsaturated C_{14}, C_{15}, and C_{16} acetates as well as Z11-14OH (van der Pers 1982). The doubly unsaturated compounds and the C_{15} acetates were never found in female pheromone gland extracts (Löfstedt et al. 1991), and thus responses to these compounds will not be discussed further. The importance of saturated acetates as phero-

mone components in some of the species was not known at the time and these are not included in the study by van der Pers (1982) but were tested later in *Yponomeuta rorrella* (see below) (Löfstedt et al. 1990).

In European ermine moth species, male antennae consist of approximately 60 segments that harbor curved sensilla trichodea ranging from 20 to 30 μm in length, and in which are housed olfactory sensory cells. In seven of the nine species, two types of sensilla with different response profiles were categorized as type I and type II. Type I sensilla are the most abundant type and their physiological response spectra are compiled in figure 13.7 (van der Pers and den Otter 1978; van der Pers 1982; Löfstedt et al. 1990). *Y. cagnagella* and *Y. irrorella* possess a third type of sensillum (type III), which has a characteristic response profile distinct from types I and II (van der Pers 1982). Typically, within each sensillum, two cells—olfactory sensory neurons (OSNs) firing with different action potential amplitudes named as Large (L), Medium (M), or Small (S) amplitude (spike) cells and with different response profiles—are activated upon odor stimulation (van der Pers 1978, 1982). Cells with the second largest spike amplitude are always called medium. Thus small spike amplitude cells are only recognized in sensilla with three cells. Recordings from *Y. plumbella* OSNs did not show spike amplitude differences, and thus it could not be determined how many OSNs were housed per sensillum in this species (van der Pers 1982). Another exception was found in *Y. rorrella* where type I sensilla consistently exhibited an additional small spike amplitude activity, suggesting the presence of a third OSN, preferentially responding to Z11-14OH. Altogether, the presence of

2–3 types of sensilla that each normally house 2 OSNs suggests that small male ermine moths typically possess at least 6 distinct types of OSNs.

The response profile of OSNs from the most abundant type of sensillum (type I) varies between species, but each OSN is usually sensitive to several stimuli (figure 13.7).

All sensillum types were found to respond to E11-14Ac and Z11-14Ac in all species but *Y. malinellus*, where no OSN responds to E11-14Ac. This is not surprising since female *Y. malinellus* produce a unique blend composed of Z11-14OH and Z9-12Ac, but Z9-12Ac was not included in the set of stimuli tested by van der Pers (1982). However, among all species tested, the strongest response to Z11-14:OH in OSN-L, in sensilla types I and II, were recorded in *Y. malinellus*.

In all species but *Y. rorrella*, responses to E11-14Ac are exclusively elicited via medium spike amplitude neurons in both types of sensilla, whereas responses to Z11-14Ac are exclusively elicited in large spike amplitude sensory neurons, indicating that a specific sensory neuron responds to each of the two most important pheromone components found across the genus.

OSNs responding to Z11-16Ac were found in *Y. cagnagella*, *Y. padella*, *Y. rorrella*, and *Y. mahalebella* (van der Pers 1982). Z9-14Ac elicited various degrees of responses in the sensillum type I in *Y. cagnagella*, *Y. malinellus*, *and Y. rorrella* (figure 13.7) and in a less abundant type of sensillum (type II, not shown in figure 13.7) in *Y. evonymella* and *Y. sedella* (van der Pers 1982). Intriguingly, no Z9-14Ac-tuned OSN was found in *Y. padella*, although Z9-14Ac was later confirmed to be a fourth pheromone component in that species, suggesting that OSNs responding to this compound exist that were not identified in earlier studies.

In *Y. rorrella*, the three types of sensory neurons were investigated in detail (Löfstedt et al. 1990). In this species the medium spike amplitude neurons in types I and II sensilla respond to Z9-14Ac and Z11-16Ac. In addition to responding to the pheromone component 14Ac, the large spike neurons respond to E6-, E9-, E11-, E12, Z6-, Z11-, and Z12-14Ac to varying extents. E6-, E12, Z6-, and Z12-14Ac are not found as pheromone components in any *Yponomeuta* species (Löfstedt et al. 1990) and this response profile is unusually broad compared to antennal responses in other small ermine moths. These results showing that a single OSN with large spike amplitude concomitantly responds to 14Ac, Z11-14Ac, and E11-14Ac differ slightly from the original study by van Der Pers (1982), reporting that like other ermine moths, Z11-14Ac and E11-14Ac were activated by distinct cells with large and medium spike amplitudes (figure 13.7). More generally, response profiles also vary within species, which can reflect functional intraspecific variation; differences in OSN action potential amplitudes may also differ between studies and individuals depending on the sensillum actually recorded from and the technique used. Differences in recorded action potentials from the tested sensillum can first be accounted for by the morphological properties of the sensillum itself. In the European corn borer, receptor cells with larger diameter have been shown to produce spikes with greater amplitudes (Hansson et al. 1994), a correlation that may raise the neuron sensitivity by allowing more receptor sites to coexist on the cell surface and has been proposed as an explanation as to why larger spike neurons also show higher spike frequencies (Cossé et al. 1995). Additionally, recordings of OSN activity in ermine moths were done by the cut sensillum tip recording technique, a technique which in contrast to tungsten electrode recordings may not have reached close enough to the source of the action potential (Olsson et al. 2010), possibly adding to variation in measurements of spike amplitudes between studies. Although we have gained a good general understanding of olfactory response profiles in *Yponomeuta*, some unknowns remain (i.e., Z9-14Ac response profiles are likely incomplete), together with subtle variations between datasets that may be both biological and technical. A comprehensive study reexamining the olfactory response profiles of *Yponomeuta* using the most recent techniques is needed.

Role of Antagonists as Enhancers of Reproductive Isolation and Interspecific Interactions

The presence of unsaturated acetates (Z11-14Ac and E11-14Ac or Z11-16Ac) is essential for male attraction in all small European ermine moth species but one (Löfstedt et al. 1986, 1991). *Yponomeuta rorrella* is the only *Yponomeuta* species—as a matter of fact the only moth—in which the primary pheromone component is a saturated acetate (14Ac). In this species, unsaturated compounds act as antagonists and the addition of as little as 1% of Z11-14Ac in synthetic pheromone blends dramatically lowers trap catches in the field (Löfstedt et al. 1986). This is a striking example of a pheromone component from one species being a strong antagonist for a closely related species. Although unsaturated 14-carbon acetates and alcohols have not be demonstrated to be part of the active blends in *Y. rorrella* and *Y. mahalebella*, males in both species possess receptor neurons responding strongly to these compounds and eliciting antagonist effects on behavior (van der Pers 1982; Löfstedt et al. 1990). In *Y. rorrella*, the Z11-14OH also acts as an antagonist and activates a specific OSN exhibiting a small spike amplitude (Löfstedt et al. 1986, 1990). Likewise the Z11-16Ac, an essential component for the attraction of *Y. padella* and the major pheromone compound in extracts of *Y. mahalebella*, acted as an antagonist for other sympatric ermine moth species in field experiments with synthetic compounds (Löfstedt and van der Pers 1985).

Male moths have a remarkable ability to distinguish between pheromone sources even when placed extremely close together (Valeur and Löfstedt 1996, and references therein). Despite this, pheromone sources have been observed to interfere with each other under field conditions (see for instance Perry and Wall 1984). In traps baited with female moths, addition of a *Y. malinellus* female to a *Y. padella* female suppressed the attraction of conspecifics in the same way as a *Y. padella* female suppressed the attraction of *Y. cagnagella* males to conspecific females (Minks et al. 1977). Intraspecific and interspecific interactions were further studied in a laboratory flight tunnel. By creating overlapping plumes of pheromones from different species, it was found that *Y. padella* had a more severe effect on *Y. cagnagella* than vice versa and that *Y. sedella* was not as strongly influenced by *Y. padella* as *Y. cagnagella*. *Y. cagnagella* appeared to be influenced by Z11-16Ac in the *Y. padella* pheromone but even more so by the "wrong" ratio of E11-/Z11-14Ac (Löfstedt 1987). Although these experiments, both the ones in the field and in the laboratory, simulated a rather artificial situation, populations of small ermine moths may locally be very high and individual females may call in close proximity to each other. Thus, there likely exists a real challenge for male small ermine moths in handling more or less overlapping plumes in the olfactory landscape.

Subtraction of Z11-14Ac or E11-14Ac fully abolishes intraspecific attractiveness of synthetic pheromone blends for six of the species and subtraction of E11-14Ac or off Z/E ratios increases heterospecific trap catches (Löfstedt and Herrebout 1988). Interestingly, subtraction of any of the five components of the synthetic pheromones of *Y. evonymella* and *Y. sedella* not only reduces their attractiveness but also causes significant attraction of several tortricid moth species. For instance, subtraction of Z11-14OH or Z11-14Ac from the *Y. sedella* pheromone attracts *Aphelia paleana* or *Dichrorampha petiverella,* respectively, whereas subtraction of E11-14Ac from the *Y. evonymella* pheromone elicits attraction of *Tortrix viridana* (Löfstedt et al. 1991). Similarly, when Z11-14Ac is absent from the *Y. irrorella* pheromone blend, it attracts *Ethmia punctella* (Ethmiidae) and *Croesia holmiana,* another tortricid (Löfstedt and Herrebout 1988).

Summary: An Emerging Model System in Research on the Role of Sex Pheromones in Speciation—Toward a New "Small Ermine Moth Project"?

The small ermine moth genus *Yponomeuta* has a worldwide distribution and includes nine species that coexist sympatrically across Europe. European species are to a large extent sympatric but have evolved unique combinations of host plants and/or daily or seasonal activity patterns and/or species-specific female sex pheromones. Cumulatively, these differences ensure efficient pre-mating isolation in their natural habitat and make the European *Yponomeuta* species complex a model system to study the role of insect–plant relationships and pheromones in reproductive isolation and speciation in moths.

However, the model system has turned out more complex to study than originally thought. Therefore, the relative importance of host plants versus pheromones and possibly other factors in population divergence, speciation, and reproductive isolation has remained difficult to determine. We conclude by summarizing some of the difficulties encountered and suggest some areas of study that would enhance our understanding of the role of plants and pheromones in this system.

Overcoming the System Limitations

First, a major limitation has been the difficulty of establishing the small ermine moths in laboratory cultures due to an obligate diapause and the need to rear larvae on their host plant. Consequently, basic studies are still needed on the pheromone composition, physiology, and behavior for some of the species including *Yponomeuta mahalebella* and *Y. gigas,* a sister species endemic to Canary Islands, about which very little is known. In the case of *Y. malinellus,* the population studied by McDonough et al. (1990) was introduced to North America, and European populations should be studied as well.

Second, we do not know whether speciation in *Yponomeuta* is any more "ongoing" than in any other taxon. Small ermine moth species originally used Celastraceae as host plants and the genus expanded throughout Europe through successive host-plant shifts to Rosaceae, Crassulaceae, and Salicaceae. The evolution of female host-plant choice and larval host acceptance were likely important drivers, yet may have involved distinct sensory and behavioral changes (e.g., Menken and Roessingh 1998). Current evidence does not support the hypothesis that these adaptations alone have been decisive in driving reproductive isolation. Among all European species, *Y. padella* is the only oligophagous taxon (*Prunus spi-*

nosa, Crataegus spp., *Sorbus aucuparia, Prunus domestica*), and we still know little of whether this oligophagous feeding pattern can be explained by larval adaptation or if the species represents a mosaic of genetically divergent populations associated with a single host. Studying model communication systems that are polymorphic and potentially undergoing current evolutionary changes may uncover the mechanisms of reproductive isolation in populations that are used as models of an early stage of speciation (Via 2009). In the case of the European small ermine moths, we still lack dating of the nodes in the phylogenetic tree, and as a result lack a time frame for the process of speciation: contemporary species in this genus might already have been formed a million years ago or, alternatively, we may face a process of "ongoing speciation." The discovery of polymorphic pheromone populations in the process of divergence or host race formation (Menken 1981) would facilitate conclusive studies.

Third, sex pheromones most likely played a significant role in the evolution of reproductive isolation in *Yponomeuta* since, except for two allochronic species, all taxa share habitats—or host plants—and overlap in periods of sexual activity. Species-specific blends and ratios of female-produced pheromones and the occurrence of pheromone components with an antagonistic behavioral effect on males of other species evidently prevent cross-attraction and hybrid formation in the field that can otherwise occur under laboratory conditions. In support of this scenario, many features of the variation in the pheromone systems of small ermine moths can likely be assigned adaptive explanations, while others cannot. Some will require further work.

Possible Areas of Future Study

(A) Evolution towards simpler pheromones with *Yponomeuta rorrella* as a model. An evolutionary mechanism that altered the function of the Δ11-desaturase (Löfstedt et al. 1986) in a common ancestor of the lineage leading to *Y. rorrella* and *Y. gigas* may explain the major shift that took place in pheromone blend composition prior to divergence of these two species from their closest ancestor with unsaturated pheromone components. Genetic changes with major effects on chain shortening, desaturation, and fatty acyl reduction have been postulated (Jurenka et al. 1994) or characterized (Roelofs et al. 2002; Lassance et al. 2010, 2013) in other species.

Y. rorrella was found to be almost completely monomorphic at some 75 enzyme loci for which its congeners exhibit a normal to high proportion of heterozygous loci (Menken 1987). The most likely explanation for this lack of variation would be a (series of) bottleneck(s) at the species origin, which would fit a scenario of saltational speciation involving relaxed selection or genetic drift on the sex pheromone during the bottleneck. *Y. gigas* also exhibits a low genetic variability (Menken 1987; Menken et al. 1992) although not as low as in *Y. rorrella.* This divergence should have taken place at least a million year ago and according to Menken (1987), it would not take more than 200,000 generations (years) to restore a normal level of genetic variation. Consequently, one or more bottlenecks in a common ancestor of the species pair *gigas/rorrella* in addition to one or more bottlenecks in the lineage giving rise to *Y. rorrella* at the time of divergence of *Y. rorrella* from its closest relative would best explain the current dearth of genetic variation (Menken 1992). The idea of speciation by genetic revolution and the passage through a bottleneck has been criticized because theoretical models show that under many assumptions the probability is low for the transition of a founder population to a new selective equilibrium reproductively isolated from the ancestral population. Nor did the reports on an aberrant number of chromosome pairs in *Y. rorrella* (29 compared to 31 in other species) (Thorpe 1929; Gershenson 1967) that would have supported the hypothesis about a genetic revolution hold up to closer examination. Studies of the karyotype in *Y. rorrella* and five other *Yponomeuta* species revealed the same haploid chromosome number in all species: 29 autosomes and a sex chromosome trivalent in females and 30 autosomes and one pair of sex chromosomes in males (Nilsson et al. 1988).

Regardless of the specific mechanism, *Y. rorrella* has obviously lost unsaturated pheromone components present in its ancestors and closest relatives. Besides alterations in biosynthetic pathways, alterations on the receiver side must have occurred involving gradual or saltational changes. Further studies should focus on *Y. rorrella* as model species. The broad olfactory receptor response of conspecific males may indicate that rare males in ancestral *Y. rorrella* populations were likely able to respond to the saturated acetate. The possible refinement of one type of OSN toward 14Ac together with the evolution of an antagonistic response to unsaturated acetates in this species would support a scenario for saltational speciation (Baker 2002) through adaptive asymmetric tracking (Phelan 1997).

(B) Molecular and functional aspects of pheromone production and reception. In light of the aforementioned observations, comparative approaches to fully elucidate the molecular and functional aspects of pheromone production and reception are still needed to contribute to a better understanding of the role of pheromones in speciation. The identification of two biosynthetic genes implicated in female signal production at the molecular level has started to shed light on some key enzymes shaping the production of multicomponent pheromones in *Yponomeuta*, but how species-specific blend ratios of alcohols and acetates are modulated in each species remains to be investigated. For this, a detailed comparative analysis of desaturase and reductase activities in the nine species is needed, together with a breakthrough in the characterization of chain shortening and acetyl-transferase enzymes. This would improve our understanding of the polygenic and functional nature of species-specific blend ratio formation among closely related species. Investigating the proximate mechanisms of pheromone reception by characterizing the number of receptors and assessing their functionality and specificity in vitro (e.g., Wanner et al. 2010; Zhang and Löfstedt 2013) will also further contribute to deciphering the evolutionary mechanisms that have shaped male-specific sensory adaptations in small ermine moths and contribute to our understanding of the potential adaptive role of male responses in the evolution of new species (Roelofs et al. 2002; Baker 2002).

(C) Ecological and evolutionary forces toward new signals. Little is known about the circumstances that have favored the maintenance and fixation of new pheromone signals. The pheromone differences observed among synchronic species with overlapping geographical distribution are consistent with reproductive character displacement. Selection for pheromonal differences, in particular divergence in Z/E11-14Ac ratios and antagonistic male responses between divergent populations, could have promoted unique communication channels to avoid interspecific hybridization and have at least facilitated coexistence of the three species (*Y. cagnagella*, *Y. plumbella*, and *Y. irrorella*) that share the European spindle tree as host plant by reducing communication interference. For other species, however, divergence by geographic isolation and drift cannot be ruled out as alternative explanations for pheromone specificity. Further evidence is therefore needed to provide conclusive answers as to which circumstances and ecological forces have favored the evolution of new signals in the genus *Yponomeuta*. Nothing is known about the evolution of the matching changes that are necessary at the level of the receiver for the evolution of a new preference. With the molecular, functional, and bioinformatic tools now available, such future comparative studies will better integrate pheromone production and reception to open the door for a continuation of the "small ermine moth project" in the (post) genomic era to understand how matching changes have come about in signal and response in *Yponomeuta* species.

Acknowledgments

We thank Drs. S. Menken and P. Roessing, as well as editors J. Allison and R. Cardé, for most valuable suggestions and helpful comments on the manuscript.

References Cited

Baker, T.C. 2002. Mechanism for saltational shifts in pheromone communication systems. *Proceedings of the National Academy of Sciences of the United States of America* 99:13368–13370.

Bjostad, L.B., Wolf, W., and W.L. Roelofs. 1987. Pheromone biosynthesis in lepidopterans: desaturation and chain-shortening. Pp. 77–120. In G.J. Blomquist and G.D. Prestwich, eds. *Pheromone Biochemistry.* New York: Academic Press.

Blomquist, G.J., Jurenka, R.A., Schal, C., and C. Tittiger. 2005. Biochemistry and molecular biology of pheromone production. Pp. 705–751. In L.L. Gilbert, K. Iatrou, and S. Gill, eds. *Comprehensive Molecular Insect Science.* San Fransisco, CA: Elsevier Academic Press.

Cossé, A., Campbell, M., Glover, T., Linn, C.J., Todd, J., Baker, T., and W.L. Roelofs. 1995. Pheromone behavioral responses in unusual male European corn borer hybrid progeny not correlated to electrophysiological phenotypes of their pheromone-specific antennal neurons. *Cellular and Molecular Life Sciences* 51:809–816.

Cox, C. 2001. The biogeographic regions reconsidered. *Journal of Biogeography* 28:511–523.

Cuperus, P. 1985. Ultrastructure of antennal sense organs of small ermine moths, *Yponomeuta* spp. (Lepidoptera: Yponomeutidae). *International Journal of Insect Morphology and Embryology* 14:179–191.

Dijkerman, H.J. 1990. Parasitoids of small ermine moths (Lepidoptera: Yponomeutidae). PhD thesis, Leiden University, the Netherlands.

Du, J.-W., Löfstedt, C., and J. Löfqvist. 1987. Repeatability of pheromone emissions from individual female ermine moths *Yponomeuta padellus* and *Yponomeuta rorellus*. *Journal of Chemical Ecology* 13:1431–1441.

Friese, G. 1960. Revision der Palaearktischen Yponomeutidae. *Beitraege zur Entomologie* 10:1–131. [In German]

Fung, S. 1989. Comparative chemical analyses of small ermine moths (Lepidoptera: Yponomeutidae) and their host plants. PhD thesis, Leiden University, the Netherlands.

Gerrits-Heybroek, E., Herrebout, W.M., Ulenberg, S., and J.T. Wiebes. 1978. Host plant preference of five species of small ermine moths (Lepidoptera:Yponomeutidae). *Entomologia Experimentalis et Applicata* 24:360–368.

Gershenson, Z. 1967. Karyotype of the willow ermine moth *Yponomeuta rorellus* Hb. (Lepidoptera: Yponomeutidae) (in Russian). *Tsitologiia i Genetika* 1:79–80.

Gershenson, Z., and S. Ulenberg. 1998. *The Yponomeutinae (Lepidoptera) of the World Exclusive of the Americas. Verhandelingen van de Koninklijke Nederlandse Akademie van Wetenschappen, Afdeling Natuurkunde Tweede Reeks*, Vol. 99, North-Holland, Amsterdam: Royal Netherlands Academy, 202pp.

Hansson, B.S., Hallberg, E., Löfstedt, C., and R. Steinbrecht. 1994. Correlation between dendrite diameter and action potential amplitude in sex pheromone specific receptor neurons in male *Ostrinia nubilalis* (Lepidoptera: Pyralidae). *Tissue & Cell* 1994:503–512.

Hendrikse, A. 1978. De lokhouding van stippelmot-wijfjes (Lep., Yponomeutidae). *Entomologische Berichte (Luzern)* 38:53–54. [In Dutch]

Hendrikse, A. 1979. Activity patterns and sex pheromone specificity as isolating mechanisms in eight species of *Yponomeuta* (Lepidoptera: Yponomeutidae). *Entomologia Experimentalis et Applicata* 25:172–180.

Hendrikse, A. 1990. Sex-pheromone communication and reproductive isolation in small ermine moths. PhD thesis, Leiden University, the Netherlands.

Hendrikse, A., Van der Laan, C.E., and L. Kerkhof. 1984. The role of male abdominal brushes in the sexual behaviour of small ermine moths (*Yponomeuta* Latr., Lepidoptera). *Mededelingen Faculteit Landbouwwetenschappen Rijksuniversiteit Gent* 49:719–726.

Herrebout, W.M., and T. van de Water. 1983. Trapping sucess in relation to height of the traps in small ermine moths (*Yponomeuta*) (Lepidoptera: Yponomeutidae). *Mededelingen Faculteit Landbouwwetenschappen Rijksuniversiteit Gent* 48(2):173–182.

Herrebout, W.M., Kuijten, P., and J. Wiebes. 1975. Stippelmotten en hun voedselplanten. *Entomologische Berichten Amsterdam* 35:84–87.

Jermy, T. 1984. Evolution of insect/host relationship. *The American Naturalist* 124:609–630.

Jurenka, R.A., Haynes, K.F., Adlof, R.O., Bengtsson, M., and W.L. Roelofs. 1994. Sex pheromone component ratio in the cabbage looper moth altered by a mutation affecting the fatty acid chain-shortening reactions in the pheromone biosynthetic pathway. *Insect Biochemistry and Molecular Biology* 24:373–381.

Kooi, R. 1988. Comparison of food-acceptance by two closely related species: *Yponomeuta malinellus* and *Yponomeuta padellus*. *Proceedings of the Koninklijke Nederlandse Akademie van Wetenschappen Series C Biologica and Medical Sciences* 91:233–242

Kooi, R. 1990. Host-plant selection and larval food acceptance by small ermine moths. PhD thesis, Leiden University, the Netherlands.

Lassance, J.-M., Groot, A.T., Liénard, M.A., Antony, B., Borgwart, C., Heckel, D.G., and C. Löfstedt. 2010. Allelic variation in a fatty-acyl reductase gene causes divergence in moth sex pheromones. *Nature* 466:486–489.

Lassance, J.-M., Liénard, M.A., Antony, B., Qian, S., Fujii, T., Tabata, J., Ishikawa, Y. and C. Löfstedt. 2013 Functional consequences of sequence variation in the pheromone biosynthetic gene pgFAR for *Ostrinia* moths. *Proceedings of the National Academy of Sciences of the United States of America* 110:3967–3972.

Liénard, M.A., and C. Löfstedt. 2010. Functional flexibility as a prelude to signal diversity? Role of a fatty acyl reductase in moth pheromone evolution *Communicative and Integrative Biology* 3:586–588.

Liénard, M.A., Hagström, Å. K., Lassance, J.-M., and C. Löfstedt. 2010. Evolution of multi-component pheromone signals in small ermine moths involves a single fatty-acyl reductase gene. *Proceedings of the National Academy of Sciences of the United States of America* 107:10955–10960.

Löfstedt, C. 1987. Behaviour of small ermine moths in overlapping pheromone plumes. Pp. 37–39. In H. Arn, ed. *Mating Disruption: Behaviour of Moths and Molecules*. West Palaearctic Regional Section (WPRS) Bulletin X/3. Paris: OILB-SROP.

Löfstedt, C., and W.M. Herrebout. 1988. Sex pheromones of three small ermine moths found on spindle tree. *Entomologia Experimentalis et Applicata* 46:29–38.

Löfstedt, C., and J.N.C. van der Pers. 1985. Sex pheromones and reproductive isolation in four european small ermine moths. *Journal of Chemical Ecology* 11:649–666.

Löfstedt, C., Herrebout, W., and J.-W. Du. 1986. Evolution of the ermine moth pheromone tetradecenyl acetate. *Nature* 323:621–623.

Löfstedt, C., Hansson, B.S., Dijkerman, H.J., and W.M. Herrebout. 1990. Behavioural and electrophysiological activity of unsaturated analogues of the pheromone tetradecyl acetate in the small ermine moth *Yponomeuta rorellus*. *Physiological Entomology* 15:47–54.

Löfstedt, C., Herrebout, W., and S.B.J. Menken. 1991. Sex pheromones and their potential role in the evolution of sex reproductive isolation in small ermine moths (Yponomeutidae). *Chemoecology* 2:20–28.

Martouret, D. 1966. Sous-famille des Hyponomeutidae. *Entomologie Appliquée à l'Agriculture II* 1:102–175. [In French]

McDonough, L. M., Davis, H. G., Smithhisler, C. L., Voerman, S., and P. S. Chapman. 1990. Apple ermine moth, *Yponomeuta malinellus* Zeller: two components of the female sex pheromone gland highly effective in field trapping tests. *Journal of Chemical Ecology* 16:477–486.

Menken, S.B.J. 1980. Allozyme polymorphism and the speciation process in small ermine moths (Lepidoptera: Yponomeutidae). PhD thesis, Leiden University, the Netherlands.

Menken, S.B.J. 1981. Host races and sympatric speciation in small ermine moths. *Entomologia Experimentalis et Applicata* 30:280–292.

Menken, S.B.J. 1987. Is the extremely low heterozygocity level in *Yponomeuta rorellus* caused by bottlenecks? *Evolution* 41:630–637.

Menken, S.B.J. 1996. Pattern and process in the evolution of insect-plant associations: *Yponomeuta* as an example. *Entomologia Experimentalis et Applicata* 80:297–305.

Menken, S.B.J., and P. Roessingh. 1998. Evolution of insect-plant associations: sensory perception and receptor modifications direct food specialization and host shifts in phytophagous insects. Pp. 145–156. In D. Howard and S.H. Berlocher, eds. *Endless forms: Species and Speciation*. New York: Oxford University Press.

Menken, S.B.J., Herrebout, W.M., and J.T. Wiebes. 1992. Small ermine moths (*Yponomeuta*): their host relations and evolution. *Annual Review of Entomology* 37:41–66.

Minks, A., Voerman, S., and W.M. Herrebout. 1977. Attractants and inhibitors of Lepidoptera: Field evaluation of pheromones and related compounds. Pp. 223–233. In N. MacFarlane, ed. *Crop Protection Agents: Their Biological Evaluation*. London: Academic Press.

Nilsson, N., Löfstedt, C., and L. Dävring. 1988. Unusual sex chromosome inheritance in six species of small ermine moths (*Yponomeuta*, Yponomeutidae, Lepidoptera). *Hereditas* 108:259–265.

Olsson, S., Kesevan, S., Groot, A., Dekker, T., Heckel, D., and B.S. Hansson. 2010. *Ostrinia* revisited: evidence for sex linkage in European corn borer *Ostrinia nubilalis* (Hübner) pheromone reception. *BMC Evolutionary Biology* 10:285.

Perry, J., and C. Wall. 1984. Short-term variation in catches of the pea moth, *Cydia nigricana*, in interacting pheromone traps. *Entomologia Experimentalis et Applicata* 36:145–149.

Peterson, S., Herrebout, W.M., and R. Kooi. 1990. Chemosensory basis of host-colonization by small ermine moth larvae. *Proceedings of the Koninklijke Nederlandse Akademie van Wetenschappen Series C: Biological and Medical Sciences* 93:287–294.

Phelan, P.L. 1997. Evolution of mate-signalling in moths: phylogenetic considerations and predictions from the assymetric tracking hypothesis. Pp. 240–256. In J.C. Choe and B.J. Crespi, eds. *The Evolution of Mating Systems in Insects and Arachnids*. Cambridge: Cambridge University Press.

Povel, G. 1984. The identification of the European small ermine moths, with special reference to the *Yponomeuta padellus*-complex (Lepidoptera, Yponomeutidae). *Proceedings of the Koninklijke Nederlandse Academie van Wetenschappen. Series C: Biological and Medical Sciences* 87:149–180.

Raijmann, L.E.L. 1996. In search for speciation: genetical differentiation and host race formation in *Yponomeuta padellus* (Lepidoptera: Yponomeutidae). PhD thesis, Amsterdam University, the Netherlands.

Roelofs, W., and L. Bjostad. 1984. Biosynthesis of lepidopteran pheromones. *Biorganic Chemistry* 12:279–298.

Roelofs, W.L., and A.P. Rooney. 2003. Molecular genetics and evolution of pheromone biosynthesis in Lepidoptera. *Proceedings of the National Academy of Sciences of the United States of America* 100:9179–9184.

Roelofs, W.L., Liu, W., Hao, G., Jiao, H., Rooney, A.P., and C.E. Jr. Linn. 2002. Evolution of moth sex pheromones via ancestral genes. *Proceedings of the National Academy of Sciences of the United States of America* 99:13621–13626.

Roessingh, P. 1989. Trail marking and following by larvae of the small ermine moth *Yponomeuta cagnagellus*. PhD thesis, Wageningen University, the Netherlands.

Roessingh, P., Hora, K.H., van Loon, J.J.A., and S.B.J. Menken. 1999. Evolution of gustatory sensitivity in *Yponomeuta* caterpillars: sensitivity to the stereo-isomers dulcitol and sorbitol is localised in a single sensory cell. *Journal of Comparative Physiology A* 184:119–126.

Roessingh, P., Hora, K., Fung, S., Peltenburg, A., and S.B.J. Menken. 2000. Host acceptance behaviour of the small ermine moth *Yponomeuta cagnagellus*: larvae and adults use different stimuli. *Chemoecology* 10:041–047.

Roessingh, P., Xu, S., and S.B.J. Menken. 2007. Olfactory receptors on the maxillary palps of small ermine moth larvae: evolutionary history of benzaldehyde sensitivity. *Journal of Comparative Physiology, A* 193:635–647.

Scoble, M. 1992. *The Lepidoptera: Form, Function and Diversity*. Oxford: Oxford University Press.

Thorpe, W. 1929. Biological races in *Hyponomeuta padella* L. *Journal of the Linnean Society Zoology* 36:621–634.

Turner, H., Lieshout, N., Van Ginkel, W.E., and S.B.J. Menken. 2010. Molecular phylogeny of the small ermine moth genus *Yponomeuta* (Lepidoptera, Yponomeutidae) in the Palaearctic. *PLoS ONE* 5(3):e9933.

Ulenberg, S. 2009. Phylogeny of the *Yponomeuta* species (Lepidoptera, Yponomeutidae) and the history of their host plant associations. *Tijdschrift voor Entomologie* 152:187–207.

Valeur, P., and C. Löfstedt. 1996. Behaviour of male oriental fruit moth, *Grapholita molesta*, in overlapping sex pheromone plumes in a wind tunnel. *Entomologia Experimentalis et Applicata* 79:51–59.

van Bruggen, A., and C. van Achterberg. 2000. In memoriam Prof. Dr. J.T. Wiebes (1931-1999), evolutionary biologist and systematic entomologist. *Zoologische Mededelingen Leiden* 74:271–282.

van der Pers, J.N.C. 1978. Responses from olfactory receptors in females of three species of small ermine moths (Lepidoptera: Yponomeutidae) to plant odours. *Entomologia Experimentalis et Applicata* 24:594–598.

van der Pers, J.N.C. 1981. Comparison of electroantennogram response spectra to plant volatiles in seven species of *Yponomeuta* and in the tortricid *Adoxophyes orana*. *Entomologia Experimentalis and Applicata* 30:181–192.

van der Pers, J.N.C. 1982. Comparison of single cell responses of antennal sensilla trichodea in the nine european small ermine moths (*Yponomeuta* spp.). *Entomologia Experimentalis et Applicata* 31:255–264.

van der Pers, J.N.C., and C.J. den Otter. 1978. Single cell responses from olfactory receptors of small ermine moths to sex-attractants. *Journal of Insect Physiology* 24:337–343.

van der Pers, J.N.C., Thomas, G., and C.J. den Otter. 1980. Interactions between plant odours and pheromone reception in small ermine moths (Lepidoptera: Yponomeutidae). *Chemical Senses* 5:369–371.

van Drongelen, W. 1979. Contact chemoreception of host plant specific chemicals in larvae of various Yponomeuta species (Lepidoptera). *Journal of Comparative Physiology* 134:265–279.

van Drongelen, W. 1980. Comparative aspects of taste receptors and hostplant selection in larvae of various *Yponomeuta* species (Lepidoptera). PhD thesis, Wageningen University, the Netherlands.

Via, S. 2009. Natural selection in action during speciation. *Proceedings of the National Academy of Sciences of the United States of America* 106:9939–9946.

Wanner, K.W., Nichols, A.S., Allen, J.E., Bunger, P.L., Garczynski, S.F., Linn, C.E., Jr., Robertson, H.M., and C.W. Luetje. 2010. Sex pheromone receptor specificity in the European corn borer moth, *Ostrinia nubilalis*. *PLoS ONE* 5(1):e8685.

Wiebes, J.T. 1976. The speciation process in the small ermine moths. *Netherlands Journal of Zoology* 26:440.

Wiegand, H. 1962. Die Deutsche Arten der Gattung *Yponomeuta* Latr. Tagungberichte 9. *Wanderversammlungen Deutscher Entomologen* 45:101–120. [In German]

Zhang, D.-D., and C. Löfstedt. 2013. Functional evolution of a multigene family: orthologous and paralogous pheromone receptor genes in the turnip moth, *Agrotis segetum*. *PLoS ONE* 8(10):e77.

CHAPTER FOURTEEN

Possible Reproductive Character Displacement in Saturniid Moths in the Genus *Hemileuca*

J. STEVEN McELFRESH and JOCELYN G. MILLAR

INTRODUCTION

Pheromone systems of *Hemileuca electra electra*,
 H. e. mojavensis, and *H. burnsi*

Pheromone systems of *Hemileuca eglanterina* and
 H. nuttalli

Pheromones of other *Hemileuca* species

REFERENCES CITED

Introduction

The process of reproductive isolation underpins the concept of a biological species because of the negative consequences of interspecific reproduction (less fit or nonviable offspring) (Dobzhansky 1937). Reproductive isolation is facilitated by a number of mechanisms that constitute ecological, behavioral, morphological, physiological, or genetic barriers to the production of viable hybrid offspring. The action of any one or any combination of these mechanisms may prohibit interspecific reproduction. In theory, discrimination among suitable and unsuitable potential mates should occur as early as possible to minimize the waste of reproductive effort and resources.

Early discrimination among suitable and unsuitable mates is particularly important for species in which adults are relatively short-lived and which expend substantial resources in each reproductive episode. Such is the case with many moth species in which nonfeeding adult males fly long distances to locate a potential mate. In addition to the risk of predation during these flights, the fact that males have limited energy resources which they cannot replenish renders all mating effort costly, and squandered mating effort particularly so. Thus, the pheromone signals that mediate the early steps of reproduction, i.e., mate location and recognition, should play a major role in pre-zygotic reproductive isolation in such lepidopteran species.

Our understanding of pheromone-mediated mate location in the Lepidoptera is based on the general premise that the pheromone blend produced by a female moth constitutes a species-specific signal to attract conspecific males. Although the compounds in a pheromone blend are often shared with congeners, the particular subset of chemicals and their ratio should provide a unique signal. Deviations from the median blend should be discouraged by the decreased ability of deviant producers to attract mates. Circumstances which allow different species to use the same or very similar blends are restrictive. Because of the high cost of mistakes (wasted reproductive effort, less fit, or nonviable hybrids), pheromone blends that are approximately equally attractive to males of more than one species should only be used by species which are rarely if ever together in space and time (sympatric and synchronic). Conversely, where there is the possibility for mistakes to occur due to sympatric and synchronic species using very similar pheromone systems, the resulting strong selection pressure should quickly result in changes in one or both pheromone systems to restore unique channels for both species. The situation can be further complicated if the two species are sympatric or synchronic over only part of their ranges, so that different populations of the same species experience different selection forces on their pheromone systems. The heterogeneous selective environments experienced by different populations of the same species thus could contribute to polymorphism of pheromone-based reproductive isolating mechanisms between populations, and eventually speciation as members of the different populations became increasingly unlikely to respond to each other. Because the selective forces operate only in the zone of overlap, the divergence of the pheromone signaling systems from the species' norms in the regions of sympatry may constitute reproductive character displacement (RCD), defined as "a pattern of greater divergence in an isolating trait in areas of sympatry between two closely related taxa than in areas of allopatry" (Howard 1993).

However, the general idea of RCD has proven to be controversial, and finding concrete examples has proven problematic because of the difficulty in deducing the direction of a process that occurs over evolutionary time from a "snapshot" of present-day observations. RCD has seldom been associated with patterns of intraspecific variation in Lepidoptera, despite the volume of studies on lepidopteran pheromones in general (summarized in Ando 2015; El-Sayed 2014. The discovery of components in sex pheromone blends which have no apparent effect on conspecific males but which deter males of other species that use similar blends has led to the conclusion that, although these compounds may initially have been biosynthetic relics or accidents, they have been secondarily selected for and maintained for the specific purpose of deterring interspecific attraction. The development of these deterrents, plus other modifications to the pheromone signaling system, may constitute examples of RCD (Cardé 1987; Löfstedt 1993; Cardé and Haynes 2004).

The information required to construct a solid argument for RCD has been outlined by Howard (1993). The primary requirement is a detailed knowledge and understanding of the mate location/identification system(s) of the study organisms. Second, there must be demonstrable variation in the system, and there must be a correlation between the geographic distribution of this variation and the presence or absence of one or more potentially interfering species, i.e., areas of sympatry and allopatry must exist for at least one of the species in the interaction. If variation is found, and it can be correlated to the presence or absence of a potentially interfering species, then it must be demonstrated that the differences can act as an isolation mechanism. Howard (1993) further suggested that the following conditions should be met:

1. There must be evidence that heterospecific matings occurred or can occur in nature.
2. There must be a cost to hybridization, to provide a selection force for change.
3. The displacement of a character must be perceptible to the opposite sex.
4. The variation in the trait must be heritable, and thus capable of responding to selection.
5. It must be demonstrable that displacement has not occurred for other reasons, such as adaptation to different ecological zones.

The search for evidence of RCD in Lepidoptera is confounded by the fact that the distributions of many species have been altered and/or dramatically expanded from their original, natural distributions by human activities, including introduction of species to new habitats, and drastic alteration or even elimination of existing habitats. In such cases, it becomes impossible to determine whether detectable changes in the pheromone systems of different populations of a species result from selection pressures from congeners, or from a host of other possible causes.

Here, we summarize the results from studies of the pheromone systems of four moth species in the genus *Hemileuca* (Lepidoptera: Saturniidae) which consists of about two dozen described species and a number of subspecies distributed throughout the United States, southern Canada, and Mexico. Their taxonomy, distributions, biologies, and host plant preferences have been summarized by Tuskes et al. (1996). Some species, such as *H. griffini*, are extremely limited in their distribution, whereas others, such as *H. maia* and *H. eglanterina*, are widely distributed over large areas that encompass several states (Tuskes et al. 1996). Whereas the potential for cross-attraction between some species is minimized by geographic and/or temporal isolation barriers, for others as many as three species may be active in the same area simultaneously (Tuskes 1984; Tuskes et al. 1996). Further competition for pheromone

channels may arise from four closely related species in the genus *Coloradia*, which share pheromone components with *Hemileuca* spp. (McElfresh and Millar 1999d; McElfresh et al. 2000).

Hemileuca moths are large, ranging in size from 45 to over 100 mm in wing span (Tuskes et al. 1996). Some species are relatively drab, such as those in the *H. tricolor* group, but most are vividly colored in black and white, often with red, yellow, or pink markings on the bodies. Unusual for moths, males of most *Hemileuca* species are diurnal. The life cycle may take from one to several years, and adults of both sexes do not feed, so adult life spans are short and virtually all adult behaviors are directed toward reproduction. Flight periods typically last for a few weeks per year.

This chapter will focus on *H. electra*, *H. burnsi*, *H. eglanterina*, and *H. nuttali*. Of particular importance to our goal of detecting reliable evidence in support of RCD was the fact that the distributions and life histories of all these species have been negligibly affected by human activities in large part because of their lack of economic importance and the generally remote natural landscapes that they inhabit. Thus, differences between populations, subspecies, and species are likely to be the result of natural selection rather than any anthropogenic disturbances. Fortunately, their large size, striking colors, and diurnal activity period have made them favorites with lepidopterists, and so there is a substantial volume of literature on their distributions and general biology (summarized in Tuskes et al. 1996).

The major components of the sex pheromones of the *Hemileuca* species studied to date include (10*E*,12*Z*)-hexadecadien-1-yl acetate (E10Z12-16Ac), (10*E*,12*Z*)-hexadecadien-1-ol (E10Z12-16OH), (10*E*,12*Z*)-hexadecadienal (E10Z12-16Ald), and (10*E*,12*E*)-hexadecadien-1-yl acetate (E10E12-16Ac) (Table 14.1). Extracts of pheromone glands of females of various species and subspecies contain a number of other compounds, including saturated and monounsaturated analogs of the diene pheromone components, but, in most cases, these components do not appear to form part of the active pheromone blends. In all cases published to date, the active blends consist of multiple components.

Pheromone Systems of *Hemileuca electra electra*, *H. e. mojavensis*, and *H. burnsi*

Hemileuca electra is distributed in a broad arc from central Baja California, Mexico, to southern Nevada and Utah, and southeast to central Arizona (figure 14.1). Populations are usually associated with the lower elevations in mountain ranges throughout the Mojave Desert and the Peninsular and Transverse mountain ranges. *H. burnsi* is found from the northern Coachella valley, California, north to the vicinity of Reno, Nevada, and as far east as central Arizona (figure 14.1). *H. electra* and *H. burnsi* are sympatric throughout the Mojave Desert. The flight seasons of both species begin in late September and in milder years extend into November.

H. electra has four recognized subspecies, two of which occur in California. The range of *H. e. mojavensis* (Hem) overlaps substantially with that of *H. burnsi* (Hb), whereas the range of *H. e. electra* (Hee) does not (figure 14.1). In addition to sharing some parts of their ranges, Hem and Hb are seasonally synchronic, and their daily activity patterns overlap; Hem males are typically active from morning to early afternoon, whereas Hb males fly from early afternoon till dusk.

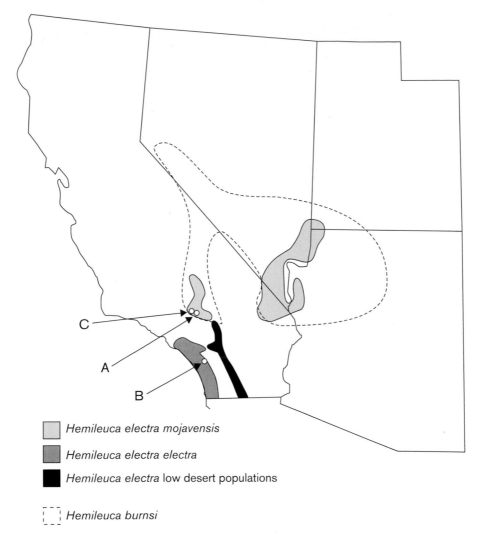

Hemileuca electra mojavensis

Hemileuca electra electra

Hemileuca electra low desert populations

Hemileuca burnsi

FIGURE 14.1 Natural ranges of *Hemileuca electra electra*, *H. e. mojavensis*, and *H. burnsi* in relation to research sites. Site A: Ord Mountains east of Hesperia, San Bernardino County, CA. Site B: 6.5 km northeast of Aguanga, Riverside County, CA. Site C: 16 km east of Phelan, San Bernardino County, CA (modified from Tuskes et al. 1996).

However, these diurnal trends probably reflect the calling periods of the respective females because males will respond to synthetic pheromone lures at times outside their natural daily activity periods. Thus, there are both temporal and spatial points of contact between the two species, but heterospecific matings do not produce fertile eggs (McElfresh and Millar 1999c).

Differential attraction of males to calling females of Hee and Hem provided the first evidence of differences in the pheromone blends of Hem and Hee as a possible consequence of the presence or absence of Hb. Thus, Hem males showed a 5:1 preference for attraction to their own females compared to Hee females. Furthermore, Hb males, which are allopatric with Hee, were strongly attracted to Hee females, whereas they were not attracted to sympatric Hem females (Tuskes 1984; Tuskes et al. 1996). Taken together, these data suggested that the presence of Hb, and the consequent potential for interference in the *H. electra* pheromone channel in one locale but not the other, may have acted selectively on one population, resulting in the pheromone blend of Hem as it is today.

The pheromone gland contents of female Hem and Hee were qualitatively the same, but detailed examination revealed that Hem extracts contained about eight times more of the minor component E10Z12-16Ald and half as much hexadecyl acetate (16Ac) as Hee extracts (Table 14.1) (McElfresh and Millar 1999c). Gland contents of Hb females were very similar to those of Hee in the proportions of E10Z12-16Ac, E10Z12-16Ald, and 16Ac, but had more E10Z12-16OH than Hee (McElfresh and Millar 2008).

As might be expected from the cross-attraction seen in field tests with live females, quantitative coupled gas chromatography-electroantennogram detection (GC-EAD) analyses showed that antennae of Hee and Hb males displayed similar trends in their responses to a panel of pheromone gland components, whereas responses from antennae of Hem males were markedly different (figure 14.2) (McElfresh and Millar 2008). In particular, Hem antennae responded most strongly to E10Z12-16OH, whereas Hee antennae responded most strongly to E10Z12-16Ac, and Hb antennae responded about equally to both compounds. Furthermore, Hem antennae were significantly more sensitive to E10Z12-16Ald than Hee and Hb antennae.

Field trials with mixtures of synthetic pheromone components supported these results (McElfresh and Millar 1999c,

TABLE 14.1

Ratios of possible pheromone components found in pheromone gland extracts of *Hemileuca* spp., as proportions
of the most abundant compound in extracts from each species

Compounds with demonstrated behavioral activity are shown in boldface type

Compounds	H. nut	H. egl SG	H. elec elec	H. elec mohav	H. burnsi	H. maia
E10Z12-16Ac	**100**	**100**	**100**	**100**	**100**	**6.3**
E10E12-16Ac	**48**	10	6.7	10	**23**	Trace
E10Z12-16OH	38	**48**	**22**	**36**	**14**	**7.4**
E10E12-16OH	18	5	2.5	15		Trace
E10Z12-16Ald	1.0	**1.1**	**0.5**	**3.5**	0.37	**100**
E10E12-16Ald						Trace
Z10-16Ac	2.5	2.1	0.4	1.0		
E/Z11-16Ac	17	11	2.2	4.0	9.1	
16Ac	160	53	**203**	104	**232**	Trace
16OH	261	58	92	111	60	11
16Ald						23

Abbreviations: *H. nut = Hemileuca nuttalli; H. egl. SG = H. eglanterina* allopatric with *H. nuttalli; H. elec elec = H. electra electra; H. elec mojav = H. electra mojavensis.*

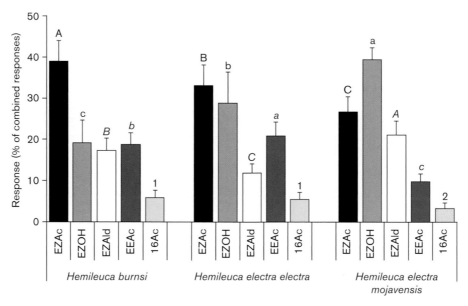

FIGURE 14.2 Antennal responses of *Hemileuca burnsi, H. electra electra*, and *H. e. mojavensis* to five
pheromone components. Bars of the same pattern surmounted by the same letter are not significantly different ($p = 0.05$).

2008). The optimal blends for both *H. electra* subspecies consisted primarily of E10Z12-16Ac, E10Z12-16OH, and E10Z12-
16Ald, but in different proportions. Both species preferred a
100:10 ratio of E10Z12-16Ac to E10Z12-16OH, but Hee was
most attracted to blends containing 0.3–1% of E10Z12-16Ald
(in relation to E10Z12-16Ac), and was inhibited by higher proportions of E10Z12-16Ald, whereas Hem preferred E10Z12-
16Ald in amounts of 10–100% that of E10Z12-16Ac. Further-

more, Hem males were indifferent to any amount of 16Ac in
blends, whereas Hee males were slightly more attracted to
blends containing a significant proportion of 16Ac.

Field trials targeting Hb showed that its preferred ratio of
E10Z12-16Ac, E10Z12-16OH, and E10Z12-16Ald was 100:10:1,
virtually identical to allopatric Hee. As with Hee, Hb males
were inhibited by higher proportions of E10Z12-16Ald,
whereas 16Ac was found to increase attraction. Unlike Hee,

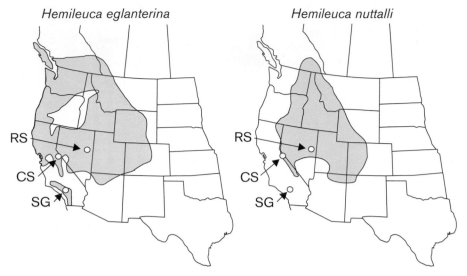

FIGURE 14.3 Natural ranges of *Hemileuca eglanterina* and *H. nuttalli*.

ABBREVIATIONS: SG: San Gabriel Mountains field site; CS: Conway Summit field site; RS: Robinson Summit field site.

SOURCE: Reprinted from McElfresh and Millar (2001).

E10E12-16Ac also increased attraction of Hb males, so that the optimal blend actually consisted of the five components E10Z12-16Ac, E10Z12-16OH, E10Z12-16Ald, E10E12-16Ac, and 16Ac in a ratio of 100:10:1:50:350. Nevertheless, blends lacking the latter two components were still sufficiently attractive to male Hb to explain their strong cross-attraction to allopatric Hee females that had been artificially moved into their range.

Overall, these results provide three pieces of evidence in support of RCD in *H. electra* and the resulting development of the Hem and Hee subspecies that exist in sympatry or allopatry with Hb, respectively. First, the pheromone gland contents of Hem and Hee females were significantly different, particularly in the amount of the crucial E10Z12-16Ald component. Second, the antennal responses of Hee and Hem males to the three critical and shared components of their pheromones were significantly different, for all three compounds. Finally, these analytical results were corroborated by the behavioral responses of males of the two subspecies in field trials, with Hee being sensitive to and inhibited by higher proportions of E10Z12-16Ald, whereas Hem males responded equally well to a wide range of proportions of E10Z12-16Ald, up to and including amounts equal to that of the main component, E10Z12-16Ac.

Pheromone Systems of *Hemileuca eglanterina* and *H. nuttalli*

The ranges of both *Hemileuca eglanterina* and *H. nuttalli* extend across much of the western United States and up into Canada, with large areas of overlap (figure 14.3) (McElfresh and Millar 2001). The flight season for *H. eglanterina* in the San Gabriel Mountains begins as early as late May in warmer years and extends through August, whereas in the eastern Sierra Nevada Mountains both *H. eglanterina* and *H. nuttalli* fly from early August through most of September. Their diurnal activity patterns overlap, with *H. eglanterina* males typically flying from morning to early afternoon, whereas *H. nuttalli* males fly

from mid to late afternoon. However, as with *H. electra electra*, *H. e. mojavensis*, and *H. burnsi*, males will fly outside their normal activity periods when presented with a pheromone lure, indicating that the natural activity periods are probably dictated by the periods when females call. Males of these species do not appear to randomly fly or patrol. Rather, when males are not in active pursuit of a pheromone plume, they perch on the upper branches of shrubs with the antennae elevated (J. Steven McElfresh, pers. obs.). Adults of the two species will mate but most crosses are infertile, and, in the one known case where the eggs did hatch, the neonates failed to feed (Collins and Tuskes 1979; Tuskes et al. 1996). *H. nuttalli* males are somewhat attracted to *H. eglanterina* females, but at close range do not exhibit the typical hovering flight exhibited by conspecific males (McElfresh and Millar 2001). However, the partial cross-attraction is asymmetric because *H. eglanterina* males are not attracted to *H. nuttalli* females.

An initial investigation of the pheromone chemistry of a population of *H. eglanterina* that is allopatric with *H. nuttalli* (SG population on figure 14.3) revealed that the pheromone gland contents were qualitatively the same as those of congeneric species, with E10Z12-16Ac being the most abundant component (Table 14.1) (McElfresh and Millar 1999a, 2001). A series of iterative field trials determined that the three components E10Z12-16Ac, E10Z12-16OH, and E10Z12-16Ald were required for attraction. Interestingly, the ratio of the two minor components to each other was much more important than the ratio of the minor components to the major component E10Z12-16Ac. Thus, blends of E10Z12-16Ac:E10Z12-16OH:E10Z12-16Ald of 100:0.3:0.03 to 100:100:10 were not significantly different in attracting male moths (McElfresh and Millar 1999a, 2001).

A parallel study of the pheromone chemistry of *H. nuttalli* found the typical pheromone gland contents for a *Hemileuca* species (Table 14.1) (McElfresh and Millar 1999b), but the gland contents by no means mirrored the optimal pheromone blend. Iterative field trials with synthetic compounds showed that the best lure for *H. nuttalli* was a two-component blend of E10Z12-16Ac and E10E12-16Ac (100:48), and that, despite the

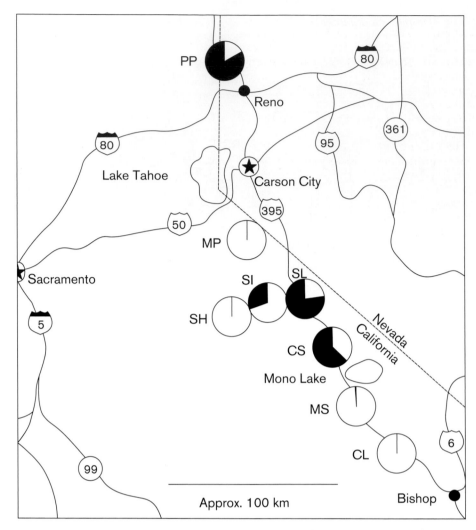

FIGURE 14.4 Results of field trials testing two discriminating lures for different *Hemileuca eglanterina* pherotypes. White portions of pie charts represent the proportion of a given population responding to the SG type lure (E10Z12-16Ac : E10Z12-16OH; E10Z12-16Ald, 100: 10: 1), black portions represent the portion responding to an RS type lure (E10Z12-16OH; E10Z12-16Ald, 100: 10).

Location designations and total number of males captured: PP = Peavine Peak, near Reno, NV, *n* = 193; MP = Monitor Pass, CA, *n* = 373; SH = Sonora Pass, upper site, *n* = 46; SI = Sonora Pass, middle site, *n* = 73; SL = vicinity of Junctions 108 and 395 below Sonora Pass, CA, *n* = 99; CS = Conway Summit, north of Lee Vining, CA, *n* = 1103; MS = south of Mono Lake, CA, *n* = 181; CL = vicinity of Crowley Lake, CA, *n* = 165.

SOURCE: Reprinted from McElfresh and Millar (2001).

considerable proportions of E10Z12-16OH in the gland contents, the alcohol was strongly inhibitory. E10Z12-16Ald, a critical component of the pheromone blends of other *Hemileuca* spp., had no significant effect at low doses, but was inhibitory at higher doses. Furthermore, it was noted that the relatively large proportions of E10E12-16Ac in *H. nuttalli* test blends inhibited the cross-attraction of *H. eglanterina*. Thus, the pheromone chemistry of *H. nuttalli* was markedly different than those of the species described above.

These two preliminary studies set the stage for a detailed comparison of the pheromone systems of *H. eglanterina* and *H. nuttalli* in areas of sympatry versus allopatry (McElfresh and Millar 2001). Efforts were focused on three major field sites (figure 14.3): the San Gabriel Mountains site (SG) where *H. nuttalli* was not present, the Conway Summit site (CS) where the edges of the ranges of the two species overlapped, and the Robinson Summit site (RS) where the two species were fully sympatric. The first trial verified that *H. eglanterina* males at the SG site

responded best to a three-component blend of E10Z12-16Ac, E10Z12-16OH, and E10Z12-16Ald (100:18:1). In stark contrast, males of the RS population of *H. eglanterina* did not require E10Z12-16Ac, with the highest trap catches occurring in traps baited with the two-component blend of E10Z12-16OH and E10Z12-16Ald (100:10). A second trial testing a range of E10Z12-16Ac doses from 0 to 333 µg per lure, with E10Z12-16OH and E10Z12-16Ald held constant at 10 and 1 µg per lure, respectively, showed that the CS population appeared to break into two groups. There appeared to be a first group which was indifferent to lower doses of E10Z12-16Ac, with no significant differences in attraction to lures containing 0–33 µg. However, a marked discontinuity in the profile of responses versus dose of E10Z12-16Ac between the 33-µg dose and the two higher doses of 100 and 333 µg suggested the presence of a second group. In contrast, the response profile of *H. eglanterina* from the SG site was markedly different, showing a monotonic increase in trap catch with increasing doses of E10Z12-16Ac.

A third trial was then conducted holding E10Z12-16Ac constant at 100 μg, and varying the doses of a fixed ratio (10:1) of E10Z12-16OH and E10Z12-16Ald. The SG population responded equally to doses of the E10Z12-16OH + E10Z12-16Ald mixture spanning more than two orders of magnitude (0.33:0.03 to 100:10), whereas trap catches of CS males increased with increasing dose of the E10Z12-16OH + E10Z12-16Ald blend.

Field trials at the RS site, where both species were fully sympatric, showed that the optimal blend for RS *H. eglanterina* males consisted of E10Z12-16OH and E10Z12-16Ald (100:3.3), with no E10Z12-16Ac. Furthermore, RS males were very sensitive to the proportions of E10Z12-16Ald to E10Z12-16OH, with deviations on either side of 3.3% aldehyde to alcohol resulting in decreased trap captures, whereas CS males responded equally well to a percentage of E10Z12-16Ald from 1% to 100% of the E10Z12-16OH dose.

E10E12-16Ac is a required pheromone component for *H. nuttalli*, and field trials with this compound revealed further subtleties in the systems. Specifically, when E10E12-16Ac was added to a generic blend of E10Z12-16Ac, E10Z12-16OH, and E10Z12-16Ald (100:10:1), *H. eglanterina* males at the CS site were inhibited by increasing doses of E10E12-16Ac, whereas allopatric SG males were not. However, when E10E12-16Ac was added to the two-component blend of E10Z12-16OH + E10Z12-16Ald (100:10) without E10Z12-16Ac, i.e., the optimized lure for populations of *H. eglanterina* with *H. nuttalli* present, males of the fully sympatric RS population were inhibited by increasing doses of E10E12-16Ac, whereas trap catches at the CS site, at the edge of the range of both species, were apparently unaffected.

With this clear evidence of differences between the responses of SG (allopatric) and RS (fully sympatric) males, and indications that the CS males might be polymorphic in their responses, discriminating blends composed of the three components E10Z12-16Ac, E10Z12-16OH, and E10Z12-16Ald (100:10:1) or the two components E10Z12-16OH + E10Z12-16Ald (100:10) were tested at an additional series of field sites with and without *H. nuttalli* present (figure 14.4). At sites at the very edge or outside the *H. nuttalli* range, *H. eglanterina* were exclusively attracted to the three-component blend, whereas at sites where both species were present, trap catches of *H. eglanterina* were split between both lures. Also, at sites where both species were present, addition of E10E12-16Ac, a crucial *H. nuttalli* pheromone component, decreased trap catches.

It was not possible to obtain sufficient numbers of females from sites with both *H. nuttalli* and *H. eglanterina* for reliable pheromone gland content analyses. However, because of the strong attraction of males to pheromone traps, it was possible to conduct detailed comparisons of the response profiles of the antennae of large numbers of males from the different populations, when challenged with a standardized 1:1:1 blend of E10Z12-16Ac, E10Z12-16OH, and E10Z12-16Ald (figure 14.5). As expected, the allopatric SG population of *H. eglanterina* showed a single response type, with the largest response to the major component of its pheromone, E10Z12-16Ac. Similarly, the fully sympatric RS population of *H. eglanterina* consisted of a single but markedly different response type, with a large response to E10Z12-16OH, a lesser response to E10Z12-16Ald, and virtually no response to E10Z12-16Ac. In contrast, *H. eglanterina* males from the CS site broke into three types: an SG type, an RS type, and a type intermediate between the two (figure 14.5). Additional testing of males from the other sites showed that there was generally a good correlation between the lure to which they had been attracted, and their

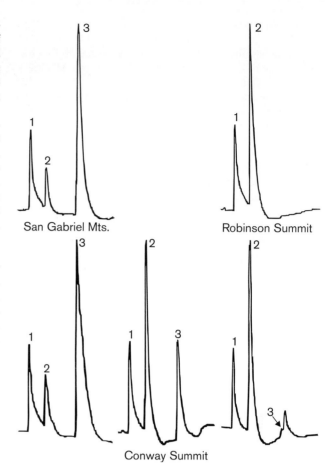

FIGURE 14.5 Antennal response types of male *Hemileuca eglanterina*. At the San Gabriel Mountains site, *H. eglanterina* is allopatric with *H. nuttalli*; at the Robinson Summit site, the two species are completely sympatric; and at the Conway Summit site, the ranges of the two species overlap. Peaks represent the responses in millivolts from antennae stimulated sequentially with 0.5 ng of (1) E10Z12-16Ald, (2) E10Z12-16OH, and (3) E10Z12-16Ac (reprinted from McElfresh and Millar 2001).

antennal response profiles. Furthermore, the antennae of males from populations of *H. eglanterina* sympatric with *H. nuttalli* responded more strongly to the *H. nuttalli* compound E10E12-16Ac than to E10Z12-16Ac, whereas the reverse was true for males from allopatric populations.

The evidence described above is consistent with RCD of *H. eglanterina* populations that are sympatric with *H. nuttalli*, as a consequence of which both species have experienced selection pressure to evolve reproductive isolating mechanisms, including unique pheromone systems. It is also crucial to note that the different *H. eglanterina* populations have not yet diverged into different species, as indicated by their reproductive compatibility and the presence of a substantial proportion of intermediate types in areas where the ranges of the two phenotypes overlap (ca. 40% of the populations at two sites). However, cross-attraction between the two extreme phenotypes is weak; for example, an SG female used as a lure at the CS site attracted 9 males with an SG antennal response profile, 27 males with an intermediate profile, but only 1 with an RS-type profile. Furthermore, these results were borne out by trapping males using synthetic lures optimized for the SG and RS phenotypes, respectively, and then checking their antennal response profiles. Males with an SG-type antennal response profile had been trapped almost exclusively in traps baited with the SG lure, and the large majority of males with RS antennal profiles had been caught in traps baited with RS lures, whereas males with intermediate antennal profiles had been caught in similar numbers in both trap types (McElfresh and Millar 2001).

In light of such weak cross-attraction between the RS and SG pherotypes, how can we explain the substantial proportions of intermediate types in some populations? One possible explanation is that the intermediates may be a viable part of the gene pool specifically because they are attracted to both of the extremes as well as to intermediate blends. That is, any possible trade-offs in fitness of the intermediates in comparison to the extreme pherotypes may be counterbalanced by having an increased pool of possible mates.

In sum, this system meets all of the requirements enumerated by Howard (1993) for a rigorous test for RCD, with the data supporting a scenario in which a single species has evolved two or more pherotypes in response to differing selection pressures on different parts of the population.

From a practical perspective, the results described above also highlight the danger of assuming that the ratio of components found in the pheromone gland represents the optimal blend, and/or that the ratio of compounds applied to a lure is representative of the ratio in which they will be released. For example, in all species, the pheromone glands contained a number of analogs of the active pheromone compounds, most of which had minimal or no activity, even though some were present in amounts substantially greater than the pheromone components. Furthermore, in most species, the optimal ratio of synthetic components applied to a lure was somewhat to very different than the ratios found in the glands. These data demonstrate that optimization of the pheromone blends of lepidopteran species requires that a series of methodical trials be conducted, in which each compound is varied individually, followed by trials in which ratios of compounds are varied in tandem.

Pheromones of Other *Hemileuca* Species

In addition to the species described above, there has been some work on pheromone systems of a number of other *Hemileuca* species. All indications are that the main components of the pheromones will adhere to the motif of (*E*10,*Z*/*E*12)-hexadecadienyl alcohols, acetates, or aldehydes. To date, the only pheromone identified for an eastern North American species, *Hemileuca maia*, was characterized as a 100:10:1 blend of E10Z12-16Ald:E10Z12-16OH:E10Z12-16Ac (McElfresh et al. 2001). Interestingly, male moths responded equally well to a broad range of relative and absolute amounts of the two minor components, as long as they were both present, analogous to examples described above. In other work by the Nature Conservancy which has not been published, using compounds supplied by our group, pheromone blends were used to survey a potentially rare and endangered *Hemileuca* species in the northeastern United States that feeds on buckbean (*Menyanthes trifoliata*: Menyanthaceae), and that so far has defied classification. Pheromone gland extracts of *H. oliviae*, which ranges from New Mexico to parts of Colorado, Oklahoma, and Texas, also have been analyzed several times,

but to date, it has not been possible to reconstruct the active blend. Fragmentary data on pheromone gland extracts, responses of male antennae to those extracts, and behavioral responses to reconstructed test blends also exist for *H. nevadensis, H. hera, H. hualapai,* and *H. neumogeni* (McElfresh and Millar, unpublished data), but in all cases, the work is not yet complete. Given the extensive overlaps in the ranges of many *Hemileuca* species, it is likely that further examples of RCD will be uncovered.

References Cited

Ando, T. 2015. Sex pheromones of moths. Available online at: http://www.tuat.ac.jp/~antetsu/LepiPheroList.htm.

Cardé, R. T. 1987. The role of pheromones in reproductive isolation and speciation in insects. Pp. 303–317. In M. D. Huettel, ed. *In Evolutionary Genetics of Invertebrate Behavior.* New York: Plenum Press.

Cardé, R. T., and K. F. Haynes. 2004. Structure of the moth pheromone communication channel. Pp. 283–332. In R. T. Cardé and J. G. Millar, eds. *Advances in Insect Chemical Ecology.* Cambridge: Cambridge University Press.

Collins, M. M., and P. M. Tuskes. 1979. Reproductive isolation in sympatric species of day-flying moths (Hemileuca: Saturniidae). *Evolution* 33:728–733.

Dobzhansky, T. 1937. Genetic nature of species differences. *The American Naturalist* 71:404–420.

El-Sayed, A. M. 2014. The Pherobase: Database of Pheromones and Semiochemicals. Available online at: http://www.pherobase.com.

Howard, D. J. 1993. Reinforcement: origin, dynamics, and fate of an evolutionary hypothesis. Pp. 46–69. In R. G. Harrison, ed. *Hybrid Zones and the Evolutionary Process.* New York: Oxford University Press.

Löfstedt, C. 1993. Moth pheromone genetics and evolution. *Philosophical Transactions of the Royal Society London, Series B* 340:167–177.

McElfresh, J. S., and J. G. Millar. 1999a. Sex pheromone of the common sheep moth, *Hemileuca eglanterina,* from the San Gabriel Mountains of California. *Journal of Chemical Ecology* 25:687–709.

McElfresh, J. S., and J. G. Millar. 1999b. Sex pheromone of Nuttall's sheep moth, *Hemileuca nuttalli,* from the eastern Sierra Nevada Mountains of California. *Journal of Chemical Ecology* 25:711–726.

McElfresh, J. S., and J. G. Millar. 1999c. Geographic variation in sex pheromone blend of *Hemileuca electra* from southern California. *Journal of Chemical Ecology* 25:2505–2525.

McElfresh, J. S., and J. G. Millar. 1999d. Sex attractant pheromone of saturniid moth, *Coloradia velda. Journal of Chemical Ecology* 25:1067–1078.

McElfresh, J. S., and J. G. Millar. 2001. Geographic variation in the pheromone system of the saturniid moth *Hemileuca eglanterina. Ecology* 82:3505–3518.

McElfresh, J. S., and J. G. Millar. 2008. Sex pheromone of the saturniid moth, *Hemileuca burnsi,* from the western Mojave Desert of California. *Journal of Chemical Ecology* 34:1115–1124.

McElfresh, J. S., X. Chen, D. W. Ross, and J. G. Millar. 2000. Sex pheromone blend of the pandora moth (Lepidoptera: Saturniidae), an outbreak pest in pine forests (Pinaceae). *The Canadian Entomologist* 132:775–787.

McElfresh, J. S., A. M. Hammond, and J. G. Millar. 2001. Sex pheromone components of the buck moth *Hemileuca maia. Journal of Chemical Ecology* 27:1409–1422.

Tuskes, P. M. 1984. The biology and distribution of California Hemileucinae (Saturniidae). *Journal of the Lepidopterists' Society* 38:281–309.

Tuskes, P. M., J. P. Tuttle, and M. M. Collins. 1996. *The Wild Silk Moths of North America: A Natural History of the Saturniidae of the United States and Canada.* Ithaca, NY: Cornell University.

The European Corn Borer *Ostrinia nubilalis*

Exotic Pest and Model System to Study Pheromone Evolution and Speciation

JEAN-MARC LASSANCE

Introduction

The European corn borer (ECB) *Ostrinia nubilalis* (Lepidoptera: Crambidae) is an economically important agricultural pest throughout much of the northern hemisphere. The ECB is highly polyphagous and colonized corn (maize, *Zea mays*) after its introduction into Europe ca. 500 years ago (Ponsard et al. 2004). The earliest records of the economic significance of this species date back to 1835 (Caffrey and Worthley 1927). The ECB attacks a variety of cereals and other crops in Europe, North Africa, and North America where it was accidentally introduced with shipments of broomcorn (*Sorghum vulgare*) imported from Central Europe and Italy between 1909 and 1914 (Caffrey and Worthley 1927). Currently, *O. nubilalis* is considered to be an important pest of corn in Europe and North America where it is distributed throughout nearly all the major corn-producing areas (Klun and Cooperators 1975; Kaster and Gray 2005; Meissle et al. 2010). Because of the potential value of pheromones for both monitoring and controlling the ECB, there has been considerable effort devoted to the study of its pheromone biology (Pélozuelo and Frérot 2007). This relatively intense focus shed light on one of the most interesting features of the ECB: the polymorphic nature of its female-produced pheromone communication system. As a result of this polymorphism and the significance of ECB as a pest, the ECB has become a model for the study of the genetics, biochemistry, and molecular biology of sexual communication systems and has provided insights into the role of sexual signals in speciation (Wyatt 2003; Coyne and Orr 2004; Smadja and Butlin 2009).

Early Work and Female Pheromone Identification

Like most moth species, ECB females release a volatile sex pheromone that attracts conspecific males. Typically, upon detection of the female-emitted sex pheromone, males initiate an oriented response upwind to locate the origin of the attractive signal. Once they arrive in the close vicinity of the signal source, males take part in a stereotyped pre-copulatory "dance" with their wings extended up and vibrating, and the genitalia expanded, followed by a clasper response (Klun 1968). The sex pheromone produced by ECB females was extracted and isolated for the first time in 1968 from 10,000 females (Klun 1968). The active pheromone component was believed to be a single component necessary and sufficient to elicit male behavioral response. The chemical identification of the molecule revealed it to be (*Z*)-11-tetradecenyl acetate (Z11-14Ac) (Klun and Brindley 1970), a fatty-acyl derivative compound that has since been found to be part of the pheromone of several hundred moth species (El-Sayed 2011). At that time, synthetic Z11-14Ac was reported to elicit a behavioral response equivalent to the one triggered by the detection of the natural pheromone when used in a close-range bioassay (Klun and Brindley 1970) or field tests (Klun and Robinson 1971). However, when geometrically pure isomers were synthesized and used in field trials, it became obvious that male ECB were only weakly attracted to traps baited with pure Z11-14Ac (Klun et al. 1973). The addition of a small amount of the geometric isomer (*E*)-11-tetradecenyl acetate (E11-14Ac) increased trap captures significantly, demonstrating the requirement of having a blend of both isomers for attraction of males that are tuned to specific proportions of

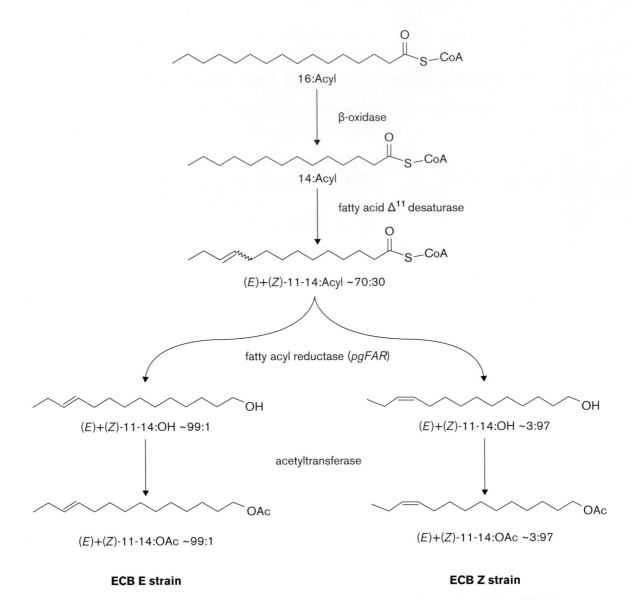

FIGURE 15.1 Biosynthetic pathway for the female *Ostrinia nubilalis* pheromone. De novo biosynthesis of (Z)- and (E)-11-tetradecenyl acetates starts from palmitoyl-CoA. One cycle of β-oxidation generates myristoyl-CoA, which is a substrate for a fatty acyl Δ11 desaturase. The (E)- and (Z)-11-tetradecenoyl moieties produced are converted into the corresponding alcohols by a fatty acyl reductase *pgFAR*. Finally, an acetyltransferase catalyzes the conversion of the fatty alcohol precursors into acetates.

Z11-14Ac and E11-14Ac (Klun et al. 1973). The results from field trials replicated in different localities in the United States revealed that large doses of the E isomer could have either a negative (Klun and Robinson 1971) or a positive (Roelofs et al. 1972; Cardé et al. 1975) effect on ECB capture rates. It was hypothesized then that the different catches corresponded to two distinctive strains of corn borer relying on different blends of the E and Z isomers (Roelofs et al. 1972; Cardé et al. 1975). Subsequent chemical analyses confirmed the existence of a pheromone polymorphism in the ECB: the so-called Z strain, with females that release and males that respond to a 97:3 Z/E11-14Ac pheromone blend, and the so-called E strain based on a 1:99 ratio of Z/E11-14Ac (Kochansky et al. 1975).

Pheromone Biosynthesis and Its Regulation

Nine abdominal segments can be distinguished in the adult female *Ostrinia nubilalis*; seven are covered with scales,

whereas the remaining terminal segments form the ovipositor. These segments remain telescoped within the preceding ones except when the female is laying eggs or protruding the pheromone gland in a typical and specialized "calling" behavior. The sex-pheromone gland is localized on the surface of the ovipositor and study of the ECB sex-pheromone gland ultrastructure revealed that the gland is formed by a single layer of hypertrophied epidermal cells that are located in the dorsal fraction of the intersegmental membrane between the 8th and the fused 9th/10th segments, thus forming a half-ring-shaped gland (Ma and Roelofs 2002). Gas chromatographic analyses of the terminal abdominal segments confirmed the presence of both the pheromone components and their precursors.

The pheromone components, E11-14Ac and Z11-14Ac, are produced de novo through the fatty acid cycle. In brief, palmitic acid is shortened to a 14-carbon intermediate through one cycle of ß-oxidation. A Δ11 fatty acid desaturase then removes two hydrogen atoms to introduce a double bond in the carbon skeleton and produces the (E)- and (Z)-11-tetradecenoyl precursors that are subsequently reduced by a fatty-acyl reductase and acetylated by an acetyltransferase to form the pheromone compounds (Roelofs et al. 1987) (figure 15.1). The fact that viable first and second filial and backcross offspring can be obtained from hybrid crosses between the strains has facilitated investigation of the genetic basis of polymorphism in the ECB communication system, notably the genetics underlying the geometric composition of the sex pheromone. It was shown that the difference in E:Z ratio is controlled primarily by a single autosomal locus with two alleles under Mendelian inheritance (Klun and Maini 1979; Roelofs et al. 1987; Dopman et al. 2004). Both strains have (E)- and (Z)-11-tetradecenoyl intermediates with a similar ~70:30 ratio (Roelofs et al. 1987; Wolf and Roelofs 1987), which indicates that the autosomal locus encodes a factor effecting one of the final steps in the biosynthesis following the production of the unsaturated precursors. Molecular investigations showed that the Δ11 fatty acid desaturase expressed in the ECB pheromone gland is responsible for the production of both geometric isomers in the aforementioned ratio (Roelofs et al. 2002). The Δ11 desaturase gene genealogy does not exhibit a pattern of variation that would be characteristic of a gene involved in the differentiation of the strains (Geiler and Harrison 2010). The selective reduction of deuterium-labeled precursors in vivo suggested that the reduction step is more likely to define the final blend of pheromone components (Zhu et al. 1996), especially since in vivo experiments demonstrated that the enzyme controlling the acetylation step seems to have low substrate specificity in *O. nubilalis* (Jurenka and Roelofs 1989; Zhu et al. 1996). Molecular characterization of *pgFAR*, the gene encoding the fatty acyl reductase responsible for the formation of fatty alcohols from the acyl precursors, confirmed that the reduction step was responsible for the observed differences in the E:Z ratio between the pheromone strains (Lassance et al. 2010). Each ECB pheromone strain expresses a different allele that encodes for a *pgFAR* with strain-specific substrate specificity that differentially reduces the *Z* and *E* precursors to yield opposite ratios in the final pheromone blend. The genotype at the *pgFAR* locus determines the phenotype for ECB female pheromone production (Lassance et al. 2010).

Sex-pheromone production in female corn borers exhibits a cyclic rhythm with peak production occurring during the scotophase and a valley with no pheromone production during the photophase (Foster 2004). The pattern changes slightly as females age, with the peak observed toward the end of the scotophase in young females, an apparent constant titer during the scotophase in 2- to 3-day-old females, and a peak in the early scotophase in older females (Foster 2004; Kárpáti et al. 2007). This pattern correlates with the mating activity as observed under laboratory conditions (Kárpáti et al. 2007). Although the timing of calling behavior of E and Z moths considerably overlap (Webster and Cardé 1982), some temporal isolation may result from differences in the 24-h rhythm of mating (Liebherr and Roelofs 1975).

Pheromone biosynthesis in female ECB is regulated by the neuropeptide PBAN (pheromone biosynthesis activating neuropeptide) (Raina et al. 1989; Ma and Roelofs 1995a). PBAN is produced in three discrete sets of neurosecretory cells in the subesophageal ganglion of adult ECB females and targets the pheromone gland after being released in the hemolymph (Ma and Roelofs 1995a, b, c). The interaction between PBAN and its receptor translates into the opening of ion channels lead-ing to an influx of extracellular calcium into the pheromone gland cells (Ma and Roelofs 1995a). In the ECB, PBAN does not activate the pheromone biosynthesis per se but rather activates *pgFAR* through an unknown mechanism, promoting the conversion of (E)- and (Z)-11-tetradecenoyl precursors into the corresponding alcohols (Eltahlawy et al. 2007).

Male Behavioral Response to the Pheromone

The detection of female pheromone by ECB males results in the following successive steps: activation, wing fanning, upwind anemotactic flight with plume-tracking maneuvers following plume contact and location of the odor source culminating in a characteristic precopulatory courtship display (Klun 1968; Glover et al. 1987; Royer and McNeil 1992).

Although ECB are capable of flying long distances (e.g., >1 km) (Showers et al. 2001), ECB adults tend to aggregate in grassy areas (Showers et al. 1976) and it is likely that males do not have to fly long distances to locate a mate. The flight tunnel assay is well suited for testing ECB because males must fly 1–3 m upwind to the chemical source and exhibit close-range behaviors including landing close to the pheromone source. Using serial blends of Z and E11-14Ac, it was demonstrated that ECB males were less responsive to ratios deviating from the natural isomeric blend and, perhaps more importantly, that ECB males exhibit limited to no attraction to the blend of the opposite strain (Glover et al. 1987; Klun and Huettel 1988; Linn et al. 1997) (figure 15.2), which corroborates results from field-trapping experiments (Klun et al. 1973 Klun and Cooperators 1975). Interestingly, males of the E and Z strains exhibit different levels of specificity for ratios. ECB Z males appear more discriminating with fewer moths orienting to off-ratio blends. Conversely, E strain males have a broad window of response, with some rare males even attracted to the blend produced by ECB Z females, which may indicate the behavioral response of E males is less canalized compared to that of Z moths (Roelofs et al. 1987; Glover et al. 1990; Linn et al. 1997) (figure 15.2). The level of discrimination among blends of E and Z11-14Ac is higher for activation and the early phase of the pheromone plume tracking than for later stages of pheromone plume tracking, including source contact (Kárpáti et al. 2013). When males experience overlapping plumes of different composition, some may eventually land on a source different than the one that elicited upwind flight, a situation that Kárpáti et al. (2013) proposed to be encountered in nature when individuals from both strains are present in the same population.

The difference in male response profile is controlled by a major sex-linked factor (*Resp*) exhibiting simple Mendelian inheritance (in Lepidoptera, males are the homogametic sex denoted ZZ, whereas females are the heterogametic sex denoted ZW) (Roelofs et al. 1987). It is not known whether the sex-linked factor *Resp* represents a single gene or a set of closely linked loci. Sex linkage of *Resp* was confirmed through the use of allozyme data and genetic mapping (Glover et al. 1990; Dopman et al. 2004, 2005). To date, the nature of *Resp* remains unknown.

The Male Olfactory System: Deciphering Pheromone Signals

Attraction of males to the female ECB pheromone is the end result of responses of peripheral and central olfactory

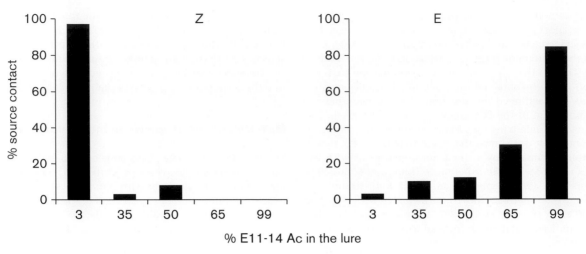

FIGURE 15.2 Flight tunnel responses for pure ECB pheromone strains exposed to different blends of Δ11-14Ac. Profiles represent the percentage of males that flew upwind and touched the pheromone source (adapted from Linn et al. 1997).

systems. The physiological and molecular basis of the behavioral differences between the ECB strains has been the focus of many studies; however, our understanding remains incomplete.

In the ECB, the olfactory sensilla occupy the ventral surface of the filiform antennae, and trichoid sensilla are the most abundant sensillar type (Cornford et al. 1973; Hallberg et al. 1994). Single-sensillum recording studies documented the existence of three types of trichoid sensilla classified based on the number of olfactory sensory neurons (OSNs) housed in them. Type A sensilla are innervated by three sensory cells responding, respectively, to the major pheromone component, the minor pheromone component, and the behavioral antagonist, (Z)-9-tetradecenyl acetate (Z9-14Ac), which is a pheromone component of other *Ostrinia* species and likely represents a pheromone component from the ancestor of the ECB (Klun and Robinson 1972; Glover et al. 1989; Ishikawa et al. 1999). Type B sensilla house OSNs that respond to the pheromone components only, whereas type C contain only one OSN that is sensitive to either the major pheromone component or the behavioral antagonist (Hansson et al. 1987; Hallberg et al. 1994). Although its adaptive significance remains to be clarified, it is worth noting that the distribution of these sensilla is uneven along the antenna: type A is the most common type in the basal antennal segments, type B sensilla are the most common in the distal antennal segments, and type C sensilla are evenly distributed throughout the antennae (Hallberg et al. 1994; Baker et al. 2012). In extracellular recordings of male antenna from the two strains, the action potential (or spike) amplitudes of OSNs responding to the pheromone components differ. In the Z strain, large-amplitude spike responses are obtained when stimulating with the major component Z11-14Ac, whereas responses to the minor component E11-14Ac come from cells characterized by a smaller amplitude spike response. Conversely, this pattern is reversed in E-strain males (Hansson et al. 1987; Hallberg et al. 1994; Cossé et al. 1995; Olsson et al. 2010). Morphological studies showed that these differences in spike amplitudes could be attributed to differences in the diameter of the pheromone-responding OSN dendrites, with large diameter OSNs displaying a large spike amplitude and vice versa (Hansson et al. 1994). In addition, the gross antennal responses measured by electroantennogram recordings are significantly different:

Z11-14Ac elicits a stronger response than E11-14Ac in Z-strain antennae, whereas the antennae of E-strain males respond approximately equally to the geometric isomers (Nagai et al. 1977; Nagai 1981; Linn et al. 1999; Kárpáti et al. 2010). These results indicate that the ECB males are exquisitely sensitive to the minor pheromone component and that it is detected via a different set of OSNs than the major component, most likely through the expression of different olfactory receptors (ORs). Recently, ORs responding to the *Ostrinia* pheromone components have been identified and functionally characterized in vitro using the *Xenopus* oocyte heterologous expression system. Some of these sex-pheromone receptors appear to be narrowly and broadly tuned, with some responding more or less specifically to E and Z11-14Ac (Miura et al. 2010; Wanner et al. 2010) (figure 15.3). Existing functional data do not suggest major differences in sensitivities between the E and Z alleles of these ORs. The pheromone strains do appear to be differentiated at three pheromone OR loci, namely *OnOR1*, *OnOR3*, and *OnOR6* (Lassance et al. 2011). The consequence of this differentiation remains unclear, but reveals that the genes are present in a part of the genome with extremely limited gene flow between the strains. Genetic mapping of these ORs revealed that they map to the Z sex chromosome, but too far away from *Resp* (the male behavioral factor) to explain the difference in male response profile (Lassance et al. 2011). Although the specificity of OSNs is mainly determined by ORs, pheromone-binding proteins (PBPs) present in the sensillar lymph might also be involved in the differential peripheral perception of species-specific pheromones, as demonstrated in *Bombyx mori* (Bombycidae) and *Heliothis virescens* (Noctuidae) (Grosse-Wilde et al. 2006, 2007). However, previous genealogical analyses included a PBP, and examination of variation at that PBP locus revealed no fixed differences between the ECB strains (Willett and Harrison 1999; Dopman et al. 2005). Further investigations are required to evaluate whether other PBP genes are significantly differentiated between the strains and play a role in the tuning of OSNs (Allen and Wanner 2011).

The genetic basis of male peripheral physiology was studied through reciprocal crosses and scoring responses in the offspring. Early studies concluded that the sensillar phenotype is determined by a single autosomal locus with three distinct phenotypes (E, Z, and intermediate) observed in the

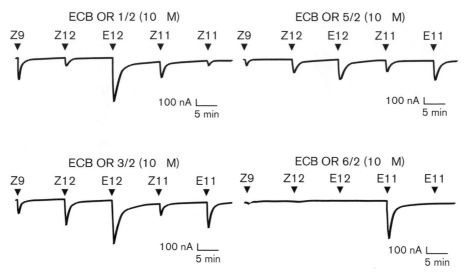

FIGURE 15.3 Functional assay for olfactory receptors from male ECB. *Xenopus* oocytes expressing the ortholog of ORCO (OnOR2) together with individual ORs were exposed to various isomers of tetradecenyl acetate used as pheromone components in *Ostrinia* species. A response is visualized by monitoring the induced current. The ORs differ in their tuning (adapted from Wanner et al. 2010).

predicted Mendelian ratio in F2 progeny and reciprocal back-crosses (Hansson et al. 1987; Roelofs et al. 1987). Because the difference in male behavioral response is sex-linked, this apparent lack of correlation between antennal and behavioral responses was surprising. A later study further emphasized the apparent lack of direct linkage between peripheral physiology and male behavioral response. Cossé et al. (1995) demonstrated that second-generation hybrids could exhibit the behavioral preference of one strain while possessing antennal electrophysiological responses representative of the other strain. Linn et al (1999) used an antennal disk transplant technique to produce males with the antennae from the other strain. Adult males with transplanted antennal disks exhibited a pheromone-mediated response in the flight tunnel toward the pheromone blend of the recipient's strain, even though they possessed antennae of the opposite race, suggesting that sex-pheromone preference in males is determined at the level of the central nervous system. Although interesting, these results should be treated with caution as the neuroanatomy and, in particular, the targeting of OSN axons are unknown in individuals with transplanted antennae. Recent reexamination of the inheritance of peripheral physiology confirmed the role of an autosomal locus but suggested the presence of an additional locus contributing to the male response phenotype (Olsson et al. 2010). Interestingly, this locus is sex-linked, exhibits E-dominance, and is involved in an epistatic interaction with the autosomal locus (Olsson et al. 2010). The discrepancies between this study and earlier investigations can be attributed to the more precise phenotyping of individual males made possible by improvements in microscopy, and electronic and computing equipment over the last 20 years. Furthermore, the gross antennal responses to the pheromone components, which is the sum electric potential of all OSNs responding to pheromone and reflect the relative abundance of both neuron types, exhibit sex-linked inheritance (Kárpáti et al. 2010). Cumulatively, these studies suggest a strong link between male antennal and behavioral response.

Antennal OSN axons project to the antennal lobe, the first-order olfactory brain area, via the antennal tract. There, OSNs make synaptic contacts with projection neurons (PNs) and local interneurons (LNs) to form glomeruli. PNs project their axons through the inner antenna-cerebral tract to the calyces of the mushroom bodies and the lateral horn of the protocerebrum. Each OSN arborizes to a single glomerulus within the antennal lobe (Anton et al. 1997). Three-dimensional high-resolution reconstructions showed that the antennal lobe of ECB males comprises 66 glomeruli, including three enlarged glomeruli forming the macroglomerular complex (MGC), which is a subgroup of glomeruli specialized in pheromone detection (Anton et al. 1997; Kárpáti et al. 2008). The MGC comprises two glomeruli receiving inputs from the pheromone-component-responsive OSNs: a large highly convoluted medial compartment folded around a smaller lateral one. The third MGC glomerulus, posterior to the two others, is dedicated to the behavioral antagonist Z9-14Ac (Anton et al. 1997; Kárpáti et al. 2008). Physiological and morphological analyses revealed that the two strains are morphologically identical and exhibit a similar functional organization: OSNs and PNs responding to the major pheromone component arborize in the large medial compartment in both strains, whereas the small lateral compartment is devoted to the detection of the minor component (Kárpáti et al. 2008; Kárpáti et al. 2010). Consequently, the strains appear to have MGC with reverse functional topologies as a given pheromone component, i.e., Z11-14Ac, elicits activity either in the large compartment (in the Z strain) or in the small compartment (in the E strain). This functional architecture of the MGC exhibits sex-linked inheritance with E-dominance (Kárpáti et al. 2010).

Kárpáti et al. (2008) suggested that the occurrence of an interchange at the level of the pheromone-sensitive OSNs without a modification of OSN axon targeting and PN arborization would explain the occurrence of these reversed functional topologies. Under such a scenario, the OSNs would project their axons to the same location in the MGC, but the ORs expressed within their dendritic membrane would be tuned to the opposite pheromone component. It is important to note that ORs do not determine axon targeting to the glomerulus in insect OSNs, neither do OSNs determine the identity of the glomerulus; rather the latter is defined by PNs

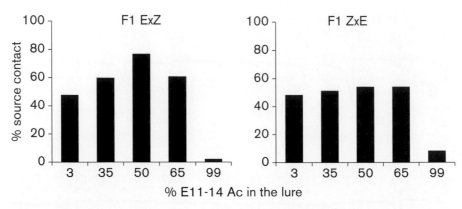

FIGURE 15.4 Flight tunnel responses for male hybrids of ECB pheromone strains from reciprocal crosses exposed to different blends of Δ11-14Ac. Profiles represent the percentage of males that flew upwind and touched the pheromone source (adapted from Linn et al. 1997).

that form "protoglomeruli" serving as template for the OSN axons (Galizia and Sachse 2010). The reorganization of the peripheral nervous system suggested by Kárpáti et al. (2008) would need a change in the regulatory region and/or in the spatial expression of transcription factors that bind to the regulatory sequence motifs. Further studies are necessary to elucidate how OR gene choice is determined in the ECB and how the E and Z strains differ in that respect. Receptor gene choice combined with variation in the OR gene repertoire might be a key element in the evolution of behavioral response in male moths. It will be important to clarify the apparent mismatch between the number of types of sensilla trichodea housing OSNs sensitive to the pheromone components and the number of pheromone ORs. Analyzing the spatial distribution of ORs on the antenna, e.g., using in situ hybridization would provide important clues to address this question and reconcile the physiological and molecular data. Finally, it remains unclear how ECB males discriminate between blends. Kárpáti et al. (2010) proposed that size of MGC glomeruli could correlate with blend preference as the ratio between the volume of the medial and lateral compartment of the MGC is similar in Z- and E-strain males (~70:30), and correlates with their opposite behavioral response profiles. Another possibility is that a particular population of blend-specific PNs are crucial for discrimination between blends (Anton et al. 1997).

To date, the molecular and physiological bases behind the divergence in male behavior in the ECB have not been fully unraveled. However, considerable progress has been made in recent years, and the availability of novel tools and approaches will facilitate cracking this intricate code.

Origin and Fate of Hybrids

Although most males are only attracted to females from their own strain, some males exhibit a broad window of response, which may result in the attraction to females of the other strain and ultimately interbreeding (presuming that some females will accept those males). There is a strong asymmetry in hybridization potential, with both male and female Z-strain individuals appearing more discriminant than E-strain males and females (Pélozuelo et al. 2007; Dopman et al. 2010). Hybrids can be found in the field (Klun and Huettel 1988; Malausa et al. 2005; Coates et al. 2013), but the ability of these hybrids to find and attract mates will ultimately determine

the importance of gene flow between the pheromone strains and the long-term impact of hybridization episodes on reproductive isolation between the strains.

Flying moths in a wind tunnel, Glover et al. (1991a) reported that a large number of hybrid males from reciprocal crosses failed to complete the behavioral sequence, i.e., taking off and orienting correctly in the pheromone plume, regardless of the blend they were exposed to. In addition, the few males exhibiting upwind flight and source contact did so to a wide range of E:Z ratios and did not exhibit a preference for any particular blend, indicating that unlike the parental types hybrids are not tuned to a specific ratio (Glover et al. 1991a; Linn et al. 1997) (figure 15.4). It is striking that hybrids showed a relative lack of response to any dose of the 99:1 E:Z ratio typically attractive of E males (Roelofs et al. 1987; Linn et al. 1997). However, many hybrid males did fly and make contact with a pheromone lure mimicking a 98:2 E:Z ratio (Linn et al. 1997), which indicates that the presence of a certain proportion of Z isomer in the blend is necessary to make a blend attractive to hybrid males.

At the peripheral level, reciprocal hybrids appear intermediate to their parents when considering both the gross antennal response and the spike amplitude of pheromone-sensitive OSNs (Hansson et al. 1987; Roelofs et al. 1987; Cossé et al. 1995; Kárpáti et al. 2010). In hybrids, responses to Z11-14Ac and E11-14Ac generate similar spike amplitudes. From a neuroanatomical point of view, hybrids have an E-strain-like functional topology with E11-14Ac- and Z11-14Ac-responding OSNs projecting to the medial and lateral glomeruli in the MGC, respectively (Kárpáti et al. 2010). Hybrids also appear intermediate to the parental strains in terms of the ratio of the two large MGC glomeruli (medial:lateral ~54:46), which may explain their relative preference for an intermediate ratio.

Hybrid females from reciprocal crosses have a pheromone blend of approximately 65:35 E:Z, which is intermediate to the parental strains (Klun and Maini 1979). This ratio corresponds to the ratio of monounsaturated intermediates produced by the desaturase in the pheromone gland. As hybrids possess and co-express both Z and E alleles of *pgFAR*, both the (*E*)-11-tetradecenoyl and (*Z*)-11-tetradecenoyl precursors would be reduced in hybrid females, and the ratio of the intermediate precursors would determine the proportion of pheromone components (Lassance et al. 2010). Only a fraction of males from either parental strain are potentially attracted by this ratio of pheromone components, making it less likely

BOX 15.1. *OSTRINIA NUBILALIS* VS. OSTRINIA *SCAPULALIS:* O BROTHER, WHO ART THOU?

The delineation of species within the genus *Ostrinia* was based on morphological characters. In particular, the genus is subdivided into three groups based on the number of lobes of the "uncus," a part of the male genitalia; *O. nubilalis* belongs to the trilobed uncus group, and so does *O. scapulalis* (Mutuura and Munroe 1970). Whereas females of both species are morphologically indistinguishable, males have either small mid-tibiae (*nubilalis*) or massive mid-tibiae (*scapulalis*). In addition, *O. nubilalis* is a pest of corn (*Zea mays* L.) and can thrive on a number of dicotyledonous plants (Ponsard et al. 2004), while *O. scapulalis* is only known to feed on dicotyledons (Mutuura and Munroe 1970). In terms of pheromone communication, the two species cannot be distinguished and *O. scapulalis* exhibits the same pheromone polymorphism as *O. nubilalis* (Huang et al. 1997, 2002). In addition, the sequences of the Δ11 desaturase and reductase genes fall within the range of variation found for the corresponding genes in *O. nubilalis* (Fukuzawa et al. 2006; Antony et al. 2009; Lassance et al. 2013). It appears as if *O. scapulalis* is the equivalent of *O. nubilalis* in the eastern Palearctic region. Several lines of evidence argue against the current version of the delineation *nubilalis–scapulalis*. In addition to pheromones not allowing to discriminate them, the inheritance of male mid-tibia morphology is due to only two Mendelian loci and appears unrelated to reproductive isolation between the taxa (Frolov et al. 2007, 2012). Recently, a taxonomic reassessment was suggested with the host plant being used as a major diagnostic trait (Frolov et al. 2007). This is based on the observation that populations feeding on corn are genetically differentiated from sympatric populations feeding on dicotyledons such as hop (*Humulus lupulus*) and mugwort (*Artemisia vulgaris*) (Bourguet et al. 2000; Martel et al. 2003; Leniaud et al. 2006; Frolov et al. 2012). These populations display strong divergent host preference, and, based on this, Frolov et al. (2007) proposed that the name *O. nubilalis* would be used for specimens feeding on corn, while *O. scapulalis* would be used for the non-corn feeders. The host races appear to use different pheromones: in France, the Z pheromone strain constituting the corn-feeding populations and the non-corn feeders using the E pheromone system (Thomas et al. 2003; Pélozuelo et al. 2004). However, this dichotomy is probably not so sharply defined and the host plant preference may be restricted to parts of the distribution range, as illustrated by individuals from the E pheromone strains thriving on corn in Eastern United States (Kochansky et al. 1975).

---

that hybrid females would attract males. In addition, hybrid females have lower mating success than pure-strain individuals in close-range mating trials (Pélozuelo et al. 2007).

Given that hybrid males do not discriminate among blends over a wide range of ratios varying from 2:98 to 98:2 E:Z and considering that hybrid females produce a suboptimally attractive blend, F1 hybridization could be considered a transient stage, thus reducing the possibility of a recurrent gene flow between the strains.

Reproductive Isolation between the Strains

The results of field-trapping experiments using synthetic sex attractants as well as chemical analyses of female effluvia documented the existence of two strains of ECB. The natural distribution of Z and E pheromone strains was established through extensive field-trapping surveys. Males at different geographic sites were found to respond to opposite isomeric blend of 11-14Ac (Cardé et al. 1975; Klun and Cooperators 1975; Kennedy and Anderson 1980; Fletcher-Howell et al. 1983; Durant et al. 1986). Although both pheromone strains have been found to occur sympatrically at several locations, the most prevalent ECB strain is the Z strain (97:3 Z/E11-14Ac), and the E strain appears to have a more restricted distribution in both Europe and North America where it is found essentially close to the sites of its introduction (Klun and Cooperators 1975; Anglade and Stockel 1984). It is presently unknown whether the strains exhibit either adaptive differences in their ability to thrive on corn or ethological differences in their host plant preference. It should be borne in mind that these surveys have usually taken place in the vicinity of cornfields, which may have biased the samplings, especially in Europe.

The relative confusion between the taxonomic status of *Ostrinia nubilalis* and *O. scapulalis* may also challenge the view of which strain is most prevalent (box 15.1).

To date, the most reliable method of discriminating between the two strains is by either the pheromone phenotype or the genotype at the pheromone-production locus *pgFAR* (Lehmhus et al. 2012; Coates et al. 2013). Morphological differences between the two strains do not exist. Differences may exist in the host plants used by the two strains (Caffrey and Worthley 1927; McLeod 1981; Straub et al. 1986; Eckenrode and Webb 1989; Bontemps et al. 2003; Thomas et al. 2003; Pélozuelo et al. 2004; Leppik and Frerot 2012), a research topic that has been relatively underexplored so far. Divergence in host preference may further promote differentiation between the pheromone types.

The two pheromone strains can produce viable hybrids in the laboratory and field. Hybridization in the field is likely facilitated by the fact that both strains aggregate in grassy areas (Showers et al. 1976; Dalecky et al. 2006). Also, the two strains can produce viable reciprocal hybrids under laboratory and field conditions (Klun and Cooperators 1975; Liebherr and Roelofs 1975; Cardé et al. 1978). However, interstrain crosses are much less frequent than intrastrain crosses under confined conditions (Liebherr and Roelofs 1975; Cardé et al. 1978; Pélozuelo et al. 2007; Dopman et al. 2010), and hybrid individuals are found at low frequency in regions where sympatric populations occur (Klun and Maini 1979; Roelofs et al. 1985; Klun and Huettel 1988; Glover et al. 1991b; Pélozuelo et al. 2004; Malausa et al. 2005; Coates et al. 2013). This indicates that assortative mating occurs, promoting and maintaining genetic differentiation between the ECB strains, and that differences in sexual communication appear to play a major role in reproductive isolation between the strains (Dopman et al. 2010).

FIGURE 15.5 The photograph shows the terminal abdominal segment of an ECB male with the hairpencils exposed. One set is located on the eighth sternite (asterisk), while another set of androconia is found on the claspers (arrowheads) (picture from Lassance and Löfstedt 2009).

Several studies have investigated molecular differentiation between the E and Z pheromone strains at a number of loci (Harrison and Vawter 1977; Cardé et al. 1978; Cianchi et al. 1980; Willett and Harrison 1999; Dopman et al. 2005; Malausa et al. 2007; Lassance et al. 2010, 2011). Although genetic differentiation exists, the two strains share an extensive amount of polymorphism at most loci examined to date. This suggests that the pheromone strains are still in an early phase of the speciation and that the ECB is a useful model system to study this evolutionary process (Smadja and Butlin 2009). The two strains of *O. nubilalis* offer excellent opportunities to unravel the origins of barriers to gene exchange that result from divergence in pheromone communication.

Identifying regions that do not exhibit shared polymorphisms and/or strain/species exclusivity has the potential to uncover the so-called speciation or barrier genes (Wu 2001). Barrier genes encode phenotypes that are directly involved in reproductive isolation and are likely to evolve under selection. So far, a limited number of loci have been shown to exhibit significant differentiation between the ECB pheromone strains, and most of these have to do with pheromone communication. These are *pgFAR* (the locus encoding a fatty acid reductase responsible for differences in pheromone production between E and Z moths [Lassance et al. 2010]) and the sex-linked markers *OnOR1*, *OnOR3* and *OnOR6* (these loci encode ORs responding to *Ostrinia* female sex-pheromone components [Wanner et al. 2010; Lassance et al. 2011]) as well as *Tpi* (*Triose-phosphate isomerase* encodes an enzyme critical

in glycolysis and efficient energy production). *Tpi* shows near exclusivity between strains and maps to a genomic region containing a major factor determining post-diapause development (*Pdd*) and ultimately voltinism differences (Roelofs et al. 1985; Glover et al. 1992; Dopman et al. 2005). However, whereas North American representatives of the pheromone strains appeared differentiated at *Tpi* (Dopman et al. 2005), differentiation at this locus is low between E and Z moths in Europe (Malausa et al. 2007; Dopman 2011). Indeed, ECB populations most likely have multiple trait differences that restrict gene flow, and some barriers may operate throughout the range, whereas others may vary geographically (Dopman et al. 2010). As such, temporal isolation caused by differences in post-diapause development may therefore operate in North America but not necessarily everywhere else.

Cumulatively, the evidence for genetic divergence between the Z and E strains is suggestive of ongoing speciation and suggests that the strains are incipient species (populations that have acquired some attributes of species rank for which the speciation process is partially complete and which intergrade only rarely when they meet).

Close-Range Interactions: Role of a Male Pheromone

Even under confined conditions where long-range behavioral barriers are suppressed, the two ECB strains do not freely interbreed, and interstrain matings are much less frequent than intrastrain matings (Liebherr and Roelofs 1975; Pélozuelo et al. 2007). This suggests that assortative mating may be the result of additional strain-specific signals active at close range that are part of the male courtship sequence. Male *Ostrinia* moths produce ultrasonic courtship songs that play a role in female acceptance (Nakano et al. 2006, 2008; Takanashi et al. 2010). However, males of both ECB strains produce similar ultrasonic courtship signals, excluding this communication modality as an isolating mechanism (Takanashi et al. 2010). In addition to acoustic communication, the use of male courtship pheromones is widespread among moths (Phelan and Baker 1987; Phelan 1997). In the ECB, males possess "hairpencils," tufts of differentiated scales, on both sides of the claspers and at the intersection between the seventh and eighth sternites (figure 15.5). The hairpencils are normally retracted inside the abdomen but are extruded during the male courtship display. The first clear evidence for the existence of a male pheromone in the ECB came from the observation that males deprived of their hairpencils have a decreased mating success (Royer and McNeil 1993). Recently, the male pheromone was demonstrated to play a crucial role in female choice and was identified in the Z strain as a mixture of (Z)-9-hexadecenyl acetate, (Z)-11-hexadecenyl acetate, (Z)-14-hexadecenyl acetate, and hexadecanyl acetate (Lassance and Löfstedt 2009). The blend produced by males of the E strain was found to be of similar composition but to lack (Z)-11-hexadecenyl acetate in most individuals, meaning that males produce distinguishable courtship signals (Lassance and Löfstedt 2009). The behavioral significance of this difference in composition remains to be demonstrated firmly, but male pheromone may allow species discrimination and reinforce reproductive isolation between the strains.

Males produce compounds that are structurally similar to those used by females, and both sexes appear to rely on the same desaturase genes to produce their pheromones (Lassance and Löfstedt 2009). Little is known about the genetic basis of female preference, but the data at hand suggest that it segregates independently from the male preference locus *Resp* (Pélozuelo et al. 2007). The existence of a close-range communication channel raises interesting questions concerning the coevolution of the two (long- and close-range) chemical communication systems, and especially on the coevolution of the production and detection systems.

Conclusion

One of the purposes of investigating the ECB chemical ecology was to provide new tools to monitor and control an invading pest. Trapping systems that rely on synthetic female pheromones as lures have been used to monitor adult activity (Pélozuelo and Frérot 2007). Nearly 50 years of research have shed light on an amazing model system. Sufficiently isolated to be considered sibling species but still capable of producing fertile offspring, the two *Ostrinia nubilalis* pheromone strains are a model system for the study of the evolution of pheromone communication and the process of speciation (Wyatt 2003; Coyne and Orr 2004; Smadja and Butlin 2009). A series of studies have elucidated the genetic bases of the polymorphism observed in this sex-pheromone communication system. While the gene responsible for the difference in female sex pheromone has been characterized, those involved in the

difference in male behavior and antennal response have not yet been identified. Recent advances in the artillery of molecular tools should help identify the molecular basis of differences in male behavior and antennal responses.

The apparent simplicity of the ECB communication system has turned out to hide considerable complexity. The recent identification of the male pheromone, a potential reproductive barrier between the strains, opens a new dimension in research on the chemical ecology of this species. Establishing the mechanisms contributing to divergence in the ECB communication channel and in particular the genetic and molecular basis of changes in pheromone production and perception holds promise for revealing general principles on the evolution of the communication systems of any given moth species.

Acknowledgments

I am indebted to Charles Linn, Jr., and Wendell Roelofs for their helpful comments, and to Christer Löfstedt for introducing me to the ECB. This book chapter is dedicated to the memory of Richard "Rick" G. Harrison and his contribution to the understanding of the European corn borer biology.

References Cited

Allen, J. E., and K. W. Wanner. 2011. Asian corn borer pheromone binding protein 3, a candidate for evolving specificity to the 12-tetradecenyl acetate sex pheromone. *Insect Biochemistry and Molecular Biology* 41:141–149.

Anglade, P., and J. Stockel. 1984. Intraspecific sex-pheromone variability in the European corn borer, *Ostrinia nubilalis* Hbn (Lepidoptera, Pyralidae). *Agronomie* 4:183–187.

Anton, S., C. Löfstedt, and B. S. Hansson. 1997. Central nervous processing of sex pheromones in two strains of the European corn borer *Ostrinia nubilalis* (Lepidoptera: Pyralidae). *Journal of Experimental Biology* 200:1073–1087.

Antony, B., T. Fujii, K. I. Moto, S. Matsumoto, M. Fukuzawa, R. Nakano, S. Tatsuki, and Y. Ishikawa. 2009. Pheromone-gland-specific fatty-acyl reductase in the adzuki bean borer, *Ostrinia scapulalis* (Lepidoptera: Crambidae). *Insect Biochemistry and Molecular Biology* 39:90–95.

Baker, T. C., M. J. Domingue, and A. J. Myrick. 2012. Working range of stimulus flux transduction determines dendrite size and relative number of pheromone component receptor neurons in moths. *Chemical Senses* 37:299–313.

Bontemps, A., D. Bourguet, L. Pélozuelo, M.-T. Bethenod, and S. Ponsard. 2003. Managing the evolution of *Bacillus thuringiensis* resistance in natural populations of the European corn borer, *Ostrinia nubilalis*: host plant, host races and phenotype of adult males at aggregation sites. *Proceedings of the Royal Society B: Biological Sciences* 271:2179–2185.

Bourguet, D., M.-T. Bethenod, C. Trouvé, and F. Viard. 2000. Host-plant diversity of the European corn borer *Ostrinia nubilalis*: what value for sustainable transgenic insecticidal *Bt* maize. *Proceedings of the Royal Society B: Biological Sciences* 267:1177–1184.

Caffrey, D. J., and L. H. Worthley. 1927. *A Progress Report on the Investigations of the European Corn Borer*, United States Department of Agriculture Bulletin 1476. Washington, DC: USDA, 155pp.

Cardé, R. T., J. Kochansky, J. F. Stimmel, A. G. Wheeler, and W. L. Roelofs. 1975. Sex pheromone of the European corn borer (*Ostrinia nubilalis*): *cis*- and *trans*-responding males in Pennsylvania. *Environmental Entomology* 4:413–414.

Cardé, R. T., W. L. Roelofs, R. G. Harrison, A. T. Vawter, P. F. Brussard, A. Mutuura, and E. Munroe. 1978. European corn borer: pheromone polymorphism or sibling species? *Science* 199:555–556.

Cianchi, R., S. Maini, and L. Bullini. 1980. Genetic distance between pheromone strains of the European corn borer, *Ostrinia nubilalis*: different contribution of variable substrate, regulatory and non regulatory enzymes. *Heredity* 45:383–388.

Coates, B.S., H. Johnson, K.-S. Kim, R.L. Hellmich, C.A. Abel, C. Mason, and T.W. Sappington. 2013. Frequency of hybridization between *Ostrinia nubilalis* E-and Z-pheromone races in regions of sympatry within the United States. *Ecology and Evolution* 3:2459–2470.

Cornford, M.E., W.A. Rowley, and J.A. Klun. 1973. Scanning electron microscopy of antennal sensilla of the European corn borer *Ostrinia nubilalis*. *Annals of the Entomological Society of America* 66:1079–1088.

Cossé, A.A., M.G. Campbell, T.J. Glover, C.E. Linn, Jr., J.L. Todd, T.C. Baker, and W.L. Roelofs. 1995. Pheromone behavioral responses in unusual male European corn borer hybrid progeny not correlated to electrophysiological phenotypes of their pheromone-specific antennal neurons. *Experientia* 51:809–816.

Coyne, J.A., and H.A. Orr. 2004. *Speciation*. Sunderland, MA: Sinauer Associates.

Dalecky, A., S. Ponsard, R.I. Bailey, C. Pélissier, and D. Bourguet. 2006. Resistance evolution to Bt crops: predispersal mating of European corn borers. *PLOS Biology* 4:e181.

Dopman, E.B. 2011. Genetic hitchhiking associated with life history divergence and colonization of North America in the European corn borer moth. *Genetica* 139:565–573.

Dopman, E.B., S.M. Bogdanowicz, and R.G. Harrison. 2004. Genetic mapping of sexual isolation between E and Z pheromone strains of the European corn borer (*Ostrinia nubilalis*). *Genetics* 167:301–309.

Dopman, E.B., L. Pérez, S.M. Bogdanowicz, and R.G. Harrison. 2005. Consequences of reproductive barriers for genealogical discordance in the European corn borer. *Proceedings of the National Academy of Sciences of the United States of America* 102:14706–14711.

Dopman, E.B., P.S. Robbins, and A. Seaman. 2010. Components of reproductive isolation between North American pheromone strains of the European corn borer. *Evolution* 64:881–902.

Durant, J.A., D.G. Manley, and R.T. Cardé. 1986. Monitoring of the European corn borer (Lepidoptera: Pyralidae) in South Carolina using pheromone traps. *Journal of Economic Entomology* 79:1539–1543.

Eckenrode, C.J., and D.R. Webb. 1989. Establishment of various European corn borer (Lepidoptera, Pyralidae) races on selected cultivars of snap beans. *Journal of Economic Entomology* 82:1169–1173.

El-Sayed, A.M. 2011. The Pherobase: database of insect pheromones and semiochemicals. Available at: http://www.pherobase.com.

Eltahlawy, H.S., J.S. Buckner, and S.P. Foster. 2007. Regulation of pheromone biosynthesis in the "Z strain" of the European corn borer, *Ostrinia nubilalis*. *Archives of Insect Biochemistry and Physiology* 65:29–38.

Fletcher-Howell, G., D.N. Ferro, and S. Butkewich. 1983. Pheromone and blacklight trap monitoring of adult European corn borer (Lepidoptera: Pyralidae) in western Massachusetts. *Environmental Entomology* 12:531–534.

Foster, S.P. 2004. Fatty acid and sex pheromone changes and the role of glandular lipids in the Z-strain of the European corn borer, *Ostrinia nubilalis* (Hübner). *Archives of Insect Biochemistry and Physiology* 56:73–83.

Frolov, A.N., P. Audiot, D. Bourguet, A.G. Kononchuk, J.M. Malysh, S. Ponsard, R. Streiff, and Y.S. Tokarev. 2012. From Russia with lobe: genetic differentiation in trilobed uncus *Ostrinia* spp. follows food plant, not hairy legs. *Heredity* 108:147–156.

Frolov, A.N., D. Bourguet, and S. Ponsard. 2007. Reconsidering the taxonomy of several *Ostrinia* species in the light of reproductive isolation: a tale for Ernst Mayr. *Biological Journal of the Linnean Society* 91:49–72.

Fukuzawa, M., X. Fu, S. Tatsuki, and Y. Ishikawa. 2006. cDNA cloning and in situ hybridization of Δ11-desaturase, a key enzyme of pheromone biosynthesis in *Ostrinia scapulalis* (Lepidoptera: Crambidae). *Journal of Insect Physiology* 52:430–435.

Galizia, C.G., and S. Sachse. 2010. Odor coding in insects. Pp. 35–70. In A. Menini, ed. *The Neurobiology of Olfaction*. Boca Raton, FL: CRC Press.

Geiler, K., and R. Harrison. 2010. A Delta11 desaturase gene genealogy reveals two divergent allelic classes within the European corn borer (*Ostrinia nubilalis*). *BMC Evolutionary Biology* 10:112.

Glover, T.J., M. Campbell, P. Robbins, and W.L. Roelofs. 1990. Sex-linked control of sex pheromone behavioral responses in European corn-borer moths (*Ostrinia nubilalis*) confirmed with TPI marker gene. *Archives of Insect Biochemistry and Physiology* 15:67–77.

Glover, T.J., M.G. Campbell, C.E. Linn, and W.L. Roelofs. 1991a. Unique sex chromosome mediated behavioral response specificity of hybrid male European corn borer moths. *Experientia* 47:980–984.

Glover, T.J., J.J. Knodel, P.S. Robbins, C.J. Eckenrode, and W.L. Roelofs. 1991b. Gene flow among three races of European corn borers (Lepidoptera: Pyralidae) in New York state. *Environmental Entomology* 20:1356–1362.

Glover, T.J., N. Perez, and W.L. Roelofs. 1989. Comparative analysis of sex-pheromone-response antagonists in three races of European corn borer. *Journal of Chemical Ecology* 15:863–873.

Glover, T.J., P.S. Robbins, C.J. Eckenrode, and W.L. Roelofs. 1992. Genetic control of voltinism characteristics in european corn borer races assessed with a marker gene. *Archives of Insect Biochemistry and Physiology* 20:107–117.

Glover, T.J., X.-H. Tang, and W.L. Roelofs. 1987. Sex pheromone blend discrimination by male moths from *E* and *Z* strains of European corn borer. *Journal of Chemical Ecology* 13:143–151.

Grosse-Wilde, E., T. Gohl, E. Bouché, H. Breer, and J. Krieger. 2007. Candidate pheromone receptors provide the basis for the response of distinct antennal neurons to pheromonal compounds. *European Journal of Neuroscience* 25:2364–2373.

Grosse-Wilde, E., A. Svatos, and J. Krieger. 2006. A pheromone-binding protein mediates the bombykol-induced activation of a aheromone receptor in vitro. *Chemical Senses* 31:547–555.

Hallberg, E., B.S. Hansson, and R.A. Steinbrecht. 1994. Morphological characteristics of antennal sensilla in the European cornborer *Ostrinia nubilalis* (Lepidoptera: Pyralidae). *Tissue and Cell* 26:489–502.

Hansson, B.S., E. Hallberg, C. Löfstedt, and R.A. Steinbrecht. 1994. Correlation between dendrite diameter and action potential amplitude in sex pheromone specific receptor neurons in male *Ostrinia nubilalis* (Lepidoptera: Pyralidae). *Tissue and Cell* 26:503–512.

Hansson, B.S., C. Löfstedt, and W.L. Roelofs. 1987. Inheritance of olfactory response to sex pheromone components in *Ostrinia nubilalis*. *Naturwissenschaften* 74:497–499.

Harrison, R.G., and A.T. Vawter. 1977. Allozyme differentiation between pheromone strains of European corn borer, *Ostrinia nubilalis* (Lepidoptera Pyralidae). *Annals of the Entomological Society of America* 70:717–720.

Huang, Y., T. Takanashi, S. Hoshizaki, S. Tatsuki, and Y. Ishikawa. 2002. Female sex pheromone polymorphism in adzuki bean borer, *Ostrinia scapulalis*, is similar to that in European corn borer, *O. nubilalis*. *Journal of Chemical Ecology* 28:533–539.

Huang, Y., S. Tatsuki, C.-G. King, S. Hoshizaki, Y. Yoshiyasu, H. Honda, and Y. Ishikawa. 1997. Identification of sex pheromone of adzuki bean borer, *Ostrinia scapulalis*. *Journal of Chemical Ecology* 23:2791–2801.

Ishikawa, Y., T. Takanashi, C.-G. Kim, S. Hoshizaki, S. Tatsuki, and Y. Huang. 1999. *Ostrinia* spp. in Japan: their host plants and sex pheromones. *Entomologia Experimentalis et Applicata* 91:237–244.

Jurenka, R.A., and W.L. Roelofs. 1989. Characterization of the acetyltransferase used in pheromone biosynthesis in moths: specificity for the Z isomer in tortricidae. *Insect Biochemistry* 19:639–644.

Kárpáti, Z., T. Dekker, and B.S. Hansson. 2008. Reversed functional topology in the antennal lobe of the male European corn borer. *Journal of Experimental Biology* 211:2841–2848.

Kárpáti, Z., B. Molnár, and G. Szőcs. 2007. Pheromone titer and mating frequency of E-and Z-strains of the european corn borer, *Ostrinia nubilalis*: fluctuation during scotophase and age dependence. *Acta Phytopathologica et Entomologica Hungarica* 42:331–341.

Kárpáti, Z., S.B. Olsson, B.S. Hansson, and T. Dekker. 2010. Inheritance of central neuroanatomy and physiology related to pheromone preference in the male European corn borer. *BMC Evolutionary Biology* 10:286.

Kárpáti, Z., M. Tasin, R.T. Cardé, and T. Dekker. 2013. Early quality assessment lessens pheromone specificity in a moth. *Proceedings of the National Academy of Sciences of the United States of America* 110:7377–7382.

Kaster, L.V., and M.E. Gray. 2005. European corn borers and western corn rootworms: old and new invasive maize pests challenge farmers on European and North American continents. *Maydica* 50:235–245.

Kennedy, G. G., and T. E. Anderson. 1980. European Corn Borer trapping in North Carolina with various sex pheromone component blends. *Journal of Economic Entomology* 73:642–646.

Klun, J. A. 1968. Isolation of a sex pheromone of the European corn borer. *Journal of Economic Entomology* 61:484–487.

Klun, J. A., and T. A. Brindley. 1970. *cis*-11Tetradecenyl acetate, a sex stimulant of the European corn borer. *Journal of Economic Entomology* 63:779–780.

Klun, J. A., and Cooperators. 1975. Insect sex pheromones: Intraspecific pheromonal variability of *Ostrinia nubilalis* in North America and Europe. *Environmental Entomology* 4:891–894.

Klun, J. A., and M. D. Huettel. 1988. Genetic regulation of sex pheromone production and response: interaction of sympatric pheromonal types of European corn borer, *Ostrinia nubilalis* (Lepidoptera: Pyralidae). *Journal of Chemical Ecology* 14:2047–2061.

Klun, J. A., and S. Maini. 1979. Genetic basis of an insect chemical communication system: the European corn borer. *Environmental Entomology* 8:423–426.

Klun, J. A., and J. F. Robinson. 1971. European corn borer: sex attractant and sex attraction inhibitors. *Annals of the Entomological Society of America* 64:1083–1086.

Klun, J. A., and J. F. Robinson. 1972. Olfactory discrimination in European corn borer and several pheromonally analogous moths. *Annals of the Entomological Society of America* 65:1337–1340.

Klun, J. A., O. L. Chapman, K. C. Mattes, P. W. Wojkowski, M. Beroza, and P. E. Sonnet. 1973. Insect sex pheromones: minor amount of opposite geometrical isomer critical to attraction. *Science* 181:661–663.

Kochansky, J., R. T. Cardé, J. Liebherr, and W. L. Roelofs. 1975. Sex pheromone of the European corn borer, *Ostrinia nubilalis* (Lepidoptera: Pyralidae), in New York. *Journal of Chemical Ecology* 1:225–231.

Lassance, J.-M., and C. Löfstedt. 2009. Concerted evolution of male and female display traits in the European corn borer, *Ostrinia nubilalis*. *BMC Biology* 7:10.

Lassance, J.-M., S. M. Bogdanowicz, K. W. Wanner, C. Löfstedt, and R. G. Harrison. 2011. Gene genealogies reveal differentiation at sex pheromone olfactory receptor loci in pheromone strains of the European corn borer, *Ostrinia nubilalis*. *Evolution* 65:1583–1593.

Lassance, J.-M., A. T. Groot, M. A. Liénard, B. Antony, C. Borgwardt, F. Andersson, E. Hedenström, D. G. Heckel, and C. Löfstedt. 2010. Allelic variation in a fatty-acyl reductase gene causes divergence in moth sex pheromones. *Nature* 466:486–489.

Lassance, J.-M., M. A. Liénard, B. Antony, S. Qian, T. Fujii, J. Tabata, Y. Ishikawa, and C. Löfstedt. 2013. Functional consequences of sequence variation in the pheromone biosynthetic gene *pgFAR* for *Ostrinia* moths. *Proceedings of the National Academy of Sciences of the United States of America* 110:3967–3972.

Lehmhus, J., G. Cordsen-Nielsen, C. Söderlind, G. Szőcs, J.-M. Lassance, J. Fodor, and A. Künstler. 2012. First records of the Z-Race of European Corn Borer *Ostrinia nubilalis* (Hübner 1796) from Scandinavia. *Journal für Kulturpflanzen* 64:163–167.

Leniaud, L., P. Audiot, D. Bourguet, B. Frerot, G. Genestier, S. F. Lee, T. Malausa, A. H. Le Pallec, M. C. Souqual, and S. Ponsard. 2006. Genetic structure of European and Mediterranean maize borer populations on several wild and cultivated host plants. *Entomologia Experimentalis et Applicata* 120:51–62.

Leppik, E., and B. Frerot. 2012. Volatile organic compounds and host-plant specialization in European corn borer E and Z pheromone races. *Chemoecology* 22:119–129.

Liebherr, J., and W. L. Roelofs. 1975. Laboratory hybridization and mating period studies using two pheromone strains of *Ostrinia nubilalis*. *Annals of the Entomological Society of America* 68:305–309.

Linn, C. E., K. Poole, A. Zhang, and W. Roelofs. 1999. Pheromone-blend discrimination by European corn borer moths with inter-race and inter-sex antennal transplants. *Journal of Comparative Physiology A, Neuroethology, Sensory, Neural, and Behavioral Physiology* 184:273–278.

Linn, C. E., M. S. Young, M. Gendle, T. J. Glover, and W. L. Roelofs. 1997. Sex pheromone blend discrimination in two races and hybrids of the European corn borer moth, *Ostrinia nubilalis*. *Physiological Entomology* 22:212–223.

Ma, P. W. K., and W. L. Roelofs. 1995a. Calcium involvement in the stimulation of sex pheromone production by PBAN in the European corn borer, *Ostrinia nubilalis* (Lepidoptera: Pyralidae). *Insect Biochemistry and Molecular Biology* 25:467–473.

Ma, P. W. K., and W. L. Roelofs. 1995b. Anatomy of the neurosecretory cells in the cerebral and subesophageal ganglia of the female European corn borer moth, *Ostrinia nubilalis* (Hübner) (Lepidoptera: Pyralidae). *International Journal of Insect Morphology and Embryology* 24:343–359.

Ma, P. W. K., and W. L. Roelofs. 1995c. Sites of synthesis and release of PBAN-like factor in the female European corn borer, *Ostrinia nubilalis*. *Journal of Insect Physiology* 41:339–350.

Ma, P. W. K., and W. L. Roelofs. 2002. Sex pheromone gland of the female European corn borer moth, *Ostrinia nubilalis* (Lepidoptera, Pyralidae): ultrastructural and biochemical evidences. *Zoological Science* 19:501–511.

Malausa, T., M.-T. Bethenod, A. Bontemps, D. Bourguet, J.-M. Cornuet, and S. Ponsard. 2005. Assortative mating in sympatric host races of the European corn borer. *Science* 308:258–260.

Malausa, T., L. Leniaud, J.-F. Martin, P. Audiot, D. Bourguet, S. Ponsard, S.-F. Lee, R. G. Harrison, and E. Dopman. 2007. Molecular differentiation at nuclear loci in French host races of the European corn borer (*Ostrinia nubilalis*). *Genetics* 176:2343–2355.

Martel, C., A. Réjasse, F. Rousset, M.-T. Bethenod, and D. Bourguet. 2003. Host-plant-associated genetic differentiation in Northern French populations of the European corn borer. *Heredity* 90:141–149.

Mcleod, D. G. R. 1981. Damage to sweet pepper in Ontario by 3 strains of European corn borer, *Ostrinia nubilalis* (Lepidoptera, Pyralidae). *Proceedings of the Entomological Society of Ontario* 112:29–32.

Meissle, M., P. Mouron, T. Musa, F. Bigler, X. Pons, V. P. Vasileiadis, S. Otto et al. 2010. Pests, pesticide use and alternative options in European maize production: current status and future prospects. *Journal of Applied Entomology* 134:357–375.

Miura, N., T. Nakagawa, K. Touhara, and Y. Ishikawa. 2010. Broadly and narrowly tuned odorant receptors are involved in female sex pheromone reception in Ostrinia moths. *Insect Biochemistry and Molecular Biology* 40:64–73.

Mutuura, A., and E. Munroe. 1970. Taxonomy and distribution of European corn borer and allied species: genus *Ostrinia* (Lepidoptera: Pyralidae). *Memoirs of the Entomological Society of Canada* 102:1–112.

Nagai, T. 1981. Electroantennogram response gradient on the antenna of the European corn borer, *Ostrinia nubilalis*. *Journal of Insect Physiology* 27:889–894.

Nagai, T., A. N. Starratt, D. G. R. Mcleod, and G. R. Driscoll. 1977. Electroantennogram responses of the European corn borer, *Ostrinia nubilalis*, to (Z)- and (E)-11-tetradecenyl acetates. *Journal of Insect Physiology* 23:591–597.

Nakano, R., Y. Ishikawa, S. Tatsuki, A. Surlykke, N. Skals, and T. Takanashi. 2006. Ultrasonic courtship song in the Asian corn borer moth, *Ostrinia furnacalis*. *Naturwissenschaften* 93:292–296.

Nakano, R., N. Skals, T. Takanashi, A. Surlykke, T. Koike, K. Yoshida, H. Maruyama, S. Tatsuki, and Y. Ishikawa. 2008. Moths produce extremely quiet ultrasonic courtship songs by rubbing specialized scales. *Proceedings of the National Academy of Sciences of the United States of America* 105:11812–11817.

Olsson, S. B., S. Kesevan, A. T. Groot, T. Dekker, D. G. Heckel, and B. S. Hansson. 2010. *Ostrinia* revisited: evidence for sex linkage in European corn borer *Ostrinia nubilalis* (Hubner) pheromone reception. *BMC Evolutionary Biology* 10:285.

Pélozuelo, L., and B. Frérot. 2007. Monitoring of European corn borer with pheromone-baited traps: review of trapping system basics and remaining problems. *Journal of Economic Entomology* 100:1797–1807.

Pélozuelo, L., C. Malosse, G. Genestier, H. Guenego, and B. Frérot. 2004. Host-plant specialization in pheromone strains of the European corn borer *Ostrinia nubilalis* in France. *Journal of Chemical Ecology* 30:335–352.

Pélozuelo, L., S. Meusnier, P. Audiot, D. Bourguet, and S. Ponsard. 2007. Assortative mating between European corn borer pheromone races: beyond assortative meeting. *PLOS ONE* 6:e555.

Phelan, P. L. 1997. Evolution of mate-signaling in moths: phylogenetic considerations and predictions from the asymmetric tracking hypothesis. Pp. 240–256. In J. C. Choe and B. J. Crespi, eds. *Mating Systems in Insects and Arachnids.* Cambridge: Cambridge University Press.

Phelan, P. L., and T. C. Baker. 1987. Evolution of male pheromones in moths: reproductive isolation through sexual selection? *Science* 235:205–207.

Ponsard, S., M. T. Bethenod, A. Bontemps, L. Pelozuelo, M. C. Souqual, and D. Bourguet. 2004. Carbon stable isotopes: a tool for studying the mating, oviposition, and spatial distribution of races of European corn borer, *Ostrinia nubilalis*, among host plants in the field. *Canadian Journal of Zoology* 82:1177–1185.

Raina, A. K., H. Jaffe, T. G. Kempe, P. Keim, R. W. Blacher, H. M. Fales, C. T. Riley, J. A. Klun, R. L. Ridgway, and D. K. Hayes. 1989. Identification of a neuropeptide hormone that regulates sex pheromone production in female moths. *Science* 244:796–798.

Roelofs, W. L., R. T. Cardé, R. J. Bartell, and P. G. Tierney. 1972. Sex attractant trapping of the European corn borer in New York. *Environmental Entomology* 1:606–608.

Roelofs, W. L., J. W. Du, X. H. Tang, P. S. Robbins, and C. J. Eckenrode. 1985. Three European corn borer populations in New York based on sex pheromones and voltinism. *Journal of Chemical Ecology* 11:829–836.

Roelofs, W. L., T. J. Glover, X.-H. Tang, I. Sreng, P. Robbins, C. Eckenrode, C. Löfstedt, B. S. Hansson, and B. O. Bengtsson. 1987. Sex pheromone production and perception in European corn borer moth is determined by both autosomal and sex-linked genes. *Proceedings of the National Academy of Sciences of the United States of America* 84:7585–7589.

Roelofs, W. L., W. Liu, G. Hao, H. Jiao, A. P. Rooney, and C. E. Linn. 2002. Evolution of moth sex pheromones via ancestral genes. *Proceedings of the National Academy of Sciences of the United States of America* 99:13621–13626.

Royer, L., and J. N. Mcneil. 1992. Evidence of a male sex pheromone in the European corn borer *Ostrinia nubilalis* (Hübner) (Lepidoptera: Pyralidae). *Canadian Entomologist* 124:113–116.

Royer, L., and J. N. Mcneil. 1993. Male investment in the European corn borer, *Ostrinia nubilalis* (Lepidoptera: Pyralidae): impact on female longevity and reproductive performance. *Functional Ecology* 7:209–215.

Showers, W. B., R. L. Hellmich, M. E. Derrick-Robinson, and W. H. Hendrix. 2001. Aggregation and dispersal behavior of marked and released European corn borer (Lepidoptera: Crambidae) adults. *Environmental Entomology* 30:700–710.

Showers, W. B., G. L. Reed, J. F. Robinson, and M. B. Derozari. 1976. Flight and sexual activity of the European corn borer. *Environmental Entomology* 5:1099–1104.

Smadja, C., and R. K. Butlin. 2009. On the scent of speciation: the chemosensory system and its role in premating isolation. *Heredity* 102:77–97.

Straub, R. W., R. W. Weires, and C. J. Eckenrode. 1986. Damage to apple cultivars by races of European corn borer (Lepidoptera, Pyralidae). *Journal of Economic Entomology* 79:359–363.

Takanashi, T., R. Nakano, A. Surlykke, H. Tatsuta, J. Tabata, Y. Ishikawa, and N. Skals. 2010. Variation in courtship ultrasounds of three *Ostrinia* moths with different sex pheromones. *PlOS One* 5:e13144.

Thomas, Y., M.-T. Bethenod, L. Pelozuelo, B. Frérot, and D. Bourguet. 2003. Genetic isolation between two sympatric host-plant races of the European corn borer, *Ostrinia nubilalis* Hübner. I. Sex pheromone, moth-emergence timing, and parasitism. *Evolution* 57:261–273.

Wanner, K. W., A. S. Nichols, J. E. Allen, P. L. Bunger, S. F. Garczynski, C. E. Linn, Jr., H. M. Robertson, and C. W. Luetje. 2010. Sex pheromone receptor specificity in the European corn borer moth, *Ostrinia nubilalis*. *PLOS ONE* 5:e8685.

Webster, R. P., and R. T. Cardé. 1982. Influence of relative-humidity on calling behavior of the female European corn-borer moth (*Ostrinia nubilalis*). *Entomologia Experimentalis et Applicata* 32:181–185.

Willett, C. S., and R. G. Harrison. 1999. Insights into genome differenciation: pheromone-binding protein variation and population history in the European corn borer (*Ostrinia nubilalis*). *Genetics* 153:1743–1751.

Wolf, W. A., and W. L. Roelofs. 1987. Reinvestigation confirms action of Δ11-desaturases in spruce budworm moth sex pheromone biosynthesis. *Journal of Chemical Ecology* 13:1019–1027.

Wu, C. I. 2001. The genic view of the process of speciation. *Journal of Evolutionary Biology* 14:851–865.

Wyatt, T. D. 2003 *Pheromones and Animal Behaviour: Communication by Smell and Taste.* Cambridge: Cambridge University Press.

Zhu, J. W., C. H. Zhao, M. Bengtsson, and C. Löfstedt. 1996. Reductase specificity and the ratio regulation of *E/Z* isomers in the pheromone biosynthesis of the European corn borer, *Ostrinia nubilalis* (Lepidoptera: Pyralidae). *Insect Biochemistry and Molecular Biology* 26:171–176.

Divergence of the Sex Pheromone Systems in "Oriental" *Ostrinia* Species

JUN TABATA and YUKIO ISHIKAWA

Introduction

The European corn borer (ECB), *Ostrinia nubilalis* (Crambidae), has long attracted the attention of chemical ecologists, particularly those interested in sex pheromones, not only because this species is a notorious agricultural pest, but also because of its sex pheromone polymorphism (Lassance 2010). The sex pheromone polymorphism in ECB suggests that speciation is ongoing and that the ECB is a useful model for studying the initial stages of speciation. As a natural extension of studies on this model insect, the pheromone communication systems of ECB congeners also have become a focus of study in the last two decades. The most recent monograph of the genus *Ostrinia* lists 20 species, with 7 reported to occur in the Far East (Mutuura and Munroe 1970). Because two new species have since been discovered in Japan (Ohno 2003a; Tabata 2010), nine species of *Ostrinia* are now known to occur in the Far East. Because of the wealth of knowledge on ECB and the diversity of Far Eastern species, the pheromone systems of oriental *Ostrinia* are model systems to study the process of moth pheromone evolution. In this chapter, we review the sex pheromones of the oriental *Ostrinia* spp., discuss sex pheromones as prezygotic reproductive isolating mechanisms, and attempt to reconstruct the evolutionary history of *Ostrinia* pheromones.

Taxonomy and Classification of The Oriental *Ostrinia* Species

The species of the genus *Ostrinia* are divided into three groups, largely based on the morphology of male genitalia (Mutuura and Munroe 1970). Group I consists of a single American species, *O. penitalis*, which is characterized by the dorsally trifid juxta and unarmed sacculus in the male genitalia. Group II consists of an assemblage of species, all of which have a dorsally spined sacculus and a simple or bifid uncus. Three species in Group II occur in the Far East—*O. palustralis* (PAL), the Far Eastern knotweed borer *O. latipennis* (LAT), and *O. ovalipennis* (OVA)—and are referred to as the palustralis group by Mutuura and Munroe (1970). The host range of members of this group is limited to a group of host plants in the Polygonaceae (Table 16.1). PAL has a trans-Palearctic distribution and feeds only on docks, *Rumex* spp. LAT, on the other hand, has a distribution limited to Northeastern China, Far East Russia, and Northern Japan, and mostly feeds on knotweeds, *Reynoutria* spp., although it is occasionally found on non-polygonaceous plants (Ohno 2000a). OVA, which is very similar to LAT, both in terms of morphology and mitochondrial DNA sequences, has the same distribution, host range, and phenology as LAT (Ohno 2003a, 2003b). These two species are classified into the latipennis subgroup (Ohno 2003b).

TABLE 16.1
Host plants of *Ostrinia* spp. inhabiting Japan

Group	Subgroup	Species	Host plant[a]	Host range[b]
Group II	–	*O. palustralis* (PAL)	*Rumex* spp. (Polygonaceae)	–
	latipennis	*O. latipennis* (LAT)	*Reynoutria sachalinensis, R. japonica* (Polygonaceae)	–[c]
		O. ovalipennis (OVA)	*Reynoutria sachalinensis, R. japonica* (Polygonaceae)	–
Group III	–	*O. furnacalis* (ACB)	*Rumex* spp., *Persicaria tinctoria* (Polygonaceae); *Helianthus annuus, Xanthium* spp., *Ambrosia* spp., *Zinnia elegans* (Asteraceae); *Capsicum annuum, Lycipersicon esculentum, Solanum melongena* (Solanaceae); *Abelmoschus esculentus, Gossypium* spp. (Malvaceae); *Phaseolus vulgaris* (Fabaceae); *Zea mays, Sorghum bicolor, Miscanthus sacchariflorus, Phragmites communis, Coix lacryma-jobi* (Graminaceae); *Iris ensata* (Iridaceae); *Zingiber officinalis* (Gingiberaceae)	+++
	scapulalis	*O. scapulalis* (SCA)	*Reynoutria sachalinensis, R. japonica, Rumex* spp., *Polygonum thunbergii, Persicaria tinctoria* (Polygonaceae); *Arctium lappa, Cirsium* spp., *Dahlia pinnata, Xanthium* spp., *Ambrosia* spp., *Artemisia* spp., *Bidens* spp., *Senecio cruentis* (Asteraceae); *Lycipersicon esculentum, Solanum melongena* (Solanaceae); *Abelmoschus esculentus, Althaea rosea* (Malvaceae); *Pelargonium* sp. (Geraniaceae); *Cannabis sativa, Humulus lupulus* (Moraceae); *Phaseolus vulgaris, Phaseolus angularis, Glycine max, Arachis hypogaea* (Fabaceae); *Zea mays, Miscanthus sacchariflorus, Phragmites communis, Coix lacryma-jobi* (Graminaceae)	+++
		O. orientalis (ORI)	*Reynoutria sachalinensis, R. japonica, Rumex* sp. (Polygonaceae); *Xanthium* spp., *Senecio cannabifolius* (Asteraceae)	++
	zealis	*O. zealis* (ZEA)	*Arctium lappa, Cirsium* spp., *Dahlia pinnata* (Asteraceae)	+
		O. zaguliaevi (ZAG)	*Petasites japonicus* (Asteraceae)	–
		O. sp. near *zaguliaevi* (SPZ)	*Farfugium japonicum* (Asteraceae)	–

a. List of host plants produced from Hattori and Mutuura (1987) with some revisions based on findings by the authors.

b. The symbols –, +, ++, and +++ indicate nearly monophagous (feeding on a single or a few kinds of plants within the same or related genus), oligophagous (feeding on a few kinds of plants within the same family), potentially polyphagous (feeding on some plants belonging to different families), and extensively polyphagous (feeding on various families of plants), respectively.

c. Ohno (2000a) lists some other host plants of *O. latipennis*, suggesting that *O. latipennis* might be potentially polyphagous.

The third group (Group III), which is considered the most derived, contains species with a trilobed uncus. Members of this group, including the ECB, have similar external morphology as well as male and female genitalia. Comprehensive studies on morphology, DNA sequences, and host plant preferences of Group III identified six *Ostrinia* species in the Far East: the Asian corn borer *O. furnacalis* (ACB), the adzuki bean borer *O. scapulalis* (SCA), the cocklebur borer *O. orientalis* (ORI), the burdock borer *O. zealis* (ZEA), the butterbur borer *O. zaguliaevi* (ZAG), and the leopard plant borer *Ostrinia* sp. near *zaguliaevi* (SPZ), which is yet to be formally described. This group is often referred to as the furnacalis group (Ishikawa et al. 1999a), and the six members are further divided into three subgroups: the furnacalis subgroup (ACB), the scapulalis subgroup (SCA and ORI), and the zealis subgroup (ZEA, ZAG, and SPZ). ACB is

widely distributed from temperate to tropical regions of Asia/ Oceania including India, China, Far East Russia, Japan, and Australia. Like ECB, ACB is a polyphagous pest that infests maize and several other crops such as cotton and ginger (Table 16.1). Molecular phylogenies based on mitochondrial DNA indicate that ACB is relatively distantly related to other members of Group III and is singly placed at the base of the Group III clade (Kim et al. 1999).

Members of the scapulalis subgroup, SCA and ORI, are most closely related to ECB (Kim et al. 1999; Frolov et al. 2007). SCA is a polyphagous species with isolated populations occurring in the Palearctic region including Eastern Europe, Southern Asia, and the Far East. In contrast, ORI occurs only in the Far East, and its main host plants are likely to be limited to the cockleburs *Xanthium* spp. and *Senecio cannabifolius* (Asteraceae)

(Ohno 2000c). Because of similarities in ecology and morphology as well as incomplete reproductive isolation between SCA and ORI, Frolov et al. (2007) synonymized these species. However, at least in Japan, the two species can be separated on the basis of some diagnostic traits, i.e., morphology, host preference, and sex pheromones, although hybridization is suggested to occur in nature (Ohno 2000b). Therefore, here we treat SCA and ORI as independent species according to the traditional taxonomy presented by Mutuura and Munroe (1970).

The zealis subgroup is characterized by larger body size and three gradually curved lobes of the uncus in contrast to pointed lobes in the other members of Group III (Hattori and Mutuura 1987). ZEA is recorded from China and Northern India in addition to the Far East (Mutuura and Munroe 1970), but ZAG and SPZ have been discovered only in the Far East (Tabata 2010). This subgroup is further characterized by specialized feeding on limited Asteraceae plants: the thistles *Cirsium* spp. and the burdock *Arctium lappa* in ZEA, the butterbur *Petasites japonicas* in ZAG, and the leopard plant *Farfugium japonicum* in SPZ (Table 16.1). The taxonomic status of SPZ is yet to be determined. Studies utilizing molecular markers have shown that this species is very closely related to ZAG, and substantial gene flow between ZAG and SPZ is observed where the two occur sympatrically, with an estimated frequency of natural hybridization of approximately 5% (Tabata 2010). Despite the occurrence of hybridization in the field, the persistence of associations of ZAG- or SPZ-specific traits, including morphology and host preference, suggests that reproductive isolating mechanisms, including sex pheromone systems, are functioning, albeit imperfectly.

Sex Pheromones of Group II, The Palustralis Group

Females of PAL emit a 99:1 mixture of (*E*)-11-tetradecenyl acetate (E11-14Ac) and (*Z*)-11-tetradecenyl acetate (Z11-14Ac) (Huang et al. 1998c). A synthetic mixture of these two components was confirmed to attract males in a wind-tunnel assay. *Ostrinia obumbratalis*, an American member of Group II, was also attracted in the field to a 1:1 mixture of E11- and Z11-14Ac (Klun and Robinson 1972). In contrast, the pheromones of LAT and OVA (latipennis subgroup) include (*E*)-11-tetradecenol (E11-14OH), an alcohol derivative of E11-14Ac (Takanashi et al. 2000; Ohno 2003b). Because all pheromone components of other *Ostrinia* moths are acetates, the alcohol pheromone in the latipennis subgroup is unique (Table 16.2). The acetate pheromones in *Ostrinia* are biosynthesized through acetylation of the corresponding alcohol precursors, which is catalyzed by acetyltransferases (Zhao et al. 1995; Zhu et al. 1996a). E11-14OH is therefore a precursor of a pheromone in *Ostrinia* moths other than LAT and OVA. LAT uses E11-14OH alone in its pheromone (Takanashi et al. 2000), indicating that the acetyltransferase is completely nonfunctional in this species. Interestingly, OVA uses a pheromone composed of a 1:9 mixture of E11-14OH and E11-14Ac (Ohno 2003b), indicating that acetyltransferase in OVA is partially functional. Since the alcohol pheromone component is only found in the latipennis subgroup, the acetyltransferase must be fully functional in all Far Eastern *Ostrinia* excluding LAT and OVA.

Another interesting issue in the pheromones of the latipennis subgroup is the lack of (*Z*)-isomers, which are commonly found in the pheromones of other *Ostrinia* species. Cloning and functional analysis of the genes coding the Δ11-desaturase, a key enzyme that introduces a double bond into pheromone molecules, revealed that the absence of the (*Z*)-isomer is due to the strict product specificity of the Δ11-desaturase in LAT (LATPG1) (Fujii et al. 2011). Moreover, *LATPG1* is not closely related to the Δ11-desaturase genes cloned from ECB and ACB (*OnuZ/E11* and *OfuZ/E11*) but rather related to cryptic and nonfunctional retroposon-linked genes of Δ11-desaturases found in *Ostrinia* genomes (*ezi*-Δ11; Xue et al. 2007, 2012). At present, biochemical and genetic backgrounds of the latipennis subgroup pheromones remain to be examined in detail. Further studies on pheromones in these relatively primitive species are essential to illustrate the whole picture of pheromone divergence in *Ostrinia*.

Sex Pheromones of Group III, The Furnacalis Group

Sex Pheromone of ACB

The sex pheromone of ACB is composed of (*E*)-12-tetradecenyl acetate (E12-14Ac) and (*Z*)-12-tetradecenyl acetate (Z12-14Ac) (Ando et al. 1980; Klun et al. 1980). Generally, the occurrence of a double bond at even-numbered positions in monoene compounds is uncommon in the sex pheromones of moths (Ando 2012). E12-14Ac and Z12-14Ac are extremely rare, having never been discovered in pheromones of insects other than ACB (El-Sayed 2012). A rare Δ14-desaturase is involved in the formation of ACB pheromones as follows. First, a double bond at the Δ14 position is introduced into palmitoyl-CoA (C16) by Δ14-desaturase, forming Δ14-hexadecenoic acyl-CoA precursors (Δ14-16Acyl) (Zhao et al. 1990). These Δ14-16Acyl precursors are subsequently chain-shortened to Δ12-14Acyl through β-oxidation. In contrast, all the other *Ostrinia* species employ E11- and Z11-14Ac, products of Δ11-desaturase as sex pheromone components (Roelofs et al. 2002; Roelofs and Rooney 2003; figure 16.1). The Δ11-desaturase converts myristoyl-CoA (C14) to Δ11-14Acyl precursors. Interestingly, the transcripts of Δ11- and Δ14-desaturase genes are detected in both ECB and ACB when analyzed by reverse transcription-PCR (Roelofs et al. 2002). Moreover, the sequences within open reading frames of these genes are very similar (Roelofs et al. 2002; Roelofs and Rooney 2003). This provides evidence for an evolutionary mechanism of pheromone divergence based on a model for subfunctionalization of diverged desaturases after a gene duplication event early in the evolution of insects (Roelofs et al. 2002). Further, an analysis of the desaturase gene family phylogeny suggests that the gene duplication event would have occurred before the order-level split of Lepidoptera, and that a gene derived from the duplication became a pseudogene and was resurrected later in a stochastic event, during the evolution of a new pheromone system. The subsequent quantitative PCR analysis has shown that differences in the control of the transcription level of the two desaturase genes can account for the selective production of Δ12-compounds or Δ11-compounds (Sakai et al. 2009). Details of the regulation of transcription remain unclear, although an epigenetic mechanism may be included (Roelofs and Rooney 2003).

Because the pheromone components emitted by ACB females are unique, males are considered to have a low risk of cross-attraction to heterospecifics. This may explain the unusual pattern of peripheral pheromone detection by Olfactory Receptor Neurons (ORNs) on male ACB antennae (Takanashi et al. 2006). Although pheromone receptor neurons are highly specific in

TABLE 16.2

Sex pheromones and their diel rhythms of emission by *Ostrinia* spp. inhabiting Japan

Group	Subgroup	Species	Pheromone component proportion (%)[a]						Pheromone amount[b]	Emission rhythm[c]	Reference
			Z9-14Ac	E11-14Ac	Z11-14Ac	E12-14Ac	Z12-14Ac	E11-14OH			
Group II	palustralis	*O. palustralis* (PAL)	–	99	1	–	–	–	+++	D4–D9	Huang et al. (1998c)
	latipennis	*O. latipennis* (LAT)	–	–	–	–	–	100	+	D4–L4	Takanashi et al. (2000)
		O. ovalipennis (OVA)	–	90	–	–	–	10	++	D4–L5	Ohno (2003b)
Group III	furnacalis	*O. furnacalis* (ACB)	–	–	–	36	64	–	++	D3–D9	Huang et al. (1998b)
	scapulalis	*O. scapulalis* (SCA)	–	99	1 (E-type)	–	–	–	++	D2–D9	Huang et al. (1997, 2002)
				3	97 (Z-type)					L1–L5	
		O. orientalis (ORI)	–	2	98	–	–	–	+	D7–L3	Fu et al. (2004)
	zealis	*O. zealis* (ZEA)	70	24	6	–	–	–	+++	D6–L3	Ishikawa et al. (1999b)
		O. zaguliaevi (ZAG)	45	5	50	–	–	–	+++	D4–D9	Huang et al. (1998a)
		O. sp. near zaguliaevi (SPZ)	19	6	75	–	–	–	+++	D3–D8	Tabata et al. (2008)

a. The proportions produced by a typical female in each species are shown.

b. Total amount of pheromone components found in extracts of pheromone glands. The symbols +, ++, and +++ indicate small (less than 5 ng/female in mean), medium (5–10 ng/female), and large (more than 10 ng/female), respectively.

c. Time when calling behavior to emit sex pheromones was observed in female moths under laboratory conditions (15L : 9D photoperiod). The letters D*x* and L*y* indicate *x* hours after light-off and *y* hours after light-on, respectively. In SCA, interpopulational variations are reported in the calling time (Ishikawa et al. 1999a).

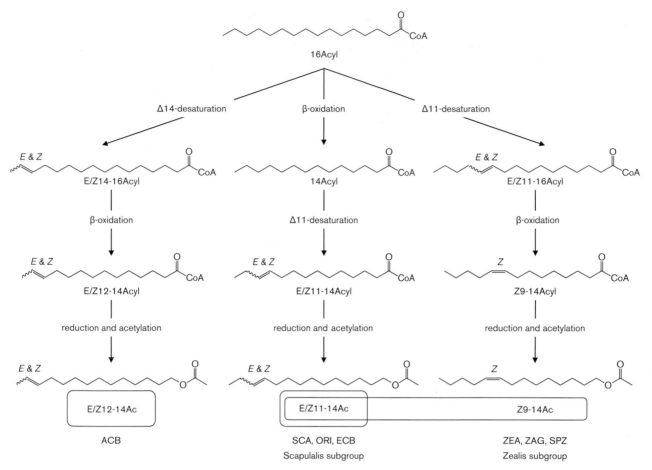

FIGURE 16.1 Biosynthetic pathways of pheromone components in *Ostrinia* Group III (the furnacalis group) adapted from Roelofs et al. (2002), Fu et al. (2005), and Tabata and Ishikawa (2011).

their response spectrum in most moths, major parts of the ORNs in ACB males are less specific and show equal responses to both (*E*)- and (*Z*)-isomers of Δ12-tetradecenyl acetate, which may increase sensitivity of their pheromone response.

The blend ratios of E12- and Z12-14Ac in ACB appear to vary among geographic populations. The proportion of (*E*)-isomer is reported to be around 50% in the Philippines (Klun et al. 1980), South China (Cheng et al. 1981), and Taiwan (Yeh et al. 1989; Kou et al. 1992). However, in Japan the proportion of (*E*)-isomer is 36–39% in northern populations but 44% in a southern population (Huang et al. 1998b). The difference is small but significant, and a family-wise analysis of the pheromone blend confirms that there are two distinct phenotypes with respect to the (*E*)-isomer proportion; a strain producing ≈46% E12-14Ac inhabits the southern part of Japan, while a strain producing ≈38% of the (*E*)-isomer seems predominant in other regions of Japan. Relatively high proportions of E12-14Ac in the southern Japanese populations are speculated to be associated with migration of ACB from Southeast Asian countries (Huang et al. 1998b).

Sex Pheromones of the Scapulalis Subgroup

The sex pheromones of SCA and ORI (the scapulalis subgroup) are composed of the same two components, E11-14Ac and Z11-14Ac, and are therefore very similar to the ECB pheromone (Huang et al. 1997; Fu et al. 2004) (Table 16.2). In addition, SCA

has ECB-like pheromone polymorphism in the blend ratio of the two components (Huang et al. 2002; Tabata et al. 2003; Takanashi et al. 2005); the E-type uses a pheromone with 99% of E11-14Ac, whereas the Z-type uses an opposite blend pheromone with 3% E11-14Ac. Their hybrid (I-type) females produce a pheromone with an intermediate blend. Moreover, the genetics of the pheromone polymorphism are identical in the SCA and ECB (Roelofs et al. 1987); the pheromone blend is under the control of an autosomal locus with two alleles (A^E/A^Z) (Takanashi et al. 2005), which are involved in the reduction step of the pheromone biosynthesis cascade (Antony et al. 2009). In ECB, these two alleles have recently been demonstrated to encode variants of a pheromone-gland-specific fatty acyl reductase (pgFAR) that show distinct substrate specificity and regulate reduction of (*E*)-11- and (*Z*)-11-tetradecenoyl pheromone precursors (Zhu et al. 1996b; Lassance et al. 2010). The *pgFAR* gene was cloned from SCA and characterized (Antony et al. 2009). Furthermore, the male SCA pheromone blend preference is controlled by one sex chromosomal locus with two alleles (Takanashi 2001), as in the case of ECB.

The sex pheromone of ORI is reported to be a 2:98 mixture of E11-14Ac and Z11-14Ac (Fu et al. 2004) and is practically identical to the Z-type SCA and ECB (Table 16.2). To date, neither the E-type nor I-type blend has been found in field-collected ORI. Crossing experiments between ORI and SCA confirmed that ORI has the same genetic architecture as Z-type SCA/ECB; an allele (A^Z) on the autosomal locus selectively produces the (*Z*)-isomer and an allele on the sex chromosomal locus determines the preference of males for a pheromone blend with a high proportion of the (Z)-isomer (Fu et al. 2005b).

Although the pheromone blend polymorphisms and their underlying genetics in the ECB and SCA are almost identical, the population-level polymorphic patterns appear to be different between the two species. For example, in a population of SCA in Japan, the hybrid type is predominant (52%), and the E-type and the Z-type are less abundant (15% and 33%, respectively) (Tabata et al. 2003). No deviation from Hardy–Weinberg expectations at equilibrium was found in the genotypic frequencies of pheromone blend types in this population. In addition, there is no indication of positive assortative mating in terms of pheromone production genotype. These may partly explain the apparently stable pheromone polymorphism in this population. Moreover, the frequencies of the three pheromone types of SCA appear to vary among populations within Japan (Takanashi et al. 2005). Conversely, either the E-type or the Z-type dominates populations of ECB or the occurrence of the hybrid I-type is rare. Klun and Huettel (1988) intensively sampled populations of the ECB in Maryland, USA, and found that Z-type females (70.2%) dominated over E-type (14.1%) and I-type (15.7%) females. There were significantly fewer hybrids (I-type) than expected under random mating and numbers strongly deviated from the Hardy–Weinberg distribution (Klun and Huettel 1988). Although the Z-type moths were dominant, they appeared not to "enjoy selective advantage"; there was no evidence that the Z-type females had more opportunities for mating than the E-type females (Bengtsson and Löfstedt 1990). This analysis indicated that the deviation in the pheromone genotype frequencies was attributed to strong assortative mating without selection, and therefore the polymorphism in this population is suggested to be in a stable state (Bengtsson and Löfstedt 1990). In New York State, the pheromone types of ECB have differences in voltinism; E-type ECB are bivoltine with flights in June and August, while Z-type ECB include a univoltine population with a flight in July as well as a bivoltine population with flights in June and August (Roelofs et al. 1985). Voltinism patterns in ECB are mainly determined by the length of post-diapause development, which is under the control of a single gene locus on the sex chromosome (Glover et al. 1992). These races of ECB differing in pheromone and voltinism often occur sympatrically, and substantial gene flow between the races is suggested by allozyme analyses (Glover et al. 1991). In contrast, in French populations of ECB, pheromone polymorphism is tightly associated with host plant preference (Thomas et al. 2003; Bontemps et al. 2004; Pélozuelo et al. 2004, 2007). The Z-type develops exclusively on maize, and the E-type develops on dicotyledonous plants such as mugworts and hops. The two types show strong assortative mating when they occur sympatrically, which prevents the formation of hybrids and leads to genetic isolation (Martel et al. 2003; Malausa et al. 2005). Because of similarities in ecology and an apparent lack of reproductive isolation between the ECB feeding on dicotyledons and SCA in the former Soviet Union, Frolov et al. (2007) proposed to treat the subpopulation of ECB feeding on dicotyledons as SCA. A recent genotyping study using a set of eight autosomal microsatellite markers supported this grouping; SCA and the dicotyledon-feeding population of ECB across Northwestern Eurasia are genetically indiscrete (Frolov et al. 2012). In contrast to the situation in Europe, to date no associations between pheromone polymorphism and host preference or voltinism have been found in populations of SCA in Japan. The reason why patterns of polymorphism are so different among species or populations is unknown.

Sex Pheromones of the Zealis Subgroup

Members of the zealis subgroup (ZEA, ZAG, and SPZ) are characterized by the use of (Z)-9-tetradecenyl acetate (Z9-14Ac), in addition to E11-14Ac and Z11-14Ac (Huang et al. 1998a; Ishikawa et al. 1999b; Tabata et al. 2008; Table 16.2). Although these three species employ the same components in pheromones, the blend ratio (Z9-14Ac:E11-14Ac:Z11-14Ac) appears to differ among the three species—70:24:6 in ZEA, 45:5:50 in ZAG, and 19:6:75 in SPZ—suggesting segregation in pheromone signals (Tabata et al. 2008). The ratio of geometric isomers (E11-14Ac and Z11-14Ac) in the ZEA pheromone (24:6) is distinct from those of the ZAG (5:50) and the SPZ (6:75) pheromones (figure 16.2). Meanwhile, the pheromones of ZAG and SPZ are distinguishable by the ratio of double-bond positional isomers (Z9-14Ac and E/Z11-14Ac), 45:55 in ZAG and 19:81 in SPZ (figure 16.2).

The biochemical basis of the production of the third component, Z9-14Ac, was studied in ZAG by using deuterium-labeled pheromone precursors (Fu et al. 2005a). The two components E11- and Z11-14Ac are biosynthesized via the following steps: C16 acid is chain-shortened to C14 acid, and then successively modified by Δ11-desaturase, reductase, and acetyltransferase, as described for ECB and SCA. The component Z9-14Ac is most likely generated via the following pathway: a double bond is first introduced into C16 acid by Δ11-desaturase to give rise to Δ11-16Acyl precursors, which are then chain-shortened to Δ9-14Acyl and successively reduced and acetylated (figure 16.1). Thus, biosyntheses of Z9- and E/Z11-14Ac share a common enzymatic step, i.e., desaturation by Δ11-desaturase (Fu et al. 2005a; Tabata et al. 2006). Interestingly, Z11-16Acyl (the precursor of Z9-14Ac) is excessively accumulated not only in the pheromone glands of the zealis subgroup species but also in the glands of ECB and SCA, which do not produce Z9-14Ac (Ma and Roelofs 2002; Fu et al. 2005a). The lack of production of Z9-14Ac in ECB and SCA can be partially ascribed to the blockage of chain-shortening from Z11-16Acyl to Z9-14Acyl (Fu et al. 2005a; Tabata and Ishikawa 2005). However, this blockage is not perfect, because a small amount of Z9-14Acyl is found in ECB and SCA (Tabata 2010).

From a series of crossing experiments between the three species of the zealis subgroup, the blend of Z9-, E11-, and Z11-14Ac is suggested to be controlled by a single major autosomal locus with three alleles ($B^{zea}/B^{zag}/B^{spz}$) (Tabata and Ishikawa 2011). The F1-hybrid females of ZAG × ZEA and SPZ × ZEA (female × male) produced pheromone blends with proportions of E11-14Ac (E/Z ratio) similar to the paternal species (ZEA), whereas the proportion of Z9-14Ac (Δ9/Δ11 ratio) was similar to the maternal species (ZAG or SPZ). When the F1 individuals were backcrossed to the parental species, the F1-type and the parent-type blends were observed at approximately a 1:1 ratio in both reciprocal crosses. The regulation of E/Z ratios and Δ9/Δ11 ratios by a common gene locus, not by independent gene loci, is consistent with the finding that Z9- and E/Z11-14Ac are generated via a common biosynthetic step (Fu et al. 2005a).

The pheromone blends of ZAG and SPZ are distinguishable by the proportion of Z9-14Ac (Δ9/Δ11 ratio). However, the proportion of Z9-14Ac in each species varies substantially and overlaps between the two species (Tabata et al. 2006, 2008) (figure 16.2). The F1 hybrids between ZAG and SPZ produce an intermediate blend, and the distribution patterns in the reciprocal crosses, ZAG × SPZ and SPZ × ZAG, are significantly different (Tabata and Ishikawa 2011). The pheromone blends

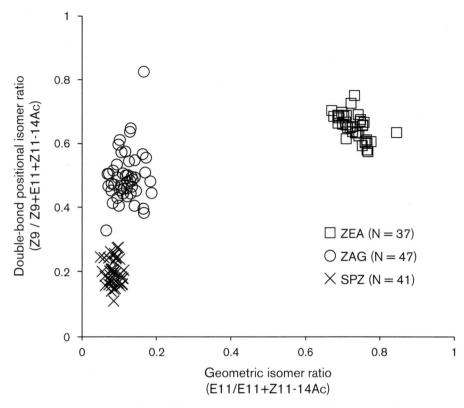

FIGURE 16.2 Sex pheromone blend differences in the zealis subgroup of *Ostrinia* Group III, which uses the three components Z9-14Ac, E11-14Ac, and Z11-14Ac. Data are shown in detail in Tabata and Ishikawa (2011).

of the F1 and backcrosses show a tendency to be more similar to the paternal characters. This inheritance pattern indicates that genes on the sex chromosome influence pheromone blend phenotypes, in addition to autosomal genes including B^{zag}/B^{spz}. Considering that females are the heterogametic (ZW) sex in Lepidoptera, at least one Z chromosome locus with two alleles in ZAG and SPZ has some effect on the pheromone blend (Tabata and Ishikawa 2011). It is unclear why evidence of sex chromosomal gene effects on the pheromone blend was not detected in the hybrids of ZEA and ZAG/SPZ. The pheromone phenotypes in ZEA and ZAG/SPZ may be canalized by the major autosomal locus ($B^{zea}/B^{zag}/B^{spz}$) with respect to control of the pheromone blend. This hypothesis is supported by the observation that in both E- and Z-types of the ECB, canalization by the major gene locus (A^E/A^Z) hides a considerable amount of underlying genetic variation in the blends of E11- and Z11-14Ac (Zhu et al. 1996a).

Putative pheromone precursors (fatty acyl pheromone analogues: FAPAs) were surveyed in three members of the zealis subgroup (Tabata and Ishikawa 2011). The amounts and proportions of FAPAs were significantly different among species, but the relative abundance of the positional isomers of tetradecenoic (C14:1) acyl was similar (Z9- > Z7- > E/Z11-14Acyl). The ratio of Z9-, E11-, and Z11-14Acyl was very different from the ratio of corresponding pheromone components (Z9-, Z11-, and E11-14Ac) in all species. In addition, a substantial amount of Z7-14Acyl, an acetate which is not used in *Ostrinia* pheromones, was commonly found. These results suggest that the production of species-specific pheromones in the zealis subgroup is due to differences in a few enzymes or genetic modifiers involved in the last two steps of pheromone biosynthesis (i.e., reduction and acetylation of fatty acyl precursors).

Because the acetyltransferase in *Ostrinia* shows very low substrate specificity (Zhao et al. 1995; Zhu et al. 1996a), reduction would be the key step for pheromone blend divergence in the zealis subgroup as well as the two-component pheromones in the scapulalis subgroup. The major autosomal locus ($B^{zea}/B^{zag}/B^{spz}$) that controls the pheromone blend in ZEA, ZAG, and SPZ is most probably involved in this biosynthetic step. As already mentioned, the ratio of FAPA geometric isomers in ECB/SCA is similar (E11- and Z11-14Acyl ≈ 7:3) regardless of pheromone type. The FAPAs are differentially reduced to yield pheromones with opposite blend ratio by two fatty acyl reductases (pgFARs) encoded in an autosomal locus (A^E/A^Z) that shows opposite substrate stereo-specificity (Lassance et al. 2010). Although the features of interspecific variation in the three-component pheromone in the *zealis* subgroup apparently differ from those of intraspecific polymorphism in the two-component pheromone in ECB/SCA, the variation may be explained by the presence of variant pgFARs reductases with different substrate specificities.

The genetics of a hypothesized saltational shift from the two-component E11/Z11-14Ac to the three-component E11/Z11-14Ac and Z9-14Ac sex pheromone was investigated by crossing ZEA × SCA (Tabata and Ishikawa 2005). The F1 females of ZEA and E-type SCA produced E11- and Z11-14Ac without Z9-14Ac in their pheromone, with a proportion of E11- and Z11-14Ac of ≈ 9:1, which is similar to the pheromone of the E-type SCA/ECB. F1 females of ZEA and Z-type SCA also produced a binary pheromone with E11- and Z11-14Ac with a ratio ≈ 6:4, which is similar to the I-type SCA/ECB. When these F1s were backcrossed to the Z-type SCA, the Z-type and the I-type pheromones were observed at approximately 1:1. These results indicate that ZEA has an allele functionally similar to A^E, which encodes the E-type reductase in ECB/SCA. When the F1 was backcrossed to ZEA, females producing pheromones with and without Z9-14Ac appeared at approximately 1:1, suggesting that the production of Z9-14Ac is also under the control of a single major gene (C^{zea}/C^{sca}). Moreover, this gene locus was shown to be tightly linked to the locus encoding the reductase (A^E/A^Z). The genetic switch for the production of Z9-14Ac may also be involved in the reduction of Z9-14Acyl to the corresponding alcohol (Z9-14OH). Indeed, pgFARs resulting from A^E/A^Z alleles in ECB did not produce detectable amounts of Z9-14OH from Z9-14Acyl in the functional assay using a yeast expression system (Lassance et al. 2010), which is consistent with the nonproduction of Z9-14Ac in ECB. Given that A^E/A^Z, $B^{zea}/B^{zag}/B^{spz}$, and C^{zea}/C^{sca} are all involved in the same biosynthetic step, they are likely to be homologues encoding variant pgFARs. The locus of A^E/A^Z is mapped on chromosome 12 in ECB (Dopman et al. 2004). Corresponding gene maps for other Ostrinia species, including ZEA, ZAG, and SPZ, and comparisons with the map of ECB are needed to verify (or invalidate) the hypothesis that A^E/A^Z, $B^{zea}/B^{zag}/B^{spz}$, and C^{zea}/C^{sca} are homologues encoding pgFARs. Recently, functional assays of *pgFARs* cloned from several Ostrinia species showed that these reductases exhibit different substrate specificities that are the direct consequences of extensive non-synonymous substitutions in their amino acid sequences (Lassance et al. 2013). This study demonstrated that substrate specificity of pgFAR and the composition of precursors (FAPAs) in the pheromone gland can explain species-specific pheromone blends in several Ostrinia species including the members of *zealis* subgroup. Cloning of pheromone-gland-specific fatty acyl reductase genes in Ostrinia species is currently in progress (Antony et al. 2009; Lassance et al. 2010). Clarification of the substrate specificity of reductases in species of the *zealis* subgroup is critical to understanding the evolution of pheromone production in Group III species.

Pheromone-Based Reproductive Isolation in The Oriental *Ostrinia* Species

Several species of oriental Ostrinia are sympatric and some even share host plants. Moreover, adult emergence of most oriental Ostrinia species occurs two to three times a year from late April to early October (Hattori and Mutuura 1987) and the emergence of several species is at least partially synchronous. Under these circumstances, differences in the timing of pheromone release and/or pheromone composition could be essential for efficient mate-finding and the prevention of the attraction of heterospecific males.

In Ostrinia, sex pheromone release and copulation mostly occur late in the scotophase under laboratory conditions (Table 16.2), with the exception of ORI in the scapulalis subgroup of Group III. Females of ORI start calling from the end of the scotophase to the beginning of photophase, which is substantially later than other Ostrinia species (Ishikawa et al. 1999a). Calling behavior may function as a prezygotic isolating mechanism among other species, particularly the Z-type SCA, which uses the same pheromone components and the same blend ratio as ORI. However, calling behaviors in the early photophase are also observed in certain populations of SCA, suggesting that calling time is not rigid (Ishikawa et al. 1999a). In fact, isolation between ORI and SCA appears

incomplete because they hybridize under laboratory conditions and produce fertile progeny (Fu et al. 2005b). Females that produce dimorphic offspring with ORI and SCA morphology have been observed in the field, indicating (suggesting) that hybridization between the two species also occurs under natural conditions (Ohno 2000b). Clearly, the sex pheromone systems in the scapulalis subgroup do not act as a complete barrier to hybridization.

Regarding the use of the same sex pheromones, the E-type SCA uses the same pheromone as PAL, a Group II species distantly related to SCA. Although these two species share the same pheromone signal, the amount of pheromone released is considerably different. The total amount of pheromone components in the pheromone gland was 37.5 ng per female in PAL (Huang et al. 1998c), compared to 6.8 ng in SCA (Huang et al. 1997). Consistent with these findings, the optimal dosage of synthetic pheromone needed to attract males of PAL in the field was 0.5 mg (Huang et al. 1998c), whereas 0.05 mg was optimal for SCA and a 0.5 mg lure failed to attract SCA (Huang et al. 1997). Furthermore, PAL and SCA are not cross-attracted, even under laboratory conditions, perhaps because of the contribution of unknown minor components.

The sex pheromone of ACB is composed of unusual components (E12- and Z12-14Ac). The use of these unusual components is likely an effective prezygotic reproductive isolating mechanism. However, rare ECB males (3–5%) that respond to the ACB pheromone exist (Linn et al. 2003). Likewise, rare males of ACB (3–4%) attracted to mixtures of E11- and Z11-14Ac, which are the pheromones of the scapulalis subgroup (Linn et al. 2007), also exist. Electrophysiological studies have illustrated the mechanisms of cross-attraction in these rare ACB and ECB males (Domingue et al. 2007a, 2007b, 2008, 2010). The ECB pheromone components (E11- and Z11-14Ac) stimulate ORNs that induce behavioral antagonism in normal ACB males, but they do not stimulate corresponding ORNs in the rare-type ACB males (Domingue et al. 2007a). Furthermore, ORNs that respond to E11- and Z11-14Ac in normal ECB males are shown to respond to the ACB pheromone components (E12- and Z12-14Ac) in the rare ECB males. Behavioral responses of rare ECB males to ACB pheromones were attributed to the similarity of the firing pattern of two different types of ORNs induced in response to ACB and ECB pheromones (Domingue et al. 2007b). Recently, Leary et al. (2012) found a mechanism that may have contributed to the shift of pheromone recognition systems in Ostrinia. ECB odorant receptor 3 (OR3), a sex pheromone receptor expressed in male ECB antennae, responds strongly to E11-14Ac but also generally to Z11-14Ac and E/Z12-14Ac. Interestingly, OR3 in ACB, which differs from ECB OR3 in the amino acid residue at position 148 (alanine in ECB while threonine in ACB), responds preferentially to E/Z12-14Ac. Discrete mutations that narrow the specificity of more broadly responsive sex pheromone receptors may have occurred in Ostrinia (Leary et al. 2012). Divergence of the Δ12-pheromone from the Δ11-pheromone could have been triggered by a saltational shift in the functionality of desaturase genes (Baker 2002), followed by an increase of "rare" males via an asymmetric tracking process (Phelan 1992, 1997a, 1997b).

Differences in the number of pheromone components may isolate members of the latipennis subgroup (LAT and OVA) and members of the scapulalis subgroup (SCA/ORI) and zealis subgroup (ZEA/ZAG/SPZ). Cross-attraction between LAT and OVA was not observed in a field trapping experiment (E11-14OH for LAT and E11-14OH + E11-14Ac for OVA), suggesting that the acetate (E11-14Ac) is essential for OVA but antagonistic to LAT (Ohno 2003b). Similarly, Z9-14Ac, included exclusively in the pheromones of the zealis subgroup, acts as an agonist in the zealis subgroup (Huang et al. 1998a; Ishikawa et al. 1999b; Tabata et al. 2008) and an antagonist in the scapulalis group (Ishikawa et al. 1999b). The "switch" in the behavioral response from preference to aversion, or from aversion to preference, is of interest, although we have almost no understanding of the neurophysiological and molecular bases of this switch.

It is noteworthy that the three species of the zealis subgroup that employ the three-component pheromone system are specialized to feed on a limited number of Asteraceae plants, whereas other members in Group III that employ the two-component system are polyphagous, feeding on a variety of host plants including Asteraceae. The increase in the complexity of the sex pheromone communication system might have facilitated host plant specialization through the formation of unique "sexual communication channels" (Löfstedt and van der Pers 1985) in the zealis subgroup. Within the zealis subgroup, prezygotic reproductive isolation between ZEA and the other two species is mediated by differences in pheromone blend ratios (figure 16.2). No cross-attraction occurred in wind-tunnel assays (Ishikawa et al. 1999b; Tabata 2010), and heterospecific mating was difficult to induce even under laboratory conditions (35% in ZAG × ZEA, 8% in SPZ × ZEA, and 0% in ZEA × ZAG/SPZ; Tabata and Ishikawa 2011). In contrast, prezygotic reproductive isolation between ZAG and SPZ is somewhat unstable, with some overlap in the proportion of the double-bond positional isomers (Z9- and E/Z11-14Ac) used by the two species. Cross-attraction is frequently observed in wind-tunnel assays (40–60%) and heterospecific mating is successfully achieved in both reciprocal crosses (30–40%; Tabata and Ishikawa 2011). Moreover, hybrids between ZAG and SPZ are not very rare (>5%) in wild populations (Tabata 2010). Despite incomplete differentiation of pheromones and a relatively high frequency of hybridization, traits that characterize each species (morphology and host preference) are firmly associated. The sex pheromone communication systems of ZEA and ZAG/SPZ may currently be best explained by the stasis model *sensu* Allison and Cardé (2008), whereas the systems in ZAG and SPZ appear to be best explained by the asymmetric tracking model.

Reconstruction of Evolutionary Histories of Pheromone Divergence in *Ostrinia*

The sex pheromones of the latipennis subgroup (LAT and OVA) in Group II are characteristic in that an alcohol is employed as a component and the geometric isomerism of double bond in the pheromone components is absolutely *E* configuration. It remains unclear whether these unique characters in this subgroup are ancestral or derived. Monophyly of Group II is not strongly supported by mitochondrial DNA-based phylogenetic analyses (Kim et al. 1999). Mutuura and Munroe (1970) inferred that LAT is relatively primitive on the bases of its morphology and host plant usage. If we assume that Group II consists of a single cluster, the acetate pheromones including E11-14Ac should have been used in the common ancestor of the genus *Ostrinia*, and the alcohol pheromones of the latipennis subgroup would be derived. In this scenario, the function of acetyltransferases was partially lost in OVA and subsequently completely lost in LAT. As a consequence, OVA and LAT began to use E11-14OH, which was formerly a precursor in the ancestral pheromones. Also in this scenario, LATPG1, a Δ11-desaturase cloned from LAT that only produces the (*E*)-isomer, is a derivative enzyme. This mutation might have acted to diminish the diversity of pheromone components, and so did not prevail in *Ostrinia*. As already mentioned, this mutation has occurred in an ancient Δ11-desaturase gene, which was generated by gene duplications and was cryptically conserved in the genome by subsequent fusion with a retroposon (Xue et al. 2007, 2012). Hence, the unique pheromone systems of the latipennis subgroup may be an "evolutionary dead end." Meanwhile, Miura et al. (2009) suggests another possibility through cloning and functional assays of orthologous olfactory receptor genes from LAT and SCA, *OlatOR1* and *OscaOR1*, respectively. These genes were shown to encode receptor proteins, which show characteristic responses to E11-14OH. Interestingly, orthologous genes have been discovered from six *Ostrinia* members that do not use E11-14OH in their pheromones. This may imply that the common ancestor of *Ostrinia* used E11-14OH as a pheromone component before species radiation (Miura et al. 2009). Further intensive phylogenetic analyses and pheromone identification in other *Ostrinia* species, particularly in putative ancestors including *O. penitalis*, *O. obumbratalis*, *O. erythrialis*, *O. marginalis*, *O. peregrinalis*, *O. kasmirica*, *O. quadripunctalis*, and *O. sanguinealis*, are necessary to test these hypotheses.

The members of Group III form a monophyletic cluster with a bootstrap value of 100% in a molecular phylogeny (Kim et al. 1999). ACB is rather distantly positioned from the other members of Group III, and is inferred to have diverged approximately 1 million years ago (Roelofs et al. 2002). The sex pheromone components E12- and Z12-14Ac involve a Δ14-desaturase and are unique to the ACB pheromone system, suggesting that they are derived from an ancestral pheromone utilizing products of Δ11-desaturase. Divergence of desaturase genes including Δ11- and Δ14-desaturases is considered to have occurred long before the divergence of *Ostrinia* species (Roelofs and Rooney 2003). Because the genes of Δ11- and Δ14-desaturases are commonly present in both ACB and ECB/SCA, alternative utilization of desaturase genes has resulted in a divergence of the ACB pheromones from the others. Acquisition of this unique pheromone system may have driven a very rapid divergence of ACB from other congeners; a mitochondrial gene sequence of ACB is clearly segregated from the others (Kim et al. 1999), suggesting low introgression of genes during the evolutionary process.

The scapulalis subgroup that uses the two-component pheromones and the zealis subgroup that uses the three-component pheromones form monophyletic clusters in a phylogeny inferred from combined mitochondrial and nuclear DNA sequences, although the zealis subgroup shows paraphyly when only mitochondrial sequences are used for the inference (Kim et al. 1999). These two subgroups appear to have diverged very recently, and it is currently unknown which of the pheromone systems is basal. The shift in the number of pheromone components between the two subgroups may be a direct result of changes in the substrate specificity of the reductase and secondarily to changes in β-oxidation enzymes, which affect the conversion of Z11-16Acyl to Z9-14Acyl precursors (Tabata and Ishikawa 2005). If the two-component pheromone is ancestral, addition of Z9-14Ac would have likely occurred in the common ancestor of the zealis subgroup. Because primitive *Ostrinia* species (Group II) do not use Z9-14Ac in their pheromones, this scenario is currently most parsimonious with

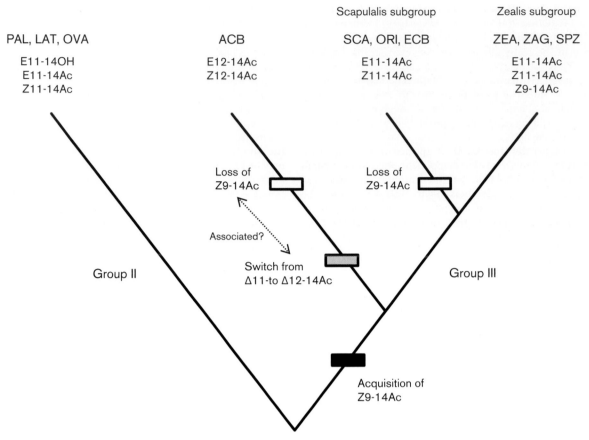

PAL, LAT, OVA	ACB	Scapulalis subgroup SCA, ORI, ECB	Zealis subgroup ZEA, ZAG, SPZ
E11-14OH E11-14Ac Z11-14Ac	E12-14Ac Z12-14Ac	E11-14Ac Z11-14Ac	E11-14Ac Z11-14Ac Z9-14Ac

Loss of Z9-14Ac

Loss of Z9-14Ac

Associated?

Group II

Switch from Δ11-to Δ12-14Ac

Group III

Acquisition of Z9-14Ac

FIGURE 16.3 Tentative reconstruction of the evolutionary history of the sex pheromone components in *Ostrinia* Group III. The tree displays only topology of the phylogenic relationships inferred from integrated data of mitochondrial and nuclear DNA sequences.

respect to current patterns of the use of Z9-14Ac. Alternatively, the common ancestor of the zealis and the scapulalis subgroups might have used the three-component pheromone. In this scenario Z9-14Ac production/perception would have evolved in an ancestral species of Group III. The use of Z9-14Ac would have been subsequently lost in derived species that use two-component pheromones, forming the current scapulalis subgroup (figure 16.3). Although Z9-14Ac is used as a pheromone component only in members of the zealis subgroup, it acts as a strong antagonist in all other pheromone communication systems of Group III (Hansson et al. 1987; Glover et al. 1989; Cossé et al. 1995; Ishikawa et al. 1999b; Takanashi et al. 2006; Linn et al. 2007). Moreover, a pheromone receptor gene (*OscaOR3*) that may account for responses to Z9-14Ac was cloned and is conserved in male antennae of all members of oriental *Ostrinia* moths (Miura et al. 2010). It is reasonable to assume that the receptor for a component used in the ancestral species would be retained even after females ceased to release this component. Reception of Z9-14Ac might have been newly acquired to avoid heterospecific attractions, but this is unlikely in ACB, which uses unique pheromone components and thus has no need to evolve additional measures for the avoidance of heterospecific females. If we assume that the three-component pheromone is an ancestral trait in Group III, the loss of Z9-14Ac should have occurred at least twice: once during the divergence of the ACB pheromone and once during the divergence of the scapulalis subgroup pheromones. The loss of Z9-14Ac from the ACB pheromone may be linked to the evolutionary shift from Δ11-14Ac to Δ12-14Ac, which includes non-

functionalization of the Δ11-desaturase, a key enzyme for the biosynthesis of Z9-14Ac (figure 16.1).

Further investigations on the molecular mechanisms of pheromone production and perception combined with biochemical and ecological studies on non-pheromonal traits may establish how pheromone channels in this intriguing genus have diverged and are maintained.

References Cited

Allison, J.D., and R.T. Cardé. 2008. Male pheromone blend preference function measured in choice and no-choice wind tunnel trials with almond moths, *Cadra cautella*. *Animal Behaviour* 75:259–266.

Ando, T. 2012. Sex pheromones of moths. Available online at: http://www.tuat.ac.jp/~antetsu/LepiPheroList.htm.

Ando, T., O. Saito, K. Arai, and N. Takahashi. 1980. (*Z*)- and (*E*)-12-tetradecenyl acetates: sex pheromone components of oriental corn borer (Lepidoptera: Pyralidae). *Agricultural Biology and Chemistry* 44:2643–2649.

Antony, B., T. Fujii, K. Moto, S. Matsumoto, M. Fukuzawa, R. Nakano, S. Tatsuki, and Y. Ishikawa. 2009. Pheromone-gland-specific fatty-acyl reductase in the adzuki bean borer, *Ostrinia scapulalis* (Lepidoptera: Crambidae). *Insect Biochemistry and Molecular Biology* 39:90–95.

Baker, T.C. 2002. Mechanism for saltational shifts in pheromone communication systems. *Proceedings of the National Academy of Sciences of the United States of America* 99:13368–13370.

Bengtsson, B.O., and C. Löfstedt. 1990. No evidence for selection in a pheromonally polymorphic moth population. *American Naturalist* 136:722–726.

Bontemps, A., D. Bourguet, L. Pélozuelo, M.-T. Bethenod, and S. Ponsard. 2004. Managing the evolution of *Bacillus thuringiensis* resistance in natural populations of the European corn borer, *Ostrinia nubilalis*: host plant, host race and pherotype of adult males at aggregation sites. *Proceedings of the Royal Society Series B: Biological Sciences* 271:2179–2185.

Cheng, Z. Q., J. C. Xiao, X. T. Huang, D. L. Chen, Q. J. Li, Y. S. He, S. R. Huang, Q. C. Luo, C. M. Yang, and T. H. Yang. 1981. Sex pheromone components isolated from China corn borer, *Ostrinia furnacalis*, Guenée (Lepidoptera: Pyralidae), (Z)- and (E)-12-tetradecenyl acetates. *Journal of Chemical Ecology* 7:841–851.

Cossé, A. A., M. G. Capbell, T. J. Glover, C. E. Linn, Jr., J. L. Todd, T. C. Baker, and W. L. Roelofs. 1995. Pheromone behavioral responses in unusual male European corn borer hybrid progeny not correlated to electrophysiological phenotypes of their pheromone-specific antennal neurons. *Experientia* 51:809–816.

Domingue, M. J., C. J. Musto, C. E. Linn, Jr., W. L. Roelofs, and T. C. Baker. 2007a. Evidence of olfactory antagonistic release as a facilitator of evolutionary shifts in pheromone blend usage in *Ostrinia* spp. (Lepidoptera: Crambidae). *Journal of Insect Physiology* 53:488–496.

Domingue, M. J., C. J. Musto, C. E. Linn, Jr., W. L. Roelofs, and T. C. Baker. 2007b. Altered olfactory receptor neuron responsiveness in rare European corn borer males (*Ostrinia nubilalis*) attracted to Asian corn borer (*O. furnacalis*) pheromone blend. *Journal of Insect Physiology* 53:1063–1071.

Domingue, M. J., C. J. Musto, C. E. Linn, Jr., W. L. Roelofs, and T. C. Baker. 2008. Olfactory neuron responsiveness and pheromone blend preference in hybrids between *Ostrinia furnacalis* and *O. nubilalis* (Lepidoptera: Crambidae). *Journal of Insect Physiology* 54:1261–1270.

Domingue, M. J., C. J. Musto, C. E. Linn, Jr., W. L. Roelofs, and T. C. Baker. 2010. Homology of olfactory receptor neuron response characteristics inferred from hybrids between Asian and European corn borer moths (Lepidoptera: Crambidae). *Journal of Insect Physiology* 56:73–80.

Dopman E. B., S. M. Bogdanowicz, and Harrison R. G. 2004. Genetic mapping of sexual isolation between E and Z pheromone strains of the European corn borer (*Ostrinia nubilalis*). *Genetics* 167:301–309.

El-Sayed, A. M. 2012. The Pherobase: Database of Pheromones and Semiochemicals. Available online at: http://www.pherobase.com.

Frolov, A. N., D. Bourguet, and S. Ponsard. 2007. Reconsidering the taxonomy of several *Ostrinia* species in the light of reproductive isolation: a tale for Ernst Mayr. *Biological Journal of the Linnean Society* 91:49–72.

Frolov, A. N., P. Audiot, D. Bourguet, A. G. Kononchuk, J. M. Malysh, S. Ponsard, R. Streiff, and Y. S. Tokarev. 2012. From Russia with lobe: genetic differentiation in trilobed uncus *Ostrinia* spp. follows food plant, not hairy legs. *Heredity* 108:147–156.

Fu, X., J. Tabata, Y. Huang, T. Takanashi, S. Ohno, H. Honda, S. Tatsuki, and Y. Ishikawa. 2004. Female sex pheromone of *Ostrinia orientalis*—throwing a light on the relationship between *O. orientalis* and the European corn borer, *O. nubilalis*. *Chemoecology* 14:175–180.

Fu, X., M. Fukuzawa, J. Tabata, S. Tatsuki, and Y. Ishikawa. 2005a. Sex pheromone biosynthesis in *Ostrinia zaguliaevi*, a congener of the European corn borer moth *O. nubilalis*. *Insect Biochemistry and Molecular Biology* 35:621–626.

Fu, X., S. Tatsuki, S. Hoshizaki, and Y. Ishikawa. 2005b. Study of the genetics of female sex pheromone production and male behavioral response in a moth, *Ostrinia orientalis*. *Entomological Science* 8:363–369.

Fujii, T., K. Ito, M. Tatematsu, T. Shimada, S. Katsuma, and Y. Ishikawa. 2011. A sex pheromone desaturase functioning in a primitive *Ostrinia* moth is cryptically conserved in congeners' genomes. *Proceedings of the National Academy of Sciences of the United States of America* 108:7102–7106.

Glover, T. J., N. Perez, and W. L. Roelofs. 1989. Comparative analysis of sex-pheromone-response antagonists in three races of European corn borer. *Journal of Chemical Ecology* 15:863–873.

Glover, T. J., J. J. Knodel, P. S. Robbins, C. J. Eckenrode, and W. L. Roelofs. 1991. Gene flow among three races of European corn borers (Lepidoptera: Pyralidae) in New York State. *Environmental Entomology* 20:1356–1362.

Glover, T. J., P. S. Robbins, C. J. Eckenrode, and W. L. Roelofs. 1992. Genetic control of voltinism characteristics in European corn

borer races assessed with a marker gene. *Archives of Insect Biochemistry and Physiology* 20:107–117.

Hansson, B. S., C. Löfstedt, and W. L. Roelofs. 1987. Inheritance of olfactory response to sex pheromone components in *Ostrinia nubilalis*. *Naturwissenschaften* 74:497–499.

Hattori, I. and A. Mutuura. 1987. Identification of Japanese species belonging to the genus *Ostrinia* with the host relationships. *Plant Protection* 41:24–31. [In Japanese]

Huang, Y., S. Tatsuki, C. Kim, S. Hoshizaki, Y. Yoshiyasu, H. Honda, and Y. Ishikawa. 1997. Identification of sex pheromone of adzuki bean borer, *Ostrinia scapulalis*. *Journal of Chemical Ecology* 23:2791–2802.

Huang, Y., H. Honda, Y. Yoshiyasu, S. Hoshizaki, S. Tatsuki, and Y. Ishikawa. 1998a. Sex pheromone of the butterbur borer, *Ostrinia zaguliaevi*. *Entomologia Experimentalis et Applicata* 89:281–287.

Huang, Y., T. Takanashi, S. Hoshizaki, S. Tatsuki, H. Honda, Y. Yoshiyasu, and Y. Ishikawa. 1998b. Geographic variation in the sex pheromone of Asian corn borer, *Ostrinia furnacalis*. *Journal of Chemical Ecology* 24:2079–2088.

Huang, Y., S. Tatsuki, C. Kim, S. Hoshizaki, and Y. Ishikawa. 1998c. Identification of the sex pheromone of *Ostrinia palustralis*. *Entomologia Experimentalis et Applicata* 86:313–318.

Huang, Y., T. Takanashi, S. Hoshizaki, S. Tatsuki, and Y. Ishikawa. 2002. Female sex pheromone polymorphism in adzuki bean borer, *Ostrinia scapulalis*, is similar to that in European corn borer, *O. nubilalis*. *Journal of Chemical Ecology* 28:533–539.

Ishikawa, Y., T. Takanashi, C. Kim, S. Hoshizaki, S. Tatsuki, and Y. Huang. 1999a. *Ostrinia* spp. in Japan: their host plants and sex pheromones. *Entomologia Experimentalis et Applicata* 91: 237–244.

Ishikawa, Y., T. Takanashi, and Y. Huang. 1999b. Comparative studies on the sex pheromones of *Ostrinia* spp. in Japan: the burdock borer, *Ostrinia zealis*. *Chemoecology* 9:25–32.

Kim, C., S. Hoshizaki, Y. Huang, S. Tatsuki, and Y. Ishikawa. 1999. Usefulness of mitochondrial COII gene sequences in examining phylogenetic relationships in the Asian corn borer, *Ostrinia furnacalis*, and allied species (Lepidoptera: Pyralidae). *Applied Entomology and Zoology* 34:405–412.

Klun, J. A., and M. D. Huettel. 1988. Genetic regulation of sex pheromone production and response: interaction of sympatric pheromonal types of the European corn borer, *Ostrinia nubilalis* (Lepidoptera: Pyralidae). *Journal of Chemical Ecology* 14:2047–2061.

Klun, J. A., and J. F. Robinson. 1972. Olfactory discrimination in the European corn borer and several pheromonally analogous moths. *Annals of the Entomological Society of America* 65:1337–1340.

Klun, J. A., B. A. Bierl-Leonhardt, M. Schwarz, L. A. Litsinger, A. T. Barrion, H. C. Chiang, and Z. Jiang. 1980. Sex pheromone of the Asian corn borer moth. *Life Science* 27:1603–1606.

Kou, R. H., H. Y. Ho, H. T. Yang, Y. S. Chow, and H. J. Wu. 1992. Investigation of sex pheromone components of female Asian corn borer, *Ostrinia furnacalis* (Hübner) (Lepidoptera: Pyralidae) in Taiwan. *Journal of Chemical Ecology* 18:833–840.

Lassance, J.-M. 2010. Journey in the *Ostrinia* world: from pest to model in chemical ecology. *Journal of Chemical Ecology* 36:1155–1169.

Lassance, J.-M., A. T. Groot, M. A. Liénard, B. Antony, C. Borgwardt, F. Andersson, E. Hedenström, D. G. Heckel, and C. Löfstedt. 2010. Allelic variation in a fatty-acyl reductase gene causes divergence in moth sex pheromones. *Nature* 466:486–489.

Lassance, J.-M., M. A. Liénard, B. Antony, S. Qian, T. Fujii, J. Tabata, Y. Ishikawa, and C. Löfstedt. 2013. Functional consequences of sequence variation in the pheromone biosynthetic gene pgFAR for *Ostrinia* moths. *Proceedings of the National Academy of Sciences of the United States of America* 110:3967–3972.

Leary, G. P., J. E. Allen, P. L. Bunger, J. B. Luginbill, C. E. Linn, Jr., I. E. Macallister, M. P. Kavanaugh, and K. W. Wanner. 2012. Single mutation to a sex pheromone receptor provides adaptive specificity between closely related moth species. *Proceedings of the National Academy of Sciences of the United States of America* 109:14081–14086.

Linn, C. E., Jr., M. O'Connor, and W. L. Roelofs. 2003. Silent genes and rare males: a fresh look at pheromone blend response specificity in the European corn borer moth, *Ostrinia nubilalis*. *Journal of Insect Science* 3:15.

Linn, C. E., Jr., C. Musto, and W. L. Roelofs. 2007. More rare males in *Ostrinia*: response of Asian corn borer moths to the sex pheromone of the European corn borer. *Journal of Chemical Ecology* 33:199–212.

Löfstedt, C., and J. N. C. van der Pers. 1985. Sex pheromones and reproductive isolation in four European small ermine moths. *Journal of Chemical Ecology* 11:649–666.

Ma, P. W. K., and W. L. Roelofs. 2002. Sex pheromone gland of the female European corn borer moth, *Ostrinia nubilalis* (Lepidoptera, Pyralidae): ultrastructural and biochemical evidences. *Zoological Science* 19:501–511.

Malausa, T., M.-T. Bethenod, A. Bontemps, D. Bourguet, J.-M. Cornuet, and S. Ponsard. 2005. Assortative mating in sympatric host races of the European corn borer. *Science* 308:258–260.

Martel, C., A. Réjasse, F. Rousset, M.-T. Bethenod, and D. Bourguet. 2003. Host-plant associated genetic differentiation in Northern French populations of the European corn borer. *Heredity* 90:141–149.

Miura, N., T. Nakagawa, S. Tatsuki, K. Touhara, and Y. Ishikawa. 2009. A male-specific odorant receptor conserved through the evolution of sex pheromones in *Ostrinia* moth species. *International Journal of Biological Sciences* 5:319–330.

Miura, N., T. Nakagawa, K. Touhara, and Y. Ishikawa. 2010. Broadly and narrowly tuned odorant receptors are involved in female sex pheromone reception in *Ostrinia* moths. *Insect Biochemistry and Molecular Biology* 40:64–73.

Mutuura, A., and E. Munroe. 1970. Taxonomy and distribution of the European corn borer and allied species: genus *Ostrinia* (Lepidoptera: Pyralidae). *Memories of the Entomological Society of Canada* 71:1–112.

Ohno, S. 2000a. A case of host expansion in the Far Eastern knotweed borer, *Ostrinia latipennis* (Lepidoptera, Crambidae, Pyraustinae). *Transactions of the Lepidopterological Society of Japan* 51:202–204.

Ohno, S. 2000b. Emergence of two nominal species, *Ostrinia scapulalis* and *O. orientalis*, from a single brood (Lepidoptera: Crambidae). *Entomological Science* 3:635–637.

Ohno, S. 2000c. New host records for *Ostrinia orientalis* (Lepidoptera, Crambidae, Pyraustinae). *Transactions of the Lepidopterological Society of Japan* 51:44–48.

Ohno, S. 2003a. A new knotweed-boring species of the genus *Ostrinia* Hübner (Lepidoptera: Crambidae) from Japan. *Entomological Science* 6:77–83.

Ohno, S. 2003b. Systematic study on the *Ostrinia latipennis* group (Lepidoptera: Crambidae), based on morphology, DNA and sex pheromones. PhD dissertation, University of Tokyo, Tokyo.

Pélozuelo, L., C. Malosse, G. Genestier, H. Guenego, and B. Frérot. 2004. Host-plant specialization in pheromone strains of the European corn borer *Ostrinia nubilalis* in France. *Journal of Chemical Ecology* 30:335–352.

Pélozuelo, L., S. Meusnier, P. Audiot, D. Bourguet, and S. Ponsard. 2007. Assortative mating between European corn borer pheromone races: beyond assortative meeting. *PLoS ONE* 2:e555.

Phelan, P. L. 1992. Evolution of sex pheromones and the role of asymmetric tracking. Pp. 563–579. In R. T. Cardé and A. K. Minks, eds. *Insect Chemical Ecology: An Evolutionary Approach*. New York: Chapman & Hall.

Phelan, P. L. 1997a. Genetics and phylogenetics in the evolution of sex pheromones. Pp. 265–314. In B. D. Roitberg and M. B. Isman, eds. *Insect Pheromone Research: New Directions*. New York: Chapman & Hall.

Phelan, P. L. 1997b. Evolution of mate-signaling in moths: phylogenetic considerations and predictions from the asymmetric tracking hypothesis. Pp. 240–256. In J. C. Choe and B. J. Crespi, eds. *Mating Systems in Insects and Arachnids*. Cambridge: Cambridge University Press.

Roelofs, W. L., and A. P. Rooney. 2003. Molecular genetics and evolution of pheromone biosynthesis in Lepidoptera. *Proceedings of the National Academy of Sciences of the United States of America* 100:9179–9184.

Roelofs, W. L., J. W. Du, X. H. Tang, P. S. Robbins, and C. J. Eckenrode. 1985. Three European corn borer populations in New York based on sex pheromones and voltinism. *Journal of Chemical Ecology* 11:829–836.

Roelofs, W. L., T. J. Glover, X. H. Tang, I. Sreng, P. S. Robbins, C. J. Eckenrode, C. Löfstedt, B. S. Hansson, and B. O. Bengtsson. 1987. Sex pheromone production and perception in European corn borer moths is determined by both autosomal and sex-linked genes. *Proceedings of the National Academy of Sciences of the United States of America* 84:7585–7589.

Roelofs, W. L., W. Liu, G. Hao, H. Jiao, A. P. Rooney, and C. E. Linn, Jr. 2002. Evolution of moth sex pheromones via ancestral genes. *Proceedings of the National Academy of Sciences of the United States of America* 99:13621–13626.

Sakai, R., M. Fukuzawa, R. Nakano, S. Tatsuki, and Y. Ishikawa. 2009. Alternative suppression of transcription from two desaturase genes is the key for species-specific sex pheromone biosynthesis in two *Ostrinia* moths. *Insect Biochemistry and Molecular Biology* 39:62–67.

Tabata, J. 2010. Studies on inter- and intra-specific variations and their mode of inheritance in sex pheromone production and response of moths. PhD dissertation, University of Tokyo, Tokyo.

Tabata, J., and Y. Ishikawa. 2005. Genetic basis to divergence of sex pheromones in two closely related moths, *Ostrinia scapulalis* and *O. zealis*. *Journal of Chemical Ecology* 31:1111–1124.

Tabata, J., and Y. Ishikawa. 2011. Genetic basis regulating the sex pheromone blend in *Ostrinia zealis* (Lepidoptera: Crambidae) and its allies inferred from crossing experiments. *Annals of the Entomological Society of America* 104:326–336.

Tabata, J., T. Takanashi, and Y. Ishikawa. 2003. Pheromone analysis of wild female moths with a PBAN C-terminal peptide injection for an estimation of assortative mating in adzuki bean borer, *Ostrinia scapulalis*. *Journal of Chemical Ecology* 29:2749–2758.

Tabata, J., S. Hoshizaki, S. Tatsuki, and Y. Ishikawa. 2006. Heritable sex pheromone blend variation in a local population of the butterbur borer moth *Ostrinia zaguliaevi* (Lepidoptera: Crambidae). *Chemoecology* 16:123–128.

Tabata, J., Y. Huang, S. Ohno, Y. Yoshiyasu, H. Sugie, S. Tatsuki, and Y. Ishikawa. 2008. Sex pheromone of *Ostrinia* sp. newly found on the leopard plant *Farfugium japonicum*. *Journal of Applied Entomology* 132:566–574.

Takanashi, T. 2001. Genetics of sex pheromone production and response in the adzuki bean borer, *Ostrinia scapulalis*. PhD dissertation, University of Tokyo, Tokyo.

Takanashi, T., S. Ohno, Y. Huang, S. Tatsuki, and Y. Ishikawa. 2000. A sex pheromone component novel to *Ostrinia* identified from *Ostrinia latipennis* (Lepidoptera: Crambidae). *Chemoecology* 10:143–147.

Takanashi, T., Y. Huang, R. Takahashi, S. Hoshizaki, S. Tatsuki, and Y. Ishikawa. 2005. Genetic analysis and population survey of sex pheromone variation in the adzuki bean borer moth, *Ostrinia scapulalis*. *Biological Journal of the Linnean Society* 84:143–160.

Takanashi, T., Y. Ishikawa, P. Anderson, Y. Huang, C. Löfstedt, S. Tatsuki, and B. S. Hansson. 2006. Unusual response characteristics of pheromone-specific olfactory receptor neurons in the Asian corn borer moth, *Ostrinia furnacalis*. *Journal of Experimental Biology* 209:4946–4956.

Thomas, Y., M.-T. Bethenod, L. Pélozuelo, B. Frérot, and D. Bourguet. 2003. Genetic isolation between host plant races of the European corn borer. I—sex pheromone, moth emergence timing and parasitism. *Evolution* 57:261–273.

Xue, B., A. P. Rooney, M. Kajiwara, N. Okada, and W. L. Roelofs. 2007. Novel sex pheromone desaturases in the genomes of corn borers generated through gene duplication and retroposon fusion. *Proceedings of the National Academy of Sciences of the United States of America* 104:4467–4472.

Xue, B., A. P. Rooney, and W. L. Roelofs. 2012. Genome-wide screening and transcriptional profile analysis of desaturase genes in the European corn borer moth. *Insect Science* 19:55–63.

Yeh, S., K. Lee, K.-T. Chang, F.-C. Yen, and J. S. Hwang. 1989. Sex pheromone components from Asian corn borer, *Ostrinia furnacalis* (Hübner) (Lepidoptera: Pyralidae) in Taiwan. *Journal of Chemical Ecology* 15:497–505.

Zhao, C.-H., C. Löfstedt, and X. Y. Wang. 1990. Sex-pheromone biosynthesis in the Asian corn borer *Ostrinia furnacalis* (II): biosynthesis of (*E*)-12-tetradecenyl and (*Z*)-12-tetradecenyl acetate involves delta-14 desaturation. *Archives of Insect Biochemistry and Physiology* 15:57–65.

Zhao, C.-H., F. Lu, M. Bengtsson, and C. Löfstedt. 1995. Substrate specificity of acetyltransferase and reductase enzyme systems used in pheromone biosynthesis by Asian corn borer, *Ostrinia furnacalis*. *Journal of Chemical Ecology* 21:1495–1509.

Zhu, J., C. Löfstedt, and B.O. Bengtsson. 1996a. Genetic variation in the strongly canalized sex pheromone communication system of the European corn borer, *Ostrinia nubilalis* Hübner (Lepidoptera; Pyralidae). *Genetics* 144:757–766.

Zhu, J., C.-H. Zhao, F. Lu, B.O. Bengtsson, and C. Löfstedt. 1996b. Reductase specificity and the ratio regulation of E/Z isomers in pheromone biosynthesis of the European corn borer, *Ostrinia nubilalis* (Lepidoptera: Pyralidae). *Insect Biochemistry and Molecular Biology* 26:171–176.

Utetheisa ornatrix (Erebidae, Arctiinae)

A Case Study of Sexual Selection

VIKRAM K. IYENGAR and WILLIAM E. CONNER

Pyrrolizidine Alkaloids: Chemical Defense and Links to Large Size

Utetheisa ornatrix, also known as the rattlebox moth, is a brightly colored erebid moth that relies on chemistry for both defense and communication, and the complex reproductive strategies of both sexes have made it a model system for studying sexual selection. *U. ornatrix* (henceforth referred to as *Utetheisa*) ranges in the Americas from the Rocky Mountains to the Atlantic coast, and from North Carolina to the northern parts of South America (Pease 1968), and can be reliably found in the dry, scrub habitat near the Archbold Biological Station in central Florida. As a larva, *Utetheisa* feeds on plants of the genus *Crotalaria* (family Fabaceae), which contain poisonous pyrrolizidine alkaloids (PAs) that play a fundamental role in the life of both sexes (Eisner and Meinwald 2003). Larvae store PAs systemically, retaining them through metamorphosis into the adult stage. At mating, the male transfers up to 11% of his body mass in a spermatophore containing sperm, nutrients, and PAs (LaMunyon and Eisner 1994). The female transmits PAs to the eggs, using not only some of the PAs she herself sequestered, but also PAs that she receives from the male via the spermatophore at mating (Dussourd et al. 1988). As a result of having PAs in their bodies, all life stages of *Utetheisa* are protected against natural enemies, the

adults and larvae against spiders (Eisner and Eisner 1991) and the eggs against ants (Hare and Eisner 1993), chrysopid larvae (Eisner et al. 2000), coccinellid beetles (Dussourd et al. 1988), and parasitoid wasps (Bezzerides 2004).

PAs are potent phagostimulants, and the presence of these compounds in their diet causes larvae to invest more time feeding, thereby growing more quickly than those on diet without PAs (del Campo et al. 2005). These data are also supported by the fact that PAs and adult body size are strongly correlated in *Utetheisa* (Conner et al. 1990). In other words, feeding on PAs during the larval period is, to some degree, responsible for the attainment of large size, which has important reproductive consequences for both sexes.

Courtship and the Role of Pheromones

The female sex attractant of *Utetheisa* is a blend of long-chained polyenes (Conner et al. 1980; Huang et al. 1983; Jain et al. 1983; Choi et al. 2007; Lim et al. 2007). The female pheromone is released through the rhythmic exposure of tubular glands and triggers the upwind flight of males at a distance (Conner et al. 1980). One of the most intriguing aspects of the female's reproductive behavior is that they respond to the pheromone of nearby female conspecifics and release more rigorously, forming "female pheromone choruses" (Lim and Greenfield 2007; Lim et al. 2007). Although there are many interesting and unexplored areas of research regarding the female pheromone, this case study will focus on *Utetheisa* as a model system for understanding the evolution of male pheromones through female mate choice.

Once in close proximity of a female, the male releases a male courtship pheromone, hydroxydanaidal (HD), that he derives chemically from the acquired PAs (figure 17.1). The chemical is more volatile than PAs, and is aired by the male from two brush-like structures (coremata) that he everts near the antennae of the female (Conner et al. 1981). Females are capable of distinguishing between males differing by less than 2 micrograms of HD, and use this pheromone as the sole criterion of choice (Iyengar et al. 2001). Mate choice mediated

Monocrotaline

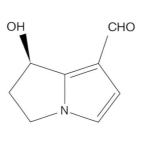

R- Hydroxydanaidal

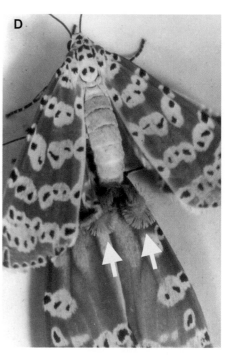

FIGURE 17.1 *Utetheisa ornatrix*: (A) Adult photographed at the Archbold Biological Station; (B) larva resting on the seed pod of *Crotalaria mucronata*, a pyrrolizidine alkaloid-containing hostplant; (C) scanning electron micrograph of the partially exposed genital coremata of a male; (D) corematal eversion captured in courtship (arrows highlight coremata). Chemical structures: (top) monocrotaline, an example of a pyrrolizidine alkaloid; (bottom) *R*-hydroxydanaidal, the male courtship pheromone.

by HD has been observed in laboratory tests, in which males with different amounts of the pheromone were offered simultaneously to females for comparative assessment (Iyengar and Eisner 1999b), and in field experiments, in which males approached females sequentially and were therefore subjected to independent assessment (Conner et al. 1981). Although PA-free (and therefore HD-free) males were not entirely rejected in these experiments, they were significantly less successful than PA-containing males. Females failed to differentiate between males of different size or PA content if these males lacked HD, but they expressed a preference for males bearing HD, even when these were size matched and PA free (Iyengar

et al. 2001). Thus, HD appears to be the only criterion on which the female bases her decision to mate.

Benefits to Female Precopulatory Mate Choice

The fitness advantages that females derive from mating with multiple males are broadly classified as direct or indirect benefits. Direct (phenotypic) benefits are the result of nongenetic quantities that have a positive impact on the survivorship and offspring production of the female such as enhanced paternal care (Davies 1992), the transmission of antipredator defensive chemicals (González et al. 1999), and the acquisition of nutrient-rich nuptial gifts (Thornhill and Alcock 1983; Gwynne 1984). Indirect (genetic) benefits are those that are experienced in the next generation in the form of increased offspring survivorship ("good genes") and mating success of sons ("sexy sons," Andersson 1994). In *Utetheisa*, females gain multiple benefits through mate choice based on the male's courtship pheromone. Laboratory studies demonstrated that females mate preferentially with males that contain higher amounts of HD and, by doing so, they also mate with males possessing larger quantities of four correlated attributes: body mass, spermatophore mass, PAs transferred in the spermatophore, and systemic content of PAs (derived from the larval diet; Conner et al. 1990; Dussourd et al. 1991). Therefore, females would derive both direct phenotypic and indirect genetic benefits by choosing males based on the amount of pheromone. By selecting an HD-rich male, females would ensure receipt of a large alkaloidal gift and consequently eggs that were better protected from predators (direct benefit; Dussourd et al. 1988), a large nutrient gift that could be used to increase fecundity by as much as 15% per mating (direct benefit; LaMunyon 1997), and genes that encode for large body size (indirect benefit; Iyengar and Eisner 1999a). Because body size is heritable, females mating with males with higher levels of HD have larger sons that are more successful in courtship (sexy sons) and larger daughters that lay more eggs (good genes; Iyengar and Eisner 1999b). The cumulative effect of both direct and indirect benefits is substantial: a female given a choice between males differing by 10% in body mass will have an estimated 25% more grandchildren by mating with the larger male (Iyengar and Eisner 1999b). The strong selection for large males in *Utetheisa* may account for why in this species, contrary to the norm for Lepidoptera (Opler and Krizek 1984), males are larger than females (LaMunyon and Eisner 1993). Interestingly, given that females profit from both genetic and nongenetic components of the male's spermatophore, females do not appear to compete for males (but see Lim and Greenfield 2007) and males do not obviously exhibit mate choice even though it would benefit them to do so (Iyengar and Eisner 2004).

The relative importance of direct phenotypic and indirect genetic benefits in the evolution of female mate choice has received considerable attention (e.g., Kirkpatrick and Barton 1997). Despite this, it is not clear whether direct or indirect selection drives female choice in *Utetheisa*. Because earlier work showed that male HD titers are correlated positively with both systemic PA content and the amount of PAs that the male transmits to the female as a gift, it was thought that females might use male HD titer as a parameter for gauging male "reproductive value" (Dussourd et al. 1991). Furthermore, because PA content also correlated with male mass (as well as the mass of the spermatophore transferred), it was hypothesized that, through assessment of male HD titer and therefore the amount released during courtship, a female could evaluate male size (Conner et al. 1990; Iyengar and Eisner 1999a). Because body size is a heritable trait with known fitness consequences (Iyengar and Eisner 1999a, 1999b), it was hypothesized that female *Utetheisa* could use pheromone titer to evaluate the phenotypic and genetic quality of individual males.

This early work described the relationship between male quality and female choice in *Utetheisa*, in virgin moths. A field study at the Archbold Biological Station in Florida revealed that while male body mass, spermatophore mass, PA transferred in the spermatophore, and systemic PA content were all positively correlated, corematal HD content was not

correlated with any of these factors (Bezzerides et al. 2005). Based on their apparent age (wing wear) and our knowledge of female mating frequency, it is likely that most of these field males were not virgins. A subsequent study attempted to reconcile the lab data on virgin males and the field data by determining whether the male pheromonal (HD) titer changes over an individual's lifetime. Male HD titer was observed to remain unchanged despite the loss of PAs and body mass with each spermatophore transfer (Kelly et al. 2012). Thus, females are choosing males based on an honest signal (HD) that represents the total amount of PAs acquired by the larva, a representation of the ability to compete for access to PA-rich seeds (likely a genetically based, heritable trait), rather than the amount of PAs transferred at mating (a phenotypic quantity). While females receive direct benefits with every mating (indeed, all males transfer *some* PA and nutrient in each spermatophore), these recent results suggest that indirect genetic benefits (and the larval acquisition of PAs) are an important evolutionary force that plays a role in driving sexual selection in *Utetheisa*.

Female Promiscuity and Postcopulatory Sperm Selection

A fundamental tenet of sexual selection has been that males maximize their reproductive success by mating as often as possible, whereas females maximize their fitness by mating selectively with one or a few mates of high quality (Williams 1966; Trivers 1972; Parker 1979). Given that one mating often provides sufficient sperm to fertilize a female's entire complement of eggs, the costs of multiple mating for a female often outweigh the benefits that she may accrue by being promiscuous (Andersson 1994). For example, in many insect species, accessory substances passed from the male to the female at mating can have numerous negative effects on female fitness (e.g., increased risk of disease transmission and energetic and opportunity costs incurred as a consequence of courtship and copulation; Thornhill and Alcock 1983; Watson et al. 1998). Despite the costs of multiple mating in females, the advent of more accurate and accessible means to assess paternity has revealed that females of many species are more promiscuous than previously thought (Birkhead and Møller 1998; Arnqvist and Nilsson 2000). In many cases where females mate multiply, researchers have found evidence that sperm from the highest quality mate is used preferentially; this postcopulatory selection leaves open the possibility that females can accrue direct benefits from multiple males while having offspring with the highest genetic quality (Birkhead and Møller 1995; Keller and Reeve 1995; Eberhard 1996; Slatyer et al. 2011).

Utetheisa females are highly promiscuous. Female mating incidence is readily established by counting the number of colla (tubular remnants of spermatophores) in the bursa (spermatophore receptacle) of females (LaMunyon and Eisner 1993). Based on such counts, previous investigators had reported mating frequencies of, on average, 4–5 per female *Utetheisa* in nature (Pease 1968). More recent field data showed colla counts to average 11 per female, and to range upward to 22 per female (Iyengar et al. 2001). These are among the highest mating frequencies recorded in Lepidoptera. Furthermore, paternity is biased among twice-mated *Utetheisa* females, and progeny are sired almost exclusively by the larger of the two males (LaMunyon and Eisner 1993). Follow-up studies (in which large males were induced to produce small spermatophores) revealed that it is not body size per se that determines paternity, but instead the relative size of the male's spermatophore that dictates the sperm used to fertilize the eggs (LaMunyon and Eisner 1994). In many males, spermatophore size is an accurate indicator of male size; thus, a female may use stretch receptors in the bursa to indirectly gauge male size and quality (Sugawara 1979).

Currently, it is unknown whether biased paternity results from cryptic female choice or sperm competition. Cryptic female choice has been demonstrated in many organisms (Eberhard 1996), and there is evidence that female *Utetheisa* control the mechanism by which one set of sperm is favored over another. LaMunyon and Eisner (1993) found that, in 70% of twice-mated females, the larger male sired *all* of the offspring. Because both males were virgins with spermatophores presumably containing some sperm, this result suggests that siring success is not proportional to the quantity of sperm transferred. When females are anesthetized to inactivate their muscles, the normal routing of sperm is inhibited because sperm never reach the spermatheca, the usual storage destination (LaMunyon and Eisner 1993). Finally, the reproductive system of female *Utetheisa* is a complex labyrinth of ducts and chambers, which leaves open the potential for sophisticated manipulation of the contents of the spermatophore (Engelmann 1970). Despite this evidence for female control, it cannot be ruled out that sperm competition (Birkhead and Møller 1998) or seminal products may bias paternity (Wolfner 1997). *Utetheisa* males, like those of most lepidopterans, possess two types of sperm (nucleated eupyrene and anucleated apyrene sperm). The presence of nonfertilizing sperm in the male ejaculate of species with promiscuous females has led many to speculate that these apyrene sperm function in postcopulatory sperm competition (Swallow and Wilkinson 2002; Curril and LaMunyon 2006). While it remains unknown precisely how the various components of the system operate, there appear to be mechanisms in place that allow *Utetheisa* females to profit directly from each partner while having the mate of highest quality fertilize the majority of her eggs. In fact, PAs from at least three different males can be found in a single egg, thus confirming a female's ability to pool resources from multiple matings to benefit offspring belonging to one sire (Bezzerides and Eisner 2002).

Utetheisa and the Sexy-Sperm Hypothesis

Female promiscuity is particularly prevalent in insects in which females receive nuptial gifts containing nutrients and other chemicals that increase egg production or offspring survival (Choe and Crespi 1997; Vahed 1998). Numerous explanations based on genetic rather than phenotypic benefits to females also have been offered to explain female promiscuity (Kokko et al. 2003). For example, multiply-mating females may obtain indirect genetic benefits by increasing their odds of finding a mate of superior genetic quality (Yasui 1998; Jennions and Petrie 2000). Alternatively, Halliday and Arnold (1987) suggested that selection for extreme male promiscuity may drive female promiscuity nonadaptively through pleiotropic gene action.

Keller and Reeve (1995) proposed another hypothesis for the evolution of female promiscuity that has received limited empirical attention. They hypothesize that female multiple mating and male sperm competition can coevolve in a runaway, or Fisherian, process (the sexually selected sperm, or

"sexy-sperm" hypothesis). It has recently been argued theoretically that ZZ/ZW genetic systems, such as those found in lepidopterans and birds, are especially conducive to Fisherian sexual selection when female preference genes lie on the Z chromosome (Reeve and Pfennig 2003; Kirkpatrick and Hall 2004). *Utetheisa* females gain nutrients, chemical protection, and higher-quality offspring through promiscuity, but the possibility that they also gain sons with more-competitive sperm, in accordance with the sexy-sperm hypothesis, has not been addressed. The sexy-sperm hypothesis involves selection generated by female choice favoring males with competitive sperm and should be promoted by Z-linkage of female preferences in *Utetheisa* (Iyengar et al. 2002). A recent study found that, although the genes for male promiscuity are inherited autosomally, the underlying genes responsible for female promiscuity are Z-linked (Iyengar and Reeve 2010). These results demonstrate a genetic pattern consistent with the sexy-sperm hypothesis, and experiments are underway to test whether increased opportunities for multiple mating will increase the sperm competitiveness of the sons resulting from such matings.

Summary

Sexual selection is an important area of behavioral ecology that explains competitive interactions that occur in both precopulatory (e.g., exaggerated male traits used as armaments to fight other males or ornaments to attract females) and postcopulatory contexts (e.g., sperm competition and cryptic female choice). *Utetheisa ornatrix* is a moth where males use larvally acquired defensive chemicals to produce a short-range pheromone used by females to assess male quality during courtship. By choosing males based on this pheromone, females receive a substantial spermatophore whose contents provide both direct phenotypic benefits (via increased nutrients and defensive PAs) and indirect genetic benefits (via genes for larger size). Given that reproduction and defense are inexorably linked via this pheromone, *U. ornatrix* is a model organism for studying all aspects of sexual selection, and there continue to be many exciting new avenues to explore regarding the reproductive strategies of both sexes in this moth.

References Cited

Andersson, M. 1994. *Sexual Selection*. Princeton, NJ: Princeton University Press.

Arnqvist, G., and T. Nilsson. 2000. The evolution of polyandry: multiple mating and female fitness in insects. *Animal Behaviour* 60:145–164.

Bezzerides, A. 2004. Phenotypic and genetic benefits of promiscuity in an arctiid moth (*Utetheisa ornatrix*). PhD dissertation, Cornell University, Ithaca, NY.

Bezzerides, A., and T. Eisner. 2002. Apportionment of nuptial alkaloidal gifts by a multiply-mated female moth (*Utetheisa ornatrix*): eggs individually receive alkaloid from more than one male source. *Chemoecology* 12:213–218.

Bezzerides, A., V. K. Iyengar, and T. Eisner. 2005. Corematal function in *Utetheisa ornatrix*: interpretation in light of data from field-collected males. *Chemoecology* 15:187–192.

Birkhead, T. R., and A. P. Møller. 1995. Extra-pair copulations and extra-pair paternity in birds. *Animal Behaviour* 49:843–848.

Birkhead, T. R., and A. P. Møller. 1998. *Sperm Competition and Sexual Selection*. San Diego, CA: Academic Press.

Choe, J. C., and B. J. Crespi. 1997. *The Evolution of Mating Systems in Insects and Arachnids*. Cambridge: Cambridge University Press.

Choi, M.-Y., H. Lim, K.C. Park, R. Adlof, S. Wang, A. Zhang, and R. Jurenka. 2007. Identification and biosynthetic studies of the hydrocarbon sex pheromone in *Utetheisa ornatrix*. *Journal of Chemical Ecology* 33:1336–1345.

Conner, W.E., T. Eisner, R.K. Vande Meer, A. Guerrero, D. Ghiringelli, and J. Meinwald. 1980. Sex attractant of an arctiid moth (*Utetheisa ornatrix*): a pulsed chemical signal. *Behavioral Ecology and Sociobiology* 7:55–63.

Conner, W.E., T. Eisner, R.K. Vander Meer, A. Guerrero, and J. Meinwald. 1981. Precopulatory sexual interaction in an arctiid moth (*Utetheisa ornatrix*): role of a pheromone derived from dietary alkaloids. *Behavioral Ecology and Sociobiology* 9:227–235.

Conner, W.E., B. Roach, E. Benedict, J. Meinwald, and T. Eisner. 1990. Courtship pheromone production and body size as correlates of larval diet in males of the arctiid moth, *Utetheisa ornatrix*. *Journal of Chemical Ecology* 16:543–552.

Curril, I.M., and C.W. LaMunyon. 2006. Sperm storage and arrangement within females of the arctiid moth *Utetheisa ornatrix*. *Journal of Insect Physiology* 52:1182–1188.

Darwin, C. 1871. *The Descent of Man, and Selection in Relation to Sex*. London: John Murray.

Davies, N.B. 1992. *Dunnock Behaviour and Social Evolution*. Oxford: Oxford University Press.

del Campo, M.L., S.R. Smedley, and T. Eisner. 2005. Reproductive benefits derived from defensive alkaloid possession in an arctiid moth (*Utetheisa ornatrix*). *Proceedings of the National Academy of Sciences of the United States of America* 102:13508–13512.

Dussourd, D.E., K. Ubik, C. Harvis, J. Resch, J. Meinwald, and T. Eisner. 1988. Biparental defensive endowment of eggs with acquired plant alkaloid in the moth *Utetheisa ornatrix*. *Proceedings of the National Academy of Sciences of the United States of America* 85:5992–5996.

Dussourd, D.E., C.A. Harvis, J. Meinwald, and T. Eisner. 1991. Pheromonal advertisement of a nuptial gift by a male moth (*Utetheisa ornatrix*). *Proceedings of the National Academy of Sciences of the United States of America* 88:9224–9227.

Eberhard, W.G. 1996. *Female Control: Sexual Selection by Cryptic Female Choice*. Princeton, NJ: Princeton University Press.

Eisner, T., and M. Eisner. 1991. Unpalatability of the pyrrolizidine alkaloid-containing moth *Utetheisa ornatrix*, and its larva, to wolf spider. *Psyche* 98:111–118.

Eisner, T., and J. Meinwald. 2003. Alkaloid-derived pheromone and sexual selection in Lepidoptera. Pp. 341–368. In G.J. Blomquist and R.G. Prestwich, eds. *Insect Pheromone Biochemistry*. Orlando, FL: Academic Press.

Eisner, T., M. Eisner, C. Rossini, V.K. Iyengar, B.L. Roach, E. Benedikt, E. and J. Meinwald. 2000. Chemical defense against predation in an insect egg. *Proceedings of the National Academy of Sciences of the United States of America* 97:1634–1639.

Engelmann, F. 1970. *The Physiology of Insect Reproduction*. Oxford: Pergamon Press.

González, A., C. Rossini, M. Eisner, and T. Eisner. 1999. Sexually transmitted chemical defense in a moth (*Utetheisa ornatrix*). *Proceedings of the National Academy of Sciences of the United States of America* 96:5570–5574.

Gwynne, D.T. 1984. Courtship feeding increases female reproductive success in bushcrickets. *Nature* 307:361–363.

Halliday, T.R., and S.J. Arnold. 1987. Multiple mating by females: a perspective from quantitative genetics. *Animal Behaviour* 35:939–941.

Hare, J.F., and T. Eisner. 1993. Pyrrolizidine alkaloid deters ant predators of *Utetheisa ornatrix* eggs: effects of alkaloid concentration, oxidation state, and prior exposure of ants to alkaloid-laden prey. *Oecologia* 96:9–18.

Huang, W., S.P. Pulaski, and J. Meinwald. 1983. Synthesis of highly unsaturated insect pheromones: (Z,Z,Z)-1,3,6,9-heneicosatetraene and (Z,Z,Z)-1,3,6,9-nonadecatetraene. *Journal of Organic Chemistry* 48:2270–2274.

Iyengar, V.K., and T. Eisner. 1999a. Heritability of body mass, a sexually selected trait, in an arctiid moth (*Utetheisa ornatrix*). *Proceedings of the National Academy of Sciences of the United States of America* 96:9169–9171.

Iyengar, V.K., and T. Eisner. 1999b. Female choice increases offspring fitness in an arctiid moth (*Utetheisa ornatrix*). *Proceedings of the National Academy of Sciences of the United States of America* 96:15013–15016.

Iyengar, V.K., and T. Eisner. 2004. Male indifference to female traits in an arctiid moth (*Utetheisa ornatrix*). *Ecological Entomology* 29:281–284.

Iyengar, V.K., and H.K. Reeve. 2010. Z-linkage of the female promiscuity genes in the moth *Utetheisa ornatrix*: support for the sexy-sperm hypothesis? *Evolution* 64:1267–1272.

Iyengar, V.K., C. Rossini, and T. Eisner. 2001. Precopulatory assessment of male quality in an arctiid moth (*Utetheisa ornatrix*): hydroxydanaidal is the only criterion of choice. *Behavioral Ecology and Sociobiology* 49:283–288.

Iyengar, V.K., H.K. Reeve, and T. Eisner. 2002. Paternal inheritance of a female moth's mating preference. *Nature* 419:830–832.

Jain, S.C., D.E. Dussourd, W.E. Conner, T. Eisner, A. Guerrero, and J. Meinwald. 1983. Polyene pheromone components from an arctiid moth (*Utetheisa ornatrix*): characterization and synthesis. *Journal of Organic Chemistry* 48:2266–2270.

Jennions, M.D., and M. Petrie. 2000. Why do females mate multiply? A review of the genetic benefits. *Biological Reviews* 75:21–64.

Keller, L., and H.K. Reeve. 1995. Why do females mate with multiple males? The sexually-selected sperm hypothesis. *Advances in the Study of Behavior* 24:291–315.

Kelly, C.A., A.J. Norbutus, A.F. Lagalante, and V.K. Iyengar. 2012. Male courtship pheromones as indicators of genetic quality in an arctiid moth (*Utetheisa ornatrix*). *Behavioral Ecology* 23:1009–1014.

Kirkpatrick, M., and N.H. Barton. 1997. The strength of indirect selection on female mating preferences. *Proceedings of the National Academy of Sciences of the United States of America* 94:1282–1286.

Kirkpatrick, M., and D.W. Hall. 2004. Sexual selection and sex linkage. *Evolution* 58:683–691.

Kokko, H., R. Brooks, M.D. Jennions, and J. Morley. 2003. The evolution of mate choice and mating biases. *Proceedings of the Royal Society B: Biological Sciences* 342:335–352.

LaMunyon, C.W. 1997. Increased fecundity, as a function of multiple mating, in an arctiid moth, *Utetheisa ornatrix*. *Ecological Entomology* 22:69–73.

LaMunyon, C.W., and T. Eisner. 1993. Postcopulatory sexual selection in an arctiid moth (*Utetheisa ornatrix*). *Proceedings of the National Academy of Sciences of the United States of America* 90:4689–4692.

LaMunyon, C.W., and T. Eisner. 1994. Spermatophore size as determinant of paternity in an arctiid moth (*Utetheisa ornatrix*). *Proceedings of the National Academy of Sciences of the United States of America* 91:7081–7084.

Lim, H., and M.D. Greenfield. 2007. Female pheromonal chorusing in an arctiid moth, *Utetheisa ornatrix*. *Behavioral Ecology* 18:165–173.

Lim, H., K.C. Park, T.C. Baker, and M.D. Greenfield. 2007. Perception of conspecific female pheromone stimulates female calling in an arctiid moth, *Utetheisa ornatrix*. *Journal of Chemical Ecology* 33:1257–1271.

Opler, P.A., and G.O. Krizek. 1984. *Butterflies East of the Great Plains*. Baltimore, MD: John Hopkins University Press.

Parker, G.A. 1979. Sexual selection and sexual conflict. Pp. 123–166. In M.S. Blum and N.A. Blum, eds. *Sexual Selection and Reproductive Competition in Insects*. New York: Academic Press.

Pease, R.W., Jr. 1968. The evolutionary and biological significance of multiple pairing in Lepidoptera. *Journal of the Lepidopterists' Society* 22:69–73.

Reeve, H.K., and D.W. Pfennig. 2003. Genetic biases for showy males: are some genetic systems especially conducive to sexual selection? *Proceedings of the National Academy of Sciences of the United States of America* 100:1089–1094.

Slatyer, R.A., B.S. Mautz, P.R. Backwell, and M.D. Jennions. 2011. Estimating genetic benefits of polyandry from experimental studies: a meta-analysis. *Biological Reviews* 87:1–33.

Sugawara, T. 1979. Stretch reception in the bursa copulatrix of the butterfly, *Pieris rapae crucivora*, and its role in behaviour. *Journal of Comparative Physiology* 130:191–199.

Swallow, J.G., and G.S. Wilkinson. 2002. The long and short of sperm polymorphisms in insects. *Biological Reviews* 77: 153–182.

Thornhill, R., and J. Alcock. 1983. *The Evolution of insect Mating Systems*. Cambridge, MA: Harvard University Press.

Trivers, R.L. 1972. Parental investment and sexual selection. Pp. 136–179. In B. Campbell, ed. *Sexual Selection and the Descent of Man*. London: Aldine Publishing Company.

Vahed, K. 1998. The function of nuptial feeding in insects: a review of empirical studies. *Biological Reviews of the Cambridge Philosophical Society* 73:43–78.

Watson, P.J., G. Arnqvist, and R.R. Stallman. 1998. Sexual conflict and the energetic costs of mating and mate choice in water striders. *The American Naturalist* 151:46–58.

Williams, G.C. 1966. *Adaptation and Natural Selection*. Princeton, NJ: Princeton University Press.

Wolfner, M.F. 1997. Tokens of love: functions and regulation of *Drosophila* male accessory gland products. *Insect Biochemistry and Molecular Biology* 27:179–192.

Yasui, Y. 1998. The genetic benefits of female multiple mating reconsidered. *Trends in Ecology and Evolution* 13:246–250.

CHAPTER EIGHTEEN

Pheromone Communication, Behavior, and Ecology
in the North American *Choristoneura* genus

PETER J. SILK and ELDON S. EVELEIGH

INTRODUCTION/OVERVIEW

SYSTEMATICS AND HOST DISTRIBUTION

HYBRIDIZATION/REPRODUCTIVE ISOLATION

SEX PHEROMONES

Chemistry
Biosynthesis
Pheromone-mediated behavior
Pheromone physiology

DETECTION AND MONITORING

MATING DISRUPTION

CONCLUSION

ACKNOWLEDGMENTS

REFERENCES CITED

Introduction/Overview

The sex pheromone communication systems of the North American coniferophagous *Choristoneura* (Lepidoptera: Tortricidae) and some angiosperm feeders in the same genus are the subject of this chapter. The role of sexual signals (pheromone chemistry and correlated behavior), ecological specialization (e.g., larval host trees), differences in life history (adult flight phenology, length of larval diapause), and morphological differentiation (adult, larval) in the reproductive isolation of these *Choristoneura* spp. have been addressed by Freeman (1953, 1967) and Powell and De Benedictis (1995a), and several reviews have summarized the sex pheromone chemistry and biology of the coniferophagous *Choristoneura* (Harvey 1985; Silk and Kuenen 1988; Powell 1995). In addition, the pheromone-mediated behaviors of *C. fumiferana*, the spruce budworm (Alford et al. 1983), and *C. occidentalis*,[1] the western spruce budworm, have also been reviewed in some detail (Alford and Silk 1983; Silk and Kuenen 1985, 1986; Sweeney et al. 1990a). This chapter will update these reviews of pheromone communication in the *Choristoneura*, with an emphasis on conifer feeders, from the perspective of (a) the systematics, distribution, and reproductive isolation, (b) the elucidation and comparison of sex phero-

mone chemistry, biochemistry, and physiology, and (c) the ecology of representative species in this group.

Coniferophagous *Choristoneura* species are found in all North American conifer forests. The distribution of most species is reasonably well established. Some species are sympatric and synchronic in parts of their range, suggesting that distinct pheromone communication channels are an important prezygotic isolating mechanism (Silk and Kuenen 1988). However, there have been recorded cases of natural hybridization between some species in areas of sympatry, including *C. occidentalis* and *C. retiniana*, the Modoc budworm (Volney et al. 1984; Powell 1995), in the west and *C. fumiferana* and *C. pinus pinus*, the jack pine budworm, in the east (DeVerno et al. 1998), indicating that isolation is not complete. In areas of sympatry among other *Choristoneura* species, data are also available to support natural hybridization (Powell 1995; Lumley and Sperling 2011a, 2011b), but, in all cases, the mechanisms of reproductive isolation are not fully understood. Pheromone differences may be an important isolating mechanism for species and useful for defining phylogenetic relationships among them (Harvey 1996a; Lumley and Sperling 2011a, 2011b).

C. fumiferana has the largest range of all the budworms (figure 18.1) and exhibits large-amplitude population oscillations (Morris 1963; Royama 1984), with episodes of extremely high densities when defoliation and tree mortality occur over large areas (Gray and MacKinnon 2006), causing serious economic losses. Similarly, populations of *C. p. pinus* (McCullough 2000)

---

[1] *C. occidentalis* is now considered a junior homonym of *C. freemani* (Gilligan et al. 2012); however, to avoid confusion, we have retained *C. occidentalis* throughout this chapter.

265

and *C. occidentalis* (Alfaro et al. 1982) also periodically increase to high densities, resulting in extensive host tree damage. During outbreaks of all three species, foliage-protection operations have been implemented (see Armstrong and Ives 1995). The study of this genus—and specifically these species—is, therefore, of considerable ecological and economic importance.

Systematics and Host Distribution

The North American coniferophagous *Choristoneura* have long been regarded as a complex of species. Over the last five decades, many studies have been conducted in efforts to delineate species on the basis of morphological (Freeman 1967; Harvey and Stehr 1967; Dang 1985, 1992; De Benedictis 1995; Lumley and Sperling 2010), behavioral (Harvey 1967; Volney et al. 1984; Silk and Kuenen 1988; Sanders and Lucuik 1992), and genetic traits (Ennis 1976; Sperling and Hickey 1994, 1995; Harvey 1996a, 1996b; Lumley and Sperling 2010), and to determine phylogenetic relationships (Harvey 1985; Dang 1992; Sperling and Hickey 1994, 1995; De Benedictis 1995; Powell and De Benedictis 1995a; Willett 2000; Lumley and Sperling 2010, 2011a, 2011b). Currently, there are eight recognized species in this coniferophagous group (Harvey 1967, 1985; Powell 1995). In addition, up to seven different subspecies are now recognized (Powell and De Benedictis 1995b; Lumley and Sperling 2011a, 2011b; Gilligan et al. 2012), for a total of 15 named "biotypes" (Volney and Fleming 2007).

Several other North American *Choristoneura* species that feed on angiosperms (Brown et al. 2008) are not included in this review, with the exception of *C. rosaceana*, the obliquebanded leafroller. However, for comparative purposes, the known sex pheromones of angiosperm-feeding species are listed in Table 18.1.

The life cycles of the coniferophagous *Choristoneura* have been extensively characterized and reviewed (see references in Silk and Kuenen 1988). In general, these budworms are univoltine and feed on the foliage, flowers, and cones of their conifer hosts; all species feed on the Pinaceae. Within this group, there are those that feed only on pines (Pinoideae), the Lambertiana complex, those that feed on spruce (Piceoideae) and true fir (Abietoideae), the Fumiferana complex, and, thirdly, those that feed on Douglas-fir (*Pseudotsuga* spp.), the Carnana complex. All species are polyphagous within their group of host trees (see Brown et al. 2008 for a current list of host trees from which *Choristoneura* species have been recorded), but host ranges are limited, so the ranges of these *Choristoneura* are defined by the distribution of their hosts within suitable climatic regions (Harvey 1985; Silk and Kuenen 1988). Only *C. fumiferana*, *C. pinus pinus*, and *C. p. maritima* are found in eastern and central North America, whereas the remaining taxa are found in western North America from the Pacific Ocean eastward through the Rocky Mountains (figure 18.1) (Powell 1980; Harvey 1985).

Hybridization/Reproductive Isolation

In geographic areas where *Choristoneura* species do not overlap (figure 18.1), species are distinct and may be distinguished on the basis of morphological and life-history traits. However, where species ranges overlap, particularly in parts of western North America (figure 18.1), reproductive isolating mechanisms (e.g., life history, adult phenology) must break down to

some extent (Lumley and Sperling 2010, 2011a, 2011b) because, as stated previously, *Choristoneura* hybrids are well documented for *C. fumiferana* and *C. pinus pinus* in eastern Canada (DeVerno et al. 1998) and between *C. occidentalis* and *C. retiniana* in western North America (Volney et al. 1984; Powell 1995). The extent of natural hybridization is unknown, but definite hybrid zones involving *C. occidentalis* and *C. retiniana* have been identified in California, Nevada, Oregon, and Utah based on morphology, life history, host tree association, and pheromone attraction (Volney et al. 1984; De Benedictis et al. 1995; Powell and De Benedictis 1995a, 1995b). This zone was also confirmed by Lumley and Sperling (2011a) using a combination of phenotypic traits and neutral markers (simple sequence repeats [SSRs] and mitochondrial DNA [mtDNA]). In addition, hybrids among all species within the *C. fumiferana* species-complex can be readily produced in the laboratory (Harvey 1997).

Qualitative and quantitative differences in sex pheromones appear to be one of the primary isolating mechanisms in western *Choristoneura* populations. Extensive studies by Powell and De Benedictis (1995a) in natural populations of *C. lambertiana*, the western pine budworm, *C. carnana*, *C. retiniana*, and *C. occidentalis* in areas south of 42°N latitude in California, Nevada, and Utah, using virgin females and synthetic pheromone, documented 100% and 99% separation of sympatric species, respectively. Morphological examination of larvae and adults collected from the same locations also found no evidence of hybridization. However, Powell and De Benedictis (1995a) found that, north of 42°N latitude, starting at the Oregon and Idaho border, there are heterogeneous populations of larvae and adults with blends of coloration patterns, host plant selection, and sex pheromones. Recently, Lumley and Sperling (2011a) studied sympatric species occurring in a geographically isolated forest "island" in western Canada. Using a combination of morphological characters (larval head color, forewing color), host plant, adult flight phenology, pheromone attraction, and neutral markers (SSRs and mtDNA), they were able to identify *C. fumiferana*, *C. occidentalis*, *C. lambertiana*, and hybrid forms of these species, with the two major species, *C. fumiferana* and *C. occidentalis*, maintaining their genomic integrity. They concluded that pheromone attraction and adult flight period were the key life-history traits involved in maintaining the genomic integrity of these *Choristoneura* species, with the hybrids having intermediate flight periods between those of *C. fumiferana* and *C. occidentalis*.

The existence of *Choristoneura* hybrids and/or cryptic species has made it difficult to find characters that can consistently delimit species throughout their entire range. Even molecular taxonomy, a method that has proven extremely useful for identifying cryptic insect species (Smith et al. 2011), has not proven useful for distinguishing *Choristoneura* species when used in the absence of other characters (Lumley and Sperling 2010). In fact, as pointed out by Lumley and Sperling (2011b), the failure to delineate *Choristoneura* species using molecular methods may indicate the existence of "widespread gene flow and/or incomplete lineage sorting amongst this group."

The fact that naturally occurring hybrids exist and have been identified in areas of species overlap suggests that sex pheromones of at least some *Choristoneura* species do not always confer complete pre-mating reproductive isolation. Although caution must be exercised when interpreting trap catch data, synthetic sex pheromones employed as chemical lures have been shown to attract related *Choristoneura* species in addition to the species for which the lures were developed. For example, the *C. fumiferana* lure attracts not only *C. fumiferana* but also

TABLE 18.1

Sex pheromone components of the North American *Choristoneura*[a]

Species	Primary[b]	Secondary	Reference
Coniferophagous species			
C. fumiferana	*E*11-14Ald		Weatherston et al. (1971)
	96:4 *E/Z*11-14Ald		Sanders and Weatherston (1976)
	95:5 *E/Z*11-14Ald	14Ald	Silk et al. (1980); Grant (1987); Alford et al. (1983)
	95/5 *E/Z*11-14Ald	*Z*11-16Ald	Silk and Kuenen (1988)
C. occidentalis	92:8*E/Z*11-14Ald		Cory et al. (1982)
	92:8 *E/Z*11-14Ald	89:11 *E/Z*11-14Ac	Silk et al. (1982);
		85:15 *E/Z*11-14OH	Alford and Silk (1983); Sweeney et al. (1990a)
C. biennis	*E/Z*11-14Ald		Sanders (1971)
	E/Z ratio unknown		Sanders et al. (1974)
C. carnana	92:8 *E/Z*11-14Ald		Liebhold et al. (1984); Liebhold and Volney (1985)
C. pinus pinus	90% (85/15) *E/Z*11-14Ac/ 10% (85/15) *E/Z*11-14OH		Silk et al. (1985) Silk et al. (1986)
C. p. maritima	90% (85/15) *E/Z*11-14Ac/10% (85/15) *E/Z*11-14OH		Liebhold and Silk (1991)
C. orae	82:9:9 *E/Z*11-14Ac & *E*11-14OH		Gray et al. (1984)
*C. retiniana**	*C. retiniana**	*E* or *Z*11-14OH	Daterman et al. (1984)
C. lambertiana	*E*11-14:Ac or *E*11-14Ald[c]		Daterman et al. (1977)
C. lambertiana subretiniana[d]	90% (60/40) *E/Z*11-14Ac + 10% (60/40) *E/Z*11-14OH		Daterman et al. (1995)
Angiosperm-feeding species			
C. parallela	94:3:3 *E/Z*11-14Ac & *Z*11-14OH		Polavarpu and Lonergan (1998); Neal et al. (1982)
C. rosaceana[e]	96:2:2 *Z/E*11-14Ac & *Z*11-14OH	*Z*11-14Ald	Hill and Roelofs (1979); El-Sayed et al. (2003)
C. conflictana	*Z*11-14Ald		Evenden and Gries (2006); Weatherston et al. (1976)
C. fractivittana[f]	95:5 *Z*11-14OH & *Z*11-14Ald		Roelofs and Comeau (1971); Weatherston et al. (1978)

a. For a complete list of these and other global *Choristoneura*, see http://www.pherobase.com.

b. *E/Z*11-14Ald = *E/Z*11-tetradecenals; *E/Z*11-14Ac = *E/Z*11-tetradecenyl acetates; *E/Z*11-14OH = *E/Z*11-tetradecen-1-ols; 14Ald = tetradecanal; *Z*11-16Ald = *Z*11-hexadecenal; *sensu* Roelofs and Cardé (1977).

c. Unspecified blend.

d. See Powell (1995).

e. Geographical variation in blend, see El-Sayed et. al. (2003) and Thomson et al. (1991).

f. Common name, the broken-banded leafroller.

* The pheromone for *C. viridis* (Daterman et al. 1977) reported in The Pherobase is not listed in the table because Brown (2006) considers *C. viridis* to be synonymous with *C. retiniana*.

C. biennis, the two-year-cycle budworm, *C. carnana*, and *C. occidentalis*; and the *C. p. pinus* lure attracts *C. lambertiana* and *C. retiniana* in addition to *C. p. pinus* (Lumley and Sperling 2011b). These results suggest that: (1) some males in these populations are less selective, enabling them to respond to a broad range of synthetic chemical blends (De Benedictis et al. 1995) such that these pheromone lures serve, at least, as long-range attractants for multiple species, and (2) this cross-attraction involves two distinct groups of species: the Abietoideae- or Piceoideae-feeding, aldehyde emitters and the Pinoideae-feeding, acetate emitters, with some overlap in the pheromone communication systems (pheromone blends) of species within each group. However, the discovery of an aldehyde component in the pheromone emissions of *C. lambertiana* (an acetate-producing, pine-feeding species)—combined with evidence that some males of *C. occidentalis* (an aldehyde-emitting species that is mainly

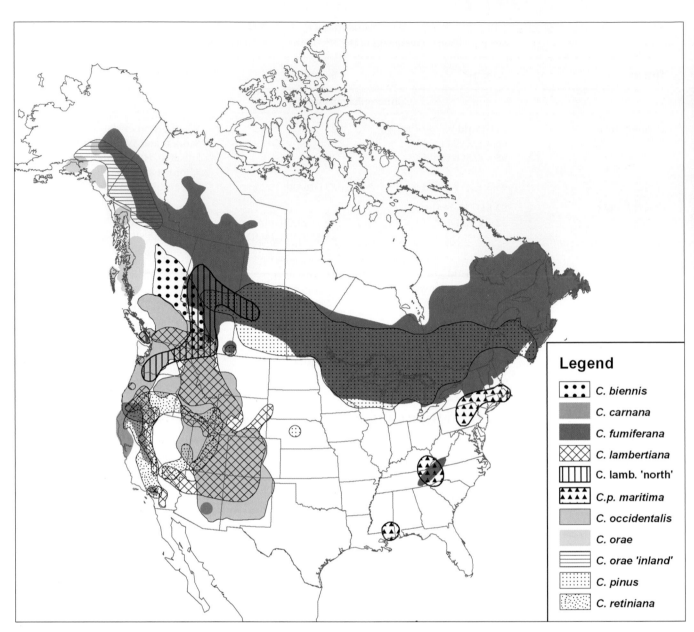

FIGURE 18.1 Distribution of coniferophagous *Choristoneura* species in North America (Lumley and Sperling 2011b; reproduced with kind permission of Lisa Lumley and Elsevier Inc.). For further details on species in the legend, see figure 4 in Lumley and Sperling (2011b).

Legend

- C. biennis
- C. carnana
- C. fumiferana
- C. lambertiana
- C. lamb. 'north'
- C.p. maritima
- C. occidentalis
- C. orae
- C. orae 'inland'
- C. pinus
- C. retiniana

found on *Pseudotsuga menziesii* but has also been found on *Pinus* sp.) were attracted to *C. lambertiana* synthetic lures to which aldehyde was added (Daterman et al. 1995)—suggests that cross-attraction and, thus, natural hybridization may also occur between sympatric species of these two groups. In fact, based on a phylogeny of the *C. fumiferana* species group constructed using mtDNA, Abietoideae feeders and Pinoideae feeders are closely related, with *C. p. pinus*, *C. biennis*, *C. occidentalis*, and *C. orae* having <1% divergence among the majority of their haplotypes that have been investigated (Sperling and Hickey 1994). This close genetic affinity among species (including sharing of mtDNA haplotypes between species), coupled with similar sex pheromones, has undoubtedly allowed hybridization and the production of viable offspring to occur in nature among some sympatric species, particularly those with similar host-feeding preferences or those occurring in mixed forest stands. As mentioned above, the extent to which hybridization

occurs and persists in nature is unknown, but hybrids are more likely to persist if they can inhabit environments that are not occupied by the parental species or where the parents have decreased fitness (Lumley and Sperling 2010). A more complete understanding of the sex pheromone chemistry, behavior, and communication systems in this group is needed to give supportive data to species assignments and, perhaps, clarify phylogenetic relationships.

Sex Pheromones

Chemistry

Sex pheromones in the *Choristoneura* are female produced, and the known sex pheromone components of the coniferophagous North American species are listed in Table 18.1. *C. fumif-*

erana, *C. occidentalis*, *C. biennis*, and *C. carnana* are aldehyde emitters, using blends of *E*- and *Z*11-tetradecenals (*E/Z*11-14Ald), whereas *C. pinus pinus*, *C. p. maritima*, *C. orae*, *C. retiniana*, *C. lambertiana*, and *C. lambertiana subretiniana* are acetate/alcohol emitters using blends of *E/Z*11-tetradecen-1-ol acetates (*E/Z*11-14Ac) and *E/Z*11-tetradecen-1-ol's (*E/Z*11-14OH). The *E/Z*11-14 acetates are the most abundant constituents in all coniferophagous *Choristoneura* for which female sex pheromone glands have been analyzed. Evidence from biochemical studies indicates that species producing aldehyde pheromones store gland-synthesized acetates, which are the direct biosynthetic precursors to the emitted aldehydes (Wiesner et al. 1979; Silk et al. 1980; Morse and Meighen 1984a, 1884b, 1986). This suggests a metabolic relationship between the acetates and aldehyde pheromones in this genus. Of the angiosperm-feeding species, *C. conflictana*, the large aspen tortrix (Evenden and Gries 2006), *C. parallela*, the spotted fireworm (Neal et al. 1982; Polavarapu and Lonergan 1998), and *C. rosaceana* (Hill and Roelofs 1979; El-Sayed et al. 2003) (Table 18.1) likely share similar biochemical pathways to their sex pheromones based on the Δ11-14Ac motif common to all.

To examine taxonomic similarities and differences among these *Choristoneura* species, alternatives to external morphological characters include analysis of pheromone communication systems (Roelofs and Brown 1982; Harvey 1985) and isozyme patterns (May et al. 1977; Harvey 1980, 1985; Stock and Castrovillo 1981; Castrovillo 1982; Willhite and Stock 1983). Trapping and male activation studies (Sanders 1971; Sanders et al. 1977) indicate similarities and differences in terms of sex pheromone chemistry and behavior; however, elucidation of complete pheromone blends, especially with respect to reproductive isolation, should indicate taxonomic affinities in this group (Harvey 1985). Population genetic studies of *C. fumiferana* have revealed few or no differences in the genetic profiles of early outbreak and declining populations (Harvey 1996b). Genetic studies among *Choristoneura* species (Harvey 1996a) using isozyme data found only small differences among the coniferophagous *Choristoneura*, with no single character permitting identification of individuals at the species or subspecies level. Secondary components (*sensu* Roelofs and Cardé 1977) (Table 18.1) may influence chemically mediated interactions within and among *Choristoneura* species (Silk et al. 1982; Alford and Silk 1983; Silk and Kuenen 1988). Given the challenges of species identification using traditional morphological and molecular traits, the role of secondary components in the reproductive isolation of *Choristoneura* species deserves further attention. The discovery of the pheromone chemistry differences and subsequent identification of their principal components have facilitated a better understanding of species relationships. For example, Powell (1995) used pheromone specificity, distribution, and morphology to separate some taxa with considerable success.

Biosynthesis

The biosynthesis of lepidopteran pheromones involves the reaction of fatty acids with two key enzymes found only in the sex pheromone glands (Roelofs and Bjostad 1984): a microsomal β-oxidation system giving limited chain shortening by two carbons and a key Δ11-desaturase system. In the budworms, the pheromones are constructed from hexadecanoate (palmitate) in a process involving limited chain shortening (β-oxidation) followed by Δ11-desaturation (Wolf and

Roelofs 1987). In *Choristoneura fumiferana* (and likely all the coniferophagous budworms), the unsaturated *E/Z*-14Ac is the microsomal gland-storage form of the aldehydic pheromone (Wiesner et al. 1979; Silk et al. 1980; Morse and Meighen 1987).

Metabolic and enzyme studies with *C. fumiferana* (Morse and Meighen 1986) allow construction of a plausible model that accounts for the biosynthesis of all known sex pheromone components in the *Choristoneura*. Following cytoplasmic fatty acid synthesis, the key steps are limited chain shortening (β-oxidation), a Δ11-desaturase (inserting the double bond (*E* or *Z*)) (Bjostad and Roelofs 1981; Roelofs and Bjostad 1984), a reduction step of acid to alcohol (a fatty alcohol reductase), and the formation of acetates by acetylation of the alcohols via an acetyl CoA:fatty alcohol acetyl transferase (Morse and Meighen 1987). All these steps occur in the microsomal fraction of the gland cells. For acetate producers, the compounds are simply transported unchanged through the gland cuticle, but for the alcohol and aldehyde producers, these compounds may be generated from the stored acetate by cuticle-bound enzymes. The alcohol is produced by hydrolysis of the acetate (alcohol esterase), and the aldehyde is produced by oxidation of the alcohol by an oxygen-dependent alcohol oxidase. This pathway model accounts for the genesis of all the *Choristoneura* pheromones reported (Table 18.1).

The sex pheromone glands of *C. fumiferana* contain both Δ11-14 and Δ11-16-fatty acids (Dunkelblum et al. 1985; Wolf and Roelofs 1987), indicating Δ11-desaturase activity; interestingly, only *Z*11-hexadecanoic acid was found, whereas both *E*- and *Z*11-tetradecenoic acids were found. [1-^{14}C]tetradecanoic acid labeled the *E/Z*11-14Ac in vivo, but [1-^{14}C]hexadecanoic and [1-^{14}C]dodecanoic acids did not. Application of [1-^{14}C]tetradecanoic acid also produced labeled *E*- and *Z*11-14:acids in a ratio close to that of the Δ11-14-acids found in the gland, and application of [U-^{14}C]hexadecanoic acid labeled both the tetradecanoic acid fraction and the *E/Z*11-14Ac components, supporting the argument that, indeed, limited chain shortening from hexadecanoic to tetradecanoic acid is the *first* step in the biosynthesis of *E/Z*11-14Ac from fatty acid precursors (Wolf and Roelofs 1987).

Direct evidence for a Δ11-desaturase in *C. fumiferana* was obtained by elegant in vitro labeling experiments using cell-free preparations from sex pheromone glands (Wolf and Roelofs 1987). Reaction of the cell-free preparation with [1-^{14}C]tetradecanoyl-CoA and [1-^{14}C]hexadecanoyl-CoA gave labeled *E/Z*11-tetradecanoic acids and *Z*11-hexadecanoic acid, respectively. With activity found only in the starting materials and *E/Z*11-14Ac, degradation and resynthesis of pheromone components are likely *not* occurring because radiolabel is not found elsewhere and being recycled in, for example, fatty acid biosynthesis and resynthesis. These data strongly support the hypothesis that *E/Z*11-14Ac are generated from hexadecanoate by limited chain shortening *followed* by Δ11-desaturation. The control of double-bond stereochemistry appears to occur in the microsomal desaturase/reductase step, these enzymes functioning in concert with the reductase modifying the isomeric ratio produced by the desaturase (Blomquist and Dillwith 1983); isomerases do not function in the pheromone biosynthesis of the Lepidoptera (Bjostad and Roelofs 1981), and so desaturase-inserted stereochemistry of double bonds is retained. Analyses of the fatty acids in the sex pheromone glands of *C. occidentalis* and *C. pinus pinus* indicate that these species use a similar pathway to synthesize *E/Z*11-14Ac (Dunkelblum et al. 1985), and, indeed, some of the identified fatty acids in all species of budworms may provide a very useful diagnostic identification tool to as yet unidentified secondary components (Silk and Kuenen 1986).

In *C. fumiferana*, *E/Z*11-14Ac are produced in a diel rhythm (Silk et al. 1980) and degraded concurrently with aldehyde release (Morse and Meighen 1984b). High amounts of an NAD-specific aldehyde dehydrogenase, associated with the membrane in *C. fumiferana* glands, have been detected (Morse and Meighen 1987). This enzyme activity also has been found on body parts of both sexes (legs, antennae, wings, and scales) where the congeneric acid is the only degradation product presumably functioning to eliminate the pheromonal signal when not required (Lonergan 1986). How species specificity in pheromone biosynthesis in this group is achieved is not certain, but it is likely through species-specific metabolic processes giving rise to specific oxygenated functional groups (aldehyde, alcohol, acetate), secondary components, geometrical isomer (*E/Z*) ratios, and release rates of single components or blends from the common unsaturated Δ11-14Ac precursor.

Pheromone-Mediated Behavior

Behavioral responses of males to the primary sex pheromone components for most of the taxa in this group have not been studied, with the exception of both *Choristoneura fumiferana* and *C. occidentalis*. Information on these two species has come from trapping studies with synthetic pheromone compounds, baiting traps with virgin females, and wind-tunnel experiments (Sanders 1978; Alford and Silk 1983; Silk and Kuenen 1985, 1986, 1988; Sweeney et al. 1990a). In addition, the pheromone response of *C. rosaceana* has been studied in the field and with wind-tunnel and electrophysiological studies in the laboratory (Evenden et al. 1999, 2000).

The primary sex pheromone components of *C. fumiferana* were identified as a blend of 95:5 *E/Z*11-14Ald (Sanders and Weatherston 1976; Silk et al. 1980), trapping males as effectively as virgin females. Addition of the saturated aldehyde, tetradecanal (14Ald), identified in effluvia (Silk et al. 1980) as ~2% of *E*11-14Ald, did not synergize trap capture in the field but caused more males to initiate flight and reach the pheromone source in wind-tunnel assays than to the primary components alone (Alford and Silk 1983; Silk and Kuenen 1986). In addition, Grant (1987) found further supporting evidence that the ternary blend of 95:5 *E/Z*11-14Ald + 14Ald also decreased the latency of males' copulatory attempts toward synthetic sources and increased the frequency of copulatory attempts. The responses were still, however, less than those produced from female effluvia, indicating an incomplete blend. Grant (1987) also found that pheromone-stimulated *C. fumiferana* males do not attempt copulation unless an appropriate release stimulus (visual, tactile, or chemo-tactile) is presented. Males did attempt copulation with filter paper impregnated with pheromone, but only when the filter paper was pinned to a rubber septum; Alford et al. (1983) reported similar results. Interestingly, conspecific wing scales from either sex applied to filter paper released copulatory attempts from pheromone-stimulated males (Grant 1987).

There is some evidence that *Z*11-hexadecenal (*Z*11-16Ald) is also a minor component in *C. fumiferana*. Analysis of the biosynthetic pathway in *C. fumiferana* (Wolf and Roelofs 1987) predicts the presence of *Z*11-16Ald, and *Z*11-16Ac was detected in the sex pheromone gland (Silk and Kuenen 1988). A 3-µg septum source of 95:5 *E/Z*11-14Ald with 3% (% of *E*11-14Ald)

*Z*11-16Ald elicited twice as many male upwind flights than the primary components alone but was still not equivalent to male responses to virgin females (Silk and Kuenen 1988). The mating behavior of the spruce budworm has been described (Sanders and Lucuik 1992), with wind-tunnel observations describing upwind flight behavior and copulatory activity of males to calling females. In addition, Palaniswamy et al. (1979) observed the behavior of male *C. fumiferana*. They concluded that males have a basal intrinsic level of sexual activity (e.g., wing fanning, copulatory attempts) that is *not* related to female presence or the pheromone, describing the role of the pheromone as enhancing this basal activity. A male always spreads its abdominal hairs (hair pencils) before attempting to mate with either sex. Female electroantennogram (EAG) responses to extracts of male abdominal tips, male wings, or airborne male volatiles were significant, perhaps revealing scent structures on wings or male abdominal tips (Palaniswamy et al. 1979). The courtship behavior of *C. rosaceana* has also been described (Curkovic et al. 2006), in which similarities to that of *C. fumiferana* (Sanders and Lucuik 1992) were noted. They demonstrated that males perform six observable and discrete steps in the courtship sequence: wing fanning, first contact, male-next-to-female, curled abdomen, genitalia engagement, and end-to-end positioning (mating). Extension of male hair pencils also was noted. There is also evidence of male mate-guarding through prolonged copulation by *C. fumiferana* (Kipp et al. 1990).

The role of a possible male-produced, close-range pheromone in *C. fumiferana* has been largely ignored. With a greater parental investment than males, females may be selective in their choice of a partner. A possible male pheromone for *C. fumiferana* located in the hair pencils (and/or possible wing glands; Palaniswamy et al. 1979) may facilitate species recognition and female mate assessment (e.g., *Ostrinia nubilalis* [Crambidae], the European corn borer [Lassance and Löfstedt 2009]). The response of female *C. fumiferana* to a putative male pheromone may act as a secondary prezygotic barrier reinforcing reproductive isolation between species. Deciphering the signals produced by males, if they exist, would help to test the hypothesis that male signals promote reproductive isolation in the *Choristoneura* genus.

The upwind flight behavior of male *C. occidentalis* moths to pheromone blends and to calling virgin females has been studied (Alford et al. 1983; Sweeney et al. 1990a). The blend of 92:8 *E/Z*11-14Ald, 89:11 *E/Z*11-14Ac, and 85:15 *E/Z*11-14OH, released together in a ratio of 2:1:1, approximating that released from a virgin female, induced a significantly higher percentage of moths to fly upwind in a wind tunnel compared with the aldehyde blend alone. However, addition of both the acetate and alcohol components (with the correct *E:Z* ratios; Table 18.1) to the primary component aldehydes in traps did not affect trap capture. The net upwind ground speed of flight in response to a PVC (0.05% 92/8 *E/Z*11-14Ald) releaser emitting at a female equivalent was lower than to virgin females, but was significantly increased by the addition of the acetate and alcohol components. These data support the hypothesis that the aldehydes are the primary components, whereas the acetate and alcohols are secondary components (*sensu* Roelofs and Cardé 1977) stimulating the precopulatory behavior of *C. occidentalis* at long range (>1 m) as well as at close range (1–2 cm).

Primary sex pheromone components for most of the remaining *Choristoneura* (Table 18.1) were identified using trap capture data as the principal assay. However, in many cases, trap capture data alone do not allow determination of the complete composition of the pheromone blend and provide only limited behavioral information. Systematic analysis (in some cases, reanalysis) of the pheromone biology of all *Choristoneura* species is needed.

Pheromone Physiology

The pheromone physiology of *Choristoneura* species is not well known, with most of our knowledge coming from studies on *C. fumiferana* and *C. rosaceana*. The diel periodicity of "calling" and pheromone production are synchronous in virgin females of *C. fumiferana* (Sanders and Lucuik 1972; Ramaswamy and Cardé 1984; Delisle et al. 1999) and of *C. rosaceana* (Delisle and Royer 1994; Delisle et al. 1999). In both species, the vast majority of females call during their first night (Delisle et al. 1999), and thus most females of these species are mated on the first night after emergence. Within 24 h after mating, pheromone production is totally suppressed (pheromonostasis) in both species, although some females may resume pheromone production at various times after mating (Delisle et al. 2000; Delisle and Simard 2002). Both juvenile hormone (JH) and neural messages modulate pheromonostasis in *C. rosaceana*, whereas control appears to be entirely neural in *C. fumiferana* (Delisle et al. 2000). As JH is involved in the migratory process (McNeil et al. 1995), the difference in JH effects on pheromone production in the two species may reflect differences in their life histories (Delisle et al. 2000); *C. rosaceana* is a nonmigratory species, whereas *C. fumiferana* can migrate over long distances after mating and laying some of their egg complement (Greenbank et al. 1980; Dobesberger et al. 1983). Because *C. fumiferana* females are able to detect and react to conspecific sex pheromone (Palaniswamy and Seabrook 1978, 1985) and show increased flight activity levels in the presence of pheromone in cage experiments (Sanders 1978), it has been suggested that high levels of sex pheromone associated with epidemic *C. fumiferana* populations may be one of the cues to initiate migration (McNeil et al. 1995; Delisle and Simard 2003). However, to our knowledge, field evidence for pheromone being a migration-triggering cue in *C. fumiferana* is lacking. Clearly, more studies are needed to unravel the complex physiological processes that govern the production of pheromones and the impact that biotic and abiotic factors and differences in life-history strategies have on these physiological processes.

Detection and Monitoring

The identification and synthesis of pheromones of the *Choristoneura* have led to the development of detection and monitoring programs against these pests. The primary goal was to use traps baited with synthetic pheromone to detect long-term variation in population trends or predict future incidence of damage. These programs may be particularly useful for *Choristoneura* species because some species defoliate trees over large tracts of forest landscapes and because other sampling procedures (larval counts, assessment of defoliation) are costly and necessitate a large number of samples for accurate estimates of abundance, especially at low population densities (Régnière and Sanders 1983). Although good correlations between numbers of males caught in pheromone-baited traps and population density in generation *n* and/or generation

n + 1 are often difficult to obtain, particularly when populations are low (Ramaswamy et al. 1982; Allen et al. 1986; Sanders 1988; Sweeney et al. 1990b), baited traps have, nevertheless, proven useful for monitoring population trends. For example, in North America, pheromone-baited traps are used to forecast spatio-temporal variation in abundance of *C. fumiferana* (Sanders 1988), *C. occidentalis* (Sweeney et al. 1990b), and *C. pinus pinus* (Clancy et al. 1980; Grant 1991).

Mating Disruption

The Canadian registration in 2007 of Hercon Disrupt SBW Micro-flakes®, a pheromone-based product for control of *Choristoneura fumiferana*, has now paved the way for large-scale trials to test pest management theories and concepts related to an early intervention strategy and population suppression (Palaniswamy et al. 1982; Sanders and Silk 1982; Silk and Kuenen 1984; Kettela and Silk 2005; Kettela et al. 2006). These data have recently been compiled and reviewed (Rhainds et al. 2012) and are based on laboratory experiments and field studies conducted mostly at moderate population densities. Application of pheromone from the ground or the air consistently reduced the orientation of males toward pheromone sources and reduced the mating success of caged or tethered wild or lab-reared females in 15 of 16 field studies. Only a limited effect on the mating success of resident wild females was found, likely because of the confounding effects of immigration of mated gravid females. No consistent difference in the density of egg masses in control and treated plots was observed, which was attributed to immigration of gravid females into treated plots. However, as lower mating success has been observed at low population densities than at moderate densities (Kipp et al. 1995), and this moth apparently depends more on short-range communication channels at high population densities (Sanders and Lucuik 1972), it has been suggested that mating disruption using synthetic pheromone would be more efficacious at relatively low population densities. Difficulties with measuring the efficacy of such an early intervention technique are associated with obtaining accurate sampling estimates (Régnière and Sanders 1983) at low population densities to forecast the onset of outbreaks as well as detecting and measuring the effects of migrating females and males into pheromone-treated areas.

Conclusion

Although the taxonomy of coniferophagous *Choristoneura* was updated by Powell in 1995, difficulties with some species remain. Many populations are taxonomically and pheromonally distinct, but in areas where the spatial and temporal distributions of some species broadly overlap, natural hybridization occurs. The mechanisms of reproductive isolation are not well understood in this complex, but as in many insect species, pheromone differences are likely important (Harvey 1996a, 1996b). For example, we might expect pheromone channels to be relatively broad in areas where there are few species but relatively narrow (more finely tuned) where several species overlap (Cardé and Baker 1984). This suggests that a closer examination of pheromone communication channels and of the amount of within- and between-species variation in blends (and also variation in male response to blends, see below) exhibited by overlapping and nonoverlapping species may facilitate the separation and/or validation of species. However, complete pheromone blends (i.e., both primary and secondary components) have not been identified in *any Choristoneura* taxon, limiting our ability to accurately answer questions about the role played by pheromones in the speciation process and to recognize cryptic species.

The successful transmission of information via a pheromone blend depends not only on the emitting female but also on the ability of the conspecific male receiver to detect and respond to the blend. A disproportionate amount of attention has been paid to the role of females in the transmission process, undoubtedly motivated by the potential use of pheromones in population monitoring and in forestry and agricultural control operations. In the few mating studies that have been conducted (e.g., *Choristoneura fumiferana* and *C. rosaceana*), the sex pheromone simply acts as a primer to "attract" potential mates, enabling males to locate females, often over long distances. Thereafter, successful mating appears to depend largely on the behavior of males and the potential involvement of other mechanical and/or chemical signals provided either by the males or females (perhaps secondary pheromone components). For example, little attention has been paid to the potential role of wing/body scales and male hair pencils as contact and/or chemical components in the mating process. Likewise, the potential involvement of visual cues or other stimuli used by either sex has also been largely ignored. More detailed studies on the sequence of behavioral events occurring after mate location by males (i.e., orientation mechanisms) have a high potential to more precisely describe the prezygotic reproductive isolating factors in *Choristoneura* species. In fact, understanding the ways in which mate choice occurs in the *Choristoneura* may be one of the keys to understanding the selective forces driving the processes of mating, reproductive isolation, population divergence, and, ultimately, speciation.

In comparison with many lepidopteran species, a considerable body of literature has accumulated on the population ecology of the economically important pest species of *Choristoneura* in North America and the factors (selective forces) that could potentially influence and mold the pheromone communication channels in these species. For example, with *C. fumiferana*, the factors influencing community structure in different forest stands (Eveleigh et al. 2007), life history and survivorship (Royama 1984), phenological development and synchrony with its host trees (Nealis and Régnière 2004), and long-range migration (Greenbank et al. 1980) are all well known. Yet, little is known about how pheromone-mediated sexual behavior shapes, or is shaped by, the population density of this species as it goes through its outbreak cycles. Thus, studies on the interaction between pheromone communication systems and population dynamics, in conjunction with genetic studies on pheromone production and male responses, are essential if we are to understand the role of pheromones, not only as a reproductive isolating mechanism, but also as a potential tool for population intervention at low densities.

Acknowledgments

We thank Amanda Keddy for providing excellent technical assistance with the figure and Maya Evenden and Krista Ryall for providing helpful comments. Special thanks are due to Lisa Lumley for providing the original copy of the figure and for a helpful critical review of an earlier version of the manuscript. We also acknowledge the assistance of our scientific editor, Caroline Simpson.

References Cited

Alfaro, R. I., G. A. van Sickle, J. Thompson, and E. Wegwitz. 1982. Tree mortality and radial growth losses caused by the western spruce budworm in a Douglas-fir stand in British Columbia. *Canadian Journal of Forest Research* 12:780–787.

Alford, A. R., and P. J. Silk. 1983. Behavioral effects of secondary components of the sex pheromone of western spruce budworm (*Choristoneura occidentalis*) free. *Journal of Chemical Ecology* 10:265–270.

Alford, A. R., P. J. Silk, M. McClure, C. Gibson, and J. Fitzpatrick. 1983. Behavioral effects of secondary components of the sex pheromone of the eastern spruce budworm *Choristoneura fumiferana* Clem. *The Canadian Entomologist* 115:1053–1058.

Allen, D. C., L. P. Abrahamson, D. A. Eggen, G. N. Lanier, S. R. Swier, R. S. Kelley, and M. Auger. 1986. Monitoring spruce budworm (Lepidoptera: Tortricidae) populations with pheromone-baited traps. *Environmental Entomology* 15:152–165.

Armstrong, J. A., and W. G. H. Ives (eds). 1995. *Forest Insect Pests in Canada.* Ottawa, ON: Natural Resources Canada, Canadian Forest Service.

Bjostad, L. B., and W. L. Roelofs. 1981. Sex pheromone biosynthesis from radiolabeled fatty acids in the red-banded leafroller moth. *Journal of Biological Chemistry* 256:7936–7940.

Blomquist, G. J., and J. W. Dillwith. 1983. Pheromones: biochemistry and physiology. Pp. 527–542. In R. G. Downer and H. Lauter, eds. *Endocrinology of Insects.* New York: Alan R. Liss Inc.

Brown, J. W. 2006. Scientific names of pest species in Tortricidae (Lepidoptera) frequently cited erroneously in the entomological literature. *American Entomologist* 52:182–189.

Brown, J. W., G. Robinson, and J. A. Powell. 2008. Food plant database of the leafrollers of the world (Lepidoptera: Tortricidae) (Version 1.0.0). Available online at: http://www.tortricidae.com/foodplants.asp.

Cardé, R. T., and T. C. Baker. 1984. Sexual communication with pheromones. Pp. 355–383. In W. J. Bell and R. T. Cardé, eds. *Chemical Ecology of Insects.* Sunderland, MA: Sinauer Associates, Inc.

Castrovillo, P. J. 1982. Interspecific and intraspecific genetic comparisons of North American spruce budworms *Choristoneura* spp. PhD dissertation, University of Idaho, Moscow, ID.

Clancy, K. M., R. L. Riese, and D. M. Benjamin. 1980. Predicting jack-pine budworm infestations in northwestern Wisconsin. *Environmental Entomology* 9:743–751.

Cory, H. T., G. E. Daterman, G. D. Daves, Jr., L. L. Sower, R. F. Shepherd, and C. J. Sanders. 1982. Chemistry and field evaluation of the sex pheromone of the western spruce budworm *Choristoneura occidentalis. Journal of Chemical Ecology* 8:339–350.

Curkovic, T., J. F. Brunner, and P. J. Landolt. 2006. Courtship behavior in *Choristoneura rosaceana* and *Pandemis pyrusana* (Lepidoptera: Tortricidae). *Annals of the Entomological Society of America* 99:617–624.

Dang, P. T. 1985. Key to adult males of conifer-feeding species of *Choristoneura* Lederer (Lepidoptera: Tortricidae) in Canada and Alaska. *The Canadian Entomologist* 117:1–5.

Dang, P. T. 1992. Morphological study of the male genitalia with phylogenetic inference of *Choristoneura* (Lepidoptera: Tortricidae). *The Canadian Entomologist* 124:7–48.

Daterman, G. E., R. G. Robbins, T. D. Eichlin, and J. Pierce. 1977. Forest Lepidoptera attracted by known sex attractants of western spruce budworms, *Choristoneura* spp. (Lepidoptera: Tortricidae). *The Canadian Entomologist* 109:875–878.

Daterman, G. E., H. T. Cory, L. L. Sower, and G. D. Daves, Jr. 1984. Sex pheromone of a conifer-feeding budworm, *Choristoneura retiniana* Walsingham. *Journal of Chemical Ecology* 10:153–160.

Daterman, G. E., L. L. Sower, R. E. Stevens, and D. G. Felin. 1995. Pheromone chemistry and response behavior of conifer-feeding *Choristoneura* in the Western United States. Pp. 69–84. In J. A. Powell, ed. *Biosystematic Studies of Conifer-Feeding Choristoneura* (Lepidoptera: Tortricidae) *in the Western United States.* Berkeley, CA: University of California Publications in Entomology.

De Benedictis, J. A. 1995. Phenetic studies of spruce budworm populations (*Choristoneura* Species). Pp. 85–150. In J.A. Powell, ed. *Biosystematic Studies of Conifer-Feeding Choristoneura* (Lepidoptera: Tortricidae) *in the Western United States*. Berkeley, CA: University of California Publications in Entomology.

De Benedictis, J.A., A.M. Liebhold, and J.A. Powell. 1995. Studies of pheromone attraction and inheritance using virgin females of *Choristoneura* species in the Central Sierra Nevada, California. Pp. 151–165. In J.A. Powell, ed. *Biosystematic Studies of Conifer-Feeding Choristoneura* (Lepidoptera: Tortricidae) *in the Western United States*. Berkeley, CA: University of California Publications in Entomology.

Delisle, J., and L. Royer. 1994. Changes in pheromone titer of oblique-banded leafroller, *Choristoneura rosaceana*, virgin females as a function of time of day, age, and temperature. *Journal of Chemical Ecology* 20:45–69.

Delisle, J., and J. Simard. 2002. Factors involved in the post-copulatory neural inhibition of pheromone production in *Choristoneura fumiferana* and *C. rosaceana* females. *Journal of Insect Physiology* 48:181–188.

Delisle, J., and J. Simard. 2003. Age-related changes in the competency of the pheromone gland and the pheromonotropic activity of the brain of both virgin and mated females of two *Choristoneura* species. *Journal of Insect Physiology* 49:91–97.

Delisle, J., J.-F. Picimbon, and J. Simard. 1999. Physiological control of pheromone production in *Choristoneura fumiferana* and *C. rosaceana*. *Archives of Insect Biochemistry and Physiology* 42:253–265.

Delisle, J., J.-F. Picimbon, and J. Simard. 2000. Regulation of pheromone inhibition in mated females of *Choristoneura fumiferana* and *C. rosaceana*. *Journal of Insect Physiology* 46:913–921.

DeVerno, L.L., G.A. Smith, and K.J. Harrison. 1998. Randomly amplified polymorphic DNA evidence of introgression in two closely related sympatric species of coniferophagous *Choristoneura* (Lepidoptera: Tortricidae) in Atlantic Canada. *Annals of the Entomological Society of America* 91:248–259.

Dobesberger, E.J., K.P. Lim, and A.G. Raske. 1983. Spruce budworm moth flight from New Brunswick to Newfoundland. *The Canadian Entomologist* 115:1641–1645.

Dunkelblum, E., S.H. Tan, and P.J. Silk. 1985. Double bond location in mono-unsaturated fatty acids by dimethyl disulfide derivatization and mass spectrometry: application to analysis of fatty acids in pheromone glands of four Lepidoptera. *Journal of Chemical Ecology* 11:265–277.

El-Sayed, A.M., J. Delisle, N. De Lury, L.J. Gut, G.J.R. Judd, S. Legrand, W.H. Reissig, W.L. Roelofs, C.R. Unelius, and R.M. Trimble. 2003. Geographic variation in pheromone chemistry, antennal electrophysiology, and pheromone-mediated trap catch of North American populations of the obliquebanded leafroller. *Environmental Entomology* 32:470–476.

Ennis, T.J. 1976. Sex chromatin and chromosome numbers in Lepidoptera. *Canadian Journal of Genetics and Cytology* 18:119–130.

Eveleigh, E.S., K.S. McCann, P.C. McCarthy, S.J. Pollock, C.J. Lucarotti, B. Morin, G.A. McDougall et al. 2007. Fluctuations in density of an outbreak species drive diversity cascades in food webs. *Proceedings of the National Academy of Sciences of the United States of America* 104:16976–16981.

Evenden, M.L., and R. Gries. 2006. Sex pheromone of the large aspen tortrix, *Choristoneura conflictana* (Lepidoptera: Tortricidae). *Chemoecology* 16:115–122.

Evenden, M.L., G.J.R. Judd, and J.H. Borden. 1999. Simultaneous disruption of pheromone communication in *Choristoneura rosaceana* and *Pandemis limitata* with pheromone and antagonist blends. *Journal of Chemical Ecology* 25:501–517.

Evenden, M.L., G.J.R. Judd, and J.H. Borden. 2000. Investigations of mechanisms of pheromone communication disruption of *Choristoneura rosaceana* (Harris) in a wind tunnel. *Journal of Chemical Ecology* 13:499–510.

Freeman, T.N. 1953. The spruce budworm, *Choristoneura fumiferana* (Clem.) and an allied new species on pine (Lepidoptera: Tortricidae). *The Canadian Entomologist* 85:121–127.

Freeman, T.N. 1967. On coniferophagous species of *Choristoneura* (Lepidoptera: Tortricidae) in North America. I. Some new forms of *Choristoneura* allied to *C. fumiferana*. *The Canadian Entomologist* 99:445–449.

Gilligan, T.M., J. Baixeras, J.W. Brown, and K.R. Tuck. 2012. T@RTS: online world catalogue of the tortricidae (Ver. 2.0). Available online at: http://www.tortricid.net/catalogue.asp.

Grant, G.G. 1987. Copulatory behavior of spruce budworm *Choristoneura fumiferana* (Lepidoptera: Tortricidae): experimental analysis of the sex pheromone and associated stimuli. *Annals of the Entomological Society of America* 80:78–88.

Grant, G.G. 1991. Development and use of pheromones for monitoring Lepidopteran forest defoliators in North America. *Forest Ecology and Management* 39:153–162.

Gray, D.R., and W.E. MacKinnon. 2006. Outbreak patterns of the spruce budworm and their impacts in Canada. *The Forestry Chronicle* 82:550–561.

Gray, T.G., K.N. Slessor, G.G. Grant, R.F. Shepherd, E.H. Holsten, and A.S. Tracey. 1984. Identification and field testing of pheromone components of *Choristoneura orae* (Lepidoptera: Tortricidae). *The Canadian Entomologist* 116:51–56.

Greenbank, D.O., G.W. Schaefer, and R.C. Rainey. 1980. Spruce budworm (Lepidoptera: Tortricidae) moth flight and dispersal: new understanding from canopy observations, radar, and aircraft. *Memoirs of the Entomological Society of Canada* 110:1–49.

Harvey, G.T. 1967. On coniferophagous species of *Choristoneura* (Lepidoptera: Tortricidae) in North America. V. Second diapause as a species character. *The Canadian Entomologist* 99:486–503.

Harvey, G.T. 1980. Sampling low density spruce budworm populations for isozyme analysis. *The Canadian Entomologist* 112:969–970.

Harvey, G.T. 1985. The taxonomy of the *Choristoneura* (Lepidoptera: Tortricidae): a review. Pp. 16–48. In C.J. Sanders, R.W. Stark, E.J. Mullins, and J. Murphy, eds. *Recent Advances in Spruce Budworms Research*. Proceedings, CANUSA Spruce Budworms Research Symposium, Bangor, Maine 1984, Natural Resources Canada, Canadian Forest Service, Ottawa, ON.

Harvey, G.T. 1996a. Genetic relationships among *Choristoneura* species (Lepidoptera: Tortricidae) as revealed by isozymes studies. *The Canadian Entomologist* 128:245–262.

Harvey, G.T. 1996b. Population genetics of the spruce budworm, *Choristoneura fumiferana* (Clem.) (Lepidoptera: Tortricidae) in relation to geographical and population density differences. *The Canadian Entomologist* 128:219–243.

Harvey, G.T. 1997. Interspecific crosses and fertile hybrids among the coniferophagous *Choristoneura* (Lepidoptera: Tortricidae). *The Canadian Entomologist* 129:519–536.

Harvey, G.T., and G. Stehr. 1967. On coniferophagous species of *Choristoneura* (Lepidoptera: Tortricidae) in North America. III. Some characters of immature forms helpful in the identification of species. *The Canadian Entomologist* 99:464–481.

Hill, A.S., and W. Roelofs. 1979. Sex pheromone components of the obliquebanded leafroller *Choristoneura rosaceana*. *Journal of Chemical Ecology* 5:3–11.

Kettela, E.G., and P.J. Silk. 2005. Development of a pheromone formulation for use in early intervention pest management strategies of the spruce budworm *Choristoneura fumiferana* (Clem.). SERG-1 Project 614, 36pp. Available online at: http://atl.cfs.nrcan.gc.ca/sprucebudworm/.

Kettela, E.G., S.E. Holmes, and P.J. Silk. 2006. Results of aerially applied Disrupt® microflakes on spruce budworm mating success, Ontario, 2005. Available online at: http://atl.cfs.nrcan.gc.ca/sprucebudworm/

Kipp, L.R., R. Ellison, and W.D. Seabrook. 1990. Copulatory mate-guarding in the spruce budworm. *Journal of Insect Behavior* 3:121–131.

Kipp, L.R., G.C. Lonergan, and W.J. Bell. 1995. Male periodicity and the timing of mating in the spruce budworm (Lepidoptera: Tortricidae): influences of population density and temperature. *Environmental Entomology* 24:1150–1159.

Lassance, J.-M., and C. Löfstedt. 2009. Concerted evolution of male and female display traits in the European corn borer, *Ostrinia nubilalis*. *BMC Biology* 7:10.

Liebhold, A.M., and P.J. Silk. 1991. Capture of *Choristoneura pinus maritima* in traps baited with *C. p. pinus* pheromone components (Lepidoptera: Tortricidae). *Journal of the Lepidopterists' Society* 45:172–173.

Liebhold, A.M., and W.J.A. Volney. 1985. Effects of attractant composition and release rate on attraction of male *Choristoneura retiniana*, *C. occidentalis* and *C. carnana* (Lepidoptera: Tortricidae). *The Canadian Entomologist* 117:447–457.

Liebhold, A.M., W.J.A. Volney, and W.E. Waters. 1984. Evaluation of cross attraction between sympatric *Choristoneura occidentalis* and

C. retiniana (Lepidoptera: Tortricidae) populations in south-central Oregon. *The Canadian Entomologist* 116:827–840.

Lonergan, G.C. 1986. Metabolism of pheromone components and analogues by cuticular enzymes of *Choristoneura fumiferana*. *Journal of Chemical Ecology* 12:483–496.

Lumley, L.M., and F.A.H. Sperling. 2010. Integrating morphology and mitochondrial DNA for species delimitation within the spruce budworm (*Choristoneura fumiferana*) cryptic species complex (Lepidoptera: Tortricidae). *Systematic Entomology* 35:416–428.

Lumley, L.M., and F.A.H. Sperling. 2011a. Life-history traits maintain the genomic integrity of sympatric species of the spruce budworm (*Choristoneura fumiferana*) group on an isolated forest island. *Ecology and Evolution* 1:119–131.

Lumley, L.M., and F.A.H. Sperling. 2011b. Utility of microsatellites and mitochondrial DNA for species delimitation in the spruce budworm (*Choristoneura fumiferana*) species complex (Lepidoptera: Tortricidae). *Molecular Phylogenetics and Evolution* 58:232–243.

May, B., D.E. Leonard, and R.L.Vardas. 1977. Electrophoretic variation and sex linkage in spruce budworm. *Journal of Heredity* 68:355–359.

McCullough, D.G. 2000. A review of factors affecting the population dynamics of jack pine budworm (*Choristoneura pinus pinus* Freeman). *Population Ecology* 42:243–256.

McNeil, J.N., M. Cusson, J. Delisle, I. Orchard, and S.S. Tobe. 1995. Physiological integration of migration in Lepidoptera. Pp. 279–302. In V.A. Drake and A.G. Gatehouse, eds. *Insect Migration: Tracking Resources through Space and Time*. Cambridge, UK: Cambridge University Press.

Morris, R.F. 1963. The dynamics of epidemic spruce budworm populations. *Memoirs of the Entomological Society of Canada* 31:1–332.

Morse, D., and E. Meighen. 1984a. Aldehyde pheromones in Lepidoptera: evidence for an acetate ester precursor in *Choristoneura fumiferana*. *Science* 226:1434–1436.

Morse, D., and E. Meighen. 1984b. Detection of pheromone biosynthetic and degradative enzymes *in vitro*. *Journal of Biological Chemistry* 259:475–480.

Morse, D., and E. Meighen. 1986. Pheromone biosynthesis and the role of functional groups in pheromone specificity. *Journal of Chemical Ecology* 12:335–351.

Morse, D., and E. Meighen. 1987. Biosynthesis of the acetate precursor of the spruce budworm sex pheromone by an acetyl CoA:fatty alcohol acetyl transferase. *Insect Biochemistry* 17:53–59.

Neal, J.W., J.A. Klun, Jr., B.A. Bierl-Leonhardt, and M. Schwarz. 1982. Female sex pheromone of *Choristoneura parallela* (Lepidoptera: Tortricidae). *Environmental Entomology* 11:893–896.

Nealis, V.G., and J. Régnière. 2004. Insect–host relationships influencing disturbance by the spruce budworm in a boreal mixedwood forest. *Canadian Journal of Forest Research* 34:1870–1882.

Palaniswamy, P., and W.D. Seabrook. 1978. Behavioral responses of the female eastern spruce budworm *Choristoneura fumiferana* (Lepidoptera: Tortricidae) to the sex pheromone of her own species. *Journal of Chemical Ecology* 4:649–655.

Palaniswamy, P., and W.D. Seabrook. 1985. The alteration of calling behavior by female *Choristoneura fumiferana* when exposed to synthetic sex pheromone. *Entomologia Experimentalis et Applicata* 37:13–16.

Palaniswamy, P., W.D. Seabrook, and R.J. Ross. 1979. Precopulatory behavior of males and perception of a potential male pheromone in spruce budworm *Choristoneura fumiferana*. *Annals of the Entomological Society of America* 72:6544–6551.

Palaniswamy, P., R.J. Ross, W.D. Seabrook, G.C. Lonergan, C.J. Wiesner, S.H. Tan, and P.J. Silk. 1982. Mating suppression of caged eastern spruce budworm moths in different pheromone atmospheres at high population densities. *Journal of Economic Entomology* 75:989–993.

Polavarapu, S., and G.C. Lonergan. 1998. Sex pheromone of *Choristoneura parallela* (Lepidoptera: Tortricidae): components and development of a pheromone lure for population monitoring. *Environmental Entomology* 27:1242–1249.

Powell, J.A. 1980. Nomenclature of nearctic conifer-feeding *Choristoneura* (Lepidoptera: Tortricidae): historical review and current status. USDA Forest Service, General Technical Report PNW-100, Pacific Northwest Forest and Range Experiment Station, Portland, OR.

Powell, J.A. (ed.). 1995. *Biosystematic Studies of Conifer-Feeding Choristoneura* (Lepidoptera: Tortricidae). University of California Publications in Entomology. 115, University of California Press, Berkeley, CA.

Powell, J.A., and J.A. De Benedictis. 1995a. Biological relationships: host tree preferences and isolation by pheromones among allopatric and sympatric populations of western *Choristoneura*. Pp. 21–68. In J.A. Powell, ed. *Biosystematic Studies of Conifer-Feeding Choristoneura* (Lepidoptera: Tortricidae) *in the Western United States*. Berkeley, CA: University of California Press.

Powell, J.A., and J.A. De Benedictis. 1995b. Evolutionary interpretation, taxonomy, and nomenclature. Pp. 217–275. In J.A. Powell, ed. *Biosystematic Studies of Conifer-Feeding Choristoneura* (Lepidoptera: Tortricidae) *in the Western United States*. Berkeley, CA: University of California Press.

Ramaswamy, S.B., and R.T. Cardé. 1984. Rate of release of spruce budworm pheromone from virgin females and synthetic lures. *Journal of Chemical Ecology* 10:1–7.

Ramaswamy, S.B., R.T. Cardé, and J.A. Witter. 1982. Relationships between catch in pheromone-baited traps and larval density of the spruce budworm, *Choristoneura fumiferana* (Lepidoptera: Tortricidae). *The Canadian Entomologist* 115:1437–1443.

Régnière, J., and C.J. Sanders. 1983. Optimal sample size for the estimation of spruce budworm (Lepidoptera: Tortricidae) populations on balsam fir and white spruce. *The Canadian Entomologist* 115:1621–1626.

Rhainds, M., E.G. Kettela, and P.J. Silk. 2012. Thirty-five years of pheromone-based mating disruption studies with spruce budworm, *Choristoneura fumiferana* (Lepidoptera: Tortricidae): a review. *The Canadian Entomologist* 144:379–395.

Roelofs, W., and L.B. Bjostad. 1984. Biosynthesis of lepidopteran pheromones. *Bioorganic Chemistry* 12:279–298.

Roelofs, W., and R.L. Brown. 1982. Pheromones and evolutionary relationships of Tortricidae. *Annual Review of Ecology and Systematics* 13:395–422.

Roelofs, W.L., and R.T. Cardé. 1977. Responses of Lepidoptera to synthetic pheromones and their analogues. *Annual Review of Entomology* 22:377–405.

Roelofs, W.L., and A. Comeau. 1971. Sex attractants in Lepidoptera. Pp. 91–114. In A.S. Tahori, ed. *Chemical Releasers in Insects, III*. New York: Gordon and Breach.

Royama, T. 1984. Population dynamics of the spruce budworm *Choristoneura fumiferana*. *Ecological Monographs* 54:429–462.

Sanders, C.J. 1971. Laboratory bioassay of the sex pheromone of the spruce budworm *Choristoneura fumiferana* (Lepidoptera: Tortricidae). *The Canadian Entomologist* 103:631–637.

Sanders, C.J. 1978. Flight and copulation of female spruce budworm in pheromone-permeated air. *Journal of Chemical Ecology* 13:1749–1758.

Sanders, C.J. 1988. Monitoring spruce budworm population density with sex pheromone traps. *The Canadian Entomologist* 120:175–183.

Sanders, C.J., and G.S. Lucuik. 1972. Factors affecting calling by eastern spruce budworm, *Choristoneura fumiferana* (Lepidoptera: Tortricidae). *The Canadian Entomologist* 104:1751–1762.

Sanders, C.J., and G.S. Lucuik.1992. Mating behavior of spruce budworm moths *Choristoneura fumiferana* (Clem.) (Lepidoptera: Tortricidae). *The Canadian Entomologist* 124:273–286.

Sanders, C.J., and P.J. Silk. 1982. Disruption of spruce budworm mating using Hercon plastic laminated flakes. CFS Information Report O-X-335. GLFRC, Sault Ste. Marie, Ontario.

Sanders, C.J., and J. Weatherston. 1976. Sex pheromone of the eastern spruce budworm (Lepidoptera: Tortricidae): optimum blend of *trans*- and *cis*-11-tetradecenal. *The Canadian Entomologist* 108:1285–1290.

Sanders, C.J., G.E. Daterman, and T.J. Ennis. 1977. Sex pheromone responses of *Choristoneura* spp. and their hybrids (Lepidoptera: Tortricidae). *The Canadian Entomologist* 109:1203–1220.

Sanders, C.J., G.E. Daterman, R.F. Shepherd, and H. Cereske. 1974. Sex attractants of two species of western spruce budworm *Choristoneura biennis* and *C. viridis* (Lepidoptera: Tortricidae). *The Canadian Entomologist* 106:157–159.

Silk, P.J., and L.P.S. Kuenen. 1984. Sex pheromones and their potential as control agents for forest Lepidoptera in eastern Canada. Pp. 35–47. In W.Y. Garner and J. Harvey, Jr., eds. *Chemical and Biological Controls in Forestry, ACS Symposium Series*, Vol. 238. Washington, DC: American Chemical Society.

Silk, P.J., and L.P.S. Kuenen. 1985. Sex pheromone chemistry and pheromone-mediated behaviour of spruce budworm moths. Pp. 393–394. In C.J. Sanders, R.W. Stark, E.J. Mullins, and J. Murphy, eds. *Recent advances in Spruce Budworms Research.* Proceedings, CANUSA Spruce Budworms Research Symposium, Bangor, Maine 1984, Natural Resources Canada, Canadian Forest Service, Ottawa, Ontario.

Silk, P.J., and L.P.S. Kuenen. 1986. Spruce budworm (*Choristoneura fumiferana*) pheromone chemistry and behavioral responses to pheromone components and analogues. *Journal of Chemical Ecology* 12:367–383.

Silk, P.J., and L.P.S. Kuenen. 1988. Sex pheromones and behavioral biology of the coniferophagous *Choristoneura. Annual Review of Entomology* 33:83–101.

Silk, P.J., L.P.S. Kuenen, E.W. Butterworth, and C.J. Sanders. 1986. The sex pheromone of the jack pine budworm (*Choristoneura pinus pinus*): identification of pheromone components and the development of a monitoring system. Jack Pine Budworm Information Exchange, Manitoba Natural Resources Publication.

Silk, P.J., L.P.S. Kuenen, S.H. Tan, W.L. Roelofs, and C.J. Sanders. 1985. Identification of sex pheromone components of jack pine budworm, *Choristoneura pinus pinus* Freeman. *Journal of Chemical Ecology* 8:159–167.

Silk, P.J., S.H. Tan, C.J. Wiesner, R.J. Ross, and G.C. Lonergan. 1980. Sex pheromone chemistry of the eastern spruce budworm, *Choristoneura fumiferana. Environmental Entomology* 9:640–644.

Silk, P.J., C.J. Wiesner, S.H. Tan, R.J. Ross, and G.G. Grant. 1982. Sex pheromone chemistry of the western spruce budworm, *Choristoneura occidentalis*, free. *Journal of Chemical Ecology* 8:351–362.

Smith, M.A., E.S. Eveleigh, K.S. McCann, M.T. Merilo, P.C. McCarthy, and K.I. van Rooyen. 2011. Barcoding a quantified food web: crypsis, concepts, ecology and hypotheses. *PLOS ONE* 6(7):e14424.

Sperling, F.A.H., and D.A. Hickey. 1994. Mitochondrial DNA sequence variation in the spruce budworm species complex (*Choristoneura*: Lepidoptera). *Molecular and Biological Evolution* 11:656–665.

Sperling, F.A.H., and D.A. Hickey. 1995. Amplified mitochondrial DNA as a diagnostic marker for species of conifer-feeding *Choristoneura* (Lepidoptera: Tortricidae). *The Canadian Entomologist* 127:277–288.

Stock, M.W., and P.J. Castrovillo. 1981. Genetic relationships among representative populations of five *Choristoneura* species: *C. occidentalis*, *C. lambertiana*, and *C. fumiferana* (Lepidoptera: Tortricidae). *The Canadian Entomologist* 113:857–865.

Sweeney, J.D., J.A. McLean, and L.M. Friskie. 1990a. Roles of minor components in pheromone-mediated behavior of western spruce budworm moths. *Journal of Chemical Ecology* 16:1517–1530.

Sweeney, J.D., J.A. McLean, and R.F. Shepherd. 1990b. Factors affecting catch in pheromone traps for monitoring the western spruce budworm, *Choristoneura occidentalis* Freeman. *The Canadian Entomologist* 122:1119–1130.

Thomson, D.R., N.P.D. Angerrilli, C. Vincent, and A.P. Gaunce. 1991. Evidence for regional difference in response of the obliquebanded leafroller (Lepidoptera: Tortricidae) to sex pheromone blends. *Environmental Entomology* 20:935–938.

Volney, W.J.A., and R. Fleming. 2007. Spruce budworm *Choristoneura* spp. biotype reactions to forest and climate characteristics. *Global Change Biology* 13:1630–1643.

Volney, W.J.A., A.M. Liebhold, and W.E. Waters. 1984. Host associations, phenotypic variation, and mating compatibility of *Choristoneura occidentalis* and *C. retiniana* (Lepidoptera: Tortricidae) populations in south-central Oregon. *The Canadian Entomologist* 116:813–826.

Weatherston, J., G.G. Grant, L.M. MacDonald, D. Frech, R.A.Werner, C.C. Leznoff, and T.M. Fyles. 1978. Attraction of various tortricine moths to blends containing *cis*-11-tetradecenal. *Journal of Chemical Ecology* 4:543–549.

Weatherston, J., J.E. Percy, and L.M. MacDonald. 1976. Field testing of *cis*-11-tetradecenal as attractant or synergist in Tortricinae. *Experientia* 32:178–179.

Weatherston, J., W.L. Roelofs, A. Comeau, and C.J. Sanders. 1971. Studies of physiologically active arthropod secretions. X. Sex pheromone of the eastern spruce budworm, *Choristoneura fumiferana* (Lepidoptera: Tortricidae). *The Canadian Entomologist* 103:1741–1747.

Wiesner, C.J., P.J. Silk, S.H. Tan, P. Palaniswamy, and J.O. Schmidt. 1979. Components of the sex pheromone gland of the eastern spruce budworm, *Choristoneura fumiferana* (Lepidoptera: Tortricidae). *The Canadian Entomologist* 111:1311.

Willett, C.S. 2000. Evidence for directional selection acting on pheromone-binding proteins in the genus *Choristoneura. Molecular and Biological Evolution* 17:553–562.

Willhite, E.A., and M.W. Stock. 1983. Genetic variation among western spruce budworm (*Choristoneura occidentalis*) outbreaks in Idaho and Montana. *The Canadian Entomologist* 115:41–54.

Wolf, W.A., and W.L. Roelofs. 1987. Reinvestigation confirms action of Δ11-desaturases in spruce budworm moth sex pheromone biosynthesis. *Journal of Chemical Ecology* 13:1019–1027.

CHAPTER NINETEEN

The Endemic New Zealand Genera
Ctenopseustis and *Planotortrix*

A Down-Under Story of Leafroller Moth Sex Pheromone Evolution and Speciation

RICHARD D. NEWCOMB, BERND STEINWENDER,
JÉRÔME ALBRE, and STEPHEN P. FOSTER

Introduction

Leafroller moths within the endemic New Zealand genera *Ctenopseustis* and *Planotortrix*, along with the Australian light-brown apple moth, *Epiphyas postvittana* (all Tortricidae), form a complex that constitutes the most economically significant group of insect pests of fruit crops in New Zealand (Wearing et al. 1991). In the 1970s and 1980s, identifications of sex pheromones of pest species proliferated worldwide, driven by the desire to use synthetic sex pheromones as integrated pest management tools for monitoring pests (McNeil 1991) as well as for directly controlling pests through environmentally benign techniques such as mating disruption (Cardé and Minks 1995). Thus, in the late 1970s/early 1980s, the identification of the sex pheromones of the economically important, endemic New Zealand leafrollers, *Ctenopseustis obliquana* (*sensu* Green and Dugdale 1982; brown-headed leafroller, BHL) and *Planotortrix excessana* (*sensu* Dugdale 1966; green-headed leafroller, GHL), was initiated. From a practical perspective, this complemented work undertaken in Australia on the identification of the sex pheromone of the Australian light-brown apple moth (Bellas et al. 1983). Although the identifications of the sex pheromones of GHL and BHL proceeded relatively smoothly, researchers at the New Zealand Department of Scientific Industrial Research (DSIR) Entomology Division in Auckland soon realized that the accepted notion of three pest leafroller species (including *E. postvittana*)

might be naïve; they found that the identified pheromone of the GHL was attractive in some regions, but not in others (Galbreath et al. 1985). As more in-depth investigations of the sex pheromones of more populations of GHL and BHL, as well as of other species in the two genera, established that this complex consisted of a number of different pheromone "types," it became clear this group was an important model for studying how differences in sex pheromonal communication contribute to species specificity in closely related and cryptic taxa. The uniqueness of this group for studies on sex pheromone evolution in Lepidoptera is highlighted by the different compounds these insects use compared to Palearctic, Nearctic, and many Gondwanan Tortricidae. Based on pheromone composition and morphology, the "two" BHL and GHL species were revised (Dugdale 1990) to recognize at least five species within *Ctenopseustis* (*C. obliquana, C. herana, C. fraterna, C. filicis,* and *C. servana*) and seven within *Planotortrix* (*P. avicenniae, P. excessana, P. flammea, P. notophaea, P. octo, P. octoides,* and *P. puffini*), including at least two species in each genus that are part of the actual leafroller pest complex.

Distribution and Pest Status

Geographically, the species within *Ctenopseustis* and *Planotortrix* vary considerably in their range across New Zealand (figure 19.1). Some, such as *C. obliquana, P. excessana,* and *P. octo,* are

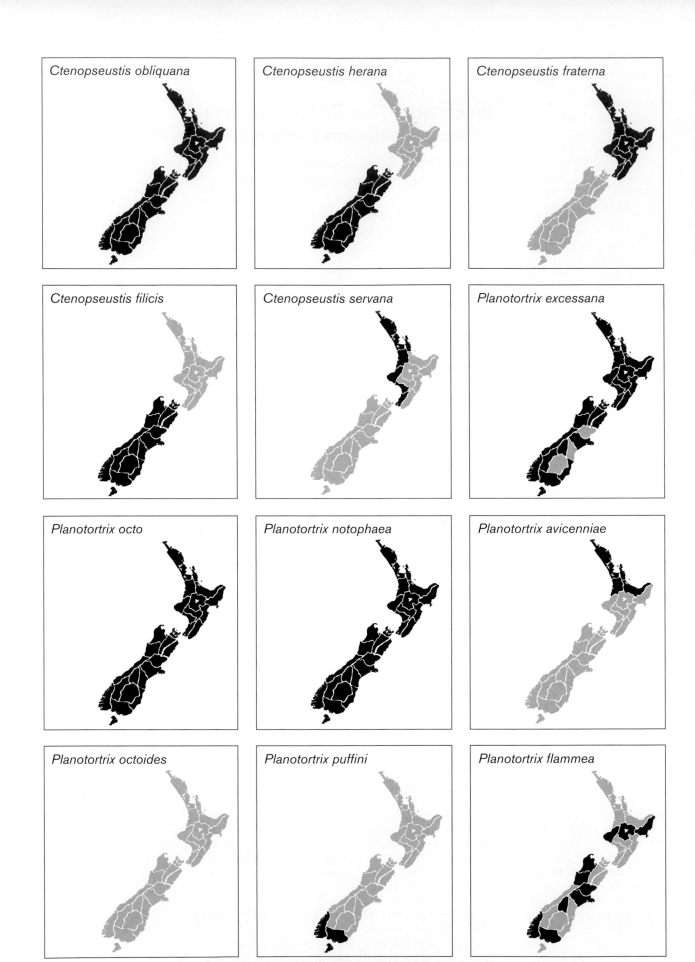

FIGURE 19.1 Distribution maps of Planotortrix and Ctenopseustis species. Dark shading refers to areas where species are commonly found. *Planotortrix octoides* is found on the Chatham Islands only, to the east of the South Island of New Zealand.

widespread throughout the country. Others are more restricted; *C. herana*, *C. filicis*, and *P. puffini* are found only on the South Island, *C. fraterna* only on the North Island, and *P. octoides* is found only on the Chatham Islands, east of the main islands of New Zealand. In addition, the group contains species with different host associations, ranging from monophagous to highly polyphagous. *C. fraterna*, *C. filicis*, *P. avicenniae*, *P. flammea*, and *P. puffini* are specialized on one or a few host plants. For example, *C. fraterna* and *C. filicis* are fern feeders, whereas *P. avicenniae* is a specialist on the mangrove *Avicennia resinifera*. *C. servana* is polyphagous on coastal angiosperms, and *C. obliquana*, *C. herana*, *P. excessana*, *P. notophaea*, *P. octo*, and *P. octoides* are all highly polyphagous.

C. obliquana, *C. herana*, *P. octo*, and *P. excessana* are the major pest species in the complex, being especially important in high value export crops such as pipfruit, summerfruit, and kiwifruit (Wearing et al. 1991). The larvae of these species not only feed on fruits of these crops, causing scarring and fruit deformation, thus reducing value, but also represent a market access problem for exporting fruit from New Zealand to overseas markets such as Japan and North America (Wearing et al. 1991). Numerous insecticide sprays are required to control these pests throughout the production season. Of the other species in the complex, only *P. notophaea* is considered an economic pest, albeit a relatively minor one of coniferous forests (Nuttall 1983).

Adult moths are typically 10–15 mm in length, with *Ctenopseustis* slightly smaller than *Planotortrix*, and males of all species smaller than conspecific females. Adult wing patterning is highly variable, both within and between species. The number of generations per year varies throughout the country, with up to three and four generations for *Ctenopseustis* and *Planotortrix species*, respectively. Both genera can overwinter as nondiapausing larvae but, on the warmer North Island especially, larvae can develop through the winter, allowing an additional generation in June or July (Wearing et al. 1991). Females can lay up to 600 eggs (Clare and Singh 1988), in batches typically on the adaxial surface of leaves (Wearing et al. 1991). Larvae emerge after 8–9 days (Clare and Singh 1988) and disperse to feed on the abaxial side of leaves, where they shelter in silken webbings (Wearing et al. 1991; Suckling and Brockerhoff 2010). The mean larval period ranges from 32 days for males to 36 days for females. Larvae typically have five or six instars (Clare and Singh 1988).

Many of the species are kept as laboratory stocks at Plant & Food Research's Insect Rearing Facility in Auckland. The larvae of *C. obliquana*, *C. herana*, *P. octo*, *P. excessana*, and *P. notophaea* are maintained individually on a general purpose diet (Singh 1983). The *C. obliquana* culture was started from larvae collected in Mt Albert, Auckland, in 1982; *C. herana* from Tai Tapu in Lincoln in 1990; *P. octo* from Canterbury in 1982; *P. notophaea* from Palmerston North in 1994; and *P. excessana* from Dunedin in 1998, with a second *P. excessana* culture derived from material collected from the North Island. Efforts are currently underway to develop laboratory colonies of *C. fraterna*, *C. filicis*, and *P. avicenniae*.

Sex Pheromones of The Green-Headed and Brown-Headed Leafrollers

Initial Identification of the Sex Pheromones

Studies identifying the sex pheromones of the GHL, *Planotortrix excessana*, and BHL, *Ctenopseustis obliquana*, commenced toward the end of the 1970s. The research team, headed by Dr. Ross Galbreath of the Entomology Division of DSIR at Auckland, started colonies of both species, with the founding insects being mated female moths collected from the Auckland area (northern North Island). Their initial sex pheromone studies found that these populations utilized unusual, for Tortricidae (Roelofs and Brown 1982), unsaturated sex pheromone components: (Z)-8-tetradecenyl acetate (Z8-14Ac) for the GHL (Galbreath et al. 1985), and an 80:20 mixture of Z8-14Ac and (Z)-5-tetradecenyl acetate (Z5-14Ac) for the BHL (Young et al. 1985). However, a subsequent rearing problem, resulting in the loss of the GHL colony, necessitated collection of new insects and led to a fortuitous discovery. Insects for the new colony were collected from Lincoln (South Island), while Galbreath also collected a single female GHL moth from Tokoroa (North Island) and reared the progeny separately from the colony originating from Lincoln. When the cross-attraction of these two populations was tested in a field cage, it was found that pheromone gland extract from the respective females only attracted males of the same population. Preliminary analysis of the pheromone gland of the Tokoroa females showed them to produce different (from the Z8-14Ac produced by the Lincoln and original Auckland colonies) unsaturated tetradecenyl acetates, which field tests with synthetic compounds allowed tentative identification as Z5-14Ac and (Z)-7-tetradecenyl acetate (Z7-14Ac) (Galbreath et al. 1985).

In late 1983, Wendell Roelofs of Cornell University arrived in New Zealand for a six-month sabbatical to work on these New Zealand leafroller moths. To coincide with his visit, a large project was initiated at DSIR Entomology Division, in which collaborators collected GHL and BHL, as well as other congeners, at various sites around New Zealand and sent live insects back to Auckland, where female moths were analyzed for pheromone content. The aims of this project were to (a) determine the distributions of the two different pheromone "types" of GHL, (b) determine whether there were any other GHL and BHL pheromone types, and (c) determine whether these taxa and their congeners used unusual pheromone components. This project also utilized the expertise of John Dugdale, an authority on New Zealand Lepidoptera at DSIR Entomology Division, to ensure that all insects were correctly identified, and to try and find consistent morphological differences among the various pheromone-characterized taxa that could assist revision of these two economically important genera.

Pheromone Survey of *Planotortrix*

The analyses of GHL females found that the two pheromone "types" identified by Galbreath et al. (1985) were widely distributed throughout New Zealand (Foster et al. 1986). Whereas females producing Z8-14Ac (termed "Type A") appeared to be fairly homogeneous in their pheromone content, females producing Z5-14Ac and Z7-14Ac consisted of two distinct phenotypes: those producing a higher proportion of Z5-14Ac and a relatively small quantity of pheromone (termed "Type B") and the other producing a higher proportion of Z7-14Ac and a relatively large quantity of pheromone (termed "Type C"; Foster et al. 1991). Subsequent studies, using selected populations of these two types from Hamilton (North Island) and Dunedin (South Island), showed that, although the populations produced significantly different ratios of pheromone components, there was overlap between some individuals in the respective populations. Furthermore, in a field-cage trial, caged females of either population attracted males from both populations equally well (Foster et al. 1989). Although these two pheromone "types" were distinct from each other in terms of pheromone production, given that they cross-attracted each other and were often found in sympatry (Foster et al. 1991), it was concluded that they represented a single species with considerable phenotypic variation (Foster et al. 1989).

Other *Planotortrix* species were found to produce one or both of Z5-14Ac and Z7-14Ac. Females of a *Planotortrix* taxon (now *P. avicenniae*) that feeds on mangrove, *Avicennia resinifera*, produce a mixture of Z5-14Ac and tetradecyl acetate ("Type M"). Field trials, using traps baited with a 1:1 synthetic mixture of this blend, caught relatively large numbers of males (Foster et al. 1987). "Type M" lures, made up of predominantly Z5-14Ac, attracted males only in mangroves, whereas "Type A" lures, containing Z8-14Ac, failed to attract *Planotortrix* species in mangrove areas (Foster et al. 1986). Later work (Foster et al. 1990) found that females also produced a small quantity of Z7-14Ac (ca 5% of that of Z5-14Ac); addition of this amount to the previously identified blend resulted in increased trap catches. Females of *P. notophaea*, a species found on conifers and some small-leaved angiosperms, produced Z7-14Ac with no detectable amount of Z5-14Ac. *P. flammea*, a species restricted to species of *Hebe*, an endemic New Zealand genus in the family Plantaginaceae, produced a roughly 50:50 mixture of Z7-14Ac:Z5-14Ac, whereas an undescribed species found on New Zealand Asteraceae in the genera *Olearia* and *Brachyglottis* species, in the lower South Island, produced a 95:5 mixture of Z7-14Ac:Z5-14Ac (Foster et al. 1990). Field trials using these respective blends, however, failed to capture males of any of these three species (Stephen P. Foster, unpublished data). A number of other *Planotortrix* species, including *P. orthropis*, *P. clarkei*, *P. spatiosa*, *P. orthocopa*, as well as the related "*Tortrix*" *flavescens*, were collected during the survey and found to produce the typically Nearctic tortricid mixture of (*Z*)-11- and (*E*)-11-tetradecenyl acetates (Z11-, E11-14Acs; Foster and Dugdale 1988).

Pheromone Survey of *Ctenopseustis*

Analyses of *Ctenopseustis obliquana* females from a collection of BHL from Tai Tapu (near Christchurch, South Island) found that the pheromone gland content was distinct from that of *C. obliquana* from Auckland analyzed by Young et al. (1985); Tai Tapu females produced Z5-14Ac and no Z8-14Ac (Foster and Roelofs 1987). Field trials using synthetic Z5-14Ac confirmed the attractiveness of Z5-14Ac to males in this region of the South Island. A further field trial, using synthetic blends formulated for this population (termed "Type II"; Z5-14Ac) and the Auckland population (termed "Type I"; consisting of both Z5-14Ac and Z8-14Ac), was conducted at Christchurch and Alexandra (South Island), the latter a site where the Auckland-type pheromone was known to be attractive to *C. obliquana* males. This trial demonstrated that male *C. obliquana* at Christchurch were attracted only to the Type II blend, whereas male *C. obliquana* from Alexandra were attracted only to the Type I blend (Foster and Roelofs 1987). A wind-tunnel trial, testing various ratios of Z5-14Ac:Z8-14Ac, demonstrated that the proportion of Type I males landing at the pheromone source dropped off steeply as the proportion of

Z5-14Ac in the blend increased, whereas the proportion of Type II males landing at the source dropped off steeply as the proportion of Z8-14Ac increased (Clearwater et al. 1991).

For females producing both Z5-14Ac and Z8-14Ac, a difference was noted between North and South Island populations. Females from the South Island (termed "Type III"; Foster et al. 1986), on average, produced blends with higher ratios of Z8-14Ac than females from the North Island (the original Type I), although ratios of some females between the two islands overlapped (Clearwater et al. 1991). A field-cage trial, using insects of an Auckland population (Type I) and those of an Alexandra population (Type III), demonstrated that Type I females attracted both types of males equally well. Type III females in the trial failed to attract significant numbers of either type of male, possibly due to the condition of the females. It was concluded that Type I and Type III female BHL represented variation in pheromone ratios within the same species (Clearwater et al. 1991). Recently, a field-trapping trial (Suckling et al. 2012) demonstrated that male *C. obliquana* in the Alexandra region were caught in higher numbers in traps baited with a 90:10 ratio (i.e., Type III blend) of Z8-14Ac:Z5-14Ac than in traps baited with a 70:30 ratio (i.e., Type I blend) of the two compounds. Thus, populations of *C. obliquana* exhibit some blend specificity to ratios of Z8-14Ac and Z5-14Ac, related to ratios produced by females from the respective populations but perhaps not enough to effect assortative mating.

Another *C. obliquana* population that feeds on *Cyathea* and *Dicksonia* tree ferns in the South and Stewart Islands was sampled, with females (termed Type IV) found to produce large quantities of (*Z*)-10-hexadecenyl acetate (Z10-16Ac; Foster et al. 1986). Limited field trapping, using sticky traps baited with synthetic Type I, Type II, or Z10-16Ac blends, showed that only traps baited with Z10-16Ac caught significant numbers of males of this undescribed species (Foster et al. 1991).

Chemical analyses of females collected as larvae, or trapping of males using blends for the respective *C. obliquana* types, showed that Type I (including Type III) was distributed throughout New Zealand, whereas Type II had a more restricted distribution in the North Island and, while found throughout most areas of the South Island that were sampled, was not found in the Central Otago region (Foster et al. 1991).

Two other species of *Ctenopseustis* were also analyzed. *C. servana* produced a mixture of Z5-14Ac and Z7-14Ac (Foster and Dugdale 1988), although traps baited with this blend failed to catch males (Stephen P. Foster, unpublished data). *C. fraterna*, a tree fern-feeding species, produces an unusual hexadecadienyl acetate, which has yet to be identified.

A Revised Taxonomy and Systematics

Before 1984, it was thought that there were just two pest species within *Ctenopseustis* and *Planotortrix*, namely *Ctenopseustis obliquana* and *Planotortrix excessana*. However, the pheromone work clearly pointed to more species within these entities. Encouraged by the pheromone findings, John Dugdale reassessed the taxonomy of the group (Dugdale 1990). He moved species that utilize E11-14Ac and Z11-14Ac as sex pheromone components into two new genera, *Apotena* and *Leucotenes*. Within the remaining genera he raised one new species in *Ctenopseustis* (*C. filicis*) and four in *Planotortrix* (*P. octo*, *P. octoides*, *P. avicenniae*, and *P. puffini*). Dugdale's 1990 manuscript provides detailed morphological descriptions,

TABLE 19.1

Sex pheromone components and host plants of leafroller species within the genera *Ctenopseustis* and *Planotortrix*

Species	"Type"	Sex pheromone components	Ratio	Host plant	References
Planotortrix					
P. excessana	B	Z5-14Ac, Z7-14Ac[1]	60:40	Polyphagous	Foster et al. (1989); Galbreath et al. (1985)
P. octo	A	Z8-14Ac, Z10-14Ac	98:2	Polyphagous	Foster et al. (1986); Galbreath et al. (1985)
P. avicenniae	M	Z5-14Ac, 14Ac	100	*Avicennia resinifera*	Foster et al. (1987)
P. octoides		Z8-14Ac	100	Polyphagous	Dugdale (1990)
P. puffini	MBS	Z5-14Ac, Z7-14Ac, Z9-14Ac	3:97:2	*Brachyglottis, Cemisia, Olearia*	Foster and Dugdale (1988)
P. flammea		Z5-14Ac, Z7-14Ac	52:48–61:39	*Hebe* spp.	Foster et al. (1990)
P. notophaea		Z7-14Ac +?	?	Polyphagous	Foster et al. (1986)
Ctenopseustis					
C. obliquana	I and II II NI	Z5-14Ac, Z8-14Ac Z5-14Ac	80:20–90:10	Polyphagous	Foster et al. (1986); Young et al. (1985)
C. herana	II SI	Z5-14Ac	100	Polyphagous	Foster and Roelofs (1987)
C. fraterna		?	?	Pteridophyta	
C. filicis	IV	Z10-16Ac	100	Pteridophyta, *Dicksonia squarrosa*	Foster and Dugdale (1988)
C. servana		Z5-14Ac, Z7-14Ac	32:68–35:65	Polyphagous on woody coastal angiosperms	Foster and Dugdale (1988); Foster et al. (1990)

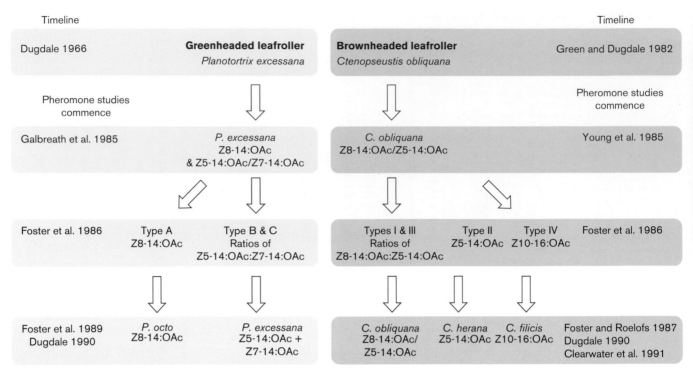

Dugdale 1966

Greenheaded leafroller
Planotortrix excessana

Brownheaded leafroller
Ctenopseustis obliquana

Green and Dugdale 1982

Pheromone studies
commence

Galbreath et al. 1985

P. excessana
Z8-14:OAc
& Z5-14:OAc/Z7-14:OAc

C. obliquana
Z8-14:OAc/Z5-14:OAc

Young et al. 1985

Pheromone studies
commence

Foster et al. 1986

Type A
Z8-14:OAc

Type B & C
Ratios of
Z5-14:OAc:Z7-14:OAc

Types I & III
Ratios of
Z8-14:OAc:Z5-14:OAc

Type II
Z5-14:OAc

Type IV
Z10-16:OAc

Foster et al. 1986

Foster et al. 1989
Dugdale 1990

P. octo
Z8-14:OAc

P. excessana
Z5-14:OAc +
Z7-14:OAc

C. obliquana
Z8-14:OAc/
Z5-14:OAc

C. herana
Z5-14:OAc

C. filicis
Z10-16:OAc

Foster and Roelofs 1987
Dugdale 1990
Clearwater et al. 1991

FIGURE 19.2 Taxonomic progression of the described greenheaded and brownheaded leafroller moths based on pheromone and morphological studies.

taxonomic keys, and distribution maps for each species. This revision took the total number of species within the 2 genera to 12: 7 in *Planotortrix* and 5 in *Ctenopseustis*. For *P. excessana*, Dugdale (1990) revised the species to include the Z5-14Ac/Z7-14Ac-producing taxa (i.e., Types B and C), while revising the Z8-14Ac-producing taxa as new species: *P. octo* for the mainland taxon and *P. octoides* for the taxon on the Chatham Islands. Type M, which is only found on mangroves and uses a predominantly Z5-14Ac blend, was revised into a separate species, *P. avicenniae*, based on both pheromone and morphological characters, whereas Type MBS was described as *P. puffini*. For *Ctenopseustis*, Dugdale (1990) revised *C. obliquana*, in accordance with the results of the pheromone studies, into three separate species: Types I and III = *C. obliquana*, Type II = *C. herana* (from the South Island), and Type IV = *C. filicis*. In this revision, he noted a further entity (Type II North Island), but did not describe it as a distinct species. Subsequently, he also noted morphological differences between lowland and montane populations of *C. fraterna* that are yet to be recognized formally (John Dugdale, unpublished data). Figure 19.2 summarizes the changes in the taxonomy of the group that occurred based on the pheromone and subsequent morphological characterization of these species.

Isozyme electrophoresis was employed by White and Lambert (1994) at the University of Auckland in the early 1990s to look at whether the newly recognized sibling species of *C. herana* and *C. obliquana*, from *Ctenopseustis*, and *P. octo*, *P. octoides*, and *P. excessana*, from *Planotortrix*, constituted robust biological species. Two monomorphic diagnostic markers were found in laboratory material (White and Lambert 1994). The β-hydroxybutyrate dehydrogenase-3 (*Hbdh-3*) locus discriminated *P. octo* and *P. octoides* (referred to as *P. excessana* Type A in the study) from *P. excessana* (referred to as *P. excessana* Types B and C in the study). In the same way, the hexokinase-1 (*Hk-1*) locus separated *C. obliquana* (*C. obliquana* Type I

and III in the study) and *C. herana* (*C. obliquana* Type II in the study) sibling species. A complementary study using a large number of field specimens from throughout New Zealand confirmed the specificity of these markers. The addition of 26 polymorphic and 4 monomorphic loci provided further evidence for a lack of gene flow between the sibling species (White and Lambert 1995). Similarly, restriction fragment length polymorphisms of the ribosomal RNA gene complex discriminated sibling species of both genera providing further evidence for genetic divergence (Sin et al. 1995).

The order of the speciation events within the group has been a major focus of investigation. By integrating studies on molecular markers with morphological and pheromone characters, some insights into speciation in the group have been obtained. Three molecular phylogenetic studies have been conducted, all utilizing regions within the mitochondrial genome that include part or all of the cytochrome oxidase genes (Newcomb and Gleeson 1998; Gleeson et al. 2000; Langhoff et al. 2009). These studies confirmed the monophyly of each genus; however, *Leucotenes coprosmae* and *Apoctena orthropis* form a group that may be sister to *Ctenopseustis* (Gleeson et al. 2000). Within *Ctenopseustis*, *C. servana* is positioned basally with respect to the other species. The relationships among the other four species are yet to be fully resolved. In *Planotortrix*, two well-separated groups emerged. The first is well resolved, and composed of *P. flammea*, *P. notophaea*, and *P. puffini*. The second group is composed of the remaining species, *P. octo*, *P. octoides*, *P. excessana*, and *P. avicenniae*, with little resolution of the relationships among them (Langhoff et al. 2009). To resolve the relationships among the sibling species that have evolved relatively recently, sequencing of further genes will be required. Notwithstanding the incomplete resolution of the phylogeny for the group, these molecular studies support to the view that the use of similar pheromone blends has evolved independently within the two genera (Newcomb and Gleeson 1998). For example, Z5-14Ac is used by both *C. herana* and *P. avicenniae*, whereas a blend of Z5-14Ac and Z7-14Ac is utilized by *C. servana* and *P. excessana*.

Sex Pheromone Biochemistry

Routes of Sex Pheromone Biosynthesis

In the early 1980s, Wendell Roelofs' research group at Cornell University was unraveling how moths made pheromone components. They identified two processes, Δ11-desaturation and cytosolic β-oxidation (two-carbon chain shortening), present in the pheromone glands of several North American moths (Bjostad et al. 1987), which were involved in producing the unsaturated carbon skeletons of the respective pheromone components (Foster, this volume). Roelofs was intrigued by the Z8-, Z7-, and Z5-14Acs produced by the endemic New Zealand leafrollers, different from the typical Δ11- or Δ9-unsaturated tetradecenyl compounds commonly produced in North American Tortricidae (Roelofs and Brown 1982), and hence took a 6-month sabbatical at DSIR Entomology Division, Auckland, that started in October 1983.

Prior to leaving for New Zealand, Roelofs requested some *Planotortrix excessana* (the type producing Z8-14Ac) and *Ctenopseustis obliquana* (the type producing Z5-14Ac and Z8-14Ac) be sent to Cornell University for analysis of the fatty acids in the pheromone gland. *P. excessana* glands contained relatively large quantities of (*Z*)-10-hexadecenoate (Z10-16:Acyl) and a smaller quantity of (*Z*)-8-tetradecenoate (Z8-14:Acyl; the immediate fatty acyl precursor of the pheromone component), whereas *C. obliquana* glands contained similar amounts of the two acids. No (*Z*)-5-tetradecenoate (Z5-14:Acyl; the immediate fatty acyl precursor of Z5-14Ac) was found in the gland of female *C. obliquana*. Both species contained large amounts of the ubiquitous (*Z*)-9-octadecenoate (Z9-18:Acyl) and traces of (*Z*)-7-hexadecenoate (Z7-16:Acyl; Löfstedt and Roelofs 1985). Based on these results, Löfstedt and Roelofs (1985) speculated that Z8-14Ac in these species must be formed by Δ10-desaturation of hexadecanoate, followed by two-carbon chain shortening of Z10-16:Acyl to Z8-14:Acyl, with Z5-14Ac being possibly formed by chain shortening of Z9-18:Acyl to Z7-16:Acyl and then to Z5-14:Acyl.

In *P. excessana* (the type later renamed *P. octo*), Δ10-desaturation was confirmed as the key step in the biosynthesis of Z8-14Ac through topical application of deuterium-labeled saturated fatty acids to the pheromone gland (Foster and Roelofs 1988). Labeled Z8-14Ac, Z10-16:Acyl, and Z8-14:Acyl were produced when labeled hexadecanoic was applied to the gland, but was not produced when labeled tetradecanoic acid was applied to the gland. The gene for the Δ10-desaturase was later isolated, sequenced, cloned into an expression vector, and functionally characterized in yeast by Roelofs' group (Hao et al. 2002).

Roelofs and Bjostad (1984) predicted that the taxa that produced Z5-14Ac and Z7-14Ac did so from ubiquitous Z9-18:Acyl (oleate) and (*Z*)-9-hexadecenoate (palmitoleate; Z9-16:Acyl), respectively. This was shown to be the case in *P. excessana* (i.e., Type B). Analysis of the fatty acids present in the gland showed the two precursor acids, Z5-14:Acyl and Z7-14:Acyl, along with a trace of Z7-16:Acyl and large quantities of Z9-18:Acyl and Z9-16:Acyl (Foster 1998); the last two acids are commonly found in pheromone glands of many moth species (Jurenka

2003). Application of labeled (*Z*)-9-octadecenoic acid to the gland yielded labeled Z5-14Ac, but not labeled Z7-14Ac, whereas application of labeled (*Z*)-9-hexadecenoic acid yielded labeled Z7-14Ac, but not labeled Z5-14Ac. Interestingly, when labeled saturated (tetradecanoic, hexadecanoic, or octadecanoic) acids were applied to the gland, no labeled pheromone components were detected, suggesting that either desaturation is a slow, perhaps rate-limiting, step and that levels of any labeled Z9-18:Acyl or Z9-16:Acyl produced were low in comparison to the large amounts of native acids in the gland, or that the relatively high concentrations of free fatty acids applied to the gland inhibited Δ9-desaturation (Foster 1998). Two genes encoding Δ9-desaturases expressed in the pheromone gland of *P. excessana* have recently been isolated and functionally characterized through collaboration with the Löfstedt group in Lund, Sweden (Albre et al. 2012).

Given the close taxonomic relationship of the two genera (Dugdale 1990), it might be expected that Z5-14Ac produced by *C. herana* and *C. obliquana* is produced by the same route as that in *P. excessana* (i.e., via chain shortening of Z9-18:Acyl). However, application of labeled (*Z*)-9-octadecenoic acid to the gland of *C. herana* failed to produce detectable levels of labeled Z5-14Ac (Foster and Roelofs 1996). In contrast, application of labeled tetradecanoic, hexadecanoic, or octadecanoic, but not dodecanoic, acids to the gland yielded labeled Z5-14Ac. Moreover, the amount of label incorporated into the pheromone component increased in the order tetradecanoic > hexadecanoic > octadecanoic, suggesting that tetradecanoic acid was the substrate for the desaturase. Application of labeled tetradecanoic acid also resulted in production of detectable amounts of labeled Z5-14:Acyl. These data imply that biosynthesis of Z5-14Ac in *C. herana* occurs through Δ5-desaturation of tetradecanoate. To date, the biosynthetic routes of Z5-14Ac and Z8-14Ac in *C. obliquana* (Type A) have not been characterized. Molecular studies on pheromone gland desaturases in the two *Ctenopseustis* species have identified genes coding for Δ10- and Δ9-desaturases (Albre et al. 2012), and most recently a Δ5-desaturase (Hagström et al. 2014). The current state of understanding of the routes of biosynthesis for the sex pheromone components used by *C. obliquana*, *C. herana*, *P. octo*, and *P. excessana*, together with the isolated and characterized desaturases, is summarized in figure 19.3.

Genetics of Sex Pheromone Biosynthesis in Females

Ctenopseustis obliquana and *C. herana* were used to investigate the genetic basis of sex pheromone production by Foster et al. (1997). Females of *C. obliquana* produce a blend of Z8-14Ac and Z5-14Ac in a ratio of about 80:20, whereas *C. herana* females produce only Z5-14Ac. F_1 females produced a pheromone mixture that was similar to the *C. obliquana* blend, whereas females from F_1 backcrosses and F_2 crosses produced a complex array of blends. A simple genetic model to explain these data could not be resolved. While F_2 offspring produced pheromone blends that seemed to suggest autosomal inheritance, the paternal backcrosses expressed phenotypes that best fitted a sex-linked model of inheritance, and the maternal backcrosses produced females that fitted neither of the two models. These results suggested that multiple genetic factors, perhaps multiple functional desaturases at different loci, are likely involved in the differences in female sex pheromone production in these species.

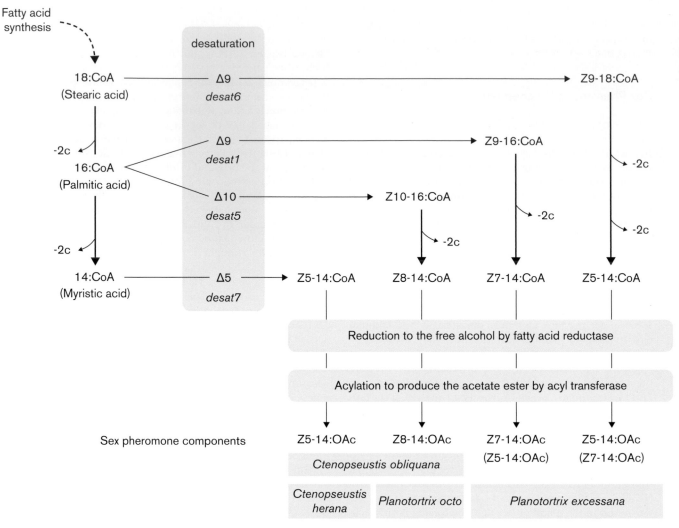

FIGURE 19.3 Schematic outlining of the likely biosynthetic routes of the sex pheromone components of *Ctenopseustis obliquana*, *C. herana*, *Planotortrix octo*, and *P. excessana* (modified from Albre et al. 2012). desat1, desat5, desat6, and desat7 correspond to the desaturase genes encoding a Δ9 desaturase with a preference for 16>18 carbon fatty acids, a Δ10-desaturase, a Δ9-desaturase with a preference for 18>16 carbon fatty acids, and a Δ5-desaturase, respectively. Chain shortening by β-oxidation is indicated by "–2C". The minor products of the two Δ9-desaturases in *P. excessana* (desat1 and desat6) are indicated in brackets. We also note that Z10-14Ac is a very minor (2%) component of the pheromone blend of *P. octo* (not shown).

More recently a similar study has been conducted with *Planotortrix octo* and *P. excessana* (Albre et al. 2013). In this case the pheromone of the F_1 hybrids and that of any backcross to *P. excessana* resembled the *P. excessana* blend, containing very little if any detectable Z8-14Ac. In F_1 hybrids, F_1 backcrosses and F_2 progeny females, the amount of Z8-14Ac was related to the level of expression of the Δ10-desaturase, *desat5*. The segregation patterns observed in the F_1 backcrosses to *P. octo* and F_2 progeny best fit a genetic model involving one major and other minor genes, with the major gene thought to be a dominant *trans*-acting repressor of *desat5* expression in the female pheromone gland.

Evolution of Sex Pheromone Biosynthesis

In the two genera, there have been changes in the use of Z8-14Ac as a component of the sex pheromones of some closely related species. In *Ctenopseustis*, Z8-14Ac is part of the sex pheromone of *C. obliquana*, but not part of the pheromone of *C. herana*; in *Planotortrix* this compound is the sole component of the sex pheromones of *P. octo* and *P. octoides*, but is not found in *P. excessana* or *P. avicenniae*. The Δ10-desaturase responsible for the production of this component has been isolated and characterized from *P. octo* (Hao et al. 2002) and an ortholog has been isolated from the pheromone gland of female *C. obliquana* (Albre et al. 2012). In these two species, the *desat5* gene that encodes the Δ10-desaturase is highly expressed in the pheromone gland of the respective females, whereas the expression of orthologs in the pheromone glands of the sibling species *P. excessana* and *C. herana* is barely detectable, but present within their respective genomes (Albre et al. 2012). These results reveal a simple switching on and off of gene expression of the relevant desaturase to produce the modified pheromone blend used by the sibling species. In *Planotortrix* the expression of *desat5* is controlled by a dominant trans-acting repressor present in *P. excessana* but absent in *P. octo*. Also, a 7-bp insertion in the promoter of desat5 in *P. octo*, which is thought to represent the binding site for an activator, is associated with high levels of expression of the gene in the pheromone gland (Albre et al. 2013).

Sex pheromone

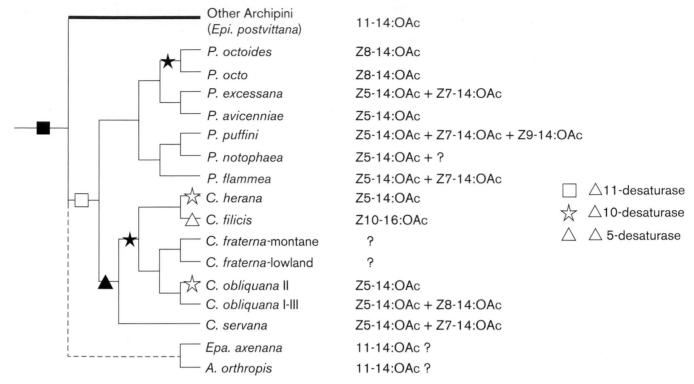

FIGURE 19.4 Hypothetical evolution of Δ5-, Δ10-, and Δ11-desaturases overlaid on a phylogeny of the Ctenopseustis/Planotortrix group based on cytochrome oxidase I gene sequence. Triangles represent Δ5-desaturase changes, boxes Δ11-desaturase changes, and stars Δ10-desaturase changes. Closed symbols are proposed gains and open symbols losses. Abbreviations: A = Apoctena; C = Ctenopseustis; Epa. = Epalxiphora; Epi. = Epiphyas; P = Planotortrix. Chemical acronyms are in the order of double bond geometry, position of unsaturation, chain length, and terminal functional group (in all cases acetate), e.g., Z5-14Ac = (Z)-5-tetradecenyl acetate.

The evolutionary events underpinning this change in Δ10-desaturase expression may represent a loss or gain of expression. Albre et al. (2012) attempted to infer the order of these evolutionary events based on the phylogenetics of the group (figure 19.2). Within *Ctenopseustis*, it appears that there has been a gain in the use of Z8-14Ac or Z10-16Ac in the evolution of the genus after the split from *C. servana*. The use of these components, presumably through the gain of expression of the Δ10-desaturase in pheromone glands, has become widespread and is used by *C. obliquana* (Z8-14Ac) and apparently by *C. filicis* (Z10-16Ac). Subsequently, *C. herana*, which only produces Z5-14Ac, appears to have lost Δ10-desaturase expression in the pheromone gland. A similar loss of Δ10-desaturase expression may also have occurred in the formation of the geographically restricted *C. obliquana* type II North Island, which also uses only Z5-14Ac in its sex pheromone. In contrast, the evolution of Δ10- and Δ8-unsaturated sex pheromone components within *Planotortrix* is likely derived from an evolutionary scenario involving a gain of expression of the Δ10-desaturase in the pheromone gland. Components such as Z5-14Ac, Z7-14Ac, and Z9-14Ac, which can be derived from the action of Δ9-desaturases, are widespread within the group and thus probably represent the ancestral pheromone blend. The use of Z8-14Ac is restricted to just two species within the genera: *P. octo* and its geographically isolated relative *P. octoides*, which is found only on the Chatham Islands. Therefore, the presence of Z8-14Ac is likely a derived condition within the genus, with Δ10-desaturase expression in the pheromone glands of *P. octo* likely a gain-of-function event probably

through the release of repression by the *P. excessana* repressor of *desat5* and the gain of the 7-bp activator binding site in the *desat5* promoter (Albre et al. 2013). The alternative scenario that all species within the genus, except for two, have lost expression of the Δ10-desaturase independently is less parsimonious, but still plausible. Simultaneously or subsequent to producing Z8-14Ac, *P. octo* must also have lost the ability to produce Z5-14Ac and Z7-14Ac.

It is more difficult to speculate on the evolution of the Δ5- and Δ9-desaturases. Certainly the use of Δ9-desaturase activity to produce Z5-, Z7-, and Z9-14Ac may be widespread within the genus *Planotortrix*. However, in *Ctenopseustis*, the Δ5-desaturase *desat7* is most likely responsible for producing Z5-14Ac in *C. herana*, if not also in *C. obliquana* and possibly *C. servana*.

There has been much discussion on the general order of evolutionary events underpinning desaturase evolution within the Lepidoptera. Roelofs and coworkers (Roelofs and Bjostad 1984; Hao et al. 2002; Roelofs and Rooney 2003) suggested that within the Lepidoptera, desaturases have evolved generally from Δ9- through Δ10- to Δ11-. Their argument was that the rarer Δ10-desaturase-derived products were found only in species possessing morphological characters associated with a primitive condition (Roelofs and Bjostad 1984; Roelofs and Wolf 1988). However more recent phylogenetic studies of the family Tortricidae reveal that the tribe Archipini, in which the genera *Ctenopseustis* and *Planotortrix* fall, is not basal within the family, but forms a large derived group of mostly cosmopolitan species (Regier et al. 2012). It seems that species within *Ctenopseustis* and *Planotortrix* may have maintained some primitive morphological characters during their evolution. Furthermore, in their analysis of the sex pheromone components of various Australian tortricid moths, Horak et al. (1988) pointed out that Δ11-unsaturated pheromone components were found in Australian tortricids that exhibit primitive morphological characters. Therefore, either the Δ11-desaturases were present early in the evolution of the Tortricidae or, perhaps, there is no direct evolutionary link between morphological and pheromonal characteristics in these Australasian moths.

The pattern of pheromone evolution within *Ctenopseustis* and *Planotortrix* seems complex (Dugdale 1997; figure 19.3). Other endemic New Zealand archipines (excluding *Planotortrix* and *Ctenopseustis*) that have been analyzed for pheromone produce compounds consistent with the action of a Δ11-desaturase (Foster and Dugdale 1988), consistent with the view that Δ11-desaturase activity is ancestral to the genera *Ctenopseustis* and *Planotortrix*. Efforts to identify "dormant" Δ11-desaturases within the preliminary genome assemblies of *C. obliquana* and *P. octo*, using the Δ11-desaturase of *Epiphyas postvittana* as a query sequence have, to date, not been successful, suggesting that an ancestor of *Ctenopseustis* and *Planotortrix* most likely lost the Δ11-desaturase gene or perhaps rapidly evolved it into a Δ10-desaturase. The latter scenario is possible, as the two types of desaturase are sisters in phylogenetic trees (Rooney 2009). This has left the two genera with a new Δ10-desaturase that is expressed in pheromone glands early in the evolution of *Ctenopseustis*, perhaps after the split of *C. servana*, and later in the evolution of *Planotortrix*, perhaps in an ancestor of *P. octo* and *P. octoides*. In addition, the two genera also harbor in their genomes a standing set of at least two Δ9-desaturases and some novel desaturases with which to construct pheromone blends (Albre et al. 2012; Hagström 2014).

Sex Pheromone Reception and Response

Genetics of Sex Pheromone Reception and Behavioral Response in Males

The genetic bases of sex pheromone perception and behavioral responses in male *Ctenopseustis obliquana* and *C. herana* have been investigated (Hansson et al. 1989; Foster et al. 1997). In the electrophysiological studies, the trichoid sensilla of *C. obliquana* males contained a large spike amplitude cell that responded strongly to the main pheromone component, Z8-14Ac, and also a small spike amplitude cell that responded weakly to the minor component, Z5-14Ac. In *C. herana*, the large spike amplitude cell responded strongly to the main pheromone component, Z5-14Ac, whereas the small spike amplitude cell responded weakly to Z8-14Ac. These differences suggested a swapping of the sensitivity of the two sensory neurons in the respective species.

Crosses between the two species produced F_1 males with an electrophysiological response similar to that of the parent *C. herana*, independent of the direction of the cross. In addition to sensilla responding to Z5-14Ac (large spike amplitude) and Z8-14Ac (low spike amplitude), sensilla that responded to neither compound were also observed; these sensilla were referred to as "atypical" (Hansson et al. 1989). Overall, the response patterns of the crosses and backcrosses best fitted a model of Z-linked inheritance with the *C. herana* type being dominant. However, as for the study on pheromone biosyn-

thesis in these species (Foster et al. 1997), the high degree of variability in the responses of the hybrids, along with the atypical receptor neurons, suggested that other genes are likely involved in the electrophysiological differences between *C. herana* and *C. obliquana*.

The behavioral responses of parent and hybrid males were tested in a wind tunnel (Foster et al. 1997). Individual males were scored for a number of behaviors, including whether they flew upwind toward blends of Z8-14Ac:Z5-14Ac (ratios of 0:100, 30:70, 70:30, and 90:10). Most *C. obliquana* parent males responded to the blends with the two highest ratios of Z8-14Ac; no *C. obliquana* males responded to the pheromone blend consisting of only Z5-14Ac. Similarly, the parent *C. herana* males flew to the blends most similar to that produced by their conspecific females, Z5-14Ac or blends low in Z8-14Ac, and few flew to blends containing greater proportions of Z8-14Ac. Males from the F_1 showed broader responses than either of the respective parents, but tended to favor blends higher in Z5-14Ac. The response patterns of the two types of F_2 males were broad, but distinct from each other. Overall the patterns of responses of the various crosses and backcrosses were similar to those of the electrophysiological experiments, with a fit to the data suggesting Z-linked inheritance of a major gene, with *C. herana*-type dominant, and other genes likely involved given the variability of the responses.

Characterization of the molecular components of the pheromone reception machinery in these moths is still in its infancy. The Australian tortricid *Epiphyas postvittana* contains three members of the Pheromone Binding Protein (PBP) family, with PBP1 and PBP3 showing male-biased antennal expression and PBP2 female-biased expression (Jordan et al. 2008). Recombinant expressed PBP1 is capable of binding the major component of the *E. postvittana* sex pheromone, E11-14Ac (Newcomb et al. 2002). Orthologs of PBP1 have also been isolated from many of the species within *Ctenopseustis* and *Planotortrix* revealing rapid rates of sequence evolution along the lineage leading to PBPs used by *C. filicis* and *C. fraterna* (Sirey 2000; Langhoff 2010). These species utilize 16 carbon acetates compared with the other species in the two genera that utilize 14 carbon acetates, suggesting that amino acid substitutions along this lineage may be involved in the ability to accommodate larger pheromone components.

The receptors responsible for the detection of the various sex pheromone components are being isolated from some of the species within the complex. Three sets of orthologous odorant receptor genes have been isolated from *C. obliquana*, *C. herana*, *P. octo*, *P. excessana*, and *P. notophaea* (Carraher et al. 2012), based on similarity to receptors found in *E. postvittana* (Jordan et al. 2009). The three odorant receptors from *E. postvittana* include the ubiquitous co-receptor Orco (EposOR2), as well as one receptor that is related to pheromone receptors from other moths (EposOR1) and another that is not (EposOR3). Sequence comparisons among the orthologs of these receptors from the *Ctenopseustis* and *Planotortrix* species reveal that OR1 is evolving rapidly compared with the other two receptors (Carraher et al. 2012). More recently a more rigorous study of the transcripts expressed in the male and female antennae of adult *E. postvittana* revealed 70 candidate odorant receptors, of which eight fall into the so-called pheromone receptor clade (Corcoran et al. 2015). Similar RNAseq studies have also been undertaken in *C. obliquana*, *C. herana*, *P. octo*, and *P. excessana* (Steinwender 2014; Steinwender et al. 2015), with the *E. postvittana* ORs used to isolate candidate ORs from these species. In *C. obliquana* and *C. herana*, Steinwender et al. (2015) identified a receptor (OR7) that responds to the pheromone component Z8-14Ac, but could not identify a receptor responding to Z5-14Ac. Interestingly, CherOR7, but not CoblOR7, was able to respond to Z7-14Ac, in addition to Z8-14Ac, suggesting that CherOR7 may be under relaxed constraint given that *C. herana* males are not behaviorally responsive to Z8-14Ac.

Concluding Remarks

In most species of moths, the female-produced sex pheromone is critical for bringing the two sexes together for reproduction. Given its key role in this process, it follows that changes in a pheromone system will profoundly influence assortative mating among individuals, leading potentially to speciation (Phelan 1992). Thus, differences in pheromone systems between closely related species of moths can provide valuable insight into the proximate mechanisms that drive speciation in the Lepidoptera (Löfstedt 1991). The discovery of sibling species complexes within the endemic New Zealand GHL and BHL complexes, along with the production of unusual pheromone components by these species and their congeners, has provided an opportunity to study speciation mechanisms, involving relatively small changes in their chemical communication systems, as well as larger-scale events across the Tortricidae and even the Lepidoptera. Much of the reason for the success of the pheromone studies in helping revise the systematics of this group was the willing uptake of the results of the pheromone studies by the systematist John Dugdale and indeed by many of the applied entomologists attempting to control pest species in this group.

Species within the genera *Ctenopseustis* and *Planotortrix* utilize various mono-unsaturated 14 (and to a lesser extent 16) carbon acetates for their pheromone components, as do most species in the tribe Archipini (Safonkin 2007). However, the novelty of this group of moths, compared to other Lepidoptera, and Tortricidae in particular, is the use of (Z)-unsaturated Δ5-, Δ7-, Δ8-, and Δ10-compounds, generally reflecting the evolution of novel (Z)-Δ5-, (Z)-Δ9-, and (Z)-Δ10-desaturases in the pheromone gland of females of the various species. Furthermore, how these novel Δ5- and Δ10-desaturase activities have arisen remains unknown and awaits further analysis of the desaturase genes from these and a wide range of closely related tortricid species.

Although species within the respective GHL and BHL complexes are cryptic and certainly closely related, the genetic differences between species in these complexes, in terms of how pheromones are produced, received, and processed, appear to be more complex than that found between the E- and Z-strains of the European corn borer, *Ostrinia nubilalis* (Haynes, this volume; Lassance, this volume; Roelofs et al. 1987). It is worth noting that, to date, the two strains of *O. nubilalis* are still considered the same species and, therefore, the pheromonal divergence between the two strains may be substantially less than that between the two *Ctenopseustis* and *Planotortrix* species. For *C. obliquana* and *C. herana*, it seems that more than one gene in each case (i.e., production, perception, and response) is responsible for the pheromonal differences between the species. Furthermore, the patterns of inheritance of the major genetic components of these traits in females (i.e., pheromone production) and males (i.e., pheromone perception and response) suggest that more than one locus underpins the evolutionary differences between the species. A similar story of one major and a number of minor

genes being responsible for pheromonal differences is also emerging for *P. octo* and *P. excessana*. At the molecular level, studies investigating the differences in desaturation between sibling species have provided evidence for the role of gene regulation in pheromone evolution, with the switching on and switching off of a Δ10-desaturase gene involved in the origin of distinct blends of sex pheromone (Albre et al. 2012, 2013). The molecular mechanisms involved in how males track changes in female pheromone blends remain to be elucidated for this group and, indeed, for most moth species.

Work on the pheromone systems of this complex of moths is continuing, addressing a number of key areas. What are the genetic differences that underpin differences between pheromone production and reception in *C. obliquana* versus *C. herana* and *P. excessana* versus *P. octo*? What are the factors that regulate Δ10-desaturase expression, and which determine the pheromone blend produced in the various species of the complex? How widespread is the use of a Δ5-desaturase versus chain shortening of oleate in this group? What are the genes that code for the molecular receptors of these unusual ligands in males? How have they or other genes changed to allow receiver evolution and how do the pheromone receptors of *Ctenopseustis* and *Planotortrix* species relate to receptors of Δ11-desaturase-derived components in other Tortricidae?

The identification of the sex pheromones of the GHL and BHL moths was driven initially by a desire to use the sex pheromones for pest management purposes. This applied problem led to the discovery of a more fundamental problem concerning the definition of species, eventually leading to fundamental studies addressing the role of sex pheromones in speciation in this group. Thus, the research on the GHL and BHL in New Zealand has been a successful example of combining applied and basic approaches to solve related problems. The widespread pheromone-based monitoring of this assortment of pests throughout various crops in New Zealand, as well as the development of specific mating disruption control systems, based on the identified sex pheromones (Suckling and Burnip 1996, 1997; Suckling et al. 2012), testifies to the impact of this research. Continuing research on this group is likely to further our understanding of sex pheromone evolution and speciation in moths.

Acknowledgments

We would like to thank the many scientists involved in pioneering this system, and especially Wendell Roelofs and John Dugdale for their vision and enthusiasm. We would also like to thank Anne Barrington for her invaluable help over the years in maintaining laboratory cultures of these insects. The authors have been supported by grants from the Allan Wilson Centre, the Marsden Fund and the Ministry of Business, Innovation and Employment in New Zealand, and the North Dakota Agricultural Experiment Station in the United States.

References Cited

Albre, J., M.A. Liénard, T.M. Sirey, S. Schmidt, L.K. Tooman, C. Carraher, D.R. Greenwood, C. Löfstedt, and R.D. Newcomb. 2012. Sex pheromone evolution is associated with differential regulation of the same desaturase gene in two genera of leafroller moths. *PLoS Genetics* 8:e1002489.

Albre, J., B. Steinwender, and R.D. Newcomb. 2013. The evolution of desaturase gene regulation involved in sex pheromone production in leafroller moths of the genus *Planotortrix*. *Journal of Heredity* 104:627–638.

Bellas, T.E., R.J. Bartell, and A. Hill. 1983. Identification of two components of the sex pheromone of the moth, *Epiphyas postvittana* (Lepidoptera: Tortricidae). *Journal of Chemical Ecology* 9:503–512.

Bjostad, L.B., W.A. Wolf, and W.L. Roelofs. 1987. Pheromone biosynthesis in lepidopterans: desaturation and chain shortening. Pp. 77–120. In G.D. Prestwich and G.J. Blomquist, eds. *Pheromone Biochemistry*. New York: Academic Press.

Cardé, R.T., and A.K. Minks. 1995. Control of moth pests by mating disruption: successes and constraints. *Annual Review of Entomology* 40:559–585.

Carraher, C., A. Authier, B. Steinwender, and R.D. Newcomb. 2012. Sequence comparisons of odorant receptors among tortricid moths reveal different rates of molecular evolution among family members. *PLOS ONE* 7:e38391.

Clare, G.K., and P. Singh. 1988. Laboratory rearing of *Ctenopseustis obliquana* (Walker) (Lepidoptera: Tortricidae) on an artificial diet. *New Zealand Journal of Zoology* 15:435–438.

Clearwater, J.R., S.P. Foster, S.J. Muggleston, J.S. Dugdale, and E. Priesner. 1991. Intraspecific variation and interspecific differences in the *Ctenopseustis obliquana* complex. *Journal of Chemical Ecology* 17:413–429.

Corcoran, J., M.D. Jordan, D. Begum, E. Hilario, A. Thrimawithana, R. Crowhurst, and R.D. Newcomb. 2015. The peripheral olfactory repertoire of the lightbrown apple moth, *Epiphyas postvittana*. *PLOS ONE* 10:e0128596.

Dugdale, J.S. 1966. A new genus for the New Zealand 'elusive *Tortrix*' (Lepidoptera: Tortricidae: Tortricinae). *New Zealand Journal of Science* 9:731–775.

Dugdale, J.S. 1990. Reassessment of *Ctenopseustis* Meyrick and *Planotortrix* Dugdale with descriptions of two new genera (Lepidoptera: Tortricidae). *New Zealand Journal of Zoology* 17:437–465.

Dugdale, J.S. 1997. Pheromone and morphology-based phylogenies in New Zealand tortricid moths. Pp. 463–472. In R.T. Carde and A.K Minks, eds. *Insect Pheromone Research: New Directions*. New York: Chapman & Hall.

Foster, S.P. 1998. Sex pheromone biosynthesis in the tortricid moth *Planotortrix excessana* (Walker) involves chain-shortening of palmitoleate and oleate. *Archives of Insect Biochemistry and Physiology* 37:158–167.

Foster, S.P., and J.S. Dugdale. 1988. A comparison of morphological and sex pheromone differences in some New Zealand Tortricinae moths. *Biochemical Systematics and Ecology* 16:227–232.

Foster, S.P., and W.L. Roelofs. 1987. Sex pheromone differences in populations of the brownheaded leafroller, *Ctenopseustis obliquana*. *Journal of Chemical Ecology* 13:623–629.

Foster, S.P., and W.L. Roelofs. 1988. Sex pheromone biosynthesis in the leafroller moth *Planotortrix excessana* by Δ10 desaturation. *Archives of Insect Biochemistry and Physiology* 8:1–9.

Foster, S.P., and W.L. Roelofs. 1996. Sex pheromone biosynthesis in the tortricid moth, *Ctenopseustis herana* (Felder & Rogenhofer). *Archives of Insect Biochemistry and Physiology* 33:135–147.

Foster, S.P., J.R. Clearwater, and S.J. Muggleston. 1989. Intra-specific variation of two components in sex pheromone gland of a *Planotortrix excessana* sibling species. *Journal of Chemical Ecology* 15:457–465.

Foster, S.P., J.R. Clearwater, S.J. Muggleston, J.S. Dugdale, and W.L. Roelofs. 1986. Probable sibling species complexes within two described New Zealand leafroller moths. *Naturwissenschaften* 73:156–158.

Foster, S.P., J.R. Clearwater, S.J. Muggleston, and P.W. Shaw. 1990. Sex pheromone of a *Planotortrix excessana* sibling species and reinvestigation of related species. *Journal of Chemical Ecology* 16:2461–2474.

Foster, S.P., J.R. Clearwater, and W.L. Roelofs. 1987. Sex pheromone of *Planotortrix* species found on mangrove. *Journal of Chemical Ecology* 13:631–637.

Foster, S.P., J.S. Dugdale, and C.S. White. 1991. Sex pheromones and the status of the greenheaded and brownheaded leafroller moths in New Zealand. *New Zealand Journal of Zoology* 17:63–74.

Foster, S.P., S.J. Muggleston, C. Löfstedt, and B. Hansson. 1997. A genetic study on pheromonal communication in two *Ctenopseustis* moths. Pp. 514–524. In R.T. Carde and A.K. Minks, eds. *Insect Pheromone Research: New Directions*. New York: Chapman & Hall.

Galbreath, R.A., M.H. Benn, H. Young, and V.A. Holt. 1985. Sex pheromone components in New Zealand greenheaded leafroller *Planotortrix excessana* (Lepidoptera: Tortricidae). *Zeitschrift für Naturforschung* 40:266–271.

Gleeson, D.M., P. Holder, R.D. Newcomb, R. Howitt, and J. Dugdale. 2000. Molecular phylogenetics of leafrollers: application to DNA diagnostics. *New Zealand Plant Protection* 53:157–162.

Green, C.J., and J.S. Dugdale. 1982. Review of the genus *Ctenopseustis* Meyrick (Lepidoptera: Tortricidae) with reinstatement of two species. *New Zealand Journal of Zoology* 9:427–435.

Hagström, Å.K., J. Albre, C. Löfstedt, and R.D. Newcomb. 2014. A novel fatty acyl desaturase from the pheromone glands of *Ctenopseustis obliquana* and *C. herana* with specific Z5-desaturase activity on myristic acid. *Journal of Chemical Ecology* 40:63–70.

Hansson, B.S., C. Lofstedt, and S.P. Foster. 1989. Z-linked inheritance of male olfactory response to sex-pheromone components in two species of tortricid moths, *Ctenopseustis obliquana* and *Ctenopseustis* sp. *Entomologia Experimentalis et Applicata* 53:137–145.

Hao, G., W. Liu, M. O'Connor, and W.L. Roelofs. 2002. Acyl-CoA Z9-and Z10-desaturase genes from a New Zealand leafroller moth species, *Planotortrix octo*. *Insect Biochemistry and Molecular Biology* 32:961–966.

Horak, M., C.P. Whittle, T.E. Bellas, and E.R. Rumbo. 1988. Pheromone gland components of some australian tortricids in relation to their taxonomy. *Journal of Chemical Ecology* 14:1163–1175.

Jordan, M.D., A. Anderson, D. Begum, C. Carraher, A. Authier, S.D. Marshall, A. Kiely et al. 2009. Odorant receptors from the light brown apple moth (*Epiphyas postvittana*) recognize important volatile compounds produced by plants. *Chemical Senses* 34:383–394.

Jordan, M.D., D. Stanley, S.D.G. Marshall, D. De Silva, R.N. Crowhurst, A.P. Gleave, D.R. Greenwood, and R.D. Newcomb. 2008. Expressed sequence tags and proteomics of antennae from the tortricid moth, *Epiphyas postvittana*. *Insect Molecular Biology* 17:361–373.

Jurenka, R. 2003. Biochemistry of female moth sex pheromones. Pp. 54–80. In G.J. Blomquist and R.C. Vogt, eds. *Insect Pheromone Biochemistry and Molecular Biology*. Amsterdam: Elsevier.

Langhoff, P. 2010. Speciation in New Zealand native leafroller moths. PhD thesis, University of Auckland, New Zealand.

Langhoff, P., A. Authier, T.R. Buckley, J.S. Dugdale, A. Rodrigo, and R.D. Newcomb. 2009. DNA barcoding of the endemic New Zealand leafroller moth genera, *Ctenopseustis* and *Planotortrix*. *Molecular Ecology Resources* 9:691–698.

Löfstedt, C. 1991. Evolution of moth pheromones. Pp. 57–73. In I. Hrdý, ed. *Chemical Ecology*. Prague: Academia.

Löfstedt, C., and W.L. Roelofs. 1985. Sex pheromone precursors in two primitive New Zealand tortricid moth species. *Insect Biochemistry* 15:729–734.

McNeil, J.N. 1991. Behavioral ecology of pheromone-mediated communication in moths and its importance in the use of pheromone traps. *Annual Review of Entomology* 36:407–430.

Newcomb, R.D., and D.M. Gleeson. 1998. Pheromone evolution within the genera *Ctenopseustis* and *Planotortrix* (Lepidoptera: Tortricidae) inferred from a phylogeny based on cytochrome oxidase I gene variation. *Biochemical Systematics and Ecology* 26:473–484.

Newcomb, R.D., T. Sirey, M. Rasam, and D.R. Greenwood. 2002. Pheromone binding proteins of *Epiphyas postvittana* are encoded at a single locus. *Insect Biochemistry and Molecular Biology* 32:1543–1554.

Nuttall, M.J. 1983. *Planotortrix excessana* (Walker), *Planotortrix notophaea* (Turner), *Epiphyas postvittana* (Walker) (Lepidoptera: Tortricidae): Greenheaded leafroller, blacklegged leafroller, light brown apple moth. Forest and Timber Insects in New Zealand No. 58, Forest Research Institute, New Zealand Forest Service, Rotorua.

Phelan, P.L. 1992. Evolution of sex pheromones and the role of asymmetric tracking. Pp. 265–314. In B.D. Roitberg and M.B. Isman, eds. *Insect Chemical Ecology: An Evolutionary Approach*. New York: Chapman & Hall.

Regier, J.C., J.W. Brown, C. Mitter, J. Baixeras, S. Cho, M.P. Cummings, and A. Zwick. 2012. A molecular phylogeny for the leaf-roller moths (Lepidoptera: Tortricidae) and its implications for classification and life history evolution. *PLOS ONE* 7:e35574.

Roelofs, W.L., and L.B. Bjostad. 1984. Biosynthesis of lepidopteran pheromones. *Bioorganic Chemistry* 12:279–298.

Roelofs, W.L., and R.L. Brown. 1982. Pheromones and evolutionary relationships of Tortricidae. *Annual Review of Ecology and Systematics* 13:395–422.

Roelofs, W.L., and A.P. Rooney. 2003. Molecular genetics and evolution of pheromone biosynthesis in Lepidoptera. *Proceedings of the National Academy of Sciences of the United States of America* 100:14599.

Roelofs, W.L., and W.A. Wolf. 1988. Pheromone biosynthesis in Lepidoptera. *Journal of Chemcial Ecology* 14:2019–2031.

Roelofs, W.L., T. Glover, X.H. Tang, I. Sreng, P. Robbins, C. Eckenrode, C. Löfstedt, B.S. Hansson, and B.O. Bengtsson. 1987. Sex pheromone production and perception in European corn borer moths is determined by both autosomal and sex-linked genes. *Proceedings of the National Academy of Sciences of the United States of America* 84:7585–7589.

Roelofs, W.L., W. Liu, G. Hao, H. Jiao, A.P. Rooney, and C.E. Linn, Jr. 2002. Evolution of moth sex pheromones via ancestral genes. *Proceedings of the National Academy of Sciences of the United States of America* 99:13621–13626.

Rooney, A.P. 2009. Evolution of moth sex pheromone desaturases. *Annals of the New York Academy of Sciences* 1170:506–510.

Safonkin, A.F. 2007. Pheromones and phylogenetic relations of leafrollers (Lepidoptera, Tortricidae). *Entomological Review* 87:1238–1241.

Sin, F.Y.T., D.M. Suckling, and J.W. Marshall. 1995. Differentiation of the endemic New Zealand greenheaded and brownheaded leafroller moths by restriction fragment length variation in the ribosomal gene complex. *Molecular Ecology* 4:253–256.

Singh, P. 1983. A general purpose laboratory diet mixture for rearing insects. *Insect Science and Application* 4:357–362.

Sirey, T.M. 2000. Pheromone binding proteins of New Zealand leafroller moths. Unpublished Masters of Science thesis, University of Auckland, New Zealand.

Steinwender, B. 2014. The evolution of sex pheromone reception in sibling species of the New Zealand endemic leafroller moth genera *Ctenopseustis* and *Planotortrix*. Unpublished PhD thesis, University of Auckland, New Zealand.

Steinwender, B., A.H. Thrimawithana, R.N. Crowhurst, and R.D. Newcomb. 2015. Pheromone receptor evolution in the cryptic leafroller species, *Ctenopseustis obliquana* and *C. herana*. *Journal of Molecular Evolution* 80:42–56.

Suckling, D.M., and E.G. Brockerhoff. 2010. Invasion biology, ecology, and management of the light brown apple moth (Tortricidae). *Annual Review of Entomology* 55:285–306.

Suckling, D.M., and G.M. Burnip. 1996. Orientation and mating disruption in *Planotortrix octo* using pheromone and inhibitor. *Entomologia Experimentalis et Applicata* 78:149–158.

Suckling, D.M., and G.M. Burnip. 1997. Orientation disruption of *Ctenopseustis herana* (Lepidoptera: Tortricidae). *Journal of Chemical Ecology* 23:2425–2436.

Suckling, D.M., G.F. McLaren, L.A. Manning, V.J. Mitchell, B. Attfield, K. Colhoun, and A.M. El-Sayed. 2012. Development of single-dispenser pheromone suppression of *Epiphyas postvittana*, *Planotortrix octo* and *Ctenopseustis obliquana* in New Zealand stone fruit orchards. *Pest Management Science* 68:928–934.

Wearing, C.H., W.P. Thomas, and J.S. Dugdale. 1991. Tortricid pests of pome and stone fruits, Australian and New Zealand species. Pp. 453–472. In L.P.S. van der Geest and H.H. Evenhuis, eds. *Tortricid Pests, Their Biology, Natural Enemies and Control.* Amsterdam: Elsevier.

White, C.S., and D.M. Lambert. 1994. Genetic differences among pheromonally distinct New Zealand leafroller moths. *Biochemical Systematics and Ecology* 22:329–339.

White, C.S., and D.M. Lambert. 1995. Genetic continuity within, and discontinuities among, populations of leafroller moths with distinct sex-pheromone. *Heredity* 75:243–255.

Young, H., R.A. Galbreath, M.H. Benn, V.A. Holt, and D.L. Struble. 1985. Sex pheromone components in New Zealand brownheaded leafroller *Ctenopseustis obliquana* (Lepidoptera: Tortricidae). *Zeitschrift für Naturforschung* 40c:262–265.

CHAPTER TWENTY

Evolution of Reproductive Isolation of
Spodoptera frugiperda

ASTRID T. GROOT, MELANIE UNBEHEND, SABINE HÄNNIGER,
MARÍA LAURA JUÁREZ, SILVIA KOST, and DAVID G. HECKEL

Introduction

Spodoptera frugiperda, the fall armyworm, is a noctuid moth occurring in North and South America with two host strains (a corn-strain and a rice-strain) identified in the 1980s (Pashley et al. 1985; Pashley 1986). These two strains were originally characterized by a polymorphism in an esterase allozyme marker and three other strain-biased protein variants in larvae collected from cornfields and rice paddies in Puerto Rico (Pashley et al. 1985; Pashley 1986). Since then, several additional strain-biased or strain-diagnostic molecular markers have been identified. The two strains differ in mitochondrial DNA sequences in the cytochrome oxidase I (COI) and NADH dehydrogenase 1 (ND1) genes (Pashley 1989; Pashley and Ke 1992; Lu and Adang 1996; Levy et al. 2002; Meagher and Gallo-Meagher 2003; Prowell et al. 2004; Nagoshi et al. 2006a; Machado et al. 2008; Juárez et al. 2012). There are also strain-biased and strain-specific amplified fragment length polymorphisms (AFLP) (McMichael and Prowell 1999; Busato et al. 2004; Prowell et al. 2004; Clark et al. 2007; Martinelli et al. 2007; Juárez et al. 2012, 2014), restriction length fragment polymorphisms (RFLP) (Lu et al. 1992), a so-called Frugiperda Rice (FR) repetitive nuclear DNA sequence, present in high copy number in the rice-strain and mostly lower copy number in the corn-strain (Lu et al. 1994; Nagoshi and Meagher 2003b;

Nagoshi et al. 2008), and nucleotide polymorphisms within the triose phosphate isomerase gene (Tpi, Nagoshi 2010).

Recently, sex pheromone differences have been found among populations of the two strains (Groot et al. 2008; Lima and McNeil 2009; Unbehend et al. 2013). However, these differences were not consistent among studies, suggesting that geographic variation may be confounded with strain-specific variation, or that pheromones may vary within strains as well. The relative importance of the pheromone differences between the two strains still needs to be established, i.e., are all pheromone compounds in the pheromone glands behaviorally important and/or are males of the two strains differentially attracted to the different pheromone blends? Since other physiological, developmental, and behavioral differences have been found among the strains (Pashley and Martin 1987; Pashley 1988b; Pashley et al. 1992, 1995; Veenstra et al. 1995; Meagher et al. 2004, 2011; Schöfl et al. 2009, 2011; Groot et al. 2010; Meagher and Nagoshi 2012), this overview integrates strain-specific variation in sexual communication (variation in the pheromone gland composition as well as variation in male response) with other possible pre-mating and postmating barriers that likely contribute to isolation of the two strains. First, we will show that the naming of the two strains is somewhat misleading, as the host specificity of the two strains is not as clear-cut as the names suggest. Then we will focus on the two types of prezygotic isolating mechanisms that have been demonstrated to differ between the two strains: (a) the diel pattern of reproductive activity and (b) pheromone signal and response traits. In addition to the pre-mating barriers, we also consider postmating barriers that may isolate the two strains. Finally, based on recent findings, we discuss a possible evolutionary scenario for the evolution of the two strains of *S. frugiperda*.

Are the Two Strains Really Host Strains?

Allozyme differences at five loci, including one apparently strain-specific esterase allele, provided the first evidence of partial genetic differentiation of populations collected from adjacent cornfields and rice fields in Puerto Rico and Louisiana (Pashley et al. 1985; Pashley 1986). Differences in mitochondrial DNA RFLP patterns were also found among these populations (Pashley 1988a). Subsequently, the same genetic differences were found in populations collected from other host plants and localities, and used to assign them to either the corn-strain or the rice-strain. The so-called corn-strain was found to infest mainly corn (i.e., maize, *Zea mays*), sorghum, (*Sorghum bicolor* subsp. *bicolor*), and cotton (*Gossypium hirsutum*), whereas the so-called rice-strain was found mostly in rice (*Oryza sativa*), sugarcane (*Saccharum officinarum*), and grasses such as Johnson grass (*Sorghum halepense*) and Bermuda grass (*Cynodon dactylon*). Genetic differentiation between these two strains has been confirmed in several regions in North and South America, using different molecular markers. The host associations of the two strains are summarized below.

Host Associations Based on Mitochondrial COI Polymorphism

Among all molecular markers available to distinguish the two strains, the most widely used target is mitochondrial DNA. For example, the two strains show differences in their COI gene and can be identified by a polymorphism in the restriction sites for SacI and AciI (both present in rice-strain and absent in the corn-strain), and for HinfI, BsmI, and MspI (all present in the corn-strain and absent in the rice-strain). The polymorphisms in SacI and MspI are used in most studies (Lu and Adang 1996; Levy et al. 2002; Meagher and Gallo-Meagher 2003; Nagoshi et al. 2006a). Based on the restriction site polymorphisms mentioned above, especially in the double digestion with SacI and MspI, the identity of the strains has been evaluated for different habitats and it has been demonstrated that the association is not always absolute.

Approximately 80% of individuals collected from corn habitats were identified as corn-strain and the remaining 20% as rice-strain (Pashley 1989; Lu and Adang 1996; Levy et al. 2002; Nagoshi et al. 2006a; Nagoshi et al. 2007b). However, exceptions from this percentage of distribution have been found as well. Prowell et al. (2004) identified samples collected from corn predominantly (i.e., 50% or more of the individuals) as rice-strain in French Guiana and in Louisiana. Nagoshi et al. (2006a) also found mostly rice-strain individuals in a sorghum field in Texas, which is considered a corn-strain habitat. In the case of larvae collected from rice fields, up to 95% of individuals have been identified as rice-strain (Nagoshi and Meagher 2003a, 2004; Machado et al. 2008; Velez-Arango et al. 2008). Recently, Juárez et al. (2012, 2014)

did not find a consistent pattern between the two strains and their respective host plants (especially, in rice habitats), when using COI markers in South American populations.

Some of these shifts in strain distributions may be due to seasonal and temporal variation in the distributions of the two strains and in the distribution of available host plants or different migration patterns of the two strains (Nagoshi et al. 2007a). For example, Nagoshi et al. (2007c) showed that the corn-strain predominated in collections from sorghum in the fall (March–June) in Brazil, but was less common in spring collections (September–November), while in Florida rice-strain larvae predominated in collections made from sorghum in the fall (September–November) and corn-strain larvae were mostly present in the spring season in sorghum (February–April). In Louisiana, Pashley et al. (1992) found that corn-strain populations were detected in the cornfields in the spring, while rice-strain populations remained at low density on various grasses until late summer when they increased in number. Together, these findings suggest that the migration pattern between the two strains may not be the same (Nagoshi and Meagher 2004).

In figure 20.1, we provide an overview of collections of the fall armyworm over a period of 27 years (from 1983 until 2010) from a number of different habitats. In 17 of 20 populations sampled from predominantly rice habitats (rice and pasture/Bermuda grass), most individuals were identified as rice-strain, whereas in 29 of 44 populations habitats (corn, cotton, and sorghum), most individuals were identified as corn-strain. Although mitochondrial markers generally show a strong correlation between strain type and host plant, in many of collections this association is lacking, especially in predominantly corn habitats (see figure 20.1).

Host Associations Based on Genome-Wide AFLP Markers

Although some studies have found a close association between the two strains and their host plants using AFLP markers (e.g., McMichael and Prowell 1999 and Busato et al. 2004 in the United States and Brazil, respectively), others have not (e.g., Martinelli et al. 2007 in Mexico, Brazil, Argentina, and the United States). Recently, we found that individuals from populations collected from corn plants from several locations in Argentina, Paraguay, and Brazil tended to cluster together and showed a high degree of homogeneity in AFLP markers (Juárez et al. 2014). This finding thus contrasts the trend found in the COI marker, where 15 out of 44 populations collected from corn (see figure 20.1) showed a significant portion of rice-strain individuals. Individuals from the populations collected from rice from several locations in Argentina and Paraguay formed three distinct groups and showed a much higher level of heterogeneity in their AFLP markers (Juárez et al. 2014). Overall, individuals collected from corn-strain habitats were clustered separately from individuals collected from rice-strain habitats, although there were some marked exceptions (Juárez et al. 2014).

Host Association Based on Mitochondrial and Nuclear Markers

Combining mitochondrial and nuclear markers with their different modes of inheritance, the rate and directionality of

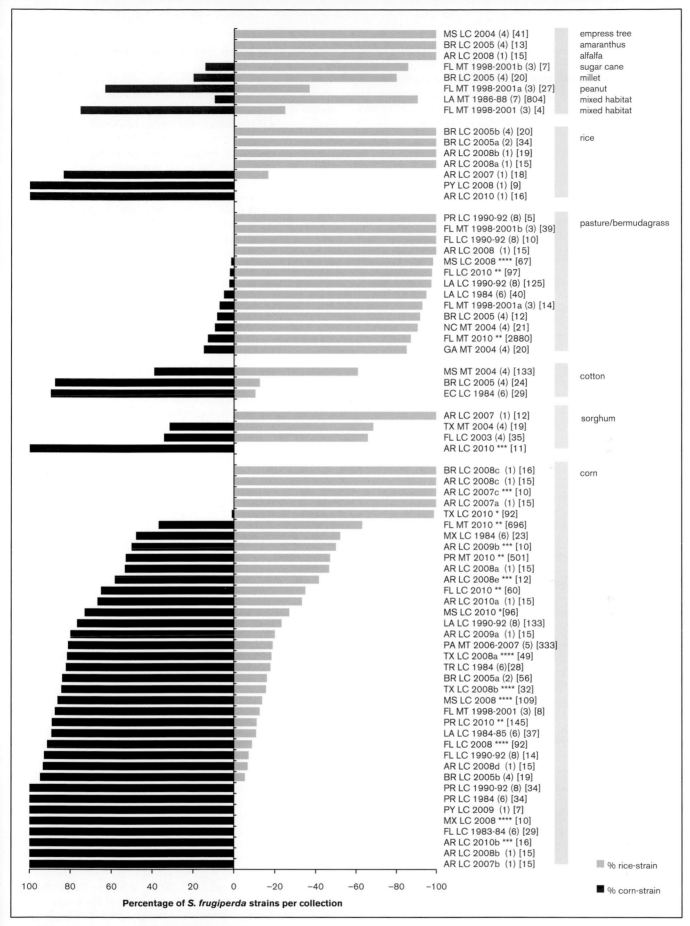

FIGURE 20.1 Distribution of *Spodoptera frugiperda* host strains in different habitats and geographic regions. Each bar shows the percentage of the identified strains per collection, based on mitochondrial markers. Habitats are indicated on the right. Numbers in [ ] indicate total number of collection.

ABBREVIATIONS: MS, Mississippi; BR, Brazil; AR, Argentina; LA, Louisiana; FL, Florida; PY, Paraguay; PR, Puerto Rico; NC, North Carolina; GA, Georgia; EC, Ecuador; TX, Texas; MX, Mexico; PA, Pennsylvania; TR, Trinidad; LC, larval collection; MT, male trapping experiments.

DATA FROM: *C. Blanco; **S. Hänniger and M. Unbehend; ***M.L. Juárez; ****G. Schöfl; (1) Juárez et al. 2012; (2) Machado et al. 2008; (3) Meagher and Gallo-Meagher 2003; (4) Nagoshi et al. 2006a; (5) Nagoshi et al. 2009; (6) Pashley 1989; (7) Pashley et al. 1992; (8) Prowell et al. 2004.

hybridization between the strains in the field can be identified. Prowell et al. (2004) analyzed populations from Louisiana, Florida, Puerto Rico, Guadeloupe, and French Guiana with different molecular markers (mitochondrial haplotype, esterase genotypes, AFLPs) and reported that 16% of the samples were potential hybrids due to discordance for at least one marker. The authors found evidence of crosses between the strains in both directions: when using mtDNA and esterase markers, 66% of the hybrids were inferred to be derived from rice-strain females mated with corn-strain males, i.e., RC hybrids, while in multi-locus comparison using the three markers, 54% of the hybrids were RC hybrids and 46% were from the reciprocal cross, i.e., CR hybrids. In addition, Prowell et al. (2004) found that these hybrids occurred mostly in the corn habitats. Similar results were found by Saldamando and Vélez-Arango (2010) with Colombian populations. In contrast, Nagoshi and Meagher (2003b) and Nagoshi et al. (2006b), using mitochondrial haplotypes and the nuclear FR tandem-repeat sequence, found 40–56% of all males collected in pheromone traps having the RC configuration, while only 3–3.5% were CR hybrids, and hybrids occurred in both corn and rice habitats.

Recently, Nagoshi (2010) identified 10 polymorphic diagnostic sites in the Z-linked (sex-linked) triose phosphate isomerase (*Tpi*) gene that can be associated with the corn- or the rice-strain of the fall armyworm (as in most Lepidoptera females are the heterogametic sex, ZW). With this marker, Nagoshi (2012) analyzed 12 populations (9 collected from corn and 3 from rice) with the COI marker, and then reanalyzed the same samples with the Tpi marker, and found that 60% and 7% of the COI-R typed individuals were Tpi-C in the corn and rice habitats, respectively (i.e., RC hybrids). The reverse constellation, COI-C and Tpi-R (i.e., CR hybrids), occurred in 8% and 22% of the COI-C typed individuals from corn and rice habitats, respectively. Like Nagoshi (2012), when we combined the COI marker with the Z-linked Tpi marker, we also found discordance between the mitochondrial and nuclear markers (43%) (Juárez et al. 2014). These configurations consisted of four different combinations: RC (30% of all hybrids), CR (7% of all hybrids), CI (20% of all hybrids), and RI (42% of all hybrids). The I stands for a Tpi-intermediate haplotype, i.e., individuals in which corn and rice SNPs were present in similar proportions or heterozygous individuals in which SNPs showed the two alternative nucleotides. The latter individuals must be hybrid males, as in Lepidoptera the females carry only one copy of the Z-linked Tpi gene. Nagoshi (2010) and Nagoshi et al. (2012) also found this intermediate configuration in a very low frequency and proposed that they may represent hybrid individuals as well.

In summary, both types of hybrids seem to occur in nature, although recent studies suggest that the RC-hybrids are more common. These hybrids are mostly found in corn habitats, while other hybrids (CR, CI, RI) are mostly found in rice habitats. Overall, the two strains seem to be predominantly found in the habitats from where they were originally described, but significant exceptions have been found with all markers used. Therefore, our preliminary conclusion is that divergence between the strains is not likely due to host plant specialization, or at least not alone. We hypothesize that an interaction between ecological and behavioral mechanisms has contributed to reproductive isolation between the two strains (Groot et al. 2010).

Behavioral Isolation Mechanism 1: Timing of Reproductive Activity

Differences in the diel pattern of mating activity between strains would create a powerful barrier to hybridization. Strain-specific differences in the timing of reproductive activity of the two strains have been consistently found, independent of the geographic origin of the strains (Pashley et al. 1992; Schöfl et al. 2009, 2011): the corn-strain is active early in the scotophase, while the rice-strain is active late in the scotophase. Schöfl et al. (2009) showed that different reproductive behaviors (calling, copulation, and oviposition) are differentially inherited and thus under complex genetic control. The coordinated timing difference between the two strains in reproductive activity and general locomotor activity suggested the involvement of the circadian clock.

When testing whether allochronic separation causes assortative mating in the laboratory, Schöfl et al. (2011) found an interaction between strain-specific timing of mating and time-independent intrinsic preferences that influenced the mating choice of both strains. Furthermore, mate choice changed over time in consecutive nights and was influenced by the timing of introduction of the mating partners (Schöfl et al. 2011). In general, females were more restricted in their mate preference than males and approximately 30% of the isolation between both strains was generated by female mate preference, suggesting the involvement of a male-specific sex pheromone that mediates close-range courtship behavior (Schöfl et al. 2011). Also, this mate-choice experiment indicates that the level of assortative mating caused by allochronic separation alone is not strong enough to cause reproductive isolation between strains.

Although the importance of differential timing of reproduction is probably not as strong as suggested by Pashley et al. (1992), the consistent timing differences between the strains, independent of the geographic origin, suggest that this behavioral difference could have a stronger influence as prezygotic isolation barrier than host plant choice. Therefore, we are tempted to argue that both strains are "timing strains" rather than "host strains."

Behavioral Isolation Mechanism 2: Variation in Sexual Communication

In the early 1990s, Pashley et al. (1992) found that males of both strains showed a slight preference for females of the same strain, 60–65% of corn- and rice-strain males being attracted to corn- and rice-strain females, respectively. These findings indicate that in addition to the differences in timing of reproduction, pheromone differences might be important for mate choice and cause assortative mating in the two strains, although Pashley et al. (1992) suggested that "pheromone chemistry may play a small role (if any) in strain separation." The sex pheromone composition of *Spodoptera frugiperda* females has been studied in different geographic regions (Mitchell et al. 1985; Tumlinson et al. 1986; Descoins et al. 1988; Batista-Pereira et al. 2006; Groot et al. 2008; Lima and McNeil 2009; Unbehend et al. 2013). While earlier studies mainly focused on the general composition of the female sex pheromone without distinguishing the two strains, later studies investigated strain-specific differences in the female pheromone composition (Groot et al. 2008; Lima and McNeil

2009; Unbehend et al. 2013). In general, the fall armyworm sex pheromone consists of the primary sex pheromone component Z9-14Ac and the critical secondary sex pheromone Z7-12Ac (Tumlinson et al. 1986; Batista-Pereira et al. 2006; Groot et al. 2008; Lima and McNeil 2009). The behavioral effect of other secondary compounds in the female gland remains unclear (Tumlinson et al. 1986; Andrade et al. 2000; Fleischer et al. 2005; Groot et al. 2008; Unbehend et al. 2013). However, twice as many males were caught when Z11-16Ac or Z9-12Ac were added to the binary blend (Fleischer et al. 2005), suggesting at least a synergistic effect of these compounds. It has been shown that corn- and rice-strain females exhibit strain-specific differences in their relative amount of Z7-12Ac (relative to the amounts of other gland compounds), as well as in the relative amount of Z9-14Ac, Z11-16Ac, and Z9-12Ac, although the type of variation found seems to vary in different geographic regions (Groot et al. 2008; Lima and McNeil 2009; Unbehend et al. 2013).

Disentangling Geographic from Strain-Specific Variation

Extractions of the pheromone glands of females from a colony, which was initiated with larvae collected in Florida, revealed that rice-strain females produce significantly higher relative amounts of Z7-12Ac and Z9-12Ac, and lower relative amounts of Z11-16Ac, than corn-strain females (Groot et al. 2008). However, laboratory rice-strain females originating from Louisiana contained lower relative amounts of the major component Z9-14Ac, as well as larger relative amounts of Z7-12Ac and Z11-16Ac, compared to laboratory corn-strain females from Louisiana (Lima and McNeil 2009). Taken together, only Z7-12Ac showed consistent strain-specific variation in females from Florida and Louisiana (Groot et al. 2008; Lima and McNeil 2009; Unbehend et al. 2013). Apparently, the selection pressure on Z7-12Ac is similar in both regions but different between the two strains. The inconsistent variation in the major sex pheromone component Z9-14Ac between the two regions suggests geographic rather than strain-specific variation. The importance of Z11-16Ac and Z9-12Ac in the attraction of fall armyworm males is not completely understood yet, but their variation suggests that these components are not under strong stabilizing selection.

Geographic variation in the strain-specific pheromone composition of females from Florida and Louisiana may be related to different haplotype profiles in Floridian and Louisianan corn-strain populations. There seem to be two main migration routes of the fall armyworm, based on haplotype patterns in the corn-strain (Nagoshi et al. 2008; Nagoshi et al. 2010). These patterns suggest an eastern migration route, i.e., populations originating from Puerto Rico and Florida move northward to Georgia, and a western migration route, i.e., populations from Texas move northeastward to Louisiana, Mississippi, Alabama, and Pennsylvania (Nagoshi et al. 2008; Nagoshi et al. 2009). If no other geographic effects influence the female pheromone, then pheromone profiles of females from Texas, Louisiana, Mississippi, Alabama, and Pennsylvania may be more similar to each other than to pheromone profiles of females from Florida, Puerto Rico, and Georgia.

In fall armyworm females from Brazil, another minor sex pheromone component, E7-12Ac, was demonstrated to be attractive to Brazilian males in the field (Batista-Pereira et al. 2006). Addition of E7-12Ac to binary blends, containing Z9-14Ac and Z7-12Ac, significantly increased the number of males captured in Brazil, i.e., from an average of 70 males per trap to an average of 100 males per trap (Batista-Pereira et al. 2006). The fact that E7-12Ac has not been found in females from Florida, Louisiana, or French Guyana (Descoins et al. 1988; Groot et al. 2008; Lima and McNeil 2009) suggests the existence of geographic variation in female pheromone production. In conclusion, the two *Spodoptera frugiperda* strains do differ in their female sex pheromone composition (Groot et al. 2008; Lima and McNeil 2009; Unbehend et al. 2013), but geographic variation seems to influence the strain-specific pheromone production. To disentangle geographic variation from strain-specific variation, additional strain-specific pheromone extractions of different populations from North and South America will be necessary.

Variation in Pheromone Composition within the Strains

In addition to strain-specific and geographic variations in the pheromone composition, pheromone differences between females of the same strain have been observed between artificially reared and field-collected corn- and rice-strain females from Florida (Unbehend et al. 2013). Females of both laboratory strains produced significantly lower relative amounts of the major pheromone component Z9-14Ac and usually higher relative amounts of Z7-12Ac, Z11-16Ac, and Z9-12Ac, compared to the field-collected females, although strain-specific pheromone variation was maintained (Unbehend et al. 2013). To estimate how much within-strain variation occurs in nature, we analysed the pheromone composition of females from seven different corn-strain families, originating from single pair matings of individuals that were collected one generation earlier from a cornfield in Florida (Marr 2009). The females of these families exhibited significant differences in their pheromone composition compared to our laboratory populations (Marr 2009). The variation of Z9-14Ac, Z7-12Ac, Z11-16Ac, and Z9-12Ac was strongly heritable and a broadsense heritability analysis showed that the variation in gland compounds within the different families is determined mainly by genetic rather than environmental effects (Marr 2009). However, the within-strain variation found in laboratory and field females, in addition to the geographic variation, indicates that laboratory rearing and environmental factors influence the pheromone composition of females. The challenge is to determine which factors may cause variation in the pheromone composition and why. Understanding the cause of variation in the pheromone composition and its genetic control will be important to understand how variation in sexual communication influences reproductive isolation and how sexual communication systems may evolve (Baker and Cardé 1979; Löfstedt 1993; Butlin and Trickett 1997; Ritchie 2007).

Male Response to Strain-Specific Pheromone

The existence of strain-specific sex pheromone blends can only contribute to differentiation between the strains if this leads to differential attraction of fall armyworm males in the field. Although several trapping experiments of *Spodoptera*

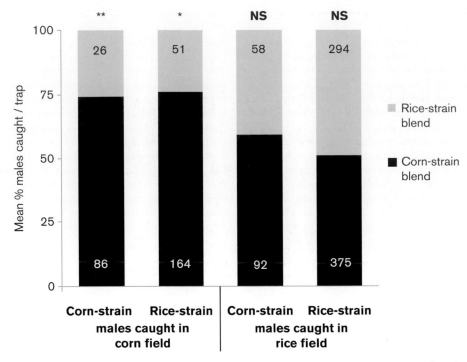

FIGURE 20.2 Mean percent of corn-strain and rice-strain males caught in sex pheromone traps baited with synthetic pheromone lures in a cornfield and a grass field in Florida. The corn-strain blend consisted of 300 μg Z9-14Ac, which was considered 100%, 6 μg (2%) Z7-12Ac, 39 μg (13%) Z11-16Ac, and 3 μg (1%) Z9-12Ac. The rice-strain blend was constructed in a similar way, only with 12 μg (4%) Z7-12Ac, 24 μg (8%) Z11-16Ac, and 6 μg (2%) Z9-12Ac. Numbers in the bars indicate total number of males caught (see Unbehend et al. 2013, for more details).

NOTE: ** $p < 0.01$; * $p < 0.05$; NS: not significant.

frugiperda males have been conducted in the field (Mitchell et al. 1985; Tumlinson et al. 1986; Meagher and Mitchell 1998; Andrade et al. 2000; Batista-Pereira et al. 2006), only one investigated strain-specific differences in the male attraction toward different pheromones (Pashley et al. 1992). In Louisiana fields containing both host plants, 60% of all rice-strain males trapped in pheromone traps were attracted to a virgin rice-strain female, while 65% of all trapped corn-strain males were caught in traps baited with virgin corn-strain females (Pashley et al. 1992). Thus, males of both strains exhibited only a slight bias toward females of their own strain in mixed habitats, suggesting that strain-specific sexual communication is a weak prezygotic isolation barrier (Pashley et al. 1992). Similarly, Lima and McNeil (2009) argued that it is quite unlikely that strain-specific sex pheromone differences alone "would be sufficient to ensure reproductive isolation of the two strains."

To evaluate whether fall armyworm males exhibit strain-specific attraction toward females of their own strain, we conducted wind tunnel choice assays and male-trapping experiments in Florida (Unbehend et al. 2013). Wind tunnel experiments without plant volatiles revealed that *S. frugiperda* males from laboratory populations show no strain-specific attraction to virgin females of their own strain. Interestingly, males of both strains were mainly influenced by the timing of female calling, and did not discriminate among calling females (Unbehend et al. 2013). However, when testing pheromone lures mimicking the pheromone gland composition of Floridian corn-strain females (i.e., 100% Z9-14Ac, 13% Z11-16Ac, 2% Z7-12Ac, 1% Z9-12Ac), 74% of all trapped corn-strain males in a cornfield were attracted to this corn-strain

lure, and only 26% to the rice-strain lure, i.e., 100% Z9-14Ac, 8% Z11-16Ac, 4% Z7-12Ac, 2% Z9-12Ac (figure 20.2). In rice fields, such a similar strain-specific attraction was not observed, and only 59% of all trapped corn-strain males were attracted to the synthetic corn-strain lure, while 41% were attracted to the rice-strain lure (figure 20.2). This result suggests that strain-specific attraction to different lures depends on the respective (volatile) environment, and hints to a synergistic effect of sex pheromones and host plant volatiles (Dekker and Barrozo, this volume). However, similar to corn-strain males, rice-strain males were also mostly attracted to the synthetic corn-strain lure in the cornfield with 76% of all trapped rice-strain males caught in traps baited with the corn-strain lure (figure 20.2). The pheromone traps that were baited with the so-called rice-strain lure (100% Z9-14Ac, 8% Z11-16Ac, 4% Z7-12Ac, 2% Z9-12Ac) did not specifically attract rice-strain males in a grass field and only 49% of all trapped rice-strain males were attracted to the rice-strain lure (Unbehend et al. 2013). Together, these results indicate that in Florida corn-strain lures are most attractive for both strains in a corn habitat, while there is no preference for a corn- or rice-strain lure in a rice habitat.

Importance of Different Pheromone Components for Male Attraction in the Field

To assess strain-specific male response toward the different pheromone components, we also evaluated the importance of single pheromone components in the attraction of corn- and rice-strain males in a corn and a grass field in Florida

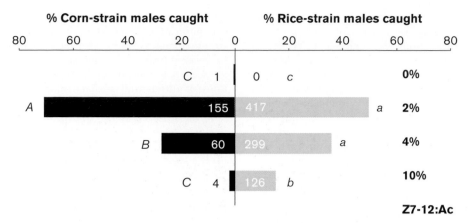

FIGURE 20.3 Strain-specific response of *Spodoptera frugiperda* males towards different doses of Z7-12Ac added to 300 µg Z9-14Ac in a cornfield and grass field in Florida. Different letters next to the bars indicate significant differences. Numbers in the bars indicate total number of males caught (see Unbehend et al. 2013, for more details).

(Unbehend et al. 2013) and in different geographic regions in North and South America (Unbehend et al. 2014). As mentioned earlier, fall armyworm males can vary in their attraction toward E7-12Ac, but show stable geographic-independent attraction toward binary blends containing Z9-14Ac and Z7-12Ac (Tumlinson et al. 1986; Andrade et al. 2000; Fleischer et al. 2005; Batista-Pereira et al. 2006; Unbehend et al. 2013). We tested different doses of the critical secondary component Z7-12Ac and found that corn-strain males had a much more pronounced optimum centered at the 2% Z7-12Ac blend, while the rice-strain male optimum was less pronounced with no discrimination between 2% and 4% (Unbehend et al. 2013; figure 20.3). This strain-specific male response is consistent with the strain-specific female pheromone production in Florida, at least in the corn-strain, because corn- and rice-strain females produce around 2% and 4% Z7-12Ac, respectively (Groot et al. 2008; Unbehend et al. 2013). These results suggest that fall armyworm corn-strain males in Florida are adapted to the strain-specific female pheromone differences in the amount of Z7-12Ac, i.e., 2% versus 4%. However, when we tested the male response toward different pheromone lures in different regions, we found that mainly corn-strain males, not rice-strain males, exhibited geographic variation in their attraction (Unbehend et al. 2014). This suggests that rice-strain males exhibit a broader response spectrum toward sex pheromone blends than corn-strain males.

The relative importance of Z11-16Ac is still unclear. In Costa Rica, the ternary blend of Z11-16Ac, Z9-14Ac, and Z7-12Ac captured marginally more males than the binary blend of Z9-14Ac and Z7-12Ac in one test and marginally fewer in another, although neither effect was statistically significant (Andrade et al. 2000). Similarly, addition of Z11-16Ac to binary blends did not significantly increase trap catches in Brazil (Batista-Pereira et al. 2006) or Florida (Tumlinson et al. 1986; Unbehend et al. 2013). However, trapping experiments in Pennsylvania suggest that the addition of Z11-16Ac, together with Z9-12Ac, enhances male attraction to Z9-14Ac and Z7-12Ac (Fleischer et al. 2005). Also, in our field experiments in Florida we found that males were differentially attracted to the two-component blends without Z11-16Ac compared to the four-component blends with Z9-12Ac and Z11-16Ac between corn- and rice-strain habitats (Unbehend et al. 2013). However, in dose-response experiments conducted in Peru, corn males

were equally attracted to blends with and without different doses of Z11-16Ac (Unbehend et al. 2014).

The compound Z9-12Ac has been reported to occur in glands of females from North and South America (Descoins et al. 1988; Batista-Pereira et al. 2006; Groot et al. 2008). In Costa Rica and Florida, fall armyworm males were attracted to traps containing only Z9-12Ac (Jones and Sparks 1979; Andrade et al. 2000). When conducting experiments where we added different relative amounts of Z9-12Ac to the binary blend of Z9-14Ac and Z7-12Ac, we found that all tertiary blends containing Z9-12Ac were similarly attractive as the binary blends without Z9-12Ac, both in corn- and rice-strain habitats in Florida (Unbehend et al. 2013). However, as pointed out above, a synergistic effect between Z9-12Ac and the other compounds cannot be excluded.

In summary, corn- and rice-strain males in Florida were mostly attracted to a corn-strain pheromone blend, at least in cornfields. Thus, there may be synergistic effects of host plant volatiles and sex pheromone components in cornfields. In grass fields, we did not find a preference for a corn- or a rice-strain pheromone blend in either strain. Strain-specific responses were found toward different doses of Z7-12Ac added to the major pheromone component Z9-14Ac, where corn-strain males were mostly attracted to 2% Z7-12Ac and rice-strain males were attracted to a wider range (2–10%). Together, these data suggest that strain-specific differences in the sexual communication of both strains do not cause assortative mating in Florida and thus are a weak prezygotic isolation barrier between the corn-strain and the rice-strain.

Level and Direction of Hybridization between the Two Strains

The fact that hybridization between the two strains can be observed in the field raises the question: are these strains in the process of divergence or convergence? RC hybrid females have been found to be less likely to mate with any kind of male (C, R, RC, or CR) and to produce a lower number of egg masses when they do mate (Pashley and Martin 1987; Whitford et al. 1988; Groot et al. 2010). Interestingly, RC hybrid males did not show this deficiency and mated readily with all types of females (C, R, and CR) (Groot et al. 2010). The fact that RC hybrid females are found to be mostly sterile in laboratory experiments seems to conflict with the field observation where mainly RC hybrids are found (see the "Host Association Based on Mitochondrial and Nuclear Markers" section). However, this contradiction makes the "reproductive problem" of RC hybrid females a perfect postzygotic isolation barrier: if the most abundant individuals are at the same time the least fertile ones, gene flow is maximally prevented at this stage. This thus indicates that these strains are in the process of divergence rather than convergence. Given the existence of RC hybrid females in the field, while in the laboratory these hybrid females are hardly able to reproduce, this hybrid incompatibility represents an essential contribution to the process of speciation between the two strains.

Possible Evolutionary Scenarios on Reproductive Isolation in the Two Strains

Since the host association of the two strains does not seem to be as strict as early studies indicated, ecological specialization based on host plant choice does not seem the most likely cause of differentiation between the two strains in *Spodoptera frugiperda*. Other factors may have influenced a host association between the strains. One of these factors may be the presence of competitors or natural enemies on the ancestral host as has been suggested for other phytophagous insects (Berlocher and Feder 2002). Pashley et al. (1995) reported that over a 2-year period, fall armyworm larval mortality caused by parasites, predators, and pathogens was higher in pastures than in cornfields. For this reason, the corn habitat may constitute a more protected environment than the rice habitats.

On the basis of the distribution of the two strains, particularly the distribution of the respective hybrids, and the behavioral differences between the two strains, we hypothesize that the rice-strain is the ancestral strain and corn-strain the derived strain (Juárez et al. 2014). Higher levels of genetic and behavioral homogeneity observed in the corn-strain than in the rice-strain suggests that the corn-strain went through a bottleneck, i.e., that the corn-strain arose from a few individuals. Additionally, in cornfields a significant portion of rice-strain individuals as well as hybrids are found, specifically RC hybrids, while in rice fields the percentage of corn-strain individuals or hybrids is generally much lower (Prowell et al. 2004; Saldamando and Vélez-Arango 2010). The observation that males of both strains are mostly attracted to a corn-strain sex pheromone blend in cornfields, while this preference is not found in rice fields, is consistent with these results. Hybrid incompatibility is between R mothers and C fathers and not vice versa, i.e., RC hybrids are incompatible with any kind of male, whereas CR hybrids produce fertile and viable offspring. Together, these findings suggest that the rice-strain is the ancestral strain and the corn-strain is the derived strain.

Conclusion

In reviewing many studies on the host plant association of the two strains, host associations do not seem to be consistent when the mitochondrial COI marker is considered. In cornfields, more rice-strain individuals seem to be found than vice versa, and RC hybrids are also mostly found in corn habitats. Thus, habitat isolation alone does not seem to be strong prezygotic isolation barrier between the corn-strain and the rice-strain. Similarly, strain-specific differences in the sexual communication system of both strains alone do not appear strong enough to cause assortative mating within strains. However, differences in diel patterns of reproductive behaviors seem to be much more consistent than host-plant associations or differential sexual communication between the strains. Since a shift in timing can immediately inhibit gene flow, the strains may be "timing strains" rather than "host strains" or "pheromone strains." Furthermore, the postmating barrier of RC hybrid female sterility seems to be most likely a key element in the divergence of these two strains.

References Cited

Andrade, R., C. Rodriguez, and A.C. Oehlschlager. 2000. Optimization of a pheromone lure for *Spodoptera frugiperda* (Smith) in Central America. *Journal of the Brazilian Chemical Society* 11:609–613.

Baker, T.C., and R.T. Cardé. 1979. Endogenous and exogenous factors affecting periodicities of female calling and male sex pheromone response in *Grapholitha molesta* (Busck). *Journal of Insect Physiology* 25:943–950.

Batista-Pereira, L.G., K. Stein, A.F. de Paula, J.A. Moreira, I. Cruz, M.D. Figueiredo, J. Perri, and A.G. Correa. 2006. Isolation, identification, synthesis, and field evaluation of the sex pheromone of the Brazilian population of *Spodoptera frugiperda*. *Journal of Chemical Ecology* 32:1085–1099.

Berlocher, S.H., and J.L. Feder. 2002. Sympatric speciation in phytophagous insects: moving beyond controversy? *Annual Review of Entomology* 47:773–815.

Busato, G.R., A.D. Grutzmacher, A.C. de Oliveira, E.A. Vieira, P.D. Zimmer, M.M. Kopp, J.D. Bandeira, and T.R. Magalhaes. 2004. Analysis of the molecular structure and diversity of *Spodoptera frugiperda* (J.E. Smith) (Lepidoptera: Noctuidae) populations associated to the corn and rice crops in Rio Grande do Sul State, Brazil. *Neotropical Entomology* 33:709–716.

Butlin, R., and A.J. Trickett. 1997. Can population genetic simulations help to interpret pheromone evolution? Pp. 548–562. In R.T. Carde, and A.K. Minks, eds. *Insect Pheromone Research: New Directions*. New York: Chapman and Hall.

Clark, P.L., J. Molina-Ochoa, S. Martinelli, S.R. Skoda, D.J. Isenhour, D.J. Lee, J.T. Krumm, and J.E. Foster. 2007. Population variation of the fall armyworm, *Spodoptera frugiperda*, in the Western Hemisphere. *Journal of Insect Science* 7:5.

Descoins, C., J.F. Silvain, B. Lalannecassou, and H. Cheron. 1988. Monitoring of crop pests by sexual trapping of males in the French West-Indies and Guyana. *Agriculture Ecosystems & Environment* 21:53–65.

Fleischer, S.J., C.L. Harding, P.E. Blom, J. White, and J. Grehan. 2005. *Spodoptera frugiperda* pheromone lures to avoid nontarget captures of *Leucania phragmatidicola*. *Journal of Economic Entomology* 98:66–71.

Groot, A.T., M. Marr, G. Schofl, S. Lorenz, A. Svatos, and D.G. Heckel. 2008. Host strain specific sex pheromone variation in *Spodoptera frugiperda*. *Frontiers in Zoology* 5:20.

Groot, A.T., M. Marr, D.G. Heckel, and G. Schofl. 2010. The roles and interactions of reproductive isolation mechanisms in fall armyworm (Lepidoptera: Noctuidae) host strains. *Ecological Entomology* 35:105–118.

Jones, R.L., and A.N. Sparks. 1979. (Z)-9-Tetradecen-1-ol acetate: a secondary sex pheromone of the fall armyworm, *Spodoptera frugiperda* (J.E. Smith). *Journal of Chemical Ecology* 5:721–725.

Juárez, M.L., M.G. Murua, M.G. Garcia, M. Ontivero, M.T. Vera, J.C. Vilardi, A.T. Groot, A.P. Castagnaro, G. Gastaminza, and E. Willink. 2012. Host association of *Spodoptera frugiperda* (Lepidoptera: Noctuidae) corn and rice strains in Argentina, Brazil, and Paraguay. *Journal of Economic Entomology* 105:573–582.

Juárez, M.L., G. Schofl, M.T. Vera, J.C. Vilardi, M.G. Murua, E. Willink, S. Hanniger, D.G. Heckel, and A.T. Groot. 2014. Population structure of *Spodoptera frugiperda* maize and rice host forms in South America: are they host strains? *Entomologia Experimentalis et Applicata* 152:182–199.

Levy, H.C., A. Garcia-Maruniak, and J.E. Maruniak. 2002. Strain identification of *Spodoptera frugiperda* (Lepidoptera: Noctuidae) insects and cell line: PCR-RFLP of cytochrome oxidase C subunit I gene. *Florida Entomologist* 85:186–190.

Lima, E.R., and J.N. McNeil. 2009. Female sex pheromones in the host races and hybrids of the fall armyworm, *Spodoptera frugiperda* (Lepidoptera: Noctuidae). *Chemoecology* 19:29–36.

Löfstedt, C. 1993. Moth pheromone genetics and evolution. *Philosophical Transactions of the Royal Society London B* 340:167–177.

Lu, Y.J., and M.J. Adang. 1996. Distinguishing fall armyworm (Lepidoptera: Noctuidae) strains using a diagnostic mitochondrial DNA marker. *Florida Entomologist* 79:48–55.

Lu, Y.J., M.J. Adang, D.J. Isenhour, and G.D. Kochert. 1992. RFLP analysis of genetic variation in North American populations of the fall armyworm moth *Spodoptera frugiperda* (Lepidoptera: Noctuidae). *Molecular Ecology* 1:199–207.

Lu, Y.J., G.D. Kochert, D.J. Isenhour, and M.J. Adang. 1994. Molecular characterization of a strain-specific repeated DNA sequence in the fall armyworm *Spodoptera frugiperda* (Lepidoptera: Noctuidae). *Insect Molecular Biology* 3:123–130.

Machado, V., M. Wunder, V.D. Baldissera, J.V. Oliveira, L.M. Fiuza, and R.N. Nagoshi. 2008. Molecular characterization of host strains of *Spodoptera frugiperda* (Lepidoptera: Noctuidae) in Southern Brazil. *Annals of the Entomological Society of America* 101:619–626.

Marr, M. 2009. Differences in pheromone composition between the two strains of the fall armyworm *Spodoptera frugiperda* (Lepidoptera: Noctuidae). Diploma thesis, 64pp. Friedrich Schiller University, Jena. Available at: http://www.clib-jena.mpg.de/theses/ice/ICE09003.pdf.

Martinelli, S., P.L. Clark, M.I. Zucchi, M.C. Silva-Filho, J.E. Foster, and C. Omoto. 2007. Genetic structure and molecular variability of *Spodoptera frugiperda* (Lepidoptera: Noctuidae) collected in maize and cotton fields in Brazil. *Bulletin of Entomological Research* 97:225–231.

McMichael, M., and D.P. Prowell. 1999. Differences in amplified fragment-length polymorphisms in fall armyworm (Lepidoptera: Noctuidae) host strains. *Annals of the Entomological Society of America* 92:175–181.

Meagher, R.L., and E.R. Mitchell. 1998. Phenylacetaldehyde enhances upwind flight of male fall armyworm (Lepidoptera: Noctuidae) to its sex pheromone. *Florida Entomologist* 81:556–559.

Meagher, R.L., and M. Gallo-Meagher. 2003. Identifying host strains of fall armyworm (Lepidoptera: Noctuidae) in Florida using mitochondrial markers. *Florida Entomologist* 86:450–455.

Meagher, R.L., and R.N. Nagoshi. 2012. Differential feeding of fall armyworm (Lepidoptera: Noctuidae) host strains on meridic and natural diets. *Annals of the Entomological Society of America* 105:462–470.

Meagher, R.L., R.N. Nagoshi, C. Stuhl, and E.R. Mitchell. 2004. Larval development of fall armyworm (Lepidoptera: Noctuidae) on different cover crop plants. *Florida Entomologist* 87:454–460.

Meagher, R.L., R.N. Nagoshi, and C.J. Stuhl. 2011. Oviposition choice of two fall armyworm (Lepidoptera: Noctuidae) host strains. *Journal of Insect Behavior* 24:337–347.

Mitchell, E.R., J.H. Tumlinson, and J.N. McNeil. 1985. Field evaluation of commercial pheromone formulations and traps using a more effective sex pheromone blend for the fall armyworm (Lepidoptera, Noctuidae). *Journal of Economic Entomology* 78:1364–1369.

Nagoshi, R.N. 2010. The fall armyworm triose phosphate isomerase (Tpi) gene as a marker of strain identity and interstrain mating. *Annals of the Entomological Society of America* 103:283–292.

Nagoshi, R.N. 2012. Improvements in the identification of strains facilitate population studies of fall armyworm subgroups. *Annals of the Entomological Society of America* 105:351–358.

Nagoshi, R.N., and R.L. Meagher. 2003a. Fall armyworm FR sequences map to sex chromosomes and their distribution in the wild indicate limitations in interstrain mating. *Insect Molecular Biology* 12:453–458.

Nagoshi, R.N., and R.L. Meagher. 2003b. FR tandem-repeat sequence in fall army-worm (Lepidoptera: Noctuidae) host strains. *Annals of the Entomological Society of America* 96:329–335.

Nagoshi, R.N., and R.L. Meagher. 2004. Behavior and distribution of the two fall armyworm host strains in Florida. *Florida Entomologist* 87:440–449.

Nagoshi, R.N., R.L. Meagher, J.J. Adamczyk, S.K. Braman, R.L. Brandenburg, and G. Nuessly. 2006a. New restriction fragment length polymorphisms in the cytochrome oxidase I gene facilitate host strain identification of fall armyworm (Lepidoptera: Noctuidae) populations in the southeastern United States. *Journal of Economic Entomology* 99:671–677.

Nagoshi, R.N., R.L. Meagher, G. Nuessly, and D.G. Hall. 2006b. Effects of fall armyworm (Lepidoptera: Noctuidae) interstrain mating in wild populations. *Environmental Entomology* 35:561–568.

Nagoshi, R.N., J.J. Adamczyk, R.L. Meagher, J. Gore, and R. Jackson. 2007a. Using stable isotope analysis to examine fall armyworm (Lepidoptera: Noctuidae) host strains in a cotton habitat. *Journal of Economic Entomology* 100:1569–1576.

Nagoshi, R.N., P. Silvie, and R.L. Meagher. 2007b. Comparison of haplotype frequencies differentiate fall armyworm (Lepidoptera: Noctuidae) corn-strain populations from Florida and Brazil. *Journal of Economic Entomology* 100:954-961.

Nagoshi, R.N., P. Silvie, R.L. Meagher, J. Lopez, and V. Machado. 2007c. Identification and comparison of fall armyworm (Lepidoptera : Noctuidae) host strains in Brazil, Texas, and Florida. *Annals of the Entomological Society of America* 100:394–402.

Nagoshi, R.N., R.L. Meagher, K. Flanders, J. Gore, R. Jackson, J. Lopez, J.S. Armstrong, G.D. Buntin, C. Sansone, and B.R. Leonard. 2008. Using haplotypes to monitor the migration of fall armyworm (Lepidoptera : noctuidae) corn-strain populations from Texas and Florida. *Journal of Economic Entomology* 101:742–749.

Nagoshi, R.N., S.J. Fleischer, and R.L. Meagher. 2009. Texas is the overwintering source of fall armyworm in central Pennsylvania: implications for migration into the northeastern United States. *Environmental Entomology* 38:1546–1554.

Nagoshi, R.N., R.L. Meagher, and D.A. Jenkins. 2010. Puerto Rico fall armyworm has only limited interactions with those from Brazil or Texas but could have substantial exchanges with Florida populations. *Journal of Economic Entomology* 103:360–367.

Nagoshi, R.N., M.G. Murua, M. Hay-Roe, M.L. Juarez, E. Willink, and R.L. Meagher. 2012. Genetic characterization of fall armyworm (Lepidoptera: Noctuidae) host strains in Argentina. *Journal of Economic Entomology* 105:418–428.

Pashley, D.P. 1986. Host-associated genetic differentiation in fall armyworm (Lepidoptera, Noctuidae): a sibling species complex. *Annals of the Entomological Society of America* 79:898–904.

Pashley, D.P. 1988a. Current status of fall armyworm host strains. *Florida Entomology* 73:227–234.

Pashley, D.P. 1988b. Quantitative genetics, development, and physiological adaptation in host strains of fall armyworm. *Evolution* 42:93–102.

Pashley, D.P. 1989. Host-associated differentiation in army worms (Lepidoptera: Noctuidae): an allozymic and mitochondrial DNA perspective. Pp. 103–114. In H.D. Loxdale, and J. Den Hollander, eds. *Electrophoretic Studies on Agricultural Pests: Systematics Association*. Oxford: Clarendon.

Pashley, D.P., and J.A. Martin. 1987. Reproductive incompatibility between host strains of the fall armyworm (Lepidoptera: Noctuidae). *Annals of the Entomological Society of America* 80:731–733.

Pashley, D.P., and L.D. Ke. 1992. Sequence evolution in mitochondrial ribosomal and ND-1 genes in Lepidoptera: implications for phylogenetic analysis. *Molecular Biology and Evolution* 9:1061–1075.

Pashley, D.P., S.J. Johnson, and A.N. Sparks. 1985. Genetic population structure of migratory moths: the fall armyworm (Lepidoptera, Noctuidae). *Annals of the Entomological Society of America* 78:756–762.

Pashley, D.P., A.M. Hammond, and T.N. Hardy. 1992. Reproductive isolating mechanisms in fall armyworm host strains (Lepidoptera, Noctuidae). *Annals of the Entomological Society of America* 85:400–405.

Pashley, D.P., T.N. Hardy, and A.M. Hammond. 1995. Host effects on developmental and reproductive traits in fall armyworm strains (Lepidoptera, Noctuidae). *Annals of the Entomological Society of America* 88:748–755.

Prowell, D.P., M. McMichael, and J.F. Silvain. 2004. Multilocus genetic analysis of host use, introgression, and speciation in host strains of fall armyworm (Lepidoptera: Noctuidae). *Annals of the Entomological Society of America* 97:1034–1044.

Ritchie, M.G. 2007. Sexual selection and speciation. *Annual Review of Ecology Evolution and Systematics* 38:79–102.

Saldamando, C.I., and A.M. Vélez-Arango. 2010. Host plant association and genetic differentiation of corn and rice strains of *Spodoptera frugiperda* Smith (Lepidoptera: Noctuidae) in Colombia. *Neotropical Entomology* 39:921–929.

Schöfl, G., D.G. Heckel, and A.T. Groot. 2009. Time-shifted reproductive behaviours among fall armyworm (Noctuidae: *Spodoptera frugiperda*) host strains: evidence for differing modes of inheritance. *Journal of Evolutionary Biology* 22:1447–1459.

Schöfl, G., A. Dill, D.G. Heckel, and A.T. Groot. 2011. Allochronic separation versus mate choice: nonrandom patterns of mating between fall armyworm host strains. *American Naturalist* 177:470–485.

Tumlinson, J.H., E.R. Mitchell, P.E.A. Teal, R.R. Heath, and L.J. Mengelkoch. 1986. Sex pheromone of fall armyworm, *Spodoptera frugiperda* (J.E. Smith): identification of components critical to attraction in the field. *Journal of Chemical Ecology* 12: 1909–1926.

Unbehend, M., S. Haenniger, G. Vásquez, M.L. Juarez, Reisig, J.N. McNeil, R.L. Meagher, D.A. Jenkins, D.G. Heckel, and A.T. Groot. 2014. Geographic variation in sexual attraction of *Spodoptera frugiperda* corn- and rice-strain males. *PLOS ONE* 9:e89255.

Unbehend, M., S. Hanniger, R.L. Meagher, D.G. Heckel, and A.T. Groot. 2013. Pheromonal divergence between two strains of *Spodoptera frugiperda*. *Journal of Chemical Ecology* 39:364–376.

Veenstra, K.H., D.P. Pashley, and J.A. Ottea. 1995. Host-plant adaptation in fall armyworm host strains: comparison of food consumption, utilization, and detoxication enzyme activities. *Annals of the Entomological Society of America* 88:80–91.

Velez-Arango, A.M., R.E. Arango, D. Villanueva, E. Aguilera, and C.I. Saldamando B. 2008. Identification of *Spodoptera frugiperda* biotypes (Lepidoptera: Noctuidae) through using mitochondrial and nuclear markers. *Revista Colombiana de Entomologia* 34:145–150.

Whitford, F., S.S. Quisenberry, T.J. Riley, and J.W. Lee. 1988. Oviposition preference, mating compatibility, and development of two fall armyworm strains. *Florida Entomology* 71:234–243.

CHAPTER TWENTY-ONE

Pheromones of Heliothine Moths

N. KIRK HILLIER and THOMAS C. BAKER

Introduction

Within noctuid moths, species in the subfamily Heliothinae
(also known as owlet moths) represent an excellent model system for examining divergence of traits associated with pheromone production, detection, and processing in closely related

species. Species of Heliothinae are ubiquitous, with many species distributed globally. Estimates of diversity have included
25–28 genera and approximately 365 species (Cho et al.
2008). Most species are seed and bud feeders, with many having the common names of "budworms" or "bollworms"
depending on their host association (Hardwick 1965, 1970;

TABLE 21.1

Heliothine spp. with documented pheromones and/or odor-mediated behavior to pheromone components (i.e., attraction to synthetic lures in the field trapping)

Species	Common name	Known distribution	Pest status	Citation(s)
Heliothis spp.				
H. belladonna	None	Western North America	Host unknown	Landolt et al. (2006)
H. maritima	Fulvous clover	Eurasia	Minor polyphagous pest	Szöcs et al. (1993)
H. maritima adaucta	Flax budworm	Japan (subspecies)	Minor polyphagous pest	Kakizaki and Sugie (1993)
H. ononis	Flax bollworm	Eurasia and western North America	Minor pest on *Linum* spp.	Steck et al. (1982)
H. peltigera	Bordered straw	Eurasia, Africa	Major polyphagous pest	Dunkelblum and Kehat (1989)
H. phloxiphaga	Darker spotted straw moth	North America	Minor polyphagous pest	Raina et al. (1986); Kaae et al. (1973)
H. subflexa	Physalis bud moth	North America, South America	Minor pest–host specialist	Vickers (2002); Heath et al. (1990); Teal et al. (1981a)
H. virescens	Tobacco budworm	Americas, Caribbean, Hawaii	Major polyphagous pest	Hendricks et al. (1989); Shaver et al. (1989); Teal and Tumlinson (1989); Ramaswamy and Roush (1986); Teal et al. (1986); Ramaswamy et al. (1985); Vetter and Baker (1983); Pope et al. (1982); Klun et al. (1980a); Mitchell et al. (1978); Roelofs et al. (1974)
Helicoverpa spp.				
H. armigera	African bollworm	Eurasia, Africa, Australia, Oceania	Major polyphagous pest	Zhang et al. (2012); Kvedaras et al. (2007); Dong et al. (2005); Kehat and Dunkelblum (1990); Nesbitt et al. (1979); Dunkelblum et al. (1980); Kehat et al. (1980); Gothilf et al. (1979); Rothschild (1978); Piccardi et al. (1977)
H. assulta	Oriental tobacco budworm	Africa, Asia, Australia, Oceania	Major polyphagous pest	Park et al. (1996); Park et al. (1994); Cork et al. (1992); Sugie et al. (1991)
H. gelotopoeon	South American bollworm	Southern South America	Major polyphagous pest	Cork and Lobos (2003)
H. punctigera	Australian bollworm	Australia	Major polyphagous pest	Rothschild (1978); Rothschild et al. (1982)
H. zea	Corn earworm	Americas	Major polyphagous pest	Descoins et al. (1988); Pope et al. (1984); Teal et al. (1984); Vetter and Baker (1984); Klun et al. (1980b)
Other genera				
Schinia bina	Bina flower moth	North America	Nonpest, feeds on selected Asteraceae	Underhill et al. (1977)
Schinia meadi	Mead's flower moth	Western North America	Host unknown	Steck et al. (1982)
Schinia mitis	Matutinal flower moth	Southeastern United States	Nonpest, specialist on *Pyrrhopappus* spp.	Mitchell (1982)
Schinia suetus	None	Western North America	Nonpest, specialist on *Lupinus* spp.	Byers and Struble (1987)

Mitter et al. 1993; Matthews 1999) (Table 21.1). In addition, the degree of host association is variable, ranging from species that are highly host specific to those that exhibit wide polyphagy, feeding on >100 plant species, or families.

Host associations are known for approximately one quarter of heliothines, of which 70% are considered largely oligophagous, with the remaining 30% being polyphagous (Cho et al. 2008) (Table 21.1). For example, *Heliothis subflexa* larvae feed exclusively on fruits of *Physalis* spp. (Solanaceae; e.g., ground-cherries, tomatillos, cape gooseberries), whereas primrose moth, *Schinia florida*, larvae are exclusive feeders on buds of evening primrose (Onagraceae: *Oenothera* spp.) (Hardwick 1958). Species in other genera, such as *Pyrrhia* and *Asidura*, maintain strict host-specialist relationships. Several heliothine species are important agricultural pests, particularly those in the genera *Heliothis* and *Helicoverpa* (Table 21.1). Based on a molecular phylogeny by Cho et al. (2008), early divergent lineages are almost exclusively host specialists, whereas species that fall into the "*Heliothis* group" represent the most extreme polyphages. This latter group includes what are more commonly referred clades of the "corn earworm complex" and "tobacco budworm group" (Cho et al. 2008).

The *Heliothis* group includes some of the most important agricultural pests worldwide, causing massive annual crop losses, particularly in the developing world. Estimates range between US$3 billion and US$7 billion in control costs per annum for the most prevalent agricultural pest species: *Helicoverpa armigera*, *H. zea*, *H. assulta*, *H. punctigera*, and *Heliothis virescens* (Fitt 1989). These species are important pests on a range of forage, oilseed, and food crops, including cotton, corn, sorghum, soybean, flax, tobacco, and tomato.

Worldwide Diversity of Heliothine Pheromone Composition

Divergence in olfactory communication is evident among heliothine species, based on shifts in the use of key components within species' pheromone blends that function to optimize attraction of conspecific males and reduce attraction of males to females of closely related species. Multicomponent pheromones have been described in all heliothine species studied to date, and their function has been supported by studies investigating pheromone gland composition, behavior, electrophysiology, field trap attraction, and odorant receptor (OR) gene expression.

Avoidance of mating mistakes with members of a wrong species during mate finding and courtship is of paramount importance for sympatric moth species, including heliothines. Male moths respond with remarkable sensitivity and selectivity to very precise multicomponent mixtures of female-emitted sex pheromone components. Deviations in blend ratios involving key components of pheromone mixtures will significantly affect the degree to which males are attracted to conspecific females and deterred from attraction to heterospecifics. Isolation is further augmented by structural specificity in genitalia, with some reports documenting irreversible genitalia "locking" between *Heliothis virescens* and *Helicoverpa zea* in the field and laboratory (Hardwick 1965; Shorey et al. 1965; Teal et al. 1981b; Stadelbacher et al. 1983). In instances where prezygotic isolation is incomplete, such as in mating of *H. virescens* and *H. subflexa*, there are distinct fitness costs: hybrid male progeny of these two species are sterile, producing largely apyrene sperm (Proshold and LaChance

1974). Attraction and subsequent successful copulation are therefore tightly linked to an individual's ability to produce (usually female), detect, and respond with appropriate behavior (male) to a conspecific versus a heterospecific signal.

With the exception of one species, all heliothines studied to date use (Z)-11-hexadecenal (Z11-16Ald) as a component of the female-produced sex pheromone (figure 21.1; Table 21.2). In most cases, Z11-16Ald is the predominant component in female effluvia. Two aldehydes used as minor components, (Z)-9-tetradecenal (Z9-14Ald) and (Z)-9-hexadecenal (Z9-16Ald), appear to be key variable components whose interchangeability dictates the species specificity of many blends. Throughout the Heliothinae, shifts in male preference for blends that include either Z9-14Ald or Z9-16Ald have apparently occurred, contributing to reduction in mating mistakes and hence to reproductive isolation and possibly speciation. Evidence also suggests isolation between *H. virescens* and *H. subflexa* is modulated by pheromone preference controlled by a single locus containing four OR genes that modulate pheromone behavioral response specificity via OR ligand selectivity (Gould et al. 2010; Vásquez et al. 2011; Wang et al. 2011).

In addition to the sex pheromone "components" in emissions produced by the pheromone glands of many moth species, there are other volatile compounds in effluvia that may decrease attraction of heterospecific males due to additional olfactory pathway antagonism that unbalances an otherwise balanced blend (Domingue et al. 2007; Baker 2008). In many cases, the effects of a heliothine species' pheromone components themselves on creating attraction depend on their relative ratio in the mixture. Sometimes an excess proportion of a minor component may decrease attraction of conspecific males by creating an excessive olfactory pathway antagonism that unbalances what should have been a balanced pheromone blend. In heliothines, such interactions have been supported by studies investigating combinatorial coding of pheromone components and pheromone-related compounds at the sensillar and antennal lobe (AL) levels (Christensen et al. 1990, 1995; Vickers et al. 1998; Vickers and Christensen 1998, 2003; Vickers 2006a,b; Hillier and Vickers 2011a). This interaction has been further proposed as "balanced" olfactory antagonism, wherein a continuum of negative and positive male behavior is modulated (Baker 2008).

Biosynthesis of Pheromone Blends in Genera *Heliothis* and *Helicoverpa*

Sequence of Action of Δ11-Desaturase and β-Oxidation Determine Composition of Many Pheromone Blends

The pheromone communication systems of heliothine moths all include various behaviorally active pheromone-component end products that are produced along pheromone biosynthetic pathways outlined by Jurenka (2003) for *Heliothis virescens* and *H. subflexa* (figure 21.2). As shown in Table 21.2, nearly all heliothine species use Z11-16Ald as the major (most abundant) sex pheromone component in their blends. The only exceptions thus far are *Helicoverpa assulta* in Asia in which Z9-16Ald is the major component, present at 10 times the abundance of Z11-16Ald (Cork et al. 1992; Park et al. 1994), and the South American species *H. gelotopoeon* in which Z11-16Ald is absent and replaced instead by saturated hexadecanal (16Ald) as the major component, with a nearly

TABLE 21.2

Heliothine female-emitted sex pheromone blends.

Pheromone components confirmed as both being produced by females and required for male attraction in laboratory and field studies are shown in headers in plain black non-italicized text. Compounds found to be produced by females but have no demonstrated behavioral effects on males or else have effects only in certain local populations are shown in headers in italicized black text. Black numbers indicate the percentages of compounds produced by females relative to Z11-16Ald (or relative to Z9-16Ald, *Heliothis assulta; or to* 16Ald, *H. gelotopoeon*). Check marks indicate trace amounts found. Green numbers indicate ranges of component percentages that have been shown to optimally evoke male attraction. Red numbers indicate heterospecific pheromone components shown to reduce attraction (behaviorally antagonistic) at the indicated percentages

Species	\	\	\	\	Compounds evidenced in chemical communication								
	14Ald	Z9-14Ald	Z9-14OH	Z9-14Ac	16Ald	Z7-16Ald	Z7-16Ac	Z9-16Ald	Z9-16OH	Z9-16Ac	Z11-16Ald	Z11-16OH	Z11-16Ac
North and South America													
H. phloxiphaga GLAND					✓			0.5			100	7.2	
H. phloxiphaga RESPONSE								0.6			100	2–3, 4	
H. subflexa GLAND	✓	✓			2.4		3.6	42.9		9.9	100	5.8	22.7
H. subflexa RESPONSE								10–50			100	1–50>	15–20
H. virescens GLAND	✓	0.4–6			25.6	✓		✓			100		
H. virescens RESPONSE		5–50									100	3–30	0.1>
H. gelotopoeon GLAND	✓				100			84					
H. gelotopoeon RESPONSE					10>			100			1>		
H. zea GLAND		0			✓	✓		1.8			100		
H. zea RESPONSE		1–3, 3>						1–15			100	0.1–10	0.1–10
*H. armigera GLAND		0.3					✓	2.5			100		
*H. armigera RESPONSE		0.3–5, 25>						3–3000			100		
Europe and Africa													
H. armigera GLAND		0.3					✓	2.5			100		
H. armigera RESPONSE		0.3–5, 25>						3–3000			100		
H. assulta GLAND					✓			100	✓	3051	6.5	✓	✓
H. assulta RESPONSE		1>			✓			100	2>	30	5.0		1.5
H. maritima GLAND					5.5			✓			100	8.1	
H. maritima RESPONSE					3–6						100	6–20	
H. peltigera GLAND	✓	14.6	✓	✓	✓	✓		✓			100	24.3	✓
H. peltigera RESPONSE		5–50									100		
Asia and Australia													
H. maritima adaucta GLAND					✓						100	24.5	
H. maritima adaucta RESPONSE											100	1–3	
H. armigera GLAND		0.3					✓	2.5			100		
H. armigera RESPONSE		0.3–5, 25>						3–3000			100		
H. assulta GLAND					✓			100	✓	3051	6.5	✓	✓
H. assulta RESPONSE		1>			✓			100	2>	30	5.0		1.5
H. punctigera GLAND		<0.5									100	25	41.7
H. punctigera RESPONSE		5									100	100>	1–10

*H. armigera is a recent introduction to Brazil (Tay et al. 2013).

SOURCES: *Heliothis maritima* (Szöcs et al. 1993); *Heliothis maritima adaucta* (Kakizaki and Sugie 2003); *Heliothis peltigera* (Dunkelblum and Kehat 1989); *Heliothis phloxiphaga* (Raina et al. 1986); *Heliothis subflexa* (Teal et al. 1981a; Klun et al. 1982; Heath et al. 1990 showed that OH is needed; Vickers 2002; Groot et al. 2007); *Heliothis virescens* (Roelofs et al. 1974; Pope et al. 1982; Vetter and Baker 1983; Teal et al. 1986; Vickers et al. 1991); *Helicoverpa armigera* (Piccardi et al. 1977; Nesbitt et al. 1979; Kehat and Dunkleblum 1990; Zhao et al. 2006; Zhang et al. 2012); *Helicoverpa assulta* (Sugie et al. 1991; Cork et al. 1992; Park et al. 1994; Zhao et al. 2006); *Helicoverpa gelotopoeon* (Cork and Lobos 2003); *Helicoverpa punctigera* (Rothschild et al. 1982); *Helicoverpa zea* (Klun et al. 1980b; Pope et al. 1984; Vetter and Baker 1984; Teal et al. 1984; Vickers et al. 1991; Fadamiro and Baker 1997; Quero and Baker 1999).

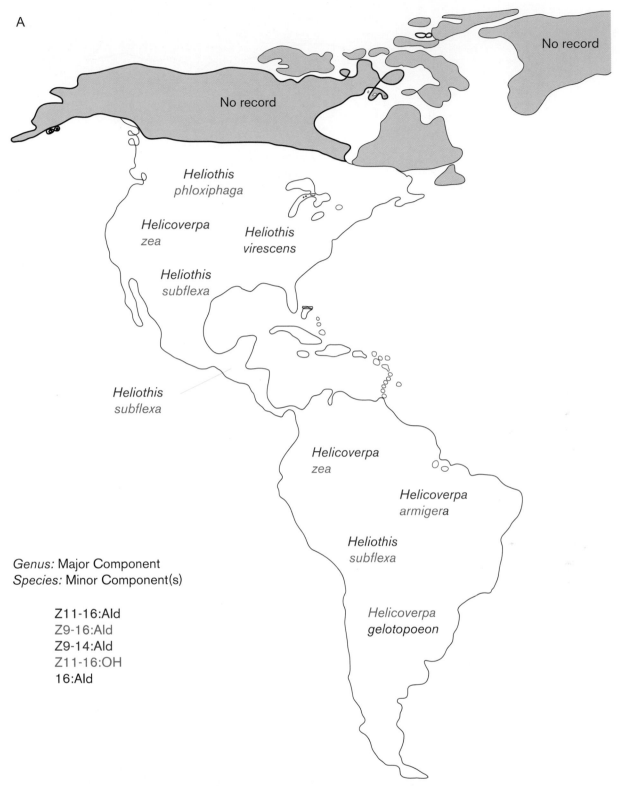

FIGURE 21.1 Worldwide distribution of *Helicoverpa* and *Heliothis* spp. and differences in their pheromone blends involving behaviorally important two-component mixtures. Genus names are color-coded according to the most abundant (major) component in the pheromone blend, as indicated in the key at the bottom of the figure. The colors depicted in each species' name denote experimentally demonstrated behaviorally important secondary components corresponding to the compounds in the color-coded key at the bottom of the figure. Some specific names are split into two colors because for those species, two secondary components are both important contributors to male behavioral response. (A) The Americas.

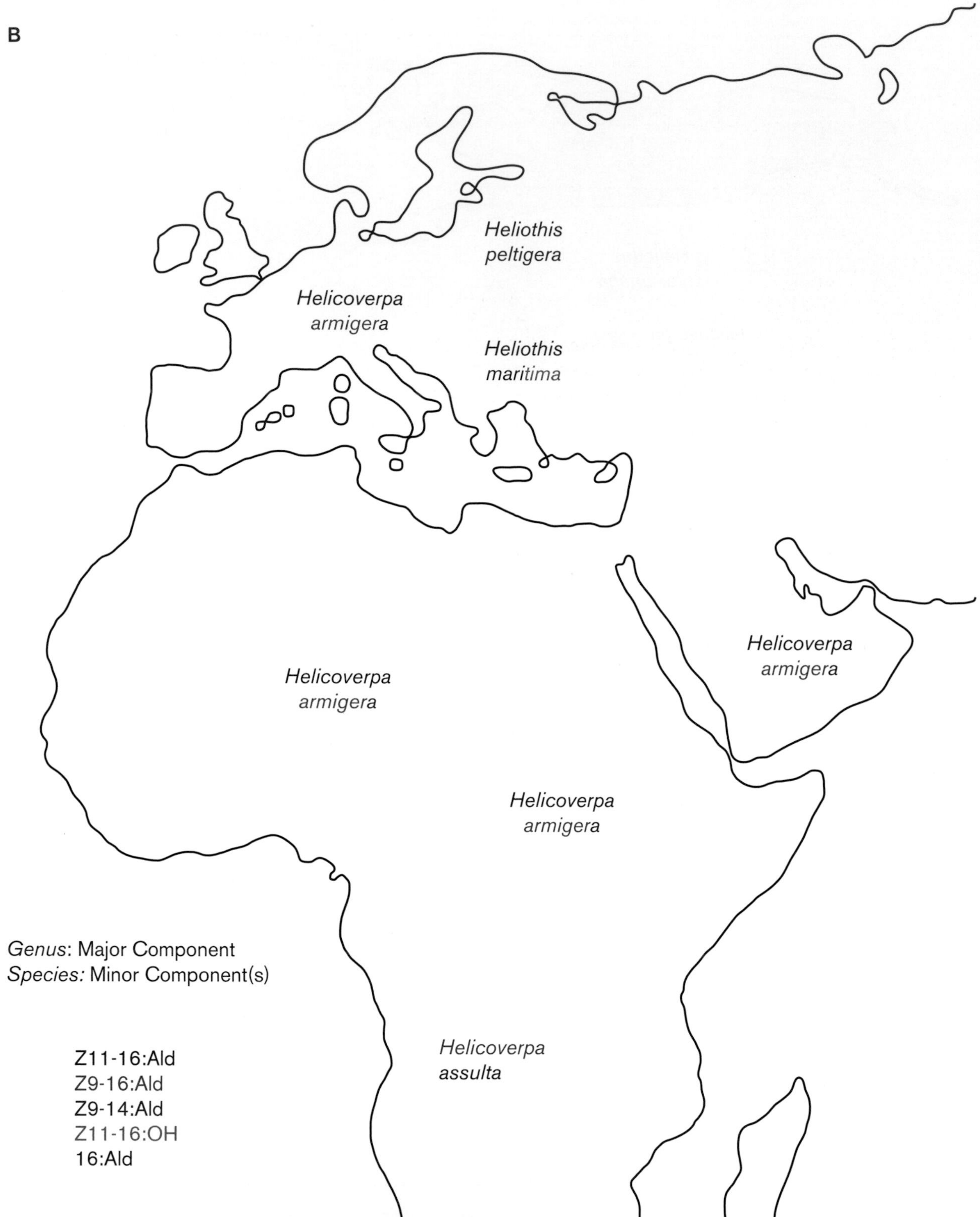

B

Heliothis
peltigera

Helicoverpa
armigera

Heliothis
maritima

Helicoverpa
armigera

Helicoverpa
armigera

Helicoverpa
armigera

Genus: Major Component
Species: Minor Component(s)

Z11-16:Ald
Z9-16:Ald
Z9-14:Ald
Z11-16:OH
16:Ald

Helicoverpa
assulta

FIGURE 21.1 *(continued)* (B) Europe and Africa.

C

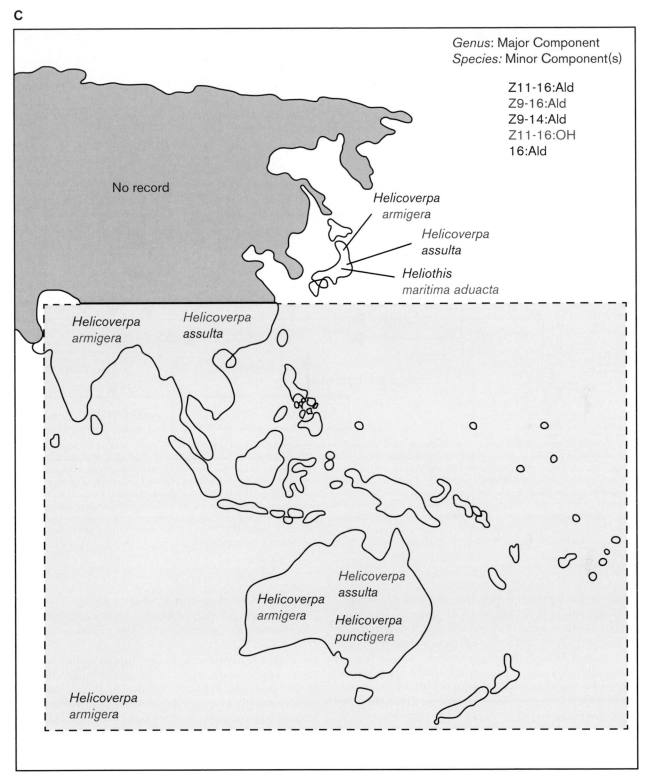

Genus: Major Component
Species: Minor Component(s)

Z11-16:Ald
Z9-16:Ald
Z9-14:Ald
Z11-16:OH
16:Ald

No record

Helicoverpa armigera

Helicoverpa assulta

Heliothis maritima aduacta

Helicoverpa armigera

Helicoverpa assulta

Helicoverpa assulta

Helicoverpa armigera

Helicoverpa punctigera

Helicoverpa armigera

FIGURE 21.1 *(continued)* (C) Australasia. Gray boxed area denotes the putative distribution of *H. armigera* throughout the island regions of Australasia.

equal amount of Z9-16Ald to comprise an unusual two-component blend (Cork and Lobos 2003).

Considering all known pheromone blends of *Helicoverpa* and *Heliothis* spp. (Table 21.2), it is clear that nearly all make use of Δ11-desaturase acting on the C16 fatty acyl substrate (16CoA) rather than on a C18 fatty acyl substrate (18CoA) (figure 21.2), to produce Z11-16Ald and the related compounds (Z)-11-hexa-

decen-1-ol acetate (Z11-16Ac) and (Z)-11-hexadecen-1-ol (Z11-16OH) (figure 21.2). A second route that results in Z9-16Ald as either a major or a minor component involves Δ11-desaturase acting on the 18CoA substrate, followed by a chain-shortening, β-oxidation step to produce the Z9-16Ald and related alcohol and acetate. This route in which chain shortening follows desaturation of the long fatty acyl group (18CoA) has been

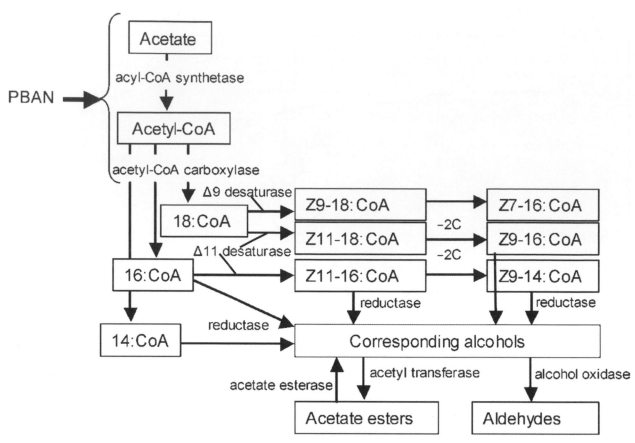

FIGURE 21.2 Pathways for biosynthesis of heliothine moth sex pheromone components. Two key enzymes explain most of the variation in blends: a Δ11-desaturase that places a (Z) double bond in the 11 position on either the C18-fatty acyl chain or the C16-fatty acyl chain, and a β-oxidation enzyme that shortens the fatty acyl chain by removing the first two carbons after the acyl group on the chain. When the Δ11-desaturase acts on the C16-fatty acyl chain, it produces Z11-16CoA that eventually gets reduced to Z11-16OH and then oxidized to Z11-16Ald. A β-oxidation step acting on Z11-16CoA ("-2C" at middle-right of figure) produces Z9-14CoA that eventually gets reduced to Z9-14OH and oxidized to Z9-14Ald, a key secondary component of *Heliothis virescens*, *H. peltigera*, and *Helicoverpa punctigera* (see figure 21.1). When the Δ11-desaturase acts on the C18-fatty acyl chain, it produces Z11-18CoA which gets chain-shortened via a β-oxidation step to Z9-16CoA ("-2C" at middle-right of figure), which is then reduced to Z9-16OH and oxidized to Z9-16Ald, the major component of *H. assulta* and *H. gelotopoeon*. Z9-16Ald is also a key minor component for *H. zea* and *H. armigera* (see figure 21.1).

SOURCE: Adapted from Jurenka (2003) and Groot et al. (2004).

favored by species that use large percentages of Z9-16Ald as a minor component (e.g., *H. subflexa)*, or predominant percentages of Z9-16Ald as a major component, (e.g., *H. assulta, H. gelotopoeon*) in their blends. Other species produce only small amounts of Z9-16Ald as pheromone components, and thus not as much 18CoA is used as the substrate for the Δ11-desaturase compared to the much larger amounts of 16CoA substrate used by these species for their blends that have Z11-16Ald as the major component and Z9-16Ald as a minor component (e.g., *H. phloxiphaga, H. armigera,* and *H. zea*). Another route involved in species using Z9-14Ald as a minor component requires that a β-oxidation chain-shortening step occur after desaturation. But here the β-oxidation step acts on the shorter, Δ11-16CoA substrate, thereby shortening Z11-16CoA down to Z9-14CoA, which is then reduced to (Z)-9-tetradecen-1-ol (Z9-14OH) and oxidized finally to Z9-14Ald (figure 21.2). The use of this post-Z11-16Ald chain-shortening step for producing Z9-14Ald is seen in *H. peltigera, H. virescens, H. punctigera,* and *H. armigera*. Rarely, some species use the saturated 16Ald as a major or minor pheromone component (e.g., *H. gelotopoeon* and *H. maritima*), and in these cases the 16CoA is reduced to the saturated hexadecanol (16OH) without the intervening step of Δ11-desaturase acting on it first, with the 16OH then being oxidized to 16Ald.

Fatty Acyl Reductase and Alcohol Oxidase Determine the Amount of Alcohol versus Aldehyde Sex Pheromone Composition of Many Blends

The terminal enzymatic steps, especially the reduction of the CoAs to the alcohols, oxidation of the alcohols to the aldehydes, as well as acetylation of the alcohols to the corresponding acetates, all can contribute significantly to the final pheromone blend composition. Fatty acyl reductases (FARs) have now been isolated from *Helicoverpa armigera, H. assulta, Heliothis virescens,* and *H. subflexa* and are shown to have the same substrate preferences across species (Hagström et al. 2012). The FARs of all four species prefer to reduce (Z)-9-tetradecenoate over (Z)-11-hexadecenoate, and the latter is reduced preferentially over (Z)-9-hexadecenoate. Thus, for all four species, the final differing species-specific blend ratios of the C14 and C16 aldehydes depend not on the differential selective activity of each species' FARs, but rather on the differing amounts of C16- and C14-acyl substrates that are available for the FARs to reduce (Hagström et al. 2012).

For all species, analyses of the amounts of Z11-16OH found in pheromone glands or shown to be emitted by female heliothines as pheromone "components" have been quite variable and a source of confusion in the literature for heliothine sex pheromones. It seems that significant amounts of Z11-16OH can always be solvent extracted from heliothine female glands, but the volatilization of Z11-16OH from the glands may be negligible. For instance, two groups found no Z11-16OH in the airborne collections from calling female *H. virescens* (Pope et al. 1982; Teal et al. 1986), but they were able to find significant percentages of it in the blends extracted from female glands. These findings of Z11-16OH in the solvent extracts were in agreement with those of several other groups (e.g., Ramaswamy et al. 1985; Ramaswamy and Roush 1986) that did not conduct airborne collections. Insight into why alcohols were not found in airborne emissions came from Teal and Tumlinson (1986), who demonstrated by applying synthetic Z9-14OH to the glands of *H. virescens* females that this compound is enzymatically converted to Z9-14Ald as it travels through the glandular epithelium to the gland's surface cuticle, suggesting the activity of a cuticular-bound oxidase. Thus, amounts of alcohols such as Z11-16OH found in gland extracts of many species likely represent only glandular biosynthetic precursors of aldehydes that are actually emitted from the gland surface after being oxidized to corresponding aldehydes by the glandular tissues.

Studies with *H. phloxiphaga* (Raina et al. 1986) further support the idea that the pheromone blend emitted can also differ quantitatively from the gland contents. When very short extraction times (<3 min) of female glands were used that probably extracted mainly the compounds present on the gland surface, the amounts of Z11-16OH in the blend were low (2.9%) and behaviorally optimal for male attraction in a wind tunnel. When extraction times of 10–15 min were used, however, that were more likely to extract pheromone-component precursors from inside the gland, the amounts of Z11-16OH increased to >4.5%, and in a wind-tunnel bioassay, this larger proportion reduced male attraction significantly (Raina et al. 1986).

As much as possible, extraction procedures should be developed that accurately estimate the actual emitted signal that has been selected for inducing male behavior, rather than the composition of the pool of precursors residing in the interior of the pheromone gland tissues. Cork et al. (1992) showed that in *H. assulta* very late in the scotophase, after the optimal calling period, the amount of the Z9-16Ald major component in the gland had diminished by more than two-thirds from its amount at the peak, whereas the amount of Z9-16OH had increased by nearly 50-fold. This finding again indicates that alcohols that can be extracted from heliothine female pheromone glands are most often biosynthetic precursors waiting to be oxidized to the aldehyde pheromone components.

In a few species, corresponding acetates to the aldehydes and alcohols, such as Z11-16Ac, have been found to be produced by females, and to cause slight but significantly elevated amounts of attraction of male conspecifics. In *H. assulta*, Z11-16Ac seems to vary geographically in its production by females (amounts found in the glands) (Cork et al. 1992; Park et al. 1994), as well as in its effect in blends containing the two main components, where it increases male *H. assulta* attraction (trap capture) in regions corresponding to its production by females (Cork et al. 1992; Park et al. 1994).

Although its effect on increasing attraction of conspecific males is small and varies according to geographic regions for both *H. assulta* in Asia and *H. subflexa* in North America, the effect of the addition of Z11-16Ac in reducing or eliminating attraction of heterospecific males in North America is pronounced. Males of species sympatric to *H. subflexa*, such as *H. zea* and *H. virescens*, are not attracted at all to calling *H. sub-*

flexa females (Lelito et al. 2008), an effect that is undoubtedly due to Z11-16Ac in the *H. subflexa* blend. Trace amounts of 0.1% or 1% Z11-16Ac added to the otherwise highly attractive blends of *H. virescens* or *H. zea* can substantially reduce attraction of the males of these two species (Fadamiro and Baker 1997; Vickers and Baker 1994, 1997; Baker et al. 1998a, 1998b). These three species are sympatric in parts of North America, and they are not temporally distinct in generational phenology or calling periodicity. It is evident that blends containing this acetate function as a means of prezygotic mating isolation, keeping heterospecific males from wasting time in long-distance orientation to females that will result in no viable progeny. At the same time, emitted blends containing this compound will for females minimize the chances that they will be harassed, with persistent courtship attempts, by heterospecific males and thus be subjected to fruitless courtships and couplings with such males. Z11-16Ac in blends thereby works at long distance to reduce male and female pre-copulatory and copulation mistakes of all types that will reduce reproductive fitness in both sexes.

Heliothine moth sex pheromone systems thus are exemplars of moth pheromone evolution via *adaptive* responses to heterospecific blends in which different species-specific blends have diverged during a reproductive character displacement process in zones of sympatry of two established species (Butlin 1987), or during a speciation event that includes reinforcement (Butlin 1987). A "species recognition" mechanism (Paterson 1985; Lambert et al. 1987) orchestrating the evolution of these pheromone blend ratios solely through normalizing (stabilizing) selection lacking adaptive responses to heterospecific pressures is not well supported. Not only is there strong behavioral aversion of heliothine males of some species to blends containing trace amounts of heterospecific components such as Z11-16Ac in areas of sympatry, but also there are olfactory sensory neurons (OSNs) on their antennae that are specifically tuned to such heterospecific components. It is hard to reconcile the presence of heterospecific-component-tuned OSNs expressed on male antenna as being anything but an adaptive natural selective response to mate finding and mating mistakes. These OSNs represent the neuronal hardwiring of males that helps them classify, at a distance, pheromone blends as being from heterospecific females and thus avoid wasting time advancing upwind in their pheromone plumes. It would be difficult to find more definitive proof for pheromones being sculpted by adaptive responses to erroneous heterospecific cross-communication than in the types of heliothine males' OSNs that are dedicated only to the detection of heterospecific pheromone components. This is not to say that blends have not also been shaped over time by the effects of normalizing selection orchestrated by females emitting very low rates of a population-mean optimal blend. Such females should attract (select for) only males that are equipped with an optimally functioning olfactory system having the greatest sensitivity and selectivity for the correct blend ratio. It appears though from abundant experimentation that very few heliothine moth species have males exhibiting such sensitivity and selectivity for only a very limited range of blends.

Hormonal Control of Pheromone Biosynthesis and Emission

Most heliothines studied to date share similar activity patterns for oviposition, feeding, and mating. In terms of calling behavior, peak pheromone production and mating receptivity by females occur approximately 2–6 h after the onset of scotophase (Shorey and Gaston 1965; Mbata and Ramaswamy 1990; Ramaswamy 1990). Furthermore, age influences pheromone production in heliothines, with peak production approximately 48 h postemergence. Peak pheromone production is modulated by several factors, including titer of juvenile hormone, octopamine (Christensen et al. 1992; Rafaeli 2009), and pheromone biosynthesis activating neuropeptide (PBAN; Raina and Klun 1984; Raina et al. 1986, 1987; Raina and Kempe 1990, 1992; for review, see Raina 1993). Finally, examination of pheromone production in *Heliothis virescens* has demonstrated that males transfer a pheromonostatic compound that inhibits the release of pheromonotropin and PBAN (Ramaswamy et al. 1995).

PBAN Activity

A thorough investigation of pheromone biosynthesis is beyond the focus of this chapter. However, a significant body of knowledge has been generated from heliothines to elucidate control of pheromone production and biosynthetic pathways. Indeed, species such as *Heliothis virescens* and *Helicoverpa zea* have been used as model systems in initial and continuing studies on such phenomena. The first evidence for neuroendocrine (peptidergic) control of female pheromone production in moths was demonstrated in *H. zea*, ultimately leading to the discovery and characterization of PBAN (Raina and Klun 1984; Raina et al. 1986, 1987; Raina and Kempe 1990, 1992; for review, see Raina 1993). Subsequent work identified the gene encoding PBAN and related peptides by using *H. zea* as a model (Jurenka et al. 1991; Davis et al. 1992; Ma et al. 1996, 1998; Choi et al. 2005). Such peptides are produced in localized regions of the brain (subesophageal ganglion, corpora cardiaca), and ganglia of the ventral nerve cord (Kingan et al. 1992, 1993; Jurenka and Rafaeli 2011). Neural and endocrine factors have been identified that may regulate PBAN production and modulate pheromone biosynthesis (Teal et al. 1989; Christensen et al. 1991; Rafaeli and Gileadi 1995). Groot et al. (2005) also demonstrated that PBAN injections in female *H. virescens* and *H. subflexa* could be used to reduce variation in production between virgin and mated females and in photophase compared to scotophase.

Several studies (again within Heliothinae) subsequently have demonstrated PBAN acts directly on the pheromone gland when the gland is stimulated in vivo or in vitro with peptide extracts (Rafaeli and Jurenka 2003). Binding occurs with epidermal cells of the pheromone gland, and G protein-coupled receptors for PBAN have been localized from *H. zea* female pheromone glands (Choi et al. 2003; Kim et al. 2008). Rate-limiting enzymatic studies in *H. armigera* suggest that PBAN specifically influences early incorporation of acetate and the activity of CoA carboxylase (Tsfadia et al. 2008; Jurenka and Rafaeli 2011). In female *H. armigera* the product malonyl-CoA is modified in series by fatty acid synthetase, Δ11-desaturase steps, and differential activity of reductases or oxidases to produce aldehydes, alcohols, and acetates (Jurenka and Rafaeli 2011). In males, the Δ11-desaturase appears to be absent, and similar steps of reduction and oxidation lead to the production of 16–18 chain-length alcohols and acetates.

Early work by Teal and Tumlinson (1989) documented enzymatic activity within pheromone glands of *H. subflexa*,

H. virescens, and *Hydraecia micacea*. By applying primary acetates and alcohols to the surface of pheromone glands, the corresponding alcohols and aldehydes were produced by female *Heliothis* spp. For heliothines, which rely on aldehydes as primary components, this step is critical: *H. micacea* did not produce aldehydes in response to alcohol application. This study provided the first evidence for oxidative steps in pheromone biosynthesis in these species.

In other moth species, PBAN activity appears more complex, affecting the activity of reductases and possibly other receptor types (Tillman et al. 1999). In *H. virescens*, PBAN influences two different stages in pheromone production, as pheromone synthesis from exogenously applied lipids is also increased (Eltahlawy et al. 2007). It remains unclear which secondary enzymatic step is influenced later in pheromone production, but it has been proposed to influence reduction of fatty acids to alcohols and aldehydes, as has been shown in other Lepidoptera (Tillman et al. 1999). In addition, heliothine species express a chain-shortening mutation that influences the production of many minor components. For example, in *H. subflexa*, Z9-16Ald is generated via Δ11 desaturation of octadecanoic acid (18OOH) followed by chain shortening, and *H. virescens* similarly uses hexadecanoic acid (16OOH) as the substrate for Δ11-desaturase followed by chain shortening to generate Z9-14Ald plus the non-chain-shortened Z11-16Ald (Choi et al. 2005).

Studies have examined the kinetics of biosynthesis in *H. virescens* to further quantify metabolites and determine flux in acetate products in pheromone production, both in photophase and scotophase (Foster and Anderson 2012). Finally, Vogel et al. (2010) have documented the transcriptome of female *H. virescens*, identifying a series of candidate genes that may influence pheromone biosynthesis. This information will provide novel strategies to examine the biosynthetic pathway within *H. virescens* and serve as a basis for interspecific variation noted in the Heliothinae.

PBAN has also been implicated in courtship pheromone biosynthesis in males, with evidence of immunoreactivity in the central nervous system of *H. armigera* (Rafaeli et al. 1991; Jurenka and Rafaeli 2011). Gene transcripts for PBAN receptors have also been documented in *H. armigera*, and RNA interference (RNAi) studies have demonstrated that production of male hairpencil compounds is influenced by PBAN (Bober and Rafaeli 2010). Given the similarity in derivative compounds used for biosynthesis in both male and female heliothines, it is perhaps not surprising to see homology in the mode of action of PBAN.

Specificity of Response by Males to Female-Emitted Compounds

The differential attraction or deterrence of males to the various combinations of pheromone gland volatiles outlined above and in Table 21.2 is integral to our understanding of how pheromone blend compositions have been shaped over evolutionary time. It is therefore best to try to understand the published data concerning male behavioral responses to different blends in the context of geographical areas of sympatry and seasonal synchrony of adults occupying the same habitats, such as in agricultural crops or surrounding vegetation.

Despite wide geographic distributions of many of these heliothine species, there is no evidence of the existence of multiple cryptic species across their ranges. For instance, an intense sampling (mitochondrial DNA from 249 individuals) of populations of *Helicoverpa armigera* around the world supports this insect's status as a single species across Africa, Asia, and Australia (Behere et al. 2007). A similar examination of *H. zea* from both North America and South America supported its status as a single species that had diverged from *H. armigera* 1.5 million years ago (Behere et al. 2007). Even though there may appear to be differences in pheromone blends and male behavioral response profiles over heliothines' wide geographic areas, these must be considered to be pheromone polymorphisms or "dialects" such as have been found in different parts of the world for the noctuid moth *Agrotis segetum* (Löfstedt et al. 1986) or for the saturniid moths in the genus *Hemileuca* (McElfresh and Millar 1999, 2001; see Allison and Cardé, Chapter 2, this volume).

For the above-mentioned *Helicoverpa* and *Heliothis* spp., there are no instances of sexual activity occurring during any hours of the diel except during scotophase. Moreover, no studies have documented significant partitioning of scotophase into species-specific sexual activity periods that might have been selected for to avoid or minimize heterospecific mate location, courtship, and copulation mistakes. Therefore, below we compare the pheromone blends of two or more species, pair by pair, for species having overlapping geographical ranges (sympatry) and overlapping adult seasonal flight periods (seasonal synchrony). Discussing such pairings may elucidate how adaptive responses to each others' pheromone blends might have helped shape their sex pheromone communication systems via divergence during a speciation event followed by reinforcement of two newly discrete communication channels, or by reproductive character displacement-related shifts in two already established species in their zones of sympatry (Butlin 1987).

Blend Specificity in North and Central America

OPTIMIZATION OF INTRASPECIFIC COMPONENT RATIOS

Species in the Americas exhibit differences in their sex pheromone blends that involve selective elimination or modulation of the use of two different enzymes, Δ11-desaturase and an enzyme that performs β-oxidation (chain shortening). The order of action of these two enzymes can also be reversed. A major selective force determining which compounds get biosynthesized for the species specificity of these pheromone blends is male preference. Although males have been selected to have a broad response spectrum to enable them to "track" (respond to) the blends of any and all conspecific females in their environment, they will be penalized by responding too broadly, that is, to heterospecific pheromone blends resulting in no offspring or hybrid progeny having reduced inclusive fitness. The degree to which such mating "mistakes" can evolutionarily sculpt male discrimination for blends having a certain composition of components at particular ratios explains why behavioral experiments are so illuminating. Such tests examine the sensitivity and specificity of male response to blends comprised of various potential synthetic pheromone components emitted in different blend ratios, and thus probe the olfactory blend integration system of a given species that has been selected for over time. We can thereby begin to understand, by determining the ranges of responsiveness that males exhibit to possible pheromone blends, the limits on blend composition that mate finding, courtship, and mating mistakes have placed on these pheromone systems.

Heliothis virescens males require the presence of Z9-14Ald and Z11-16Ald to fly upwind to and locate a source of sex pheromone. Of the known pheromone blends of North American heliothines, *H. virescens* is the only species that uses Z9-14Ald, and this component involves the chain shortening of Z11-16CoA to create the Z9-14CoA needed for Z9-14Ald. *H. virescens* males respond very well in wind-tunnel assays to percentages of Z9-14Ald in blends loaded on dispensers that ranged from 5% to 50% Z9-14Ald (Vickers et al. 1991). In field-trapping studies, Groot et al. (2010a) found that *H. virescens* males were trapped in equivalently optimal numbers when 1–10% Z9-14Ald was loaded onto dispensers with Z11-16Ald, with significant numbers of males still captured to lures loaded with 25% Z9-14Ald. The airborne percentages of Z9-14Ald relative to Z11-16Ald will of course be much higher due to the 14-carbon aldehyde's higher volatility. Pope et al. (1982) measured percentages of Z9-14Ald emitted by individual females and found that only 3 of 40 females emitted over 10% Z9-14Ald in Z11-16Ald (10%, 11%, and 14%). The rest of the females' emitted percentages varied between 7% and 2% with a mean of 5.04% Z9-14Ald. Thus, it is the females that have a narrow variance in emission of Z9-14Ald in their blends, and the experimentally demonstrated male acceptance of quite broad ratios of Z9-14Ald in Z11-16Ald seems to indicate that neither males nor females have imposed any kind of strong stabilizing selection on this two-component blend.

In addition to Z11-16Ald, *H. subflexa* males require the emission of large proportions of both Z11-16OH and Z9-16Ald in the blend to be attracted (Vickers 2002). Z11-16Ac is a fourth component, increasing male attraction when added to the Z11-16Ald, Z11-16OH, and Z9-16Ald three-component blend (Groot et al. 2009a). Emission of three pheromone-gland-constituent acetates, including Z11-16Ac, along with the above-mentioned three requisite components (Vickers 2002), had been reported to be more important for male attraction in the eastern (North Carolina) compared to the western (Mexico) part of this species' range (Groot et al. 2007). However, these field-trapping tests showed that there was a significant increase in trap catch in *both* regions due to the addition of just Z11-16Ac to this three-component blend, with no further increase when the two remaining acetates were also added (Groot et al. 2007). Therefore, this suggests not only is there no significant geographic variation in response to this four-component blend, but also that Z11-16Ac must now be considered to be a fourth pheromone component of this species' pheromone. Its effect on increasing trap catch was consistent in all tests in all regions. Groot et al. (2007) then compared differences in the ratios of increase in trap catch in both regions due to the addition of Z11-16Ac (or the acetates together) to the three main components compared to the three main components alone. When this trap catch ratio analysis was performed, the increase in trap catch in North Carolina compared to the blend lacking Z11-16Ac was declared to be significantly greater (approximately four-fold) than the increase in Mexico (approximately twofold). However, attaching statistical significance to such a ratio difference in magnitude of trap catch between two regions is problematic because in moth field-trapping pheromone blend experiments, the magnitude of increase in trap catch, even from the same locale but from different fields, in response to equally extra-component-fortified blends can vary greatly for many different reasons (cf. Baker and Cardé 1979). The shape of the trap capture response profile may vary (i.e., ratio of captures of suboptimal blends relative to optimal blends) according to test location and population density. A further consideration is that the lure compositions that were tested were based on gland extracts (Groot et al. 2007), which in heliothine moths may vary considerably from the gland volatiles' actual emission ratios (Pope et al. 1982, 1984).

Z11-16Ac is always present in female pheromone gland extracts whenever (Z)-7-hexadecen-1-ol acetate (Z7-16Ac) and (Z)-9-hexadecen-1-ol acetate (Z9-16Ac) are found there, and thus Z11-16Ac should always be present in females' volatile emissions whenever either of these other two acetates are emitted. Hence, there is no known natural situation in which Z7-16Ac can affect male behavior on its own as a female-emitted pheromone component, and it does not matter that Z7-16Ac was shown in one of the three field tests to increase trap catch (as did Z11-16Ac in that same test) when added, as the only acetate, to the three-component requisite blend (Groot et al. 2007). Furthermore, regarding Z9-16Ac, in no case was this acetate shown to have any effect on behavior when added by itself to the three-component blend lacking the other acetates. Therefore, again, evidence suggests that Z11-16Ac must be considered to be the only proven fourth component of the *H. subflexa* blend.

The range of ratios of both Z11-16OH and Z9-16Ald relative to the Z11-16Ald major component that results in optimal *H. subflexa* male behavioral responses is quite broad. Ratios of Z11-16OH that result in optimal male attraction in wind-tunnel assays ranged from 1% to 50% of the Z11-16Ald amount, with no diminution of attractiveness (Vickers 2002). In field-trapping studies using just these three components, the ratio requirements for Z11-16OH seemed much narrower but that is because greater than 10% Z11-16OH was not tried (Groot et al. 2007). Similar to the wind-tunnel results of Vickers (2002), either 1% or 10% Z11-16OH significantly increased trap catch compared to treatments lacking Z11-16OH (Groot et al. 2007). Interestingly, when the fourth component, Z11-16Ac, was present in the blend in these field trials, a slightly broader range of percentages of Z11-16OH, from 1% to 25%, resulted in equivalently high increased levels of trap catch compared to blends lacking Z11-16OH (Groot et al. 2007). In wind-tunnel assays, the percentage of Z9-16Ald in the blend that resulted in optimal attraction ranged from 10% to 50% (Vickers 2002).

Helicoverpa zea males require only a small amount of Z9-16Ald in the blend with Z11-16Ald to evoke optimal attraction, although a small amount of Z9-14Ald can substitute for the Z9-16Ald (Table 21.2; Vickers et al. 1991), despite the fact that *H. zea* females do not synthesize or emit detectable amounts of Z9-14Ald (Pope et al. 1984). The range of percentages of Z9-16Ald that evoke optimal attraction when blended with Z11-16Ald is from 1% to 15% (Klun et al. 1980b; Vetter and Baker 1984; Vickers et al. 1991), even though Z9-16Ald makes up only approximately 1–2% of *H. zea* female effluvia (Pope et al. 1984). Males do not require emission of any other compounds, including Z7-16Ald or 16Ald (Vetter and Baker 1984), even though both of these compounds are emitted by *H. zea* females (Pope et al. 1984) and are found in gland extracts (Klun et al. 1980b).

H. phloxiphaga males have a critical requirement for optimal male attraction: a very small amount of Z9-16Ald (no more than 0.6%) for optimal male attraction as well as only a small amount of Z11-16OH (2–3%), both admixed with Z11-16Ald

(Table 21.2). When Z11-16OH is deleted from the blend, or when it exceeds 4%, male attraction is severely reduced (Raina et al. 1986). The requirement for a precise percentage of Z11-16OH (2–3%) to be blended with Z11-16Ald seen in *H. phloxiphaga* is unique among the heliothines (Raina et al. 1986).

BEHAVIORAL ANTAGONISM TO HETEROSPECIFIC BLENDS AND BLEND COMPONENTS

Infield-trapping experiments using large-diameter Hartstack wire-screen traps (Hartstack et al. 1979), female *Heliothis subflexa* and *H. virescens* were both able to attract significant numbers of *Helicoverpa zea* males (Groot et al. 2009a), and also to attract small numbers of each other's males (Groot et al. 2006, 2009a). In these experiments, males did not have to orient all the way to the calling female pheromone source; they only needed to approach within approximately 25 cm of the female to be captured. Wind-tunnel experiments testing cross-attraction of male *H. virescens*, *H. subflexa*, and *H. zea* all the way to the calling females themselves showed that males of none of the three species were able to lock onto the plumes and fly all the way upwind to arrive at heterospecific females (Lelito et al. 2008). The reasons for this inability in males of all three species originates with blends emitted by heterospecific females that contain behaviorally antagonistic components or else are comprised of unbalanced, behaviorally antagonistic blend ratios deficient in one or more essential components. *H. virescens* males are not attracted to *H. subflexa* females because although these females do emit a small amount of Z9-14Ald, the large percentages of Z11-16Ac and Z11-16OH emitted by *H. subflexa* females are antagonistic to *H. virescens* male attraction. The addition of just 0.1% Z11-16Ac or ≥3% Z11-16OH to the optimal *H. virescens* blend is sufficient to reduce male *H. virescens* attraction significantly (Vetter and Baker 1983; Vickers and Baker 1997). *H. virescens* males are not attracted to *H. zea* females because they do not emit any Z9-14Ald, and for *H. virescens* Z9-16Ald cannot substitute behaviorally for the absence of Z9-14Ald (Vickers et al. 1991). Similarly, *H. virescens* males would not be attracted to *H. phloxiphaga* females for the same reason as for *H. zea* above. In addition, the 3% Z11-16OH emitted by the *H. phloxiphaga* females would be behaviorally antagonistic to *H. virescens* males (Vetter and Baker 1983).

Female *H. virescens* probably do not attract *H. subflexa* males because they do not emit the large percentage of Z9-16Ald or any Z11-16OH, both of which are required for response by *H. subflexa* males (Vickers 2002). Male *H. subflexa* likewise will not be attracted to *H. zea* females due to the lack of emission of Z11-16OH (Pope et al. 1984), as well as the tiny amount of Z9-16Ald (1–2%) emitted by *H. zea* females. These low levels of Z9-16Ald would be inadequate to attract *H. subflexa* males (Vickers 2002). Similarly, *H. subflexa* males would likely not be attracted to *H. phloxiphaga* females because the percentages of Z9-16Ald and Z11-16OH (0.4% and 6.7%, respectively) they produce in their glands (Raina et al. 1986) are too small to evoke upwind flight to the source in *H. subflexa* males (Vickers 2002).

H. zea males are not attracted to *H. virescens* females because even though a small percentage of Z9-14Ald can substitute for the Z9-16Ald of *H. zea* females, the amount emitted by *H. virescens* females will exceed that amount and be behaviorally antagonistic (Table 21.2; Vickers et al. 1991). Also, the addition of small percentages of Z9-16Ald to the Z9-14Ald that *H. virescens* females typically emit has been shown to antagonize flight to nearly complete suppression of male attraction (Shaver et al. 1982). *H. zea* males are not attracted to *H. subflexa* females because the percentage of Z11-16OH produced by *H. subflexa* females is antagonistic to their upwind flight (Quero and Baker 1999; Quero et al. 2001). Finally, *H. zea* males should not be attracted to *H. phloxiphaga* females because the percentage of Z11-16OH they produce (2–3%) will be antagonistic to *H. zea* male upwind flight (Quero and Baker 1999; Quero et al. 2001). However, a small amount (5%) of cross-attraction of *H. zea* males in the field (cross-trapping in large cone traps) was found using both *H. phloxiphaga* calling females (Kaae et al. 1973; Raina et al. 1986) and the *H. phloxiphaga* synthetic blend (Raina et al. 1986). The percentage of Z11-16OH in the blend emitted by *H. phloxiphaga* females does not appear to be enough to completely prevent cross-attraction of *H. zea* males, although the amount of cross-attraction cannot be assessed without knowing how many *H. zea* males would have been trapped in response to *H. zea* females if these females had been used as the proper control treatments in these same experiments. Note that it is unlikely that *H. zea* males would occupy the same habitat at the same time of the year as *H. phloxiphaga* females.

For *H. phloxiphaga*, the requirement for a precise percentage of Z11-16OH (2–3%) in the blend may explain why cross-attraction of male *H. phloxiphaga* to females of other sympatric heliothine moths does not occur (Raina et al. 1986). Thus, *H. phloxiphaga* males will not be attracted to *H. virescens* (Pope et al. 1982; Teal et al. 1986) or to *H. zea* (Pope et al. 1984) females due to their lack of emission of Z11-16OH, or to *H. subflexa* females because the large percentages of Z11-16OH emitted by *H. subflexa* females will be behaviorally antagonistic to *H. phloxiphaga* male attraction (Raina et al. 1986).

HETEROSPECIFIC FEMALE PHEROMONE PLUME INTERFERENCE VIA OVERLAPPING PLUMES?

The effect of communication interference on mate attraction by calling females has been measured by placing two calling heterospecific females on either side of a calling conspecific female (Lelito et al. 2008). Male orientation behavior was measured to quantify the impact of potentially antagonistic pheromone plume strands of heterospecific females interleaved with conspecific plume strands. In five out of six possible combinations, the close positioning of two heterospecific interfering females next to one conspecific female generating overlapping plumes caused a significant, 50–80%, reduction of attraction of males to within 10 cm of their female. The one exception was of *Helicoverpa zea* males. Attraction of *H. zea* males to a calling *H. zea* female, although significantly reduced by the presence of two calling *Heliothis subflexa* females, was not at all reduced by the presence of two calling *H. virescens* females (Lelito et al. 2008). The degree of coincident, completely mixed heterospecific and conspecific plume strands in this experimental setup was measured via a four-channel EAG as being approximately 50% mixed.

GEOGRAPHIC VARIATION IN THE PHEROMONE BLENDS OF *HELIOTHIS SUBFLEXA* AND *H. VIRESCENS*

The female-produced pheromone blends in the *Heliothis virescens* and *H. subflexa* pheromone communication systems have been reported to vary significantly across a geographical range in North America (Groot et al. 2009a). Unfortunately, Groot et al. (2009a) did not actually measure any variation in

the *pheromone* of this species, because measurements were not reported for variation in the major component, Z11-16Ald, to determine how it may have covaried with any of the minor sex pheromone components (nor were the compositions of the emitted mixtures reported). In all the gland extractions for both species, only the large number of behaviorally inert minor gland extract constituents, plus the few actual minor sex pheromone components known to be emitted by females of each species, were analyzed. The actual quantities of these compounds were not reported, but rather, only their ratios of relative abundance were calculated. Again, to quantify geographic variation in the behaviorally active pheromone blends of these two species, the amount of covariance of Z11-16Ald abundance with the abundances of minor pheromone components needed to be reported, but it was not. We thus do not know how, or whether, these pheromone blends varied geographically. Moreover, the degree of ratio variation of even the behaviorally active actual minor pheromone components will have been influenced by the amounts of some of the more abundant, behaviorally inert constituents in the gland extracts. Because all results were reported as percentages of all minor/inert components, fluctuations in the amounts of these compounds, such as 16Ald for *H. virescens* and Z9-16Ac, 16Ald, and Z9-16OH for *H. subflexa*, will have confounded even these gland-constituent ratio results in unknown ways, and the actual *emitted* blend ratios in still other indiscernible ways.

SEASONAL "TEMPORAL VARIATION" IN FEMALE PHEROMONE BLENDS

In addition to varying geographically, the sex pheromone blends of *Heliothis subflexa* and *H. virescens* have been reported to vary temporally (i.e., among years) (Groot et al. 2009a). Again, unfortunately, variation in "chemical communication" that was mentioned in the title of Groot et al. (2009a) was not measured, because chemical communication in these species involves the complete sex pheromone blend, with Z11-16Ald being the essential component in both species. The abundance of Z11-16Ald was not reported, and thus it is difficult to determine how it may have covaried with the abundances of behaviorally active minor components. Minor gland constituents, including minor pheromone components, are all behaviorally inert in any blends without the inclusion of Z11-16Ald; thus, the seasonal variation in behaviorally inert compounds was all that was analyzed (Groot et al. 2009a). Finally, because ratios and not amounts were reported, fluctuations in the amounts of non-pheromonal component in gland extracts will have confounded the ratios of even the known, behaviorally active minor pheromone components in the glands in unknown ways.

The "positive assortative" attraction of North Carolina *H. subflexa* and *H. virescens* males in the field to females collected from North Carolina rather than from Texas was implicated as being due to differences in the pheromone blends of the females that had been collected from these two regions (Groot et al. 2009a). However, the lower trap captures of North Carolina male *H. subflexa* and *H. virescens* in response to Texas females than to North Carolina females could be due to many other factors than the quality of these females' sex pheromone blends. For example, Texas females may inherently call less frequently than North Carolina females and thus have a smaller total attraction period during the night compared to North Carolina females. Unfortunately, the reciprocal field tests were not conducted in Texas to see whether this is true, and that perhaps North Carolina females would attract more males in Texas as well as in North Carolina.

"PHENOTYPIC PLASTICITY" IN FEMALE BLEND COMPOSITION FOLLOWING EXPOSURE TO HETEROSPECIFIC PHEROMONE PLUMES?

Groot et al. (2010b) exposed female *Heliothis subflexa* to the synthetic pheromone blends of *H. virescens* and *H. subflexa* over three continuous days and reported that the gland compositions of female *H. subflexa* seemed to change after they had been exposed to the *H. virescens* heterospecific pheromone blend, but not after they were exposed to the conspecific *H. subflexa* blend. Females that exhibited a blend shift showed a slight, but significantly, elevated percentage of Z11-16Ac and the two other acetates in their gland extracts. Groot et al. (2010b) proposed this phenotypic plasticity as a mechanism for females to increase their mating success in the presence of higher density populations of *H. virescens* that might otherwise have increased the chances of *H. subflexa* females cross-attracting *H. virescens* males and incurring reduced fitness.

There are several difficulties in trying to assess the validity of these results, not the least of which is that all gland-extracted compounds were reported as "normalized" values to Z9-16OH (Groot et al. 2010b). First, the titers of Z9-16OH extracted, and their variation from gland to gland, were never shown, nor was the calculation method for normalizing the abundances. If Z9-16OH was present in very small amounts and used as a denominator in calculations, then very slight variations in the titers of Z9-16OH could create large variations in the normalized amounts of other compounds. Regardless, it is not possible from the information given to calculate back to find the actual titers of pheromone components that were extracted or to determine whether *the pheromone*, i.e., the *blend ratios of behaviorally active components*, did actually vary significantly between the differently exposed groups of females. Some of the increases in gland constituents deemed "significant," such as those of Z7-16Ac and Z9-16Ac after heterospecific pheromone exposure, appear to be vanishingly small and no larger than those from the females exposed to the conspecific blend (Groot et al. 2010b).

Second, the exposure of a group of females to the heterospecific *H. virescens* blend versus their blank control was done as a single lengthy cohort over several weeks during an earlier time period than was the exposure of the second several-weeks-long cohort of females to the *H. subflexa* blend versus their blank control. The difference in the apparent increase of "acetates" in the former group to that of the latter group might have been due to some unknown differences in the quality of the *H. subflexa* females being tested during these earlier weeks versus those tested during the later sets of weeks.

Third, it should be noted that following *H. subflexa* females' preexposure to the heterospecific *H. virescens* blend, the normalized amount of Z11-16Ald major component extracted from these females also was seen to increase to seemingly as large a degree as that of Z11-16Ac. This calls into question whether the *pheromone blend* (e.g., the ratios of behaviorally active components including Z11-16Ald) in these females actually changed significantly, even though the amount of increase in the acetates was the authors' focus. Unfortunately, we are not given enough information to be able to calculate whether the actual abundances of the *pheromone blend* com-

ponents changed (or not) because we were not shown how these normalized data were obtained.

Fourth, the Groot et al. (2010b) experiments were performed in the laboratory by using prolonged, 3-day exposures of females located in close proximity (in the upwind portion of the aeration cylinder) to sources of synthetic pheromone blends. These were extremely long-duration, chronic exposure levels that would never occur in nature (3-day exposure as adults and 1–7 days as pupae). Calling female moths' pheromone plumes do not persist for more than a few tens of seconds in the field. Pheromone researchers over decades have found it virtually impossible to locate a calling female of any species under natural conditions in the field because they usually become mated so quickly that only moths already *in copula* are found. In addition, such brief natural occurrences of heterospecific female plume emission in nature will be made even more fleeting due to the plumes' meanderings caused by shifts in wind direction.

Finally, the percentages of minor components in heliothine moth gland extracts have been known to vary considerably due to extracts being prepared over periods differing by only a few hours during scotophase (Pope et al. 1984; Park et al. 1996), or due to the glands being extracted for different durations, even if only by a few minutes (Raina et al. 1986). Also, the compounds present in a solvent extract from a pheromone gland, which may include biosynthetic precursors, are not necessarily reflective of the blend volatilizing into the air from the gland surface.

Blend Specificity in South America

BEHAVIORAL ANTAGONISM TO HETEROSPECIFIC BLENDS AND BLEND COMPONENTS

In South America, the same three types of non-cross-attractive heterospecific interactions between *Helicoverpa zea*, *Heliothis virescens*, and *H. subflexa* as in North America (see previous sections above) should be occurring. In Uruguay, Argentina, and Chile, however, there is another species, *H. gelotopoeon*, a pest of local field crops, that is sympatric and synchronic with these first three species (Cork and Lobos 2003). Females of *H. gelotopoeon* emit an approximately 1:1 ratio of 16Ald and Z9-16Ald, and they do not emit any Z11-16Ald (Table 21.2). In fact, addition of ≥1% Z11-16Ald to the 16Ald-plus-Z9-16Ald blend significantly antagonizes conspecific male attraction (Cork and Lobos 2003). The optimal conspecific blend involves only Z9-16Ald, with 10–100% 16Ald admixed with it. Females will be completely unattractive to males of the other three species, all of which require Z11-16Ald in their blends. Also, male *H. gelotopoeon* will be deterred from being attracted to females of the other species due to the predominance of Z11-16Ald as the major component in their blends. Thus, both females and males of *H. gelotopoeon* should be prevented from making mating mistakes with these other three species in their environment.

Blend Specificity in Europe and Asia

OPTIMIZATION OF INTRASPECIFIC COMPONENT RATIOS

Male *Helicoverpa armigera* require the presence of Z9-16Ald as a secondary component in their female sex pheromone blends

(Kehat et al. 1980; Kehat and Dunkelblum 1990). At least 3% Z9-16Ald relative to Z11-16Ald must be present in a two-component blend to be optimally attractive to *H. armigera* males, with ≥24% causing a reduction in male attraction (Table 21.2; Kehat et al. 1980). In addition, whereas it had previously been thought that *H. armigera* was a species that did not use Z9-14Ald as part of its blend, Zhang et al. (2012) found that very small percentages (0.3–5.0%) of Z9-14Ald can contribute to doubling the trap capture of males when it is added to the Z11-16Ald-plus-Z9-16Ald blend. The component Z9-14Ald is found at a level of approximately 0.3% in female pheromone gland volatiles (Zhang et al. 2012), so it must be considered to be a pheromone component that contributes to increased male attraction in this species (Zhang et al. 2012). Indeed, further support for this compound as a sex pheromone component for *H. armigera* comes from the studies of Rothschild (1978) who showed that the addition of 2–5% Z9-14Ald to Z11-16Ald in Australia increased male trap catch significantly.

H. assulta is only one of two known *Helicoverpa* or *Heliothis* species in which Z11-16Ald is not the major pheromone component. Here, Z9-16Ald is the major component, with Z11-16Ald contributing to optimal male attraction when present at only 5–10% of the Z9-16Ald (Table 21.2). It has been suggested that *H. assulta* is a polymorphic pheromone species with geographical variation in the most effective blends in different regions. Males in Thailand were trapped optimally when the percentage of Z11-16Ald in a blend of Z11-16Ald plus Z9-16Ald was 13% of the amount of Z9-16Ald, whereas males in China and Korea were trapped optimally when Z11-16Ald was 2–5% (Cork et al. 1992). Too much Z11-16Ald in the blend reduces *H. assulta* male response significantly; no males were found to fly upwind to and locate sources emitting the *H. armigera* blend ratio of 3% Z9-16Ald/97% Z11-16Ald. With the optimal *H. assulta* blend ratio of Z9-16Ald to Z11-16Ald, addition of the Z11-16Ac and Z9-16Ac female gland constituents in Korea significantly increased male response (trap catch), indicating activity as two minor sex pheromone components in the *H. assulta* pheromone communication system in Korea (Cork et al. 1992). However, in China the addition of acetates to the blend reduced male trap catch and the addition of the acetates to the Thailand aldehyde blend had no effect on trap catch of males there (Cork et al. 1992).

Males of *Heliothis peltigera* are attracted optimally in field tests to a two-component blend of Z11-16Ald plus Z9-14Ald, with anywhere from 20% to 50% Z9-14Ald added as the minor component. The percentage of Z9-14Ald relative to Z11-16Ald extracted from female pheromone glands was observed to be 13–17% (Table 21.2; Dunkelblum and Kehat 1989). In the wind tunnel, Z9-14Ald blended with Z11-16Ald at both 5% and 50% of the amount of Z11-16Ald elicited equivalently high levels of upwind flight, source contact, and copulatory attempts. Addition of 30% Z11-16OH to the binary blend, mirroring the 24% found in female glands, evoked no significant changes in these categories of behavior (Dunkelblum and Kehat 1989).

H. maritima has been designated as having many subspecies, and among the seven such subspecies are *H. m. hungarica* in Hungary and *H. m. aduacta* in Japan. Males of *H. m. hungarica* in Hungary have been shown to respond optimally to the only two ratios that were tested that approximate the 100:8.1:5.5 ratio of Z11-16Ald/Z11-16OH/16Ald gland constituents shown to be produced by Hungarian females (Szőcs et al. 1993). Two blend ratios, 100:20:6 and 100:6:3, of these three compounds were shown to have an equivalently high

level of male attraction in a wind tunnel (Table 21.2). A ratio similar to these was also used successfully in field-trapping tests. Binary blends were not tested against either in the field or in the wind tunnel against the ternary blends, so it is not clear whether both Z11-16OH and 16Ald are pheromone components or only one of them is. Addition of a fourth compound, Z9-16Ald, that had been isolated in trace amounts from *H. m. hungarica* female glands, helped increase attraction in the wind tunnel when added at 0.1% Z11-16Ald, but in field-trapping experiments this compound had no effect (Szőcs et al. 1993). It may be tentatively labeled also as being a pheromone component, based on the wind-tunnel results.

Females of *H. m. aduacta* in Japan produce ratios of Z11-16Ald/Z11-16OH/16Ald (100:24.5:2.4) that are quite similar to those produced by the Hungarian (subspecies) females (Kakizaki and Sugie 2003). Field-trapping tests showed that binary blends of 1% Z11-16OH relative to Z11-16Ald attracted and captured males, whereas each compound alone did not. In several experiments, trap capture was highest when the percentage of Z11-16OH was in the range of 1–5%. The addition of either 16Ald or Z9-16Ald to the binary blend had no effect on male trap capture (Kakizaki and Sugie 2003).

It may be concluded from work on Japanese *H. maritima* that *H. maritima* in Japan definitely, and in Hungary most likely, use Z11-16Ald and Z11-16OH as sex pheromone components. There is no strong evidence for 16Ald or Z9-16Ald being behaviorally active in Hungary and no evidence at all that they are active in Japan. No ranges of Z11-16OH were tried in binary blends in Hungary, and so we do not know whether the 1–5% Z11-16OH used in Japan is a different optimal ratio than the 6–20% tested in Hungary. Therefore, there may be no strong differences in Japan and Hungry in the binary blends used by these two "subspecies."

BEHAVIORAL ANTAGONISM TO HETEROSPECIFIC BLENDS AND BLEND COMPONENTS

The most geographically widespread species, *Helicoverpa armigera*, might be expected to exert the biggest influence on heterospecific pheromone blend specificity across Europe and Asia. In both Hungary and Japan, *Heliothis maritima* does not use Z9-16Ald as a minor component, using Z11-16OH instead; therefore, its blend does not attract *H. armigera* males, which require Z9-16Ald. *H. maritima*, conversely, would not be attracted to *H. armigera* females because they do not emit significant percentages of Z11-16OH that *H. maritima* males need to be attracted.

In the Mediterranean region, *H. peltigera* females produce a large percentage (15%) of Z9-14Ald as their secondary pheromone component, a percentage that is antagonistic to *H. armigera* male attraction (Dunkelblum and Kehat 1989). Percentages of Z9-14Ald that are this high, coupled with the large percentage of Z11-16OH produced by *H. peltigera* females, which is not a pheromone component of this species, were shown to effectively eliminate attraction of *H. armigera* males (Dunkelblum and Kehat 1989). Conversely, *H. peltigera* males will not be attracted to *H. armigera* females because they emit too little Z9-14Ald (0.3%; Zhang et al. 2012) to be effective for *H. peltigera* attraction, as shown by the wind-tunnel studies of Dunkelblum and Kehat (1989) in which 1% Z9-14Ald was ineffective in causing upwind flight of *H. peltigera* males.

Male *H. peltigera* should not be attracted to female *H. maritima* because they do not produce detectable amounts of Z9-14Ald. It is not clear whether *H. maritima* males would be attracted to *H. peltigera* females because the latter produce significant amounts of Z11-16OH, a secondary pheromone component used by *H. maritima*; however, *H. peltigera* females also produce Z9-14Ald that might be antagonistic to *H. maritima* male attraction. This hypothesis would need to be confirmed experimentally, as would the amounts actually emitted by both *H. peltigera* and *H. maritima* females, not just the amounts extracted from glands.

The species *H. assulta* and *H. armigera* are sympatric over large areas of Asia and may present an interesting study in the evolution of blend specificity as a consequence of selection to reduce mating mistakes. In wind-tunnel cross-attraction studies, Ming et al. (2007) showed that calling *H. assulta* and *H. armigera* females were completely unattractive to *H. armigera* and *H. assulta* males, respectively. It had previously been shown that *H. armigera* males could be cross-attracted to the synthetic two-component *H. assulta* blend of 97% Z9-16Ald plus 3% Z11-16Ald but that *H. assulta* males were not attracted to the *H. armigera* blend of 97% Z11-16Ald plus 3% Z9-16Ald (Zhao et al. 2006). Therefore, Ming et al. (2007) conjectured that because *H. armigera* males were not attracted to calling *H. assulta* females, other compounds emitted by *H. assulta* females might be responsible for antagonizing *H. armigera* attraction. It is possible that these additional antagonistic compounds might be the corresponding acetates, such as Z11-16Ac that were found to be produced by *H. assulta* females from populations in China (Cork et al. 1992).

However, Wu et al. (2013) later demonstrated in wind-tunnel experiments with two-component synthetic blends of Z9-16Ald plus Z11-16Ald that ratio differences alone could explain the lack of cross-attraction of *H. armigera* males to *H. assulta* females. It may still be possible that the emission of other compounds such as Z9-16Ac and Z11-16Ac in the *H. assulta* gland volatiles (Cork et al. 1992) might explain the lack of *H. armigera* attraction to calling *H. assulta* females. This possibility needs to be investigated in experiments using synthetic mixtures and moths of carefully chosen geographic origin.

No additional compounds need to be involved to explain the absence of cross-attraction of *H. assulta* males to *H. armigera* females. Differences in the ratio of Z9-16Ald to Z11-16Ald between the species are enough to prevent cross-attraction (Zhao et al. 2006; Ming et al. 2007). However, *H. armigera* females also have been shown to produce small amounts (0.3%) of Z9-14Ald as part of their sex pheromone blend (Zhang et al. 2012), and between 0.1% and 1% Z9-14Ald added to the *H. assulta* two-component blend was shown to be behaviorally antagonistic to male *H. assulta* attraction in the field (Boo et al. 1995). Thus, Z9-14Ald might contribute to further reducing the chance of male *H. assulta* attraction to *H. armigera* females.

It should be expected that *H. assulta* males will, across their geographic range, not be attracted to any of the other heliothine species whose pheromone blends are thus far known, because of their preference alone for the skewed ratio of predominantly Z9-16Ald compared to Z11-16Ald. The same should be true in the other direction: males of no other species should be attracted to female *H. assulta* due to this unique Z9-16Ald-to-Z11-16Ald ratio. In addition, the requirement by *H. peltigera* for Z9-14Ald as a secondary component would prevent these males from being attracted to *H. assulta* females. *H. maritima* does not use Z9-16Ald as a secondary component; it uses Z11-16OH instead. Because *H. assulta* females do not emit Z11-16OH, this is yet a further reason (in addition to the atypical Z9-16Ald-to-Z11-16Ald ratio) why *H. maritima* males should not be cross-attracted to *H. assulta* females.

Blend Specificity in Australia

It was reported that the "native" budworm of Australia, *Helicoverpa punctigera*, uses a sex pheromone blend ratio of 100:100:5 of Z11-16Ald/Z11-16Ac/Z9-14Ald, respectively (Table 21.2; Rothschild et al. 1982). In exhaustive field trials, this ratio resulted in optimal trap capture of males. In addition to the above-mentioned three components, Z11-16OH was found to be a constituent of female pheromone glands at levels of ≥15% relative to Z11-16Ald, but its inclusion in synthetic blends did not affect trap capture levels.

It is highly unlikely that *H. punctigera* males would be attracted to *H. assulta* females due to the predominance of Z9-16Ald in that species' blend, plus the lack of emission of Z9-14Ald and low level of emission of Z11-16Ac (Cork et al. 1992). Attraction of *H. assulta* males to *H. punctigera* females would be low due to the need for Z9-16Ald to predominate in a blend (Zhao et al. 2006; Wu et al. 2013). It should be expected that *H. punctigera* males will not be attracted to *H. armigera* females due to the latter females' lack of emission of Z11-16Ac. *H. armigera* males likewise will not be attracted to *H. punctigera* females due to the negligible amounts of Z9-16Ald in *H. punctigera* female glands, and their excessive amounts of Z11-16Ac and Z11-16OH, the latter of which has been shown to reduce upwind flight and source location by *H. armigera* males when present at ≥5% relative to Z11-16Ald in the *H. armigera* two-component blend of Z11-16Ald plus Z9-16Ald (Kehat and Dunkelblum 1990).

Hybridization Studies Reveal Heritable Features of Pheromone Production and Attraction

Heliothine Hybrids in Nature and in the Laboratory

Despite evidence of optimized sex pheromone blends for individual species, there is some evidence that heterospecific matings might occasionally occur among heliothines in the field. Cross-attraction among sympatric populations of several heliothine species have been documented in field-trapping studies (Hardwick 1965; Klun et al. 1980a; Cork et al. 1992; Cork and Lobos 2003; Wang et al. 2005; Groot et al. 2006, 2009a). Any cross-attraction among heliothines resulting in heterospecific copulations has high negative fitness consequences. For instance, *Helicoverpa zea* males have been shown to be attracted to female *Heliothis virescens* and *H. subflexa* in the field (Groot et al. 2006, 2009a), but without the prospect for viable progeny. In the laboratory, *H. zea* males paired with *H. virescens* females experience irreversible locking of genitalia (Shorey et al. 1965). Likewise, *H. virescens* and *H. subflexa* are able to hybridize, but there is a fitness cost due to the sterility of male hybrids, and also possibly from the emission of less attractive pheromone blends by hybrid female offspring (Laster 1972; Proshold and LaChance 1974; Laster et al. 1976; Teal and Oostendorp 1995). In these two species and in *H. zea*, the presence of either Z9-14Ald or Z9-16Ald in the blend is critical to modulating attraction in possible zones of sympatry where such heterospecific mating mistakes might occur (figure 21.1). Male attraction to blends containing either of these components appears to be dictated by how broadly or narrowly tuned OSNs are in "B-type" *H. subflexa* and *H. virescens* trichoid sensilla (Cossé et al. 1998) on their antennae (Baker et al. 2004, 2006; Gould et al. 2010; see sections below on olfactory architectures of hybrids and parental species). With regard to other components that can vary, the addition of Z11-16OH to a mixture is critical for *H. subflexa* male attraction; conversely, the presence of Z11-16Ac will typically suppress approach by heterospecific males such as *H. virescens* or *H. zea* (Vickers and Baker 1997; Fadamiro et al. 1999; Vickers 2002).

The existence of these mechanisms for pre-mating isolation suggests that hybridization of species such as *H. virescens* and *H. subflexa* in nature is rare. However, hybridization in the field cannot be dismissed because these two species can be readily hybridized in the laboratory (Laster 1972; Teal and Oostendorp 1995; Groot et al. 2004). The presence of hairpencil organs in male heliothines, their active display during courtship, and the behavioral activity of courtship pheromones in these species provide evidence that there have likely been significant levels of heterospecific cross-attraction in the past to long-distance female-emitted sex pheromone. The cross-attraction will have resulted in selection for choosy females that require males to identify themselves as conspecific via olfactory cues emitted during the males' courtship displays (Birch et al. 1990; Hillier and Vickers 2011b). Such displays, if heritable, may additionally have contributed to the process of runaway sexual selection, in which increasingly discriminating females select for the best, most reproductively "fit" males that can produce the optimal quality and quantities of olfactory signals (Birch et al. 1990; Hillier and Vickers 2004; Hillier and Vickers 2011b).

Heritable Features Affecting Pheromone Blend Biosynthesis

Phenotypic variation in female sex pheromone mixture production and male attraction is modulated in the Heliothinae through autosomal inheritance (i.e., not on sex chromosomes). Wang and colleagues (Wang et al. 2005, 2008; Zhao et al. 2005, 2006; Ming et al. 2007; Wang 2007; Zhang et al. 2010) have investigated species isolation, fitness costs, and heritable features of hybridization in *Helicoverpa armigera* and *H. assulta*. These species produce different ratios of Z11-16Ald and Z9-16Ald as primary and secondary sex pheromone components (100:2.1 for *H. armigera* and 1739:100 for *H. assulta*; Wang and Dong 2001; Wang et al. 2005). Hybrid females produce these components in a ratio similar to, but slightly higher than, *H. armigera* (100:4). This difference is likely due to increased use of both palmitic (C16) and stearic (C18) acids as substrates for Δ9-desaturase and Δ11-desaturase, respectively, for producing Z9-16Ald, instead of only stearic acid for Δ11-desaturase in *H. armigera* to get to Z9-16Ald. Of course, Δ11-desaturase would use palmitic acid as the substrate in both *H. armigera* and the hybrids for them to create the Z11-16Ald end product. Biosynthesis of these components suggests polygenic determinism in these hybrids (Wang et al. 2005), similar to what has been found in studies of hybrid and backcross progeny of *Heliothis virescens* and *H. subflexa* (see next paragraph).

H. virescens and *H. subflexa* hybrids were studied initially to test the feasibility of the sterile insect technique in controlling *H. virescens* populations (Proshold and LaChance 1974). Variation in sex pheromone biosynthesis has been well researched in hybrid and backcross studies of *H. virescens* and *H. subflexa* (Groot et al. 2006, 2009b). Studies using amplified fragment length polymorphism marker mapping and backcross families of these species isolated a series of quantitative trait loci (QTLs) that are closely linked with either *H. virescens* or *H. sub-*

flexa pheromone blend phenotypes. Six QTLs that influence pheromone biosynthesis have been identified in these species (Sheck et al. 2006; Groot et al. 2009b, 2013).

In comparison to *H. subflexa* females, *H. virescens* pheromone glands contain more 16Ald, which is *not* a pheromone component of either species; concentrations of this compound are higher in backcross progeny possessing at least one copy of chromosome 24 from *H. virescens* (Groot et al. 2009b). Presence of *H. virescens* chromosome 13 likewise increases the production of the Z9-14Ald pheromone component of *H. virescens*.

Finally, relative acetate production in backcross progeny is also dependent on the presence, absence, and interaction of chromosomes 4 and 22 from *H. virescens*. Presence of either chromosome from *H. virescens* significantly decreases acetate production and is also linked with increased production of the *H. subflexa* pheromone component Z11-16OH (Groot et al. 2006). This finding suggested the presence of a gene encoding acetyl transferase on this QTL. However, candidate genes for acetyl- and acyl-transferases have not been located on this QTL, suggesting rather that activity is modulated by a transcription factor on this chromosome that underlies this variation (Groot et al. 2009b). Subsequent work by Groot et al. (2009b) experimentally confirmed that there was intense interspecific selection against the *H. subflexa* females exhibiting phenotypes that they had created that had shown enhanced acetate production. In cage trials using female *H. subflexa* with introgressed QTLs from *H. virescens* to produce low amounts of acetates, male *H. virescens* mated significantly more frequently with these females than those with normal acetate production (Groot et al. 2006).

These studies confirm that production of multicomponent pheromone blends in these species are autosomally inherited, that they can be proportionally influenced by the chromosomal copies present, that QTLs from single chromosomes may influence multiple components, and that epistatic interactions may occur between QTLs on different chromosomes. Such traits directly influence the degree to which conspecific males are attracted, the mating success of individuals, and the fitness of progeny.

Heritable Features Affecting Male Response Specificity to Sex Pheromone Blends

Hybrids of *Heliothis virescens* and *H. subflexa* have been used to explore how the specificity of sex pheromone behavioral response might be orchestrated by their peripheral and central olfactory pathways (Baker et al. 2006; Vickers 2006a,b). A key determinant for response specificity of male heliothines and hybrids is the type of OR that is expressed on OSNs that are housed in trichoid sensilla on male heliothine moth antennae and are sensitive to pheromone components (figure 21.3).

COUPLED QTL AND MALE PHEROMONE RECEPTION STUDIES

As with biosynthetic variation in the females, ORs of heliothine moths represent heritable features that are known to modulate attraction of male progeny to selected pheromone blends. Heritability of pheromone blend preference has been investigated in *Heliothis subflexa* and *H. virescens* by performing behavioral preference tests on hybrids combined with single-cell recordings from hybrid and parental type males

Heliothis virescens sensilla trichodea

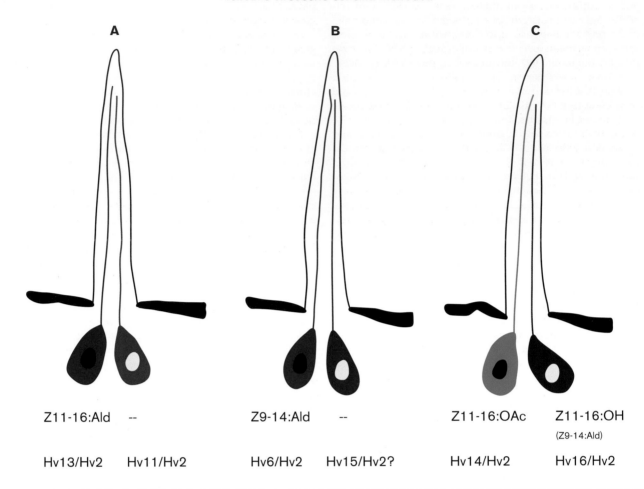

A **B** **C**

Z11-16:Ald -- Z9-14:Ald -- Z11-16:OAc Z11-16:OH
(Z9-14:Ald)

Hv13/Hv2 Hv11/Hv2 Hv6/Hv2 Hv15/Hv2? Hv14/Hv2 Hv16/Hv2

	HvOR13	HvOR11	HvOR6	HvOR15	HvOR14	HvOR16
Z11-16:Ald	⬤		•			•
Z9-14:Ald	⊖		⬤		⦷	⦷
Z11-16:OAc					⊞	•
Z11-16:OH			•			⬤
Z9-16:Ald			•			

FIGURE 21.3 Relationship between OSNs co-compartmentalized within A-, B-, and C-type sensilla of *Heliothis virescens* and the ligand sensitivities of the ORs (Wang et al. 2011) that have been shown via in situ hybridization studies to be expressed on them (Große-Wilde et al. 2007; Baker 2009; Krieger et al. 2009). Adapted from Wang et al. (2011). (Top) Optimal sensitivities of OSNs that reside in different sensilla to the pheromone-related compounds Z11-16Ald, Z9-14Ald, Z11-16OAc, and Z11-16OH, as shown in single-cell recording studies (Berg et al. 1995; Baker et al. 2004). On each of these OSNs, the expression of the ORs Hv13, Hv11, Hv6, Hv14, and Hv16, along with Orco (Hv2) co-expressed with all the ORs, is shown, as was demonstrated using in situ hybridization (Große-Wilde et al. 2007; Baker 2009; Krieger et al. 2009). The expression of Hv15 in B-type sensilla on the companion OSN that expresses Hv6 has not been resolved, hence the question mark. (Bottom) Response spectra of the ORs to possible pheromone component ligands, summarized from the work of Krieger et al. (2004, 2009) and Große-Wilde et al. (2007). The size of the circles represents the magnitude of the response of the ORs to compounds presented at a concentration of 10−4 M. Filled circles represent 150–200 nA, horizontally striped circles represent 100–150 nA, and vertically striped circles represent 50–100 nA (adapted from Wang et al. 2011).

(Baker et al. 2006; Vickers 2006a,b). Other studies have combined extensive QTL and candidate gene analyses of pheromone ORs, their functionalization (ascribing their response specificity to a range of candidate pheromone-component ligands), and neuroanatomical studies definitively locating certain ORs populating OSNs within stereotypical trichoid sensilla (Große-Wilde et al. 2007; Gould et al. 2010; Wang et al. 2011; Vásquez et al. 2013). In the latter studies, several *H. virescens* ORs (HvOR6, HvOR14, HvOR15, and HvOR16) were mapped to a single chromosome, Hv/Hs-C27 (Gould et al. 2010). By introgressing Hs-C27 into an *H. virescens* backcrossed background, it was discovered that minor DNA sequence differences between orthologs of *H. subflexa* and *H. virescens* for this chromosome (and presumably Hv/HsOR6) determine attraction and ORN receptivity to either Z9-14Ald or Z9-16Ald (Gould et al. 2010). These studies corroborated the collaborative neuroethological studies of hybrids by Baker et al. (2006) and Vickers (2006a,b), all of which provided evidence that male response can be strongly influenced by shifts in the expression levels of one or a few OR genes and affect the tuning curves of OSNs to create novel behavioral responses to new pheromone components and blends. Furthermore, through such hybrid and backcross studies, the framework for potential speciation was established via novel trait emergence by introgression of pheromone ORs and OSNs in the background of closely related species (Gould et al. 2010).

BASIC ARCHITECTURE OF THE HELIOTHINE MALE MOTH PHEROMONE OLFACTORY SYSTEM

The OSNs of heliothine moths respond quite narrowly and specifically with action potentials to one pheromone component of their conspecific pheromone blend. Their response specificity is determined by the OR that is expressed on them, such as is shown for the OSNs of *Heliothis virescens* (figure 21.3). The axons of pheromone-sensitive OSNs such as these for *H. virescens*, as in other moth species, project to pheromone-component-specific knots of neuropil called glomeruli in the AL of the brain (figure 21.4). Second-order neurons, called projection interneurons (PNs), exit each glomerulus after synapsing within them and send pheromone-component stimulation along their axons to higher centers of the brain (figures 21.5 and 21.6). In the heliothine moths, the specificities of response of PNs to particular pheromone components have been well worked out, as have their pathways to, and arborization locations in, higher neuropils such as the mushroom body and lateral protocerebrum (see more concerning architectures of higher-order neuronal pathways in the next sections). Some of these PN pathways synapse with integrative circuits in these higher centers and create the odor sensation "pheromone," whereas other PN integrations result in an odor that is behaviorally antagonistic (figures 21.5 and 21.6).

COORDINATED BEHAVIORAL AND SINGLE-CELL STUDIES OF HYBRID *HELIOTHIS VIRESCENS* AND *H. SUBFLEXA*

Some major findings from the collaborative wind-tunnel behavioral and single-cell studies of parental- and hybrid-type males of *Heliothis virescens* and *H. subflexa* (Baker et al. 2006; Vickers 2006a) revealed how the olfactory pathways of hybrids can shift their tuning profiles to allow broadened behavioral responsiveness to new combinations of components. The key changes in behavioral attraction responses and the concomitant changes in tuning profiles of the OSNs in hybrid males of *H. subflexa* and *H. virescens* are as follows and complement the extensive QTLs and OR-gene expression neuroanatomical studies cited above.

1. Z11-16OH Is Required for Attraction of Hybrid Heliothis virescens × H. subflexa *Males*

The response of *Heliothis virescens* × *H. subflexa* hybrid males to different blends of the two species' pheromone components was found to be more similar to the parental *H. subflexa* response type than to the *H. virescens* type (Vickers 2006a). Unlike the *H. virescens* parental type, but similar to *H. subflexa* males, hybrid males all required the presence of Z11-16OH in whatever blend mixture was tested for upwind flight to any of the blends (Vickers 2006a). This is consistent with peripheral neurophysiological recordings that found that nearly all of the OSNs responding to Z11-16OH in the "C-type" sensilla of hybrid males retained fidelity to only Z11-16OH, similar to *H. subflexa* males but unlike C-type OSNs in *H. virescens* males (Baker et al. 2006). This would mean that the Z11-16OH-dedicated OSN line of *H. subflexa* to the anteromedial (AM) glomerulus that is related to positive upwind flight to pheromone was retained in hybrid males, identical to that in *H. subflexa* parental-type males (figure 21.5B). Had OSNs in hybrid males been constructed more like those of *H. virescens* (figure 21.5A), the line to this glomerulus would have switched to now be tuned to Z11-16Ac with a minor responsiveness to Z9-14Ald, and the Z11-16OH-tuned OSNs would now be arborizing in the ventromedial (VM) glomerulus (figure 21.5A). Recordings from PNs (Vickers 2006b) confirmed that the AM glomerulus in hybrids retained its responsiveness to Z11-16OH (figure 21.5C), just as in parental-type *H. subflexa* PNs arborizing in the AM glomerulus (Vickers 2006b) (figure 21.5B). This result with the PNs shows that the AM is the target glomerulus of the Z11-16OH-tuned OSNs found in hybrids (Baker et al. 2006) (figure 21.5C), just as it is in parental-type *H. subflexa* males (Lee et al. 2006b).

2. Z9-14Ald Can Substitute for Z9-16Ald in Attracting Hybrid Heliothis virescens × H. subflexa *Males*

A second major finding was that in hybrid males, Z9-14Ald could substitute behaviorally for Z9-16Ald in any of the blends (Vickers 2006a); this flexibility does not exist in *Heliothis subflexa* males (Vickers 2002, 2006a). Similarly, Z9-16Ald could not substitute for Z9-14Ald in *H. virescens* males (Vickers et al. 1991; Vickers 2006a). Therefore, this ability to substitute Z9-16Ald for Z9-14Ald and still get good attraction in hybrid males was surprising, and correlated with single-cell recordings from hybrid B-type sensilla in hybrids (Baker et al. 2006). Hybrid B-type OSNs exhibited *equal* sensitivity to *both* Z9-14Ald and Z9-16Ald (figure 21.5C), which was quite unusual compared to the complete inactivity of B-type OSNs to Z9-16Ald in *H. virescens* (Baker et al. 2004, 2006) (figure 21.5A) and only slight co-responsiveness of B-type OSNs of *H. subflexa* to Z9-14Ald in combination with high responsiveness to Z9-16Ald (Baker et al. 2004, 2006) (figure 21.5B). Recordings from PNs exiting the dorsomedial (DM) glomerulus of hybrid males corroborated the OSN tuning shift; the PNs of hybrid males exiting the DM glomerulus were now equally responsive to Z9-14Ald and Z9-16Ald (Vickers 2006b) (figure 21.5C), whereas in parental *H. subflexa* they were only responsive to Z9-16Ald (Vickers and Christensen 2003) (figure 21.5B) and in *H. virescens* only responsive to Z9-14Ald (Vickers et al. 1998) (figure 21.5A).

Baker et al. (2006) had speculated that the tuning shift in B-type sensillar OSNs may have been caused by the

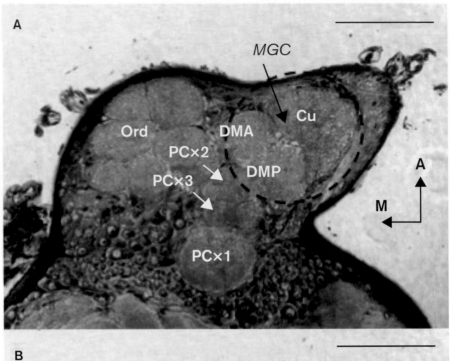

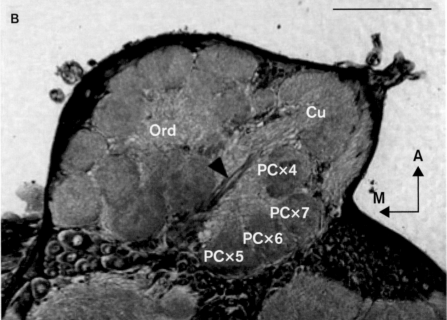

FIGURE 21.4 Horizontal (top-down) view of sections of the right antennal lobe of the brain of a *Helicoverpa zea* male. The pinkish-stained areas are the knots of neuropil called glomeruli, and the small blue-stained ovoids are the cell bodies of olfactory sensory neurons, local interneurons, and projection neurons.

A Section taken from approximately 70 μm from the most dorsal surface showing the macroglomerular complex (MGC) (dashed circled area) with its three glomeruli: cumulus (Cu) and dorsomedial posterior (DMP) and dorsomedial anterior (DMA) glomeruli of the pheromone-component-related MGC. This section also shows the location of PCx1, PCx2, and PCx3 glomeruli immediately posteriomedial to the three MGC glomeruli. PCx1 is the arborization destination of the second OSN co-compartmentalized with the Z11-16Ald–responding OSN in A-type trichoid sensilla. The second OSN that targets PCx1 is unresponsive to all odorants that have been tested on it, and as such no ligand can be assigned to it. Ord designates ordinary glomeruli that receive general odorant-related OSN inputs in the antennal lobe.

B A more ventral section taken at a depth of approximately 130 μm from the dorsal surface of the same animal, still showing the Cu of the MGC, but now other PCx glomeruli—PCx4, PCx5, PCx6, and PCx7—are apparent. A confluent bundle of fibers from interneurons of the MGC and PCx is indicated by the arrowhead.

ABBREVIATIONS: A, anterior; M, medial; MGC, macroglomerular complex; Cu, cumulus; DMA, dorsomedial anterior; DMP, dorsomedial posterior; Ord, ordinary glomeruli; PCx1, posterior complex glomerulus 1; PCx2, posterior complex glomerulus 2; PCx3, posterior complex glomerulus 3; PCx4, posterior complex glomerulus 4; PCx5, posterior complex glomerulus 5; PCx6, posterior complex glomerulus 6; PCx7, posterior complex glomerulus 7.

Scale bars = 100 μm.

SOURCE: Adapted from Lee et al. (2006a).

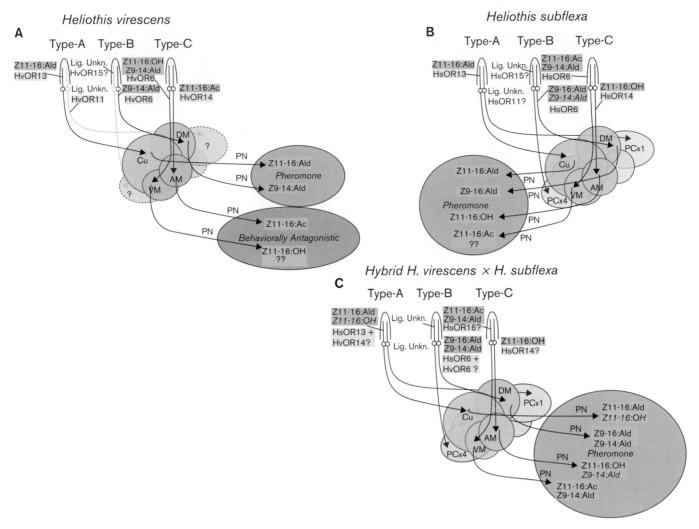

FIGURE 21.5 Neuroanatomical depictions of the known olfactory pathways of male *Heliothis virescens*, *H. subflexa*, and *H. virescens* × *H. subflexa* hybrids. Sensilla are represented by black-outlined conical figures, and OSNs are represented by the black, gray, or blue lines with cell bodies (tiny circles) within the sensilla having axons that project (arrows) to arborize in glomeruli in the MGC (brown or pink ovoid figures) or to the posterior complex (PCx; gray or blue ovoid figures). The targeting of particular glomeruli by pheromone-component-specific OSNs has been shown definitively by cobalt backfilling in *H. virescens* (A) (Berg et al. 1998) and *H. subflexa* (B) (Lee et al. 2006b). For the hybrids (C), the targeting of glomeruli by OSNs has not been demonstrated by cobalt backfilling, but it is assumed to be as depicted due to the demonstrated response profiles and arborization locations of PNs leaving the MGC glomeruli (Vickers 2006b). The known tuning profiles of OSNs are shown in gray rectangles. The ORs of *H. virescens* having tuning profiles matching particular OSNs and that have also been characterized via in situ hybridization studies as being expressed on particular OSNs (Große-Wilde et al. 2007; Krieger et al. 2009) are shown in yellow rectangles (A). Orthologous ORs of *H. subflexa* (Vásquez et al. 2011) (yellow rectangles in B) are placed on respective OSNs according to the known tuning profiles of OSNs (Baker et al. 2004) shown to reside in particular sensilla (Lee et al. 2006b). Placement of ORs of hybrids (yellow rectangles in C) on particular OSNs has been deduced from the OSN tuning profiles of hybrids found by Baker et al. (2006). OSNs having no known ligand are designated as such with the abbreviation, "Lig. Unkn." and have no rectangle. PCx1 and PCx4 arborization destinations of the axons of particular "no-ligand" OSNs of *H. subflexa* from A- and B-type sensilla are shown as blue axons due to their proven anatomies from cobalt staining. Because of the proven inputs after cobalt staining, PCx1 and PCx4 are likewise designated with blue ovoid figures. Conversely, PCx OSN arborization destinations of "no-ligand" OSNs of *H. virescens* are shown as gray ovoids, and their OSN inputs as gray axons following the careful reassessment of the stained OSNs shown in Berg et al. (1998) by Lee et al. (2006b). In particular, Lee et al. (2006b) strongly suggested that the co-localized unknown-ligand OSNs from *H. virescens* A- and B-type sensilla that had previously been designated by Berg et al. as arborizing in "ordinary" glomeruli actually are targeting PCx1 and PCx4 of this species. PNs are shown leaving their known arborization locations within component-specific glomeruli and projecting to protocerebral centers such as the mushroom body or the inferior lateral protocerebrum (see Chapter 10) to be integrated as "pheromone" (green ellipses) or as a "behaviorally antagonistic" odor feature (lavender ellipse). PN tuning profiles and PN arborization locations within particular MGC glomeruli have been determined through neuroanatomical studies of PNs in *H. virescens* (Berg et al. 1998; Vickers et al. 1998; Vickers and Christensen 2003), *H. subflexa* (Vickers and Christensen 2003), and hybrid *H. virescens* × *H. subflexa* (Vickers 2006b). These studies show a predominantly linear relationship of pheromone-component-specific OSN-to-glomerulus-to-PN pathways in these two parental species and their hybrids (Berg et al. 1998; Vickers and Christensen 2003; Vickers 2006b). Graphic scheme adapted from Lee et al. (2006b) which was also adapted and employed by Berg et al. (2014).

ABBREVIATIONS: Cu, cumulus; AM, anteromedial; VM, ventromedial; DM, dorsomedial; PCx1, posterior complex glomerulus 1; PCx4, posterior complex glomerulus 4; PN, projection neuron.

SOURCE: Adapted from Lee et al. (2006b).

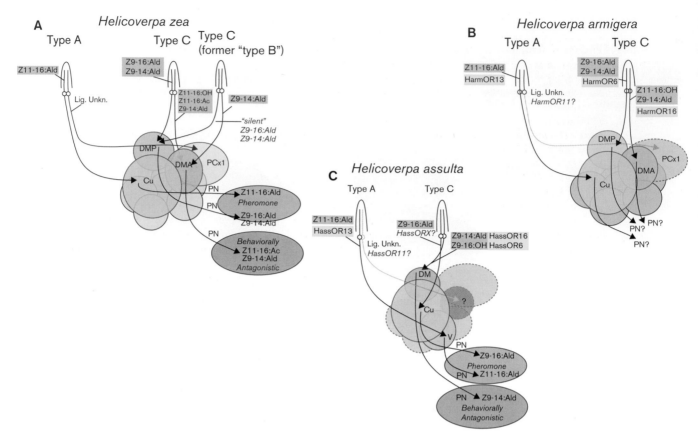

FIGURE 21.6 Neuroanatomical depictions of the known olfactory pathways of male *Helicoverpa zea*, *H. assulta*, and *H. armigera*. Sensilla are represented by black-outlined conical figures, and OSNs are represented by the black, gray, or blue lines with cell bodies (tiny circles) within the sensilla having axons that project (arrows) to arborize in glomeruli in the MGC (brown or pink ovoid figures) or to the posterior complex (PCx; gray or blue ovoid figures). The optimal pheromone-component-related odorant molecules for the OSNs (gray rectangles) and OSN co-localization relationships within A- or C-type sensilla have all been proven using single sensillum recordings for *H. zea* (Cossé et al. 1998; Lee et al. 2006a), *H. assulta* (Berg and Mustaparta 1995; Berg et al. 2005; Wu et al. 2013), and *H. armigera* (Wu et al. 2013; except for the non-Z9-16Ald-responding OSN in the C-type sensillum whose profile is merely conjectured here). OSNs having no known ligand are designated as such with the abbreviation "Lig. Unkn." or with the word "silent," and have no rectangle. For *H. armigera* and *H. assulta*, the names of ORs that have been shown to have the particular indicated tuning profile (Liu et al. 2013; Jiang et al. 2014) and to be expressed on particular OSNs based on OSNs' particular tuning profiles (Berg and Mustaparta 1995; Berg et al. 2005; Wu et al. 2013) are shown in yellow rectangles. No ORs are labeled as residing on any OSNs for *H. zea* because none have been characterized. The targeting of particular glomeruli by component-specific OSNs has been shown by cobalt backfilling in *H. zea* (Lee et al. 2006a) and *H. assulta* (Berg et al. 1998). For *H. armigera*, the targeting of glomeruli by OSNs has not been demonstrated by cobalt backfilling, but it is assumed to be as depicted due to the tuning profiles of OSNs (Wu et al. 2013) and *H. armigera*'s pheromonal similarity to *H. zea*. The PCx1 arborization destination of the "no-ligand" OSN of *H. zea* from A-type sensilla was proven via cobalt backfilling (blue axon from the OSN from the A-type sensillum) and is designated as a blue ovoid figure, as are the other *H. zea* PCx glomeruli. The OSN arborization destination of the no-ligand OSN (gray axon) from the A-type sensilla of *H. assulta* into a PCx is shown in gray and was done (Lee et al. 2006b) by reassessing older figures and data from Berg et al. (2005). The gray axon from the co-localized OSN in the A-type sensillum of *H. armigera* arborizing in PCx1 of this species is just suggested here due to *H. armigera*'s pheromonal similarity to *H. zea*. PNs are shown leaving their known arborization locations within component-specific glomeruli and projecting to protocerebral centers such as the mushroom body or the inferior lateral protocerebrum (see figure 21.3C) to be integrated as "pheromone" (green ellipses) or as a "behaviorally antagonistic" odor feature (lavender ellipse). PN tuning profiles and PN arborization locations within particular MGC glomeruli have been determined through neuroanatomical studies of PNs in *H. zea* (Vickers et al. 1998) and *H. assulta* (Berg et al. 2005). No PN analyses have yet been performed in *H. armigera*, as indicated by the question marks for this species. These studies of the PNs of *H. zea* and *H. assulta* (Vickers et al. 1998; Berg et al. 2005) again demonstrate the predominantly linear relationship of pheromone-component-specific OSN-to-glomerulus-to-PN pathways in heliothine moths (graphic scheme adapted from Lee et al. 2006b, also employed by Berg et al. 2014).

ABBREVIATIONS: Cu, cumulus; DMA, dorsomedial anterior; DMP, dorsomedial posterior; DM, dorsomedial; V, ventral; PCx1, posterior complex glomerulus 1; PN, projection neuron.

SOURCE: Adapted from Lee et al. (2006a).

co-expression of two different ORs on the same sensillum, with one OR being tuned to Z9-14Ald and the other OR tuned to Z9-16Ald. Although such co-expression of multiple pheromone OR types on a single OSN has been demonstrated in moths (Koutroumpa et al. 2014), another possibility is that a completely different and new OR might now be being expressed in the membranes of OSN dendrites of the hybrids. These different alleles would code for ORs having significantly different affinities for different ranges of pheromone-component ligands.

A striking example of this broadened acceptance of ligands comes from the work of Leary et al. (2012). They showed in the *Ostrinia nubilalis* and *O. furnicalis* sex pheromone receptor systems that a change in a single amino acid in the third transmembrane domain (of seven) of an OR that usually is optimally responsive to (*E*)-11-tetradecenyl acetate (E11-14Ac) plus several other pheromone-related ligands now severely restricts its tuning profile to respond only to (*Z*)-12-tetradecenyl acetate (Z12-14Ac) and (*E*)-12-tetradecenyl acetate (E12-

14Ac) (Leary et al. 2012). The fact that an OR, "HarmOR6," has now been found in *H. armigera* that is equally stimulated by either Z9-14Ald or Z9-16Ald (Liu et al. 2013) supports the idea that there may be a type of OR in heliothine moths amenable to a tuning shift similar to that occurring in *O. nubilalis* and *O. furnicalis*. This OR would be exhibiting a shifted, broadened acceptance of ligands due to a minor amino acid substitution, and an OR similar to HarmOR6 might be being expressed in hybrid *H. virescens* × *H. subflexa* B-type OSNs and explain their tuning to both Z9-14Ald and Z9-16Ald.

3. Addition of Z11-16Ac Does Not Adversely Affect Attraction of Hybrid Heliothis virescens × H. subflexa *Males*

A third major finding in *Heliothis virescens* × *H. subflexa* hybrid males was that the addition of Z11-16Ac to the optimal hybrid blend did not affect attraction of the hybrid males significantly (Vickers 2006a), a finding that is quite different than the antagonistic effect on upwind flight that even 0.1% or 1.0% Z11-16Ac has on *H. virescens* males when added to its two-component blend (Vickers and Baker 1997). It was clear from OSN recordings that in hybrids, the OSN that previously would have responded only to Z11-16Ac in parental type *H. virescens* now responded to Z9-14Ald as well (figure 21.5C). This arrangement is more like the parental-type *H. subflexa* OSN in C-type sensilla (figure 21.5B), except in hybrids the sensitivity to Z9-14Ald relative to Z11-16Ac in this OSN had been elevated (figure 21.5C). This pattern is yet a third way that the hybrids exhibit a more *H. subflexa*-like behavior and olfactory pathway phenotype than a parental *H. virescens* phenotype (Baker et al. 2006).

The behavior and OSN tuning profiles for hybrid C-type sensilla are borne out in the responsiveness of PNs of hybrid males that were recorded from and responded to Z11-16Ac, Z11-16OH, and Z9-14Ald (Vickers 2006b) (figure 21.4C). The PNs responsive to Z11-16OH did not respond to Z9-14Ald, a lack of Z9-14Ald-responsiveness that is similar to that of the PNs usually found in *H. subflexa* (Vickers and Christensen 2003). These Z11-16OH-responding PNs of hybrid males arborized in the AM glomerulus of the macroglomerular complex (MGC), again just like those of parental *H. subflexa* (Vickers and Christensen 2003) (figure 21.5C). This type of PN in *H. virescens* would have arborized in the VM glomerulus and would have responded to high doses of Z9-14Ald as well as Z11-16OH (Christensen et al. 1995; Vickers et al. 1998), mirroring the response profiles of *H. virescens* OSNs in C-type sensilla that were tuned to Z11-16OH (Baker et al. 2004, 2006) (figure 21.5A).

Furthermore, the Z11-16Ac-responding PNs in the hybrid males arborized in the VM glomerulus, exactly like Z11-16Ac-responding PNs of *H. subflexa* males (Vickers and Christensen 2003) (figures 21.5B,C) and unlike those of *H. virescens* males, which arborize in the AM glomerulus (Christensen et al. 1995) (figure 21.5A). These hybrid PNs exhibited an elevated responsiveness to Z9-14Ald compared to the parental-type PNs of *H. subflexa*, which are tuned to Z11-16Ac (Vickers and Christensen 2003), and again mirrors the OSN response profiles found in these same hybrid males (Baker et al. 2006) (figure 21.5C).

DIFFERENTIAL EXPRESSION OF ORs ON OSNs DETERMINES OSN RESPONSE PROFILES AND BEHAVIORAL RESPONSE SPECIFICITY

Studies with hybrids between *Heliothis virescens* and *H. subflexa* showed behavioral response shifts in hybrids to different blends of components that corresponded to shifts in OSN tuning curves as well as to shifts in PN tuning profiles and associated MGC arborization locations. All of these changes in the response spectra of hybrids illustrate how important OSN tuning profiles are in determining the subsequent neuronal activities, ending with behavior. Furthermore, the determinants of OSN tuning profiles are the ligand-specific response profiles of the ORs that are expressed on OSN dendritic membranes. For each OSN type, the selectors of which OR out of all the many possible ORs will populate its dendritic membrane are the presumed transcription factors that choose one *OR* gene to be expressed on an OSN over all the others (Ray et al. 2007; Fujii et al. 2011).

Years of work have gone into de-orphanizing putative pheromone-component-tuned ORs that are expressed on the dendrites of heliothine OSNs. These OR response profiles have been characterized using heterologous expression systems, the first and most widespread of which is the *Xenopus* oocyte system (see Baker and Hansson, this volume), in which the ORs are expressed in the oocyte membrane and pheromone-related compounds are solubilized to flow over the oocyte in the aqueous bath to contact the oocyte membrane. With the *Xenopus* system, the magnitude of the change in membrane conductance as the potential ligands contact the ORs bound to the membrane is measured electrophysiologically (Wang et al. 2011). A second heterologous expression system used in studying heliothine moth OR tuning profiles is the *Drosophila* "empty neuron" system in which the OR is expressed in the *Drosophila* OSN and its activity is monitored electrophysiologically (Dobritsa et al. 2003; Hallem et al. 2004; Hallem and Carlson 2004, 2006). A third heterologous expression system that has been used successfully with helothine moths (Große-Wilde et al. 2007) is a Flp-In-System coupled with a human embryonic kidney 293 (HEK293) cell line stably carrying a mouse $G\alpha_{15}$ gene; this system is named Flp-In T-Rex293/$G\alpha_{15}$. With this HEK293 cell line expression system, the magnitude of the change in membrane conductance is measured via calcium imaging; this imaging involves monitoring increases in fluorescence due to a calcium-sensitive dye responding to calcium release by the cells when they are contacted by the prospective ligands (Große-Wilde et al. 2007; Krieger et al. 2009).

Several of these heterologous expression studies showed clear tuning profiles of putative *H. virescens* pheromone-component-responsive ORs that corresponded with previously performed single-cell neurophysiological assays and to neuro-anatomical tracings of OSN axons to specific MGC glomeruli (Große-Wilde et al. 2007; Krieger et al. 2009; Wang et al. 2011; Liu et al. 2013; Vásquez et al. 2013; Jiang et al. 2014). Importantly, in situ hybridization studies, electrophysiological recordings, and cobalt backfilling of OSNs from single sensilla showed definitively that certain pairs of ORs could be characterized as being expressed on co-localized OSNs within the same sensilla. Thus, specific ORs could be mapped to their sensillar locations on stereotypical pairs of OSNs, and the axonal projections to AL glomeruli could be surmised (Baker et al. 2004, 2006; Lee et al. 2006a,b; Große-Wilde et al. 2007; Baker 2009; Krieger et al. 2009; Vásquez et al. 2011; Wang et al. 2011; Jiang et al. 2014) (figure 21.3).

These studies culminated in a definitive map of OR expression on certain OSNs for the *H. virescens* sex pheromone olfaction system (figure 21.3). The OR that is expressed on the OSN in *H. virescens* "A-type" sensilla that respond optimally to Z11-16Ald is HvOR13 (previously known as HR13; Große-Wilde et al. 2007; Krieger et al. 2009). The response spectrum of the

HvOR13 expressed in the *Drosophila* empty neuron system was shown to be very specific for Z11-16Ald (Vásquez et al. 2013), and as such mirrors the OSN responsive to Z11-16Ald in A-type sensilla (Berg et al. 1995; Baker et al. 2004). The OSN in A-type sensilla that is co-localized with the Z11-16Ald-responsive OSN was shown in in situ hybridization studies (Krieger et al. 2009) to have HvOR11 (previously known as HR11) expressed on its dendritic membranes, but as yet there is no known ligand for this OR that excites the corresponding OSN (Große-Wilde et al. 2007; Wang et al. 2011) (figures 21.3 and 21.5A).

The C-type sensilla of *H. virescens* have two known co-localized OSNs whose electrophysiological response profiles mirror those of two ORs, HvOR14 and HvOR16 (Große-Wilde et al. 2007; Krieger et al. 2009). HvOR14 (previously known as HR14) was shown to be optimally responsive to Z11-16Ac, whereas HvOR16 (previously known as HR16) was responsive to both Z11-16OH and Z9-14Ald, just like the response profiles of the two OSNs in C-type sensilla recorded in vivo (Baker et al. 2004, 2006; Große-Wilde et al. 2007; Baker 2009; Krieger et al. 2009) (figures 21.3 and 21.5A). In situ hybridization studies had shown these two ORs to be expressed in co-localized OSNs (Große-Wilde et al. 2007). The OSNs expressing these Z11-16Ac-tuned ORs had been shown to project to the AM glomerulus of the MGC (Berg et al. 1998) with PNs exiting the AM, confirming this pathway for reporting the presence of Z11-16Ac to protocerebral centers (Vickers et al. 1998; Vickers and Christensen 2003) (figure 21.5A). OSNs expressing the Z11-16OH-plus-Z9-14Ald-tuned ORs have been shown to project to the VM glomerulus (Berg et al. 1998), projections that presumably would be substantiated with further neuroanatomical studies of PNs, because VM is the only MGC glomerulus that has had no PNs neuroanatomically characterized and Z11-16OH- and Z9-14Ald-responding PNs are the only remaining unrepresented types of PNs for *H. virescens* (Vickers et al. 1998; Vickers and Christensen 2003; Vickers 2006b).

After years of uncertainty, it was finally shown definitively that in B-type sensilla, the OSN responsive to Z9-14Ald expresses HvOR6 (Wang et al. 2011) (previously known as HR6). It is still unknown what the co-localized OSN expresses, but it is suggested that it might be HvOR15 (Wang et al. 2011) (previously known as HR15). As with HvOR11, there is currently no known ligand for HvOR15. It will take double-staining in situ hybridization studies to determine whether in fact HvOR15 and HvOR6 are expressed on co-localized OSNs in B-type sensilla of *H. virescens*. We now know at least that the HvOR6 must be expressed on the Z9-14Ald-responsive OSN in B-type sensilla, but we do not know about the companion OSN (figures 21.4A and 21.6). Studies of PNs substantiate the conclusion that the DM glomerulus receives input from Z9-14Ald-responding OSNs in *H. virescens*, because this is the arborization location of PNs responding to Z9-14Ald, but to no other ligands in *H. virescens* (Vickers et al. 1998; Vickers and Christensen 2003).

The OSN in B-type sensilla that responds to Z9-14Ald in *H. virescens* and Z9-16Ald in *H. subflexa* (figures 21.5A and 21.5B) seems to be the OSN that determines behavioral response specificity in the two species, yet allows a broadened response in hybrids to both Z9-16Ald and Z9-14Ald (figure 21.5C). Orthologous ORs in *H. subflexa* have been isolated using QTL techniques (Gould et al. 2010), with the expression of HsOR6, along with HsOR14, HsOR16, HsOR13, HsOR11, and HsOR15, being implicated as orthologs to the *H. virescens* ORs of corresponding sequence similarity. However, the functionalization of *H. subflexa* ORs measuring their actual responses to ligands has not been performed.

In reviewing the data on the intermediate type of responsiveness of the B-type OSN of hybrids that responded equally well to both Z9-14Ald and Z9-16Ald, Baker et al. (2006) conjectured that perhaps this new, dual responsiveness to both components was due to a co-expression of two different types of ORs on the dendrites of a single type of OSN, with one OR being tuned to Z9-14Ald and the other OR to Z9-16Ald. Vásquez et al. (2011) examined the relative expression levels of HvOR6 and HsOR6 in *H. subflexa* and *H. virescens* and concluded that HvOR6 and HsOR6 are not differentially expressed in male antennae; therefore, interspecific sequence differences between these two genes must explain the species-specificity of their tuning on their B-type OSNs and the resultant differential responsiveness to their pheromone blends. Therefore, the possibility exists that a co-expression of HvOR6 and HsOR6 on hybrid B-type OSNs as conjectured by Baker et al. (2006) might explain the co-responsiveness of these OSNs to both Z9-14Ald and Z9-16Ald and the mutual interchangeability of Z9-14Ald and Z9-16Ald with regard to successful upwind flight behavior to the hybrid sex pheromone blend.

Heliothine Moth Pheromone Olfactory Pathways Common across Species

The underlying circuitry of the pheromone olfactory systems of moths is quite similar across all species (see Baker and Hansson, this volume), and this similarity holds true especially for *Heliothis* and *Helicoverpa* spp. that have been thoroughly examined neuroanatomically. Because the pheromone olfactory pathways ascending from the AL to higher neuropils are so similar across heliothines, the major changes that can be expected to be at work in changing the specificity of behavioral response by male heliothines will be at the periphery. OSNs can be selected to express different ORs and change the types of pheromone compounds that can be registered by the existing circuitry platform of the olfactory system. Because the target glomeruli of OSNs of insects are not known to change when new ORs are expressed on them (Dobritsa et al. 2003; Hallem et al. 2004; Goldman et al. 2005), the malleability of pheromone-component reporting by OSNs to glomeruli causing changes in behavioral response profiles will rest predominantly with any shifts in expression of ORs on OSNs. Males can be expected to "track" shifting female-emitted blends (Phelan 1992) by being selected to have new and different ORs expressed on their antennal OSNs. The action potentials of these OSNs in response to now-different pheromone-component ligands will be received and integrated as representing the same pheromone blend as before due to the unvarying, underlying synaptic circuitry. Such shifts in OSN tuning will be what optimizes conspecific mate finding and minimizes heterospecific attraction and mating mistakes.

The general theme across all of the heliothine species that have been intensely looked at is one of balanced olfactory antagonism (Baker 2008) that is orchestrated first by OSNs that are optimally responsive to each of the species' own conspecific pheromone components. These OSN inputs are carried by OSN axons to their sex pheromone-component-specific glomerular targets in the MGC of the AL (see Baker and Hansson, this volume) (figure 21.4A). The balance of excitations across the MGC glomeruli will determine the pheromone blend sensation that is registered by protocerebral neurons after they receive inputs from PNs carrying information out of the AL to the protocerebrum (see Baker and Hansson, this volume).

In heliothines, there are either three or four glomeruli that comprise the MGC. In the genus *Helicoverpa*, there are three MGC glomeruli that receive pheromone-component-related inputs from pheromone-component-tuned OSNs: the "cumulus" or largest glomerulus in all three species, a second "dorsomedial anterior" (DMA) glomerulus in *H. zea* and *H. armigera*, which is called "ventral" in *H. assulta*, and a third glomerulus called "dorsomedial posterior" (DMP) in *H. zea* and *H. armigera*, and just "dorsomedial" in *H. assulta*) (Vickers et al. 1998; Berg et al. 2002, 2005, 2014) (figure 21.4B). In the genus *Heliothis*, there are four MGC glomeruli, including the cumulus, the DM, AM, and VM (Vickers et al. 1998; Berg et al. 2002; Vickers and Christensen 2003) (figure 21.5). In both *Heliothis* and *Helicoverpa* spp., there is an unusual "posterior complex" of glomeruli, immediately behind the MGC (figure 21.3), some of whose glomeruli receive inputs from thus far unexcitable OSNs whose ligands have yet to be discovered. The OSNs arborizing in posterior complex glomeruli are co-compartmentalized in A-type trichoid sensilla with OSNs responsive to Z11-16Ald and in B-type sensilla with OSNs responsive to Z9-14Ald or Z9-16Ald (Lee et al. 2006a,b; Baker 2009) (figures 21.4–21.6).

The correct ratio of conspecific sex pheromone components results in an optimal ratio of excitatory action potential inputs to the target glomeruli, and these excitations, usually still component specific even after the MGC AL level, are then conducted by PNs to integrative neuropils in the protocerebrum, including the calyces of the mushroom body (MBC) and the inferior lateral protocerebrum (ILPC) (see Baker and Hansson, this volume). Incorrect sex pheromone-component ratio inputs to the glomeruli of the MGC (i.e., too little or too much of any of the components), or else the presence of heterospecific female pheromone components in the plume strands reported to other MGC glomeruli, can result in a suboptimal balance of MGC glomerular activity that now results in poor behavioral response and little or no attraction.

A-, B-, and C-Type Sensilla House Stereotypically Paired, Differentially Tuned, Olfactory Sensory Neurons

The differentially tuned pheromone-component-responsive OSNs of heliothine pheromone olfactory systems reside in trichoid sensilla on the male antennae (Almaas and Mustaparta 1990, 1991; Almaas et al. 1991; Vickers et al. 1991; Wu 1993; Cossé et al. 1998). Sensilla are named according to the responsiveness of the OSNs that are housed in them. For instance, OSNs tuned to Z11-16Ald have been deemed to reside in A-type sensilla (Cossé et al. 1998; Baker et al. 2004), along with a second OSN for which no ligand has yet been found (figures 21.5 and 21.6). Across all the species examined thus far, except for *Helicoverpa assulta*, the A-type sensilla are the most abundant and are correlated with Z11-16Ald being the most abundant sex pheromone component in the blend (Baker et al. 2004). A-type sensilla for these species include >70% of the trichoid sensilla and thus 70% of the pheromone-component-tuned OSNs in these species are Z11-16Ald-responding OSNs (Berg et al. 1995; Cossé et al. 1998; Baker et al. 2004; Lee et al. 2006a,b; Wu et al. 2013). *H. assulta* differs from these species that use Z11-16Ald as their major component, yet it follows the rule of greater abundance of OSNs tuned to the major component because >80% of the OSNs of *H. assulta* are tuned to its major pheromone component Z9-16Ald (Berg et al. 2005; Wu et al. 2013) and reside in C-type sensilla (Cossé et al. 1998; Lee et al. 2006a,b) (figure 21.6).

In the C-type sensilla of four of the five closely examined species (e.g., all but *Heliothis subflexa*), one of the two OSNs is always tuned to behaviorally antagonistic compounds emitted by other species. For instance, a single C-type OSN is tuned to both Z9-14Ald and Z9-16OH in *H. assulta* (Berg et al. 1995, 2005, 2014) and to Z11-16Ac, Z11-16OH, and Z9-14Ald in *H. zea* (Cossé et al. 1998). The two OSNs in C-type sensilla of *H. virescens* are tuned to Z11-16Ac and Z11-16OH, respectively (Baker et al. 2004, 2006; Baker 2009; Krieger et al. 2009). Although no direct electrophysiological recordings have been made from the second OSN in C-type sensilla of *H. armigera* (the first OSN of the pair has been shown electrophysiologically to be tuned to Z9-16Ald; Wu et al. 2013), the second OSN has been labeled as being tuned to behavioral antagonists Z11-16OH and Z9-14Ald, based on OR de-orphanization results (Liu et al. 2013; Jiang et al. 2014) (figure 21.6).

In three of these four species, the second OSN in C-type sensilla is tuned to one of the minor conspecific pheromone components, e.g., to Z9-14Ald in *H. zea* and *H. armigera* (Cossé et al. 1998; Wu et al. 2013), or to the major conspecific component Z9-16Ald in *H. assulta* (Berg et al. 1995, 2005, 2014).

The exception for C-type sensilla having at least one OSN that is tuned to a heterospecific antagonist is *H. subflexa*. In this species, no heterospecific behavioral antagonists have been found to date. One of the *H. subflexa* C-type OSNs is tuned to its third sex pheromone component Z11-16OH (Baker et al. 2004, 2006), with the second co-localized OSN tuned to Z11-16OAc (figure 21.5). This compound has been shown in field tests to function as a fourth sex pheromone component across the entirety of its geographical range (Groot et al. 2007).

For *H. virescens* and *H. subflexa*, B-type sensilla house an OSN tuned either to strictly Z9-14Ald (*H. virescens*) or to Z9-16Ald with slight responsiveness to Z9-14Ald (*H. subflexa*) (Baker et al. 2004, 2006). This type of OSN in B-type sensilla is critical for reporting the presence of these two species' minor sex pheromone components, either Z9-14Ald (*H. virescens*) or Z9-16Ald (*H. subflexa*). In both species, there is a second OSN co-localized in B-type sensilla that has not yet been found to respond to any ligands (Berg et al. 1998; Lee et al. 2006b) (figure 21.5).

Lee et al. (2006a) concluded that for a third species, *H. zea*, their initial designation of the presence of a B-type sensillum needed to be reconsidered. The two OSNs in the purportedly B-type sensilla that were recorded from and stained were observed to be anatomically C-type sensilla in their glomerular targets (Lee et al. 2006a). They found that OSNs in these sensilla projected their axons to the DMP and DMA glomeruli (see next section) just like the OSNs of C-type sensilla, with one OSN responding only to Z9-14Ald (figure 21.6). For some reason in these few sensilla, the large-spiking OSN that usually would have responded to Z9-16Ald, Z9-14Ald, or both seemed to have become silent. In addition, the small spiking OSN in this sensillum had become unresponsive to Z11-16OH and Z11-16Ac, remaining responsive only to Z9-14Ald (Lee et al. 2006a). The conclusion that the "B-type" sensilla of *H. zea* are actually C-type, with some of the C-type OSNs having become totally or partially silenced in response to some ligands, makes sense when comparing *H. zea* sensilla to those of other *Helicoverpa* species. For instance, to date no B-type sensilla have been recorded from and characterized for either *H. armigera* or *H. assulta*; there are only A- and C-type sensilla in these two species (Berg et al. 1995, 2005, 2014; Wu et al. 2013; Jiang et al. 2014) (figure 21.6).

A-, B-, and C-Type Olfactory Receptor Neurons Project to Stereotypical Antennal Lobe Glomeruli

For all of the heliothine spp. whose olfactory systems have been thoroughly examined, each species' major pheromone-component inputs from antennal OSNs tuned to the major component arborize in the largest glomerulus, the cumulus (Hansson et al. 1991), of the MGC (Berg et al. 1998, 2005, 2014; Lee et al. 2006a,b) (Cu in figures 21.4–21.6). The cumulus sits immediately at the base of the antenna, at the confluence of all the axons from the antenna entering the AL through the antennal nerve. For most species, the cumulus-arborizing OSNs are tuned to Z11-16Ald, but for *Helicoverpa assulta* they are tuned to Z9-16Ald (Berg et al. 2005) (figure 21.6). Minor component-tuned OSNs arborize in smaller companion glomeruli that usually are either dorsomedial (DM or DMP in figures 21.5 and 21.6), anteriomedial (AM or DMA in figures 21.5 and 21.6), or ventral (VM or V in figures 21.5 and 21.6) to the cumulus (Berg et al. 1998, 2005; Lee et al. 2006a,b).

It seems that the contributions of the activities within these small glomeruli vary between species with regard to the overall antagonistic balance of the signal that is integrated in protocerebral neuropils. Rather than considering that different glomeruli represent neuronal activities having a positive or negative odor "valence" (Knaden et al. 2012), we prefer to consider the activity, for instance of Z11-16Ac stimulation within the AM glomerulus of *Heliothis virescens*, as having extra gain (strength) compared to the same activity in the AM glomerulus of *H. subflexa*. Thus, although the activity in this glomerulus in *H. virescens* appears to have a negative valence due to its negative effect on behavior, and the activity in the corresponding glomerulus in *H. subflexa* seems to have a positive valence due to its positive behavioral effect, the antagonistic influence of Z11-16Ac in *H. virescens* might be viewed as being extra strong and creating an extra-heavy weight on the overall blend-balance between all the glomeruli and thus results in behavioral antagonism. Similarly, in *H. subflexa* the Z11-16OH-related activity within the AM glomerulus can be considered as exerting a moderate antagonism that serves as a perfect counterweight to balance the antagonistic strengths of the synaptic activities within the Cu and the DM glomeruli, thus contributing to balanced olfactory antagonism and male attraction to the blend. In both *H. virescens* and *H. subflexa*, inputs to the DM glomerulus by either Z9-14Ald from OSNs in the B-type sensilla of *H. virescens* or Z9-16Ald from the OSNs of the B-type sensilla of *H. subflexa* could be contributing an appropriate amount of antagonistic counterweight to the antagonistic inputs of other glomeruli in these species, resulting in balanced olfactory antagonism and male attraction (Baker 2008).

The PNs whose pheromone-related response profiles and neuroanatomies have been characterized for *H. virescens*, *H. subflexa* (figures 21.5A and 21.5B), *H. zea*, and *H. assulta* (figures 21.6A and 21.6C) have been found to project along linear, component-specific pathways, emerging from the same MGC glomeruli that are visited by OSNs tuned to the same major and minor pheromone components of these species that the PNs are tuned to (Vickers et al. 1998; Vickers and Christensen 2003; Zhao and Berg 2010). This theme of one-component-to-one-glomerulus-and-PN linear design is typical of heliothine pheromone olfactory systems (Vickers et al. 1998; Vickers and Christensen 2003; Zhao and Berg 2010), but it is not typical of moth sex pheromone olfactory systems in general (see Baker and Hansson, this volume).

ORs of other heliothine moth species have now begun to be functionalized, and data concerning their tuning profiles mirror OR homology with regard to their sequence similarity versus their ligand responsiveness. The orthologs HarmOR6 and HassOR6 have recently been expressed in *Xenopus* heterologous expression systems and provided a first glimpse of their tuning profiles in response to possible ligands (Liu et al. 2013; Jiang et al. 2014). HarmOR13 was found to be tuned as strongly to Z11-16Ald as its ortholog HvOR13 is, and it shares a 91% sequence identity with HvOR13 (Liu et al. 2013; Jiang et al. 2014). However, the *H. armigera* ortholog to HvOR6 (HarmOR6) in *H. virescens* has a distinctly different tuning profile, despite having an 88% sequence identity to HvOR6 (Liu et al. 2013). HarmOR6 responded equally well to Z9-16Ald and Z9-14Ald, whereas HvOR6 was specifically tuned to Z9-14Ald. This HarmOR6 profile fits well with the behavioral and female pheromone emission data showing that Z9-14Ald is a third pheromone component in this species (Zhang et al. 2012). This is consistent with the congruence between OR response profiles and their respective OSNs' tuning profiles in *H. virescens* and *H. subflexa*. There was increased male *H. armigera* behavioral responsiveness (trap catch) when both Z9-14Ald and Z9-16Ald were added into blends containing Z11-16Ald (Zhang et al. 2012).

Jiang et al. (2014) found that the *H. assulta* ortholog HassOR6 was optimally responsive to Z9-16OH, with lower, and equivalent, levels of reactivity to both Z9-16Ald and Z9-14Ald. For *H. assulta*, it is not clear why there should be equal responsiveness to Z9-16Ald and Z9-14Ald on this type of pheromone-component-tuned OSN (Wu et al. 2013), because Z9-14Ald is highly behaviorally antagonistic (Boo et al. 1995), as is Z9-16OH (Cork et al. 1992). However, Z9-16Ald is the most abundant of the two pheromone components and should predominate in the stimulation of HassOR6 on this OSN, Furthermore, because the behavioral antagonists Z9-14Ald and Z9-16OH are detected by HassOR16 on the other co-localized C-type OSN (Berg et al. 2005), it should not matter if either of these compounds also coincidentally stimulate HassOR6 on the large-spiking C-type OSN (Jiang et al. 2014). It is clear that work on the complete functionalization coupled with studies of in situ hybridization of these HassORs and HarmORs has only just begun. More research is needed to characterize fully their response profiles and to map them to their resident OSNs to determine patterns of co-localization in sensilla. Further work on these *H. armigera* and *H. assulta* ORs (along with OR11 and OR15 in all heliothines) also needs to be performed to reconcile some of these seemingly conflicting and confusing early results. It would be particularly helpful to have different research groups use the same panel of prospective pheromone-component odorants in their experiments.

Courtship Pheromones

Given their relevance for agricultural applications, such as monitoring programs, much research has focused on female sex pheromones of the Heliothinae. However, male-produced courtship pheromones are also quite prevalent within this group and influence mating encounters between males and females. The majority of studies have focused on documenting the behavioral relevance and impact of such compounds on females (and in some instances males) during a courtship bout (Agee 1969; Teal and Tumlinson 1989; Cibrian-Tovar and Mitchell 1991; Hillier and Vickers 2004).

The pre-copulatory behavior of heliothines has been documented in several species, along with the impact of various factors on such behavior, including geographic region (Colvin et al. 1994), host plant distribution and moth age (Kvedaras et al. 2000), disruption of courtship with pheromone components or ultrasound (Callahan 1958; Agee 1969; Huang et al. 1997), and general ethology (Hendricks and Tumlinson 1974; Mitchell et al. 1974; Mitchell 1976; Teal et al. 1981b). Similar to many other Lepidoptera, male Heliothinae release volatile courtship pheromones from eversible structures (hairpencils) associated with the distal eighth abdominal segment (Birch et al. 1990). Hairpencil structures, closely associated with male claspers, disperse pheromones when everted during courtship. In different species, the effects of these pheromones have been proposed to either increase female receptivity to courting males, attract females to males, induce or arrest female calling, arrest female movement to facilitate copulation, or inhibit approach of competing males (Birch 1974; Baker 1981; Dong et al. 2005; Hillier and Vickers 2004, 2007; Jurenka and Rafaeli 2011). Overall, these compounds seem to modulate mating and courtship, offering a secondary pre-copulatory (prezygotic) barrier to mating mistakes beyond species-specific female pheromone production and male attraction.

A similar general pattern of courtship behavior has been documented in multiple heliothine species (*Heliothis virescens*, *H. subflexa*, *Helicoverpa zea*, and, to a lesser degree, *H. armigera* and *H. assulta*) (Agee 1969; Teal et al. 1981b, 1986; Cibrian-Tovar and Mitchell 1991; Hillier and Vickers 2004, 2011b; Ming et al. 2007). Behaviors may be separated in to pre-courtship behaviors (calling by females and activation or orientation by males) and courtship behaviors (hairpencil display, abdominal extension, clasping attempts) (Teal et al. 1989; Hillier and Vickers 2004, 2011b). Generally, a male will approach a stationary calling female from downwind. On arrival, the male will typically tap his antennae on her abdomen, near her ovipositor. The male then moves adjacent to the female and exposes his hairpencils, followed shortly thereafter by curling his abdomen toward the female and attempting clasping and copulation. The female typically either moves or flies away from the male, or she curls her own abdomen to accept the copulation attempt. There is considerable variation among and within species regarding the length of time spent conducting a given behavior within the sequence, and there are likely various modalities of feedback (visual, chemical, tactile, acoustic) that are likely used during courtship.

Teal et al. (1981b) documented the composition of such male-produced compounds from *H. virescens* as primarily being a combination of 16–18 chain-length acetates, alcohols, and carboxylic acids. Interestingly, these compounds share similarity to pheromone components of female Heliothinae, suggesting some common biosynthetic pathways, and perhaps functional homology in reception between sexes. However, to date, there has been no evidence of desaturase activity (commonly found in biosynthetic pathways for female-produced sex pheromones), from either the examination of male hairpencil gland extracts or hairpencil airborne emissions. Furthermore, gene transcripts and immunoassays have indicated the presence of PBAN in male *H. armigera* (Hirsch 1991; Rafaeli 2009; Ma et al. 1998). RNAi receptor studies in *H. armigera* also demonstrated that PBAN influences the production of male pheromone components, stimulating similar fatty acid biosynthetic pathways (Bober and Rafaeli 2010; for review, see Jurenka and Rafaeli 2011).

The composition, ratio, and concentration of male pheromone components can differ dramatically among species. For example, *H. subflexa* and *H. virescens* have opposite ratios of hexadecan-1-ol (16OH) to hexadecyl acetate (16Ac), and the concentration found in *H. subflexa* males is often substantially (20–100 times) lower than that for *H. virescens* males (Teal and Oostendorp 1995). Teal and Oostendorp (1995) also found that in hybrids and backcross progeny between *H. virescens* and *H. subflexa*, the ratio of 16OH/16Ac production was determined by dominant autosomal inheritance of *H. subflexa* alleles. Irrespective of cross direction, hairpencil pheromone titers of hybrid males were quantitatively and qualitatively similar to those of *H. subflexa* adults. In backcrosses between F1 hybrids and either *H. virescens* or *H. subflexa*, the composition and ratio of hairpencil components 16OH and 16Ac varied with cross direction, producing phenotypes similar to each species or intermediate phenotypes between species (Teal and Oostendorp 1993, 1995). Results suggest that the production of hairpencil pheromone by hybrids and backcrosses is under dominant, sex-linked control of alleles on the *H. subflexa* Z chromosome (Teal and Oostendorp 1993, 1995; Teal and Tumlinson 1997). Despite this, the morphology of hairpencils seems to be dictated by sex-linked inheritance from the male (Z) *H. virescens* sex chromosome (Teal and Oostendorp 1993).

Jacobson et al. (1984) reported quantities of 1 mg/male of Z9-14Ald in hairpencil extracts from *H. virescens*. Furthermore, this compound was proposed by Jacobson et al. (1984) to repel males during a courtship bout. However, subsequent studies on *H. virescens* and *H. subflexa* have not isolated Z9-14Ald from male glands, and further have not shown the presence of large quantities of any unsaturated hydrocarbons, suggesting that desaturases may not be present in the male pheromone glands of these species (Teal and Tumlinson 1989; Teal and Oostendorp 1993, 1995; Hillier and Vickers 2004). Finally, Huang et al. (1997) documented a similar complement of compounds (saturated 14–18 chain-length acetates, alcohols, and corresponding carboxylic acids) in the hairpencil gland extracts and headspace of *H. armigera*. Huang et al. (1997) also documented that the titers of hairpencil compounds peak 2–5 days after emergence and that titers are typically highest during scotophase.

The detection and behavioral effects of courtship pheromones have been investigated in multiple species, but perhaps most extensively in *H. virescens*. Male and female *H. virescens* share two short sensillar types that house olfactory receptor neurons that have been shown to selectively respond to both 16Ac and octadecyl acetate (18Ac), or to both 16OH and octadecan-1-ol (18OH) from among a selection of behaviorally relevant compounds (Hillier et al. 2006; Hillier and Vickers 2007). In *H. virescens*, OSNs from these sensilla selectively stain glomeruli near the entrance of the AL; the OSNs responding to acetate project to glomerulus 24 (adjacent and medial to the MGC for males), and to the possibly homologous glomerulus 59 in females (adjacent and medial to the large female glomeruli [LFG]) (Hillier and Vickers 2007; Hillier et al. 2007). OSNs responding to 16OH and 18OH project to glomerulus 41, adjacent and ventral to the LFG (Hillier and Vickers 2007). The proximity of these glomeruli to the MGC, along with similarity in position in males and females, suggests that there may be a "map" in the organization of the AL based upon molecular structure. Another option is that there is further behavioral segregation of pheromone and non-pheromone odor processing with processing of male hairpencil components and female sex pheromone components in a similar region, thus representing a co-localization of conspecific odorant processing, irrespective of the producing sex.

In female *H. virescens*, hairpencil compounds induce quiescence during a courtship bout, thereby increasing male success in mating by preventing females from moving away (and increasing female receptivity) (Teal et al. 1981b; Hillier and Vickers 2004). Furthermore, the efficacy of hairpencil extracts to induce quiescence in females is species specific, potentially linked to differential ratios of acetates and alcohols in the hairpencil composition of related species. Consequently, females can distinguish conspecific and heterospecific males by hairpencil composition, an important feature, as there are costs associated with mating mistakes (e.g., irreversible locking of genitalia or inviable male progeny) (Hardwick 1965; Goodpasture et al. 1980; Stadelbacher et al. 1983). It is possible that the initial choosiness by females for signals indicative of conspecific males may be the initial step in subsequent runaway female choice sexual selection. Females will then become even more discriminating and proceed to evaluate the quality and reproductive fitness of conspecific males according to the quality and quantity of their hairpencil volatiles, which will lead to increasingly amplified and specific courtship pheromone signals (Birch et al. 1990; Hillier and Vickers 2004, 2011b).

Moreover, a series of experiments comparing the mating behavior and success of interspecific mating trials between *H. subflexa* and *H. virescens* confirmed that hairpencil composition differentially influences courtship success between these species (Hillier and Vickers 2011b). Hairpencil-ablated males were more successful mating with the opposing species, provided females were stimulated with an artificial odor source with male conspecific pheromone. This effect was much more pronounced in trials involving male *H. virescens*. *H. subflexa* males were significantly more successful clasping and mating with both *H. virescens* and *H. subflexa* females in the absence of hairpencils, suggesting that there are fundamental differences in each species' requirements for these compounds to increase female receptivity or quiescence.

Field-cage studies with *H. virescens* suggest that female calling (and associated pheromone release) are inhibited by exposure to 50-male-equivalents of hairpencil extracts or exposure to 2-day-old virgin males (Hendricks and Shaver 1975). In *H. armigera*, an opposing effect has been found, as onset of female calling behavior may also be slightly influenced by saturated acids found within male hairpencil glands (percentage of calling females and duration of calling do not appear to be influenced) (Dong et al. 2005).

Wind-tunnel studies with male *H. virescens* also suggest 16Ac and 18Ac inhibit upwind male flight toward a synthetic female sex pheromone blend (Hillier and Vickers 2007). In *H. armigera*, saturated alcohols have also been tested as potential inhibitors of male approach, but they did not significantly influence male behavior (whereas addition of 5% Z11-16OH to an attractive blend inhibited approach) (Huang et al. 1997). Induction of quiescence in females and inhibition of approach by conspecific males may facilitate copulation and reduce competition from other suitors, ultimately increasing mating success. However, it remains unclear what the costs are for males to continue to orient upwind and attempt mating, despite the presence of conspecific males releasing courtship pheromone. Further research is required to reveal the potential behavioral roles of these compounds in courtship behavior.

Conclusions

The role of pheromones in male and female moth reproductive behaviors has doubtlessly been most comprehensively documented in the heliothine moths. Well-studied species of the genera *Heliothis* and *Helicoverpa* share biosynthetic pathways that produce similar compounds, but they exhibit considerable variation in pheromone composition between allopatric and sympatric populations around the world. In particular, a discrete balance is maintained between female pheromone production of key conspecific attractant blends (often including heterospecific antagonists) and male attraction to blends containing key compounds within a critical range of ratios. Field studies, laboratory behavioral tests, chemical analyses, and hybridization and crossing studies have provided a wealth of information on the probable evolution of pheromone communication in this group. Recent advances in neurophysiology, genetic analyses, and molecular biology offer great promise to continue to expand on and unlock key mechanisms that drive selection and diversification of these pheromone chemical signals and receptors.

References Cited

Agee, H. R. 1969. Mating behavior of bollworm moths. *Annals of the Entomological Society of America* 62:1120–1122.

Almaas, T. J., and H. Mustaparta. 1990. Pheromone reception in tobacco budworm moth, *Heliothis virescens*. *Journal of Chemical Ecology* 16:1331–1347.

Almaas, T. J., and H. Mustaparta. 1991. *Heliothis virescens*: response characteristics of receptor neurons in sensilla trichodea type 1 and type 2. *Journal of Chemical Ecology* 17:953–972.

Almaas, T. J., T. A. Christensen, and H. Mustaparta. 1991. Chemical communication in heliothine moths. I. Antennal receptor neurons encode several features of intra-and interspecific odorants in the male corn earworm moth *Helicoverpa zea*. *Journal of Comparative Physiology A* 69:249–258.

Baker, T. C. 2008. Balanced olfactory antagonism as a concept for understanding evolutionary shifts in moth sex pheromone blends. *Journal of Chemical Ecology* 34:971–981.

Baker, T. C. 2009. Nearest neural neighbors: moth sex pheromone receptors HR11 and HR13. *Chemical Senses* 34:465–468.

Baker, T. C. and R. T. Cardé. 1979. Analysis of pheromone-mediated behaviors in male *Grapholitha molesta*, the oriental fruit moth (Lepidoptera: Tortricidae). *Environmental Entomology* 8:956–968.

Baker, T. C., Nishida, R., and W. L. Roelofs. 1981. Close-range attraction of female oriental fruit moths to herbal scent of male hairpencils. *Science* 214:1359–1361.

Baker, T. C., A. A. Cossé, and J. L. Todd. 1998a. Behavioral antagonism in the moth *Helicoverpa zea* in response to pheromone blends of three sympatric heliothine moth species is explained by one type of antennal neuron. *Annals of the New York Academy of Sciences* 855:511–513.

Baker, T. C., H. Y. Fadamiro, and A. A. Cossé. 1998b. Moth uses fine tuning for odour resolution. *Nature* 393:530.

Baker, T. C., A. A. Cossé, S. G. Lee, J. L. Todd, C. Quero, and N. J. Vickers. 2004. A comparison of responses from olfactory receptor neurons of *Heliothis subflexa* and *Heliothis virescens* to components of their sex pheromone. *Journal of Comparative Physiology A* 190:155–165.

Baker, T. C., C. Quero, S. A. Ochieng', and N. J. Vickers. 2006. Inheritance of olfactory preferences II. Olfactory receptor neuron responses from *Heliothis subflexa* × *Heliothis virescens* hybrid male moths. *Brain, Behavior and Evolution* 68:75–89.

Behere, G. T., W. T. Tay, D. A. Russell, D. G. Heckel, B. R. Appleton, K. R. Kranthi, and P. Batterham. 2007. Mitochondrial DNA analysis of field populations of *Helicoverpa armigera* (Lepidoptera: Noctuidae) and of its relationship to *H. zea*. *BMC Evolutionary Biology* 7:117.

Berg, B.G., and H. Mustaparta. 1995. The significance of major pheromone components and interspecific signals as expressed by receptor neurons in the oriental tobacco budworm moth, *Helicoverpa assulta*. *Journal of Comparative Physiology A* 177:683–694.

Berg, B.G., J.H. Tumlinson, and H. Mustaparta. 1995. Chemical communication in heliothine moths. IV. Receptor neuron responses to pheromone compounds and formate analogues in the male tobacco budworm moth *Heliothis virescens*. *Journal of Comparative Physiology A* 177:527–534.

Berg, B.G., T.J. Almaas, J.G. Bjaalie, and H. Mustaparta. 1998. The macroglomerular complex of the antennal lobe in the tobacco budworm moth *Heliothis virescens*: specified subdivision in four compartments according to information about biologically significant compounds. *Journal of Comparative Physiology A* 183:669–682.

Berg, B.G., C.G. Galizia, R. Brandt, and H. Mustaparta. 2002. Digital atlases of the antennal lobe in two species of tobacco budworm moths, the oriental *Helicoverpa assulta* (male) and the American *Heliothis virescens* (male and female). *Journal of Comparative Neurology* 446:123–134.

Berg, B.G., T.J. Almaas, J.G. Bjaalie, and H. Mustaparta. 2005. Projections of male-specific receptor neurons in the antennal lobe of the oriental tobacco budworm moth, *Helicoverpa assulta:* a unique glomerular organization among related species. *Journal of Comparative Neurology* 486:209–220.

Berg, B.G., X.-C. Zhao, and G. Wang. 2014. Processing of pheromone information in related species of heliothine moths. *Insects* 5:742–761. doi:10.3390/insects5040742.

Birch, M.C. 1974. Aphrodisiac pheromones in insects. Pp. 115–134. In M.C. Birch, ed.. *Pheromones*. Amsterdam: North Holland.

Birch, M.C., G.M. Poppy, and T.C. Baker. 1990. Scents and eversible scent structures of male moths. *Annual Review of Entomology* 35:25–58.

Bober, R., and Rafaeli, A. 2010. Gene-silencing reveals the functional significance of pheromone biosynthesis activating neuropeptide receptor (PBAN-R) in a male moth. *Proceedings of the National Academy of Sciences of the United States of America* 107:16858–16862.

Boo, K.S., K.C. Park, D.R. Hall, A. Cork, B.G. Berg, and H. Mustaparta. 1995. (Z)-9-Tetradecenal: a potent inhibitor of pheromone-mediated communication in the oriental tobacco budworm moth, *Helicoverpa assulta*. *Journal of Comparative Physiology A* 177:695–699.

Butlin, R. 1987. Species, speciation, and reinforcement. *The American Naturalist* 130:461–464.

Byers, J.R., and D.L. Struble. 1987. Monitoring population levels of eight species of noctuids with sex-attractant traps in southern Alberta, 1978–1983: specificity of attractants and effect of target species abundance. *The Canadian Entomologist* 119:541–556.

Callahan, P.S. 1958. Behavior of the imago of the corn earworm, *Heliothis zea* (Boddie), with special reference to emergence and reproduction. *Annals of the Entomological Society of America* 51:271–283.

Cho, S., A. Mitchell, C. Mitter, J. Regier, M. Matthews, and R.O.N. Robertson. 2008. Molecular phylogenetics of heliothine moths (Lepidoptera: Noctuidae: Heliothinae), with comments on the evolution of host range and pest status. *Systematic Entomology* 33:581–594.

Choi, M.Y., Fuerst, E.J., Rafaeli, A., and R. Jurenka. 2003. Identification of a G protein-coupled receptor for pheromone biosynthesis activating neuropeptide from pheromone glands of the moth *Helicoverpa zea*. *Proceedings of the National Academy of Sciences of the United States of America* 100:9721–9726.

Choi, M.Y., A. Groot, and R.A. Jurenka. 2005. Pheromone biosynthetic pathways in the moths *Heliothis subflexa* and *Heliothis virescens*. *Archives of Insect Biochemistry and Physiology* 59:53–58.

Christensen, T.A., S.C. Geofrion, and J.G. Hildebrand. 1990. Physiology of interspecific chemical communication in *Heliothis* moths. *Physiological Entomology* 15:275–283.

Christensen, T.A., H. Itagaki, P.E. Teal, R.D. Jasensky, J.H. Tumlinson, and J.G. Hildebrand. 1991. Innervation and neural regulation of the sex pheromone gland in female *Heliothis* moths. *Proceedings of the National Academy of Sciences of the United States of America* 88:4971–4975.

Christensen, T.A., H.K. Lehman, P.E.A. Teal, H. Itagaki, J.H. Tumlinson, and J.G. Hildebrand. 1992. Diel changes in the presence and physiological actions of octopamine in the female sex-pheromone glands of heliothine moths. *Insect Biochemistry and Molecular Biology* 22:841–849.

Christensen, T.A., H. Mustaparta, and J.G. Hildebrand. 1995. Chemical communication in heliothine moths. VI: Parallel pathways for information processing in the macroglomerular complex of the male tobacco budworm moth *Heliothis virescens*. *Journal of Comparative Physiology A* 177:545–557.

Cibrian-Tovar, J., and E.R. Mitchell. 1991. Courtship behavior of *Heliothis subflexa* (Gn.) (Lepidoptera: Noctuidae) and associated backcross insects obtained from hybridization with *H. virescens* (F.). *Environmental Entomology* 20:419–426.

Colvin, J., R.J. Cooter, and S. Patel. 1994. Laboratory mating behavior and compatibility of *Helicoverpa armigera* (Lepidoptera: Noctuidae) originating from different geographical regions. *Journal of Economic Entomology* 87:1502–1506.

Cork, A., and E.A. Lobos. 2003. Female sex pheromone components of *Helicoverpa gelotopoeon*: first heliothine pheromone without (Z)-11-hexadecenal. *Entomologia Experimentalis et Applicata* 107:201–206.

Cork, A., K.S. Boo, E. Dunkelblum, D.R. Hall, K. Jee-Rajunga, M. Kehat, E. Kong Jie, K.C. Park, P. Tepgidagarn, and L. Xun. 1992. Female sex pheromone of oriental tobacco budworm, *Helicoverpa assulta* (Guenée) (Lepidoptera: Noctuidae): identification and field testing. *Journal of Chemical Ecology* 18:403–418.

Cossé, A.A., J.L. Todd, and T.C. Baker. 1998. Neurons discovered in male *Helicoverpa zea* antennae that correlate with pheromone-mediated attraction and interspecific antagonism. *Journal of Comparative Physiology A* 182:585–594.

Davis, M.T., V.N. Vakharia, J. Henry, T.G. Kempe, and A.K. Raina. 1992. Molecular cloning of the pheromone biosynthesis-activating neuropeptide in *Helicoverpa zea*. *Proceedings of the National Academy of Sciences of the United States of America* 89:142–146.

Descoins, C., Silvain, J.F., Lalanne-Cassou, B., and H. Cheron. 1988. Monitoring of crop pests by sexual trapping of males in Guadeloupe and Guyana. *Agriculture, Ecosystems & Environment* 21:53–56.

Dobritsa, A.A., W. Van der Goes van Naters, C.G. Warr, R.A. Steinbrecht, and J.R. Carlson. 2003. Integrating the molecular and cellular basis of odor coding in the *Drosophila* antenna. *Neuron* 37:827–841.

Domingue, M.J., C.J. Musto, C.E. Linn, Jr., W.L. Roelofs, and T.C. Baker. 2007. Evidence of olfactory antagonistic inhibition as a facilitator of evolutionary shifts in pheromone blend usage in *Ostrinia* spp. (Lepidoptera: Crambidae). *Journal of Insect Physiology* 53:488–496.

Dong, W.X., B.Y. Han, and J.W. Du. 2005. Inhibiting the sexual behavior of female cotton bollworm *Helicoverpa armigera*. *Journal of Insect Behavior* 18:453–463.

Dunkelblum, E., and M. Kehat. 1989. Female sex pheromone components of *Heliothis peltigera* (Lepidoptera: Noctuidae). Chemical identification from gland extracts and male response. *Journal of Chemical Ecology* 15:2233–2245.

Dunkelblum, E., Gothilf, S., and M. Kehat. 1980. Identification of the sex pheromone of the cotton bollworm, *Heliothis armigera*, in Israel. *Phytoparasitica* 8:209–211.

Eltahlawy, H., J.S. Buckner, and S.P. Foster. 2007. Evidence for two-step regulation of pheromone biosynthesis by the pheromone biosynthesis-activating neuropeptide in the moth *Heliothis virescens*. *Archives of Insect Biochemistry and Physiology* 64:120–130.

Fadamiro, H.Y., and T.C. Baker. 1997. *Helicoverpa zea* males (Lepidoptera: Noctuidae) respond to the intermittent fine structure of their sex pheromone plume and an antagonist in a flight tunnel. *Physiological Entomology* 22:316–324.

Fadamiro, H.Y., A.A. Cossé, and T.C. Baker. 1999. Fine-scale resolution of closely spaced pheromone and antagonist filaments by flying male *Helicoverpa zea*. *Journal of Comparative Physiology A* 185:131–141.

Fitt, G.P. 1989. The ecology of *Heliothis* species in relation to agroecosystems. *Annual Review of Entomology* 34:17–52.

Foster, S.P., and K.G. Anderson. 2012. Synthetic rates of key stored fatty acids in the biosynthesis of sex pheromone in the moth *Heliothis virescens*. *Insect Biochemistry and Molecular Biology* 42:865–872.

Fujii, T., T. Fujii, S. Namiki, H. Abe, T. Sakurai, A. Ohnuma, R. Kanzaki, S. Katsuma, Y. Ishikawa, and T. Shimada. 2011. Sex-linked transcription factor involved in a shift of sex-pheromone preference in the silkmoth *Bombyx mori*. *Proceedings of the*

National Academy of Sciences of the United States of America 108:18038–18043.

Goldman, A.L., W. Van der Goes van Naters, D. Lessing, C.G. Warr, and J.R. Carlson. 2005. Coexpression of two functional odor receptors in one neuron. *Neuron* 45:661–666.

Goodpasture, C., R.D. Richard, D. Martin, and M. Laster. 1980. Sperm cell abnormalities in progeny from interspecific crosses between *Heliothis virescens* and *H. subflexa*. *Annals of the Entomological Society of America* 73:529–532.

Gothilf, S., Kehat, M., Dunkelblum, E., and M. Jacobson. 1979. Efficacy of (Z)-11-hexadecenal and (Z)-11-tetradecenal as sex attractants for *Heliothis armigera* on two different dispensers. *Journal of Economic Entomology* 72:718–720.

Gould, F., M. Estock, N.K. Hillier, B. Powell, A.T. Groot, C.M. Ward, J.L. Emerson, C. Schal, and N.J. Vickers. 2010. Sexual isolation of male moths explained by a single pheromone response QTL containing four receptor genes. *Proceedings of the National Academy of Sciences of the United States of America* 107:8660–8665.

Große-Wilde, E., T. Gohl, E. Bouché, H. Breer, and J. Krieger. 2007. Candidate pheromone receptors provide the basis for the response of distinct antennal neurons to pheromonal compounds. *European Journal of Neuroscience* 25:2364–2373.

Groot, A.T., C. Ward, J. Wang, A. Pokrzywa, J. O'Brien, J. Bennett, R.G. Santangelo, C. Schal, and F. Gould. 2004. Introgressing pheromone QTL between species: towards an evolutionary understanding of differentiation in sexual communication. *Journal of Chemical Ecology* 30:2495–2514.

Groot, A.T., Y. Fan, C. Brownie, R.A. Jurenka, F. Gould, and C. Schal. 2005. Effect of PBAN on pheromone production by mated *Heliothis virescens* and *Heliothis subflexa* females. *Journal of Chemical Ecology* 31:15–28.

Groot, A.T., J.L. Horovitz, J. Hamilton, R.G. Santangelo, C. Schal, and F. Gould. 2006. Experimental evidence for interspecific directional selection on moth pheromone communication. *Proceedings of the National Academy of Sciences of the United States of America* 103:5858–5863.

Groot, A.T., R.G. Santangelo, E. Ricci, C. Brownie, F. Gould, and C. Schal. 2007. Differential attraction of *Heliothis subflexa* males to synthetic pheromone lures in eastern US and western Mexico. *Journal of Chemical Ecology* 33:353–368.

Groot, A.T., M.L. Estock, J.L. Horovitz, J. Hamilton, R.G. Santangelo, C. Schal, and F. Gould. 2009a. QTL analysis of sex pheromone blend differences between two closely related moths: insights into divergence in biosynthetic pathways. *Insect Biochemistry and Molecular Biology* 39:568–577.

Groot, A.T., O. Inglis, S. Bowdridge, R.G. Santangelo, C. Blanco, J.D. López, Jr., A.T. Vargas, F. Gould, and C. Schal. 2009b. Geographic and temporal variation in moth chemical communication. *Evolution* 63:1987–2003.

Groot, A.T., C.A. Blanco, A. Claßen, O. Inglis, R.G. Santangelo, J. Lopez, D.G. Heckel, and C. Schal. 2010a. Variation in sexual communication of the tobacco budworm, *Heliothis virescens*. *Southwestern Entomologist* 35:367–372.

Groot, A.T., A. Classen, H. Staudacher, C. Schal, and D.G. Heckel. 2010b. Phenotypic plasticity in sexual communication signal of a noctuid moth. *Journal of Evolutionary Biology* 23:2731–2738.

Groot, A.T., H. Staudacher, A. Barthel, O. Inglis, G. Schöfl, R.G. Santangelo, S. Gebauer-Jung et al. 2013. One quantitative trait locus for intra- and interspecific variation in a sex pheromone. *Molecular Ecology* 22:1065–1080.

Hagström, A.K., M.A. Liénard, A.T. Groot, E. Hedenström, and C. Löfstedt. 2012. Semi-selective fatty acyl reductases from four heliothine moths influence the specific pheromone composition. *PLOS ONE* 7:e37230.

Hallem, E.A., and J.R. Carlson. 2004. The odor coding system of Drosophila. *Trends in Genetics* 20:453–459.

Hallem, E.A., and J.R. Carlson. 2006. Coding of odors by a receptor repertoire. *Cell* 125:143–160.

Hallem, E.A., M.G. Ho, and J.R. Carlson. 2004. The molecular basis of odor coding in the *Drosophila* antenna. *Cell* 117:965–979.

Hansson, B.S., T.A. Christensen, and J.G. Hildebrand. 1991. Functionally distinct subdivisions of the macroglomerular complex in the antennal lobe of the male sphinx moth *Manduca sexta*. *Journal of Comparative Neurology* 312:264–278.

Hardwick, D.F. 1958. Taxonomy, life history, and habits of the elliptoid-eyed species of *Schinia* (Lepidoptera: Noctuidae), with notes on the Heliothidinae. *Memoirs of the Entomological Society of Canada* 90:5–116.

Hardwick, D.F. 1965. The corn earworm complex. *Memoirs of the Entomological Society of Canada* 97(Suppl 40):5–247.

Hardwick, D.F. 1970. The biological status of *Heliothis stombleri*. *Canadian Entomologist* 102:339–341.

Hartstack, A.W., J.A. Witz, and D.R. Buck. 1979. Moth traps for the tobacco budworm. *Journal of Economic Entomology* 72:519–522.

Heath, R.R., E.R. Mitchell, and J.C. Tovar. 1990. Effect of release rate and ratio of (Z)-11-hexadecen-1-ol from synthetic pheromone blends on trap capture of *Heliothis subflexa* (Lepidoptera: Noctuidae). *Journal of Chemical Ecology* 16:1259–1268.

Hendricks, D.E., and T.N. Shaver. 1975. Tobacco budworm: male pheromone suppressed emission of sex pheromone by the female. *Environmental Entomology* 4:555–558.

Hendricks, D.E., and J.H. Tumlinson. 1974. A field cage bioassay system for testing candidate sex pheromones of the tobacco budworm. *Annals of the Entomological Society of America* 67:547–552.

Hendricks, D.E., B.A. Leonardt, and T.N. Shaver. 1989. Development of optimized blends of two sex pheromone components impregnated in PVC dispensers for tobacco budworm bait. *Southwestern Entomologist* 14:17–25.

Hillier, N.K., and N.J. Vickers. 2004. The role of heliothine hairpencil compounds in female *Heliothis virescens* (Lepidoptera: Noctuidae) behavior and mate acceptance. *Chemical Senses* 29:499–511.

Hillier, N.K., and N.J. Vickers. 2007. Physiology and antennal lobe projections of olfactory receptor neurons from sexually isomorphic sensilla on male *Heliothis virescens*. *Journal of Comparative Physiology A* 193:649–663.

Hillier, N.K., and N.J. Vickers. 2011a. Mixture interactions in moth olfactory physiology: examining the effects of odorant mixture, concentration, distal stimulation, and antennal nerve transection on sensillar responses. *Chemical Senses* 36:93–108.

Hillier, N.K., and N.J. Vickers. 2011b. Hairpencil volatiles influence interspecific courtship and mating between two related moth species. *Journal of Chemical Ecology* 37:1127–1136.

Hillier, N.K., C. Kleineidam, and N.J. Vickers. 2006. Physiology and glomerular projections of olfactory receptor neurons on the antenna of female *Heliothis virescens* (Lepidoptera: Noctuidae) responsive to behaviorally relevant odors. *Journal of Comparative Physiology A* 192:199–219.

Hillier, N.K., D. Kelly, and N.J. Vickers. 2007. A specific male olfactory sensillum detects behaviorally antagonistic hairpencil odorants. *Journal of Insect Science* 7:4.

Huang, Y., S. Xu, X. Tang, Z. Zhao, and J. Du. 1997. Male orientation inhibitor of cotton bollworm: inhibitory effects of alcohols in wind-tunnel and in the field. *Insect Science* 4:173–181.

Jacobson, M., V.E. Adler, and A.H. Baumhover. 1984. A male tobacco budworm pheromone inhibitory to courtship. *Journal of Environmental Science and Health A* 19:469–476.

Jiang, X.-J., H. Guo, C. Di, S. Yu, L. Zhu, L.-Q. Huang, and C.-Z. Wang. 2014. Sequence similarity and functional comparisons of pheromone receptor orthologs in two closely related *Helicoverpa* species. *Insect Biochemistry and Molecular Biology* 48:63–74.

Jurenka, R.A. 2003. Biochemistry of female moth sex pheromones. Pp. 53–80. In G.J. Blomquist and R. Vogt, eds. *Insect Pheromone Biochemistry and Molecular Biology*. Amsterdam: Elsevier.

Jurenka, R., and A. Rafaeli. 2011. Regulatory role of PBAN in sex pheromone biosynthesis of heliothine moths. *Frontiers in Endocrinology* 2:46.

Jurenka, R.A., E. Jacquin, and W.L. Roelofs. 1991. Control of the pheromone biosynthetic pathway in *Helicoverpa zea* by the pheromone biosynthesis activating neuropeptide. *Archives of Insect Biochemistry and Physiology* 17:81–91.

Kaae, R.S., H.H. Shorey, S.U. McFarland, and L.K. Gaston. 1973. Sex pheromones of Lepidoptera. XXXVII. Role of sex pheromones and other factors in reproductive isolation among ten species of Noctuidae. *Annals of the Entomological Society of America* 66:444–448.

Kakizaki, M., and H. Sugie. 2003. Sex pheromone of the flax budworm, *Heliothis maritima adaucta* Butler (Lepidoptera: Noctuidae). *Applied Entomology and Zoology* 38:73–78.

Kehat, M., and E. Dunkelblum. 1990. Behavioral responses of male *Heliothis armigera* (Lepidoptera: Noctuidae) moths in a flight tunnel to combinations of components identified from female sex pheromone glands. *Journal of Insect Behavior* 3:75–83.

Kehat, M., S. Gothilf, E. Dunkelblum, and S. Greenberg. 1980. Field evaluation of female sex pheromone components of the cotton bollworm, *Heliothis armigera. Entomologia Experimentalis et Applicata* 27:188–193.

Kim, Y.J., R.J. Nachman, K. Aimanova, S. Gill, and M.E. Adams. 2008. The pheromone biosynthesis activating neuropeptide (PBAN) receptor of *Heliothis virescens*: identification, functional expression, and structure–activity relationships of ligand analogs. *Peptides* 29:268–275.

Kingan, T.G., M.B. Blackburn, and A.K. Raina. 1992. The distribution of pheromone-biosynthesis-activating neuropeptide (PBAN) immunoreactivity in the central nervous system of the corn earworm moth, *Helicoverpa zea. Cell and Tissue Research* 270:229–240.

Kingan, T.G., P.A. Thomas-Laemont, and A.K. Raina. 1993. Male accessory gland factors elicit change from 'virgin' to 'mated' behaviour in the female corn earworm moth *Helicoverpa zea. Journal of Experimental Biology* 183:61–76.

Klun, J.A., B.A. Bierl-Leonhardt, J.R. Plimmer, A.N. Sparks, M. Primiani, O.L. Chapman, and G.H. Lee. 1980a. Sex pheromone chemistry of the female tobacco budworm moth, *Heliothis virescens. Journal of Chemical Ecology* 6:177–183.

Klun, J. A., Plimmer, J. R., Bierl-Leonhardt, B. A., Sparks, A. N., Primiani, M., Chapman, O. L., Lee, G.H., and G. Lepone. 1980b. Sex pheromone chemistry of female corn earworm moth, *Heliothis zea. Journal of Chemical Ecology* 6:165–175.

Klun, J. A., Leonhardt, B. A., Lopez, J. D., and L. E. Lachance. 1982. Female *Heliothis subflexa* (Lepidoptera: Noctuidae) sex pheromone: chemistry and congeneric comparisons. *Environmental Entomology* 11:1084–1090.

Knaden, M., A. Strutz, J. Ahsan, S. Sachse, and B.S. Hansson. 2012. Spatial representation of odorant valence in an insect brain. *Cell Reports* 1:392–399.

Koutroumpa, F.A., Z. Kárpáti, C. Monsempes, S.R. Hill, B.S. Hansson, E. Jacquin-Joly, J. Krieger, and T. Dekker. (2014) Shifts in sensory neuron identity parallel differences in pheromone preference in the European corn borer. *Frontiers in Ecology and Evolution* 2:65. doi:10.3389/fevo.2014.00065.

Krieger, J., E. Grosse-Wilde, T. Gohl, Y.M.E. Dewer, K. Raming, and H. Breer. 2004. Genes encoding candidate pheromone receptors in a moth (*Heliothis virescens*). *Proceedings of the National Academy of Sciences of the United States of America* 101:11845–11850.

Krieger, J., I. Gondesen, M. Forstner, T. Gohl, Y. Dewer, and H. Breer. 2009. HR11 and HR13 receptor-expressing neurons are housed together in pheromone-responsive sensilla trichodea of male *Heliothis virescens. Chemical Senses* 34:469–477.

Kvedaras, O.L., P.C. Gregg, and A.P. Del Socorro. 2000. Techniques used to determine the mating behaviour of *Helicoverpa armigera* (Hübner) (Lepidoptera: Noctuidae) in relation to host plants. *Australian Journal of Entomology* 39:188–194.

Kvedaras, O.L., A.P. Del Socorro, and P.C. Gregg. 2007. Effects of phenylacetaldehyde and (Z)-3-hexenyl acetate on male response to synthetic sex pheromone in *Helicoverpa armigera* (Hübner) (Lepidoptera: Noctuidae). *Australian Journal of Entomology* 46:224–230.

Lambert, D.M., B. Michaux, and C.S. White. 1987. Are species self-defining? *Systematic Zoology* 36:196–205.

Landolt, P.J., C.L. Smithhisler, R.S. Zack, and L. Camelo. 2006. Attraction of *Heliothis belladonna* (Henry and Edwards) to the sex pheromone of the corn earworm moth, *Helicoverpa zea* (Boddie) (Lepidoptera: Noctuidae). *Journal of the Kansas Entomological Society* 79:303–308.

Laster, M.L. 1972. Interspecific hybridization of *Heliothis virescens* and *H. subflexa. Environmental Entomology* 1:682–687.

Laster, M. L., Martin, D. F., and D. W. Parvin, Jr. 1976. Potential for suppressing tobacco budworm (Lepidoptera: Noctuidae) by genetic sterilization. *Technical Bulletin of the Mississippi State University Agriculture and Forestry Experimental Station.*

Leary, G.P., J.E. Allen, P.L. Bunger, J.B. Luginbill, C.E. Linn, Jr., I.E. Macallister, and K.W. Wanner. 2012. Single mutation to a sex pheromone receptor provides adaptive specificity between closely related moth species. *Proceedings of the National Academy of Sciences of the United States of America* 109:14081–14086.

Lee, S.G., M.A. Carlsson, B.S. Hansson, J.T. Todd, and T.C. Baker. 2006a. Antennal lobe projection destinations of *Helicoverpa zea* male olfactory receptor neurons responsive to heliothine sex pheromone components. *Journal of Comparative Physiology A* 192:351–363.

Lee, S.G., N.J. Vickers, and T.C. Baker. 2006b. Glomerular targets of *Heliothis subflexa* male olfactory receptor neurons housed within long trichoid sensilla. *Chemical Senses* 31:821–834.

Lelito, J.P., A.J. Myrick, and T.C. Baker. 2008. Interspecific phero-mone-plume interference among sympatric heliothine moths: a wind tunnel test using live, calling females. *Journal of Chemical Ecology* 34:725–733.

Liu, Y., C. Liu, K. Lin, and G. Wang. 2013. Functional specificity of sex pheromone receptors in the cotton bollworm *Helicoverpa armigera. PLOS ONE* 8:e62094.

Löfstedt, C., J. Löfqvist, B.S. Lanne, J.N.C. van der Pers, and B.S. Hansson. 1986. Pheromone dialects in European turnip moths *Agrotis segetum. Oikos* 46:250–257.

Ma, P.W.K., W.L. Roelofs, and R.A. Jurenka. 1996. Characterization of PBAN and PBAN-encoding gene neuropeptides in the central nervous system of the corn earworm moth, *Helicoverpa zea. Journal of Insect Physiology* 42:257–266.

Ma, P.W., D.C. Knipple, and W.L. Roelofs. 1998. Expression of a gene that encodes pheromone biosynthesis activating neuropeptide in the central nervous system of corn earworm, *Helicoverpa zea. Insect Biochemistry and Molecular Biology* 28:373–385.

Matthews, M. 1999. *Heliothine Moths of Australia: A Guide to Pest Bollworms and Related Noctuid Groups.* Melbourne: CSIRO.

Mbata, G.N., and S.B. Ramaswamy. 1990. Rhythmicity of sex pheromone content in female *Heliothis virescens*: impact of mating. *Physiological Entomology* 15:423–432.

McElfresh, J.S., and J.G. Millar. 1999. Geographic variation in sex pheromone blend of *Hemileuca electra* from southern California. *Journal of Chemical Ecology* 25:2505–2525.

McElfresh, J.S., and J.G. Millar. 2001. Geographic variation in the pheromone system of the saturniid moth *Hemileuca eglanterina. Ecology* 82:3505–3518.

Ming, Q.L., Y.H. Yan, and C.Z. Wang. 2007. Mechanisms of premating isolation between *Helicoverpa armigera* (Hübner) and *Helicoverpa assulta* (Guenée) (Lepidoptera: Noctuidae). *Journal of Insect Physiology* 53:170–178.

Mitchell, E. R. 1982. Attraction of *Schinia mitis* males to southern armyworm females. *The Florida Entomologist* 65:291.

Mitchell, E. R., Tumlinson, J. H., and A. H. Baumhover. 1978. *Heliothis virescens*: attraction of males to blends of (Z)-9-tetra-decen-1-ol formate and (Z)-9-tetradecenal. *Journal of Chemical Ecology* 4:709–716.

Mitter, C., R.W. Poole, and M. Matthews. 1993. Biosystematics of the Heliothinae (Lepidoptera: Noctuidae). *Annual Review of Entomology* 38:207–225.

Nesbitt, B.F., P.S. Beevor, D.R. Hall, and R. Lester. 1979. Female sex pheromone components of the cotton bollworm, *Heliothis armigera. Journal of Insect Physiology* 25:535–541.

Park, K.C., A. Cork, K.S. Boo, and D.R. Hall. 1994. Biological activity of female sex pheromone of the oriental tobacco budworm, *Helicoverpa assulta* (Guenee) (Lepidoptera: Noctuidae): electroan-tennography, wind tunnel observation and field trapping. *Korean Journal of Applied Entomology* 33:26–32.

Park, K.C., A. Cork, and K.S. Boo. 1996. Intrapopulational changes in sex pheromone composition during scotophase in oriental tobacco budworm, *Helicoverpa assulta* (Guenée) (Lepidoptera: Noctuidae). *Journal of Chemical Ecology* 22:1201–1210.

Paterson, H.E.H. 1985. The recognition concept of species. Pp. 21–29. In E.S. Vrba, ed. *Species and Speciation. Transvaal Museum Monograph No. 4.* Pretoria, South Africa: Transvaal Museum.

Phelan, P.L. 1992. Evolution of sex pheromones and the role of asymmetric tracking. Pp. 265–314. In B.D. Roitberg and M.B. Isman, eds. *Insect Chemical Ecology.* New York: Chapman & Hall.

Piccardi, P., A. Capizzi, G. Cassani, P. Spinelli, E. Arsura, and P. Massardo. 1977. A sex pheromone component of the Old World bollworm *Heliothis armigera. Journal of Insect Physiology* 23:1443–1445.

Pope, M.M., L.K. Gaston, and T.C. Baker. 1982. Composition, quantification, and periodicity of sex pheromone gland volatiles from individual *Heliothis virescens* females. *Journal of Chemical Ecology* 8:1043–1055.

Pope, M.M., L.K. Gaston, and T.C. Baker. 1984. Composition, quantification, and periodicity of sex pheromone volatiles from individual *Heliothis zea* females. *Journal of Insect Physiology* 30:943–945.

Proshold, F.I., and L.E. LaChance. 1974. Analysis of sterility in hybrids from interspecific crosses between *Heliothis virescens* and *H. subflexa*. *Annals of the Entomological Society of America* 67:445–449.

Quero, C., and T.C. Baker. 1999. Antagonistic effect of (Z)-11-hexadecen-1-ol on the pheromone-mediated flight of *Helicoverpa zea* (Boddie) (Lepidoptera: Noctuidae). *Journal of Insect Behavior* 12:701–709.

Quero, C., H.Y. Fadamiro, and T.C. Baker. 2001. Responses of male *Helicoverpa zea* to single pulses of sex pheromone and behavioural antagonist. *Physiological Entomology* 26:106–115.

Rafaeli, A. 2009. Pheromone biosynthesis activating neuropeptide (PBAN): regulatory role and mode of action. *General and Comparative Endocrinology* 162:69–78.

Rafaeli, A., and C. Gileadi. 1995. Modulation of the PBAN-stimulated of pheromonotropic activity in *Helicoverpa armigera*. *Insect Biochemistry and Molecular Biology* 25:827–834.

Rafaeli, A., and R.A. Jurenka. 2003. PBAN regulation of pheromone biosynthesis in female moths. Pp. 107–136. In G.J. Blomquist and R.G. Vogt, eds. *Insect Pheromone Biochemistry and Molecular Biology*. Oxford: Elsevier.

Rafaeli, A., J. Hirsch, V. Soroker, B. Kamensky, and A.K. Raina. 1991. Spatial and temporal distribution of pheromone biosynthesis-activating neuropeptide in *Helicoverpa* (*Heliothis*) *armigera* using RIA and in vitro bioassay. *Archives of Insect Biochemistry and Physiology* 18:119–129.

Raina, A.K. 1993. Neuroendocrine control of sex pheromone biosynthesis in Lepidoptera. *Annual Review of Entomology* 38:329–349.

Raina, A.K., and Kempe, T.G. 1990. A pentapeptide of the C-terminal sequence of PBAN with pheromonotropic activity. *Insect Biochemistry* 20:849–851.

Raina, A.K., and T.G. Kempe. 1992. Structure activity studies of PBAN of *Helicoverpa zea* (Lepidoptera: Noctuidae). *Insect Biochemistry and Molecular Biology* 22:221–225.

Raina, A.K., and J.A. Klun. 1984. Brain factor control of sex pheromone production in the female corn earworm moth. *Science* 225:531–533.

Raina, A.K., J.A. Klun, J.D. Lopez, and B.A. Leonhardt. 1986. Female sex pheromone of *Heliothis phloxiphaga* (Lepidoptera: Noctuidae): chemical identification, male behavioral response in the flight tunnel, and field tests. *Environmental Entomology* 15: 931–935.

Raina, A.K., H. Jaffe, J.A. Klun, R.L. Ridgway, and D.K. Hayes. 1987. Characteristics of a neurohormone that controls sex pheromone production in *Heliothis zea*. *Journal of Insect Physiology* 33:809–814.

Ramaswamy, S.B. 1990. Periodicity of oviposition, feeding, and calling by mated female *Heliothis virescens* in a field cage. *Journal of Insect Behavior* 3:417–427.

Ramaswamy, S.B., and R.T. Roush. 1986. Sex pheromone titers in females of *Heliothis virescens* from three geographical locations (Lepidoptera: Noctuidae). *Entomologia Generalis* 12:19–23.

Ramaswamy, S.B., S.A. Randle, and W.K. Ma. 1985. Field evaluation of the sex pheromone components of *Heliothis virescens* (Lepidoptera: Noctuidae) in cone traps. *Environmental Entomology* 14:293–296.

Ramaswamy, S.B., R.A. Jurenka, C.E. Linn, Jr., and W.L. Roelofs. 1995. Evidence for the presence of a pheromonotropic factor in hemolymph and regulation of sex pheromone production in *Helicoverpa zea*. *Journal of Insect Physiology* 41:501–508.

Ray, A., W. van der Goes van Naters, T. Shiraiwa, and J.R. Carlson. 2007. Mechanisms of odor receptor gene choice in *Drosophila*. *Neuron* 53:353–369.

Roelofs, W.L., A.S. Hill, R.T. Cardé, and T.C. Baker. 1974. Two sex pheromone components of the tobacco budworm moth, *Heliothis virescens*. *Life Sciences* 14:1555–1562.

Rothschild, G.H.L. 1978. Attractants for *Heliothis armigera* and *H. punctiger*. *Journal of the Australian Entomological Society* 17:389–390.

Rothschild, G.H.L., B.F. Nesbitt, P.S. Beevor, A. Cork, D.R. Hall, and R.A. Vickers. 1982. Studies of the female sex pheromone of the native budworm, *Heliothis punctiger*. *Entomologia Experimentalis et Applicata* 31:395–401.

Shaver, T.N., J.D. Lopez, Jr., and A.W. Hartstack, Jr. 1982. Effects of pheromone components and their degradation products on the response of *Heliothis* spp. to traps. *Journal of Chemical Ecology* 8:755–762.

Shaver, T.N., D.E. Hendricks, and J.D. Lopez. Jr. 1989. Influence of (Z)-11-hexadecen-1-ol on field performance of *Heliothis virescens* pheromone in a PVC dispenser as evidenced by trap capture. *Journal of Chemical Ecology* 15:1637–1644.

Sheck, A.L., Groot, A.T., Ward, C.M., Gemeno, C., Wang, J., Brownie, C., Schal, C., and F. Gould. 2006. Genetics of sex pheromone blend differences between *Heliothis virescens* and *Heliothis subflexa*: a chromosome mapping approach. *Journal of Evolutionary Biology* 19:600–617.

Shorey, H.H., and L.K. Gaston. 1965. Sex pheromones of noctuid moths. V. Circadian rhythm of pheromone-responsiveness in males of *Autographa californica*, *Heliothis virescens*, *Spodoptera exigua*, and *Trichoplusia ni* (Lepidoptera: Noctuidae). *Annals of the Entomological Society of America* 58:597–600.

Shorey, H.H., L.K. Gaston, and J.S. Roberts. 1965. Sex pheromones of noctuid moths. VI. Absence of behavioral specificity for the female sex pheromones of *Trichoplusia ni* versus *Autographa californica*, and *Heliothis zea* versus *H. virescens* (Lepidoptera: Noctuidae). *Annals of the Entomological Society of America* 58:600–603.

Stadelbacher, E.A., M.W. Barry, A.K. Raina, and J.R. Plimmer. 1983. Fatal interspecific mating of two *Heliothis* species induced by synthetic sex pheromone. *Experientia* 39:1174–1176.

Steck, W., Underhill, E.W., and M.D. Chisholm. 1982. Structure-activity relationships in sex attractants for North American noctuid moths. *Journal of Chemical Ecology* 8:731–754.

Sugie, H., Tatsuki, S., Nakagaki, S., Rao, C.B.J., and Yamamato, A. 1991. Identification of the sex pheromone of the oriental tobacco budworm, *Heliothis assulta* (Guenee) (Lepidoptera: Noctuidae). *Applied Entomology and Zoology* 26:151–153.

Szŏcs, G., A. Raina, M. Tóth, and B.A. Leonhardt. 1993. Sex pheromone components of *Heliothis maritima*: chemical identification, flight tunnel and field tests. *Entomologia Experimentalis et Applicata* 66:247–253.

Tay, W.T., M.F. Soria, T. Walsh, D. Thomazoni, P. Silvie, G.T. Behere, C. Anderson, and S. Downes. 2013. A brave new world for an old world pest: *Helicoverpa armigera* (Lepidoptera: Noctuidae) in Brazil. *PlOS ONE* 8:e80134.

Teal, P.E.A., and A. Oostendorp. 1993. Interspecific hybridization between *Heliothis virescens* and *H. subflexa* (Lepidoptera: Noctuidae) affects the presence and structure of hairpencil glands of males. *Annual Review of Entomology* 86:322–326.

Teal, P.E.A., and A. Oostendorp. 1995. Production of pheromone by hairpencil glands of males obtained from interspecific hybridization between *Heliothis virescens* and *H. subflexa* (Lepidoptera: Noctuidae). *Journal of Chemical Ecology* 21:59–67.

Teal, P.E.A., and J.H. Tumlinson. 1986. Terminal steps in pheromone biosynthesis by *Heliothis virescens* and *H. zea*. *Journal of Chemical Ecology* 12:353–366.

Teal, P.E.A., and J.H. Tumlinson. 1989. Isolation, identification, and biosynthesis of compounds produced by male hairpencil glands of *Heliothis virescens* (F.) (Lepidoptera: Noctuidae). *Journal of Chemical Ecology* 15:413–427.

Teal, P.E.A., and J.H. Tumlinson. 1997. Effects of interspecific hybridization between *Heliothis virescens* and *Heliothis subflexa* on the sex pheromone communication system. Pp. 535–547. In R.T. Cardé and A.K. Minks, eds. *Insect Pheromone Research: New Directions*. New York: Chapman & Hall.

Teal, P.E.A., R.R. Heath, J.H. Tumlinson, and J.R. McLaughlin. 1981a. Identification of a sex pheromone of *Heliothis subflexa* (Gn.) (Lepidoptera: Noctuidae) and field trapping studies using different blends of components. *Journal of Chemical Ecology* 7:1011–1022.

Teal, P.E.A., J.R. McLaughlin, and J.H. Tumlinson. 1981b. Analysis of the reproductive behavior of *Heliothis virescens* (F.) under laboratory conditions. *Annals of the Entomological Society of America* 74:324–330.

Teal, P.E.A., J.H. Tumlinson, J.R. McLaughlin, R. Heath, and R.A. Rush. 1984. (Z)-11-Hexadecen-1-ol: a behavioral modifying chemical present in the pheromone gland of female *Heliothis zea* (Lepidoptera: Noctuidae). *Canadian Entomologist* 116:777–779.

Teal, P.E.A., J.H. Tumlinson, and R.R. Heath. 1986. Chemical and behavioral analyses of volatile sex pheromone components released by calling *Heliothis virescens* (F.) females (Lepidoptera: Noctuidae). *Journal of Chemical Ecology* 12:107–126.

Teal, P.E.A., J.H. Tumlinson, and H. Oberlander. 1989. Neural regulation of sex pheromone biosynthesis in *Heliothis* moths. *Proceedings of the National Academy of Sciences of the United States of America* 86:2488–2492.

Tillman, J.A., S.J. Seybold, R.A. Jurenka, and G.J. Blomquist. 1999. Insect pheromones—an overview of biosynthesis and endocrine regulation. *Insect Biochemistry and Molecular Biology* 29:481–514.

Tsfadia, O., A. Azrielli, L. Falach, A. Zada, W.L. Roelofs, and A. Rafaeli. 2008. Pheromone biosynthetic pathways: PBAN-regulated rate-limiting steps and differential expression of desaturase genes in moth species. *Insect Biochemistry and Molecular Biology* 38:552–567.

Underhill, E.W., Chisholm, M.D., and W. Steck. 1977. Olefinic aldehydes as constituents of sex attractants for noctuid moths. *Environmental Entomology* 6:333–337.

Vásquez, G.M., P. Fischer, C.M. Grozinger, and F. Gould. 2011. Differential expression of odorant receptor genes involved in the sexual isolation of two *Heliothis* moths. *Insect Molecular Biology* 20:115–124.

Vásquez, G.M., Z. Syed, P.A. Estes, W.S. Leal, and F. Gould. 2013. Specificity of the receptor for the major sex pheromone component in *Heliothis virescens*. *Journal of Insect Science* 13:160.

Vetter, R.S., and T.C. Baker. 1983. Behavioral responses of male *Heliothis virescens* in a sustained-flight tunnel to combinations of seven compounds identified from female glands. *Journal of Chemical Ecology* 9:747–749.

Vetter, R.S., and T.C. Baker. 1984. Behavioral responses of male *Heliothis zea* moths in sustained flight-tunnel to combinations of four compounds identified from female sex pheromone gland. *Journal of Chemical Ecology* 10:193–202.

Vickers, N.J. 2002. Defining a synthetic pheromone blend attractive to male *Heliothis subflexa* under wind tunnel conditions. *Journal of Chemical Ecology* 28:1255–1267.

Vickers, N.J. 2006a. Inheritance of olfactory preferences I. Pheromone-mediated behavioral responses of *Heliothis subflexa* × *Heliothis virescens* hybrid male moths. *Brain, Behavior and Evolution* 68:63–74.

Vickers, N.J. 2006b. Inheritance of olfactory preferences. III. Processing of pheromonal signals in the antennal lobe of *Heliothis subflexa* × *Heliothis virescens* hybrid male moths. *Brain Behavior & Evolution* 68:90–108.

Vickers, N.J., and T.C. Baker. 1994. Reiterative responses to single strands of odor promote sustained upwind flight and odor source location by moths. *Proceedings of the National Academy of Sciences of the United States of America* 91:5756–5760.

Vickers, N.J., and T.C. Baker. 1997. Chemical communication in heliothine moths. VII. Correlation between diminished responses to point-source plumes and single filaments similarly tainted with a behavioral antagonist. *Journal of Comparative Physiology A* 180:523–536.

Vickers, N.J., and T.A. Christensen. 1998. A combinatorial model of odor discrimination using a small array of contiguous, chemically defined glomeruli. *Annals of the New York Academy of Sciences* 855:514–516.

Vickers, N.J., and T.A. Christensen. 2003. Functional divergence of spatially conserved olfactory glomeruli in two related moth species. *Chemical Senses* 28:325–338.

Vickers, N.J., T.A. Christensen, H. Mustaparta, and T.C. Baker. 1991. Chemical communication in heliothine moths. III. Flight

behavior of male *Helicoverpa zea* and *Heliothis virescens* in response to varying ratios of intra-and interspecific sex pheromone components. *Journal of Comparative Physiology A* 169:275–280.

Vickers, N.J., T.A. Christensen, and J.G. Hildebrand. 1998. Combinatorial odor discrimination in the brain: attractive and antagonist odor blends are represented in distinct combinations of uniquely identifiable glomeruli. *Journal of Comparative Neurology* 400:35–56.

Vogel, H., A.J. Heidel, D.G. Heckel, and A.T. Groot. 2010. Transcriptome analysis of the sex pheromone gland of the noctuid moth *Heliothis virescens*. *BMC Genomics* 11:29.

Wang, C. 2007. Interpretation of the biological species concept from interspecific hybridization of two *Helicoverpa* species. *Chinese Science Bulletin* 52:284–286.

Wang, C., and J. Dong. 2001. Interspecific hybridization of *Helicoverpa armigera* and *H. assulta* (Lepidoptera: Noctuidae). *Chinese Science Bulletin* 46:489–491.

Wang, G., G.M. Vásquez, C. Schal, L.J. Zwiebel, and F. Gould. 2011. Functional characterization of pheromone receptors in the tobacco budworm *Heliothis virescens*. *Insect Molecular Biology* 20:125–133.

Wang, H.-L., Zhao, C.-H., and C.-Z. Wang. 2005. Comparative study of sex pheromone composition and biosynthesis in *Helicoverpa armigera*, *H. assulta* and their hybrid. *Insect Biochemistry and Molecular Biology* 35:575–583.

Wang, H.-L., Q.L. Ming, C.H. Zhao, and C.-Z. Wang. 2008. Genetic basis of sex pheromone blend difference between *Helicoverpa armigera* (Hübner) and *Helicoverpa assulta* (Guenée) (Lepidoptera: Noctuidae). *Journal of insect physiology* 54: 813–817.

Wu, C.-I.. 1993. Responses from sensilla on the antennae of male *Heliothis armigera* to its sex pheromone components and analogs. *Acta Entomologica Sinica* 36:385–389.

Wu, H., C. Hou, L.-Q. Huang, F.-S. Yan, and C.-Z. Wang. 2013. Peripheral coding of sex pheromone blends with reverse ratios in two *Helicoverpa* species. *PLOS ONE* 8:e70078.

Zhao, X.C., and B.G. Berg. 2010. Arrangement of output information from the 3 macroglomerular units in the heliothine moth *Helicoverpa assulta*: morphological and physiological features of male-specific projection neurons. *Chemical Senses* 35: 511–521.

Zhao, X.-C., J.-F. Dong, Q.-B. Tang, Y.-H. Yan, I. Gelbic, J.J.A. Van Loon, and C.-Z. Wang. 2005. Hybridization between *Helicoverpa armigera* and *Helicoverpa assulta* (Lepidoptera: Noctuidae): development and morphological characterization of F_1 hybrids. *Bulletin of Entomological Research* 95:409–416.

Zhao, X.-C., Y.-H. Yan, and C.-Z. Wang. 2006. Behavioral and electrophysiological responses of *Helicoverpa assulta*, *H. armigera* (Lepidoptera: Noctuidae), their F_1 hybrids and backcross progenies to sex pheromone component blends. *Journal of Comparative Physiology A* 192:1037–1047.

Zhang, D.D., K.Y. Zhu, and C.-Z. Wang. 2010. Sequencing and characterization of six cDNAs putatively encoding three pairs of pheromone receptors in two sibling species, *Helicoverpa armigera* and *Helicoverpa assulta*. *Journal of Insect Physiology* 56: 586–593.

Zhang, J.-P., C. Salcedo, Y.-L. Fang, R.-L. Zhang, and Z.-N. Zhang. 2012. An overlooked component: (Z)-9-tetradecenal as a sex pheromone in *Helicoverpa armigera*. *Journal of Insect Physiology* 58:1209–1216.

PART THREE

CHAPTER TWENTY-TWO

Monitoring for Surveillance and Management

D. M. SUCKLING

Introduction

The identification of pheromones of many economically important moth pests has revolutionized their management. Pheromone-baited traps have been used in a wide variety of ways for pest management, including seasonal phenology, population estimation, and decision support, as well as early detection and delimitation of invasive species. All sectors affected by moth pests have benefited from these technologies. Moth sex pheromone traps are now widely deployed, and have contributed significantly to sustainable pest management. New developments, including lures and traps for additional species, the use of pheromones in the biological control of weeds, and self-reporting camera traps linked with geographical information systems, are providing exciting opportunities for the expansion of the use of pheromones for surveillance and monitoring of pest populations, as well as detection and delimitation of new invaders.

Trapping Moths with Pheromones

The identification and availability in the early 1970s of the first pheromones for agricultural and forestry pests (Roelofs and Arn 1968; Roelofs et al. 1971; Bierl et al. 1972; Roelofs et al. 1973; Weatherston et al. 1974) was quickly followed by the realization that insect traps baited with pheromones might be able to be used in a range of ways for monitoring, and in some cases direct control (see Cork, this volume; Evenden, this volume). In several countries, pheromone-baited traps have been used with thresholds for determining the timing of insecticide sprays in orchards to manage the codling moth (Riedl and Croft 1974; Rock et al. 1978; Wearing and Charles 1978; Alford et al. 1979) and several leafrollers (Knight and Hull 1989a; Minks et al. 1995; Wearing 1995). Trapping systems have also been developed for pests of field crops (Macaulay 1977; Drapek et al. 1990), cotton (Foster et al. 1977), forestry (Sweeney et al. 1990), vegetables (Nakasuji and Kiritani 1978; Hoffmann et al. 1986), as well as stored products pests (Mullen et al. 1991; Trematerra 2012).

Traps for male moths are typically very simple, consisting of a pheromone lure and a retention system. This is usually a sticky surface for smaller moths, but can involve a funnel leading to a chamber with volatile insecticide for larger species (Howse et al. 1998). A wide range of trap designs is available; however, the simple delta trap is most commonly used, probably in part because its design allows air currents to pass through the trap and generate a favorable plume structure. Traps are normally made of plastic or weatherproof cardboard. The sticky bases can be made of a wide variety of materials, including paper-covered materials, coated with an inexpensive, polybutene-based glue.

The lures typically are formulated to match or surpass the attractiveness of females, but this is not always achieved, particularly if the synthetic blend varies from the natural product because of missing compounds or incorrect ratios. The female moth calls (releases pheromone) for a shorter interval than the duration of male responsiveness, and this disparity can result in higher catch from synthetic lures than females even if the synthetic blend is incomplete. In many reported

337

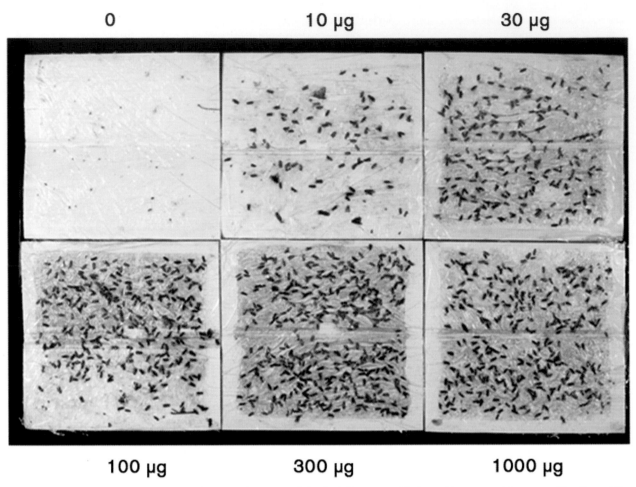

0   10 µg   30 µg

100 µg   300 µg   1000 µg

FIGURE 22.1 Catch on sticky bases of citrus flower moth (*Prays nephelomima*) (Yponomeutidae) to rubber septa with a range of loadings of the single component synthetic sex pheromone, (*Z*)-7-tetradecenal (Gibb et al. 2005). This compound is reported to attract all *Prays* species tested so far. Mass trapping has been developed against this species (Jamieson et al. 2008).

cases of pheromone identifications, female moths are not used as a positive control, or, if they are, trap catches can be too low to be meaningful. The absence of a female moth-baited positive control can be due to difficulties in obtaining sufficient numbers of live insects. However, if attraction of male moths to synthetic lures is lower than to wild female moths, this can lead to underestimation of pest populations, so it is useful to check this assumption. It is possible to exceed the catch achieved with females, and this too has ramifications. For example, in the assessment of mating disruption a higher catch of males to synthetic lures compared with traps with virgin female moths can lead to an underestimation of levels of control. Ideally, catch to a range of doses is determined during the pheromone identification and development (figure 22.1), which enables recommendations of suitable lures for future trapping programs (Gibb et al. 2005). Higher loadings usually have larger active spaces and offer greater sensitivity of detection, but can increase the risk of sampling an area outside the crop of interest.

That pheromone traps can capture moths at very low density is an incredibly useful attribute, especially when combined with their typically high species-specificity. Additionally, the ease of counting adult moths on a sticky base compared to sampling other life stages (e.g., larvae which are often cryptic and hard to count, especially when rare) is one obvious advantage of pheromone-baited traps. It is essential to determine the relationship between larval and adult densities, to facilitate interpretation of adult trap catch in terms of F_1 larval population levels. However, there can be issues of interpretation of what trap catch actually means, because of factors such as changes in trap efficiency through competition from wild female moths which are changing in density (Riedl et al. 1976), effects of weather (Adams et al. 1995), as well as trap configuration or placement (Drapek et al. 1990) and capture efficiency (Lingren et al. 1980). Interpretation is complicated if the effects of these factors on trap catch are nonlinear. Ideally, traps should perform with constant efficiency over time, in both attraction and capture efficiency. However, some release substrates such as rubber septa show first-order release kinetics (see below), with a logarithmic decrease in release of pheromone over time (Mullen et al. 1991), which requires more frequent rebaiting. Furthermore, if moths are large or very numerous, they can saturate traps (Elkinton 1987), thereby reducing possible correlations between catch and ambient population densities, unless traps are checked frequently. Thus, non-saturating traps are preferred in pest management where large numbers are captured. Temperature can also affect retention of the insects on glue. Bird or wasp predation of captured target moths can also be a problem.

TABLE 22.1

Typical applications of moth sex pheromones in integrated pest management, weed biological control, and biosecurity

Application of trapping	Purpose
IPM	Phenology of emergence and seasonal flights
	Density estimation
	Threshold for decisions on management interventions
	Sampling for insecticide resistance
	Species complex identification
	Sampling for timing of scouting for other life stages
	Direct control through mating disruption, mass trapping, or lure-and-kill systems
Weed biological control	Sampling for phenology and synchrony with host
	Establishment of new agent in a country
	Estimation of minimum release size for establishment
Biosecurity	Detection of an exotic species
	Establish the distribution limits of an invader
	Monitoring effectiveness of intervention efforts
	Demonstration of eradication

Trapping Requirements and Interpretation

Traps can be used in a wide range of ways by researchers exploring basic and applied questions, as well as operationally by growers and advisors (Table 22.1). Pheromone-baited traps can be used to document changes in pest density (assuming that catch is closely related to density), to determine seasonal flight patterns, presence in an area, distribution, and to establish pest action thresholds. Pheromone traps can also offer an effective tool for assessment of the best timing and level of efficacy of intervention tactics such as insecticides, mating disruption, or the sterile insect technique.

Ideally, lures are designed to reproduce the ratio of chemical components and the emission rate of calling females. There are many different lure dispensers in use, although rubber septa have been used successfully for many years. This usage was supported by analysis of release rates for a range of pheromone structures (Heath et al. 1986; McDonough et al. 1989). More recently, commercialized lures have used plastic fibers, polyethylene, PVC, urethane, and laminated plastic flakes in an attempt to stabilize release rates to zero order or constant release.

Trap design is also very important for effective monitoring, because moth behavior near traps can be influenced by visual cues (Foster and Muggleston 1993). Ideally, trap efficiency should not be affected by previous male catch (Rumbo 1993), competition with changing density of calling female moths (Riedl and Croft 1974; Riedl et al. 1976), or changes in attractiveness of the lure over time. Trap placement is also important, including height above the ground (Bohnenblust et al. 2011) or height within or above a growing crop (Kondo and Tanaka 1991; De Lame and Gut 2006; Kovanci et al. 2006). The insects should be caught and still be present when the trap is checked, which is often done weekly. The catch should also have some relationship with the actual insect population density, if it is to be used as a quantitative sampling tool. This is important in the application of traps to assess population suppression under tactics including mating disruption. The best trapping systems are sensitive, have constant trapping efficiency, are inexpensive, and easy to use. Consistency in trapping protocol is also important to enable comparisons between years, management options, and locations. Data are usually converted to catch per trap per day, to correct for varying time intervals between collections.

Pheromones and other attractants are now known for a large number of species (El-Sayed 2016), and many are available for purchase as lures that are ready to use, or as individual synthetic compounds that can be mixed and loaded onto lures. For this reason, increased use of pheromones has greatly supported the development of Integrated Pest Management (IPM), where a range of tactics are used together to reduce pest damage. Although pheromone trap catches have proven to be very useful for guiding pest management decisions in many different crops, their use is expanding making it difficult to estimate their global use (Witzgall et al. 2010). However, in addition to the factors above, diligent monitoring of catches and a certain amount of experience in interpreting the collected data are both critical for success. Developing an understanding of the relationship between trap catches and either crop damage or the abundance of the damaging stage of the insect is also essential.

Trapping without capture of moths is a primary tool for indicating the efficacy of mating disruption in fruit trees (McLaren et al. 1998; Stelinski et al. 2005), nuts (Higbee and Burks 2008), as well as crops such as blackcurrants (Suckling et al. 2005a), and stored products (Trematerra 2012). Conversely, trap catch can indicate the loss of effect of pheromone under mating disruption and therefore the need to reapply a pheromone or other treatment to prevent the realization of losses at harvest (Pringle et al. 2003).

Estimating Pest Density

Prediction of absolute pest density is complex, and consequently most pest management researchers report information about changes in relative density over time or space. For example, the temporal distribution of trap catches may indicate the start, middle, or end of an adult male flight period. This type of seasonal phenology information can then be used in various ways, such as to predict egg laying or hatch, enabling growers to target insecticide applications against larval stages. It may be necessary to consider the effects of protandry (earlier emergence of males), depending on species. Such information can be used with phenological models based on heat unit accumulation to provide decision support for estimating the correct timing of key life-history events and management interventions (Riedl and Croft 1974; Solomon 1978; Bailey 1980; Macaulay et al. 1985; Knight and Light 2005; Henneberry and Naranjo 2008; Knutson and Muegge 2010). In other cases, catch may be sufficient to show that the pest density is high enough to warrant intervention with insecticides (Dhawan and Sidhu 1984). In forestry, traps can serve as an early warning system of pest build-up (Sanders 1996; Daterman et al. 2004) or the need for intervention (Gargiullo et al. 1985; Ravlin 1991). Trap catch can also be used to determine the need and timing of other sampling methods, such as commencing larval scouting for diamondback moth *Plutella xylostella* (Plutellidae) (Walker et al. 2003) or pink bollworm *Pectinophora gossypiella* (Gelechiidae) (Henneberry and Naranjo 2008).

Trap Active Space

Trap performance in the field is affected by many factors (Houseweart et al. 1981) and trap active space is an important concept (Linn et al. 1987). It is affected by the threshold for perception of the pheromone. Wall and Perry (1987) reported extensive investigations into the application of traps for pea moth *Cydia nigricana* (Tortricidae), in which trap competition and other effects were demonstrated. Trap efficiency changes when multiple traps are present, due to competition. This has a spatial component and is predictable but often overlooked.

The active space of pheromone lures has seldom been estimated (Roelofs 1978; Elkinton and Cardé 1980; Baker and Roelofs 1981), but this information is actually very useful for interpretation of trapping results. The release rate of a semiochemical lure that attracts flying insects has a specific effective attraction radius that corresponds to the lure's orientation response strength and serves as a standardized method for comparing responses to lures that is independent of population density (Byers 2008). However, it should be noted that the idea of a fixed attraction radius or "active space" of a trap does not imply that any moth that flies within that space will be captured. In reality, there will be a pronounced fall-off of capture probability with distance, even inside that area where the insect may be able to sense and respond to the attractant. Capture efficiency, which concerns the number of attracted insects that are actually caught, also needs to be taken into account. The chapter on orientation covers the concept of active space in more detail (see Cardé, this volume).

Correlation of Catch with Damage

A good correlation between catch and population density can provide the basis for the development of action thresholds for insecticides or other management interventions (Madsen et al. 1974; Riedl and Croft 1974; Knight and Hull 1989b; Damos and Savopoulou-Soultani 2010). However, some pests with many hosts may not have good correlations between catch and subsequent crop damage because of populations from alternative hosts outside the crop. In this situation, it is important to identify the source of the trapped insects, because this information may lead to other tactics for management (Rogers et al. 2003; Trematerra 2012).

Trapping Exotic Species for Detection and Delimitation

Merely having a lure can increase the chance of a successful eradication more than 20-fold (Tobin et al. 2014). Traps baited with pheromones (i.e., synthetic pheromones or female moths) or (rarely) other attractants can be highly effective for the detection of incipient populations of exotic species, potentially well before any pest damage or other population stage becomes apparent. The chances of successful intervention are greatest while the distribution of an invader is still limited and amenable to area-wide control. Early detection is an essential prerequisite for successful eradication, and pheromone trapping has a proven track record in eradication and response (Brockerhoff et al. 2010; Tobin et al. 2014; Kean et al. 2016). Pheromone-baited traps have been used in eradication programs with species of moths in 10 families (Table 22.2). Once an invasive species has been detected, pheromone-baited traps are generally used to delimit the affected area and to monitor population trends at the same time.

To determine the rate of spread of invasive species such as gypsy moth, traps have been the principal means for detection, delimitation, and population estimation. The Slow-the-Spread program against the European gypsy moth (with flightless females) has consistently been the biggest user of pheromone traps for many years (Sharov et al. 2002), with 90,000 traps deployed in 2010 (V. Mastro, pers. comm.). The large, comprehensive, multiagency containment and exclusion program uses traps to monitor the expansion of the European strain from New England across many other states to the south and west. The current distribution covers 860,000 km² in North America (Bigsby et al. 2011). Traps have provided valuable information to managers, allowing for the accurate targeting of aerial treatments of sex pheromone to disrupt mate finding or insecticides to slow the spread of this moth through North America. Trap catch data have been used in a variety of ways, including to help identify the mode of spread (Bigsby et al. 2011). A further 200,000 traps are deployed annually throughout un-infested areas of the United States to detect incipient populations arising from human transportation. This early detection system has enabled treatment of small areas while avoiding significant disruption or inconvenience to the public. Extensive detection programs also target the Asian gypsy moth, a defoliator mainly of oaks, and other broadleaved trees with flight-capable females. Both types of gypsy moth are captured with the same traps, and distinguished by genetic methods. Similar detection programs aimed at gypsy moth are carried out in New Zealand (Ross 2005), Aus-

TABLE 22.2

Species of 21 moths from 10 families that have been trapped using synthetic pheromones as part of incursion responses for eradication or containment (Kean et al. 2016)

Species	Common name	Family	Number of incursion responses
Ostrinia nubilalis	European corn borer	Crambidae	1
Hyphantria cunea	Fall webworm	Erebidae, Arctiinae	1
Lymantria dispar dispar	European gypsy moth	Erebidae, Lymantriinae	73
Lymantria dispar asiatica	Asian gypsy moth	Erebidae, Lymantriinae	18
Euproctis chrysorrhoea	Brown-tail moth	Erebidae, Lymantriinae	2
Lymantria umbrosa	Hokkaido gypsy moth	Erebidae, Lymantriinae	1
Tuta absoluta	Tomato leafminer	Gelechiidae	2
Pectinophora gossypiella	Pink bollworm	Gelechiidae	1
Dendrolimus pini	Pine tree lappet moth	Lasiocampidae	1
Helicoverpa armigera	Corn earworm	Noctuidae	2
Spodoptera litura	Tropical armyworm	Noctuidae	1
Uraba lugens	Gum-leaf skeletonizer	Nolidae	1
Thaumetopoea pityocampa	Pine processionary moth	Notodontidae	2
Thaumetopoea processionea	Oak processionary moth	Notodontidae	2
Duponchelia fovealis	European pepper moth	Pyralidae	3
Cactoblastis cactorum	Cactus moth	Pyralidae	2
Opogona sacchari	Banana moth	Tineidae	1
Cydia pomonella	Codling moth	Tortricidae	9
Epiphyas postvittana	Light-brown apple moth	Tortricidae	4
Grapholita molesta	Oriental fruit moth	Tortricidae	1
Lobesia botrana	European grapevine moth	Tortricidae	1

tralia (Anonymous 2009), and Canada (Régnière et al. 2009). Other species have been investigated also (Elkinton et al. 2010).

Monitoring with pheromone traps has been essential for decision support in the area-wide eradication campaign against pink bollworm across several states. It has enabled evaluation of program success in real time and between seasons (Henneberry 2007; Henneberry and Naranjo 2008). The sterile insect release program has been operating in the United States since the late 1960s (Henneberry and Naranjo 2008), and today sterile insect release is integrated with mating disruption and Bt cotton where it provides additional value from managing the risk of resistance evolving (Tabashnik et al. 2010). However, there is a problem of separating marked released sterile insects from wild insects in pheromone traps. The ability to separate these males is indispensable for determining the performance of factory-reared sterile moths, and for estimating ratios of sterile to wild moths. A solution was recently demonstrated in the field with transgenic insects (Simmons et al. 2011) that were genetically modified to contain a highly visible fluorescent protein marker. Trapped insects can be quickly scanned for calco red

dye and identified as wild or sterile, including F_1 (wild × sterile) males with inherited sterility (low irradiation dose produces some viable F_1 adults but no F_2 offspring). The F_1 males had no dye but were identifiable by an inserted fluorescent gene construct, while wild insects lack any markers.

The Canadian codling moth program similarly has used pheromone traps to provide evidence of the successful reduction of populations (Bloem et al. 2007), as well as to assess the quality and performance of sterile insects (Judd et al. 2011). The program operates with a combination of lower risk insecticides, mating disruption, and sterile insects to suppress the pest, and uses a web-based geographical information systems approach (www.oksir.org). The use of traps for showing progress with eradication of codling moth was also used in Brazil, where host tree removal was a key operational feature, but progress toward the goal was monitored using traps (Kovaleski and Mumford 2007).

Delimitation is also conducted using pheromone traps after the detection of an incursion by the public or a trapping program. After a single male Hokkaido gypsy moth was caught in a pheromone trap in Hamilton, New Zealand, in March 2003

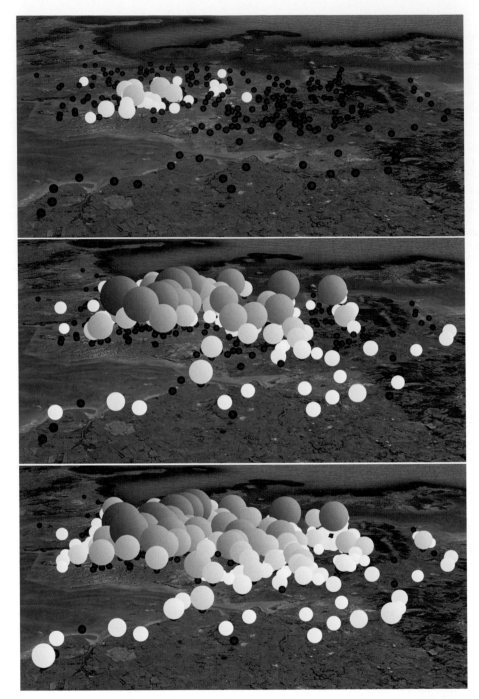

FIGURE 22.2 Catch of gum leaf skeletonizer, *Uraba lugens* (Nolidae), as it spread out of the trapping grid over three successive generations in the city of Auckland, New Zealand, in 2003–2004 (top to bottom). Circle size is proportional to catch but the smallest red circles denote traps that caught no males (after Suckling et al. 2005b). A Google Earth method enables file sharing of such data within programs (Suckling et al. 2014).

(Armstrong and Ball 2005), aerial spraying of *Bacillus thuringiensis kurstaki* (Btk) commenced in October 2003 and was completed after eight treatments in November 2003 (Ross 2005). A total of 1700 pheromone traps at 50 m spacings in the area where the first moth was trapped failed to capture any more males. In May 2005, the Technical Advisory Group concluded that gypsy moth could be declared eradicated from Hamilton (or at least not present), based on no sign of activity for two generations after aerial treatment with Btk. Pheromone trapping for the fall webworm *Hyphantria cunea* (Erebidae, Arctiinae) (El-Sayed et al. 2005a) also supported a similar eradication in Auckland, New Zealand, using ground application of Btk.

A trapping program for gum leaf skeletonizer, *Uraba lugens* (Nolidae), used the newly identified pheromone (Gibb et al. 2008) to delimit the population in the city of Auckland, New Zealand, but unfortunately the rate of spread of the insect exceeded the limits of a grid of 1500 traps placed at 500-m spacing, and, in three generations (figure 22.2), it successfully established (Suckling et al. 2005b). The pheromone was also used to sample the insect in Tasmania, in order to delimit the local population, develop parameters, and derive a model of its potential eventual distribution in New Zealand (Kriticos et al. 2007).

The 2008 discovery in California of a single male false codling moth *Thaumatotibia leucotreta* (Tortricidae) in one of 911 pheromone traps was followed by no further detections, with the conclusion that it had failed to establish (Gilligan et al. 2011). The recommendation of setting 36 traps baited with pheromone per 2.59 km² (700 m spacing) at or near the original infestation site and in the adjacent 20.72 km² was accompanied by the need for visual inspections of hosts as a backup (USDA 2010).

For the goal of eradication, the number and density of traps for delimiting and monitoring the population are likely to be much greater than for detection trapping. If there is no known long-range attractant, it may be possible to use caged female moths for this purpose. This was done during the eradication of the white spotted tussock moth *Orgyia thyellina* (Erebidae, Lymantriinae) (Hosking et al. 2003) and painted apple moth (*Teia anortoides*) (Erebidae, Lymantriinae) in New Zealand (Suckling et al. 2005), and initially in a slow-the-spread program for cactus moth, *Cactoblastis cactorum* (Pyralidae) in the United States, until the pheromone was identified (Heath et al. 2006).

Epiphyas postvittana, the light-brown apple moth (Tortricidae), a native to Australia (Geier and Briese 1981; Fowler et al. 2009), was identified from a black light collection in California in 2007 (Brown et al. 2010). Because it was determined to be a high-risk pest by the California Department of Agriculture after it was discovered (Johnson et al. 2007), and presented a market access barrier for fruit and nursery exports to noninfested countries and other parts of the United States, a major ($74M USD) response from the federal and state levels was triggered (Suckling and Brockerhoff 2010), with more than 50,000 pheromone traps deployed. Surprisingly, in 2005 trapping was conducted without any detection, but by April 2007 it was surprisingly widespread in areas near Santa Cruz, California (Suckling et al. 2014). The trapping program like the moth population rapidly expanded, with more than 100 staff involved in pheromone trap operations across the state at one time.

More recently, detection of European grape vine moth *Lobesia botrana* (Tortricidae) in northern California vineyards triggered the deployment of 248 pheromone traps in 2009, insecticides and mating disruption, with quarantine restrictions on plant movement (Varela et al. 2010). A trapping program has been deployed to determine and manage the quarantine areas.

There are some other US examples of trapping efforts that have ranged from moderate sized (e.g., winter moth *Operophtera brumata* [Geometridae] in New England, leek moth *Acrolepiopsis assectella* [Acrolepiidae] in New York) to very large geographically (*Duponchelia fovealis*) (Crambidae), but these trapping programs have only delineated the insects' range in very broad terms and were not used in formulating a regulatory response (D. R. Lance, pers. comm.).

Problems with Traps and Trapping Programs

The advent of convenient trapping systems has almost certainly led to a reduction in use of alternative more costly sampling systems for other life stages, such as eggs or larvae. This reliance on pheromone traps carries the risk that the trap capture data may be misleading, for various reasons including nonlinearity or an unknown relationship between catch and actual wild population size.

For detection programs, there is a risk of complacency in the assumption that zero catch indicates the absence of a pest. In the case of painted apple moth, recapture of sterile males in female-baited traps placed at 500-m spacing averaged 5%. This information enabled the development of a model to determine the probability of zero catch, indicating successful eradication (Kean and Suckling 2005). This approach has also been used with fruit flies (Lance and Gates 1994) and tsetse fly (Barclay and Hargrove 2005), but is not widespread. To interpret the meaning of null capture, some knowledge of the "detectability" of the species must be available. Otherwise, there is a risk of declaring successful eradication when a breeding population remains.

One issue is the downwind projection of active space of traps and the spacing between traps in a trapping grid, because traps placed at considerable distances apart may fail to catch any insects until either the population increases or individuals disperse close enough to traps to be captured. This is another reason why an understanding of the concept of trap active space is important. Sampling effort is a compromise based on effort and the cost-benefit ratio, where it is critical to fully understand the objectives. For eradication, where the stakes can be high and there is uncertainty in trap grid efficacy, it may be appropriate to continue trapping for several generations or even years after the last capture. Several interacting factors can cause trap capture to tail off with increasing inter-trap distance, including how the pheromone disperses from the trap (which can be a function of landscape, surface roughness, and wind speed), and how well the moth can sense and respond to the plume.

Sometimes, synthetic pheromones fail to meet expectations for trapping. This was the case with the painted apple moth, because the compounds were too unstable, with a half-life estimated at about 12 hours (El-Sayed et al. 2005b). Lures stored for long periods can become unattractive if behaviorally active contaminants, such as inhibitory geometric isomers, are present as a result of chemical breakdown. Detection programs for exotic species should be supported by careful quality assurance of lure efficacy, ideally involving some form of behavioral bioassay or field test.

Traps for some species catch related nontarget species (Weber and Ferro 1991), which then require separation and identification. This is particularly a problem for biosecurity trapping in locations where congeneric insects occur. This problem has prevented the use of multiple species trapping through the combination of lures in traps. In addition, some combinations were inhibitory, which creates the risk of reduced surveillance efficacy (Schwalbe and Mastro 1988). The potential of simultaneously monitoring multiple insect threats is being examined for New Zealand forests and other ecosystems. No interference in catch resulted when lures for up to four species were added to the same trap (Brockerhoff et al. 2012).

The simultaneous application of pheromone for mating disruption and surveillance may not be compatible. Ironically, although the human psyche appears to favor positive results, in this case empty traps (i.e., negative results) indicate successful application of the technology and progress toward eradication or at least suppression. Similarly, growers using traps in pheromone-based IPM may become lax in checking traps when catches are low. This points to the need for alternative sampling systems, such as the female moth attractants developed for codling moth (Light et al. 2001; Landolt et al. 2007). Sampling using a lure that is independent of the disruptant blend offers a solution, since monitoring can continue unaffected.

Other Applications of Monitoring Using Pheromones

Pheromones and pheromone traps have been developed for various applications in IPM and biosecurity for a range of pest species (e.g., Table 22.2), including assessment of insecticide resistance in light-brown apple moth (Suckling et al. 1985), codling moth (Riedl et al. 1985), pink bollworm (Haynes et al. 1987), tufted apple bud moth *Platynota idaeusalis* (Tortricidae) (Knight and Hull 1989a), and leaf miners *Phyllonorycter elmaella* (Gracillariidae) (Shearer and Riedl 1994). Uses of traps in IPM and to support mating disruption have been well canvased (Witzgall et al. 2010). Presence of regular catches in pheromone traps can indicate ineffective application of mating disruption by growers (Suckling and Shaw 1995). This is a technology which can offer resistance management at the same time (Suckling et al. 1990).

There is also potential for use of pheromones in biocontrol of weeds, as was demonstrated for gorse pod moth *Cydia succedana* (Tortricidae) (Suckling et al. 1999), gorse shoot moth *Agonopterix ulicetella* (Oecophoridae) (Suckling et al. 2000), *Xubida infusella* (Pyralidae) (Stanley et al. 2000), *Croesia zimmermani* (Tortricidae), and *Schreckensteinia festaliella* (Heliodinidae) (Suckling et al. 2006). Traps can be used to investigate the establishment, dispersal, and distribution of insects after release, as well as phenological synchrony of host and herbivore (Hill and Gourlay 2002). Similarly, pheromones can be used in arthropod biological control for comparing phenology and distribution of the natural enemy with the pest, as shown with simultaneous pheromone trapping for codling moth and an egg parasitoid (Suckling et al. 2002).

New technologies are likely to emerge around pheromones and trapping systems. Automatic traps that time stamp and count the moth catch have been tested (Kondo et al. 1994; Schouest and Miller 1994), and can offer information about conditions suitable for male flight. One interesting development concerns the use of cameras from cellular telephones with traps (Guarnieri et al. 2011). This type of technology stands to have a high impact offering real-time monitoring for the first time, and there are many developments underway. Although this technology has been proposed for use in IPM, it seems likely that its use may be most economically feasible for detection of high-impact exotic pests, where the first detection of a new moth incursion would be the easiest and most valuable task. Together with multiple species detection trapping, these developments can greatly enhance surveillance.

Summary

Pheromones continue to be identified and developed into trapping systems for a wide range of species of pest management and biosecurity concern. The number of applications for pheromones continues to expand and technology is being added to trapping systems to reduce labor costs. Greater knowledge about pheromone interactions between species would be useful, since this could expand the simultaneous use of lures in traps for multiple species. The role of pheromones in large surveillance trapping programs appears to be expanding, for detecting or delimiting invasive species, beyond the exemplar case of gypsy moth in the United States.

Such programs normally use centralized collation of trapping information to look at patterns of spread. In IPM, the number of traps used for decision support is very large, with many products and suppliers of traps and lures.

Some niche uses, such as the use of pheromones in classical weed biological control, where they can help to track newly released natural enemies and measure synchrony with the target weed, have been demonstrated. There remain important gaps in knowledge for many species, especially relating to the estimation of the wild population size from the number of moths captured, and the relationship between catch and damage. More work is needed on trap-active space and trap efficiency in order to better interpret catches and improve forecasting of population growth and potential for damage. The continued discovery and development of IPM applications for moth pheromones seems assured.

Acknowledgments

This chapter was funded by the New Zealand government (Better Border Biosecurity, www.b3nz.org) and completed during a fellowship supported by the Organisation for Economic Cooperation and Development (Cooperative Research Programme: Biological Resource Management for Sustainable Agricultural Systems) at the European Biological Control Laboratory (USDA-ARS), Montpellier. Dave Lance (USDA APHIS) and Garry Hill (NZIPFR) provided helpful comments on the manuscript. Tom Sullivan kindly prepared the Google Earth file.

References Cited

Adams, C.J., C.A. Beasley, and T.J. Henneberry. 1995. Effects of temperature and wind speed on pink bollworm (Lepidoptera: Gelechiidae) moth captures during spring emergence. *Journal of Economic Entomology* 88:1263–1270.

Alford, D.V., P.W. Carden, E.B. Dennis, J.H. Gould, and J.D.R. Vernon. 1979. Monitoring codling and tortrix moths in United Kingdom apple orchards using pheromone traps. *Annals of Applied Biology* 91:165–178.

Anonymous. 2009. Field guide to exotic pests and diseases: Asian gypsy moth. Department of Agriculture, Fisheries and Forestry 2009. Available online at: http://www.daff.gov.au/aqis/quarantine/pests-diseases/forests-timber/asian-gypsy-moth.

Armstrong, K.F., and S.L. Ball. 2005. DNA barcodes for biosecurity: invasive species identification. *Philosophical Transactions of the Royal Society B: Biological Sciences* 360:1813–1823.

Bailey, P. 1980. Oriental fruit moth in South Australian peach orchards: monitoring moth activity and abundance and estimating first egg hatch. *Zeitschrift für Angewandte Entomologie* 89:377–386.

Baker, T.C., and W.L. Roelofs. 1981. Initiation and termination of oriental fruit moth male response to pheromone concentrations in the field. *Environmental Entomology* 10:211–218.

Barclay, H.J., and J.W. Hargrove. 2005. Probability models to facilitate a declaration of pest-free status, with special reference to tsetse (Diptera: Glossinidae). *Bulletin of Entomological Research* 95:1–11.

Bierl, B.A., M. Beroza, and C.W. Collier. 1972. Isolation, identification, and synthesis of the gypsy moth sex attractant. *Journal of Economic Entomology* 65:659–664.

Bigsby, K.M., P.C. Tobin, and E.O. Sills. 2011. Anthropogenic drivers of gypsy moth spread. *Biological Invasions* 13:2077–2090.

Bloem, S., A. McCluskey, R. Fugger, S. Arthur, S. Wood, and J. Carpenter. 2007. Suppression of the codling moth *Cydia pomonella* in British Columbia, Canada Using an area-wide integrated approach with an SIT component. Pp. 591–601. In M.J.B. Vreysen, A.S. Robinson, and J. Hendrichs, eds. *Area-Wide Control of Insect*

Pests: from Research to Field Implementation. Dordrecht, The Netherlands: Springer.

Bohnenblust, E., L.A. Hull, G. Krawczyk, and N.K. Joshi. 2011. Capture of two moth species in traps placed at different heights in sex pheromone treated apple orchards. *Journal of Entomological Science* 46:223–231.

Brockerhoff, E.B., A. Liebhold, B. Richardson, and D.M. Suckling. 2010. Eradication of invasive forest insects: concept, methods, costs and benefits. *New Zealand Journal of Forestry* 40:S117–S135.

Brockerhoff, E.G., D.M. Suckling, A. Roques, H. Jactel, M. Brancoe, A.M. Twidle, V. Mastro, and M.O. Kimberley. 2012. Improving the efficiency of lepidopteran pest detection and surveillance: constraints and opportunities for multiple-species trapping. *Journal of Chemical Ecology* 39:50–58.

Brown, J.W., M.E. Epstein, T.M. Gilligan, S.C. Passoa, and J.A. Powell. 2010. Biology, identification and history of the light-brown apple moth *Epiphyas postvittana* (Walker) (Lepidoptera: Tortricidae: Archipini) in California: an example of the importance of local faunal surveys to document the establishment of exotic insects. *American Entomologist* 56:34–43.

Byers, J. 2008. Active space of pheromone plume and its relationship to effective attraction radius in applied models. *Journal of Chemical Ecology* 34:1134–1145.

Damos, P., and M. Savopoulou-Soultani. 2010. Population dynamics of *Anarsia lineatella* and their relation to crop damage in Northern Greece IPM peach orchards: towards the development of EIL. *IOBC/WPRS Bulletin* 54:31–32.

Daterman, G.E., J.M. Wenz, and K.A. Sheehan. 2004. Early warning system for Douglas-fir tussock moth outbreaks in the western United States. *Western Journal of Applied Forestry* 19:232–241.

De Lame, F.M., and L.J. Gut. 2006. Effect of monitoring trap and mating disruption dispenser application heights on captures of male *Grapholita molesta* (Busck; Lepidoptera: Tortricidae) in pheromone and virgin female-baited traps. *Environmental Entomology* 35:1058–1068.

Dhawan, A.K., and A.S. Sidhu. 1984. Assessment of capture threshold of pink-bollworm moths for timing insecticidal applications on *Gossypium hirsutum* Linn. *Indian Journal of Agricultural Sciences* 54:426–433.

Drapek, R.J., L.B. Coop, B.A. Croft, and G.C. Fisher. 1990. Studies of trap and lure combinations and field placement in sweet corn. *Southwestern Entomologist* 15:63–69.

El-Sayed, A.M. 2016. The Pherobase: Database of insect pheromones and semiochemicals. Available online at: www.pherobase.com/ (accessed April 26, 2016).

El-Sayed, A.M., A.R. Gibb, and D.M. Suckling. 2005a. Chemistry of the sex pheromone gland of the fall webworm, *Hyphantria cunea*, discovered in New Zealand. *New Zealand Plant Protection* 58:31–36.

El-Sayed, A.M., A.R. Gibb, D.M. Suckling, B. Bunn, S. Fielder, D. Comeskey, L.A. Manning et al. 2005b. Identification of sex pheromone components of the painted apple moth: a tussock moth with a thermally labile pheromone component. *Journal of Chemical Ecology* 31:621–646.

Elkinton, J.S. 1987. Changes in efficiency of the pheromone-baited milk carton trap as it fills with male gypsy moths (Lepidoptera: Lymantriidae). *Journal of Economic Entomology* 80:754–757.

Elkinton, J.S., and R.T. Cardé. 1980. Distribution, dispersal, and apparent survival of male gypsy moths as determined by capture in pheromone-baited traps. *Environmental Entomology* 9:729–737.

Elkinton, J.S., G.H. Boettner, M. Sremac, R. Gwiazdowski, R.R. Hunkins, J. Callahan, S.B. Scheufele et al. 2010. Survey for winter moth (Lepidoptera: Geometridae) in Northeastern North America with pheromone-baited traps and hybridization with the native bruce spanworm (Lepidoptera: Geometridae). *Annals of the Entomological Society of America* 103:135–145.

Foster, R., R. Staten, and E. Miller. 1977. Evaluation of traps for pink bollworm. *Journal of Economic Entomology* 70:289–291.

Foster, S.P., and S.J. Muggleston. 1993. Effect of design of a sex-pheromone-baited delta trap on behavior and catch of male *Epiphyas postvittana* (Walker). *Journal of Chemical Ecology* 19:2617–2633.

Fowler, G., L. Garrett, A. Neeley, R. Magarey, D. Borchert, and B. Spears. 2009. *Economic Analysis: Risk to U.S. Apple, Grape, Orange and Pear Production from the Light Brown Apple Moth,* Epiphyas postvittana *(Walker).* Raleigh, NC: USDA-APHIS. Available online at: http://www.aphis.usda.gov/plant_health/plant_pest_info /lba_moth/downloads/lbameconomicanalysis.pdf (accessed January 28, 2014).

Gargiullo, P.M., W.C. Berisford, and J.F. Godbee. 1985. Prediction of optimal timing for chemical control of the Nantucket pine tip moth, *Rhyacionia frustrana* (Comstock) (Lepidoptera: Tortricidae), in the Southeastern Coastal Plain. *Journal of Economic Entomology* 78:148–154.

Geier, P., and D. Briese. 1981. The light-brown apple moth, *Epiphyas postvittana* (Walker): a native leafroller fostered by European settlement. Pp. 131–155. In R. Kitching and R. Jones, eds. *The Ecology of Pests.* Melbourne, Australia: CSIRO.

Gibb, A.R., L.E. Jamieson, D.M. Suckling, P. Ramankutty, and P.S. Stevens. 2005. Sex pheromone of the citrus flower moth *Prays nephelomima*: pheromone identification, field trapping trials, and phenology. *Journal of Chemical Ecology* 31:1633–1644.

Gibb, A.R., D.M. Suckling, S. Fielder, B. Bunn, L.E. Jamieson, M.L. Larsen, G.H. Walter, and D.J. Kriticos. 2008. Major sex pheromone components of the Australian gum leaf skeletonizer *Uraba lugens*: (10E,12Z)-hexadecadien-1-yl acetate and (10E,12Z)-hexadecadien-1-ol. *Journal of Chemical Ecology* 34:1125–1133.

Gilligan, T.M., M.E. Epstein, and K.M. Hoffman. 2011. Discovery of false codling moth, *Thaumatotibia leucotreta* (Meyrick), in California (Lepidoptera: Tortricidae). *Proceedings of the Entomological Society of Washington* 113:426–435.

Guarnieri, A., S. Maini, G. Molari, and V. Rondelli. 2011. Automatic trap for moth detection in integrated pest management. *Bulletin of Insectology* 64:247–251.

Haynes, K.F., T.A. Miller, R.T. Staten, W.G. Li, and T.C. Baker. 1987. Pheromone trap for monitoring insecticide resistance in the pink bollworm moth (Lepidoptera: Gelechiidae): new tool for resistance management. *Environmental Entomology* 16:84–89.

Heath, R.R., P.E.A. Teal, J.H. Tumlinson, and L.J. Mengelkoch. 1986. Prediction of release ratios of multicomponent pheromones from rubber septa. *Journal of Chemical Ecology* 12:2133–2143.

Heath, R.R., P.E.A. Teal, N.D. Epsky, B.D. Dueben, S.D. Hight, S. Bloem, J.E. Carpenter et al. 2006. Pheromone-based attractant-for males of *Cactoblastis cactorum* (Lepidoptera: Pyralidae). *Environmental Entomology* 35:1469–1476.

Henneberry, T.J. 2007. Integrated systems for control of the pink bollworm *Pectinophora gossypiella* in cotton. Pp. 567–579. In M.J.B. Vreysen, A.S. Robinson, and J. Hendrichs, eds. *Area-Wide Control of Insect Pests: From Research to Field Implementation.* Dordrecht, The Neatherlands: Springer.

Henneberry, T.J., and S.E. Naranjo. 2008. Integrated management approaches for pink bollworm in the southwestern United States. *Integrated Pest Management Reviews* 3:31–52.

Higbee, B.S., and C.S. Burks. 2008. Effects of mating disruption treatments on navel orangeworm (Lepidoptera: Pyralidae) sexual communication and damage in almonds and pistachios. *Journal of Economic Entomology* 101:1633–1642.

Hill, R.L., and A.H. Gourlay. 2002. The introduction and establishment of *Cydia succedana* as a biological control agent for gorse (*Ulex europaeus*) in New Zealand. *Biological Control* 25:173–186.

Hoffmann, M.P., L.T. Wilson, F.G. Zalom, and L. McDonough. 1986. Lures and traps for monitoring tomato fruitworm. *California Agriculture* 40:17–18.

Hosking, G., J. Clearwater, J. Handiside, M. Kay, J. Ray, and N. Simmons. 2003. Tussock moth eradication: a success story from New Zealand. *International Journal of Pest Management* 49:17–24.

Houseweart, M.W., D.T. Jennings, and C.J. Sanders. 1981. Variables associated with pheromone traps for monitoring spruce budworm populations (Lepidoptera: Tortricidae). *The Canadian Entomologist* 113:527–537.

Howse, P.E., I.D.R. Stevens, and O.T. Jones. 1998. *Insect Pheromones and Their Use in Pest Management.* London: Chapman & Hall.

Jamieson, L.E., D.M. Suckling, and P. Ramankutty. 2008. Mass trapping of *Prays nephelomima* (Lepidoptera: Yponomeutidae) in citrus orchards: optimizing trap design and density. *Journal of Economic Entomology* 101:1295–1301.

Johnson, M.W., C. Pickel, L.L. Strand, L.G. Varela, C.A. Wilen, M.P. Bolda, M.L. Flint, W.K. Frankie Lam, and F.G. Zalom. 2007. Light brown apple moth in California: quarantine, management, and potential impacts. UC Statewide Integrated Pest Management Program. Oakland: University of California Agriculture and Natural Resources. Available online at: http://www.aphis.usda

.gov/plant_health/plant_pest_info/lba_moth/downloads
/LBAM_IPM_UCDavis.pdf (accessed October 20, 2011).

Judd, G.J.R., S. Arthur, K. Deglow, and M.G.T. Gardiner. 2011. Operational mark-releaser-recapture field tests comparing competitiveness of wild and differentially mass-reared codling moths from the Okanagan–Kootenay sterile insect program. *The Canadian Entomologist* 143:300–316.

Kean, J.M., and D.M. Suckling. 2005. Estimating the probability of eradication of painted apple moth from Auckland. *New Zealand Plant Protection* 58:7–11.

Kean, J.M., D.M. Suckling, N.J. Sullivan, P.C. Tobin, D.C. Lee, L.D. Stringer, R. Flores Vargas et al. 2016. Global eradication and response database. Available online at: http://b3.net.nz/gerda /index.php (accessed May 26, 2016)

Knight, A.L., and L.A. Hull. 1989a. Use of sex pheromone traps to monitor azinphosmethyl resistance in tufted apple bud moth (Lepidoptera: Tortricidae). *Journal of Economic Entomology* 82:1019–1026.

Knight, A.L., and L.A. Hull. 1989b. Predicting seasonal apple injury by tufted apple bud moth (Lepidoptera: Tortricidae) with early-season sex pheromone trap catches and brood I fruit injury. *Environmental Entomology* 18:939–944.

Knight, A.L., and D.M. Light. 2005. Timing of egg hatch by early-season codling moth (Lepidoptera: Tortricidae) predicted by moth catch in pear ester- and codlemone-baited traps. *The Canadian Entomologist* 137:728–738.

Knutson, A.E., and M.A. Muegge. 2010. A degree-day model initiated by pheromone trap captures for managing pecan nut casebearer (Lepidoptera: Pyralidae) in Pecans. *Journal of Economic Entomology* 103:735–743.

Kondo, A., and F. Tanaka. 1991. Pheromone trap catches of the rice stem borer moth, *Chilo suppressalis* (Walker) (Lepidoptera Pyralidae) and related trap variables in the field. *Applied Entomology and Zoology* 26:167–172.

Kondo, A., T. Sano, and F. Tanaka. 1994. Automatic record using camera of diel periodicity of pheromone trap catches. *Japanese Journal of Applied Entomology and Zoology* 38:197–199.

Kovaleski, A., and J. Mumford. 2007. Pulling out the evil by the root: the codling moth *Cydia pomonella* eradication programme in Brazil. Pp. 581–590. In M.J.B. Vreysen, A.S. Robinson, and J. Hendrichs, eds. *Area-Wide Control of Insect Pests: From Research to Field Implementation.* Dordrecht, The Netherlands: Springer.

Kovanci, O.B., C. Schal, J.F. Walgenbach, and G.G. Kennedy. 2006. Effects of pheromone loading, dispenser age, and trap height on pheromone trap catches of the oriental fruit moth in apple orchards. *Phytoparasitica* 34:252–260.

Kriticos, D.J., K.J.B. Potter, N.S. Alexander, A.R. Gibb, and D.M. Suckling. 2007. Using a pheromone lure survey to establish the native and potential distribution of an invasive Lepidopteran, *Uraba lugens. Journal of Applied Ecology* 44:853–863.

Lance, D.R., and D.B. Gates. 1994. Sensitivity of detection trapping systems for Mediterranean fruit flies (Diptera: Tephritidae) in Southern California. *Journal of Economic Entomology* 87:1377–1383.

Landolt, P.J., D.M. Suckling, and G.J.R. Judd. 2007. Positive interaction of a feeding attractant and a host kairomone for trapping the codling moth, *Cydia pomonella* (L.). *Journal of Chemical Ecology* 33:2236–2244.

Light, D.M., A.L. Knight, C.A. Henrick, D. Rajapaska, B. Lingren, J.C. Dickens, K.M. Reynolds et al. 2001. A pear-derived kairomone with pheromonal potency that attracts male and female codling moth, *Cydia pomonella* (L.). *Naturwissenschaften* 88:333–338.

Lingren, P., J. Burton, W. Shelton, and J. Raulston. 1980. Night vision goggles: for design, evaluation, and comparative efficiency determination of a pheromone trap for capturing live adult male pink bollworms. *Journal of Economic Entomology* 73:622–630.

Linn, C.E., M.G. Campbell, and W.L. Roelofs. 1987. Pheromone components and active spaces—what do moths smell and where do they smell it? *Science* 237:650–652.

Macaulay, E.D.M. 1977. Field trials with attractant traps for timing sprays to control pea moth. *Plant Pathology* 26:179–188.

Macaulay, E.D.M., P. Etheridge, D.G. Garthwaite, A.R. Greenway, C. Wall, and R.E. Goodchild. 1985. Prediction of optimum spraying dates against pea moth, *Cydia nigricana* (F.), using pheromone traps and temperature measurements. *Crop Protection* 4:85–98.

Madsen, H.S., A.C. Myburgh, D.J. Rust, and I.P. Bosman. 1974. Codling moth (Lepidoptera: Olethreutidae): correlation of male

sex attractant trap captures and injured fruit in South African apple and pear orchards. *Phytophylactica* 6:185–187.

McDonough, L.M., D.F. Brown, and W.C. Aller. 1989. Insect sex pheromones: effect of temperature on evaporation rates of acetates from rubber septa. *Journal of Chemical Ecology* 15:779–790.

McLaren, G.F., J.A. Fraser, and D.M. Suckling. 1998. Mating disruption for the control of leafrollers on apricots. *New Zealand Journal of Crop and Horticultural Science* 26:259–268.

Minks, A.K., P. Vandeventer, J. Woets, and E. Vanremortel. 1995. Development of thresholds based on pheromone trap catches for control of leafroller moths in apple orchards: a first report. *Proceedings of the Section Experimental and Applied Entomology of the Netherlands Entomological Society* 6:125–132.

Mullen, M.A., H.A. Highland, and F.H. Arthur. 1991. Efficiency and longevity of two commercial sex-pheromone lures for Indian meal moth and almond moth (Lepidoptera Pyralidae). *Journal of Entomological Science* 26:64–68.

Nakasuji, F., and K. Kiritani. 1978. Estimating the control threshold density of the tobacco cutworm *Spodoptera litura* (Lepidoptera: Noctuidae) on a corm crop, taro, by means of pheromone traps. *Protection Ecology* 1:23–32.

Pringle, K.L., D.K. Eyles, and L. Brown. 2003. Trends in codling moth activity in apple orchards under mating disruption using pheromones in the Elgin area, Western Cape Province, South Africa. *African Entomology* 11:65–75.

Ravlin, F.W. 1991. Development of monitoring and decision-support systems for integrated pest management of forest defoliators in North America. *Forest Ecology and Management* 39:3–13.

Régnière, J., V. Nealis, and K. Porter. 2009. Climate suitability and management of the gypsy moth invasion into Canada. *Biological Invasions* 11:135–148.

Riedl, H., and B.A. Croft. 1974. A study of pheromone trap catches in relation to codling moth (Lepidoptera: Olethreutidae) damage. *The Canadian Entomologist* 106:525–537.

Riedl, H., B.A. Croft, and A.J. Howitt. 1976. Forecasting codling moth phenology based on pheromone trap catches and physiological-time models. *The Canadian Entomologist* 108:449–460.

Riedl, H., A. Seaman, and F. Henrie. 1985. Monitoring susceptibility to azinphosmethyl in field populations of the codling moth (Lepidoptera: Tortricidae) with pheromone traps. *Journal of Economic Entomology* 78:692–699.

Rock, G.C., C.C. Childers, and H.J. Kirk. 1978. Insecticide applications based on Codlemone trap catches vs. automatic schedule treatments for codling moth control in North Carolina apple orchards. *Journal of Economic Entomology* 71:650–653.

Roelofs, W.L. 1978. Threshold hypothesis for pheromone perception. *Journal of Chemical Ecology* 4:685–699.

Roelofs, W.L., and H. Arn. 1968. Sex attractant of the red-banded leafroller moth. *Nature* 219:513.

Roelofs, W.L., A. Comeau, A.M. Hill, and G. Milicevic. 1971. Sex attractant of the codling moth: characterization with electroantennogram technique. *Science* 174:297–299.

Roelofs, W.L., J. Kochansky, R.T. Cardé, H. Arn, and S. Rauscher. 1973. Sex attractant of the grape vine moth, *Lobesia botrana. Mitteilungen der Schweizerischen Entomologischen Gesellschaft* 46:71–73.

Rogers, D.J., J.T.S. Walker, I.C. Moen, F. Weibel, P.L. Lo, and L.M. Cole. 2003. Understory influence on leafroller populations in Hawke's Bay organic apple orchard. *New Zealand Plant Protection* 56:168–173.

Ross, M.G. 2005. Response to a gypsy moth incursion within New Zealand. Paper read at Proceedings of the IUFRO Working Parties D7 & D8 Conference: Forest Diversity and Resistance to Native and Exotic Pest Insects, 2004, Hanmer Springs, New Zealand.

Rumbo, E.R. 1993. Interactions between male moths of the lightbrown apple moth, *Epiphyas postvittana* (Walker) (Lep.: Tort.), landing on synthetic sex pheromone sources in a wind-tunnel. *Physiological Entomology* 18:79–86.

Sanders, C.J. 1996. Pheromone traps for detecting incipient outbreaks of the spruce budworm, *Choristoneura fumiferana* (Clem.). NODA/NFP Technical Report TR-32.

Schouest, L., Jr., and T. Miller. 1994. Automated pheromone traps show male pink bollworm (Lepidoptera: Gelechiidae) mating response is dependent on weather conditions. *Journal of Economic Entomology* 87:965–974.

Schwalbe, C.P., and V.C. Mastro. 1988. Multispecific trapping techniques for exotic pest detection. *Agriculture, Ecosystems and Environment* 21:43–51.

Sharov, A.A., D. Leonard, A.M. Liebhold, E.A. Roberts, and W. Dickerson. 2002. "Slow the Spread": a national program to contain the gypsy moth. *Journal of Forestry* 100:30–35.

Shearer, P.W., and H. Riedl. 1994. Comparison of pheromone trap bioassays for monitoring insecticide resistance of *Phyllonorycter elmaella* (Lepidoptera: Gracillariidae). *Journal of Economic Entomology* 87:1450–1454.

Simmons, G.S., A.R. McKemey, N.I. Morrison, S. O'Connell, B.E. Tabashnik, J. Claus, G. Fu et al. 2011. Field performance of a genetically engineered strain of pink bollworm. *PLoS ONE* 6:e24110.

Solomon, M.E. 1978. Relationships between pheromone trap catches of codling moths, damage to fruit, and seasonal heat sums. Mitteilungen aus der Biologischen Bundesanstalt für Land- und Forstwirtschaft Berlin-Dahlem, Berlin, pp.36–37.

Stanley, J.N., M.H. Julien, E.R. Rumbo, A.J. White, and N.R. Spencer. 2000. Post release monitoring of *Xubida infusella* (Lep.: Pyralidae): an example of using pheromones for the early detection of establishing populations of biological control agents. Paper read at Proceedings of the X International Symposium on Biological Control of Weeds, 4–14 July 1999, Montana State University, Bozeman, Montana, USA.

Stelinski, L.L., J.R. Miller, and L.J. Gut. 2005. Captures of two leafroller moth species (Lepidoptera: Tortricidae) in traps baited with varying dosages of pheromone lures or commercial mating-disruption dispensers in untreated and pheromone-treated orchard plots. *The Canadian Entomologist* 137:98–109.

Suckling, D.M., and E.G. Brockerhoff. 2010. Invasion biology, ecology, and management of the light brown apple moth (Tortricidae). *Annual Review of Entomology* 55:285–306.

Suckling, D.M., and P.W. Shaw. 1995. Large-scale trials of mating disruption of lightbrown apple moth in Nelson, New Zealand. *New Zealand Journal of Crop and Horticultural Science* 23:127–137.

Suckling, D.M., D.R. Penman, R.B. Chapman, and C.H. Wearing. 1985. Pheromone use in insecticide resistance surveys of lightbrown apple moths (Lepidoptera: Tortricidae). *Journal of Economic Entomology* 78:204–207.

Suckling D.M., P.W. Shaw, J.G.I. Khoo, V. Cruickshank. 1990. Resistance management of lightbrown apple moth, *Epiphyas postvittana* (Lepidoptera: Tortricidae) by mating disruption. *New Zealand Journal of Crop and Horticultural Science* 18:89–98.

Suckling, D., R. Hill, A. Gourlay, and P. Witzgall. 1999. Sex attractant-based monitoring of a biological control agent of gorse. *Biocontrol Science and Technology* 9:99–104.

Suckling, D.M., A.R. Gibb, H. Gourlay, P. Conant, C. Hirayama, R. Leen, and G. Szocs. 2000. Sex attractant for the gorse biological control agent *Agonopterix ulicetella* (Oecophoridae). *New Zealand Plant Protection* 53:66–70.

Suckling, D.M., A.R. Gibb, G.M. Burnip, and N.C. Delury. 2002. Can parasitoid sex pheromones help in insect biocontrol? A case study of codling moth (Lepidoptera: Tortricidae) and its parasitoid *Ascogaster quadridentata* (Hymenoptera: Braconidae). *Environmental Entomology* 31:947–952.

Suckling, D., A. Gibb, G. Burnip, C. Snelling, J. De Ruiter, G. Langford, and A. El-Sayed. 2005a. Optimization of pheromone lure and trap characteristics for currant clearwing, *Synanthedon tipuliformis*. *Journal of Chemical Ecology* 31:393–406.

Suckling, D.M., A.R. Gibb, P.R. Dentener, D.S. Seldon, G.K. Clare, L. Jamieson, D. Baird, D.J. Kriticos, and A.M. El-Sayed. 2005b. *Uraba lugens* (Lepidoptera: Nolidae) in New Zealand: pheromone trapping for delimitation and phenology. *Journal of Economic Entomology* 98:1187–1192.

Suckling, D.M., J. Charles, D. Allan, A. Chaggan, A. Barrington, G.M. Burnip, and A.M. El-Sayed. 2005c. Performance of irradiated *Teia anartoides* (Lepidoptera: Lymantriidae) in urban Auckland, New Zealand. *Journal of Economic Entomology* 98:1531–1538.

Suckling, D.M., A.R. Gibb, T. Johnson, and D.R. Hall. 2006. Examination of sex attractants for monitoring weed biological control agents in Hawaii. *Biocontrol Science and Technology* 16:919–927.

Suckling, D.M., L.D. Stringer, D.B. Baird, R.C. Butler, T.E.S. Sullivan, D.R. Lance, and G.S. Simmons. 2014. Light brown apple moth (*Epiphyas postvittana*) (Lepidoptera: Tortricidae) colonization of California. *Biological Invasions* 16:1851–1863. doi:10.1007/s10530-013-0631-8.

Sweeney, J.D., J.A. McLean, and R.F. Shepherd. 1990. Factors affecting catch in pheromone traps for monitoring the western spruce budworm, *Choristoneura occidentalis* Freeman. *The Canadian Entomologist* 122:1119–1130.

Tabashnik, B.E., M.S. Sisterson, P.C. Ellsworth, T.J. Dennehy, L. Antilla, L. Liesner, M. Whitlow et al. 2010. Suppressing resistance to Bt cotton with sterile insect releases. *Nature Biotechnology* 28:1304–1307.

Tobin, P.C., J.M. Kean, D.M. Suckling, D.G. McCullough, D.A. Herms, and L. Stringer. 2014. Determinants of successful arthropod eradication programs. *Biological Invasions* 16:401–414. doi:10.1007/s10530-013-0529-5.

Trematerra, P. 2012. Advances in the use of pheromones for stored-product protection. *Journal of Pest Science* 85: 285–289.

USDA. 2010. *New Pest Response Guidelines False Codling Moth* Thaumatotibia leucotreta. Edited by UDSA APHIS. Riverdale, MD: Department of Agriculture, Plant Protection and Quarantine, Emergency and Domestic Programs. Available online at: http://www.aphis.usda.gov/import_export/plants/manuals/online_manuals.shtml (accessed February 13, 2012).

Varela, L.G., R.J. Smith, M.L. Cooper, and R.W. Hoenisch. 2010. European grapevine moth, *Lobesia botrana*, in Napa Valley vineyards. *Practical Winery and Vineyard* 30:1–5. Available online at: http://www.practicalwinery.com/marapr10/moth1.htm.

Walker, G.P., A.R. Wallace, R. Bush, F.H. Macdonald, and D.M. Suckling. 2003. Evaluation of pheromone trapping for prediction of diamondback moth infestations in vegetable brassicas. *New Zealand Plant Protection* 56:180–184.

Wall, C., and J.N. Perry. 1987. Range of action of moth sex-attractant sources. *Entomologia Experimentalis et Applicata* 44:5–14.

Wearing, C.H. 1995. A recommended spray programme for leafroller and codling moth control in Central Otago apple orchards. Proceedings of the Forty Eighth New Zealand Plant Protection Conference, Angus Inn, Hastings, New Zealand, August 8–10, pp. 111–116.

Wearing, C.H., and J.G. Charles. 1978. Integrated control of apple pests in New Zealand. 14. Sex pheromone traps to determine applications of azinphos-methyl for codling moth control. Proceedings of the Thirty-First New Zealand Weed and Pest Control Conference, Devon Motor Lodge, New Plymouth, August 8–10 , pp. 229–235.

Weatherston, J., L.M. Davidson, and D. Simonini. 1974. Attractants for several male forest Lepidoptera. *The Canadian Entomologist* 106:781–782.

Weber, D.C., and D.N. Ferro. 1991. Nontarget noctuids complicate integrated pest management monitoring of sweet corn with pheromone traps in Massachusetts. *Journal of Economic Entomology* 84:1364–1369.

Witzgall, P., P. Kirsch, and A. Cork. 2010. Sex pheromones and their impact on pest management. *Journal of Chemical Ecology* 36:80–100.

Pheromones as Management Tools

Mass Trapping and Lure-and-Kill

ALAN CORK

BACKGROUND

Pheromone blends and controlled release formulations

Trapping systems and killing agents

Spacing between point sources

Method and timing of application, duration of control period

CURRENT STATUS OF ATTRACT-AND-CONTROL SYSTEMS

Cotton and oilseed pests

Vegetable pests

Fruit tree pests

Sugarcane pests

Rice pests

Storage pests

DISCUSSION

REFERENCES CITED

Background

To the uninitiated, terms such as mass trapping, lure-and-kill (El-Sayed et al. 2006), and lure-and-infect (Klein and Lacey 1999) appear to describe the same technique. In essence an attractant is employed to increase population density of a pest species at a location where they can be controlled. In the context of integrated pest management (IPM), these techniques utilize an insects' natural response to an attractant that could stimulate a combination of senses. The most widely applied cues in control programs are olfactory, covering the whole range of semiochemicals from host-plant attractants to aggregation pheromones. In the case of moths, the lure is typically based on a female sex pheromone because of their high level of species-specificity, the low concentration needed to elicit a behavioral response, and nontoxic nature to nontarget organisms (vertebrates in particular). The primary difference among these techniques is the method of "killing" the target population. In mass trapping, insects are confined in a receptacle and killed either with a toxicant or naturally by heat exhaustion and/or dehydration, typically using a sticky trap. Similarly, lure-and-infect involves inoculation of trapped insects with an infectious microorganism. Infected insects are then allowed to escape back into the environment so that they can cross-infect conspecifics to spread the inoculum and establish an epizootic in the pest population. Lure-and-kill typically involves an insect making contact with a substrate that contains a conventional killing agent such as a synthetic insecticide but, importantly, does not require the use of a trap. In the context of this chapter, "kill" means to render male moths reproductively inca-

pacitated. Thus, it is not necessarily important to physically kill a moth but rather to prevent or delay a male moth from mating and all the techniques are seen as variants on that theme involving two distinct phases that result from "attraction and kill" and lead to "control" or "attract-and-control" systems.

The idea of luring or attracting insects to a source where they can be killed is by no means new, especially for Lepidoptera. Ancient civilizations were well aware of the power of a candle flame to both attract and kill moths. The method of attraction is well understood. Moths navigate using the sunlight reflected by the moon at night as a fixed reference point. They become confused by stronger terrestrial light sources which result in them adopting a spiraling course that inevitably leads them to the light source. If the light is emitted by a flame, then on reaching the light source they are killed. Light traps utilize the same principle but replace a flame with an electric light that is incorporated into a trapping system. This standard entomological tool is primarily used for studying insect populations, and in recent years researchers have claimed success for light traps in the control of moth populations, most notably for use with rice stem borers (Zhu et al. 2007). Similarly, light traps together with other non-pesticide methods have been used to control red hairy caterpillar, *Amsacta albistriga* (Erebidae, Arctiinae), in Andhra Pradesh, India, on castor, groundnut, sesame, sorghum, and pigeon pea on over 18,000 ha between 1989 and 1993 (Ramanjaneyulu et al. 2009). More recently, a solar-powered UV light source has been elegantly combined with a female sex pheromone-baited water trap to produce the Ferolite trap (Russell IPM, United Kingdom). This trap is effective at catching both male and female moths but utilizes the pheromone bait to attract male moths from a greater distance than would be possible with a light trap alone. The downside of light-based systems is their lack of species-specificity and relatively high cost.

There are a number of parameters associated with attract-and-control systems that are related to the use of traps for monitoring moth populations (Suckling, this volume). The two approaches vary insofar as monitoring populations require only a representative sample of a population to be trapped, whereas attract-and-control systems require large proportions of an adult population to be attracted to a source, particularly where males are the responding sex. Thus, for example, trap designs for each purpose may well vary, at least in their capacity. Similarly, in common with mating disruption (Minks and Cardé 1988), the "natural" blend of compounds in a semiochemical should be used to achieve the highest level of control in attract-and-control systems.

To be effective there are a number of parameters in attract-and-control systems for Lepidoptera that should be fully optimized including:

- Pheromone blends and controlled release formulations;
- Trapping systems and killing agents;
- Spacing between point sources;
- Method and timing of application, duration of control period.

This chapter will discuss the optimization of these parameters to achieve the highest level of attraction of receptive members of a target population with pheromones and the methods available to effectively kill attracted moths. In addition the chapter will further consider the potential scope, limitations, and current application of such approaches in lepidopteran pest management.

In attract-and-control systems, by far the most research effort has been applied to developing effective female sex pheromone lures. In particular, the structural elucidation of natural attractants and semiochemicals has been extraordinarily successful over the 50 years since the identification of the first lepidopteran female sex pheromone, bombykol, by Butenandt and Hecker (1961). Indeed, there seems no let-up in the pace of chemical identification of ever more semiochemicals. Helpfully the chemical composition of most of these semiochemicals has been catalogued and made freely available on the internet (Pheronet, Pherobase). However, the presumption that the chemical composition of a pheromone is absolute, and that having identified a species of pest in a particular location, this chemical blend can be deployed without further thought, is overly simplistic. There are now a number of well-studied examples demonstrating variation in the chemical composition of pheromones among (Löfstedt et al. 1986; Cork et al. 1992) and within (Roelofs et al. 1987) populations, providing an important insight into evolutionary mechanisms of speciation (see Allison and Cardé, this volume; Cardé and Haynes 2004) but potentially adding complexity to the successful commercialization of attract-and-control systems.

Pheromone Blends and Controlled Release Formulations

Fortunately, for most species of Lepidoptera where the composition of the sex pheromone has been analyzed, optimiza-

tion of the composition of the semiochemical is usually a part of the identification process. Modeling mass trapping parameters Yamanaka (2007) identified lure efficiency to be particularly important for achieving high levels of mating suppression. The optimal blend typically reflects the composition of chemicals identified in the insect pheromone gland, or volatiles released from them. Arguably the blend of compounds released by the insect constitutes the pheromone, but because of the differential evaporation rates of compounds, which are related to their molecular weight and polarity, the blend found in a gland may better reflect the blend required to produce an effective synthetic lure. However, this is not always the case since some species, most notably Heliothentines, store pheromone components as alcohols which are subsequently oxidized to the bioactive aldehydes as they are released from the pheromone glands (Teal and Tumlinson 1986). Importantly, the extraction and analysis of sex pheromone glands from virgin female moths in the period of time when they are releasing a pheromone usually yields the highest quantity of pheromone. The quantity and quality of sex pheromone in a sample are particularly important for identifying essential minor components of that semiochemical.

The behavioral function of some compounds identified in a female sex pheromone gland remains uncertain, notably those that are found to have no effect on trap catch and those that actually reduce trap catch over a range of doses tested. Thus, it might be argued that some compounds are precursors of pheromone components. For example, aliphatic straight-chained primary alcohols are typically considered to be precursors of aldehydes. Compounds may be biologically redundant for evolutionary reasons, but in reality this may simply reflect our lack of knowledge about how sex pheromones function. We typically assume that a sex pheromone "message" only contains information for a conspecific male searching for a receptive female. Some compounds may inhibit attraction of heterospecific males (Szocs et al. 1990). For example, while working with the sibling species *Earias insulana* (Nolidae) and *E. vittella* in Pakistan, Cork et al. (1988) established that the EZ-isomer of the common major sex pheromone component, (*E,E*)-10,12-hexadecadienal, mediated signal specificity between *E. insulana* and *E. vittella*. The presence of the EZ-isomer enabled attraction of *E. insulana* and absence resulted in the attraction of *E. vittella*. Before the identification of the *E. vittella* sex pheromone, lures to attract *E. insulana* using only the EE-isomer had been used successfully on cotton in Israel and Egypt. The reasons why the lures were effective in Egypt and Israel are still uncertain, but may either reflect a difference in composition of the *E. insulana* pheromone in different locations or the presence of the EZ-isomer as an impurity in the synthetic blend, perhaps due to isomerization of the EE-isomer under field conditions.

Thus, the substitution of a synthetic chemical for a naturally produced compound can bring with it new complications in the form of impurities associated with the method of synthesis, method of storage, and exposure to field conditions. Researchers with access to good-quality chemistry benefit from the use of high-purity compounds. However, many commercially available compounds are at best typically <95% pure. The most common contaminants are geometric or optical isomers of pheromone components, perhaps where a selective hydrogenation has resulted in the production of a mixture of (*Z*)- and (*E*)-isomers of a compound. The presence of these additional compounds may or may not influence the behavior of a target species, as seen with the *Earias* spp.

This, however, needs to be established by testing the effect of introducing different amounts of the anticipated impurity into the blend in a fully replicated behavioral bioassay. Ideally such research should be undertaken with a natural population of the species and, preferably, under field conditions. In an elegant study of pheromones of two related species, Hall et al. (1993) demonstrated that when a natural population of *Alabama argillacea* (Erebidae) was presented with both *S* and *R* enantiomers of the major pheromone component, 9-methylnonadecane, no male moth catches were obtained, although the natural homochiral *S* compound alone was highly attractive. In contrast, a racemic mixture of the related 7-methylheptadecane, the major sex pheromone component of *Anomis texana* (Erebidae), caught as many *A. texana* as the natural *S* enantiomer. This is not to say that the unnatural isomers were not detected by male moths, indeed EAG studies indicated that they were, but merely that under the conditions of the field trials, no discernible difference in trap catch, as a proxy for male attraction, was observed.

Thus, while the "natural blend" is considered the best attractant for attract-and-control systems, this ideal is usually difficult to realize. Nevertheless, by employing high doses of a sex pheromone, lures can be produced that can out-compete a virgin female moth, at least on the basis of averaged trap catch. As we will see later, this does not necessarily mean that synthetic lures have to be more attractive than a female moth; even if they are, this may not prevent a male from locating a female moth, but employing a high density of traps will increase the probability of a male being trapped before mating. Thus, both the relative density of traps to virgin females and the relative attractiveness of synthetic and natural sex pheromones at any particular time can influence the level of mating success of a pest species in a crop.

Lures should remain highly competitive with virgin females, not only over a single night, but for the entire period during which the lures are deployed. Ideally a lure should release the components of a sex pheromone blend in the same ratio and rate throughout the entire time it is deployed in the field.

A crop may be susceptible to attack by the larvae of a pest species during a specific period of time and the trapping period for adults will either overlap with the larval damaging period, if there are no distinct generations of the species, or may be quite separate from the larval damaging period if there are distinct generations. Thus, for example, a monophagous species that specializes on a host that produces one crop of seeds or fruits in a year may well have distinct generations and require control at specific stages in crop development. Alternatively an oligophagous species that feeds on a number of crop stages, such as cereal stem borers, have distinct generations in the vegetative as well as reproductive stages of Graminaceae. In contrast, host species that continually produce shoots and fruits on reaching maturity, such as the eggplant (*Solanum melongena*, Solanaceae), may have pest species, such as the eggplant borer, *Leucinodes orbonalis* (Crambidae), which has overlapping generations continually attacking both young stems and fruit. Similarly, pickleworms, *Diaphania* spp. (Crambidae), are known to attack young leaves of cucurbits as well as fruit. Other species, such as some heliothentines, can be highly mobile, polyphagous pests that lay individual eggs in crops such as maize and cotton from the flowering stage onwards and are characterized by overlapping generations. Unless the timing of mating is well understood with reference to crop development and phenology, traps may

need to be deployed prophylactically and possibly over larger areas than a single crop field to achieve control.

The majority of synthetic lures utilize simple polymers to control release, and emission from these materials is governed primarily by temperature. As a consequence, most of the behaviorally active compounds released occur during the photophase. However, most male moths respond to female sex pheromones during the scotophase when temperatures, and hence release rates, are lower. Perhaps more importantly, release rates are strongly influenced by seasonable variation in nighttime temperatures. Typically, a lure designed and tested for use in temperate conditions may release sex pheromone too rapidly under tropical conditions, reducing the effective field longevity of the lure, although possibly increasing catch during the period of higher release rate. The problem is more acute where a crop is grown in both summer and winter conditions at the same location. Thus, lures developed to address high pest population pressures experienced in vegetables (e.g., diamondback moth) during the summer months, for example, may well be ineffective on winter crops at the same location because of reduced release rates of pheromone components. This problem might be addressed by modifying the dispenser, but such products have not been commercially developed.

The question of whether suitably optimized lures may be equally effective for attract-and-control systems in winter crops when population density is low and summer crops when population density is high may ultimately depend on the package of technologies utilized and scale of operation. Thus, in a winter crop, mass trapping of the target species alone may be enough to achieve acceptable levels of control. In the same crop, but under summer conditions, more than one life stage of the pest may need to be controlled and other pest species become economically important. The latter requiring additional measures to be incorporated into an IPM package and an area-wide approach to control adopted instead of control in a single crop field.

The choice of dispenser for a sex pheromone is critical to obtaining optimal release of pheromone components. Working with *L. orbonalis*, Cork et al. (2001) found that the relative attractiveness of lures containing the same pheromone components was dependent on the polymer used in the dispenser. Under wind-tunnel conditions, at 27°C, the half-lives of the two pheromone components, (*E*)-11-hexadecenyl acetate (E11-16Ac) and (*E*)-11-hexadecen-1-ol (E11-16OH), were approximately 52 and 22 days, respectively, in low density polyethylene vials, suggesting a field life of at least one month. Nevertheless, as the alcohol was released at twice the rate of the acetate, the actual amount of E11-16OH in the blend released would initially have been significantly higher than the 1% present in the lure. Once exhausted of E11-16OH, the relative attractiveness of the lure would be considerably diminished. In contrast, only 8% of E11-16Ac was released from rubber septa after 40 days under wind-tunnel conditions, at 27°C. Field data from winter field studies in West Bengal supported this interpretation, with traps baited with lures prepared from rubber septa catching significantly fewer male moths than the same pheromone blend and dose dispensed from low-density polyethylene vials (Cork et al. 2001).

The release rates of n-alkyl and n-alkenyl acetates and related alcohols from industry standard natural rubber septa are well understood, following the exhaustive study conducted by McDonough et al. (1989). Importantly, they determined the values based on the actual amount of compound released and not the residual compound remaining in the

dispenser. This is particularly important for n-alkenyl aldehydes that McDonough (1991) suggested were not entirely stable in rubber septa and as a result some of the loss attributed
to evaporation was in fact due to chemical degradation.

The release rates of compounds from natural rubber septa
were found to be first order, related to the amount of material
remaining. The half-lives of compounds in rubber septa can
be calculated from Equation (1) where ΔH is the heat of evaporation, R is the gas constant, T absolute temperature, and γ_0
a constant:

$$t_{1/2} = \frac{\Delta H}{RT} + \gamma_0 \qquad (1)$$

The data clearly showed that temperature and molecular
weight (which is related to rate of evaporation) have a profound and quantifiable effect on release rate. Thus, for a given
temperature decreasing the molecular weight by one methylene group ($-CH_2-$) almost triples release rate, while decreasing
the temperature by 5°C can almost half a release rate. The
half-lives of aldehydes were found to be very similar to those
of acetates containing two fewer methylene groups. Thus, the
half-lives of 14:Ac and 16:Ald at 30°C were 84.2 and 86.9 days,
respectively. This result was to be expected, because the
molecular weights and polarity of the compounds are similar,
resulting in both compounds having similar ΔH values.
McDonough et al. (1989) calculated the half-lives of hexadecenyl acetates to be 2900 and 200 days at 15 and 35°C, respectively. This may explain the poor trap catches of *L. orbonalis*
in northern India during winter months compared to summer catches further south with similar levels of infestation.

For sex pheromones composed of compounds with similar
physicochemical properties, designing lures with predictable
field efficiency and durability is relatively straightforward.
Where components of a female sex pheromone have very different physicochemical properties, maintaining the release of
a constant blend is more complicated. For example, *Chilo auricilius* (Crambidae) which has a female sex pheromone is composed of an 8:4:1 blend of (*Z*)-8-tridecenyl acetate, (*Z*)-9-tetra-
decenyl acetate, and (*Z*)-10-pentadecenyl acetate (Nesbitt
et al. 1986). Nevertheless, because of the work by McDonough
et al. (1989) and others, this process is predictable and the initial blend can be modified to produce a lure that maintains its
attractiveness in the field for a significant period of time.

In a study of gossyplure (1:1 blend of (Z,Z)-7,11,
hexadecadienyl acetate and (Z,E)-7,11,-hexadecadienyl acetate)
release rates from commercial suppliers in India (Alan Cork
et al., unpublished data), a wide range of release rates were
obtained (figure 23.1), varying between 0.3 and 4.5 mg over a
six-week period in a wind tunnel at 27°C. The results partially
reflected different loadings used in the lures as well as differences in the polymer or blend of polymers used to fabricate the
lures, as is apparent from the percentage release data (figure
23.2). Polymers commonly used included silicone rubber, natural rubber composed of polyisoprene molecules, and nitrile
rubber synthesized from a copolymer of acrylonitrile and
butadiene. Other commonly available polymers included
styrene-butadiene copolymer and "triblock" copolymers, typically composed of styrene-butadiene-styrene (SBR) and
styrene-isoprene-styrene (SIS) rubbers that have relatively
high polarity. Such polymers might be expected to be potentially useful for sex pheromones, given that they would slow
the release of alcohols compared to the equivalent acetate for
example compared to nonpolar polyethylene dispensers. One

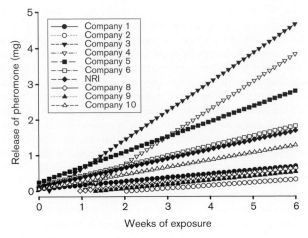

FIGURE 23.1 Cumulative release of gossyplure from commercial
lures, in a wind tunnel at 27°C.

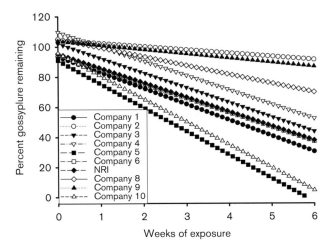

FIGURE 23.2 Percent release of gossyplure from commercial lures, in
a wind tunnel at 27°C.

of the problems experienced by the pheromone companies
studied was uncertainty over the polymeric composition of
materials they purchased for fabricating lures and batch-to-
batch variation. The lack of research on controlled release
technology is reflected in the current adherence of most
researchers to the standard "natural rubber septa" even though
the polymer is by no means the most suitable substrate for
releasing all female lepidopteran sex pheromone blends.

Trapping Systems and Killing Agents

For most purposes a trap is used to catch and retain insects.
Typically a trap is composed of two chambers; the first houses
the lure and ideally provides easy access for the insect and
release of the pheromone. The second chamber contains either
a killing agent or mechanism for preventing escape. Many
commonly used traps utilize plastic funnels with lids in which
the lure is placed directly above the center of the funnel. Such
omnidirectional traps permit moths to orient to the trap from
any direction with minimal disturbance to the plume structure (Wall 1989). Moths in close proximity to the lure are disoriented and fall through the funnel into the second chamber

often composed of a plastic bag where they die from heat exhaustion and/or dehydration. A volatile insecticide such as malathion, DDVP (2-dichlorovinyl dimethyl phosphate) or a related organophosphate (1,2-dibromo-2,2-dichloroethyl dimethyl phosphate) may be incorporated to increase the rate of kill, thereby reducing the possibility of escape.

Single chamber traps are often cheaper to fabricate, but to be effective the lure must be located close to the killing agent. Water, with or without a wetting agent such as the surfactant Triton, or locally available detergents are the most commonly used killing agent. The International Atomic Energy Authority recommends the addition of 10% propylene glycol to water for trapping fruit flies (Anonymous 2003) but this has not yet been widely adopted for Lepidoptera. The use of mineral oils (for example, castor oil) to reduce evaporation and surface tension is more widely adopted. Alternatively, water can be replaced entirely with engine oil (Huber and Hoffman 1979) but trapped insects are poorly preserved and information on species present, for example, may be lost. The use of water or oils as killing agents requires considerable effort to maintain and while the costs of such activity may be prohibitive in developed countries they are acceptable in developing countries, especially in high value crops grown on a small scale. Polybutene adhesives are popular substrates for killing moths. They have low mammalian toxicity, and are inexpensive given their wide application as polymer extenders, plasticizers, and wetting agents. Combined with low molecular weight polyethylenes they produce tacky gels that can be applied as thin films onto boards, and used as liners in the base of traps. Nevertheless, tacky films used in delta and wing traps result in traps with finite capacity (Sanders 1986). Such limitations may be acceptable for monitoring insect populations, but not attract-and-control systems, unless the insects are micro-Lepidoptera and population densities low. Water and dry funnel traps are recommended for attract-and-control because of their higher capacity and the latter's low maintenance costs.

Spacing between Point Sources

Having developed an attractive, long-lasting lure and a low-cost efficient trap for capturing male moths, the question remains whether sufficient male moths can be trapped to have an effect on mating and ultimately the damage caused by their progeny. Before such questions can be answered it is important to establish parameters associated with the deployment of traps that will ensure that catch per unit time and area are maximized.

For most annual field crops the highest catches are obtained at or just above crop height (Cork et al. 2003; Byers 2008). This information is readily obtained by placing traps at a range of heights within a crop in replicated trials. The situation may vary with orchard crops where some moth species prefer to fly within a canopy and not above (Tyson et al. 2006). Intriguingly, higher trap catches have been obtained with *Tuta absoluta* (Gelechiidae) when pheromone traps were placed at or close to ground level compared to 1 or 1.5 m above even with crop heights in excess of 2 m (Ferrara et al. 2001; Lobos et al. 2013). Care needs to be taken over how such trials are conducted. Youm et al. (1997) working with millet stem borer, *Coniesta ignefusalis* (Crambidae), found that when traps were placed individually between 0.1 and 2 m above the ground, the highest number of male moths were caught in traps 0.1 m above ground level in crops of between 0.13 and 1.92-m height. However, when male moths were given a choice of traps at different heights located at the same position (i.e., vertical transects) there was no significant difference in the trap catches.

The density of traps within a crop is critical for mass trapping. A high density of traps might be seen as advantageous but can lead to trap interference (Trematerra 1993; McMahon et al. 2010). Interference is thought to be minimized if trap spacing is at least two times the "attractive" radii (Dodds and Ross 2002). High trap densities are expensive to maintain. The actual density used may reflect a compromise between the density that delivers the highest catch per unit area and that which is most cost-effective. Because insect density within a crop is heterogeneous and varies over time, methods for determining the most efficacious trap density are far from straightforward, and may require extensive field trials. In one of the few such studies, Jamieson et al. (2008) optimized trap density in replicated trials of 3, 10, 30, 100, and 300 traps/ha to show that for the citrus pest *Prays nephelomima* (Yponomeutidae), 100 traps/ha provided good levels of control, for that trap and lure system. The approach adopted by Cork et al. (2005a) was to set out a typical matrix of traps with 10-m spacing. Within that matrix, groups of four traps were selected for counting trap catch data (four traps per 100 m²). By adding or subtracting traps within that space and rearranging inter-trap distances to maximize the distance between them, a range of densities was created (three to six traps per 100 m²) and estimates of catch as a function of trap density established for the target species. For *Leucinodes orbonalis*, there was no significant difference in the number of moths caught per unit area between four and six traps per 100 m². Thus, a density of four traps per 100 m² (100 traps/ha) was selected for mass trapping *L. orbonalis* using a water trap with polyethylene vial dispensers containing 2 mg of pheromone. In contrast, Gururaj et al. (2001) working with yellow rice stem borer, *Scirpophaga incertulas* (Crambidae) in India, found that 20 traps/ha was optimal for control but used a higher pheromone loading (5 mg, rubber septa) and different trap design (funnel trap). The trap density selected should periodically be reviewed as pest population, scale of operation, and improvements in trap design and lure performance, all work to reduce the number of traps needed to achieve economically acceptable levels of control.

Method and Timing of Application, Duration of Control Period

Hand application of point sources, as used in attracticides, can be time-consuming and hence expensive to apply in the field, although farmers save the cost and maintenance associated with traps used in mass trapping. The time taken to apply attracticides is comparable to mating disruption techniques given that workers have to move similar distances as they apply treatments, although the number of point sources may be fewer for mating disruption. Mass trapping is more labor intensive when the installation of traps requires the erection of structures to support them, as would be the case in cereal crops, notably rice.

Attract-and-control systems are designed to target the adult of a lepidopteran species and yet it is the larval stage that is damaging. Thus, application has to be undertaken before the pest is established in a crop. This could be decided on the basis of pheromone monitoring trap catches or applied prophylactically, depending on the value of a crop, and economic threshold of damage. In the case of groundnut leaf miner, *Aproaerema modicella* (Gelechiidae), application is recommended a few days after sowing, and certainly before the first leaves are visible (A. R. Prasad, personal communication). Similarly, control of rice stem borers is required in the nursery onwards, and yet mass trapping is normally only initiated within 14 days of transplanting to target the second generation of adults that attack at the reproductive phase: control in the vegetative stage then being assured by application of granular insecticide in the nursery. Experience has shown that in long duration crops, where pests have overlapping generations, application can be effective at any time during the fruit development stage, although for maximum efficiency mass trapping should be applied throughout the cropping period to prevent the initial build-up of pest populations. Attract-and-control systems should be developed in collaboration with farmers to ensure that the recommended timing and method of application is appropriate within their production schedule to ensure good adoption (Cork 1998).

Nevertheless, mass trapping can be effective even when applied late season. Working in Bangladesh, Cork et al. (2003) assessed the impact of mass trapping of *Leucinodes orbonalis* in eggplant on two physiological stages of development: a mature crop and a young crop still in the vegetative stage. Insecticides were applied every two days to both farmer practice plots and IPM plots. However, in the IPM plots mass trapping and removal of infested shoots were also imposed. In both farmer practice and pheromone treatment plots, trap catches were comparable for at least four weeks after application but then reduced significantly in mass trapped plots. The delayed change in moth catches presumably reflected the emergence of adults from pupae already present in the crop prior to the implementation of mass trapping. In contrast, the level of fruit damage, relative to control plots, significantly decreased in less than two weeks of the application of IPM tactics despite high initial fruit damage in the mature crop. Low fruit numbers in the immature crop in the initial stages of the trial obscured the effects of mass trapping, but while the mass trapped plots remained at an average of 20% fruit damage over the trial period, damage levels increased significantly to 45% over the same period in the insecticide-treated, farmers' practice plots.

Having gained farmers' confidence in mass trapping, they were persuaded to allow some plots to be treated with IPM practices alone. Weekly fruit harvesting began 10 weeks after transplanting and 6 weeks after installing the IPM regime. From the first fruit harvest, damage caused by larvae of *L. orbonalis* was significantly less in IPM + insecticide application plots compared to insecticide-treated plots, as shown previously. However, IPM-treated plots in the complete absence of insecticides showed significantly lower levels of fruit damage than IPM + insecticide. The difference in fruit damage between IPM plots with or without insecticide was later attributed to the increased impact of natural enemies, notably *Trathala flavoorbitalis* (Ichneumonidae), which only emerged from infested fruit collected in IPM plots and were subsequently shown to cause between 20% and 44% parasitism in IPM fields over a 12-month period in a range of locations (Rahman et al. 2008). The effect of mass trapping on the ability of adult *L. orbonalis* to mate, as expressed by subsequent reduction in larval fruit damage levels, was particularly significant, given that the trials were conducted on replicated 0.5-ha plots. Increasing the plot size would be expected to

improve the levels of control achieved in mass trapped plots, but plots of 0.5 ha were deliberately chosen to reflect the typical field size cultivated in the region.

More recently in Bangladesh, yields have been further improved by incorporating the larval parasitoid, *Bracon hebetor* (Braconidae) and the egg parasitoid, *Trichogramma chilonis* (Trichogrammatidae) into the IPM program. This resulted in crops with typical fruit damage levels of less than 10%. Similarly, grafted plants with root stock resistant to bacterial wilt, *Ralstonia solanacearum* (Ralstoniaceae) (Bletsos et al. 2003), have been incorporated into the IPM package to further extend the value of the approach to farmers and likelihood of adoption (Cork 1998).

Current Status of Attract-and-Control Systems

Cotton and Oilseed Pests

In cotton, early work conducted on *Spodoptera littoralis* in Israel (Teich et al. 1979, 1985) resulted in mass trapping being optimized with 1.7 traps/ha and the adoption of a dry funnel trap incorporating diazinon as the killing agent. The technique achieved 40–50% reduction in egg mass numbers and 20–25% savings in insecticide use compared to conventional control methods. This success led to an increase in area treated from 2500 ha in 1975 to 20,000 ha in 1981 (Shani 1982). Similar results were obtained in Egypt with 50% and 77% reductions in egg masses being achieved using 3 and 5 traps/ha, respectively, in 500-ha trials. In subsequent work a trap density of 9 traps/ha was found to give the maximum catch/unit area, which would ultimately be expected to result in better control of mating and hence lower infestation rates in treatment plots (Campion and Nesbitt 1981). Work on the related species *S. litura* (Noctuidae) in Japan, India, and China suggested that mass trapping was efficacious at controlling pest populations (Nemoto et al. 1980; Pawar and Prasad 1983; Tang and Su 1988). In particular, Krishnaiah (1986) reported that in 240 ha of black gram, *Vigna mungo*, mass trapping with 10 traps/ha was more effective than 5 traps/ha and that both trap densities produced significant reductions in damage levels compared to controls. This approach has been adapted and promoted by the National Centre for Integrated Pest Management in India (Singh et al. 2004) as part of an IPM package for farmers in a range of crops. Results from a recent survey of pheromone producers in India (Alan Cork, unpublished data) has shown that production of *S. litura* lures remained essentially constant over the past eight years: between 500,000 and 600,000 lures per annum. This is despite their use in cotton being essentially discontinued because of the widespread adoption of *Bt* cotton (Deguine et al. 2008). Nevertheless, while mass trapping can have an impact on populations of *S. litura*, it seems most likely that combinations of component IPM technologies controlling more than one life stage will produce the most sustainable methods of control in the future (Zhong 2009).

De Souza et al. (1992) evaluated eight insecticides for use in a lure-and-kill attracticide for *S. littoralis* and found that the pyrethroid lambda-cyhalothrin had the greatest toxicity, and, perhaps more importantly, the highest knockdown rates within 30 min of a 10-s exposure to leaves sprayed with 500 μg/ml (AI). Initial field trials in Egypt suggested that lambda-cyhalothrin gave excellent levels of control of male *S. littoralis* for up to 14 days. However, subsequent field trials by Downham et al. (1995) using 500 point sources/ha of PVC-resin and micro-encapsulated pheromone dispensers treated with lambda-cyhalothrin suggested that, although mating levels and trap catch in treatment plots were significantly reduced compared to controls for up to 24 nights, there was no treatment-related reduction in egg masses. They concluded that mating suppression was due primarily to disruption, i.e., delay in mating caused by the presence of the pheromone point sources, and not increased mortality arising from contact with the insecticide. Subsequent laboratory examination of field exposed PVC lures, suggested that a lack of contact with the insecticide sources may have been due to the production of trace amounts of the inhibitor (Z)-9-tetradecen-1-ol caused by de-acetylation of (Z)-9-tetradecenyl acetate under field conditions.

Early research suggested that mass trapping can be effective for controlling pink bollworm, *Pectinophora gossypiella* (Gelechiidae) (Huber et al. 1979). However, the lower maintenance costs, and higher consistency of data obtained with mating disruption products, meant that mass trapping techniques were not developed further. In contrast mating disruption was refined and adopted on a large scale for a number of years, notably in the United States and up to 400,000 ha treated per annum in Egypt (Witzgall et al. 2010). Widespread adoption of transgenic *Bt* cotton has overshadowed further development of pheromone-based control products. It remains to be seen whether concerns about the long-term viability of *Bt* transgenics in regions where refuges are not adopted (Huang et al. 2011) are realized and low end-of-season *Bt* titers compromise the ability of the technology to control *P. gossypiella*. Certainly research interest did continue for some time in mass trapping *P. gossypiella* (Ahmad and Attique 1993; Mafra-Neto and Habib 1996; Nassef et al. 1999) and the early promise of attracticides (Haynes et al. 1986; Hofer and Angst 1995) in lure-and-kill systems may yet produce viable control products for the future. Ideally more benign insecticides than permethrin should be incorporated into the systems to protect non-target predaceous insects (Butler and Las 1983).

Control of heliothentines was the main driving force for developing *Bt* cotton, given their ability to develop resistance to a wide range of insecticides. The extraordinary success of *Bt* cotton in India (Choudhary and Gaur 2010) might have been expected to result in a reduction in the production of pheromone traps for these species but in fact production increased from 830,000 in 2004 to 6.7 million in 2010–2011 (Alan Cork, unpublished data). As with *S. litura* the current production of lures is associated with a shift in emphasis from use in cotton to pulse IPM. The introduction of *Bt* cotton has resulted in extraordinary reductions in overall populations of *Helicoverpa armigera* (Noctuidae) across a range of other crops in India (Deguine et al. 2008) which has enabled researchers to reconsider the potential of IPM component technologies, such as mass trapping, for control. Interestingly, the only semiochemical-based control product to be commercialized in Australian cotton is Magnet, an attracticide not based on a pheromone, but rather an artificial five-component blend of green-leaf and floral volatiles (Gregg et al. 2010). The reason why this product has succeeded compared to mating disruption using the female sex pheromone (Chamberlain et al. 2000; Britton 2005) is presumably because both male and female *H. armigera* are attracted to the odor-bait and killed.

Aproaerema modicella (Gelechiidae) is a serious pest of groundnut and soybean in Southeast Asia where the larvae typically cause up to 50% yield loss (Hall et al. 1993). Farmers are reliant on insecticides for control (Shanower et al. 1993), though even with 11–12 applications a season insecticides can be ineffective. The Indian Institute of Chemical Technology has developed a mass trapping procedure based on 18–22 delta traps/ha and lures (3-mg dose, polyethylene vials) replaced every 3 weeks until harvest, 110–120 days. Total cost of mass trapping was estimated to be cost-effective (25 US$/ha), with significant gains in yield and insecticide use reduced to two to three applications per season for control of other pests. Demonstration trials to promote mass trapping to farmers have been conducted on 650 ha/annum for the past three years (A. R. Prasad, personal communication).

Vegetable Pests

Diamondback moth, *Plutella xylostella* (Plutellidae), is perhaps the most damaging pest of cruciferous crops in the world. Talekar and Shelton (1993) estimated the global costs of control at 1 billion US$. IPM-based approaches are difficult to implement in many peri-urban areas where cruciferous vegetables are grown throughout the year because of the high reproductive rate of *P. xylostella*, typically 16–17 generations per year. However, with increasing concern about the high levels of resistance to insecticides (Zhao and You 2001) considerable efforts have been made to develop pheromone-based control techniques for *P. xylostella* and, in particular, mass trapping. Uncertainty about the appropriate sex pheromone blend (Chisholm et al. 1983) to use has resulted in researchers developing best-bet options for use in their own regions, although a compromise blend of (Z)-11-hexadecenyl acetate, (Z)-11-hexadecenal, and (Z)-11-hexadecen-1-ol in a 50:50:1 ratio is now widely commercialized. Trap design can have a profound effect on catch and plastic bucket traps with water containing detergent located just above the crop canopy have been found to be particularly effective. Importantly, the lure should be located close to the water surface; Wang et al. (2004) recommend 1 cm above the water surface. Similarly, pheromone loading is important for maximum catch; septa loaded with between 10 and 100 µg being recommended (Chisholm et al. 1983; Chen et al. 1990) and lures requiring replacement after no more than 30 days. Mass trapping of male moths with female sex pheromone lures has clearly been demonstrated to have a significant impact on population density of subsequent larval populations (Reddy and Guerrero 2000a) but in many cases the level of control demonstrated has not been sufficient to merit complete reliance on this single technique. Some workers prefer to combine mass trapping with judicious applications of soft insecticides such as *Bt* when populations exceed Economic Threshold Levels (ETL) (Wang et al. 2004). More recently, combinations of host odors and female sex pheromone have been developed that enhance attraction of male *P. xylostella* and catch significant numbers of female conspecifics (Reddy and Guerrero 2000b), in particular blends of pheromone plus (Z)-3-hexenyl acetate, (Z)-3-hexen-1-ol, and allyl isothiocyanate (Dai et al. 2008). Similarly, yellow-colored sticky traps were found to be more attractive to male and female *P. xylostella* than other colors tested (Sivapragasam and Saito 1986). The combination of improved lure attractancy and trap efficiency raise the possibility of achieving more acceptable levels of control by mass trapping in the future. In the United States, the cabbage caterpillar complex, composed of *P. xylostella*, *Trichoplusia ni* (Noctuidae), and *Pieris rapae* (Pieridae) (Shelton et al. 1982), is normally controlled together by farmers. Small-scale experiments have demonstrated that low-to-moderate populations of the three species can be controlled using an attracticide (Last Call, IPM Development Company, OR, United States) containing the pheromones of *P. xylostella* and *T. ni* but not when high populations of *P. rapae* were present in the spring season. Interestingly, pheromone trap catches were not suppressed in plots treated with attracticide, suggesting that the killing agent, permethrin, controlled the population rather than interruption of pheromone communication (Maxwell et al. 2006).

Tuta absoluta is an example of an invasive pest that has had a profound effect on the sustainability of tomato production in South America, Europe, and North Africa. Chemical control of this multivoltine species whose larvae mine the fruit and leaves of solanaceous crops is difficult to achieve. Since its introduction into Spain from South America in 2006 the pest has spread throughout the Mediterranean basin and Near East, threatening glasshouse production in Northern Europe. Because biological control is used extensively in Europe in greenhouse crops, this has encouraged the development of complementary IPM approaches for control. European farmers increasingly favor systematic releases of *Macrolophus pygmaeus* (Miridae) and *Trichogramma achaeae* (Trichogrammatidae) in greenhouses and screened houses together with mass trapping. The latter may involve between 20 and 25 pheromone traps/ha and UV traps recommended in packing areas. Most lures are comprised of the major component of the sex pheromone (*E,Z,Z*)-3,8,11-tetradecatrien-1-yl acetate alone (Svatos et al. 1996) because addition of the minor component, (*E,Z*)-3,8-tetradecadienyl acetate (Griepink et al. 1996), does not appear to increase catch significantly. In South America mass trapping is recommended before planting, followed by soil clearance, application of imidacloprid in irrigation water 10 days after planting and spinosad application as needed, and destruction of crop residues immediately after last harvest (Junco and Herrero 2008). Trapping of *T. absoluta* has been reported to significantly lower damage (Yucra Equize 2002; Lobo Pinheiro 2005; Montserrat 2009; Mohamed and Siam 2011). Working in open-field plantations Lobos et al. (2013) demonstrated that at 30–40 traps/ha, leaf damage was significantly reduced compared to conventional insecticide treatments even when adult male trap catches were >35 per trap per day and that repeated treatments in successive years further reduced population levels. Interestingly, Núñez et al. (2009) demonstrated that in an enclosed glasshouse monitoring trap catches were reduced by more than 92% for a period of 106 days, when lures (0.2 mg, rubber septa) were placed out at densities of 16 and 32/ha whether or not they were placed in traps, suggesting that at those densities the lures were releasing sufficient pheromone to achieve conditions for mating disruption, rather than mass trapping.

Fruit Tree Pests

Early research on codling moth, *Cydia pomonella* (Tortricidae), suggested that mass trapping was ineffective for control. In particular, a study by Proverbs et al. (1975) found that larval damage over a three-year period in a 14 ha orchard increased from 0.00075% to 0.033% with related increases in male moth catches using a trap density of 34 traps/ha but only 1 mg of pheromone per lure. The reasons for this apparent failure are uncertain but, given the lack of control plots and replication, may simply reflect seasonal variation. Given the high level of control required by growers, the polygamous nature of this male moth, and success of mating disruption, the prospects and rationale for developing cost-effective and sustainable mass trapping systems appear limited at this time. Nevertheless, mating disruption is expensive, requiring typically 140 g/ha/season of (*E,E*)-8,10-dodecadien-1-ol (codlemone) and, in apple-producing countries with low labor costs, mass trapping may be an economically viable alternative (Pawar and Tuhan 1985).

More recently attention has shifted to the development of attract-and-kill products for *C. pomonella* that typically incorporates 0.16% pheromone with 6% permethrin (Sirene CM®). Using tethered females and pheromone trap catches to assess the ability of males to locate females and mate, the formulation was found to have a field longevity of 5–7 weeks, requiring only 2 applications to give season-long control (Charmillot et al. 2003) with fruit damage below the ETL of 1% in 14 out of 15 trials. Losel et al. (2002) substituted the pyrethroid cyfluthrin in the attracticide and was able to achieve control using 4500 sources/ha with 100-µl droplets containing 1 mg/g of codlemone and 40 mg/g of cyfluthrin. Similar levels of control have been achieved by substituting an insect growth regulator, fenoxycarb, for the pyrethroid. Field trials confirmed that autosterilization contributed to the level of control, with effective transfer of fenoxycarb to female moths while mating (Charmillot et al. 2002). As with mass trapping, the density of point sources was important to achieve control, with applications of typically 3000 50-µl droplets/ha considered optimal. Importantly, the authors only used between 0.12 and 0.35 g/ha of codlemone, representing a significant cost saving over mating disruption.

Meng et al. (1985) conducted mass trapping trials with 15 traps/ha on 4000 and 5400 ha of pear orchards in 1981 and 1982, respectively, to control light-brown apple moth *Epiphyas postvittana* (Tortricidae). Comparing the results with insecticide-treated controls there were between 74.2% and 82.8% reductions in female mating and 79% increase in egg parasitism in the pheromone-treated plots. Importantly, fruit damage was reduced by 50–73% compared to the insecticide-treated plots. More recently, Suckling and Brockerhoff (1999) demonstrated that an attracticide (Sirene®) was also effective at suppressing trap catches of *E. postvittana* in 0.3-ha plots by up to 96% four days after application. By comparing caged and non-caged toxicant they were able to demonstrate that 50% of the suppression in monitoring trap catch was due to the effects of insecticide, suggesting that mating disruption contributed significantly to the level of control achieved. However, further work is needed to establish whether the technology can achieve control.

Oriental fruit moth, *Grapholita molesta* (Tortricidae), has been the subject of a renewed interest in attract-and-control, most notably using lure-and-kill which offers reduced pheromone application rates compared to mating disruption and reduced application costs compared to mass trapping. In apple orchards Evenden and McLaughlin (2004a, 2004b) demonstrated that a commercial attracticide formulation (Last Call) significantly reduced trap catches in pheromone-baited traps at 1500 and 3000 droplets/ha, but more importantly only sentinel females in untreated portions of tree canopies were mated. Mate finding behavior was equally disrupted by formulations with and without insecticide, suggesting that mating disruption was more important than the effect of the toxicant on mating success. Evenden et al. (2005) tested the behavioral effect of an attracticide in a wind tunnel and found that sublethal exposure to the attracticide significantly reduced mating success but not on exposure to the formulation without insecticide. Evenden and McLaughlin (2005) studied the behavioral responses of male *G. molesta* and *C. pomonella* to an attracticide containing sex pheromones of both species. They concluded that male *G. molesta* were more attracted to the combined pheromone blend in field and wind-tunnel experiments, and spent longer at the droplets, enhancing the probability of killing the moths. The authors considered that Last Call OFM released a suboptimal amount of (*Z*)-8-dodecen-1-ol but that addition of codlemone in the combined formulation compensated for this. However, trap catches of male *C. pomonella* to the combined attractant in field trials were significantly reduced compared to attracticide droplets containing codlemone alone.

Evidence that the efficacy of mass trapping is dependent on adult population density comes from early work on spotted tentiform leaf miner, *Phyllonorycter blancardella* (Gracillariidae) (Trimble and Hagley 1988) using baits containing (*E*)-10-dodecenyl acetate in Multi-Pher-II traps. They found that larval damage was lower in mass trapped plots compared to check plots in a trial conducted on an experimental orchard, but in commercial orchards where trap catches were 91 times higher mass trapping was ineffective. The low level of control of *P. blancardella* observed in pheromone-treated plots may reflect the use of a suboptimal attractant, as Gries et al. (1993) found that the addition of (*E,E*)-4,10-dodecadienyl acetate enhanced attraction of the pheromone, although the compound was not confirmed in more recent GC-MS studies of the sex pheromone using SPME collections (Mozuraitis et al. 1999) and gland extracts (El-Sayed et al. 2005).

Preliminary field trials to develop mass trapping systems for apple clearwing moth, *Synanthedon myopaeformis* (Sesiidae), were conducted by Önucar and Ulu (1999) between 1994 and 1998 in which mass trapping with one trap/tree was found to reduce pupal numbers by 57% when trap catches ranged between 200 and 1200 moths/week. In a more comprehensive study with *S. myopaeformis*, Trematerra (1993) compared trap designs, pheromone loading, trap density, trap location in trees, and the effect of wind direction on trap catches between 1986 and 1991, and recommended 12 traps/ha for mass trapping.

Surrounding a centrally placed monitoring trap with either lures (mating disruption) or traps (mass trapping) Teixeira et al. (2010) were able to demonstrate that as the number of traps or lures per ha increased monitoring trap catch of peach tree borer, *S. exitiosa* (Sesiidae), lesser peach tree borer, *S. pictipes* (Sesiidae), and American plum borer, *Euzophera semifuneralis*, decreased. By comparing data from mating disruption and mass trapping trials they were able to demonstrate that traps were between 1.9 and 4.4 times more effective in reducing moth catch than dispensers alone, at the densities tested, and that competitive attraction was the mechanism responsible for disrupting monitoring trap catches. This hypothesis has been further tested by Aurelian (2011) who compared the level of catches in open assessment yellow unitraps (1-mg pheromone, grey halobutyl rubber septa dispensers) in plots treated with either open or closed traps (sealed with aluminum stoppers) at 25 traps/ha and found that catches in open assessment traps in plots containing open traps decreased compared to plots treated with closed traps. This result clearly demonstrated that the mechanism of action in the mass trapping plots was due to male moth disorientation and competitive attraction (i.e., mating disruption) and not simply removal of male moths by trap capture.

On other fruit tree crops preliminary studies designed to optimize parameters required for mass trapping have been conducted on *S. tipulifornis* (Sesiidae) on black currants, *G. funebrana* (Tortricidae) on plums (Koltun and Yarchakovskaya 2006), and *Rondotia menciana* (Bombycidae) on mulberry (Dai et al. 1988).

As early as 1976, field trials to control citrus flower moth, *Prays citri* (Yponomeutidae) were conducted in Israel by Sternlicht et al. (1990). Utilizing 120 traps/ha, significant reductions in larval infestations of lemon flowers were observed in treated plots provided the traps were operational throughout the year. Commercial field trials from 1980 to 1983 confirmed these results in over 30 ha of lemon trees using rubber septa lures loaded with 0.5 mg of pheromone replenished every four months. The sex pheromone, (*Z*)-7-tetradecenal (Z7-14Ald), first identified by Nesbitt et al. (1977), has proven to be a common pheromone for a large number of economically important *Prays* spp. (Gibb et al. 2005). The technique was found to be significantly cheaper than using three to six applications of insecticides/year and by 1989 almost all lemon growers in Israel were using the sex pheromone for mass trapping (Bakke and Lie 1989).

More recent work by Gibb et al. (2005) suggested the prospects for control of the related species *P. nephelomima* in citrus by mass trapping were good. Given the small size of the moth, (even biweekly catches of 300 per trap did not saturate a delta trap), the simplicity of the pheromone (Z7-14Ald) and that rubber septa loaded with 0.3 mg had a field life of six weeks, the economics of mass trapping as part of an IPM program would certainly be compelling. Indeed, Jamieson et al. (2008) subsequently optimized trap density in replicated trials of 3, 10, 30, 100, and 300 traps/ha to show that trap catch and number of infested flowers were reduced as trap density increased. Importantly, rindspot damage caused by larval feeding on fruit was significantly reduced from 45% to 16% as trap densities increased from 3 to 100 traps/ha.

Vang et al. (2011) working on *P. endocarpa* (Yponomeutidae), a pest on pomelo, *Citrus grandis* (Rutaceae), in Vietnam showed that although the pheromone is composed of three compounds, Z7-14Ald, (*Z*)-7-tetradecenyl acetate, and (*Z*)-7-tetradecen-1-ol, in a ratio of 10:3:10, lures baited with 0.5 mg of Z7-14:Ald alone were sufficient to control the pest in mass trapping trials to a level equivalent to that achieved with insecticides. However, the trial was only conducted on a plot of 0.1 ha and employed the equivalent of 200 traps/ha.

Working on tea tussock moth, *Euproctis pseudoconspersa* (Erebidae, Lymantriine) in southern China, Wang et al. (2005) were able to demonstrate significantly reduced levels of mating in 9 out of 12 sample dates and egg densities reduced by up to 51% compared to untreated plots. The research was conducted over a two-year period using 32-cm diameter white funnel traps located 90 cm above ground level, containing lures with 1.5-mg pheromone and deployed at a density of 25 traps/ha.

A number of groups have studied the possibility of using mass trapping for control of the goat moth, *Cossus cossus* (Cossidae), by optimizing the sex pheromone and dispenser type (Bratti et al. 1988) and number of traps needed per ha (Baronio et al. 1992b). A preliminary field trial was conducted by Pasqualini et al. (1985), but to date no significant application of the technology has been forthcoming. Similarly, considerable preliminary research has been conducted on mass trapping of *Thaumetopoea pityocampa* (Thaumetopoeidae) (Baronio et al. 1992a) building on earlier work by Cuevas et al. (1983) on poplar. The clearwing moth *Paranthrene tabaniformis* (Sesiidae) has also been the subject of considerable efforts to develop a mass trapping control technology (Du et al. 1984, 1985; Moraal et al. 1993).

The sex pheromone of the cocoa pod borer, *Conopomorpha cramerella* (Gracillariidae), is composed of two hexadecatrienyl acetates and their related alcohols (Beevor et al. 1986). The cost of using the pheromone for control by mating disruption was considered to be prohibitively expensive by Beevor et al. (1993). However, encouraged by the apparent trapping out of populations during pheromone optimization trials and the fact that the adults were weak flyers, they developed and tested a mass trapping technique that employed between 4 and 16 traps/ha in Sabah, East Malaysia. Traps were composed of two corrugated plastic squares to form a sandwich, with a sticky insert on the base. Polyethylene vial dispensers loaded with 0.05 or 0.1 mg of pheromone had field lives of up to four weeks and were deployed 0.5 m above the top of the tree canopies to maximize catch. In a 200-ha field trial conducted in 1988, Beevor et al. (1993) demonstrated that during the first two weeks of the trial, trap catches declined in the mass trapped area from a total of 3242 moths to 390 *C. cramerella* moths/night. Damage to harvested cocoa pods in the mass trapped area concomitantly decreased steadily during the 52 weeks of the trial from 450 to 230 pods/ha/week. In contrast, damage increased from 150 to 190 pods/ha/week in the control plot and remained constant at 100 pods/ha/week in another control plot over the same period. The success of the trial encouraged the introduction of estate-wide (2800 ha) mass trapping and other cultural control measures (e.g., bagging susceptible cocoa pods), which resulted in further significant reductions in damaged cocoa pods in all plots to 50 pods/ha/week. Ultimately the pheromone-based technology was not adopted even though a 5% reduction in damage resulting from the treatment would be sufficient to make mass trapping cost-effective (Beevor et al. 1993).

Sugarcane Pests

A wide range of sugarcane borer sex pheromones have been identified (Cork and Hall 1998). Initial research was conducted by David et al. (1985) on control of internode borer, *Chilo sacchariphagus indicus* (Crambidae) by mass trapping employing 25 water traps/ha in 5 and 10 ha scale trials during 1984 and 1985. Ten 0.004-ha subplots were sampled for damage and yield data in both trials demonstrating significantly increased average yields of cane sugar; 564 and 514 kg/subplot in mass trapped areas compared to 473 and 390 kg/subplot, respectively, in the control plots (Beevor et al. 1990). Later work on early shoot borer, *C. infuscatellus* (Crambidae) (Govindachari et al. 1992; Narasimhan 1995), confirmed that mass trapping could reduce dead heart formation from 45% to less than 14% with concomitant increases in yield of up to 30% more than farmers' practice.

Rice Pests

Rice can be infested by a complex of stem borers, from the families Crambidae and Noctuidae. The species present vary with location and season (Kamal et al. 2006). Early work on control of these pest species focused on the development and application of formulations for control of *Scirpophaga incertulas*, by mating disruption (Cork et al. 2008). Even though the technique was found to be highly efficacious (Cork et al. 1998), it was not cost-effective, given the relatively low value of the harvested rice, or acceptable to risk-averse farmers (Cork 1998). In order to try to reduce the cost of pheromone-based control, Gururaj et al. (2001) developed a mass trapping system using sleeve traps at a density of 20 traps/ha and pheromone lures (2 mg a.i., rubber septa) that had field longevity of approximately 20 days. In multilocation trials, the technique was found to be as effective as conventional insecticides, keeping infestations below the ETL of 10% dead hearts (DH) and 5% white heads (WH) in the vegetative and reproductive phases of the crop, respectively. Similar results were obtained in Bangladesh and the technique recommended for use in the transplanted Aman season rice crop (Cork et al. 2005b). In India, commercial production of *S. incertulas* lures increased from 20,000 in 2004–2005 to 308,000 in 2010–2011 (Alan Cork, unpublished data) suggesting that mass trapping is becoming increasingly accepted as an effective method of control.

Building on earlier work by Sheng et al. (2000) and Su et al. (2003) in China, Jiao et al. (2005) found that mass trapping with female sex pheromone of the striped rice stem borer, *Chilo suppressalis* (Crambidae) in trial plots of 80 and 100 ha (30 water traps/ha) over two years significantly reduced male populations by 84% on average and both DH and WH damage. Importantly, they established that the percentage of mated females was significantly reduced compared to controls over a two-year period (e.g., on average 77% of females were mated in control plots and 41% in mass trapped plots), although DH and WH damage levels in control plots were below economic levels.

Storage Pests

Karg and Suckling (1999) suggested that mass trapping would be most efficacious in protected areas in which the influx of adults is minimized, such as warehouses or other storage facilities. While this could be true, the presence of a semiochemical-baited trap in a storage facility may well result in the introduction of a pest into the facility, particularly if the bait was attractive to females, such as carob-based attractants (cf. Wakefield et al. 2005). Similarly, re-infestation of storage facilities after fumigation with methyl bromide is rapid (Trematerra 1991; Süss et al. 1996), suggesting that flour storage and milling facilities are by no means isolated from external pest populations (Doud and Phillips 2000). Trematerra and Gentile (2010) addressed this issue by placing traps both inside and outside an 11,500-m³ flour mill to assess the potential of mass trapping for control of Mediterranean flour moth, *Ephestia kuehniella* (Pyralidae) over a five-year period using funnel traps baited with 2 mg of (Z,E)-9,12-tetradecadienyl acetate (ZE9,12-14Ac) on natural rubber septa. Traps were located 2.5 m above floor level and 3.5 m from walls every 270 m³ inside the storage facility, and eight traps located outside the mill adjacent to areas that were contaminated with wheat residues such as the loading area and silos based on results from earlier work (Trematerra and Battanini 1987; Trematerra and Capizzi 1991). Their results demonstrated that populations could be significantly reduced inside the mills (>90% compared to initial infestation levels) and maintained at low levels when mass trapping was supplemented by spot treatments of insecticide and vigorous cleaning of equipment and facilities. However, before such IPM approaches to storage pest control can be adopted, ETLs need to be developed for insect infestations in commercial flour mills.

In toxicity tests of attracticide gels (Last Call) all blends containing permethrin (3–18% wt/wt) caused significant reductions in mating and increases in mortality of male Indian meal moth, *Plodia interpunctella* (Pyralidae) (Nansen and Phillips 2004). Of the two concentrations of the attractant, ZE9,12-14Ac, a loading of 0.16% (0.024 μg in a 0.015 g droplet) was found to be the most attractive in wind-tunnel bioassays of the attracticide with no evidence of repellency to permethrin. They were also able to demonstrate efficacy of the optimized attracticide in small warehouse rooms, provided the moth density did not exceed one adult pair per 11.3 m³.

Discussion

Lepidopteran pest management by means of attract-and-control techniques has considerable appeal for practitioners. Placing traps in a crop or storage facility to control by mass trapping has an immediate reassuring impact on producers. This psychological advantage is reenforced when dead insects accumulate in traps. Lure-and-kill and attracticide approaches offer users the reassurance that target pests will be killed by the insecticide applied, unlike mating disruption where male moths are not killed. The efficacy of such approaches can be optimized and several excellent examples now exist to demonstrate the validity of such an approach being adopted for use in commercial settings. Nevertheless, questions remain about the cost-effectiveness of the technology compared to other approaches and the robustness of the technology when presented with high population densities of target pest species. Attract-and-control approaches work most effectively as part of IPM programs where other stages of the target species are controlled by complementary bioagents, not least because lepidopteran larvae and not the adults are the damaging stage.

The actual mechanisms by which attract-and-control systems function are generally considered to be simple, involving primarily attraction to a source where the targets can be "killed." Recent research would suggest that this is not the case, with "high" semiochemical source densities evoking effects on responding insects more reminiscent of mating disruption, notably including false trial following. Contentiously, Yamanaka (2007) argues that improving capture efficiency of traps does not improve mating suppression, although improving lure efficiency does. There is certainly an upper limit to the dose that pheromones can be released from traps, if they are to remain attractive to male moths (e.g., Baker et al. 1981). However, application dose does not necessarily reflect release; as discussed above, release is highly dependent on the molecular structure of sex pheromone components. Thus, application dose and source density (e.g., traps or attracticide point sources) per hectare may vary with pheromone structure, formulation, insect population density, flight potential, and crop stage. While the elimination of more adult moths from a population should achieve a greater level of control (Byers 2007), elimination or removal need not necessarily involve killing insects, but merely delaying mating, as indeed mating disruption strives to achieve (Mori and Evenden 2013).

Combining sex pheromones of more than one species in attract-and-control systems has not been found to be particularly effective and using separate traps for each target species in a crop would be prohibitively expensive. Nevertheless, there are also significant opportunities to enhance the efficacy of lures by including kairomones that are attractive to both sexes (Camelo et al. 2007; Gregg et al. 2010) and possibly attract a range of pest species in a single crop (Tóth et al. 2010).

With improvements in killing agents, whether by trap design or toxicant, the variants of attract-and-control techniques continue to challenge the current focus on mating disruption and where the economics are favorable, they will prevail. However, unless and until the mechanisms which drive these behavioral responses are better understood, progress will remain slow.

References Cited

Ahmad, Z., and M. R. Attique. 1993. Control of pink bollworm with gossyplure in the Punjab, Pakistan. *International Organisation for Biological and Integrated Control of Noxious Animals and Plants, West Palaearctic Region Section Bulletin* 39:141–148.

Anonymous. 2003. *Trapping Guidelines for Area-Wide Fruit Fly Programmes*. Vienna: International Atomic Energy Agency, p. 16.

Aurelian, V. M. 2011. Semiochemical-based mass trapping of the apple clearwing moth *Synanthedon myopaeformis* (Borkhausen) (Lepidoptera: Sesidae). MSc thesis, University of Alberta, Canada.

Baker, T. C., W. Meyer, and W. L. Roelofs. 1981. Sex pheromone dosage and blend specificity of response by oriental fruit moth males. *Entomologia Experimentalis et Applicata* 30:269–279.

Bakke, A., and R. Lie. 1989. Mass trapping. Pp. 67–87. In A. R. Jutsum and R. F. S. Gordon, eds. *Insect Science in Plant Protection*. London: John Wiley and Sons, Ltd.

Baronio, P., N. Baldassari, and D. Scaravelli. 1992a. Quantitative development of a population of *Thaumetopoea pityocampa* (Den. and Schiff.) (Lepidoptera, Thaumetopoeidae) assessed by the mass trapping technique. *Frustula Entomologica* 15:1–9.

Baronio, P., E. Pasqualini, G. Faccioli, and G. Pizzi. 1992b. Determination of the number of delta-traps per hectare to optimize mass trapping of *Cossus cossus* L. (Lepidoptera: Cossidae). *Bollettino dell'Istituto di Entomologia della Università Bologna* 46:223–228.

Beevor, P. S., A. Cork, D. R. Hall, B. F. Nesbitt, R. K. Day, and J. D. Mumford. 1986. Components of the female sex pheromone of the cocoa pod borer, *Conopomorpha cramerella*. *Journal of Chemical Ecology* 12:1–23.

Beevor, P. S., H. David, and O. T. Jones. 1990. Females sex pheromones of *Chilo* spp. (Lepidoptera: Pyralidae) and their development in pest control applications. *Insect Science and its Application* 11:787–794.

Beevor, P. S., J. D. Mumford, S. Shah, R. K. Day, and D. R. Hall. 1993. Observations on pheromone-baited mass trapping for control of cocoa pod borer, *Conopomorpha cramerella*, in Sabah, East Malaysia. *Crop Protection* 12:134–140.

Bletsos, F., C. Thanassoulopoulos, and D. Roupakias. 2003. Effect of grafting on growth, yield, and verticillium wilt of eggplant. *Hortscience* 38:183–186.

Bratti, A., C. Malavolta, S. Maini, E. Pasqualini, and A. Capizzi. 1988. Comparative trials of sex attractant and dispenser types for *Cossus cossus* L. (Lepidoptera: Cossidae). *Bollettino dell'Istituto di Entomologia della Università Bologna* 42:179–192.

Britton, D. R. 2005. Using sex pheromone to control male *Helicoverpa armigera* Hübner (Lepidoptera: Noctuidae) in cotton. PhD thesis, University of New England, Australia, 233pp.

Butenandt, A., and E. Hecker. 1961. Synthese des Bombykols, des Sexuallockstoffes des Seidenspinners, und seiner geometrischen Isomeren. *Angewandte Chemie* 73:349–353. [In German]

Butler, G. D., Jr., and A. S. Las. 1983. Predaceous insects: effect of adding permethrin to the sticker used in gossyplure applications. *Journal of Economic Entomology* 76:1448–1451.

Workshop on Current Approaches to Pheromone Technology. Chennai, India: SPIC Science Foundation.

Nassef, M.A., A.M. Hamid, and W.M. Watson. 1999. Mass-trapping of pink bollworm with gossyplure. *Alexandria Journal of Agricultural Research* 44:327–334.

Nemoto, H., K. Takahashi, and A. Kubota. 1980. Reduction of the population density of *Spodoptera litura* (F.) (Lepidoptera: Noctuidae) using a synthetic sex pheromone. I. Experiment in taro field. *Japanese Journal of Applied Entomology and Zoology* 24:211–216.

Nesbitt, B.F., P.S. Beevor, D.R. Hall, R. Lester, M. Sternlicht, and S. Goldenberg. 1977. Identification and synthesis of the female pheromone of the citrus flower moth, *Prays citri. Insect Biochemistry* 7:355–359.

Nesbitt, B.F., P.S. Beevor, A. Cork, D.R. Hall, H. David, and V. Nandagopal. 1986. The female sex pheromone of the sugarcane stalk borer, *Chilo auricilius*: identification of four components and field trials. *Journal of Chemical Ecology* 12:1377–1387.

Núñez, P., A. Zignago, J. Paullier, and S. Núñez. 2009. Sex pheromones to control tomato moth *Tuta absoluta* (Meyrick) (Lep., Gelechiidae). *Agrociencia Uruguay* 13:20–27.

Önucar, A., and O. Ulu. 1999. Investigations on the possibility of mass-trapping technique for the control of apple clearwing moth, *Synanthedon myopaeformis* (Borkh.) (Lep. Sesiidae) in Aegean Region. *Bitki Koruma Bülteni* 39:115–125.

Pasqualini, E., F. Gavioli, P. Baronio, C. Malavolta, G. Campadelli, and S. Maini. 1985. Study on the possibility of implementing the mass-trapping method for *Cossus cossus* L. (Lep. Cossidae). *Bollettino dell'Istituto di Entomologia della Università Bologna* 39:187–199.

Pawar, A.D., and J. Prasad. 1983. Field evaluation of a synthetic sex pheromone "Litlure" as an attractant for males of *Spodoptera litura* (F.) (Lepidoptera: Noctuidae). *Indian Journal of Plant Protection* 11:108–109.

Pawar, A.D., and N.C. Tuhan. 1985. Codling moth (Lepidoptera: Olethreutidae): suppression by male removal with sex pheromone traps in Ladakh, Jammu and Kashmir. *Indian Journal of Entomology* 47:226–229.

Proverbs, M.D., D.M. Logan, and J.R. Newton. 1975. Study to suppress codling moth (Lepidoptera: Olethreutidae) with sex pheromone traps. *The Canadian Entomologist* 107:1265–1269.

Rahman, M.A., S.N. Alam, M.Z. Alam, and M.M. Hossain. 2008. Host preference of *Trathala flavoorbitalis* on brinjal shoot and fruit borer and rice leaffolder. *Journal of Biopesticides* 1:92–97.

Ramanjaneyulu, G.V., M.S. Chari, T.A.V.S. Raghunath, Z. Hussain, and K. Kuruganti. 2009. Non-pesticidal management: learning from experiences. Pp. 543–573. In R. Peshin and A.K. Dhawan, eds. *Integrated Pest Management: Innovation-Development Process*, Vol. 1. Berlin: Springer Sciences + Business Media B.V.

Reddy, G.V.P., and A. Guerrero. 2000a. Pheromone-based integrated pest management to control the diamondback moth *Plutella xylostella* in cabbage fields. *Pest Management Science* 56:882–888.

Reddy, G.V.P., and A. Guerrero. 2000b. Behavioral responses of the diamondback moth to green leaf volatiles of *Brassica oleracea* subsp. *capitata. Journal of Agricultural Food Chemistry* 48:6025–6029.

Roelofs, W., T. Glover, X.H. Tang, I. Sreng, P. Robbins, C. Eckenrode, C. Löfstedt, B.S. Hansson, and B.O. Bengtsson. 1987. Sex pheromone production and perception in European corn borer moths is determined by both autosomal and sex-linked genes. *Proceedings of the National Academy of Sciences of the United States of America* 84:7585–7589.

Sanders, C.J. 1986. Evaluation of high-capacity, nonsaturating sex pheromone traps for monitoring population densities of spruce budworm (Lepidoptera: Tortricidae). *The Canadian Entomologist* 118:611–619.

Shani, A. 1982. Field studies and pheromone application in Israel. 3rd Israeli Meeting on Pheromone Research, 4 May, Ben-Gurion University of the Negev, Beer-Sheva, Israel, pp. 18–22.

Shanower, T.G., J.A. Wightman, and A.P. Gutierrez. 1993. Biology and control of groundnut leaf miner (*Aproaerema modicella* Deventor) (Lepidoptera: Gelechiidae). *Crop Protection* 12:3–10.

Shelton, A.M., J.T. Andaloro, and J. Barnards. 1982. Effects of cabbage looper, imported cabbageworm, and diamondback moth on fresh market and processing cabbage. *Journal of Economic Entomology* 75:742–745.

Sheng, C.F., F.A. Yang, Y.B. Wei, C.Q. Zhu, and Y.W. Xiong. 2000. Field trials for mass trapping of rice stem borer *Chilo suppressalis* by sex pheromone. *Plant Protection* 26:4–5.

Singh, A., R.K. Tanwar, A.J. Tamhankar, and O.M. Bambawale. 2004. NCIPM initiatives to promote pheromones in IPM. Pp. 56–58. In A. Cork, P.K. Jayanth, S. Narasimhan, eds. *Enabling Small and Medium Enterprises to Promote Pheromone-Based Pest Control Technologies in South Asia.* Kent, UK: DFID Crop Protection Programme, R8304. Available online at: www.researchintouse.com

Sivapragasam, A., and T. Saito. 1986. A yellow sticky trap for the diamondback moth *Plutella xylostella* (L.) (Lepidoptera: Yponomeutidae). *Applied Entomology and Zoology* 21:328–333.

Sternlicht, M., I. Barzakay, and M. Tamin. 1990. Management of *Prays citri* in lemon orchards by mass trapping of males. *Entomologia Experimentalis et Applicata* 55:59–68.

Su, J.W., W.J. Xuan, C.F. Sheng, and F. Ge. 2003. The sex pheromone of rice stem borer, *Chilo suppressalis* in paddy fields: suppressing effect of mass trapping with synthetic sex pheromone. *Chinese Journal of Rice Science* 17:171–174.

Suckling, D.M., and E.G. Brockerhoff. 1999. Control of light brown apple moth (Lepidoptera: Tortricidae) using an attracticide. *Journal of Economic Entomology* 92:367–372.

Süss, L., D.P. Locatelli, and R. Marrone. 1996. Possibilities and limits of mass trapping and mating disruption techniques in the control of *Ephestia kuehniella* (Zell.) (Lepidoptera, Phycitidae). *Bollettino di Zoologia Agraria e di Bachicoltura* 28:77–89.

Svatos, A., A.B. Attygalle, G.N. Jham, R.T.S. Frighetto, E.F. Vilela, D. Saman, and J. Meinwald. 1996. Sex pheromone of tomato pest *Scrobipalpuloides absoluta* (Lepidoptera: Gelechiidae). *Journal of Chemical Ecology* 22:787–800.

Szocs, G., L.A. Miller, W. Thomas, R.A. Vickers, G.H.L. Rothschild, M. Schwarz, and M. Tóth. 1990. Compounds modifying male responsiveness to main female sex pheromone component of the currant borer (*Synanthedon tipuliformis* Clerk) (Lepidoptera: Sesiidae) under field conditions. *Journal of Chemical Ecology* 16:1289–1306.

Talekar, N.S., and A.M. Shelton. 1993. Biology, ecology, and management of the diamondback moth. *Annual Review of Entomology* 38:275–301.

Tang, L.C., and T.H. Su. 1988. Field trials of the synthetic sex pheromone of *Spodoptera litura* (F.). 1. Mass trapping of males. *Chinese Journal of Entomology* 8:11–22.

Teal, P.E.A., and J.H. Tumlinson. 1986. Terminal steps in pheromone biosynthesis by *Heliothis virescens* and *H. zea. Journal of Chemical Ecology* 12:353–366.

Teich, I., S. Neumark, M. Jacobson, J. Klug, A. Shani, and R.M. Waters. 1979. Mass trapping of males of Egyptian cotton leafworm *Spodoptera littoralis* and large scale synthesis of prodlure. Pp. 343–350. In F.J. Ritter, ed. *Chemical Ecology, Odour Communication in Animals.* Amsterdam: Elsevier/North Holland Biomedical Press.

Teich, I., A. Shani, and J.T. Klug. 1985. The role of mass trapping of the Egyptian cotton leafworm (*Spodoptera littoralis* Boisd.) in its integrated pest control. *Journal of Environmental Science and Health* 20:943–955.

Teixeira, L.A.F., J.R. Miller, D.L. Epstein, and L.J. Gut. 2010. Comparison of mating disruption and mass trapping with Pyralidae and Sesiidae moths. *Entomologia Experimental et Applicata* 137:176–183.

Tóth, M., I. Szarukán, B. Dorogi, A. Gulyás, P. Nagy, and Z. Rozgonyi. 2010. Male and female noctuid moths attracted to synthetic lures in Europe. *Journal of Chemical Ecology* 36:592–598.

Trematerra, P. 1991. Population dynamic of *Ephestia kuehniella* Zeller in a flour mill: three years of mass trapping. Pp. 1435–1443. In F. Fleurat-Lessard and P. Ducom, eds. *Proceedings of the 5th International Working Conference on Stored-Product Protection.* Bordeaux: Imprimerie du Me'doc.

Trematerra, P. 1993. On the possibility of mass-trapping *Synanthedon myopaeformis* BKh. (Lep., Sesiidae). *Journal of Applied Entomology* 115:476–483.

Trematerra, P., and F. Battanini. 1987. Control of *Ephestia kuehniella* Zeller by mass-trapping. *Journal of Applied Entomology* 104:336–340.

Trematerra, P., and A. Capizzi. 1991. Attracticide method in the control of *Ephestia kuehniella* Zeller: studies on effectiveness. *Journal of Applied Entomology* 111:451–456.

Trematerra, P., and P. Gentile. 2010. Five years of mass trapping of *Ephestia kuehniella* Zeller: a component of IPM in a flour mill. *Journal of Applied Entomology* 134:149–156.

Trimble, R.M., and E.A.C. Hagley. 1988. Evaluation of mass trapping for controlling the spotted tentiform leafminer, *Phyllonorycter blancardella* (Fabr.) (Lepidoptera: Gracillariidae). *The Canadian Entomologist* 120:101–107.

Tyson, R., H. Thistlewood, and G.J.R. Rudd. 2006. Modelling dispersal of sterile male codling moths, *Cydia pomonella* across orchard boundaries. *Ecological Modelling* 205:1–12.

Vang, L.V., N.D. Do, L.K. An, P.K. Son, and T. Ando. 2011. Sex pheromone components and control of the citrus pock caterpillar, *Prays endocarpa*, found in the Mekong Delta of Vietnam. *Journal of Chemical Ecology* 37:134–140.

Wakefield, M.E., G.P. Bryning, L.E. Collins, and J. Chambers. 2005. Identification of attractive components of carob volatiles for the foreign grain beetle, *Ahasverus advena* (Waltl) (Coleoptera: Cucujidae). *Journal of Stored Product Research* 41:239–253.

Wall, C. 1989. Monitoring and spray timing. Pp. 39–67. In A.R. Jutsum and R.F.S. Gordon, eds. *Insect Pheromones in Plant Protection*. London: John Wiley and Sons, Ltd.

Wang, X.P., V.T. Le, Y.L. Fang, and Z.N. Zhang. 2004. Trap effect on the capture of *Plutella xylostella* (Lepidoptera: Plutellidae) with sex pheromone lures in cabbage fields in Vietnam. *Applied Entomology and Zoology* 39:303–309.

Wang, Y., F. Ge, X. Liu, F. Feng, and L. Wang. 2005. Evaluation of mass-trapping for control of tea tussock moth *Euproctis pseudoconspersa* (Strand) (Lepidoptera: Lymantriidae) with synthetic sex pheromone in south China. *International Journal of Pest Management* 51:289–295.

Witzgall, P., P. Kirsch, and A. Cork. 2010. Sex pheromones and their impact on pest management. *Journal of Chemical Ecology* 36:80–100.

Yamanaka, T. 2007. Mating disruption or mass trapping? Numerical simulation analysis of a control strategy for lepidopteran pests. *Population Ecology* 49:75–86.

Youm, O., P.S. Beevor, L.J. McVeigh, and A. Diop. 1997. Effect of trap height and spacing in relation to crop height on catches of the millet stemborer, *Coniesta ignefusalis* males. *Insect Science and its Application* 17:235–240.

Yucra Equize, E. 2002. Densidad de trampas de feromonas para la captura de la polilla del tomate, *Tuta absoluta*, Meyrick. Pp. 17–82. In *Tesis de Grado*, Santa Cruz, Bolivia: Universidad Autonoma Gabriel Rene Moreno. [In Spanish]

Zhao, H.L., and M.S. You. 2001. Advances in research on the insecticide resistance of diamondback moth *Plutella xylostella* (L.) and its management. *Entomological Journal of East China* 10:82–88.

Zhong, Z.S. 2009. A review on control of tobacco caterpillar, *Spodoptera litura*. *Chinese Bulletin of Entomology* 46:354–361.

Zhu, Z.R., J. Cheng, W. Zuo, X.W. Lin, Y.R. Guo, Y.P. Jiang, X.W. Wu, et al. 2007. Integrated management of rice stem borers in the Yangtze delta, China. Pp. 373–382. In M.J.B. Vreyson, A.S. Robinson, and J. Hendrichs, eds. *Area-Wide Control of Insect Pests*. Vienna: International Atomic Energy Agency.

CHAPTER TWENTY-FOUR

Mating Disruption of Moth Pests in Integrated Pest Management

A Mechanistic Approach

MAYA EVENDEN

Introduction

The strict reliance of most species of moths on female-produced pheromones for long-range attraction of mates, and the relative stability of the signaling system, makes production of and response to sex pheromones ideal targets for management of lepidopteran pests. Synthetic copies of lepidopteran sex pheromones are most widely used in integrated pest management (IPM) to bait traps to attract moth pests in managed ecosystems (Miller et al. 2010; Witzgall et al. 2010). Capture of male moths in pheromone-baited traps can indi-

cate the presence of a target pest in the cropping area or monitor the density of an established pest population (Witzgall et al. 2010), facilitating subsequent decisions on pest control strategies. Sex pheromones are less widely used to directly control pest populations, but several tactics have been developed and deployed for the direct control of moth pests in a variety of managed ecosystems (Witzgall et al. 2010). To date, mating disruption is the most developed pheromone-based technology for direct control of moth pests and commercial efforts have been successful primarily against pests of perennial horticultural crops (Jones et al. 2009a), invasive species

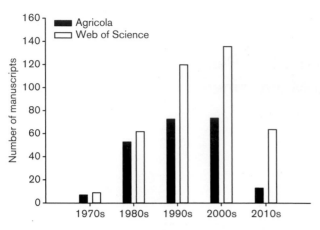

FIGURE 24.1 Number of manuscripts per decade recovered from Agricola and Web of Science databases with the term "mating disruption" in the title of the manuscript. Manuscripts include experimental studies on Lepidoptera only; reviews, patents, and conference proceedings are not included.

at low population densities (Sharov et al. 2002), and pest populations that cannot be otherwise managed with insecticides due to resistance or lack of efficacious registered products.

The idea of using synthetic sex pheromone as an atmospheric treatment to control insect pests was proposed by Babson (1963) as a method to eradicate the gypsy moth, *Lymantria dispar* (Erebidae), and by Wright (1963), as a selective method to control insect pests. Field trials in which pheromone treatment disrupted mate-finding behavior were first conducted against the cabbage looper, *Trichoplusia ni* (Noctuidae) (Gaston et al. 1967; Shorey et al. 1967). Terms such as pheromone-mediated communication disruption have been used to refer to this pest management tactic (Shorey 1977; Cardé 1990), but the term mating disruption is most commonly used, because the ultimate aim is to prevent insects from mating by disabling the pair-formation process. Following the inaugural studies illustrating the potential for mating disruption, studies on the development and implementation of mating disruption have increased (figure 24.1), and this approach is now used commercially to control a number of moth pests (Witzgall et al. 2010). In most instances, mating disruption is used as a component piece in a broader IPM program (Sarfraz et al. 2006; Witzgall et al. 2010). The most successful mating disruption programs are those in which the mechanisms of interference of the pair formation process by pheromone application have been studied and are exploited (Miller et al. 2006a; Miller and Gut 2015). Further field evaluation of mating disruption mechanisms (Miller et al. 2006b; Byers 2007) will likely reveal that mating disruption is achieved through several mechanisms acting together in a given system (Cardé and Minks 1995; Cardé 2007) (figure 24.2). The prevalence of a given mechanism will depend on the pheromone formulation (Minks and Cardé 1988; Gut et al. 2004), the method of pheromone release into the crop (Bartell 1982; Cardé and Minks 1995), and the insect's biology (Gut et al. 2004) and behavior (Cardé and Minks 1995).

Earlier reviews on population control of lepidopteran pests by mating disruption include larger treatments of the general use of semiochemicals in IPM (Foster and Harris 1997; Gut et al. 2004; Baker and Heath 2005; Witzgall et al. 2010) or target the research conducted on mating disruption of specific key pests (Witzgall et al. 2008; Ioriatti et al. 2011; Rhainds et al. 2012) or in specific agroecosystems (Phillips 1997;

Suckling 2000; Tamhankar et al. 2000). Reviews that broadly treat the use of mating disruption to control moth pests and to understand the mechanisms by which pheromone treatment interferes with the pair formation process include the following: Bartell (1982), Minks and Cardé (1988), Cardé (1990, 2007), Cardé and Minks (1995), Sanders (1997), and Miller and Gut (2015). The goal of the current review is to highlight recent research on mating disruption mechanisms evoked by pheromone treatment and the factors that impact these mechanisms and therefore the potential for population control by mating disruption. The review aims to understand mating disruption as a component of an IPM program and how it can be combined with other pest management tactics to achieve management of moth pest populations in different managed ecosystems. Potential areas for increased research and implementation of mating disruption are evaluated.

Mechanisms of Mating Disruption

Theory

Most research to understand the mechanisms by which treatment with synthetic pheromone interferes with moth mate finding, mating, and subsequent oviposition has focused on the interruption of male moth responsiveness to female-produced pheromone. Early researchers hypothesized several mechanisms by which pheromone treatment might interfere with moth pair formation, but theoretical and empirical testing of a mechanistic framework (Miller et al. 2010; Miller and Gut 2015) has lagged behind. Shorey (1977) suggested three mechanisms by which synthetic pheromone could influence male moths: (1) sensory adaptation of chemoreceptors on the male moths' antennae; (2) habituation of the central nervous system; and (3) confusion due to competition among sources of synthetic and natural pheromone. Bartell (1982) discussed four possible mechanisms evoked by pheromone treatment: (1) neural effects including both sensory adaptation and habituation of the central nervous system; (2) false-trail following, similar to Shorey's (1977) confusion mechanism, whereby males pursue plumes generated by synthetic pheromone in competition with female-produced plumes; (3) inability of responding males downwind of calling females to distinguish naturally produced plumes from a synthetic pheromone-laden background, also termed camouflage (Cardé 1981); and (4) alteration of behavior caused by an imbalance in the pattern of sensory input, e.g., when only one component of a multicomponent pheromone is released. Cardé (1990) suggested that an imbalance in sensory input was caused by either adaptation of antennal neurons or long-term habituation to one component of the insects' pheromone blend. Another potential mechanism of mating disruption includes a shift in the rhythm of response to pheromone so that periods of receptivity by the males and calling by females are offset (Cardé et al. 1998). These possible mechanisms of mating disruption are not mutually exclusive (Cardé and Minks 1995) and most likely multiple mechanisms interact (figure 24.2) to suppress mate-finding behavior in pheromone-treated crops (Cardé 2007).

Recent theoretical research on the mechanisms of mating disruption has grouped the previously described mechanisms into two broad categories: competitive and noncompetitive (Barclay and Judd 1995; Miller et al. 2006a; Miller and Gut 2015). Mating disruption by competitive mechanisms occurs

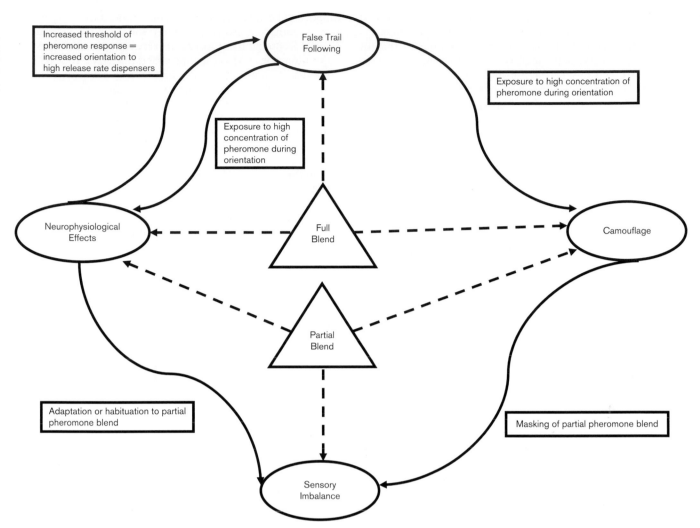

FIGURE 24.2 Mating disruption mechanisms ◯ that influence male moth behavior under treatment with attractive full pheromone blends or less attractive partial pheromone blends △. Dashed arrows indicate mechanisms elicited by the different blends. Solid arrows indicate how mechanisms may interact to effect male moth behavior as suggested in each ☐ placed between mechanisms ◯.

by competitive attraction, or false-trail following (Bartell 1982), in which males spend time and energy orienting to synthetic pheromone dispensers instead of calling females. The models describe an increase in mating disruption with pheromone dispenser density under competitive attraction that is nonlinear and asymptotically approaches, but never reaches, complete interruption of mate-finding behavior (Miller et al. 2006a). Simulation models reveal that the percentage of females that mate in a cropping system under mating disruption by competitive attraction is affected by the product of the effective attraction radius of the pheromone dispenser and the density of dispensers in the treated area (Byers 2007). Models predict that mating disruption via competitive attraction is sensitive to the population density of the target pest (Barclay and Judd 1995; Miller et al. 2006a; Byers 2007). The degree to which mating disruption declines with moth population density depends on the relative attractiveness of pheromone dispensers and calling females to male moths (Miller et al. 2006a). Model assumptions tested by a reanalysis of data from mating disruption field studies from the literature (Miller et al. 2006b) show that competitive attraction is the main mechanism that evokes mating disruption in the majority of case studies. Theoretical predictions of

mating disruption by competitive attraction were experimentally corroborated using the codling moth, *Cydia pomonella* (Tortricidae), in a large-cage study (Miller et al. 2010).

Models describing mating disruption via noncompetitive mechanisms result in different predictions than those generated for mating disruption by competitive attraction. The efficacy of mating disruption should be independent of pest population density under noncompetitive mechanisms (Miller et al. 2006a). If camouflage is the main mechanism, models predict that orientation to calling females should decrease linearly with dispenser density as pheromone coverage approaches 100%, instead of the nonlinear fashion predicted by competitive attraction (Miller et al. 2006a). The efficacy of mating disruption by camouflage is predicted to decrease with non-oriented movement of moths in and out of the camouflaging plumes provided by synthetic pheromone dispensers (Miller et al. 2006a; Byers 2007). Early hypotheses suggested that camouflage of calling females would be best achieved with pheromone blends that most closely mimic the natural female-produced blend (Cardé 1990). Moths may be less likely to move out of camouflaging synthetic plumes if the blend released evokes orientation behavior. This suggests that mating disruption by camouflage would be enhanced by competitive attraction (Miller et al. 2006a) (figure 24.2). Models emulating mating disruption as a result of neurophysiological effects or male moth desensitization predict that disruption efficacy should increase with pheromone release rate in the treated area (Barclay and Judd 1995; Miller et al. 2006a). Although pest population density should not affect mating disruption caused by neurophysiological effects, the sex ratio of active adults is predicted to influence treatment efficacy and species with high recruitment rates (males per female) and survivorship may be difficult to disrupt by this mechanism (Barclay and Judd 1995).

False-Trail Following

The importance of false-trail following as a mechanism for mating disruption by competitive attraction has been empirically tested in a number of ways. Several researchers have directly assessed the capacity of male moths to orient to synthetic pheromone dispensers in mating disruptant-treated plots. Orientation of male pink bollworm moths, *Pectinophora gossypiella* (Gelechiidae), to pheromone dispensers applied to cotton was videotaped in a wind tunnel positioned over two cotton rows in pheromone-treated and untreated fields. This approach revealed that male moths orient to and land on or near dispensers even in a pheromone-treated background, supporting false-trail following as one of the mating disruption mechanisms in this system (Cardé et al. 1998). The orientation behavior of four orchard-inhabiting tortricid moths toward high-dosage synthetic pheromone dispensers was observed in pheromone-treated and control orchards. Pheromone treatment of the orchard reduced the visitation rate of moths to the sentinel dispenser, but orientation within 100 cm of the dispenser occurred in three of the four species in the control orchard and two of the four species in the pheromone-treated orchard (Stelinski et al. 2004a). Capture of male red clover casebearer moths, *Coleophora deauratella* (Coleophoridae), on sticky cards positioned adjacent to high-dosage pheromone dispensers releasing the full attractive pheromone blend of this species (Evenden et al. 2010) in red clover, *Trifolium pratense*, revealed that false-trail following to dispensers occurs in pheromone-treated fields. Invocation of false-trail following is dependent on the attractiveness of the pheromone blend, as similarly positioned sticky cards adjacent to dispensers releasing the less attractive major component alone did not capture a significant number of moths (Mori and Evenden 2014). Videotape analysis of male peach tree borers, *Synanthedon exitiosa* (Sesiidae), shows that orientation approaches to pheromone dispensers are more brief in mating-disruption-treated orchards compared to untreated control orchards with >50% of approaches lasting <1 s in the mating-disruption-treated plot. This observation shows that false-trail following to synthetic dispensers occurs in this system, but the time spent interacting with the dispenser varies with dispenser density (Teixeira et al. 2010). The variation in the time spent by males at synthetic dispensers will be captured in the "retentiveness" term in models of mating disruption by competitive attraction (Miller et al. 2010). Capture or direct observation of moths at pheromone dispensers may be an underestimation of the importance of the false-trail following mechanism in mating disruption systems, as orientation behavior may occur far downwind of the dispenser, but due to adaptation (Baker et al. 1989) or habituation to pheromone the male moth becomes arrested in flight (Cardé and

Minks 1995) before contacting the dispenser. Direct observation of moths to assess false-trail following could be particularly misleading if the dispenser density is high and the pest population density is low to moderate (Miller et al. 2006a).

False-trail following can also be tested by comparing the efficacy of mating disruption achieved by dispensers releasing an attractive pheromone blend to that of dispensers releasing less attractive partial blends or blends of pheromone with behavioral antagonists. If this approach is used to test for false-trail following, it is important to first establish that partial pheromone blends and blends that include a pheromone antagonist do not in fact elicit upwind-oriented flight by the target moth. In several instances, attraction of male moths to dispensers releasing pheromone plus an antagonist has been reported (Suckling and Angerilli 1996; Witzgall et al. 1999; Stelinski et al. 2004a), suggesting that false-trail following could occur to blends containing antagonists.

Interestingly, most studies that test for false-trail following by comparing blends of various levels of attractiveness conclude that there is no difference in mating disruption efficacy between attractive and less attractive pheromone blends. Disruption of *Planotortrix octo* (Tortricidae) mate-finding communication was achieved equally in small-plot orchard trials with the attractive pheromone and a partial pheromone combined with a pheromone component of a sympatric species that acts as a pheromone antagonist to *P. octo*. These formulations both had greater disruptive efficacy than the antagonist alone. As the partial pheromone with the antagonist is not attractive, the authors concluded that false-trail following was not necessary to achieve disruption of this species (Suckling and Burnip 1996). The Guatemalan potato moth, *Tecia solanivora* (Gelechiidae), is equally disrupted by formulations releasing their attractive pheromone and a generic off-ratio blend. Wind-tunnel experiments show similar, albeit low, rates of upwind-oriented flight to dispensers releasing both blends, suggesting that some degree of false-trail following could occur to the off-ratio blend (McCormick et al. 2012). The pea moth, *Cydia nigricana* (Tortricidae), is attracted to fresh dispensers releasing its pheromone (*E, E*)-8, 10-dodecadienyl acetate in pheromone-treated pea fields, indicating that false-trail following occurs soon after pheromone treatment. However, isomerization of the pheromone to contain >6% of a combination of unattractive isomers occurs as early as two days post application and results in moth dispersal away from the pheromone-treated crop (Bengtsson et al. 1994; Witzgall et al. 1996). Despite the change in blend composition and the resultant effect on moth behavior, mate-finding communication is suppressed throughout the entire season, even in the absence of false-trail following to the unattractive blend (Bengtsson et al. 1994; Witzgall et al. 1996).

This approach has been tested thoroughly with the obliquebanded leafroller, *Choristoneura rosaceana* (Tortricidae), a secondary pest of pome fruits in North America. The four-component pheromone blend attractive to obliquebanded leafrollers in western North America (Vakenti et al. 1988) was compared with less attractive two- and three-component partial blends as mating disruptants over a range of release rates and dispenser densities. At the threshold pheromone release rate of 1.3 mg ha^{-1} h^{-1} released from either 1000 or 250 dispensers ha^{-1}, there was no difference in the level of disruption of mating of tethered sentinel females in plots treated with the attractive four-component blend and the less attractive two-component blend (Evenden et al. 1999a). These results, in conjunction with those in which a pheromone antagonist was incorporated into the four-component blend (Evenden et al. 1999b), suggest that false-trail following is not necessary to achieve mating disruption in this species. However, evidence for disruption by competitive attraction of eastern populations of the obliquebanded leafroller comes from a reanalysis (Miller et al. 2006b) of experiments that tested the disruption efficacy of high-dosage polyethylene tube dispensers at four different densities (Lawson et al. 1996). The straight-line relationship between dispenser density and the inverse of moth capture in assessment traps (Miller et al. 2006b) indicates competition for pheromone sources or false-trail following as a mating disruption mechanism under the experimental conditions reported. The conflicting results in these studies illustrate that, even within the same species, different mating disruption mechanisms can be evoked most likely due to differences in insect density, population differences in pheromone communication, mating disruption formulations, pheromone release rate, and environmental factors.

Camouflage

Camouflage is a noncompetitive mechanism by which the female-produced plume is obscured by that of the synthetic dispensers, so that at some distance downwind of the calling female her plume is masked (Cardé 1990). The efficacy of disruption by camouflage will depend on the uniformity of pheromone coverage and the concentration of synthetic pheromone as it compares to the pheromone release rate of the female (Cardé 2007). Camouflage of the female's signal may be best achieved by a synthetic pheromone that closely mimics the natural pheromone blend (Cardé 1990). However, because camouflage is considered as a noncompetitive mechanism in theoretical assessments of mating disruption mechanisms, synthetic plumes should not evoke orientation to pheromone dispensers (Miller et al. 2006a). In reality, orientation to dispensers that also cause camouflage of female plumes is likely and may actually enhance the effect of camouflage by retaining male moths in the plume produced by the synthetic dispensers (figure 24.1).

Empirical evidence of camouflage under field conditions is scant, but there are examples of this mechanism resulting from treatment with either attractive or unattractive pheromone formulations. Pheromone plumes from monitoring traps targeting the pink bollworm were camouflaged by pheromone released from high-dosage polyethylene dispensers releasing the attractive "gossyplure" blend, a 49:51 ratio of (Z,Z)- and (Z,E)-7,11-hexadecadienyl acetate. Monitoring traps were positioned at the downwind end of a wind tunnel positioned over two cotton rows in pheromone-treated and control fields. Trap capture of released pink bollworm males was almost eliminated when air from the pheromone-treated field was pulled into the tunnel at concentrations high enough to camouflage the plumes emanating from the monitoring traps. In untreated control fields, camouflage of plumes from monitoring traps was achieved by placement of dispensers on cotton plants on one of the two rows covered by the field wind tunnel. In this case, only the plume from the monitoring trap positioned within the same row as the dispensers was camouflaged and resulted in reduced trap capture of released pink bollworm males (Cardé et al. 1998). This experiment provided evidence that camouflage likely acts on a local

level and is most effective directly downwind of synthetic dispensers (Cardé et al. 1998).

The use of high-release-rate pheromone lures to bait assessment traps in pheromone-treated crops is one way that researchers can test for camouflage as a mating disruption mechanism. Early experimentation in cotton treated with mating disruption to control the pink bollworm showed that high-release-rate lures were "apparent" to male moths in both treated and control plots, whereas low release rate lures were attractive only in untreated plots. The concentration of pheromone released from mating disruption dispensers was great enough to mask the plumes emanating from the low release but not high-release-rate lures (Doane and Brooks 1981). Schwalbe and Mastro (1988) suggested that camouflage of female-produced plumes contributes to communication disruption of dense populations of gypsy moths as a result of high application rates of racemic disparlure, $(7R, 8S)$-cis-7,8-epoxy-2-methyloctadecane, composed of both the attractive (+) enantiomer and the (–) enantiomer that acts as a pheromone antagonist (Miller et al. 1977). Camouflage was assumed to be the mechanism causing disruption because male moth capture in traps baited with lures releasing a low dose of the attractive(+)-disparlure was readily reduced by 95% in plots treated with low application rates of racemic disparlure (5 g ha^{-1}), as compared to non-treated control plots, whereas moth response to high-dose (+)-disparlure lures was reduced by only 27% in these same plots, suggesting that the plumes emanating from low-dose lures were better camouflaged than those from high-dose lures. Camouflage of plumes from high-dose (+)-disparlure lures was achieved by high application rates of 500 g ha^{-1} of racemic disparlure (Schwalbe and Mastro 1988). These same findings could also be interpreted as a shift in threshold of pheromone responsiveness (figure 24.2) caused by neurophysiological effects of adaptation or habituation of pheromone processing. Similarly, camouflage was considered to be the major mechanism of mating disruption of the Guatemalan potato moth using an unattractive off-ratio blend (Bosa et al. 2006; McCormick et al. 2012), although the role of neurophysiological effects could not be ruled out in these studies.

Other studies have tested for the possibility of camouflage using wind-tunnel bioassays in which homogenous application of pheromone is feasible. The major pheromone component of two sympatric tortricid leafrollers, the obliquebanded leafroller and the three-lined leafroller *Pandemis limitata* (Tortricidae), was incorporated into a sprayable, microencapsulated formulation and applied to the upwind end of a wind tunnel. Male moths were less likely to orient to calling females in the treated wind tunnel, but disruption was more effective against the three-lined leafroller than the obliquebanded leafroller (Judd et al. 2005a), a species that is known to be difficult to disrupt (Gut et al. 2004). The mechanism of disruption was most likely camouflage of female-produced plumes because the formulation is not attractive to either species, and spray application results in uniform coverage at pheromone release rates too low to cause disruption by neurophysiological effects in these species (Stelinski et al. 2003a). Microencapsulated formulations result in a more uniform permeation of the environment than point dispensers (Sanders 1997) and are considered to be more likely to invoke camouflage than hand-applied dispensers that produce discrete pheromone plumes for false-trail following. However, there is evidence that male moths will orient to clumps of microcapsules in both wind-tunnel (Stelinski et al. 2005a) and field bioassays (Knight and Larsen 2008).

Neurophysiological Effects

The neurophysiological effects of male moth exposure to pheromone released from mating disruption dispensers have been divided into two main categories: adaptation of antennal pheromone receptors and habituation of central nervous system processing of pheromone signals. This division is somewhat artificial (Cardé 2007) as processing of pheromone signals relies on both peripheral and central nervous system function and both mechanisms can result in male moth inability to locate pheromone or a shift in response to natural sources of pheromone. Long-lasting adaptation of antennal receptors to pheromone as a result of atmospheric pheromone treatment may protect the central nervous system from habituation (Stelinski et al. 2003a), so these two neurophysiological effects of pheromone exposure are inextricably linked.

It has long been recognized that exposure to either continuous or pulsed sources of pheromone influences subsequent processing of pheromone signals and male moth behavior (Bartell and Lawrence 1973; Kuenen and Baker 1981). In most cases, researchers report a reduction in electrophysiological or behavioral response after pheromone exposure as measured by electroantennogram (EAG) or wind-tunnel bioassays, respectively. In a few instances, exposure to pheromone appears to sensitize male moths so that subsequent response to low doses of pheromone is enhanced (Anderson et al. 2003, 2007; Stelinski et al. 2004b). Enhanced responsiveness following pheromone preexposure is not the result of increased sensitivity of pheromone receptors on the antennae (Stelinski et al. 2004b; Anderson et al. 2007) but is correlated with an increase in sensitivity of the interneurons in the primary olfactory center of the brain in *Spodoptera littoralis* (Noctuidae) exposed to female-produced pheromone. Interestingly, brief exposures of male obliquebanded leafrollers to high doses of synthetic pheromones from reservoir-type mating disruption dispensers induce increased responsiveness to an attractive pheromone source in a wind-tunnel assay, as compared to that of naïve males. The physiological trait of enhanced sensitivity to low doses of pheromone following pheromone exposure may be one reason the obliquebanded leafroller is difficult to control with pheromone-based mating disruption (Stelinski et al. 2004b).

Pheromone exposure does not completely desensitize pheromone processing in many male moths but can instead lead to a shift in the threshold of pheromone responsiveness. Pheromone preexposure can cause a reduction in subsequent response to the female-produced pheromone plume or to low release rates of synthetic pheromone, but can also invoke responsiveness to high doses of synthetic pheromone that may normally not attract male moths. Preexposure of the almond moth, *Cadra cautella* (Pyralidae), to various doses of a pheromone spray in a wind tunnel induced a shift in responsiveness so that male moths were no longer responsive to normally attractive low-dose pheromone lures but oriented to high-dose lures to which naïve males were not attracted (Mafra-Neto and Baker 1996). Electrophysiological measurements of antennae from male red clover casebearer moths show that antennae from moths exposed to high doses of pheromone are less responsive to low-dose pheromone stimuli and more responsive to high-dose pheromone stimuli than naïve moths, suggesting that adaptation of pheromone receptors elevates the threshold of pheromone response (Mori and Evenden 2014).

There is little evidence that neurophysiological effects contribute much to mating disruption under field conditions. This is most likely because the atmospheric concentration of pheromone that male moths are exposed to in mating-disruption-treated crops (Koch et al. 2002) is far below the threshold concentration that has been shown experimentally to induce neurophysiological effects (Rumbo and Vickers 1997; Stelinski et al. 2003a). However, there does appear to be a correlation between the degree to which male moths are desensitized by high doses of pheromone and the ease by which they can be controlled by mating disruption (Gut et al. 2004). Long-lasting adaptation of peripheral nervous system pheromone receptors (Stelinski et al. 2003b) may protect moths from central nervous system habituation to pheromone. Long-lasting adaptation of antennal receptors in response to exposure of high concentrations of airborne pheromone has been studied in several orchard-inhabiting tortricid moths (Tortricidae) (Stelinski et al. 2003b, 2005b). Moth species that are not highly susceptible to long-lasting adaptation, such as the oriental fruit moth *Grapholita molesta* (Tortricidae) (Rumbo and Vickers 1997; Stelinski et al. 2003b) and the redbanded leafroller *Argyrotaenia velutinana* (Tortricidae) (Stelinski et al. 2003b), appear to be readily controlled by pheromone-based mating disruption. Conversely, species that demonstrate some level of long-lasting adaptation following pheromone exposure, including the codling moth (Judd et al. 2005b; Stelinski et al. 2005b), the obliquebanded leafroller (Stelinski et al. 2003b), and *Pandemis pyrusana* (Tortricidae) (Stelinski et al. 2005b), are relatively more difficult to control by pheromone-based mating disruption (Stelinski et al. 2005b). The susceptibility to peripheral nervous system adaptation varies by species but also with pheromone concentration and duration of the preexposure period (Stelinski et al. 2003a). Although the obliquebanded leafroller exhibits long-lasting adaptation after preexposure to pheromone above 500 pg ml^{-1} of air (Stelinski et al. 2003a, 2003b), there was no adaptation of antennal receptors after males were exposed to microencapsulated pheromone releasing ~8 pg cm^{-1} min^{-1} (Wins-Purdy et al. 2008). Preexposure to microencapsulated pheromone did result in significant reduction in orientation of male obliquebanded leafrollers to calling females up to 1 h following pheromone exposure, suggesting that central nervous system habituation occurs in this difficult-to-disrupt species (Wins-Purdy et al. 2008).

Some studies have tried to link the occurrence of neurophysiological effects measured in the laboratory to moth behavior in pheromone-treated plots in the field. European grape vine moths, *Lobesia botrana* (Tortricidae), preexposed to pheromone released from a mating disruption dispenser (0.3 mg h^{-1} for 3 and 8 h) in the laboratory exhibit antennal adaptation and reduced responsiveness to pheromone-baited traps positioned in an untreated vineyard for one night following exposure. Moths conditioned to pheromone in the field for 8 h in a 200 m^2 vineyard plot treated with 1 dispenser per 5 m^2 exhibited no subsequent decline in response to pheromone-baited traps positioned in an untreated vineyard (Schmitz et al. 1997). Similarly, codling moth males preexposed to the high airborne concentration of 35 ng ml^{-1} of pheromone in the laboratory have reduced antennal and behavioral response to pheromone tested in a wind-tunnel bioassay for 1 and 4 h following exposure, respectively. Placement of male codling moths in an orchard treated with 1000 rope dispensers each releasing 20 µg h^{-1} of pheromone did not affect subsequent response to pheromone 30 min after moths were removed from the orchard. This suggests that neurophysiological effects invoked by pheromone treatment in the field are not the result of exposure to atmospheric concentrations of pheromone but would require moths to be in close proximity to high-dosage pheromone dispensers (Judd et al. 2005b). Support for this hypothesis comes from a comparison of neurophysiological effects resulting from exposure of male obliquebanded leafrollers to pheromone dispensers in an orchard environment. Close proximity of male moths to high-dose dispensers resulted in long-lasting adaptation in which EAG response to pheromone stimuli was reduced for >5 min following pheromone exposure, whereas moths positioned 2 m from the nearest dispenser in a pheromone-treated plot did not display any level of adaptation (Stelinski et al. 2003a).

Sensory Imbalance

Sensory imbalance is a mechanism that can be invoked when an off-ratio pheromone blend or single pheromone component is used in the mating disruption formulation (figure 24.2). Early researchers (Bartell 1982; Flint and Merkle 1983) originally envisioned this to affect male moths through a disruption in sensory input caused by treatment with partial blends rendering the female-produced signal as unrecognizable. This definition was adopted by Miller et al. (2006b) who stressed that treatment with the off-ratio blend could interfere with the male moth's ability to perceive or recognize the female-produced signal rather than receive the pheromone molecules on odorant receptors. Other authors suggest that the mating disruption mechanism of sensory imbalance can also be a manifestation of the neurophysiological effects (figure 24.2) caused by the off-ratio blend that influence the males' ability to receive and process the entire pheromone signal necessary for orientation (Flint and Merkle 1984; Cardé 1990).

There are several studies that provide field evidence for sensory imbalance as a mating disruption mechanism when an off-ratio pheromone blend is used. Flint and Merkle (1984) treated cotton plots with either gossyplure (1:1 ratio of (Z,Z)- and (Z,E)-7,11-hexadecadienyl acetates) or with the (Z,Z) isomer alone and monitored capture of pink bollworm males in assessment traps baited with lures releasing varying ratios of the two isomers. Mating disruption was achieved by both formulations; however, in plots treated with a high dose of the (Z,Z) isomer alone, male moths were not captured in traps baited with the normally attractive 1:1 ratio of the two isomers, but their response was shifted so that orientation to traps releasing a 9:1 ratio of the (Z,Z):(Z,E) occurred. The shift in response to a blend with a higher percentage of the (Z,Z) isomer was attributed to adaptation or habituation to this compound as a result of background treatment with the (Z,Z) isomer alone (Flint and Merkle 1984). Lapointe et al. (2009) used response surface modeling and mixture-amount designs to test various blends of the two pheromone components, (Z,Z,E)-7,11,13-hexadecatrienal and (Z,Z)-7,11-hexadecadienal, as mating disruptants of the citrus leafminer, *Phyllocnistis citrella* (Gracillariidae). Disruption of communication to traps baited with the natural 3:1 combination of the two isomers was highest in plots treated with the pure triene. This finding could be the result of a shift in optimal response to pheromone through neurophysiological effects or reception of an off-ratio signal that the sensory system is unable to process (Lapointe et al. 2009). Conclusive evidence of sensory imbalance in this system would require replicated field trials

using the triene component alone as the mating disruption formulation with assessment traps baited with varying ratios of the two components.

Shift in Pheromone Response

An additional proposed effect of pheromone-based mating disruption is the induction of a shift in the timing of pheromone responsiveness that decouples female and male components of mate-finding behavior. The omnipresence of pheromone in mating-disruption-treated crops may cause male moths to respond to pheromonal cues before females call (Cardé and Minks 1995). Few studies have directly addressed this possibility in the field, probably due to the difficulty of observing moth flight behavior, especially at night. Videotape analysis of released pink bollworm males in field wind tunnels showed that pheromone responsiveness was advanced by 2–3 h as a result of pheromone exposure (Cardé et al. 1998). A recent field study explicitly tested the potential of this mechanism in mating disruption plots targeting the diurnally active peach tree borer. Videotape analysis of male moth approaches to various pheromone stimuli in pheromone-treated and untreated peach orchards showed that diel response to pheromone was not shifted as a result of background pheromone treatment (Teixeira et al. 2010).

Effects of Pheromone Treatment on Female Behavior

Most research probing the mechanisms behind mating disruption has focused on the effect of pheromone treatment on male moths. As females of some moth species can autodetect (den Otter et al. 1978) the pheromone that they produce, there is potential that mating disruption could also impact female behavior. Electrophysiological analyses illustrate female moths in the Tortricidae (den Otter et al. 1978; Palaniswamy and Seabrook 1978; DeLury et al. 2005; Stelinski et al. 2006; Gökçe et al. 2007), Noctuidae (Light and Birch 1979; Ljungberg et al. 1993; Groot et al. 2005; Sadek et al. 2012), Yponomeutidae (van der Pers and den Otter 1978), Sesiidae (Pearson and Schal 1999), and Erebidae, Arctiinae (Schneider et al. 1998) can detect their sex pheromone. Autodetection of pheromone may provide female moths with a way to detect intraspecific competition and elicit subsequent behavioral changes in dispersal (Sanders 1987; DeLury et al. 2005), calling (Weissling and Knight 1996; Stelinski et al. 2006; Gökçe et al. 2007), mating (Kuhns et al. 2012), oviposition (Gökçe et al. 2007), and aggregation (Birch 1977).

Change in female behavior as a result of pheromone exposure is species-specific and has been best studied in tortricid moths. Exposure to pheromone under lab conditions increased flight activity of female spruce budworm *Choristoneura fumiferana* (Tortricidae) measured electrostatically in a Faraday cage (Sanders 1987). In contrast, flight activity of the three-lined leafroller was reduced in pheromone-treated portions compared to clean-air portions of a wind tunnel (DeLury et al. 2005). In the field, increased flight activity in response to pheromone may lead to dispersal away from pheromone-treated areas. Pheromone exposure induces calling behavior in some tortricid females (Weissling and Knight 1996) but has no effect (DeLury et al. 2005) or reduces (Gökçe et al. 2007) calling propensity in others. The timing of calling can be advanced (Palaniswamy and Seabrook 1985; Stelinski et al. 2006) or delayed (Gökçe et al. 2007) in tortricid moths as a result of pheromone treatment. A shift in the timing of female pheromone release as a result of pheromone treatment under field conditions could desynchronize female and male mate-finding behavior. Pheromone preexposure of females reduced subsequent copulation success observed in small mating arenas in two of four species of tortricid moths tested (Kuhns et al. 2012). Such an effect of pheromone preexposure on female copulatory behavior could increase the effectiveness of mating disruptant treatments in the field. Egg laying is reduced under pheromone treatment for some species (Gökçe et al. 2007) but not others (Stelinski et al. 2006). There are few studies, however, that examine the effect of mating disruption on female moth behavior in the field. Observations of obliquebanded leafroller females positioned in pheromone-treated versus control plots over two nights revealed that calling behavior was delayed in pheromone-treated plots and the overall time spent calling was reduced (Evenden 1998). Similar responses to pheromone treatment in the laboratory were later reported for this species (Gökçe et al. 2007). Recent field research on the effect of mating disruption on female European grapevine moths showed that pheromone treatment increased female movement and reduced the time spent calling (Harari et al. 2015). Much more research is needed under field conditions to determine how female moth behavior is modified as a result of pheromone treatment and whether any behavioral change will enhance or detract from the success of mating disruption.

Delayed Mating

Mating disruption may not always interfere directly with mating of moth pests, but sometimes causes a delay of mating that reduces offspring production and contributes to population control. Modeling of pest populations treated with pheromone-based mating disruption shows that the reproductive rate is reduced regardless of moth fertility when pheromone treatment delays but does not eliminate mating (Barclay and Judd 1995). This theoretical prediction is supported by empirical studies: a recent meta-analysis conducted on 24 moth species found mating delay reduces female fecundity, fertility, and pre-oviposition period but increases moth longevity (Mori and Evenden 2013). The effect of delayed mating on moth fitness has been well studied in the laboratory, but there are few studies demonstrating delayed mating under mating disruption treatment in the field (Rice and Kirsch 1990; Knight 1997; Fadamiro et al. 1999). Mating was delayed by at least 3 days in female codling moths released into pheromone-treated orchards compared to those released into untreated control orchards. As a 2-day mating delay in codling moth females results in a 40% reduction in the production of viable offspring, delayed mating under mating disruption will impact population density (Knight 1997). Knight (2007) showed that the proportion of mated female codling moths captured in pear ester kairomone-baited traps was similar in pheromone-treated and control orchards but multiple mating was reduced by pheromone treatment. It was estimated that pheromone treatment caused a reduction of female fertility through direct interference of mating (8%); reductions in multiple mating (7%); and a delay of mating (40%) that significantly impacted subsequent population levels (Knight 2007). Jones et al. (2008) note that codling moth fitness is reduced when mating is delayed for a short period, whereas a longer delay is required to reduce the fitness of the obliquebanded leafroller, a species that is known to be difficult to control by pheromone-based mating disruption. Perhaps species whose fitness is greatly impacted by delayed mating may be more vulnerable to control by mating disruption and, in this way, delayed mating could be considered an additional mechanism by which pheromone treatment interferes with population growth (Mori and Evenden 2013). More field studies are required to determine the degree to which delayed mating contributes to population control under pheromone-based mating disruption of different moth species.

Mating Disruption Formulations

Multiple mechanisms can be evoked by various mating disruption formulations. Which mechanisms and their comparative importance will depend on the type of formulation used and the attractiveness of the pheromone released. Different mechanisms of mating disruption were evoked by different formulations targeting the oriental fruit moth tested under the same experimental conditions. Dispensers releasing low levels of pheromone disrupted oriental fruit moth by false-trail following, whereas noncompetitive mechanisms were evoked as a result of treatment with high-release-rate dispensers (Reinke et al. 2014).

Mating disruption formulations must protect pheromone from degradation by UV and oxygen and provide adequate release of active ingredients for the required period of time (Heuskin et al. 2011). Release rate of semiochemicals from most formulations is dependent on the diffusion speed of the compound through the matrix of the dispenser and the evaporation from its surface. Characteristics of the dispenser including the matrix type, the distribution of pheromone throughout the dispenser, and the shape and size of the dispenser will influence the diffusion speed of pheromone through the dispenser (Heuskin et al. 2011). Once diffusion through the dispenser has occurred, evaporation of the molecule will depend on the chemical characteristics of the pheromone compounds (Gut et al. 2004) and the environmental conditions. Pheromone release from most formulations increases with ambient temperature which results in loss of active ingredient during the day when temperatures are high but when most moth pests are inactive (Witzgall et al. 2010). Mating disruption formulations can be divided into several categories with inherintly different characteristics. Major ways in which formulations differ include the following: (i) the matrix from which pheromone is released; (ii) pheromone release rate per dispenser; (iii) application procedure; (iv) pheromone distribution in the crop; (v) longevity of the formulation in the cropping season; and (vi) the mating disruption mechanisms evoked by the application. In some crop-pest systems many mating disruption formulations have been tested, whereas in others only one or few formulations are practical.

Reservoir-Type Dispensers

Reservoir-type dispensers are the most commonly used commercial type of mating disruption formulation. Despite their commonly used name, most reservoir-type pheromone dispensers have pheromone released through a solid matrix (Heuskin et al. 2011) of plastic or synthetic polymers (Gut et al.

2004). Reservoir-type pheromone dispensers are characterized by gradual release of the pheromone at rates of micrograms (McDonough et al. 1992; Knight et al. 1995) per hour. Although the goal is for dispensers to have a zero-order release kinetic or constant release rate, most reservoir type dispensers cannot maintain a constant release (Heuskin et al. 2011). Release of pheromone from the commonly used polyethylene rope dispensers is faster during the first three weeks of deployment in the field followed by a slower loss (McDonough et al. 1992), whereas PVC-resin dispensers achieve a first-order release kinetic of constant release that is dependent on the amount of pheromone applied per dispenser (Cork et al. 2008). In addition to the overall release rate of pheromone, the flux of pheromone dictates male moth response to pheromone (Baker and Heath 2005). Field EAG recordings show that application of polyethlyene rope dispensers to orchard plots results in large-scale fluctuations in pheromone concentration in the treated area (Karg and Suckling 1997), but this fluctuation did not impact the efficacy of mating disruption against the light-brown apple moth, *Epiphyas postvittana* (Tortricidae). Reservoir-type dispensers are applied by hand and due mainly to labor costs are dispensed at densities of 250–1000 ha^{-1} (Gut et al. 2004). These formulations can be designed with two reservoirs that each release a different species' pheromone blend. Application of dual reservoir dispensers can result in the disruption of more than one pest species in a given agroecosystem (Judd and Gardiner 2004; Il'ichev et al. 2007). Newer reservoir-type formulations incorporate biodegradeable dispensers to avoid additional labor costs associated with removal of dispensers from the managed ecosystem (Anfora et al. 2008).

Because pheromone release rate from individual reservoir-type dispensers is several magnitudes higher than a calling female (Sanders 1997), treatment with these formulations should result in camouflage and neurophysiological effects as mechanisms of mating disruption (Cardé 2007). The importance of false-trail following as a mechanism evoked by reservoir-type dispensers is difficult to assess if trail following occurs far downwind of high-release-rate dispensers. False-trail following directly to reservoir dispensers occurs in some systems (Stelinski et al. 2004a; Mori and Evenden 2014) and works in concert with neurophysiological effects (Miller et al. 2010) (figure 24.2).

Female-Equivalent Dispensers

Female-equivalent dispensers also release pheromone from a solid matrix but are designed to release pheromone at rates similar to a calling female (Sanders 1997) to promote the attractiveness of the formulation to male moths. Common female-equivalent dispensers include flake and hollow fiber dispensers. Because of the relatively low release rate from each individual point source, many more dispensers must be applied per hectare (~10,000) than the higher release rate reservoir-type dispensers. The advantage of these types of formulations is that they can be aerially or manually sprayed onto crops, but application still requires specialized equipment (Cardé 2007). Although female-equivalent formulations were among the first to be registered for commercial use (Baker and Heath 2005), their use has been limited mainly to forestry applications (Sharov et al. 2002; Rhainds et al. 2012) where application by air is necessary.

A major issue of sprayable female-equivalent formulations is a lack of adhesion of dispensers to the crop canopy, particu-

larly following rain (Stelinski et al. 2008a). Atmospheric concentration of gypsy moth pheromone measured by portable EAGs in forest plots treated with flake dispensers with and without a sticker adjuvant was higher in plots where flakes remained on the foliage due to the addition of the sticker. Mating success of female gypsy moths was negatively correlated with atmospheric pheromone concentration (Thorpe et al. 2007a). The nonbiodegradable female-equivalent dispensers can build up in the environment and unintentially impact managed ecosystems and pest management activities. Flake dispensers applied for gypsy moth control in plots in both Wisconsin and Virginia prolong the efficacy of mating disruption based on reduced capture of males in pheromone-baited traps one year after application compared to that in control plots. Although increased temporal efficacy of the mating disruption treatment might be considered beneficial to the management of gypsy moth, it can also interfere with population density assessments obtained by male moth capture in pheromone-baited traps in the year following application (Onufrieva et al. 2013).

Female-equivalent dispensers promote competitive attraction via false-trail following (Stelinski et al. 2008a). Based on the release rate from individual dispensers (Golub et al. 1983), the concentration of atmospheric pheromone would not be high enough to induce significant neurophysiological effects (Stelinski et al. 2003a). An even distribution of pheromone throughout the crop canopy should promote camouflage of female-produced plumes (Cardé 2007).

Wax Emulsion Formulations

In an effort to design a mating disruption formulation that could be sprayed onto the crop, protect the pheromone compounds from degradation but biodegrade at the end of the season, paraffin wax was developed as a pheromone release matrix (Atterholt et al. 1998). Wax emulsion formulations can be considered as female-equivalent dispensers if the wax droplets are small and release similar amounts of pheromone to females. Automated application of wax formulations (Stelinski et al. 2007a) could be adopted to dispense various differently sized droplets which remain active in the field for different periods of time. The release rate of pheromones from wax emulsion formulations targeting several tortricid moths (Stelinski et al. 2005c; Epstein et al. 2006; Jenkins and Isaacs 2008) followed first-order release characteristics. The longer chain length pheromone of the citrus leafminer was released from one gram dollops of a wax emulsion formulation at a rate of 70 µg day^{-1} over the first week post application and a constant release rate of 6.4 µg day^{-1} for the next 15 weeks. This slow release rate provided season-long disruption of mating with two spray applications (Stelinski et al. 2010). Like most female-equivalent dispensers, a high density of wax droplets is required to achieve disruption. An application of 27,300 wax droplets ha^{-1} was more effective as a mating disruptant against the oriental fruit moth than lower densities (820–8200 ha^{-1}) tested (Stelinski et al. 2005c). An additional advantage of the wax emulsion formulations is that color can be incorporated into the formulation. Kwon (2014) showed that diurnally active apple clearwing moths, *Synanthedon myopaeformis* (Sesiidae), are more attracted to wax droplets that are black than the standard gray-colored formulation or any other color tested.

Mechanisms evoked by wax emulsion formulations will differ by the percent of pheromone in the formulation, and the size and density of droplets applied and by the pest-crop system. False-trail following was suggested as the main mechanism elicited by wax droplets in small plot trials against the oriental fruit moth. This conclusion was based on direct observation of male moths orienting close to droplets, increased efficacy of disruption with increased droplet density, and an overall low release rate that would be unlikely to camouflage female pheromone plumes (Stelinski et al. 2005c). In contrast, male citrus leafminers were not attracted to wax droplets formulated with various concentrations of the full and partial pheromone blends of this species (Stelinski et al. 2010), suggesting a noncompetitive mechanism is the main mechanism of mating disruption using wax droplets, as has been illustrated with other mating disruption dispensers used against this species (Stelinski et al. 2008b).

Microencapsulated Formulations

The convenience and economics of adoption of conventional ground and air spray equipment to apply pheromone formulations led to the early investigation of pheromone encapsulation in microscopic polymer capsules to disrupt mating of forest (Cameron et al. 1974), orchard (Cardé et al. 1975), and vineyard (Taschenberg and Roelofs 1976) moth pests. Application of microencapsulated mating disruption formulations results in many millions of capsules dispensed per hectare providing a fairly homogenous distribution of pheromone in the treated area (Cardé 2007). Microencapsulated formulations can be tank mixed with other agrochemicals such as fertilizers or fungicides and thereby reduce overall labor costs associated with application of pheromone. Adoption of microencapsulated formulations, however, has been hampered by the short life span (2–4 weeks) of the formulation in comparison with other dispenser technologies. Microencapsulated formulations characteristically release a large initial pulse of pheromone which is followed by a sharp decline in airborne pheromone concentration 7–10 days after application. Pheromone concentration in the air and release from foliage remains at low levels for 4 weeks after application (Polavarapu et al. 2001). An increase in the concentration of pheromone in the microcapsules does not measureably increase the effectiveness of the formulation as compared to multiple applications of microcapsules with low pheromone loading (Polavarapu et al. 2001; Stelinski et al. 2007b).

The short period of effectiveness of microencapsulated formulations under field conditions may in part be due to a lack of rainfastness and/or photodegradation of the capsules on the crop foliage (Waldstein and Gut 2004). Greater than 10 mm of rain results in a significant reduction in the number of microcapsules of three different generations of microencapsulated pheromone formulations retained on apple foliage (Waldstein and Gut 2004). Retention of microcapsules on the crop foliage may be particularly important to the treatment efficacy as contact with the pheromone-treated surface appears to be important to adequately disrupt mate-finding behavior (Judd et al. 2005a). Various types of sticker adjuvants have been added to microencapsulated pheromone formulations in an effort to better retain microcapsules on foliage and increase the rainfastness of microencapsulated pheromone formulations. Addition of a pine-resin sticker to a microencapsulated pheromone formulation targeting the oriental fruit moth increased the effectiveness of the formulation but not its longevity. In contrast, the same sticker did not

improve efficacy of microencapsulated pheromone formulations targeting three other orchard pests (Stelinski et al. 2007b). A direct comparison of three different sticker types under simulated rainfall conditions on the foliage of apple, pear, and walnut showed that a latex sticker had greater potential to increase retention of microcapsules than either a pine-resin or polyvinyl-polymer stickers, but the effect was dependent on leaf type and morphology (Knight et al. 2004). Addition of horticultural oil to a microencapsulated pheromone formulation targeting the obliquebanded leafroller did not increase the efficacy or longevity of the formulation (Wins-Purdy et al. 2007).

Microencapsulated pheromone formulations release a relatively diffuse cloud of pheromone rather than distinct plumes released by other formulation types. It follows that disruption of male mate finding by microencapsulated pheromone formulations should stem from noncompetitive mechanisms such as camouflage and neurophysiological effects that are effective at high pheromone concentrations (Miller et al. 2006a). Although initial airborne concentrations of pheromone released from microencapsulated formulations (100–400 ng m–3) reduce mating of male moths in several systems, an increase in the concentration of the active ingredient in the formulation does not enhance effectiveness in field trials (Polavarapu et al. 2001; Stelinski et al. 2007b). Trials using a microencapsulated formulation loaded with a partial pheromone blend against two orchard tortricids disrupted mate-finding behavior for three weeks in the field but only for 2–9 days in a wind tunnel treated with the formulation at different pheromone concentrations. It was postulated that male moths need to contact the foliage treated with the microcapsules for the formulation to be effective (Judd et al. 2005a). Subsequent studies show that surface exposure to the formulation has a strong negative effect on mate-finding behavior of the obliquebanded leafroller but 75 s after removal from the treated surface antennal adaptation was not observed. These studies suggest that habituation of the central nervous system may be an important mechanism caused by unattractive microencapsulated formulations if moths are in contact with the treated substrate (Wins-Purdy et al. 2008). There is a possibility that some false-trail following can occur to microencapsulated pheromone formulations as orientation by male codling moths to leaves treated with high densities of microcapsules has been observed in wind-tunnel studies (Knight and Larsen 2008). Wind-tunnel assays have documented initial disorientation of male codling moths caused by camouflage or neurophysiological effects followed by false-trail following to clumps of ~30 microcapsules, 2–6 days after application (Stelinski et al. 2005a). It is not known if orientation to microcapsules on leaves occurs in treated orchards and if spray application can be manipulated to increase this effect.

Aerosol Dispensers

The early work of Farkas et al. (1974) illustrated that pheromone dispensers could be widely spaced (200–400 m) in the cropping area and still provide communication disruption if the amount of pheromone released per dispenser was very high (40–640 µg min-1). This information was later exploited by researchers in the development of dispensers made up of pressurized cannisters from which pheromone is emmitted in an aerosol spray (Shorey et al. 1996; Baker et al. 1997; Isaacs et al. 1999). The advantages of this type of dispenser are that the release rate of pheromone and the timing of pheromone release can be programmed to the specific requirements of the pest-crop system. This eliminates emission and waste of pheromone during the day when moth pests are not active. Pheromone components are also protected from photodegradation within the cannister (Isaacs et al. 1999). Newer versions of these dispensers are equipped with light and temperature detectors to fine-tune pheromone release based on the pheromone biology of the target pest. Although aerosol dispensers represent an expensive upfront cost (Suckling et al. 2007), there are savings in labor associated with their application, as only ~1–10 dispensers are deployed per hectare (Cardé 2007) and placement of cannisters around the perimeter of the cropping area is adequate to provide disruption in some systems (Shorey et al. 1996). An economic analysis conducted in pear orchards in California illustrated that the use of aerosol dispensers for mating disruption of codling moth resulted in a $352 ha^{-1} savings as compared to conventional insecticide-based management. Because of the upfront costs of aerosol dispensers, there were minimal savings in the first two years post adoption of the aerosol dispensers, but savings were realized after three years of use (Elkins et al. 2005).

Successful communication disruption with aerosol dispensers occurs with pheromone release rates ranging from 0.9 mg ha^{-1} day^{-1} (Shorey et al. 1996) to 254 mg ha^{-1} day^{-1} (Shorey and Gerber 1996). High-release-rate dispensers can be positioned around the perimeter of the cropping area and still provide protection throughout the crop. Communication disruption of moth pests in field, vineyard, and orchard, 1 and 4 ha plots treated with aerosol dispensers around the plot perimeter, was as effective as evenly distributed aerosol dispensers, as long as the overall pheromone release rate was maintained at 0.9 mg ha^{-1} day^{-1}. The perimeter treatment, however, broke down when plot sizes of 16 ha were compared (Shorey et al. 1996). Perimeter treatment with aerosol dispensers is particularly useful for mating disruption treatment in crops such as cranberry where deployment of dispensers in the interior of the field can result in crop damage (Fadamiro et al. 1998). The distribution and concentration of pheromone released from widely spaced aerosol dispensers will depend on the crop canopy structure and the wind speed. EAG recordings in an orchard setting illustrated that light-brown apple moth males can detect pheromone 5–40 m downwind of a single aerosol dispenser (Suckling et al. 2007). It is likely that continuous rerelease of pheromone from the surrounding foliage or from the dispenser itself (Fadamiro et al. 1998; Suckling et al. 2007) contributes to the efficacy of pheromone-based mating disruption using aerosol dispensers. Captures of *Platynota stultana* (Tortricidae) were reduced by 84% compared to those in control plots the night following removal of aerosol dispensers from treated grape and peach plots, illustrating that pheromone concentrations were high enough to disrupt behavior after dispenser removal (Shorey et al. 1996). These findings prompted the development of a modified high-release-rate dispenser in which passive pheromone release instead of active aerosol pulses is maintained (Baker and Heath 2005).

The high concentration of pheromone released by aerosol dispensers and re-released by the dispenser and/or surrounding crop foliage should impact male moths through neurophysiological effects and camouflage of female-produced plumes (Cardé 2007). If an attractive pheromone is released by aerosol dispensers, false-trail following far downwind of the dispenser will occur and increase the likelihood of male moths experiencing adaptation or habituation as a result of pheromone exposure (Suckling et al. 2007) (figure 24.2). To enhance moth exposure to pheromone from widely spaced, high-release-rate dispensers, an attractive pheromone blend should be used in the formulation (Baker and Heath 2005).

Other Formulations

Other mating disruption formulations have been proposed and active research continues in their development. Baker and Heath (2005) suggested that research on high-emission-rate, widely spaced dispensers should focus on the development of dispensers that generate a large plume-strand flux, rather than overall high release rates. Development of asymmetrical dispensers of either a planar or a cylindrical design would result in increased pheromone flux of individual strands of pheromone released by the dispenser and would have a greater impact on responding male moths. To maximally enrich the pheromone signal released from the dispenser the long axis of the dispenser should align along the wind line (Baker and Heath 2005).

Recent research on the use of insects as vectors of pheromone dissemination, in what has been termed "mobile mating disruption," represents a promising new avenue for mating disruption, particularly in sensitive urban areas for the control of invasive pests (Suckling et al. 2011). Sterile Mediterranean fruit flies, *Ceratitis capitata* (Diptera: Tephritidae), treated with microencapsulated formulations of the light-brown apple moth pheromone and released at densities of 1000 flies per hectare disrupted mate-finding behavior of the light-brown apple moth for several days following the release of treated flies. The short period of activity is a major issue in this approach and regular releases of the pheromone-treated insect vector, as would occur in a Sterile Insect Release program, would need to be coupled with this approach (Suckling et al. 2011). Alternatively, the target insect pest could act as the pheromone vector of its own pheromone. Electrostatic powders can be treated with sex pheromone and adhere to the cuticle of the insect (Nansen et al. 2007). Treatment of European grape vine moths with pheromone-treated electrostatic powder reduced the electrophysiological and behavioral response of the treated moths to sex pheromone sources in comparison to non-treated control moths (Nansen et al. 2007). Therefore, pheromone treatment of male moths with electrostatic powder would reduce mating attempts of the vectors while also disrupting the mate-finding behavior of untreated conspecific males. This approach would likely be most effective using autodissemination techniques in which male moths treat themselves with pheromone-laden electrostatic powder at bait stations positioned in the cropping system (Baxter et al. 2008) and would likely need to be incorporated into a broader IPM system to achieve population control.

Chemical Characteristics of Formulations

Pheromone Blend

A long-standing question in mating disruption research has been whether or not the natural pheromone blend is necessary to achieve mating disruption. As a blend closely mimicking the female blend would be most likely to elicit the greatest number of mating disruption mechanisms at the lowest release rate, it would follow that mating disruption would be most efficacious using the natural pheromone blend (Minks and Cardé 1988). Theoretical treatment of data from various empirical mating disruption studies (Miller et al. 2006b) indeed suggests that mating disruption in the majority of the cases studied is achieved principally by competitive attraction which would require an attractive pheromone. The pheromone blend used in a given mating disruption formulation will influence the mechanism by which mating disruption principally occurs (figure 24.2).

In addition to the empirical studies that directly compare differentially attractive pheromone blends as mating disruptants that were discussed in the false-trail following section above, there are several examples of successful applications of mating disruption using incomplete pheromone blends. Interestingly, mating disruption using partial pheromone blends has been quite successful for moth species with a pheromone comprised of only two pheromone components, both of which make up a large (>20%) part of the pheromone signal. The citrus leaf miner pheromone consists of a 3:1 ratio of (Z,Z,E)-7,11,13-hexadecatrienal and (Z,Z)-7,11-hexadecadienal and both components are necessary to elicit attraction (Lapointe et al. 2009). Mating disruption can be achieved by the attractive full pheromone blend (Stelinski et al. 2008b) or either component alone (Lapointe et al. 2009), although the full attractive pheromone blend has a longer period of effectiveness in the field. Similarly, the pink bollworm pheromone has a two-component pheromone comprised of a 50:50 blend of (Z,Z) and (Z,E)-7,11 hexadecadienyl acetates and mating disruption can be achieved through treatment with the entire attractive blend or either isomer alone (Flint and Merkle 1983). The effectiveness of the partial blends against these two species maybe due to the invocation of the sensory imbalance mechanism in the pink bollworm (Flint and Merkle 1984) and citrus leafminer (Lapointe et al. 2009). Sensory imbalance is a mechanism that can only be invoked by a partial pheromone blend (figure 24.2).

Partial blends have also been used as mating disruptants against species for which most of the pheromone is comprised of a single major component (>60%) with several minor components that are often critical to the attractiveness of the pheromone constituting the remainder of the signal. In these cases, disruption can usually only be achieved with a partial blend that includes the major component (Evenden et al. 1999c). Due to the constraints on moth pheromone biosynthesis, closely related species often share the same major pheromone component. This allows pest managers to target more than one species using a partial pheromone blend that contains the shared major component of multiple sympatric species in a managed ecosystem. This approach can reduce the overall cost of management of the pest complex. Several species of leafroller pests of pome fruits (Tortricidae) in western North America all use the major pheromone component (Z)-11-tetradecenyl acetate (Z11-14Ac). The potential for multispecies mating disruption targeting 2–3 of these species with a partial blend containing Z11-14Ac has been tested with polyethylene tube reservoir-type dispensers (Deland et al. 1994), groups of female-equivalent fiber dispensers (Evenden et al. 1999a, 1999c), and microencapsulated pheromone (Judd et al. 2005a). Although the efficacy of the formulations releasing the partial pheromone blend varied depending on the leafroller species, mate-finding behavior of all species was signif-

icantly diminished as a result of treatment. As the major component is less attractive than the full pheromone blend, competitive attraction may have a diminished role in mating disruption using these partial blends. Neurophysiological effects are some of the main mechanisms that cause mating disruption using partial blends (Evenden et al. 1999a), especially when the major component is released from a microencapsulated formulation (Judd et al. 2005a; Wins-Purdy et al. 2008). Sensory imbalance is less likely to occur when the disruptant is a component that constitutes the majority of the pheromone blend. The fruittree leafroller, *Archips argyrospila* (Tortricidae), however, is attracted to pheromone-baited traps with a higher concentration of Z11-14Ac in plots treated with a 93:7 mixture of (Z):(E)-11-tetradecenyl acetates (Deland et al. 1994), indicating that mating disruption might be in part due to a shift in pheromone response caused by sensory imbalance.

Pheromone Characteristics

A blend of 2–7 individual compounds make up most lepidopteran pheromone signals. The chemical characteristics of these compounds including molecular weight and functional groups will dictate how these compounds are delivered into and persist in the environment (Gut et al. 2004). Long-chain moth pheromone molecules (18–21 carbon atoms) are more persistent in the cropping environment than short-chain pheromones (12–14 carbon atoms) which might contribute to the successful disruption of moths with long-chain pheromone signals (Gut et al. 2004). Pheromone released from mating disruption formulations is in the vapor state but can condense and be adsorbed onto solid surfaces such as foliage in the crop canopy. The molecular weight and functional groups of the pheromone molecule dictate the likelihood of pheromone adherence to foliage. The "stickiness" of the pheromone molecule increases with carbon chain length and oxygenation of the hydrocarbons (Gut et al. 2004). This effect would be expected to be great for pheromones of peach tree borers (18 carbon atoms), gypsy moth (19 carbon atoms) and Douglas-fir tussock moth, *Orgyia pseudotsugata* (Erebidae) (21 carbon atoms). Pheromones that are likely to adhere to the foliage can result in a buildup of pheromone in the canopy and contribute to successful mating disruption (Gut et al. 2004). This effect can result in prolonged mating disruption that is efficacious for greater than one season. Application of the high dose of 72 g ha^{-1} of pheromone applied in PVC beads from the air completely interfered with male Douglas-fir tussock moth orientation to feral females and no egg masses were found in treated plots during the year of application. The PVC beads continued to emit pheromone that impacted male moth behavior up to 2 years after treatment, which highlights the long-lasting effect of the 21-carbon chain length pheromone, (Z)-6-heneicosen-11-one (Gray and Hulme 1995). The mating disruption effect of a laminate flake formulation releasing racemic disparlure, (Z)-7,8-epoxy-2-methyloctadecane, at 37.5 g ha^{-1} against the gypsy moth lasted for 2 years. Trap capture of male gypsy moth was significantly reduced during the season of application and the subsequent two seasons, whereas female mating success was reduced during the season of application and for one additional year (Thorpe et al. 2007b). Although prolonged activity of mating disruption formulations may be considered beneficial, the lingering effects can also interfere with population monitoring using pheromone-baited traps.

Pheromone Antagonists

Some chemical signals that evolved primarily as moth pheromone components also function secondarily as antagonistic semiochemicals to prevent interspecific mating through partitioning the chemical communication channel (Evenden et al. 1999d). The potential for compounds that inhibit pheromone response to be used as mating disruptants against moth pests was recognized by researchers early on, but pheromone antagonists alone have not been effective mating disruptants (Kaae et al. 1974; Daterman et al. 1975). A combination of pheromone components and antagonists could result in mating disruption by causing species to emigrate from treated areas. Male pea moths are repelled from plots treated with pheromone-based mating disruption after isomerization of pheromone components to antagonistic isomers (Bengtsson et al. 1994).

The obliquebanded and three-lined leafrollers occur sympatrically and synchronically in pome fruit orchards in western Canada. Both species share the same major component, Z11-14Ac, and the minor pheromone component of the three-lined leafroller, (Z)-9-tetradecenyl acetate (Z9-14Ac), also acts as an antagonist to pheromone response in the obliquebanded leafroller. Application of Z9-14Ac as a mating disruptant in small plots did not influence mate-finding success in either moth species as compared to non-treated control plots (Evenden et al. 1999c). Although atmospheric treatment with Z9-14Ac did not interfere with conspecific mate location, it did induce heterospecific mate location in both directions (Evenden et al. 1999d). A combination of Z9-14Ac with Z11-14Ac disrupted mate finding of both species of leafroller but not to a greater degree than the shared major component alone (Evenden et al. 1999b).

Pheromone Analogs

Pheromone analogs or mimics are modified compounds that are structurally related to pheromones (Renou and Guerrero 2000) and evoke a behavioral response in the target pest (Cardé 2007). The use of pheromone analogs in pest management is attractive if the analog is cheaper or easier to produce and/or more stable than the natural pheromone. Despite broad research on the discovery, synthesis, and biological response to pheromone analogs (Renou and Guerrero 2000), few have been tested as mating disruptants. Pheromone analogs of aldehyde-based moth pheromone components have been investigated due to the cost of synthesis and instability of aldehyde pheromones. Application of a formate pheromone analog (Z)-9-tetradecen-1-ol formate disrupted pheromone communication of both corn earworms, *Helicoverpa zea*, and tobacco budworms, *Heliothis virescens* (Noctuidae), in small plots of tobacco (Mitchell et al. 1975). The formate analog is a structural mimic of the 14- and 16-carbon aldehyde compounds that are components of both species' pheromone. A hydrocarbon (Z)-1,12-heptadecadiene was tested as a pheromone analog of (Z)-11-hexadecanal to disrupt mating of *H. zea* and *H. virescens* in tobacco. Significant disruption of male moth capture in female-baited traps occurred only for *H. zea*, illustrating that pheromone analog activity is species specific (Carlson and McLaughlin 1982). The formate pheromone analog (Z,Z)-9,11-tetradecadien-1-ol formate achieved mating disruption activity against the navel orangeworm, *Amyelois transitella* (Pyralidae), but not to a greater degree

than the main pheromone component, (Z,Z)-11,13-hexadecandienal (Landolt et al. 1982). More recently, a chain length pheromone analog successfully disrupted sexual communication of the invasive guava moth, *Coscinoptcha improbana* (Carposinidae), in New Zealand (Suckling et al. 2013). The pheromone analog is also a pheromone component of the related peach fruit moth *Carposina sasakii* (Carposinidae). Guava moth olfactory neurons tuned to its own pheromone components also respond to the longer chain length analog. In mating disruption trials using the fruit moth pheromone, sexual communication of the guava moth was disrupted for 470 days in macademia nut orchards. The chain length analog is already registered as a mating disruptant in New Zealand and the longer chain length of the analog in comparison to the guava moth pheromone may increase the longevity of the treatment in the field (Suckling et al. 2013).

Pheromone and Plant Volatiles

Volatile chemical cues released by plants are often used as cues for herbivorous insects for host location and acceptance. In some instances, host plant volatiles also synergize the response of male moths to conspecific pheromone signals. The addition of host plant volatiles to attractive pheromone blends increases male moth trap capture for both corn earworm and codling moth (Light et al. 1993). A combination of host volatiles and sex pheromone produces a synergistic effect on the firing rate of olfactory receptor neurons in comparison to the pheromone alone in corn earworm (Ochieng et al. 2002). In the context of pheromone-based mating disruption, traps baited with host plant volatiles provide a way to monitor pest populations with noncompetitive lures. Pear ester, ethyl (E, Z)-2,4-decadienoate, is attractive to female codling moths (Light et al. 2001) and enhances attraction of male codling moths when combined with codlemone (Knight et al. 2005). Pear ester has been adopted as one way to monitor codling moth in pheromone-treated orchards (Knight and Light 2005), as it is an attractive lure that is not masked by codlemone released from the dispensers positioned in the treated orchard.

Recent work has evaluated the potential of combining codlemone with pear ester in mating disruption formulations to improve the efficacy of codling moth control (Knight et al. 2012; Stelinski et al. 2013) in so-called high-performance mating disruption (Stelinski et al. 2007c). It was anticipated that the combination of pear ester with codlemone would enhance the effectiveness of pheromone-based mating disruption by competitive attraction and provide improved control at high codling moth densities. However, using metrics such as trap capture in synthetic and virgin-female-baited traps as well as fruit damage, PVC dispensers loaded with codlemone or a blend of codlemone and pear ester were similarly effective (Knight et al. 2012). Two different kairomones, pear ester and (E)-β-farnesene, released alone and in conjunction with codlemone were tested as mating disruptants of codling moth. Both kairomones alone disrupted pheromone communication of codling moths and (E)-β-farnesene treatment also reduced mating of sentinel virgin females. Co-deployment of either kairomone with pheromone did not enhance the effectiveness of pheromone alone as a mating disruptant (Stelinski et al. 2013). The biological activity of kairomones as mating disruptants suggests that further study on using these compounds alone or in combination with pheromone treatment is warranted.

Distribution of Pheromone in the Cropping Environment

Dispenser Distribution

Mating disruption formulations inherently differ in the number of pheromone point sources that are distributed in the cropping area from several per hectare for aerosol dispensers to millions per hectare for microcapsule applications. Positioning of the chosen mating disruption formulation within the cropping environment can affect the efficacy of the formulation. While maintaining the overall pheromone application rate constant, Epstein et al. (2006) varied the number of point sources of pheromone release from reservoir-type dispensers targeting the codling moth in small orchard plots. The percent disruption of male codling moth orientation to pheromone-baited traps increased with an increase in density of pheromone release sites from 10 to 1000 ha^{-1}. Accordingly, fruit damage was significantly reduced only in plots treated with 1000 dispensers ha^{-1} evenly distributed throughout the plot (Epstein et al. 2006). Increased mating disruption efficacy with dispenser density is not limited to reservoir-type dispensers. Mating disruption against the light-brown apple moth increased with dispenser density of a reservoir-type dispenser and an emulsifiable wax-based formulation. A high number of pheromone release sites results in an atmosphere in which the insect will frequently encounter synthetic pheromone filaments and is correlated with high levels of mating disruption (Suckling et al. 2012a). Even distribution of 500 ha^{-1} reservoir-type dispensers targeting the grape vine moth in vineyards in Europe resulted in a uniform level of ambient pheromone within the treated plot as measured with field EAGs. Pheromone concentration dropped dramatically outside of the treated plots (Karg and Sauer 1995). For this reason, additional pheromone is often applied around the crop margins (Gut et al. 2004) or as part of an area-wide contiguous treatment (Calkins and Faust 2003).

The main mating disruption mechanism elicited by a given formulation and the biology of the target pest will dictate the optimal dispenser density in a given system. Female gypsy moths are flightless and cannot disperse. Mating disruption is achieved through aerial application of flake dispensers. Although the goal of application is even distribution of pheromone throughout the tree canopy, disruption also can be achieved with areas of clean air between swaths of pheromone application (Tcheslavskaia et al. 2005a) and at low moth densities on isolated trees in open landscapes (Onufrieva et al. 2008). Reduced coverage of flake dispensers can save on the costs of application time and product formulation. A 29% cost savings in pheromone and application costs was achieved with intentional coverage gaps in the application of a wax formulation targeting the citrus leafminer. Mating disruption efficacy was slightly reduced (4%) when a coverage gap of 4 rows for every 10 treated rows was applied (Lapointe et al. 2014). For systems in which mating disruption is achieved through competitive attraction, dispensers could be positioned such that their effective attractive radii (Byers 2007) overlap to ensure a maximum number of false trails and minimal clean air within the cropping area. More research is needed to understand factors that influence the effective attractive radii of different dispenser types in various mating disruption systems.

Dispenser Height

Most research conducted on vertical placement of mating disruption formulations in the cropping area has focused on mating disruption of tree crop and vineyard pests because of the relatively deep canopies of these crops. Theoretically, placement of dispensers in parts of the canopy where females release pheromone should result in the greatest level of mating disruption, as airborne pheromone concentrations are highest at the height where dispensers are applied (Karg and Sauer 1995). Disruption of codling moth males to synthetic pheromone-baited traps and tethered virgin females was greatest when dispensers were positioned in both the upper and the mid-canopy of 4–5 m tall apple trees. This treatment was significantly more effective than the industry standard placement of dispensers in the upper canopy alone (Epstein et al. 2011). In contrast, disruption of orientation of male oriental fruit moths to pheromone-baited traps was not influenced by the height in the tree canopy (low, high, throughout) where droplets of an emulsified wax pheromone formulation (De Lame and Gut 2006) or a microencapsulated formulation were applied (Stelinski et al. 2007b). Airborne pheromone concentrations differ with the height of dispensers positioned in vineyards. Plots treated with mating disruption dispensers targeting the grape vine moth at low and mid-canopy have higher ambient pheromone concentrations than those in which the dispensers are positioned high in the canopy (Sauer and Karg 1998). This could be due to an increased wind speed toward the top of the canopy that acts to disperse pheromone out of the crop.

Crop Canopy

It has long been known that atmospheric pheromone can be adsorbed onto plant foliage and then rereleased at concentrations that are behaviorally relevant to male moths (Wall et al. 1981). Apple leaves exposed to a constant stream of pheromone released from two reservoir-type pheromone dispensers for >3 days were attractive to male light-brown apple moths when used as lures in traps for three days following pheromone exposure (Karg et al. 1994). The importance of crop foliage to the efficacy of pheromone-based mating disruption is less well understood but release of pheromone from foliage should act to distribute pheromone throughout the treated area and may be particularly important when a low density of high-release-rate pheromone point sources are deployed. Some researchers have made use of field EAG measurements to assess the atmospheric concentration of pheromone at different stages of foliar development in the crop (Sauer and Karg 1998) or after the removal of pheromone dispensers from the cropping environment (Koch et al. 2009). In vineyards treated with dispensers releasing the 12-carbon diene acetate pheromone of the European grape vine moth, atmospheric pheromone concentration was higher in the summer when foliage was fully developed compared to the spring (Sauer and Karg 1998). This may be the result of rerelease of pheromone from foliage or a reduction of wind in the plot due to foliage development. Pheromone concentration in summer in a defoliated vineyard after a hail storm was significantly lower than summer measurements in a vineyard with leaves and did not differ from measurements taken in the spring. These findings suggest that foliage buffers and distributes pheromone and may play a more important role than tempertature in seasonal differences of pheromone concentration in mating disruption plots (Karg and Sauer 1997). In contrast, atmospheric concentration of gossyplure following removal of reservoir-type dispensers targeting the pink bollworm from 0.4 ha plots of cotton declined to levels undetectable by field EAG 1–10 h following dispenser removal. The small canopy (30–150 cm) of cotton plants may not contribute significantly to pheromone distribution in treated plots. Disruption of orientation of male light-brown apple moths to pheromone-baited traps continued at significant levels for three nights following removal of reservoir-type dispensers from apple trees (10 per tree). This finding suggests that the pheromone rereleased from foliage either masks the plume eminating from the trap or causes neurophysiological effects that impairs the males' ability to orient to the trap (Suckling et al. 1996).

Environmental Factors

Environmental conditions partly dictate how pheromone is distributed in mating disruption plots and therefore the efficacy of mating disruption. The principal environmental factors that affect pheromone release and movement in the plot include temperature and wind. Precipitation can also influence mating disruption efficacy primarily through dislodging sprayable formulations from foliage (Waldstein and Gut 2004). In addition to influencing pheromone distribution, these conditions will also influence insect behavior in the agroecosystem.

There is a direct impact of temperature on the release of pheromone from mating disruption dispensers. Pheromone release increases with ambient temperature from all dispenser types except for aerosol dispensers. Although this results in high levels of pheromone in the cropping area under warm air temperatures, much of this pheromone may be wasted if moths are only active at night (Witzgall et al. 2010). Despite a reduced release rate of pheromone through the season, atmospheric pheromone concentration in the treated area can actually increase later in the growing season. This is primarily the result of crop phenology, as a developed crop canopy can retain pheromone through buffering the effect of the wind on pheromone dispersion (Sauer and Karg 1998). In addition, leaves can adsorb and rerelease pheromone (Karg et al. 1994), contributing to uniform coverage of the cropping area with pheromone. High temperatures in mid-season may result in more pheromone released into the treated area despite an overall reduction in release rate over time (Karg and Sauer 1997). There is a higher ambient pheromone concentration in vineyards treated with mating disruption targeting the European grape vine moth in summer than in spring despite overall lower pheromone release rates from depleted dispensers later in the season. Maximum ambient concentrations of pheromone correlate with the warmest days of the season (Karg and Sauer 1997). Pheromone loss from dispensers as a result of high temperatures will also reduce the longevity of the formulation throughout the season and may necessitate that the initial pheromone loading per dispenser vary across growing regions with disparate temperature regimes (Gut et al. 2004). Ambient temperature can also influence the distribution of pheromone in the treated area. In the canopy of an almond orchard, stratification of temperature with height in the canopy promotes vertical mixing of pheromone plumes

and could influence the optimal placement of pheromone dispensers (Girling et al. 2013).

Wind speed and direction have a major impact on the distribution of pheromone in the mating-disruption-treated area. A Langarian model developed to predict the distribution of pheromone in a mating-disruption-treated orchard revealed that mean wind speeds have a major effect on the ambient concentration of pheromone within the canopy. Pheromone concentration is lower under high wind conditions (Suckling et al. 1999). Moderate to high wind speeds lead to vertical movement of pheromone plumes within the cropping environment. In an orchard environment, codling moth pheromone was contained within the orchard canopy under low wind conditions, but could be detected up to 6 m above the canopy in the presence of moderate wind speeds (Milli et al. 1997). Horizontal movement of pheromone out of the plot occurs under high winds and can lead to a depletion of pheromone concentration at the edge of the treated area, causing decreased mating disruption and control, often referred to as an "edge effect." Movement of pheromone out of the plot can also have implications for the assessment of mating disruption products. Milli et al. (1997) measured ambient pheromone concentrations similar to that in a treated orchard 60 m downwind of the treated plot and recommended that control plots be positioned a minimum of 500 m upwind from pheromone-treated areas.

The Effect of the Target Pest Biology on the Efficacy of Mating Disruption

Population Density

The density of the target pest population can be one of the most important criteria dictating the success of mating disruption, but this effect varies with the moth pest and the mechanisms by which mating disruption acts. In general, the efficacy of mating disruption is negatively correlated with the population density of the target pest, especially when mating disruption is achieved by competitive attraction (Miller et al. 2006a). An increase in density of reservoir-type dispensers and pheromone release rate increases the efficacy of mating disruption of the pink bollworm (Kehat et al. 1999), suggesting that mating disruption occurs, at least in part, by competitive attraction (Cardé et al. 1998). Although the percent disruption of pink bollworm populations is similar regardless of insect density, the absolute efficacy of the treatment declines with population density (Kehat et al. 1999). In other systems in which pheromone release rate does not influence mating disruption efficacy, high insect population densities can still reduce control. The efficacy of mating disruption of *Pandemis* leafroller moths (Knight and Turner 1999) and spotted tentiform leafminer, *Phyllonorycter blancardella* (Gracillariidae) (Trimble and Tyndall 2000), in apple orchards did not increase with pheromone dispenser density or release rate, suggesting that noncompetitive mechanisms may be disrupting mate-finding behavior. Nevertheless, mating disruption of both species declines with increased moth population density. Population thresholds above which mating disruption should not be applied will differ for each crop-insect system and need to be established to promote the economic use of pheromone-based mating disruption. There are a few moth species that appear to be susceptible to control by pheromone-based mating disruption even at high population densities. These species include those that are classified as "easy-to-disrupt," including the oriental fruit moth, the tomato pinworm, *Keiferia lycopersicella* (Gelechiidae), and the peach-tree borer (Gut et al. 2004). High levels of communication disruption (>99%) of the red clover casebearer moth were achieved at extremely high population densities (>6500 moths captured per week in control plots) using reservoir-type pheromone dispensers. Further, this level of disruption was achieved at three different dispenser densities, suggesting that noncompetitive mechanisms are involved in the control of such dense populations (Mori and Evenden 2014).

Population density is most likely to influence the efficacy of mating disruption for species that exhibit cyclical increases in population density, like many lepidopteran forest defoliators. Mating disruption is applied to at least 200,000 ha of forested land in the United States each year (Witzgall et al. 2010) to target low population densities of gypsy moth in the "slow-the-spread" program. Mating disruption is only efficacious against low population densities of gypsy moth (Webb et al. 1990) and can therefore only be applied in areas at the leading edge of gypsy moth invasion where moths are present at low population densities. Webb et al. (1988, 1990) estimate that gypsy moth populations with ≤15 egg masses ha⁻¹ can be controlled by mating disruption applied at a minimum of 50 g AI ha⁻¹. Later studies showed that much less pheromone (15 g AI ha⁻¹) can be used at the low population densities targeted in the slow-the-spread program (Tcheslavskaia et al. 2005b). At high population densities, or when flightless female gypsy moths are concentrated on a few host trees, the proximity of female to male gypsy moths dictates the success of mating disruption. The effect of female proximity to male gypsy moths in mating-disruption-treated plots was tested with released sentinel moths. Mating of sentinel gypsy moth females was suppressed by mating disruption treatment if male moths were positioned at least 1 m from females on a different tree. When males and and females were positioned <5 cm apart, pheromone treatment did not influence mating success (Onufrieva et al. 2008). Under the constant stimulation of pheromone in mating-disruption-treated forests, diurnally active male gypsy moths may use cues other than olfaction for close-range mate location (Charlton and Cardé 1990), which negates the effect of pheromone treatment at high population densities. At extremely high moth densities, the pheromone released by all the females can cause a level of natural mating disruption. This effect has been illustrated in theoretical models (Pearson et al. 2004) and empirically for forest defoliators (Jones et al. 2009b; Evenden et al. 2015).

The impact of pest population density on the efficacy of mating disruption may be increased if moths are aggregated in their distribution pattern throughout the cropping area. Mating disruption of the European grape vine moth was effective when applied to both high and low population densities if the moths were evenly distributed throughout the vineyard. When insects were aggregated, mating disruption was less effective at both population densities. Mating disruption of dispersed insects maintained damage at low levels (2.5–3%), whereas high levels of damage were incurred (10–17%) when insects were aggregated at either density (Schmitz et al. 1995). The distance between males and females will dictate the probability that males will encounter the pheromone plume released by the female moth, and this appears to be an important factor limiting the success of mating disruption in several systems (Howell et al. 1992; Schmitz et al. 1995; Onufrieva et al. 2008).

Female Moth Dispersal

It has long been recognized that the dispersal capacity of female moths can impact the success of pheromone-based mating disruption as a management tactic (Cardé and Minks 1995). Females that mate outside and then immigrate into the pheromone-treated area to oviposit can contribute to failure of the mating disruption technique. Moths with polyphagous feeding habits as larvae may be more difficult to control by mating disruption than moth pests with specialized feeding habits on the crop to be protected. Eclosion and mating on alternate plant hosts outside of the pheromone-treated cropping area followed by movement of females into a pheromone-treated cropping area for ovipositon can create edge effects whereby there is a greater presence and activity level of moths at the perimeter than center of the treated area (Hsu et al. 2009). This, in addition to larger gaps in pheromone coverage at the plot edge, can lead to higher than acceptable damage at the edge of mating-disruption-treated plots. Placement of additional dispensers around the perimeter of the treated area (Knight 2004) or widespread adoption of mating disruption for the target pest following an "area-wide" approach (Calkins and Faust 2003) can help to mitigate damage at the crop edge. Byers and Castle (2005) incorporated dispersal and population growth of a polyphagous insect pest in simulation models to compare area-wide control versus control by individual producers using mating disruption. Direct comparison of the resulting models showed that the area-wide approach would provide better control at a lower cost than asynchronous treatment of individual cropping areas (Byers and Castle 2005).

Female moth dispersal may also be altered by pheromone treatment itself. Female moths in many families are able to detect the pheromone they produce and exposure to pheromone elicits changes in moth dispersal in some species (Sanders 1987; Pearson and Meyer 1996; DeLury et al. 2005). A step-time model that incorporates female movement into models of mate finding by male moths indicates that female movement in all directions except downwind of a competitor increases the likelihood of moth mating success over remaining stationary (Pearson et al. 2004). This suggests that if females are stimulated to move as a result of perceiving pheromone in a mating-disruption-treated environment, their mating success may be increased. In an extreme case in which female moths disperse out of the treated area, females may attract mates in clean air and then return to the treated area to lay eggs (Pearson et al. 2004).

Adult Moth Longevity and Voltinism

Adult longevity and the length of the adult eclosion period vary greatly among different species of moths and will influence the ease by which their mating behavior can be disrupted with pheromone treatment. Theoretically, a short-lived, univoltine species in which moths emerge over a confined time period should be easier to target with pheromone-based mating disruption than a long-lived or multivoltine species, as the formulation will only need to be active in the cropping area for a short time. As has been previously discussed, crop and environmental effects play a large role in the efficacy of season-long mating disruption. The overall release rate of pheromone from most dispenser types will decline with time in the season. This effect is most noticeable in formulations that are based on small amounts of pheromone per point source such as sprayable formulations that may require multiple applications through the growing season to control long-lived or multivoltine species (Stelinski et al. 2007b).

There are also biological effects of the target pest that can influence the success of mating disruption throughout the season. In multivoltine species, the population density can vary with generation and have a direct effect on the efficacy of mating disruption (Stelinski et al. 2007d). Lifespan and fecundity of females can differ by generation and in response to varying levels of nutrition as crop phenology changes across the season (Fitzpatrick and Troubridge 1993). These differences in life history traits across generations can be particularly important when the effectiveness of pheromone treatment relies on delaying mating in a portion of the population. A delay in mating caused by pheromone-based mating disruption can differentially impact moths from different generations. Delayed mating of the blackheaded fireworm, *Rhopobota naevana* (Tortricidae), resulted in a greater reduction in fecundity of females from the spring than summer generation, suggesting that pheromone-based mating disruption could have more impact in the spring (Fitzpatrick 2006). In contrast, the fecundity of the overwintering codling moth generation is less affected by delayed mating than it is in the subsequent summer generations. The greater effect of delayed mating on fecundity reduction later in the season is driven by heat unit accumulations as delayed mating also reduces fecundity to a greater extent in warm parts of the codling moth range. These results suggest that control of codling moth by mating disruption can be achieved more easily under warm than cool conditions because the effect of delayed mating on population reduction is driven by physiological time (Jones and Wiman 2012). The differential impact of delayed mating on different moth generations remains to be confirmed using empirical field testing.

Male Moth Pheromone Response

Gut et al. (2004) consider the male moth's response to pheromone to be the most important biological characteristic that dictates the success of mating disruption in moths. Male moths vary in their response profile to pheromone over a range of doses. Some species are sensitive to low doses of pheromone and exhibit an upper threshold of response (Baker et al. 1989) above which moth orientation is arrested. Other species are not arrested at high pheromone concentrations and can orient to pheromone released at rates far higher than that released by calling females (Cardé et al. 1998). Still other species do not exhibit an explicit dose-response and are able to orient to pheromone across a wide range of release rates (Evenden et al. 2010). This variation in response to pheromone may dictate differences in the mechanisms of mating disruption that impact individual species and the ease at which different species can be disrupted (Gut et al. 2004). Mating disruption of the oriental fruit moth, a relatively easy species to disrupt, occurs by competitive attraction to female-equivalent pheromone dispensers and by noncompetitive mechanisms in response to high-release-rate reservoir-type dispensers (Reinke et al. 2014). This differential effect of pheromone treatment based on pheromone release rate probably occurs because the oriental fruit moth becomes arrested in response to high pheromone release rates (Baker et al. 1989). Susceptibility to more than one mating disruption mechanism may be one reason that the oriental fruit moth is rela-

tively easy to control with pheromone-based mating disruption, even at high population denisties.

Male moth pheromone response also dictates their susceptibility to the neurophysiological effects that can be caused by mating disruption treatment. Although exposure to pheromone usually results in a subsequent decrease in pheromone responsiveness that contributes to mating disruption, some species have enhanced sensitivity to low doses of pheromone following pheromone exposure. This has been found in species that are relatively difficult to control with pheromone-based mating disruption (Gut et al. 2004), such as the obliquebanded leafroller (Stelinski et al. 2004b). Moths that are difficult to control with mating disruption also readily exhibit long-lasting adaptation as a result of pheromone exposure; this may serve to "protect" the central nervous system from habituation (Stelinski et al. 2003b; Judd et al. 2005b; Stelinski et al. 2005b). Species in which long-lasting adaptation does not occur are relatively easier to disrupt (Rumbo and Vickers 1997; Stelinski et al. 2003b).

Incorporation of Mating Disruption into Integrated Pest Management

Most mating disruption research has attempted to develop this management tactic as a "stand-alone" tool. This approach can be successful when mating disruption targets the key pest at low population densities in a managed ecosystem and additional pest management is less important. In many agroecosystems, mating disruption is more commonly a component in a suite of management tools applied against several pests. The combination of pheromone formulations with other management tactics can increase the efficacy of the formulation on the target pest, help to prevent the development of resistance to any one given management tactic, and contribute to management of more than one insect pest.

Resistance to Pheromone-Based Mating Disruption

Long-term application of pheromone-based mating disruption that effectively controls the target pest population could exert selection pressure that alters the pheromone-based communication system and contributes to the evolution of resistance to mating disruption (Cardé and Minks 1995). Several studies have followed populations exposed to pheromone-based mating disruption over time and document changes in the communication system that result in differential mate-finding ability. In a field cage experiment, treatment with the most commonly produced pheromone blend of the cabbage looper moth over four generations acted to preserve an uncommon mutant pheromone-producing genotype in the population. Likely the mutant females that release a different pheromone were more apparent to male moths in an atmosphere treated with the common wild-type pheromone blend and were more successful in securing a mate (Evenden and Haynes 2001). Selection caused by pheromone treatment can also result in females that release more pheromone (Shani and Clearwater 2001). Females with a stronger signal would be more apparent to males in a pheromone-treated atmosphere which could be one mechanism leading to the evolution of resistance to mating disruption. Calling females are likely to be more apparent to male moths in an atmosphere treated with a mating disruption formulation that releases only a partial pheromone blend as compared to a pheromone blend that closely mimics the female's natural blend. Resistance to mating disruption is predicted to be most likely with repeated use of partial pheromone blends as the mating disruptant (Cardé 2007).

The only documented field case of resistance to mating disruption is in response to treatment with a formulation releasing a single pheromone component, Z11-14Ac, against the smaller tea tortrix, *Adoxophyes honmai* (Tortricidae), in tea plantations in Japan (Mochizuki et al. 2002). The efficacy of mating disruption of the small tea tortrix declined after 14–16 years of continuous pheromone treatment, as measured by pheromone-baited trap catch and larval density in treated plantations. Other potential causes of control failure were ruled out as the pheromone formulation was effective in regions that had not been previously treated with the single-component formulation. Mating disruption of the resistant population was restored as a result of treatment with a formulation releasing the full pheromone blend (Mochizuki et al. 2002). The pheromone released by resistant smaller tea tortrix females contained significantly more of two of the pheromone components than females from a susceptible lab population. This slightly altered female blend could make females more apparent to searching males in an atmosphere treated with the single pheromone component but resistant females released less pheromone than susceptible females (Tabata et al. 2007). Resistant males had a much broader response profile to tested pheromone blends than susceptible males and a portion of resistant males responded to blends lacking components crucial to attraction of susceptible males. Males capable of responding to off-ratio pheromone blends might be more successful at finding females in tea plantations treated with the partial pheromone than susceptible males with a narrowly tuned response (Tabata et al. 2007).

Mating Disruption and Insecticides

In agroecosystems in which activity by a pest complex results in economically important damage, mating disruption has been used as one tactic that contributes to overall arthropod management in the system. Incorporation of mating disruption into an overall IPM strategy is often initiated as a result of the development of insecticide-resistant populations or legislation for restricted insecticidal inputs. Integration of mating disruption, insecticides, and other tactics has promoted "reduced-risk" pest management that lowers insecticidal input into the system. In a number of cases, when mating disruption with limited insecticide input is directly compared to conventional management with insecticides alone, the inclusion of mating disruption in the program results in better management and reduced insecticidal input. Reduced-risk pest management programs in peach orchards in the mid-Atlantic United States (Atanassov et al. 2002) and Ontario, Canada (Trimble et al. 2001), used mating disruption to control the key pest, the oriental fruit moth. Insecticides were applied to target the overwintering generation and mating disruption was used against the subsequent summer generations. Capture of moths in bait traps and samples of shoot tip and fruit damage in both reduced-risk and conventional peach blocks that treated all generations with insecticide showed that both approaches equally reduced oriental fruit moth populations and the reduced-risk program resulted in a lower insecticidal input. Similarly, a combination of mating disruption with reduced insecticidal application was as effective as conventional insecticidal treatment of pink bollworm

in cotton. In this case, the addition of mating disruption reduced insecticidal input at flowering and fruiting when beneficial insects are most abundant (Critchley et al. 1991).

In other studies, the combination of mating disruption with insecticides has not enhanced overall control of the target pest. Mating disruption alone controlled codling moth populations at low population densities in apple orchards in Australia but at high densities neither mating disruption alone nor mating disruption with an insecticidal spray targeting the overwintering generation provided adequate control (Vickers et al. 1998). The addition of pheromone-based mating disruption to insecticidal management of the obliquebanded leafroller on apple did not increase control of larval feeding on shoots or fruit (Trimble and Appleby 2004). The obliquebanded leafroller is not highly susceptible to control by mating disruption (Gut et al. 2004) and any marginal effect of mating disruption treatment did not enhance the control provided by insecticides alone (Trimble and Appleby 2004).

Another approach is to combine the pheromone and insecticide into a single formulation called an attracticide. Lepidopteran attracticides were developed to improve control by mating disruption at high population densities and to reduce the amount of pheromone required in the formulation. Most lepidopteran attracticides combine low concentrations of pheromone in female-equivalent dispensers with a pyrethroid insecticide that acts on contact. Toxicity of the formulation can result in direct removal of males from the population or sublethal effect that decreases subsequent orientation to pheromone (Haynes et al. 1986; Evenden et al. 2005). As moths are exposed to the insecticide only through contact with the formulation, attracticides must release the full attractive pheromone at a rate and dispenser density that will promote contact with the dispensers. Several studies have compared the efficacy of pheromone formulations with and without the incorporation of an insecticide. In some cases, approximately 50% of the observed communication disruption was a "mating-disruption effect" of pheromone alone, and the additional 50% was due to removal of males from the population from insecticide exposure (Charmillot et al. 1996; Suckling and Brockerhoff 1999). The insecticide component of an attracticide targeting the oriental fruit moth did not enhance disruption of male moth orientation or reduce mating with sentinel females in small plots (Evenden and McLaughlin 2004). These findings suggest that male moths are not reaching the formulation and may be arrested during upwind flight to the source. Proximity of the insecticide and pheromone components of an attracticide formulation may influence the exposure of the target pest to the insecticide and affect the efficacy of the formulation. Physical separation of the toxicant from the pheromone source promotes greater contact with the insecticide-treated surface by male codling moths than when the two components are mixed in the same formulation (Huang et al. 2013).

Mating Disruption and Biological Control

Because implementation of mating disruption into an IPM system can result in reduced insecticide application, it may also enhance biological control in the cropping area. The widespread use of codling moth mating disruption on >200,000 ha worldwide (Witzgall et al. 2010) has promoted investigation into the diversity and abundance of beneficial arthropods found in mating-disruption-treated orchards. The density and diversity of predatory ground beetles were compared in large blocks (3–4 ha) of apple orchards treated with mating disruption or conventional neuroactive insecticides. Blocks treated with mating disruption targeting the codling moth supported more ground beetles (Coleoptera: Carabidae) than the conventionally managed blocks. In total, 75% of the most commonly captured ground beetles (*Pterostichus* spp.) were recovered from samples collected in the mating-disruption-treated plots and several of these species were predators of codling moth (Epstein et al. 2001). The density of predatory arboreal arthropods was higher in apple orchards treated with mating disruption than conventionally managed orchards. In these same orchards, predation of sentinel codling moth eggs was higher in the mating-disruption-treated orchards late but not early in the season (Knight et al. 1997). In a subsequent study, arboreal and understory dwelling spider abundance was higher in organically managed orchards than in orchards treated with mating disruption and reduced insecticide input for codling moth control or in conventionally managed orchards, suggesting that spiders either are extremely sensitive to synthetic insecticides or require a long population recovery period (Miliczky et al. 2000). Although mating disruption targeting codling moth benefited natural enemy populations, 3 years after its adoption there was no appreciable increase in biological control of secondary pests of apple (Knight 1995).

Reduced risk management of peaches with limited insecticide application and mating disruption of oriental fruit moth promoted beneficial arthropod fauna that was more abundant than in conventionally managed orchards (Atanassov et al. 2003). The mortality rate of oriental fruit moth sentinel eggs positioned in peaches under the two management regimes was higher in the reduced-risk than the conventional orchards and was attributed mainly to predation (Atanassov et al. 2003). In cotton, counts of arthropods on 10 plants per field at 4 sample periods throughout the growing season were compared in fields under conventional or mating disruption management for the pink bollworm in Egypt (Boguslawski and Basedow 2001). Most predatory arthropods were present in low densities and did not differ in abundance on cotton grown under the different management regimes. Secondary heteropteran pests were more abundant in the conventionally managed cotton fields than in the organic production using mating disruption to control the pink bollworm, whereas spiders were more commonly found on cotton plants in the organic production system (Boguslawski and Basedow 2001). This study suggests that mating disruption of the key pest of a crop production system can promote biological control of secondary pests in the system.

Although the use of pheromone-based mating disruption and the subsequent reduction of insecticidal input into the cropping system can result in enhanced biological control, there is also the potential that pheromone treatment itself may influence the behavior of nontarget beneficial arthropods. Some parasitoids use sex pheromones of their moth host as kairomones to assist in location of moth eggs (Lewis et al. 1982; Boo and Yang 2000). Application of fiber dispensers releasing the pheromone of the corn earworm caused an increase in parasitism of corn earworm eggs by *Trichogramma* in the treated cotton plots (Lewis et al. 1982), suggesting that mating disruption may directly increase parasitism by recruitment of parasitoids into the area. Alternatively, mating disruption could negatively impact beneficial nontarget arthropods in the system by disruption of kairomonal response or through reduction in host availability in the treated area. Measurement of abundance of the most common parasitoid of the European pine shoot moth, *Rhyacionia buoliana* (Tor-

tricidae), before and after mating disruption treatment illustrated a decline in abundance of the larval parasitoid *Glypta zozanae* (Hymenoptera: Ichneumonidae) after mating disruption treatment. Because it is unlikely that larval parasitoids use the moth sex pheromone as a kairomone, the lower abundance of the parasitoid following mating disruption may be an indirect effect of reduced host density that promoted parasitoid dispersal out of the patch (Niwa and Daterman 1989).

Mating Disruption and Horticultural Oil

Application of mating disruption formulations to the crop remains one of the most costly aspects of this technology and limits its adoption in IPM in some systems. Formulations that can be tank mixed with compatible insecticides, fungicides, or fertilizers and applied with conventional spray equipment would reduce costs associated with pheromone application and would promote use of mating disruption. Microencapsulated pheromone formulations can be applied with conventional spray equipment, but to date the release rate properties of these formulations (Albajes et al. 2002) and their poor rainfastness (Waldstein and Gut 2004) have limited their longevity and efficacy under field conditions. Recent research has investigated the possibility of horticultural oil as an adjuvant to microencapsulated pheromone formulations (Wins-Purdy et al. 2007). Horticultural oils are used as adjuvants for many pesticides (Zabkiewicz 2002) and as stand-alone pesticides primarily against soft-bodied insects and insect eggs in perennial tree fruit crops (Fernandez et al. 2005). Horticultural oil is compatible with microencapsulated pheromone, but it does not increase the efficacy of mating disruption against the obliquebanded leafroller in apple (Wins-Purdy et al. 2007). In apple plots treated with microencapsulated pheromone with and without horticultural oil, mate-finding communication was similarly and significantly disrupted as compared to that in water-treated control plots for 42 days post application. The percent of males that were disrupted by the formulation with and without oil gradually decreased over the 42-day period and the formulation lost effectiveness by 49 days post application (Wins-Purdy et al. 2007). Accompanying wind-tunnel and electrophysiological studies suggested that mating disruption with a microencapsulated formulation releasing only the major pheromone component of this species, Z11-14Ac, with or without horticultural oil was the result of noncompetitive mechanisms. The proximity of male obliquebanded leafroller moths to the pheromone-treated foliage is crucial in maintaining the effectiveness of the formulation that appears to be mainly caused by habituation of the central nervous system (Wins-Purdy et al. 2008). Horticultural oil can penetrate insect tissues and exhibits neurotoxicity in other tortricid moths (Taverner et al. 2001). These physical properties of light horticultural oils may facilitate the physiological uptake of pheromone into the male moth nervous system. This, in addition to the pesticidal properties of horticultural oil on obliquebanded leafroller eggs (Wins-Purdy et al. 2009), may help to promote the adoption of horticultural oil adjuvants to sprayable pheromones for difficult-to-disrupt species like the obliquebanded leafroller.

Mating Disruption and the Sterile Insect Technique

Sterile insect technique is a pest management tactic that is used to eradicate a key pest in an agroecosystem by government-regulated release of sterilized insects into the area. The effectiveness of the sterile insect technique relies on initial low population densities of the target pest and the competitiveness of sterile insects to locate and mate with insects in the wild population. A combined strategy of sterile insect technique with pheromone-based mating disruption seems like an incompatible pairing, as the success of the sterile insect technique relies on successful mate location. There are, however, several examples of successful integration of these two pest management tactics. The first study to combine these two tactics showed that mating disruption alone or in combination with the sterile insect technique could control pink bollworm moths over large cotton-growing areas of the SW United States so that populations were undetectable in some areas (Staten et al. 1997). This approach has also been tested and used in area-wide control of codling moth in British Columbia, Canada. Initial problems associated with implementation of the sterile insect release program against the codling moth in the montane fruit-growing regions of British Columbia included reducing the population density of wild insects so that a 40:1 ratio of released sterile:wild moths could be maintained. This was particularly problematic in the spring when sterile moths were less competitive than wild moths (Bloem et al. 1998) and in organic orchards in which insecticides could not be applied to lower population densities (Judd and Gardiner 2005). A 5-year study conducted in organic and conventional orchards similarly treated with sterile insects showed that the addition of mating disruption and tree banding in organic orchards was as or more effective than supplementation of sterile insect release with insecticides in conventional orchards (Judd and Gardiner 2005). Reduction of codling moth populations as measured by overwintering larval populations and male moth capture in pheromone-baited traps declined more rapidly over the 5-year period in orchards in which sterile insect release was supplemented by mating disruption and tree banding than in orchards supplemented with insecticide (Judd and Gardiner 2005). This integrated approach provided codling moth population reduction in both organic and conventional orchards so that area-wide management of codling moth could be achieved with release of sterile insects. Mating disruption was also successfully combined with release of partially sterile codling moths that received less gamma radiation than fully sterile moths and were more competitive with wild moths in the field. A synergistic effect of mating disruption with release of partially sterile insects on fruit damage at harvest could not be illustrated, likely because population densities of codling moth were below 0.1% in all treated plots. Mating disruption and release of partially sterile moths alone and together resulted in significantly less damage at harvest as compared to untreated control plots (Bloem et al. 2001). The success of combining mating disruption and sterile insect release against the codling moth may be driven by the fact that much of the population reduction caused by pheromone-based mating disruption is the result of a delay in mating (Knight 2007) rather than complete interference of mating in this species.

Summary/Discussion

The adoption of pheromone-based mating disruption as part of an IPM program often occurs because of a breakdown in control of a key pest in an agroecosystem. Insecticide resistance (Cardé 2007) and government regulation to reduce the use of neuroactive insecticides (Gut et al. 2004) have played key roles in increasing the commercial use of mating disrup-

tion to over 750,000 ha worldwide in 2010 (Witzgall et al. 2010). Application of mating disruption to a wider acreage in managed ecosystems will require the following: (1) continued identification and formulation of insect semiochemicals (Witzgall et al. 2010); (2) further understanding of the mechanisms by which mating disruption works against different species in various cropping systems with various formulations (Cardé 2007) including the effect of pheromone treatment on female behavior and mating delay; (3) development of efficacious and cost-effective formulations and methods of pheromone application; (4) increased cooperation among producers and government regulators to implement area-wide pheromone treatment; and (5) a favorable economic environment for the incorporation of mating disruption into IPM in various cropping systems.

Because mating disruption is a species-specific pest management tactic, crops that have a single key moth pest may be the most suitable environments for the rapid incorporation of mating disruption into an overall IPM strategy. Mating disruption that targets multiple moth pests in a single agroecosystem can be achieved by a common formulation of pheromone components used by more than one related species (Chamberlain et al. 1992; Deland et al. 1994; Gronning et al. 2000; Suckling et al. 2012b) or dispensers that are modified to release more than one pheromone blend (Judd and Gardiner 2004; Il'ichev et al. 2007; Stelinski et al. 2009). This approach deserves further attention, especially if reduced insecticide use as a result of control of the key pest by mating disruption results in release of secondary sympatric moth pests from control.

Implementation of mating disruption on a commercial scale is skewed by crop type. In addition to the composition of the pest complex, other biological, environmental, and economic conditions seem to promote the adoption of mating disruption in certain managed ecosystems over others. The majority of studies on the application and underlying mechanisms of mating disruption have been conducted on moth pests of tree fruit production (figure 24.3). There are multiple reasons for the success of mating disruption in tree fruit agroecosystems. Tree fruits produced for the fresh fruit market are high-value crops. The economics of pheromone application for codling moth, the key pest of apple production worldwide, has improved in the last two decades. An initial economic analysis that compared pest management of codling moth with pheromone versus insecticide revealed that mating disruption was more expensive by $188.22 ha^{-1} in the early 1990s and was only a viable option under low pest population densities (Williamson et al. 1996). Reduction in the cost of codlemone production (Witzgall et al. 2010) and area-wide application of codling moth mating disruption (Calkins and Faust 2003; McGhee et al. 2011) have removed the economic uncertainty previously associated with control of codling moth by mating disruption. A recent economic analysis shows that growers realize savings of $55–65 ha^{-1} with application of area-wide mating disruption of codling moth over insecticidal control. Savings are achieved through better levels of control and reduced fruit injury compared to orchards treated with insecticides and reduced input costs associated with insecticidal sprays (McGhee et al. 2011). Savings are even greater in pears when pheromone treatment is applied by aerosol dispensers. An estimated savings of $352 ha^{-1} compared to conventional insecticide-based management was achieved after 3 years of pheromone application with aerosol dispensers (Elkins et al. 2005).

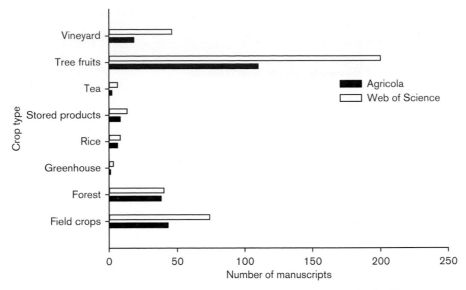

FIGURE 24.3 Number of manuscripts by crop type recovered from Agricola and Web of Science databases with the term "mating disruption" in the title of the manuscript. Manuscripts include experimental studies on Lepidoptera only; reviews, patents, and conference proceedings are not included.

Biological features of orchards also promote the use of mating disruption as an important part of IPM in tree fruit production systems. Tree fruits are a perennial cropping system to which cultural control measures such as crop rotation and alteration of planting or harvest dates cannot be applied. Because some cultural control options are not available to tree fruit growers, other low-impact pest management tactics, such as mating disruption, may be adopted out of necessity. Tree crops also provide a large canopy that can reduce wind (Milli et al. 1997), trap pheromone in the cropping area, and promote the success of mating disruption. A buildup of pheromone in the tree canopy in combination with the waxy cuticle of tree leaves will cause adsorption and rerelease of pheromone by the plant itself (Karg et al. 1994). However, not all tree canopies enhance the effectiveness of mating disruption. In large tree canopies (such as nut crops) penetration of the canopy with pheromone can be difficult, and detailed study is required to understand pheromone movement in relation to moth activity (Girling et al. 2013). The reduced canopy of high-density orchard plantings may also reduce the efficacy of mating disruption against tree fruit pests. The biology of the lepidopteran pest complex in temperate tree fruit production systems also promotes high levels of adoption of mating disruption. The key lepidopteran pests of tree fruit production are internal fruit feeders in the family Tortricidae (figure 24.4) that cause direct damage to the harvested product. The feeding habits of these pests make control necessary and difficult as the damaging larval stage is shielded from control by contact insecticidal sprays. A pest management tactic that targets the adult stage of the insect before oviposition occurs is favorable when larvae are difficult to control.

Just as mating disruption has been more developed in particular managed ecosystems (figure 24.3), some moth pests may be better targets for control by mating disruption than others (figure 24.4). The available literature suggests that pheromone biology of the target pest can influence the probability of successful mating disruption. Moth pheromones that consist of long-chain pheromone components with high levels of oxidation of hydrocarbon subunits are well retained in the environment (Gut et al. 2004). More moth pests in the Geometridae, Erebidae, and Noctuidae that use sex pheromones with chain lengths of 17–23 carbon atoms (Millar 2000) should be targeted for control by mating disruption. More basic research on the threshold response of males to sex pheromone is needed as mating disruption seems to be more effective against moth species that are sensitive to low pheromone doses but become arrested to high doses (Gut et al. 2004).

Moth mobility and larval host range can also influence the success of mating disruption. Mating disruption should be most effective against species in which the female moth is not very mobile and the larvae are feeding specialists on the crop to be protected. Area-wide mating disruption is successful in management of lepidopteran pests of forest (Sharov et al. 2002), tree fruit (Brockerhoff et al. 2012), and field crop (Staten et al. 1997) ecosystems and helps to ameliorate mating disruption failure as a result of immigration of mobile, mated females from outside the cropping system. Because the economics of area-wide management are also better than mating disruption in isolated crop plantings (McGhee et al. 2011), this approach should be promoted by producer organizations and government agencies in the adoption of commercial pheromone-based mating programs.

Population density of the target pest can be the most important biological characteristic that impacts the success of mating disruption. Mating disruption enhances Allee effects that prevent the establishment of small populations (Brockerhoff et al. 2012). For this reason, mating disruption should be developed to combat the spread of invasive species before populations become established in the introduced range. Aerially applied pheromone used in area-wide management against low densities of gypsy moth has successfully slowed the westward range expansion of this polyphagous, invasive species in the United States (Sharov et al. 2002). Adoption of a similar program against expanding populations of gypsy moth in Canada should be considered. Aerial application of

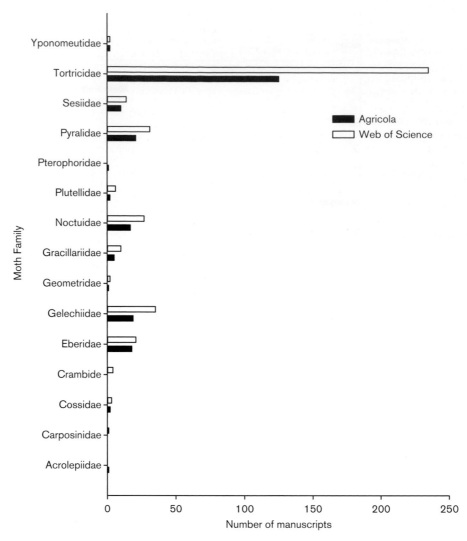

FIGURE 24.4 Number of manuscripts by moth family recovered from Agricola and Web of Science databases with the term "mating disruption" in the title of the manuscript. Manuscripts include experimental studies on Lepidoptera only; reviews, patents, and conference proceedings are not included.

pheromone to pine forests in New Zealand significantly reduced trap capture of the polyphagous light-brown apple moth, suggesting that area-wide application of pheromone could be used to eradicate this species in its invasive range in California (Brockerhoff et al. 2012). Use of mating disruption to control or eradicate invasive species will require knowledge of the identity of the invasive species and its pheromone biology. A sensitive monitoring tool (such as pheromone-baited traps) is required to identify invasion before populations either become established or become too high or widely dispersed to be managed by mating disruption. Access to large amounts of pheromone that can be formulated for aerial application also could be a limiting step in adoption of mating disruption for control of invasive species.

Incorporation of mating disruption into IPM programs will require a greater understanding of how mating disruption works (Miller and Gut 2015) and a better refinement of how it can be combined synergistically with other pest management tactics. Dissemination of pheromone on sterile insects (Suckling et al. 2011) is an example of synergy of pest management tactics that in the future could also be combined with autodis-

semination of insect pathogens (Baverstok et al. 2010). Genetic manipulation of plants to release alarm pheromones of aphids has already been achieved (Pickett et al. 2014) and recent research has achieved transient expression of a multicomponent moth pheromone in the leaves of *Nicotiana benthamiana* (Ding et al. 2014). It is conceivable that similarly modified crop plants that release moth pheromones could disrupt moth mating without the addition of synthetic pheromone. In addition to the regulatory hurdles that would accompany this approach, pheromone-producing crop plants could actually attract the target pest to the cropping area and increase population densities to levels that could not be controlled by mating disruption. For this reason, genetic manipulation of plants to produce pheromones should probably target unattractive partial pheromones that disrupt moth behavior through noncompetitive mechanisms. Before such a cost-intensive approach is taken, ways in which noncompetitive mechanisms can be enhanced in mating disruption systems should be addressed. Therefore, the long-standing question of whether attractive pheromone blends make the best mating disruptants (Minks and Cardé 1988) is still relevant and deserves further study.

Acknowledgments

My interest and passion for pheromone-based mating disruption was fostered in the laboratories of Drs. John Borden and Gary Judd in the mid-1990s. I continue to collaborate on pheromone-based mating disruption with Dr. Judd and much of my current thinking on the topic is the result of lively discussions with him and the students we co-supervise. Recent research on mating disruption in my laboratory is the result of the hard work of Dr. Boyd Mori. Dr. Mori's work links mating disruption to mechanism in a little-studied field crop pest, the red clover casebearer moth. Dr. Mori contributed suggestions to an earlier version of this manuscript. Continuous funding from the Natural Science and Engineering Research Council (NSERC) of Canada for my laboratory at the University of Alberta has made contribution of this chapter possible.

References Cited

Albajes, R., M. Konstantopoulou, O. Etchepare, M. Eizguirre, B. Frerot, A. Sans., F. Krokos, A. Ameline, and B. Mazomenos. 2002. Mating disruption of the corn borer *Sesamia nonagrioides* (Lepidoptera: Noctuidae) using sprayable formulations of pheromone. *Crop Protection* 21:217–225.

Anderson, P., M.M. Sadek, and B.S. Hansson. 2003. Pre-exposure modulates attraction to sex pheromone in a moth. *Chemical Senses* 28:285–291.

Anderson, P., B.S. Hansson, U. Nilsson, Q. Han, M. Sjoholm, N. Skals, and S. Anton. 2007. Increased behavioral and neuronal sensitivity to sex pheromone after brief odor experience in a moth. *Chemical Senses* 32:483–491.

Anfora, G., M. Baldessari, A. de Cristofaro, G. Germinara, C. Ioriatti, F. Reggiori, S. Vitagliano, and G. Angeli. 2008. Control of *Lobesia botrana* (Lepidoptera: Tortricidae) by biodegradable Ecodian sex pheromone dispensers. *Journal of Economic Entomology* 101:444–450.

Atanassov, A., P.W. Shearer, G. Hamilton, and D. Polk. 2002. Development and implementation of a reduced risk peach arthropod management program in New Jersey. *Journal of Economic Entomology* 95:803–812.

Atanassov, A., P.W. Shearer, and G. Hamilton. 2003. Peach pest management programs impact beneficial fauna abundance and *Grapholita molesta* (Lepidoptera: Tortricidae) egg parasitism and predation. *Environmental Entomology* 32:780–788.

Atterholt, C.A., J.M. Krochta, R.E. Rice, and M.J. Delwiche. 1998. Study of biopolymers and paraffin as potential controlled-release carriers for insect pheromones. *Jouranl of Agricultural and Food Chemistry* 46:4429–4434.

Babson, A.L. 1963. Eradicating the gypsy moth. *Science* 142:447–448.

Baker, T.C., and J.J. Heath. 2005. Pheromones: function and use in insect control. Pp. 406–459. In L. Gilbert, K. Iatrou, and S. Gill, eds. *Comprehensive Molecular Insect Science*. Amsterdam: Elsevier.

Baker, T.C, B.S. Hansson, C. Löfstedt, and J. Löfqvist. 1989. Adaptation of male moth antennal neurons in a pheromone plume is associated with cessation of pheromone-mediated flight. *Chemical Senses* 14:439–448.

Baker, T.C., T. Dittl, and A. Mafra-Neto. 1997. Disruption of sex pheromone communication in the blackheaded fireworm in Wisconsin cranberry marshes by using MSTRS devices. *Journal of Agricultural Entomology* 14:449–457.

Barclay, H.J., and G.J.R. Judd. 1995. Models for mating disruption by means of pheromone for insect pest control. *Researches on Population Ecology* 37:239–247.

Bartell, R.J. 1982. Mechanisms of communication disruption by pheromone in the control of Lepidoptera: a review. *Physiological Entomology* 7:353–364.

Bartell, R.J., and L.A. Lawrence. 1973. Reduction in responsiveness of males of *Epiphyas postvittana* (Lepidoptera) to sex pheromone following previous brief pheromonal exposure. *Physiological Entomology* 19:845–855.

Baverstok, J., H.E. Roy, and J.K. Pell. 2010. Entomopathogenic fungi and insect behaviour: from unsuspecting hosts to targeted vectors. *Biocontrol* 55:89–102.

Baxter, I.H., N. Howard, C.G. Armsworth, L.E.E. Barton, and C. Jackson. 2008. The potential of two electrostatic powders as the basis for an autodissemination control method of *Plodia interpunctella* (Hübner). *Journal of Stored Products Research* 44:152–161.

Bengtsson, M., G. Karg, P.A. Kirsch, J. Löfqvist, A. Sauer, and P. Witzgall. 1994. Mating disruption of pea moth *Cydia nigricana* F. (Lepidoptera:Tortricidae) by a repellent blend of sex pheromone and attraction inhibitors. *Journal of Chemical Ecology* 20:871–887.

Birch, M.C. 1977. Response of both sexes of *Trichoplusia ni* (Lepidoptera: Noctuidae) to virgin females and to synthetic pheromone. *Ecological Entomology* 2:99–104.

Bloem, S., K.A. Bloem, and A.L. Knight. 1998. Assessing the quality of mass-reared codling moths (Lepidoptera: Tortricidae) by using field release-recapture tests. *Journal of Economic Entomology* 91:1122–1130.

Bloem, S., K.A. Bloem, J.E. Carpenter, and C.O. Calkins. 2001. Season-long releases of partially sterile males for control of codling moth (Lepidoptera: Tortricidae) in Washington apples. *Environmental Entomology* 30:763–769.

Boguslawski, C., and T. Basedow. 2001. Studies in cotton fields in Egypt on the effects of pheromone mating disruption on *Pectinophora gossypiella* (Saund.) (Lep., Gelechiide), on the occurrence of other arthropods, and on yields. *Journal of Applied Entomology* 125:327–331.

Boo, K.S., and J.P. Yang. 2000. Kairomones used by *Thrichogramma chilonis* to find *Helicoverpa assulta* eggs. *Journal of Chemical Ecology* 26:359–375.

Bosa, C.F., A.M. Cotes, P. Osorio, T. Fukumoto, M. Bengtsson, and P. Witzgall. 2006. Disruption of pheromone communication in *Tecia solanivora* (Lepidoptera: Gelechiidae): flight tunnel and field studies. *Journal of Economic Entomology* 99:1245–1250.

Brockerhoff, E.G., D.M. Suckling, M. Kimberly, B. Richardson, G. Coker, S. Gous, J.L. Kerr, et al. 2012. Aerial application of pheromones for mating disruption of an invasive moth as a potential eradication tool. *PLOS ONE* 7:e43767.

Byers, J.A. 2007. Simulation of mating disruption and mass trapping with competitive attraction and camouflage. *Environmental Entomology* 36:1328–1338.

Byers, J.A., and S.J. Castle. 2005. Areawide models comparing synchronous versus asynchronous treatments for control of dispersing insect pests. *Journal of Economic Entomology* 98:1763–1773.

Calkins, C.O., and R.J. Faust. 2003. Overview of areawide programs and the program for suppression of codling moth in the western USA directed by the United Strates Department of Agriculture-Agricultureal Research Service. *Pest Management Science* 59:601–604.

Cameron, E.A., E.F. Knipling, M. Beroza, and C.P. Schwalbe. 1974. Disruption of gypsy moth mating with microencapsulated disparlure. *Science* 4128:972–973.

Cardé, R.T. 1981. Disruption of long-distance pheromone communication in the oriental fruit moth: camouflaging the natural aerial trails from females? Pp. 385–398. In E.R. Mitchell, ed. *Management of Insect Pests with Semiochemicals*. New York: Plenum Press.

Cardé, R.T. 1990. Principles of mating disruption. Pp. 47–71. In R.L. Ridgway, R.M. Silverstein, and M.N. Inscoe, eds. *Behavior-Modifying Chemicals for Insect Management: Applications of Pheromones and other Attractants*. New York: Marcel Dekker.

Cardé, R.T. 2007. Using pheromones to disrupt mating of moth pests. Pp. 122–169. In M. Kogan, ed. *Perspectives in Ecological Theory and Integrated Pest Management*. New York: Cambridge University Press.

Cardé, R.T., and A.K. Minks. 1995. Control of moth pests by mating disruption: successes and constraints. *Annual Review of Entomology* 40:559–585.

Cardé, R.T., K. Trammel, and W.L. Roelofs. 1975. Disruption of sex attraction of the redbanded leafroller (*Argyrotaenia velutinana*) with microencapsulated pheromone components. *Environmental Entomology* 4:448–450.

Cardé, R.T., R.T. Staten, and A. Mafra-Neto. 1998. Behaviour of pink bollworm males near high-dose, point sources of pheromone in field wind tunnels: insights into mechanisms of mating disruption. *Entomologia Experimentalis et Applicata* 89:35–46.

Carlson, D. A., and J. R. McLaughlin. 1982. Diolefin analog of a sex pheromone component of *Heliothis zea* active in disrupting mating communication. *Experientia* 38:309–310.

Chamberlain, D. J., M. Rafique, M. I. Arif, M. R. Attique, B. R. Critchley, and D. G. Campion. 1992. Use of multi-component pheromone formulation for control of cotton bollworms (Lepidoptera: Gelechiidae and Noctuidae) in Pakistan. *Bulletin of Entomological Research* 82:449–458.

Charlton, R. E., and R. T. Cardé. 1990. Orientation of male gypsy moths, *Lymantria dispar* (L.), to pheromone sources: the role of olfactory and visual cues. *Journal of Insect Behavior* 3:443–469.

Charmillot, P. J., D. Pasquier, A. Scalco, and D. Hofer. 1996. Essais de lutte contre le carpocapse *Cydia pomonella* L. par un procédé attracticide. *Mitteilungen der Schweizerischen Entomologischen Gesellschaft* 69:431–439.

Cork, A., K. De Souza, D. R. Hall, O. T. Jones, E. Casagrande, K. Krishnaiah, and Z. Syed. 2008. Development of PVC-resin-controlled release formulation for pehromones and use in mating disruption of yellow rice stem borer, *Scirpophaga incertulas*. *Crop Protection* 27:248–255.

Critchley, B. R., D. J. Chamberlain, D. G. Campion, M. R. Attique, M. Ali, and A. Ghaffar. 1991. Integrated use of pink bollworm pheromone formulations and selected conventional insecticides for the control of the cotton pest complex in Pakistan. *Bulletin of Entomological Research* 81:371–378.

Daterman, G. E., J. G. Doyle Daves, and R. G. Smith. 1975. Comparison of sex pheromone versus an inhibitor for disruption of pheromone communication in *Rhyacionia buoliana*. *Environmental Entomology* 4:944–946.

De Lame, F. M., and L. J. Gut. 2006. Effect of monitoring trap and mating disruption dispenser application heights on captures of male *Grapholita molesta* (Busck; Lepidoptera: Tortricidae) in pheromone and virgin female-baited traps. *Environmental Entomology* 35:1058–1068.

Deland, J.-P., B. R. Roitberg, and G. J. R. Judd. 1994. Disruption of pheromone communication in three sympatric leafroller (Lepidoptera:Tortricidae) pests of apple in British Columbia. *Environmental Entomology* 23:1084–1090.

DeLury, N. C., G. J. R. Judd, and M. G. T. Gardiner. 2005. Antennal detection of sex pheromone by female *Pandemis limitata* (Robinson) (Lepidoptera: Tortricidae) and its impact on their calling behaviour. *Journal of Entomological Society of British Columbia* 102:3–11.

den Otter, C. J., H. A. Schuil, and S. van Oosten. 1978. Reception of host-plant odours and female sex pheromone in *Adoxophyes orana* (Lepidoptera: Tortricidae): electrophysiology and morphology. *Entomologia Experimentalis et Applicata* 24:570–578.

Ding, B.-J., P. Hofvander, H.-L. Wang, T. P. Durrett, S. Stymne, and C. Löfstedt. 2014. A plant factory for moth pheromone production. *Nature Communications* 5:3353. doi:10.1038/ncomms4353.

Doane, C. C., and W. Brooks. 1981. Research and development of pheromones for insect control with emphasis on the pink bollworm. Pp. 285–303. In E. R. Mitchell, ed. *Management of Insect Pests with Semiochemicals*. New York: Plenum Press.

Elkins, R. B., R. L. DeMoura, and K. M. Klonsky. 2005. Cost of production of transitioning from conventional codling moth control to aerosol-released mating disruption ("puffers") in pears. *Acta Horticulturae* 671:559–563.

Epstein, D., R. S. Zack, J. F. Brunner, L. J. Gut, and J. J. Brown. 2001. Ground beetle activity in apple orchards under reduced pesticide management regimes. *Biological Control* 21:97–104.

Epstein, D. L., L. L. Stelinski, T. Reed, J. R. Miller, and L. J. Gut. 2006. Higher densities of distributed pheromone sources provide disruption of codling moth (Lepidoptera: Tortricidae) superior to that of lower densities of clumped sources. *Journal of Economic Entomology* 99:1327–1333.

Epstein, D. L., L. L. Stelinski, J. R. Miller, M. J. Grieshop, and L. J. Gut. 2011. Effects of reservoir dispenser height on efficacy of mating disruption of codling moth (Lepidoptera: Tortricidae) in apple. *Pest Management Science* 67:975–979.

Evenden, M. L. 1998. Semiochemical-based disruption of mate-finding behaviour in *Choristoneura rosaceana* (Harris) and *Pandemis limitata* (Robinson) (Lepidoptera: Tortricidae) in British Columbia orchards. PhD thesis, Simon Fraser University, Burnaby, Canada.

Evenden, M. L., and K. F. Haynes. 2001. Potential for the evolution of resistnce to pheromone-based mating disruption tested using two pheromone strains of the cabbage looper, *Trichoplusia ni*. *Entomologia Experimentalis et Applicata* 100:131–134.

Evenden, M. L., and J. R. McLaughlin. 2004. Factors influencing the effectiveness of an attracticide formulation against the oriental fruit moth, *Grapholita molesta*. *Entomologia Experimentalis et Applicata* 112:89–97.

Evenden, M. L., G. J. R. Judd, and J. H. Borden. 1999a. Pheromone-mediated mating disruption of *Choristoneura rosaceana*: is the most attractive blend really the most effective? *Entomologia Experimentalis et Applicata* 90:37–47.

Evenden, M. L., G. J. R. Judd, and J. H. Borden. 1999b. Simultaneous disruption of pheromone communication in *Choristoneura rosaceana* and *Pandemis limitata* with pheromone and antagonist blends. *Journal of Chemical Ecology* 25:501–517.

Evenden, M. L., G. J. R. Judd, and J. H. Borden. 1999c. Mating disruption of two sympatric, orchard-inhabiting tortricids, *Choristoneura rosaceana* and *Pandemis limitata* (Lepidoptera: Tortricidae), with pheromone components of both species' blends. *Journal of Economic Entomology* 92:380–390.

Evenden, M. L., G. J. R. Judd, and J. H. Borden. 1999d. A synomone imparting distinct sex pheromone communication channels for *Choristoneura rosaceana* (Harris) and *Pandemis limitata* (Robinson) (Lepidoptera: Tortricidae). *Chemoecology* 9:73–80.

Evenden, M. L., J. R. McLaughlin, and D. Czokajlo. 2005. Effects of exposure to pheromone and insecticide constituents of an attracticide formulation on reproductive behavior of oriental fruit moth (Lepidoptera: Tortricidae). *Journal of Economic Entomology* 98:334–341.

Evenden, M. L., B. A. Mori, R. Gries, and J. Otani. 2010. Sex pheromone of the red clover casebearer moth, *Coleophora deauratella*, an invasive pest of clover in Canada. *Entomologia Experimentalis et Applicata* 137:255–261.

Evenden, M. L., B. A. Mori, K. D. Sjostrom, and J. Roland. 2015. Forest tent caterpillar, *Malacosoma disstria* (Lepidoptera: Lasiocampidae), mate-finding behavior is greatest at intermediate population densities: implications for interpretation of moth capture in pheromone-baited traps. *Frontiers in Ecology and Evolution* 3:78. doi:10.3389/fevo.2015.00078.

Fadamiro, H. Y., A. A. Cossé, T. Dittl, and T. C. Baker. 1998. Suppression of mating by blackheaded fireworm (Lepidoptera: Tortricidae) in Wisconsin cranberry marshes by using MSTRS devices. *Journal of Agricultural Entomology* 15:377–386.

Fadamiro, H. Y., A. A. Cossé, and T. C. Baker. 1999. Mating disruption of European corn borer, *Ostrinia nubilalis* by using two types of sex pheromne dispensers deployed in grassy aggregation sites in Iowa cornfields. *Journal of Asia-Pacific Entomology* 2:121–132.

Farkas, S. R., H. H. Shorey, and L. K. Gaston. 1974. Sex pheromones of Lepidoptera: the use of widely separated evaporators of looplure for the disruption of pheromone communication in *Trichoplusia ni*. *Environmental Entomology* 3:876–877.

Fernandez, D. E., E. H. Beers, J. F. Brunner, M. D. Doerr, and J. E. Dunley. 2005. Effects of seasonal mineral oil applications on the pest and natural enemy complexes of apple. *Journal of Economic Entomology* 98:1630–1640.

Fitzpatrick, S. M. 2006. Delayed mating reduces fecundity of blackheaded fireworm, *Ropobota naevana*, on cranberry. *Entomologia Experimentalis et Applicata* 120:245–250.

Fitzpatrick, S. M., and J. T. Troubridge. 1993. Fecundity, number of diapause eggs, and egg size of successive generations of the blackheaded fireworm (Lepidoptera: Tortricidae) on cranberries. *Environmental Entomology* 22:818–823.

Flint, H. M., and J. R. Merkle. 1983. Pink bollworm (Lepidoptera: Gelechiidae): communication disruption by pheromone composition imbalance. *Journal of Economic Entomology* 76:40–46.

Flint, H. M., and J. R. Merkle. 1984. The pink bollworm (Lepidoptera: Gelechiidae): alteration of male response to gossyplure by release of its component Z,Z, isomer. *Journal of Economic Entomology* 77:1099–1104.

Foster, S. P., and M. O. Harris. 1997. Behavioral manipulation methods for insect pest-management. *Annual Review of Entomology* 42:123–146.

Gaston, L. K., H. H. Shorey, and C. A. Saario. 1967. Insect population control by the use of sex pheromones to inhibit orientation between the sexes. *Nature* 213:1155.

Girling, R. D., B. S. Higbee, and R. T. Cardé. 2013. The plume also rises: trajectories of pheromone plumes issuring from point sources in an orchard canopy at night. *Journal of Chemical Ecology* 39:1150–1160.

Gökçe, A., L.L. Stelinski, L.J. Gut, and M.E. Whalon. 2007. Comparative behavioral and EAG responses of female oblique-banded and redbanded leafroller moths (Lepidoptera: Tortricidae) to their sex pheromone components. *European Journal of Entomology* 104:187–194.

Golub, M., J. Weatherston, and M.H. Benn. 1983. Measurement of release rates of gossyplure from controlled release formulations by mini-airflow method. *Journal of Chemical Ecology* 9:323–333.

Gray, T.G., and M.A. Hulme. 1995. Mating disruption of Douglas-fir tussock moth one and two years after application of pheromone. *Journal of the Entomological Society of British Columbia* 92:101–105.

Gronning, E.K., D.M. Borchert, D.G. Pfeiffer, C.M. Felland, J.F. Walgenbach, L.A. Hull, and J.C. Killian. 2000. Effect of specific and generic sex attractant blends on pheromone trap captures of four leafroller species in mid-Atlantic apple orchards. *Journal of Economic Entomology* 93:157–164.

Groot, A., C. Gemeno, C. Brownie, F. Gould, and C. Schal. 2005. Male and female antennal responses in *Heliothis virescens* and *H. subflexa* to conspecific and heterospecific sex pheomone compounds. *Environmental Entomology* 34:256–263.

Gut, L.J., L.L. Stelinski, D.R. Thomson, and J.R. Miller. 2004. Behaviour-modifying chemicals: prospects and constraints in IPM. Pp. 73–120. In O. Koul, G. Dhaliwal, and G. Cuperus, eds. *Integrated Pest Management: Potential, Constraints, and Challenges.* New York: CABI Publishing.

Harari, A.R., T. Zahavi, and H. Steinitz. 2015. Female detection of the synthetic sex pheromone contributes to the efficacy of mating disruption of the European grapevine moth, *Lobesia botrana. Pest Management Science* 71:316–22. doi:10.1002/ps.3830.

Haynes, K.F., W.G. Li, and T.C. Baker. 1986. Control of the pink bollworm moth (Lepidoptera: Gelechiidae) with insecticides and pheromones (attracticide): lethal and sublethal effects. *Journal of Economic Entomology* 79:1466–1471.

Heuskin, S., F.J. Verheggen, E. Haubruge, and J.-P. Wathelet. 2011. The use of semiochemical slow-release devices in integrated pest management strategies. *Biotechnology Agronomy Society and Environment* 15:459–470.

Howell, J.F., A.L. Knight, T.R. Unruh, D.F. Brown, J.L. Krysan, C.R. Sell, and P.A. Kirsch. 1992. Control of codling moth in apple and pear with sex pheromone-mediated mating disruption. *Journal of Economic Entomology* 85:918–925.

Hsu, C.L., A.M. Agnello, and W.H. Reissig. 2009. Edge effects in the directionally biased distribution of *Choristoneura rosaceana* (Lepidoptera: Tortricidae) in apple orchards. *Environmental Entomology* 38:433–441.

Huang, J., Gut, L.J., and J.R. Miller. 2013. Separating the attractant from the toxicant improves attact-and-kill of codling moth (Lepidoptera: Tortricidae). *Journal of Economic Entomology* 106:2144–2150.

Il'ichev, A.L., L.J. Gut, and D.G. Williams. 2007. Dual pheromone dispenser for combined control of codling moth *Cydia pomonella* L. and oriental fruit moth *Grapholita molesta* (Busck) (Lep., Tortricidae) in pears. *Journal of Applied Entomology* 131:368–378.

Ioriatti, C., G. Anfora, M. Tasin, A. De Cristofaro, P. Witzgall, and A. Lucchi. 2011. Chemical ecology and management of *Lobesia botrana* (Lepidoptera: Tortricidae). *Journal of Economic Entomology* 104:1125–1137.

Isaacs, R., M. Ulczynski, B. Wright, L.J. Gut, and J.R. Miller. 1999. Performance of the microsprayer, with application for pheromone-mediated control of insect pests. *Environmental Entomology* 92:1157–1164.

Jenkins, P.E., and R. Isaacs. 2008. Mating disruption of *Paralobesia viteana* in vineyards using pheromone deployed in SPLAT-GBM wax droplets. *Journal of Chemical Ecology* 34:1089–1095.

Jones, V.P., and N.G. Wiman. 2012. Modelling the interaction of physiological time, seasonal weather patterns, and delayed mating on population dynamics of codling moth *Cydia pomonella* (L.) (Lepidoptera: Tortricidae). *Population Ecology* 54:421–429.

Jones, V.P., N.G. Wiman, and J.F. Brunner. 2008. Comparison of delayed female mating on reproductive biology of codling moth and obliquebanded leafroller. *Environmental Entomology* 37:679–685.

Jones, V.P., T.R. Unruh, D.R. Horton, N. Mills, J.F. Brunner, E.H. Beers, and P.W. Shearer. 2009a. Tree fruit IPM programs in the Western United States: the challenge of enhancing biological control through intensive management. *Pest Management Science* 65:1305–1310.

Jones, B.C., J. Roland, and M.L. Evenden. 2009b. Development of a combined sex pheromone-based monitoring system for *Malacosoma disstria* Hübner (Lepidoptera: Lasiocampidae) and *Choristoneura conflictana* (Walker) (Lepidoptera: Tortricidae). *Environmental Entomology* 38:459–471.

Judd, G.J.R., and M.G.T. Gardiner. 2004. Simultaneous disruption of pheromone communication and mating in *Cydia pomonella, Choristoneura rosaceana* and *Pandemis limitata* (Lepidoptera: Tortricidae) using Isomate-CM/LR in apple orchards. *Journal of the Entomological Society of British Columbia* 101:3–13.

Judd, G.J.R., and M.G.T. Gardiner. 2005. Towards eradication of codling moth in British Columbia by complimentary actions of mating disruption, tree banding and sterile insect technique: five-year study in organic orchards. *Crop Protection* 24:718–733.

Judd, G.J.R., N.C. DeLury, and M.G.T. Gardiner. 2005a. Examining disruption of pheromone communication in *Choristoneura rosaceana* and *Pandemis limitata* using microencapsulated (Z)-11-tetradecenyl acetate applied in a laboratory flight tunnel. *Entomologia Experimentalis et Applicata* 114:35–45.

Judd, G.J.R., M.G.T. Gardiner, N.C. DeLury, and G. Karg. 2005b. Reduced antennal sensitivity, behavioural response, and attraction of male codling moths, *Cydia pomonella*, to their pheromone (E,E)-8-10-dodecadien-1-ol following various pre-exposure regimes. *Entomologia Experimentalis et Applicata* 114:65–78.

Kaae, R.S., H.H. Shorey, L.K. Gaston, and H.H. Hummel. 1974. Sex pheromones of Lepidoptera: disruption of pheromone communication in *Trichoplusia ni* and *Pectinophora gossypiella* by permeation of the air with nonpheromone chemicals. *Environmental Entomology* 3:87–89.

Karg, G., and A.E. Sauer. 1995. Spatial distribution of pheromone in vineyards treated for mating disruption of the grape vine moth *Lobesia botrana* measured with electroantennograms. *Journal of Chemical Ecology* 21:1299–1314.

Karg, G., and A.E. Sauer. 1997. Seasonal variation of pheromone concentration in mating disruption trials against European grape vine moth *Lobesia botrana* (Lepidoptera: Tortricidae) measured by EAG. *Journal of Chemical Ecology* 23:487–501.

Karg, G., and D.M. Suckling. 1997. Polyethylene dispensers generate large-scale temporal fluctuations in pheromone concentration. *Environmental Entomology* 26:896–905.

Karg, G., D.M. Suckling, and S.J. Bradley. 1994. Absorption and release of pheromone of *Epiphyas postvittana* (Lepidoptera: Tortricidae) by apple leaves. *Journal of Chemical Ecology* 20:1825–1841.

Kehat, M., L. Anshelevich, D. Gordon, M. Harel, L. Zilberg, and E. Dunkelblum. 1999. Effect of density of pheromone sources, pheromone dosage and population pressure on mating of pink bollworm, *Pectinophora gossypiella* (Lepidoptera: Gelechiidae). *Bulletin of Entomological Research* 89:339–345.

Knight, A.L. 1995. The impact of codling moth (Lepidoptera: Tortricidae) mating disruption on apple pest management in Yakima Valley, Washington. *Journal of the Entomologial Society of British Columbia* 92:29–38.

Knight, A.L. 1997. Delay of mating of codling moth in pheromone disrupted orchards. *International Society of Biological Control WPRS Bulletin* 20:203–206.

Knight, A.L. 2004. Managing codling moth (Lepidoptera: Tortricidae) with an internal grid of either aerosol puffers or dispenser clusters plus border applications of individual dispensers. *Journal of the Entomological Society of British Columbia* 101:69–77.

Knight, A.L. 2007. Multiple mating of male and female codling moth (Lepidoptera: Tortricidae) in apple orchards treated with sex pheromone. *Environmental Entomology* 36:157–164.

Knight, A.L., and T.E. Larsen. 2008. Creating point sources for codling moth (Lepidoptera: Tortricidae) with low-volume sprays of a microencapsulated sex pheromone formulation. *Environmental Entomology* 37:1136–1144.

Knight, A.L., and D.M. Light. 2005. Sasonal flight patterns of codling moth (Lepidoptera: Tortricidae) monitored with pear ester and codlemone-baited traps in sex pheromone-treated apple orchards. *Environmental Entomology* 34:1028–1035.

Knight, A.L., and J.E. Turner. 1999. Mating disruption of *Pandemis* spp. (Lepidoptera: Tortricidae). *Environmental Entomology* 28:81–87.

Knight, A.L., J.F. Howell, L.M. McDonough, and M. Weiss. 1995. Mating disruption of codling moth (Lepidoptera: Tortricidae) with polyethylene tube dispensers: determining emission rates and distribution of fruit injuries. *Journal of Agricultural Entomology* 12:85–100.

Knight, A.L., J.E. Turner, and B. Brachula. 1997. Predation on eggs of codling moth (Lepidoptera: Tortricidae) in mating disrupted and conventional orchards in Washington. *Journal of the Entomological Society of British Columbia* 94:67–74.

Knight, A.L., T.E. Larsen, and K.C. Ketner. 2004. Rainfastness of microencapsulated sex pheromone formulation for codling moth (Lepidoptera: Tortricidae). *Journal of Economic Entomology* 97:1987–1992.

Knight, A.L., R. Hilton, and D.M. Light. 2005. Monitoring codling moth (Lepidoptera: Tortricidae) in apple with blends of ethyl (*E, Z*)-2, 4-decadienoate and codlemone. *Environmental Entomology* 34:598–603.

Knight, A.L., L.L. Stelinski, V. Hebert, L.J. Gut, D.M. Light, and J.F. Brunner. 2012. Evaluation of novel semiochemical dispensers simultaneously releasing pear ester and sex pheromone for mating disruption of codling moth (Lepidoptera: Tortricidae). *Journal of Applied Entomology* 136:79–86.

Koch, U.T., A.M. Cardé, and R.T. Cardé. 2002. Calibration of an EAG system to measure concentration of pheromone formulated for mating disruption of the pink bollworm moth, *Pectinophora gossypiella* (Saunders) (Lepidoptera: Gelechiidae). *Journal of Applied Entomology* 126:431–435.

Koch, U.T., W. Luder, U. Andrick, R.T. Staten, and R.T. Cardé. 2009. Measurement by electroantennogram of airborne pheromone in cotton treated for mating disruption of *Pectinophora gossypiella* following removal of pheromone dispensers. *Entomologia Experimentalis et Applicata* 130:1–9.

Kuenen, L.P.S., and T.C. Baker. 1981. Habituation versus sensory adaptation as the cause of reduced attraction following pulsed and constant sex pheromone pre-exposure in *Trichoplusia ni*. *Journal of Insect Physiology* 27:721–726.

Kuhns, E.H., K. Pelz-Stelinski, and L.L. Stelinski. 2012. Reduced mating success of female tortricid moths following intense pheromone auto-exposure varies with sophistication of mating system. *Journal of Chemical Ecology* 38:168–175.

Kwon, J.J. 2014. Development of a pheromone-based attract and kill formulation with visual cues to target the diurnally active apple clearwing moth, *Synanthedon myoaeformis* (Borkhausen), (Lepidoptera: Sesiidae). MSc thesis, University of Alberta, Alberta, Canada.

Landolt, P.J., C.E. Curtis, J.A. Coffelt, K.W. Vick, and R.E. Doolittle. 1982. Field trials of potential navel orangeworm mating disruptants. *Journal of Economic Entomology* 75:547–550.

Lapointe, S.L., L.L. Stelinski, T.J. Evens, R.P. Niedz, D.G. Hall, and A. Mafra-Neto. 2009. Sensory imbalance as mechanism of orientation disruption in the leafminer *Phyllocnistis citrella*: elucidation by multivariate geometric designs and response surface models. *Journal of Chemical Ecology* 35:896–903.

Lapointe, S.L., L.L. Stelinski, C.P. Keathley, and A. Mafra-Neto. 2014. Intentional coverage gaps reduce cost of mating disruption for *Phyllocnistis citrella* (Lepidoptera: Gracillariidae) in citrus. *Journal of Economic Entomology* 107:718–726.

Lawson, D.S., W.H. Reissig, A.M. Agnello, J.P. Nyrop, and W.L. Roelofs. 1996. Interference with the mate-finding communication system of the obliquebanded leafroller (Lepidoptera: Tortricidae) using synthetic sex pheromones. *Environmental Entomology* 25:895–905.

Lewis, W.J., D.A. Nordlund, R.C. Gueldner, P.E.A. Teal, and J.H. Tumlinson. 1982. Kairomones and their use for management of entomophagous insects. XIII. Kairomonal activity for *Trichogramma* spp. of abominal tips, excretion, and a synthetic sex pheromone blend of *Heliothis zea* (Boddie) moths. *Journal of Cheimcal Ecology* 8:1323–1331.

Light, D.M. and M.C. Birch. 1979. Electrophysiological bsis for the behvioural response of male and female *Trichoplusia ni* to synthetic female pheromone. *Journal of Insect Physiology* 25:161–167.

Light, D.M., R.A. Flath, R.G. Buttery, F.G. Zalom, and R.E. Rice. 1993. Host-plant green leaf volatiles synergize the synthetic sex pheromones of the corn earworm and codling moth (Lepidoptera). *Chemoecology* 4:145–152.

Light, D.M., Knight, A.L., Henrick, C.A., Rajapaska, D., B. Lingren, J.C. Dickens, K.M. Reynolds et al. 2001. A pear-derived kairomone with pheromonal potency that attracts male and female codling moth, *Cydia pomonella* (L.). *Naturwissenschaften* 88:333–338.

Ljungberg, H., P. Anderson, and B.S. Hansson. 1993. Physiology and morphology of pheromone-specific sensilla on the antennae of male and female *Spodoptera littoralis* (Lepidoptera: Noctuidae). *Journal of Insect Physiology* 39:253–260.

Mafra-Neto, A., and T.C. Baker. 1996. Elevation of pheromone response threshold in almond moth males pre-exposed to pheromone spray. *Physiological Entomology* 21:217–222.

McCormick, A.L.C., M. Karlsson, C.F.B. Ochoa, M. Proffit, M. Bengtsson, M.V. Zuluag, T. Fukumoto, C. Oehlschlager, A.M.C. Prado, and P. Witzgall. 2012. Mating disruption of Guatamalan potato moth *Tecia solanivora* by attractive and non-attractive pheromone blends. *Journal of Chemical Ecology* 38:63–70.

McDonough, L.M., W.C. Aller, and A.L. Knight. 1992. Performance characteristics of a commercial controlled-release dispenser of sex pheromone for control of codling moth (*Cydia pomonella*) by mating disruption. *Journal of Chemical Ecology* 18:2177–2189.

McGhee, P.S., D. Epstein, and L.J. Gut. 2011. Quantifying the benefits of areawide pheromone mating disruption programs that target codling moth (Lepidoptera: Tortricidae). *American Entomologist* 57:94–100.

Miliczky, E.R., C.O. Calkins, and D.R. Horton. 2000. Spider abundance and diversity in apple orchards under three insect pest management programmes in Washington State, USA. *Agricultural and Forest Entomology* 2:203–215.

Millar, J.G. 2000. Polyene hydrocarbons and epoxides: a second major class of lepidopteran sex attractant pheromones. *Annual Review of Entomology* 45:575–604.

Miller, J.R., and L.J. Gut. 2015. Mating disruption for the 21st century: matching technology with mechanism. *Environmental Entomology* 44:427–453.

Miller, J.R., K. Mori, and W.L. Roelofs. 1977. Gypsy moth field trapping and electroantennogram studies with pheromone enantiomers. *Journal of Insect Physiology* 23:1447–1453.

Miller, J.R., L.J. Gut, F.M. de Lamé, and L.L. Stelinski. 2006a. Differentiation of competitive vs. non-competitive mechanisms mediating disruption of moth sexual communication by point sources of sex pheromone (Part 1): theory. *Journal of Chemical Ecology* 32:2089–2114.

Miller, J.R., L.J. Gut, F.M. de Lamé, and L.L. Stelinski. 2006b. Differentiation of competitive vs. non-competitive mechanisms mediating disruption of moth sexual communication by point sources of sex pheromone (Part 2): case studies. *Journal of Chemical Ecology* 32:2115–2143.

Miller, J.R., P.S.McGhee, P.Y.Siegert, C.G. Adams, J. Huang, M.J. Grieshop, and L.J. Gut. 2010. General principles of attraction and competitive attraction as revealed by large-cage studies of moths responding to sex pheromone. *Proceedings of the National Academy of Sciences of the United States of America* 107:22–27.

Milli, R., U.T. Koch, and J.J. de Kramer. 1997. EAG measurement of pheromone distribution in apple orchards treated for mating disruption of *Cydia pomonella*. *Entomologia Experimentalis et Applicata* 82:289–297.

Minks, A.K., and R.T. Cardé. 1988. Disruption of pheromone communication in moths: is the ntural blend really most efficacious? *Entomologia Experimentalis et Applicata* 49: 25–36.

Mitchell, E.R., M. Jacobson, and A.H. Baumhover. 1975. *Heliothis* spp.: disruption of pheromonal communication with (*Z*)-9-tetradecen-1-ol formate. *Environmental Entomology* 4:577–579.

Mochizuki, F., T. Fukumoto, H. Noguchi, H. Sugie, T. Morimoto, and K. Ohtani. 2002. Resistance to a mating disruptant composed of (*Z*)-11-tetradecenyl acetate in the smaller tea tortrix, *Adoxophyes honmai* (Yasuda) (Lepidoptera: Tortricidae). *Applied Entomology and Zoology* 37:299–304.

Mori, B.A., and M.L. Evenden. 2013. When mating disruption does not disrupt mating: fitness consequences of delayed mating in moths. *Entomologia Experimentalis et Applicata* 146:50–65.

Mori, B.A., and M.L. Evenden. 2014. Efficacy and mechanisms of communication disruption of the red clover casebearer moth

(*Coleophora deauratella*) with complete and partial pheromone formulations. *Journal of Chemical Ecology* 40:577–589.

Nansen, C., K.M. MacDonald, C.D. Rogers, M. Thomas, G.M. Poppy, and I.H. Baxter. 2007. Effects of sex pheromone in electrostatic powder on mating behaviour by *Lobesia botrana* males. *Journal of Applied Entomology* 131:303–310.

Niwa, C.G., and G.E. Daterman. 1989. Pheromone mating disruption of *Rhyacionia zozana* (Lepidoptera: Tortricidae): influence on the associated parasite complex. *Environmental Entomology* 18:570–574.

Ochieng, S.A., K.C. Park, and T.C. Baker. 2002. Host plant volatiles synergize responses of sex pheromone-specific olfactory receptor neurons in male *Helicoverpa zea. Journal of Comparative Physiology A* 188:325–333.

Onufrieva, K.S., K.W. Thorpe, A.D. Hickman, D.S. Leonard, V.C. Mastro, and E.A. Roberts. 2008. Gypsy moth mating disruption in open landscapes. *Agricultural and Forest Entomology* 10:175–179.

Onufrieva, K.S., K.W. Thorpe, A.D. Hickman, D.S. Leonard, E.A. Roberts, and P.C. Tobin. 2013. Persistence of gypsy moth pheromone, disparlure in the environment in various climates. *Insects* 4:104–116.

Palaniswamy, P., and W.D. Seabrook. 1978. Behavioral responses of female eastern spruce budworm *Choristoneura fumiferana* (Lepidoptera, Tortricidae) to the sex pheromone of her own species. *Journal of Chemical Ecology* 4:649–655.

Palaniswamy, P., and W.D. Seabrook. 1985. The alteration of calling behaviour by female *Choristoneura fumiferana* when exposed to synthetic sex pheromone. *Entomologia Experimentalis et Applicata* 37:13–16.

Pearson, G.A., and J.R. Meyer. 1996. Female grape root borer (Lepidoptera: Sesiidae) mating success under synthetic sesiid sex pheromone treatment. *Journal of Entomological Science* 3:323–330.

Pearson, G.A., and C. Schal. 1999. Electroantennogram responses of both sexes of grape root borer (Lepidoptera: Sesiidae) to synthetic female sex pheromone. *Environmental Entomology* 28:943–946.

Pearson, G.A., S. Dillery, and J.R. Meyer. 2004. Modeling intrasexual competition in a sex pheromone system: how much can female movement affect female mating success? *Journal of Theoretical Biology* 231:549–555.

Phillips, T.W. 1997. Semiochemicals of stored-product insects: research and applications. *Journal of Stored Products Research* 33:17–30.

Pickett, J.A., G.I. Aradottir, M.A. Birkett, T.J.A. Bruce, A.M. Hooper, C.A.O.Midega, H.D. Jones, et al. 2014. Delivering sustainable crop protection systems via the seed: exploiting natural constitutive and inducible defence pathways. *Philosophical Transactions of the Royal Society B Biological Sciences* 369:20120281.

Polavarapu, S., G. Lonergan, H. Peng, and K. Neilsen. 2001. Potential for mating disruption of *Sparganothis sulfureana* (Lepidoptera: Tortricidae) in cranberries. *Journal of Economic Entomology* 94:658–665.

Reinke, M.D., P.Y. Siegert, P.S. McGhee, L.J. Gut, and J.R. Miller. 2014. Pheromone release rate determines whether sexual communication of oriental fruit moth is disrupted competitively vs. non-competitively. *Entomologia Experimentalis et Applicata* 150:1–6.

Renou, M., and A. Guerrero. 2000. Insect parapheromones in olfaction research and semiochemical-based pest control strategies. *Annual Review of Entomology* 48:605–630.

Rhainds, M., E.G. Kettela, and P.J. Silk. 2012. Thirty-five years of pheromone-based mating disruption studies with *Choristoneura fumiferana* (Clemens) (Lepidoptera: Tortricidae). *The Canadian Entomologist* 144:379–395.

Rice, R.E., and P.A. Kirsch. 1990. Mating disruption of the oriental fruit moth in the United States. Pp. 193–211. In R. Ridgeway, R. Silverstein, and M. Inscoe, eds. *Behavior-Modifying Chemicals for Insect Management: Applications of Pheromones and other Attractants.* New York: Marcel Dekker.

Rumbo, E.R., and R.A. Vickers. 1997. Prolonged adaptation as possible mating disruption mechanism in Oriental fruit moth, *Cydia* (=*Grapholita*) *molesta. Journal of Chemical Ecology* 23:445–457.

Sadek, M.M., G. von Wowern, C. Löfstedt, W.-Q. Rosen, and P. Anderson. 2012. Modulation of the temporal pattern of calling behavior of female *Spodoptera littoralis* by exposure to sex pheromone. *Journal of Insect Physiology* 58:61–66.

Sanders, C.J. 1987. Flight and copulation of female spruce budworm in pheromone-permeated air. *Journal of Chemical Ecology* 13:1749–1758.

Sanders, C.J. 1997. Mechanisms of mating disruption in moths. Pp. 333–346. In R.T. Cardé and A.K. Minks, eds. *Insect Pheromone Research: New Directions.* New York: Chapman & Hall.

Sarfraz, R.M., M.L. Evenden, B.A. Keddie, and L.M. Dosdall. 2006. Pheromone-mediated mating disruption: a powerful tool in insect pest management. *Outlooks on Pest Management* 17:36–45.

Sauer, A.E., and G. Karg. 1998. Variables affecting pheromone concentration in vineyards treated for mating disruption of grape vine moth *Lobesia botrana. Journal of Chemical Ecology* 24:289–302.

Schmitz, V., R. Roehrich, and J. Stockel. 1995. Disruption mechanisms of pheromone communication in the European grape moth *Lobesia botrana* (Lep., Tortricidae) II. Influence of population density and the distance between insects for males to detect the females in atmosphere impregnated by pheromone. *Journal of Applied Entomology* 119:303–308.

Schmitz, V., M. Renou, R. Roehrich, J. Stockel, and P. Lecharpentier. 1997. Disruption mechanisms of pheromone communication in the European grape moth *Lobesia botrana* Den & Schiff. III. Sensory adaptation and habituation. *Journal of Chemical Ecology* 23:83–95.

Schneider, D., S. Schulz, E. Priesner, J. Ziesmann, and W. Francke. 1998. Autodetection and chemistry of female and male pheromone in both sexes of the tiger moth *Panaxia quadripunctaria. Journal of Comparative Physiology* 182:153–161.

Schwalbe, C.P., and V.C. Mastro. 1988. Gypsy moth mating disruption: dosage effects. *Journal of Chemical Ecology* 14:581–588.

Shani, A., and J. Clearwater. 2001. Evasion of mating disruption in *Ephestia cautella* (Walker) by increased pheromone production relative to that of undisrupted populations. *Journal of Stored Products Research* 37:237–252.

Sharov, A.A., D. Leonard, A.M. Liebhold, E.A. Roberts, and W. Dickerson. 2002. Slow the spread: a national program to contain the gypsy moth. *Journal of Forestry* 100:30–35.

Shorey, H.H. 1977. Manipulation of insect pests of agricultural crops. Pp. 353–367. In H.H. Shorey and J.J. McKelvey, Jr., eds. *Chemical Control of Insect Behavior: Theory and Application.* Toronto: John Wiley and Sons.

Shorey, H.H., and R.G. Gerber. 1996. Use of puffers for disruption of sex pheromone communication of codling moths (Lepidoptera: Tortricidae) in walnut orchards. *Environmental Entomology* 25:1398–1400.

Shorey, H.H., L.K. Gaston, and C.A. Saario. 1967. Sex pheromones of noctuid moths. XIV. Feasibility of behavioral control by disruption of pheromone communication in cabbage loopers. *Journal of Economic Entomology* 60:1541–1545.

Shorey, H.H., C.B. Sisk, and R.G. Gerber. 1996. Widely separated pheromone release sites for disruption of sex pheromone communication in two species of Lepidoptera. *Environmental Entomology* 25:446–454.

Staten, R.T., O. El-Lissy, and L. Antilla. 1997. Successsful area-wide program to control pink bollworm by mating disruption. Pp. 383–396. In R.T. Cardé and A.K. Minks, eds. *Insect Pheromone Research: New Directions.* New York: Chapman & Hall.

Stelinski, L.L., L.J. Gut, and J.R. Miller. 2003a. Concentration of air-borne pheromone required for long-lasting adaptation in the obliquebanded leafroller, *Choristoneura rosaceana. Physiological Entomology* 28:97–107.

Stelinski, L.L., J.R. Miller, and L.J. Gut. 2003b. Presence of long-lasting peripheral adaptation in oblique-banded leafroller, *Choristoneura rosaceana* and absence of such adaptation in redbanded leafroller, *Argyrotaenia velutinana. Journal of Chemical Ecology* 29:405–423.

Stelinski, L.L., L.J. Gut, A.V. Pierzchala, and J.R. Miller. 2004a. Field observations quantifying attraction of four tortricid moth species to high-dosage pheromone rope dispensers in untreated and pheromone-treated apple orchards. *Entomologia Experimentalis et Applicata* 113:187–196.

Stelinski, L.L., L.J. Gut, K.J. Vogel, and J.R. Miller. 2004b. Behaviors of naive vs. pheromone-exposed leafroller moths in plumes from high-dosage pheromone dispensers in a sustained-flight wind tunnel: implications for mating disruption of these species. *Journal of Insect Behavior* 17:533–554.

Stelinski, L.L., L.J. Gut, K.C. Ketner, and J.R. Miller. 2005a. Orientational disruption of codling moth, *Cydia pomonella* (L.) (Lep., Tortricidae), by concentrated formulations of microencapsulated pheromone in flight tunnel assays. *Journal of Applied Entomology* 129:481–488.

Stelinski, L. L., L. J. Gut, and J. R. Miller. 2005b. Occurrence and duration of long-lasting peripheral adaptation among males of three species of economically important totricid moths. *Annals of the Entomological Society of America* 98:580–586.

Stelinski, L. L., L. J. Gut, R. E. Malllinger, D. Epstein, T. P. Reed, and J. R. Miller. 2005c. Small plot trials documenting effective mating disruption of oriental fruit moth by issuing high densities of wax-drop pheromone dispensers. *Journal of Economic Entomology* 98:1267–1274.

Stelinski, L. L., A. L. Il'ichev, and L. J. Gut. 2006. Antennal and behavioral responses of virgin and mated oriental fruit moth (Lepidptera: Tortricidae) females to their sex pheromone. *Annals Entomological Society of America* 99:898–904.

Stelinski, L. L., J. R. Miller, R. Ledebuhr, and L. J. Gut. 2007a. Mechanized applicatior for large-scale field deployment of paraffin-wax dispensers of pheromone for mating disruption in tree fruit. *Journal of Economic Entomology* 99:1705–1710.

Stelinski, L. L., P. McGhee, M. Haas, A. L. Il'ichev, and L. J. Gut. 2007b. Sprayable microencapsulated sex pheromone formulations for mating disruption of four tortricid species: effects of application height, rate, frequency and sticker adjuvant. *Journal of Economic Entomology* 100:1360–1369.

Stelinski, L. L., L. J. Gut, P. McGee, and J. R. Miller. 2007c. Towards high performance mating disruption of codling moth. *International Society of Biological Control wrps Bulletin* 30:115–122.

Stelinski, L. L., J. R. Miller, R. Ledebuhr, P. Siegert, and L. J. Gut. 2007d. Season-long mating disruption of *Grapholita molesta* (Lepidoptera: Tortricidae) by one machine application of pheromone in wax drops (SPLAT-OFM). *Journal of Pest Science* 80:109–117.

Stelinski, L. L., P. McGhee, M. Grieshop, J. F. Brunner, and L. J. Gut. 2008a. Efficacy and mode of action of female-equivalent dispensers of pheromone for mating disruption of codling moth. *Agricultural and Forest Entomology* 10:389–397.

Stelinski, L. L., J. R. Miller, and M. E. Rogers. 2008b. Mating disruption of citrus leafminer mediated by a noncompetitive mechanism at a remarkably low pheromone release rate. *Journal of Chemical Ecology* 34:1107–1113.

Stelinski, L. L., A. L. Il'ichev, and L. J. Gut. 2009. Efficacy and release rate of reservoir pheromone dispensers for simultaneous mating disruption of codling moth and oriental fruit moth (Lepidoptera: Tortricidae). *Journal of Economic Entomology* 102:315–323.

Stelinski, L. L., S. L. Lapointe, and W. L. Meyer. 2010. Season-long mating disruption of citrus leafminer, *Phyllocnistis citrella* Stainton, with an emulsified wax formulation of pheromone. *Journal of Applied Entomology* 134:512–520.

Stelinski, L. L., L. J. Gut, and J. R. Miller. 2013. An attempt to increse efficacy of moth mating disruption by co-releasing pheromones with kairomones and to understand possible underlying mechanisms of this technique. *Environmental Entomology* 42:158–166.

Suckling, D. M. 2000. Issues affecting the use of pheromones and other semiochemicals in orchards. *Crop Protection* 19:677–683.

Suckling, D. M., and N. P. D. Angerilli. 1996. Point source distribution affects pheromone spike frequency and communication disruption of *Epiphyas postvittana* (Lepidoptera: Tortricidae). *Environmental Entomology* 25:101–108.

Suckling, D. M., and E. G. Brockerhoff. 1999. Control of light brown apple moth (Lepidoptera: Tortricidae) using an attracticide. *Journal of Economic Entomology* 92:367–372.

Suckling, D. M., and G. M. Burnip. 1996. Orientation disruption of *Planotortrix octo* using pheromone or inhibitor blends. *Entomologia Experimentalis et Applicata* 78:149–158.

Suckling, D. M., G. Karg, and S. J. Bradley. 1996. Apple foliage enhances mating disruption of light-brown apple moth. *Journal of Chemical Ecology* 22:325–341.

Suckling, D. M., S. R. Green, A. R. Gibb, and G. Karg. 1999. Predicting atmospheric concentration of pheromone in treated apple orchards. *Journal of Chemical Ecology* 25:117–139.

Suckling, D. M., J. M. Daly, X. Chen, and G. Karg. 2007. Field electroantennogram and trap assessments of aerosol pheromone dispensers for disrupting mating in *Epiphyas postvittana*. *Pest Management Science* 63:202–209.

Suckling, D. M., B. Woods, V. J. Mitchell, A. Twidle, I. Lacey, E. B. Jang, and A. R. Wallace. 2011. Mobile mating disruption of light-brown apple moths using pheromone-treated sterile Mediterranean fruit flies. *Pest Management Science* 67: 1004–1014.

Suckling, D. M., E. G. Brockerhoff, L. D. Stringer, R. C. Butler, D. M. Campbell, L. K. Mosser, and M. F. Cooperband. 2012a. Communication disruption of *Epiphyas postvittana* (Lepidoptera: Tortricidae) by using two formulations at four point source densities in vineyards. *Journal of Economic Entomology* 105:1694–1701.

Suckling, D. M., G. F. McLaren, L.-A. M. Manning, V. J. Mitchell, B. Attfield, K. Colhoun, and A. El-Sayed. 2012b. Development of single-dispenser pheromone suppression of *Epiphyas postvittana*, *Planotortrix octo* and *Ctenopseustis obliquana* in New Zealand stone fruit orchards. *Pest Management Science* 68:928–934.

Suckling, D. M., J. J. Dymock, K. C. Park, R. H. Wakelin, and L. E. Jamieson. 2013. Communication disruption of guava moth (*Coscinoptcha improbana*) using a pheromone analog based on chain length. *Journal of Chemical Ecology* 39:1161–1168.

Tabata, J., H. Noguchi, Y. Kainoh, F. Mochizuki, and H. Sugie. 2007. Sex pheromone production and perception in the mating disruption-resistant strain of the smaller tea leafroller moth, *Adoxophyes honmai*. *Entomologia Experimentalis et Applicata* 122:145–153.

Tamhankar, A. J., R. T. Gahukar, and T. P. Rajendrum. 2000. Pheromones in the management of major lepidopterous and coleopterous pests of cotton. *Integrated Pest Management Reviews* 5:11–23.

Taschenberg, E. F., and W. L. Roelofs. 1976. Pheromone communication disruption of the grape berry moth with micoencapsulated and hollow fiber systems. *Environmental Entomology* 5:688–691.

Taverner, P. D., R. V. Gunning, P. Kolesik, P. T. Bailey, A. B. Inceoglu, B. Hammock, and R. T. Rousch. 2001. Evidence for direct neural toxicity of a "light" oil on the peripheral nerves of light brown apple moth. *Pesticide Biochemistry and Physiology* 69:153–165.

Tcheslavskaia, K. S., C. C. Brewster, K. W. Thorpe, A. A. Sharov, D. S. Leonard, and E. A. Roberts. 2005a. Effects of intentional gaps in spray coverage on the efficacy of gypsy moth mating disruption. *Journal of Applied Entomology* 129:475–480.

Tcheslavskaia, K. S., K. W. Thorpe, C. C. Brewster, A. A. Sharov, D. S. Leonard, R. C. Reardon, V. C. Mastro, P. Sellers, and E. A. Roberts. 2005b. Optimization of pheromone dosage for gypsy moth mating disruption. *Entomologia Experimentalis et Applicata* 115:355–361.

Teixeira, L. A., M. J. Grieshop, and L. J. Gut. 2010. Effect of dispenser density on timing and duration of approaches by peach tree borer. *Journal of Chemical Ecology* 36:1148–1154.

Thorpe, K. W., J. van der Pers, D. S. Leonard, P. Sellers, V. C. Mastro, R. E. Webb, and R. C. Reardon. 2007a. Electroantennogram mesurements of atmospheric pheromone concentration after aerial and ground application of gypsy moth mating disruptants. *Journal of Applied Entomology* 131:146–152.

Thorpe, K. W., K. S. Tcheslavskaia, P. C. Tobin, L. M. Blackbum, D. S. Leonard, and E. A. Roberts. 2007b. Persistant effects of aerial applictions of disparlure on gypsy moth: trap catch and mating success. *Entomologia Experimentalis et Applicata* 125:223–229.

Trimble, R. M., and M. E. Appleby. 2004. Comparison of efficacy of programs using insecticide and insecticide plus mating disruption for controlling obliquebanded leafroller in apple (Lepidoptera: Tortricidae). *Journal of Economic Entomology* 97:518–524.

Trimble, R. M., and C. A. Tyndall. 2000. Disruption of mating in the spotted tentiform leafminer (Lepidoptera: Gracillaiidae) using synthetic sex pheromone. *The Canadian Entomologist* 132:107–117.

Trimble, R. M., D. J. Pree, and N. J. Carter. 2001. Integrated control of Oriental fruit moth (Lepidoptera: Tortricidae) in peach orchards using insecticide and mating disruption. *Journal of Economic Entomology* 94:476–485.

Vakenti, J. M., A. P. Gaunce, K. N. Slessor, G. G. S. King, S. A. Allan, H. F. Madsen, and J. H. Borden. 1988. Sex pheromone components of the oblique-banded leafroller, *Choristoneura rosaceana* in the Okanagan Valley of British Columbia. *Journal of Chemical Ecology* 14:605–621.

van der Pers, J. N. C., and C. J. den Otter. 1978. Single cell responses from olfactory receptors of small ermine moths to sex-attractants. *Journal of Insect Physiology* 24:337–343.

Vickers, R. A., W. G. Thwaite, D. G. Williams, and A. H. Nicholas. 1998. Control of codling moth in small plots by mating disruption: alone and with limited insecticide. *Entomologia Experimentalis et Applicata* 86:229–239.

Waldstein, D. E., and L. J. Gut. 2004. Effects of rain and sunlight on oriental fruit moth (Lepidoptera: Tortricidae) pheromone microcapsules applied to apple foliage. *Journal of Agricultural and Urban Entomology* 21:117–128.

Wall, C., D. M. Sturgeon, A. R. Greenway, and J. N. Perry. 1981. Contamination of vegetation with synthetic sex attractant released from traps for the pea moth, *Cydia nigricana*. *Entomologia Experimentalis et Applicata* 30:111–115.

Webb, R. E., K. M. Tatman, B. A. Leonhardt, J. R. Plimmer, V. K. Boyd, P. G. Bystrak, C. P. Schwalbe, and L. W. Douglass. 1988. Effect of aerial application of racemic disparlure on male trap catch and female mating success of gypsy moth (Lepidoptera: Lymantriidae). *Journal of Economic Entomology* 81:268–273.

Webb, R. E., B. A. Leonhardt, J. R. Plimmer, K. M. Tatman, V. K. Boyd, D. L. Cohen, C. P. Schwalbe, and L. W. Douglass. 1990. Effect of racemic disparlure released from grids of plastic ropes on mating success of gypsy moth (Lepidoptera: Lymantriidae) as influenced by dose and population density. *Journal of Economic Entomology* 83:910–916.

Weissling, T. J., and A. L. Knight. 1996. Oviposition and calling behavior of codling moth (Lepidoptera: Tortricidae) in the presence of codlemone. *Annals Entomological Society of America* 89:142–147.

Williamson, E. R., R. J. Folwell, A. L. Knight, and J. F. Howell. 1996. Economics of employing pheromones for mating disruption of the codling moth, *Carpocapsa pomonella*. *Crop Protection* 15:473–477.

Wins-Purdy, A. H., G. J. R. Judd, and M. L. Evenden. 2007. Disruption of pheromone communication of *Choristoneura rosaceana* (Lepidoptera: Tortricidae) using microencapsulated sex pheromones formulated with horticultural oil. *Environmental Entomology* 36:1189–1198.

Wins-Purdy, A. H., G. J. R. Judd, and M. L. Evenden. 2008. Mechanisms of pheromone communication disruption in *Choristoneura rosaceana* exposed to microencapsulated (*Z*)-11-tetradecenyl acetate formulated with and without horticultural oil. *Journal of Chemical Ecology* 34:1096–1106.

Wins-Purdy, A. H., C. M. Whitehouse, G. J. R. Judd, and M. L. Evenden. 2009. Effect of horticultural oil on oviposition behaviour and egg survival in the obliquebanded leafroller (Lepidoptera: Tortricidae). *The Canadian Entomologist* 141:86–94.

Witzgall, P., M. Bengtsson, G. Karg, A.-C. Bäckman, L. Streinz, P. A. Kirsch, Z. Blum, and J. Löfqvist. 1996. Behavioral observations and measurements of aerial pheromone in mating disruptionn trial against pea moth *Cydia nigricana* F. (Lepidoptera: Tortricidae). *Journal of Chemical Ecology* 22:191–206.

Witzgall, P., A.-C. Bäckman, M. Svensson, U. T. Koch, F. Rama, A. El-Sayed, J. Brauchli, H. Arn, M. Bengtsson, and J. Löfqvist. 1999. Behavioral observations of codling moth, *Cydia pomonella*, in orchards permeated with synthetic pheromone. *BioControl* 44:211–237.

Witzgall, P., L. L. Stelinski, L. J. Gut, and D. R. Thomson. 2008. Codling moth management and chemical ecology. *Annual Review of Entomology* 53:503–522.

Witzgall, P., P. A. Kirsch, and A. Cork. 2010. Sex pheromones and their impact on pest management. *Journal of Chemical Ecology* 36:80–100.

Wright, R. H. 1963. Chemical control of chosen insects. *New Scientist* 20:598–600.

Zabkiewicz, J. A. 2002. Enhancement of pesticide activity by oil adjuvants. Pp. 52–61. In G. Beattie, D. Watson, M. Stevens, D. Rae, and R. Spooner-Hart, eds. *Spray Oils beyond 2000*. Sydney: University of Western Sydney.

INDEX

Note: Page number followed by (b), (f) and (t) indicates box, figure and table, respectively.

visual motion resolution, of moths, 146
visual systems, of insects, 146

wallflower (WF) hypothesis, 13(f), 15–16, 19
wax emulsion formulations, 373–74
weed biological control, use of sex phero-mone in, 339(t)
Wichmann, J. K., 5
Wiebes, Jacobus, 211
Wilson, E. O., 71–72

wind-tunnel bioassays, 17, 26–27, 37, 357, 359, 370
    camouflage using, 369
Wittig reaction, 4
Wood, Dave, 3
Wüthrich, Kurt, 130

xenobiotics, plant-derived, 131
Xenopus laevis, 13
Xenopus oocyte system, 128–29, 133, 236

Yponomeuta evonymella, 217
Yponomeuta gigas, 213, 221
Yponomeuta mahalebella, 221
Yponomeuta malinellus, 211, 213, 216, 218, 220–21
Yponomeuta padellus, 32, 213
Yponomeuta rorrella, 215(f), 220

Zea mays, 233, 292
Zeiraphera diniana, 33